VLSI MICRO- and NANOPHOTONICS

Science, Technology, and Applications

VLSI MICRO- and NANOPHOTONICS

Science, Technology, and Applications

Edited by

El-Hang Lee • Louay A. Eldada
Manijeh Razeghi • Chennupati Jagadish

CRC Press
Taylor & Francis Group
Boca Raton   London   New York

CRC Press is an imprint of the
Taylor & Francis Group, an **informa** business

CRC Press
Taylor & Francis Group
6000 Broken Sound Parkway NW, Suite 300
Boca Raton, FL 33487-2742

© 2011 by Taylor and Francis Group, LLC
CRC Press is an imprint of Taylor & Francis Group, an Informa business

---

Library of Congress Cataloging-in-Publication Data

VLSI micro- and nanophotonics : science, technology, and applications / editors, El-Hang Lee ... [et al.].
    p. cm.
    Includes bibliographical references and index.
    ISBN 978-1-57444-729-3 (acid-free paper)
    1. Microelectromechanical systems--Technological innovations. 2. Nanophotonics--Technological innovations. 3. Integrated circuits--Very large scale integration. I. Lee, El-Hang. II. Title.

TK7875.V57 2010
621.39'5--dc22

2010021867

Visit the Taylor & Francis Web site at
http://www.taylorandfrancis.com

and the CRC Press Web site at
http://www.crcpress.com

Contents

PART I Introduction

PART II Scientific and Engineering Issues

PART III Integrated Photonic Technologies: Micro-Ring Structures

v

PART IV   Integrated Photonic Technologies: Photonic Crystals and Integrated Circuits

PART V   Integrated Photonic Technologies: Plasmonics and Integration

PART VI   Integrated Photonic Technologies: Quantum Devices and Integration

PART VII   Integrated Photonic Technologies: Planar Lightwave Circuits

PART VIII   Integrated Photonic Technologies: Optical-Printed Circuit Boards

PART IX   Technologies for Emerging Applications: Renewable Energy Generation

PART X   Technologies for Emerging Applications: Photonic DNA Computing

PART XI   Technologies for Emerging Applications: Sensing Applications

Preface

This book is about an advanced technical field we have termed "VLSI micro/nanophotonics." It is not about what is generally known as "VLSI micro/nanoelectronics." VLSI stands for very-large-scale integration. From a historical perspective, we may recall that VLSI micro/nanoelectronics played a central role in the progress of twentieth-century information technology. With the phenomenal growth of photonics in information technology in the last several decades and well into this millennium, the question to be asked is which field can further play a pivotal role in the advancement of twenty-first century information technology. Surely, the answer to this question can be controversial. However, most people would agree that one vivid phenomenon that the world has witnessed in recent years is that, while mega-scale long-distance optical cables are replacing century-old copper cables, optical wires and devices are increasingly replacing metal-based electrical wires and devices on the micro/nanoscale in information systems, breathing new life into the information technology arena.

The concept of VLSI micro/nanophotonics may be a bold proposition at the present stage especially in light of the well-established VLSI micro/nanoelectronics. However, it is becoming an increasingly important concept in modern times as the demand for larger capacity information technology rapidly rises. Further, as the components of VLSI micro/nanophotonics can be compact, lightweight, and portable, and can consume low power, be wearable, and environment friendly (similarly to components of VLSI micro/nanoelectronics), they can be in compliance with global issues such as green technology, environmental technology, and biotechnology, now and in the future. VLSI micro/nanophotonics is thus becoming a new frontier of exploration in the new millennium.

The new demand for this technology entails technological and scientific issues, and evolutionary and revolutionary challenges, which we discuss in this book. These include issues such as miniaturization, interconnection, and integration of photonic devices on a micron, submicron, and nanometer scale. While many associated issues are the subject of continuing study of mega-scale optical fiber communication technology, new needs require these issues be dealt with at the micro/nanoscale. Throughout the course of this book, we closely follow the experiences of VLSI micro/nanoelectronics technology as well as knowledge from mega-scale optical technology so that they can provide insight from which micro/nanophotonics can benefit.

This book is intended only as an introduction to the field. There were many additional subjects that could have been covered, such as in-depth photonic circuit theories; however, this book cannot represent the complete story of this emerging field. It can only be a small compendium of explorations and advances in the field. In light of the phenomenal growth and development in the technologies that make up the field of micro/nanoscale photonic devices, it is notable that there is very little effort to develop integration technologies for such devices. Thus, from the point of view of integration of micro/nanoscale photonic devices, the publication of this book is an important and necessary first step to share an early stage perspective in this field.

Many prominent scholars and practitioners from around the world were invited to contribute to this book, and while a significant number participated, the works of some others could not be included, not only due to limitations in space, but also due to other commitments currently preventing us from including some aspects of this technology. This will be reflected in future books of this series.

We hope this book will be a good starter reference for those who are interested in this technical field. The book covers both basic and practical aspects of the technology. Thus, it will be useful not only for students and scholars but also for practitioners, explorers, and industrialists in diverse fields. We hope that readers can provide criticism and insight so that our collective thinking can advance this technology to a more mature one, not only for our generation but also for generations to come.

Finally, and most importantly, we would like to express our gratitude to all the contributing authors. Without their expertise, extraordinary enthusiasm, willing spirit, participation, dedication, support, and inspiration, this book would not have been possible. We are truly grateful to them for their extraordinary effort.

El-Hang Lee
Louay Eldada
Manijeh Razeghi
Chennupati Jagadish

Acknowledgments

The vision, conception, planning, thoughts, editing, writing, and consolidation of this book required lots of inspiration, encouragement, and support from a number of enlightened and very wise individuals. It has also taken a lot of time and effort from our family members and loved ones, without whom this book would not have been possible. We would like to express our sincere thanks and gratitude to them.

Professor Lee wishes to thank Professor John B. Fenn (Professor Emeritus, Yale University, Nobel Laureate, chemistry, 2002, Professor Richard K. Chang (Henry Ford II Professor Emeritus, Yale University, applied physics), Professor N. Bloembergen, Harvard University, Nobel Laureate, physics, 1981), and other professors at Yale University and Seoul National University, who have given him a solid educational foundation and inspiration for a lifetime pursuit of science; many of his former mentors and colleagues, including Professor Steven L. Bernasek (Princeton University), Professor George A. Rozgonyi (North Carolina State University), Professor Lionel C. Kimerling (Massachusetts Institute of Technology), Professor David A. B. Miller (Stanford University), Dr. Alastair M. Glass (Deputy Minister, Research and Innovation, Ontario, Canada), Dr. Thomas P. Pearsall (European Photonics Industry Consortium), Dr. Sang Hyun Kyong (former Minister of Information and Communication, South Korea), Dr. Seung Taek Yang (former Minister of Information and Communication, South Korea), and Professor Un-Chul Paek (GIST Chair Professor), who, since the days of AT&T Bell Laboratories, have given him immeasurable support and nurturing throughout his career; and tens and hundreds of his colleagues in South Korea and from around the world whose names cannot be mentioned here due to limited space. Professor Lee also thanks his dear parents, Professor Pyung Sup Lee (Ewha) and Aae Y. Lee; his wife, Professor Namsoo Chang (Ewha); and his children, David and Jennifer, for their loving support.

Dr. Louay Eldada wishes to express special thanks to Larry Bossidy (former CEO, Honeywell, former vice chairman, General Electric) who instilled in him the fundamentals for successful development and commercialization of advanced technology in a competitive global economy; he also wishes to thank Dr. Richard M. Osgood, Jr. (Professor, Columbia University), Dr. James T. Yardley (Professor, Columbia University; former director, Honeywell), Dr. Thomas M. Connelly, Jr. (executive vice president and chief innovation officer, DuPont), Dr. E. James Prendergast (executive director & COO, IEEE; former CTO, DuPont Electronics and Communications), and Dr. Arno Penzias (Nobel Laureate, physics, 1978; former chief scientist, Bell Labs) for the years of exciting innovation together. Dr. Eldada also thanks his mother Adela, his sister Rania, his wife Katharina, and his daughter Seraina for their loving support.

Professor Jagadish wishes to express gratitude to his late parents, C. Dharma Rao and C. Vimala Kumari, for their love and encouragement; sincere thanks to his teachers C. Sambi Reddy and the late V. Srinivasa Rao for his early education and their support; and special thanks to his wife Vidya, daughter Laya, sister Jhansi, and brother-in-law Pattabhi for their love and support.

Editors

Professor El-Hang Lee graduated from Seoul National University with a BSEE (summa cum laude) in 1970 and received his MS, MPhil, and PhD in applied physics from Yale University in 1973, 1975, and 1977, respectively, under the guidance of Professor John B. Fenn (Yale, Nobel Laureate, chemistry, 2002) and Professor Richard K. Chang (Henry Ford II Professor, former student of Professor N. Bloembergen, Harvard, Nobel Laureate, physics, 1981). Professor Lee subsequently conducted teaching, research, and management in the fields of semiconductor physics, materials, devices, optoelectronics, photonics, and optical communication at Yale University, Princeton, MEMC, AT&T, ETRI (Electronics and Telecommunications Research Institute; vice president), and KAIST (Korea Advanced Institute of Science and Technology). In 1999, Professor Lee joined INHA University, where he has been a distinguished university professor, founder of the School of Information and Communication Engineering, dean of the Graduate School of Information Technology, founding director of the micro/nano-PARC (Photonics Advanced Research Center), and founding director of the OPERA (Optics and Photonics Elite Research Academy) National Research Center of Excellence for VLSI Photonics. Professor Lee served as vice president of the Optical Society of Korea and as the founding president of IEEE/LEOS-Korea, founding director of SPIE-Korea, and founding editor of the international version of *ETRI Journal*. Professor Lee is an author and coauthor of more than 280 international journal papers and 840 worldwide conference and proceeding papers. He has edited several proceedings and has written one book. He has delivered nearly 150 plenary, keynote, and invited talks in international conferences and institutions. Professor Lee also holds over 130 international patents, and has served over 100 times as the conference chair, organizer, committee member, and international advisor for international conferences. He is an elected fellow of APS (American Physical Society), IEEE, OSA (Optical Society of America), SPIE (Society of photographic Instrumentation Engineers, International Society for Optical Engineering), IET (formerly IEE, United Kingdom), KPS, OSK, and IEEK, and a life fellow of KAST (Korean Academy of Science and Technology). Professor Lee is a recipient of more than 15 national and international awards, including the National Medal of Mok-Ryun Civic Order, National Medal of Science, and King Sejong Prize (all from the president of Korea); Grand Science Award; IEEE Third Millennium Medal (IEEE); Best Research Award (Korea Federation of the Academic and Professional Societies); and Distinguished Lecturer Award (IEEE). He is currently serving as the editor in chief of *IEEE Photonics Technology Letters*.

Dr. Louay Eldada is the chief technology officer of HelioVolt Corporation, Austin, Texas, where he leads the development and manufacture of compound semiconductor-based photovoltaic modules and systems. He holds a PhD from Columbia University, specializing in optoelectronic devices, modules, and systems. Dr. Eldada started his professional career with Honeywell International, where he founded the Telecom Photonics venture and directed its research and development arm for six years. The group's success led to its acquisition by Corning Inc., where he continued to manage technical development. After leaving Corning Inc., Dr. Eldada founded Telephotonics Inc., a start-up company, where he took on the responsibilities of chief technical officer and vice president of engineering for the development and manufacture of innovative organic–inorganic highly integrated optoelectronic integrated circuits and systems. The commercial success of Telephotonics Inc. in two years after its launch led to its acquisition by E. I. du Pont de Nemours and Company (DuPont)®, where Dr. Eldada served as chief technical officer and vice president of engineering at DuPont Photonic Technologies for six years. Dr. Eldada has published 220 technical papers, books, and book chapters. He has also organized and chaired 150 conferences; presented plenary, keynote, and invited talks at 160 conferences worldwide; garnered 47 technical awards; and holds 44 patents.

Professor Manijeh Razeghi (fellow, IEEE) received her Doctorat d'État es Sciences Physiques from the Université de Paris, France, in 1980. After receiving her doctoral degree, she joined Thomson-CSF (Orsay, France) as a senior research scientist and then served as the head of the Exploratory Materials Laboratory from 1980 to 1991. While at Thomson, Dr. Razeghi developed and implemented modern metalorganic chemical vapor deposition (MOCVD) epitaxial growth for entire compositional ranges of III–V compound semiconductors and heterostructures. She realized the first InP quantum wells and superlattices and demonstrated the marvels of quantum mechanics in the low-dimensional world. Her pioneering work on InP-based compound semiconductors culminated with the publication of *The MOCVD Challenge Volume 1: A Survey of GaInAsP-InP for Photonic and Electronic Applications* in 1989. She joined Northwestern University, Evanston, Illinois, as a Walter P. Murphy Professor and Director of the Center for Quantum Devices in the fall of 1991, where she created the undergraduate and graduate programs in solid-state engineering. She is one of the leading scientists in the field of semiconductor science and technology, pioneering in the development and implementation of major modern epitaxial techniques. Her current research interest is in nanoscale optoelectronic quantum devices. She has authored or coauthored more than 1000 papers, 12 books, and more than 30 book chapters. She holds 30 U.S. patents and has given more than 1000 invited and plenary talks. She received the IBM Europe Science and Technology Prize in 1987, the Achievement Award from the Society of Women Engineers (SWE) in 1995, the R.F. Bunshah Award in 2004, and many best paper awards. Dr. Razeghi is an elected fellow of SWE (1995), SPIE (2000), International Engineering Consortium (IEC) (2003), OSA (2004), APS (2004), Institute of Physics (IOP) (2005), Institute of Electrical and Electronics Engineers (IEEE) (2005), and Materials Research Society (MRS) (2008).

Professor Chennupati Jagadish was born and educated in India and worked in India and Canada, prior to moving to Australia in 1990. He is currently a federation fellow, professor, and head of the Semiconductor Optoelectronics, Nanotechnology and Photovoltaics Group in the Research School of Physical Sciences and Engineering, Australian National University. He is also the convenor of the Australian Research Council Nanotechnology Network, the director (ACT Node) of the Australian National Fabrication Facility, and the president of the IEEE Nanotechnology Council (NTC). Professor Jagadish serves on editorial boards of 12 international journals in addition to being a founding editor of the *Online Journal of Nanotechnology*, editor of *IEEE Electron Device Letters*, editor of *Progress in Quantum Electronics*, and editor of the *Journal of Semiconductor Technology and Science*. He served as an associate editor of the *IEEE/OSA Journal of Lightwave Technology* (2003–2008) and as an associate editor of the *Journal of Nanoscience and Nanotechnology* (2000–2005). He advises many high-tech companies in Australia and overseas, and has collaborated with scientists from 20 countries. His research interests include quantum dots, nanowires, quantum dot lasers, quantum dot photodetectors, quantum dot photonic integrated circuits, photonic crystals, THz photonics, plasmonics, and third-generation photovoltaics. He has published more than 590 research papers (390 journal papers), holds 5 U.S. patents, and has coauthored and coedited a book. He has also edited 12 conference proceedings and guest edited special issues of 5 journals. He has chaired many international conferences and has served on many professional society committees. He is the recipient of the 2000 Institute of Electrical and Electronics Engineers, Inc. (IEEE) Millennium Medal and has received distinguished lecturer awards from both the IEEE Lasers and Electro-Optics Society and the IEEE Electron Devices Society; he has also received the

Peter Baume Award from the Australian National University (ANU). Recently, he received the prestigious Australian Laureate Fellowship from the Australian Government. He is a fellow of the Australian Academy of Science, the Australian Academy of Technological Sciences and Engineering, IEEE, the American Physical Society, the American Association of Advancement of Science, the Optical Society of America, the SPIE (International Society for Optical Engineering), the Electrochemical Society, the American Vacuum Society, the Institute of Physics (United Kingdom), the Institute of Nanotechnology (United Kingdom), the Institute of Engineering and Technology (United Kingdom), and the Australian Institute of Physics.

Contributors

Vilson Rosa Almeida
Instituto Tecnologico de Aeronautica
Sao Paulo, Brazil

and

Instituto de Estudos Avancados
Universidade de Sao Paulo
Sao Paulo, Brazil

Elisa Antolín
Instituto de Energía Solar
Escuela Técnica Superior de Ingenieros de
    Telecomunicación
Universidad Politécnica de Madrid
Madrid, Spain

Ulaş Kemal Ayaz
Department of Mechanical Engineering
Southern Methodist University
Dallas, Texas

Oliver Benson
Department of Physics
Humboldt University of Berlin
Berlin, Germany

Pierre Berini
School of Information Technology and
    Engineering
University of Ottawa
Ottawa, Ontario, Canada

and

Department of Physics
University of Ottawa
Ottawa, Ontario, Canada

Peter Blood
School of Physics and Astronomy
Cardiff University
Cardiff, Wales, United Kingdom

R. Chatterjee
Optical Nanostructures Laboratory
Center for Integrated Science and Engineering
Columbia University
New York, New York

Mario Dagenais
Department of Electrical and Computer
    Engineering
University of Maryland
College Park, Maryland

Daoxin Dai
Center for Optical and Electromagnetic Research
Zhejiang University
Zhejiang, China

Pierre-Yves Delaunay
Center for Quantum Devices
Northwestern University
Evanston, Illinois

Louay Eldada
HelioVolt Corporation
Austin, Texas

Lan Fu
Department of Electronic Materials Engineering
Research School of Physical Sciences and
    Engineering
The Australian National University
Canberra, Australian Capital Territory, Australia

F. Y. Gardes
Advanced Technology Institute
University of Surrey
Guildford, United Kingdom

S. Götzinger
Laboratory of Physical Chemistry
Eidgenössische Technische Hochschule Zurich
Zurich, Switzerland

Elmar Griese
Department of Electrical Engineering and
    Informatics
University of Siegen
Siegen, Germany

Jian-jun He
Center for Optical and Electromagnetic Research
Zhejiang University
Zhejiang, China

Sailing He
Center for Optical and Electromagnetic Research
Zhejiang University
Zhejiang, China

and

Division of Electromagnetic Engineering
Royal Institute of Technology
Stockholm, Sweden

C. Jagadish
Department of Electronic Materials Engineering
Research School of Physical Sciences and
    Engineering
The Australian National University
Canberra, Australian Capital Territory, Australia

S. Kawanishi
NTT Basic Research Laboratories
Atsugi, Japan

S. Kocaman
Optical Nanostructures Laboratory
Center for Integrated Science and Engineering
Columbia University
New York, New York

Sanjay Krishna
Electrical and Computer Engineering
    Department
Center for High Technology Materials
University of New Mexico
Albuquerque, New Mexico

E. Kuramochi
NTT Basic Research Laboratories
Atsugi, Japan

Adnan Kurt
Teknofil Ltd.
Istanbul, Turkey

El-Hang Lee
Graduate School of Information Technology
Inha University
Incheon, South Korea

Hyun-Shik Lee
Graduate School of Information Technology
Inha University
Incheon, South Korea

Antonio Luque
Instituto de Energía Solar
Escuela Técnica Superior de Ingenieros de
    Telecomunicación
Universidad Politécnica de Madrid
Madrid, Spain

Antonio Martí
Instituto de Energía Solar
Escuela Técnica Superior de Ingenieros de
    Telecomunicación
Universidad Politécnica de Madrid
Madrid, Spain

G. Z. Mashanovich
Advanced Technology Institute
University of Surrey
Guildford, United Kingdom

A. Mazzei
Department of Physics
Humboldt University of Berlin
Berlin, Germany

Ryan McClintock
Center for Quantum Devices
Northwestern University
Evanston, Illinois

J. F. McMillan
Optical Nanostructures Laboratory
Center for Integrated Science and Engineering
Columbia University
New York, New York

Omer G. Memis
Bio-Inspired Sensors and Optoelectronics Lab
Electrical Engineering and Computer Science
    Department
Northwestern University
Evanston, Illinois

L. de S. Menezes
Departamento de Fisica
Universidade Federal de Pernambuco
Recife, Brazil

M. M. Milosevic
Advanced Technology Institute
University of Surrey
Guildford, United Kingdom

Adam Mock
Department of Electrical
    Engineering—Electrophysics
University of Southern California
Los Angeles, California

Hooman Mohseni
Bio-Inspired Sensors and Optoelectronics Lab
Electrical Engineering and Computer Science
    Department
Northwestern University
Evanston, Illinois

S. Mokkapati
Department of Electronic Materials Engineering
Research School of Physical Sciences and
    Engineering
The Australian National University
Canberra, Australian Capital Territory, Australia

Bijan Movaghar
Center for Quantum Devices
Northwestern University
Evanston, Illinois

Binh-Minh Nguyen
Center for Quantum Devices
Northwestern University
Evanston, Illinois

M. Notomi
NTT Basic Research Laboratories
Atsugi, Japan

John O'Brien
Department of Electrical
    Engineering—Electrophysics
University of Southern California
Los Angeles, California

Yusuke Ogura
Department of Information and Physical
    Sciences
Graduate School of Information Science and
    Technology
Osaka University
Osaka, Japan

Roberto Ricardo Panepucci
Electrical and Computer Engineering
    Department
Florida International University
Miami, Florida

David V. Plant
Department of Electrical and Computer
    Engineering
McGill University
Montreal, Quebec, Canada

C. E. Png
Institute of High Performance Computing, A*Star
Singapore, Singapore

Manijeh Razeghi
Center for Quantum Devices
Northwestern University
Evanston, Illinois

G. T. Reed
Advanced Technology Institute
University of Surrey
Guildford, United Kingdom

V. Sandoghdar
Laboratory of Physical Chemistry
Eidgenössische Technische Hochschule Zurich
Zurich, Switzerland

Henning Schröder
Fraunhofer Institute for Reliability and
    Microintegration
Berlin, Germany

Joshua D. Schwartz
Department of Electrical and Computer
    Engineering
McGill University
Montreal, Quebec, Canada

Ali Serpengüzel
Microphotonics Research Laboratory
Department of Physics
Koç University
Istanbul, Turkey

A. Shinya
NTT Basic Research Laboratories
Atsugi, Japan

Peter M. Smowton
School of Physics and Astronomy
Cardiff University
Wales, United Kingdom

Jun Song
Center for Optical and Electromagnetic Research
Zhejiang University
Zhejiang, China

Christopher J. Stanford
Department of Electrical and Computer
    Engineering
University of Maryland
College Park, Maryland

H. H. Tan
Department of Electronic Materials Engineering
Research School of Physical Sciences and
    Engineering
The Australian National University
Canberra, Australian Capital Territory, Australia

T. Tanabe
NTT Basic Research Laboratories
Atsugi, Japan

Jun Tanida
Department of Information and Physical
    Sciences
Graduate School of Information Science and
    Technology
Osaka University
Osaka, Japan

H. Taniyama
NTT Basic Research Laboratories
Atsugi, Japan

Thomas Vandervelde
Department of Electrical and Computer
    Engineering
Tufts University
Medford, Massachusetts

C. W. Wong
Optical Nanostructures Laboratory
Center for Integrated Science and Engineering
Columbia University
New York, New York

Tian Yang
Department of Electrical
    Engineering—Electrophysics
University of Southern California
Los Angeles, California

X. Yang
Optical Nanostructures Laboratory
Center for Integrated Science and Engineering
Columbia University
New York, New York

Yiğit Ozan Yılmaz
Department of Physics and Optical Science
University of North Carolina
Charlotte, North Carolina

G. Zumofen
Laboratory of Physical Chemistry
Eidgenössische Technische Hochschule Zurich
Zurich, Switzerland

I

Introduction

Introduction: A Preamble

El-Hang Lee

I.1 Introduction

The world is changing rapidly. We are now living in the information age. The whole world is networked by electrical wires, optical wires, and free-space electromagnetic waves [1]. Terabits of information flow through networks to transmit information from one corner of the world to the other, almost instantly. On the other hand, information of such magnitude is stored in, and retrieved from, tiny silicon chips that have ever-increasing capacities. The quest to store ever-larger information on the smallest possible chips and the quest to transmit such information to the longest possible distance in the shortest period through wired networks is never ending and is expected to continue until the fullest potential is realized. Information technology explores every possible means to increase information capacity, starting from the production of information to the consumption of information; it is also becoming faster, larger, more intelligent, humanized, personally oriented, multimedia savvy, and environment friendly, and is consuming less power. This trend has become more pronounced in recent years given the rising concerns on energy consumption and environmental issues.

Information capacity can be maximized by maximizing the multiplexing of time division, frequency division, polarization division, space division, code division, and others, using both electronics technology and photonics technology. Although very-large-scale integrated (VLSI) photonics technology can utilize most of these techniques, it deals particularly with the concept of space division. While the century-old, long-distance copper cables are continually being replaced with optical cables to meet the increasing demand for information capacity, a new challenge is now presenting itself, which is to solve the information bottleneck in the small areas that are in the centimeter to nanometer range. This book aims to systematically explore and utilize the miniaturization and integration of micro/nano-photonic devices and components to their maximum. It tries to bring together various disciplines of science and engineering to address the issues of miniaturizing and integrating micro- and nanoscale photonic devices, circuits, and networks for high-density photonic integration technology leading to VLSI photonics.

The challenges facing micro- and nanoscale utilization of space for information technology were laid down by Richard Feynman in the late 1950s [2]. The challenge was whether one could store a complete encyclopedia set on the tip of a clothespin. At this time, even semiconductor electronics technology was in its germinating stage. But the information storage technology using semiconductor-based electronics technology has made a lot of progress toward these goals. Many people now agree that as long as information technology progresses, the human dream to achieve the highest possible level of information storage capacity will not end. Then, the next logical question that inevitably entails this pursuit is how to get information in and out most effectively and efficiently from a device of such a small volume. After all, the scientific community has been delivering and obtaining information from small objects, such as atoms and molecules, by way of optical means called optical spectroscopy. The challenge here is to devise means to perform similar tasks using micro/nanoscale photonics on a board or on a chip. Figure I.1 shows a conceptual schematic diagram that incorporates these views. It depicts a large green-colored board on which wires and devices of micro/nanoscale are interconnected and integrated. There

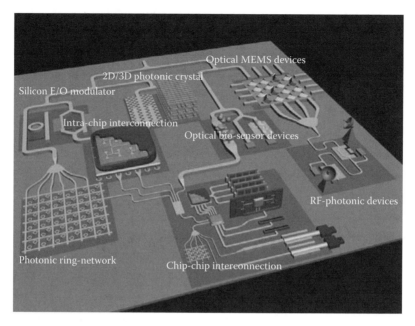

FIGURE I.1 A conceptual schematic diagram of an O-PCB on which VLSI-level micro/nanophotonic chips of diverse functions are interconnected and integrated by optical wires.

are wires of diverse thicknesses and devices of diverse functions. This diagram symbolically represents what this book addresses. Throughout this book, scientific and engineering solutions are sought and discussed to realize a vision of this nature.

Conceptually, this can be pursued by way of miniaturization, interconnection, and integration of devices of diverse functions. In order to realize photonic integration, two specific and practical concepts have been proposed [3–9]: the concept of "optical-printed circuit board (O-PCB)" and the concept of "VLSI photonic chip." This book examines the viability of both the O-PCB and the VLSI photonic integration on micron and submicron scales and their utilization for specific and practical applications. The generic approaches to the design and construction of O-PCBs are described in this book mainly for the purpose of building a platform for VLSI photonic circuits. The historical and global background that gave birth to O-PCB and VLSI photonics have already been well documented [3–9].

However, even before these concepts were developed, there have been historical efforts to realize photonic integration by way of micro-optics or integrated optics since the 1960s and 1970s [10,11]. These developments then evolved into a variety of disciplines like optoelectronic integration, integrated optoelectronics, optical integration, photonic integration, and integrated photonics, which continue to be subjects of intense study even to date [12–18]. O-PCBs and VLSI photonic chips are part of these continuing efforts and are intended as platforms to interconnect and integrate micro/nanoscale discrete devices for diverse functions and applications. This book discusses the scientific and engineering principles and methodologies to realize O-PCB and VLSI photonic technologies. Both the generic approaches and application-specific approaches have been considered.

I.2 Technological Approaches and Challenges

The process of optical and photonic integration can be conducted either in the form of monolithic integration or hybrid integration. From a historical perspective, both these methods have been pursued independently as well as interdependently. Although monolithic integration is well developed in

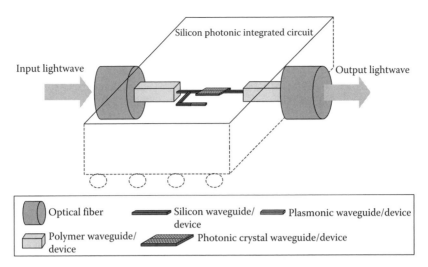

FIGURE I.2 A schematic diagram illustrating the interconnection of dissimilar micro/nanophotonic wires and devices on a board through which lightwave signal would travel from one end of an input optical fiber to the opposite end of an output optical fiber.

semiconductor-based integration technology, hybrid integration has not been well developed yet due to difficulties in interconnection and integration of devices in micro/nanoscale. While monolithic integration is better suited for most of integration applications, hybrid integration cannot be ignored. There are many technical problems to consider for hybrid integration even for monolithically prepared VLSI chips. It is highly beneficial to learn and utilize whatever technology is available from the already well-established electronics technology.

Figure I.2 shows a schematic diagram illustrating the interconnection of dissimilar micro/nanophotonic wires and devices on a board through which lightwave signal would travel from one end of an input optical fiber to the opposite end of an output optical fiber. The photonic devices may include photonic wires and devices made of polymer, silicon, silica, or other semiconductor materials, photonic crystal devices, plasmonic devices, micro-ring devices, or metamaterials, in differing shapes, sizes, materials, and characteristics. The mismatch problems have to be addressed for each pair of wires and devices to be interconnected. The issues include power mismatch, mode mismatch, polarization mismatch, size mismatch, shape mismatch, material mismatch, mechanical mismatch, thermal mismatch, and many other possible mismatches, including path mismatch, alignment mismatch, and angle mismatch. Examples of devices to be paired for interconnection and the mismatches thereof may include mismatches between the cylindrical optical fibers and rectangular polymer waveguides; between polymer waveguides and silicon waveguides; between rectangular polymer/silicon waveguides and photonic crystal waveguides and devices; between rectangular polymer/silicon waveguides and silicon nanowires; between rectangular polymer/silicon waveguides and plasmonic wires/devices; between slot waveguides and dielectric nanowires; and between plasmonic wires/devices and various nanoscale photonic, electronic, or bio devices. The mismatch issues have to be considered for monolithic integration too. Some of the chapters describe the efforts to analyze these problems and suggest some solutions.

In treating the integration issues, it would be useful to consider scaling rules for photonic miniaturization and integration. VLSI microelectronics technology has been mainly guided by two scaling rules: one for miniaturization and the other for integration (the so-called Moore's law). In VLSI photonics, however, no scaling rules have been established; the field requires systematic study, although there have been sporadic efforts [8]. Scaling rules will be helpful for the design of micro/nanophotonic wires and devices both in miniaturization and integration.

I.3 Building Blocks of VLSI Photonics

As briefly mentioned in the section above, a variety of basic building blocks have been developed and it would be useful to integrate them for a variety of practical applications. Some of the building blocks being pursued in recent years include nanowires [19,20], micro-ring resonator devices [21,22], photonic crystals [23–29], surface plasmon polaritons [30–36], silicon photonics [37], and metamaterials [38,39], paving the way for microscale to nanoscale integration.

Normally, in micron-size devices, the propagation of light through a waveguide is governed by the total internal reflection between two different dielectrics and is thus limited in the device size by the diffraction limit. However, structures like photonic crystals or plasmonics show promise of propagating light beyond the diffraction limit. As the propagation of lightwaves by these two means is not governed by the total internal reflection, they can realize light confinement in the regime below the diffraction limit and can open ways to the miniaturization of photonic devices in the nanoscale regime.

Silicon for high-density photonic integration has attracted much interest since the early 1980s [37]. Silicon-based photonics offers great potential for low-cost manufacturing of photonics because standard semiconductor device fabrication tools and processes can also be used to fabricate photonic devices and circuits. Silicon, however, has some limitations to overcome. The extremely low electro-optic coefficient, for example, usually makes the device too long and big. Also, silicon has limitations in light emission due to its nature of indirect electronic bandgap. To overcome these shortcomings, various structures, such as micro-ring resonators, photonic crystals, and strained silicon structures, have been proposed.

Micro-ring resonators consist of a ring-type curved waveguide with one or two straight waveguides coupled to it. Micro-ring resonators have attracted much attention in recent years for their potential use as novel optical components such as optical filters, switches, routers, and lasers. The advantage of a micro-ring resonator device is that it can be fabricated for VLSI photonic systems by cascading the resonators in various ring sizes ranging from several microns to several tens of microns in radii.

Photonic crystals are known to have capability to control the characteristics of the lightwave propagation, such as the path, speed, dispersion, and nonlinearity, and, as such, have been utilized for many functional devices such as optical filters, switches, routers, and lasers, allowing small feature size of functional devices and leading to potentially VLSI photonic systems. Also, lightwaves can be made to slow down or even to stop in the vicinity of photonic band edge due to anomalous dispersive relation, which has been studied to pursue optical switches and modulators, showing potentials for new functionalities.

Surface plasmon (SP) has emerged in the fields of optics, photonics, and bioscience after the pioneering work in the 1950s [30]. Unlike conventional total internal refection traveling at the interface between two dielectric materials, plasmonic waves, traveling along the dielectric interface between a dielectric material and a metallic material, is considered not limited by the diffraction limit, thereby having the potential to realize subwavelength photonic devices. Many studies have been devoted to reveal its mode confinement below the wavelength. Various functionalities, using slot structures of the surface plasmonics and directional coupling and Mach–Zehnder interference, have been reported. However, the strong confinement of lightwave in the SP structure is known to entail an extremely high propagation loss. One of the most practical issues in the SP is, therefore, to develop means of low propagation loss and strongly confined SP mode generation. One possible approach to minimize the propagation loss is to integrate a dielectric nanowire with SP. The coupling between the SP and a dielectric waveguide can allow effective subwavelength transmission in nonmetallic regions. A metal–insulator–metal structure, for example, can allow surface plasmon polaritons to travel over long distances, like tens of millimeters, with a strong mode confinement.

Another promising candidate is the quantum dot. Quantum dots have discrete energy spectra that depend on their sizes. Quantum dots can emit lightwaves of discrete colors, thereby allowing ways to make devices of diverse functions. This book discusses the characteristics of quantum dots of diverse materials and the ways to utilize them and their integrated devices for various functional applications.

In terms of material, currently, as well as in the past, the most commonly used materials for micro/nanophotonic integration include silica, silicon, polymers, compound semiconductors, metamaterials, metals, and other materials. Silica has been used for planar lightwave circuits, such as arrayed waveguide gratings, while polymer has been used as wires for board-level photonic integration. In recent years, silicon has been given much attention because it is a material that can utilize and can take advantage of the already well-established silicon-based electronic technology. Compound semiconductor materials are expected to be continually useful for photonics because they are also already well known and well established. Metamaterials have also been given much attention in recent years because of their potential applications in diverse engineering fields. Other materials that would increase lightwave confinement, and have low loss, high nonlinearity, high electro-optic coefficient, ease of processing, low cost, low power, and high volume processing would also continually be explored.

I.4 Potential Applications

From a historical perspective, VLSI electronics technology has penetrated virtually every corner of electronic applications. In the same manner, VLSI photonic technologies need to be made to penetrate whatever application areas they can. The applications can potentially be limited by the cost of the technologies. However, there are some areas where the technology will be required to be put into practice regardless of the cost of the technology.

The proposed VLSI photonic circuits can not only help electrical VLSI circuits, but they can also provide a platform to allow integration of diverse devices of information technology, biotechnology, and nanotechnology that can bring about new functions as applicable for telecommunication, data communication, information processing, aerospace, defense, transportation, environment, biomedical, and manufacturing systems. In telecommunication areas, the need for VLSI photonics may arise in the applications for high-speed and high-volume optical transmission, switching, network, routing, wavelength division multiplexing, polarization division multiplexing, and other active and passive functional components for wired and wireless applications. In the information processing areas, the need for VLSI photonics may arise in the applications for high-speed and high-volume information storage, imaging, monitoring, sensing, quake detection, security, data acquisition/analysis, and others. Biomedical applications may include diagnostics, data analysis, precision medical examination, medical imaging and processing, and information transmission. The applications in the transportation and aerospace technologies may include exploration, data acquisition, automatic control, guidance, image processing, and others. The applications in the energy technology may include energy harvesting, energy storage, energy distribution, smart and intelligent control of energy, and associated technologies.

In recent years, there have been efforts to develop a basis of photonic circuit theories. Electronics technology and VLSI electronic circuit technology have well-established theories on circuits and networks. However, photonics technology and especially VLSI photonic integrated circuit technology have not had the opportunity to develop theories on the micro/nanophotonic circuits and networks. In recent years, the development of photonic circuit theories has assumed some importance [40–42]. This book does not discuss this subject but it is an area to be addressed as the level of integration diversity and complexity rises.

I.5 Summary

We have briefly reviewed and described, from a historical and global perspective of information technology, how the concept of VLSI photonic technology has originated . In the context of these perspectives, we have described why the high-density photonic integration in micro/nanoscale is becoming an increasingly important subject in information technology. We have explained ways to approach the technology of micro/nanophotonic integration. We discussed the scientific and engineering issues and challenges for the miniaturization and integration of photonic wires and devices in micro/nanoscale, including scaling rules. We also discussed the building blocks of VLSI photonic integration and related materials thereof. Finally, we discussed the potential applications of VLSI photonic circuits. VLSI electronic circuits, which originated more than half a century ago and which prevailed over the then strong vacuum electronics technology, have penetrated almost every corner of modern-day electronics and information technology. VLSI photonics technology, when fully matured, can also progress similarly, taking information technology to new heights. This book aims to take the initial steps to that end by working out solutions to scientific and technological problems.

References

1. Welcome to the wired world—What the networked society means to you, your business, your country, and the globe. TIME, Special Report, February 3, 1997.
2. R. Feynman, There's plenty of room at the bottom, *American Physical Society, Annual Meeting*, California Institute of Technology, Pasadena, CA, December 29, 1959.
3. E. H. Lee, S. G. Lee, O. Beom-Hoan, and S. G. Park, Optical printed circuit board (O-PCB): A new platform toward VLSI micro/nano-photonics? *IEEE/LEOS, Summer Topical Meeting, Optical Interconnects and VLSI Photonics*, San Diego, CA, June 29–30, 2004.
4. E. H. Lee, S.G. Lee, O. Beom-Hoan, and S.G. Park, Optical printed circuit board (O-PCB): A new platform toward VLSI micro/nano-photonics? *IEEE-LEOS Newsletter*, 18(5), October 2004.
5. E. H. Lee, S. G. Lee, and O. Beom-Hoan, Microphotonics: Physics, technology, and an outlook toward the 21st century, *SPIE Proceedings*, 4580, 263, 2001.
6. E. H. Lee, S. G. Lee, and O. Beom-Hoan, VLSI microphotonics: Issues, challenges, and prospects, *SPIE Proceedings*, 4652, 1, 2002.
7. E. H. Lee, S. G. Lee, O. Beom-Hoan, and S. G. Park, Miniaturization and integration of micro/nano-scale photonic devices: Scientific and technical issues, *SPIE Proceedings*, 5356, 1–12, 2004.
8. E.-H. Lee, S.-G. Lee, O. Beom-Hoan, M.-Y. Jeong, K.-H. Kim, and S.-H. Song, Fabrication and integration of VLSI micro/nano-photonic circuit board, *Microelectronic Engineering*, 83, 1767, 2006.
9. H.-S. Lee, S.-M. An, Y. Kim, D.-G. Kim, J.-K. Kang, Y.-W. Choi, S.-G. Lee, O. Beom-Hoan, and E.-H. Lee, Fabrication of a 2.5 Gbps × 4 channel optical micro-module for O-PCB application, *Microelectronic Engineering*, 83, 1347–1351, 2006.
10. S. E. Miller, Integrated optics: An introduction, *Bell System Technical Journal*, 48, 2059, 1969.
11. D. Marcuse, Ed., *Integrated Optics*, IEEE Press, New York, 1973.
12. R. G. Hunsperger, *Integrated Optics: Theory and Technology*, Springer Verlag, Berlin, Germany, 1984.
13. T. L. Koch and U. Koren, Semiconductor photonic integrated circuit, *Journal of Quantum Electronics*, 27, 641, 1991.
14. D. A. B. Miller, Physical reasons for optical interconnection, *International Journal of Optoelectronics*, 11, 155, 1997.
15. E. Towe, Ed., *Heterogeneous Optoelectronic Integration*, SPIE Press, Bellingham, WA, 2000.
16. J. Takahara and T. Kobayashi, Low-dimensional optical waves and nano-optical circuits, *Optics and Photonics News*, 15(54), 54, 2000.
17. MIT Microphotonics Center, *Annual Report, Bringing New Technology to Light*, MIT Press, Cambridge, MA, 2000.
18. Annual Report, Inha University, OPERA (Optics and Photonics Elite Research Academy) Research Center for VLSI Photonic Integration, Incheon, South Korea.
19. C. M. Lieber, Nanoscale science and technology: Building a big future from small things, *MRS Bulletin*, 28, 486–491, 2003.
20. H.-G. Park, C. J. Barrelet, Y. Wu, B. Tiani, F. Qian, and C. M. Lieber, A wavelength-selective photonic-crystal waveguide coupled to a nanowire light source, *Nature Photonics*, 2, 622–626, 2008.
21. B. Little, S. Chu, J. Foresi, G. Steinmeyer, E. Thoen, H. A. Haus, E. P. Ippen, L. Kimerling, and W. Greene, Microresonators for integrated optical devices, *Optics and Photonics News*, 9, 32–33, December 1998.

22. R. K. Chang and Y. L. Pan, Linear and non-linear spectroscopy of microparticles: Basic principles, new techniques and promising applications, *Faraday Discussion*, 137, 9–36, 2008.
23. E. Yablonovitch and T. J. Gmitter, Photonic band structure: The face-centered-cubic case, *Physics Review Letters*, 63, 1950, 1989.
24. T. D. Happ, M. Kamp, and A. Forchel, Photonic crystal tapers for ultracompact mode conversion, *Optics Letters*, 26(14), 1102–1104, 2001.
25. A. Talneau, Ph. Lalanne, M. Agio, and C. M. Soukoulis, Low-reflection photonic-crystal taper for efficient coupling between guide sections of arbitrary widths, *Optics Letters*, 27(17), 1522–1524, 2002.
26. J. P. Hugonin, P. Lalanne, T. P. White, and T. F. Krauss, Coupling into slow-mode photonic crystal waveguides, *Optics Letters*, 32(18), 2638–2640, 2007.
27. V. Berger, Nonlinear photonic crystals, *Physical Review Letters*, 81, 4136–4139, 1998.
28. Y. A. Vlasov, M. O'Boyle, H. F. Hamann, and S. J. McNab, Active control of slow light on a chip with photonic crystal waveguides, *Nature*, 438, 65–69, 2005.
29. M. Soljacic and J. D. Joannopoulos, Enhancement of nonlinear effects using photonic crystals, *Nature Materials*, 3, 211–219, 2004.
30. R. H. Ritchie, Plasma losses by fast electrons in thin films, *Physical Review*, 106, 874–881, 1957.
31. E. Betzig, J. K. Trautman, T. D. Harris, J. S. Weiner, and R. L. Kostelak, Breaking the diffraction barrier: Optical microscopy on a nanometric scale, *Science*, 251, 1468, 1991.
32. W. L. Barnes, A. Dereux, and T. W. Ebbesen, Surface plasmon subwavelength optics, *Nature*, 424, 824–829, 2003.
33. M. Hochberg, T. Baehr-Jones, C. Walker, and A. Scherer, Integrated plasmon and dielectric waveguides, *Optics Express*, 12, 5481–5486, 2004.
34. J. R. Krenn and J. C. Weeber, Surface plasmon polaritons in metal stripes and wires, *Philosophical Transactions of the Royal Society of London Series A*, 362, 739, 2004.
35. W. Nomura, M. Ohtsu, and T. Yatsui, Nanodot coupler with a surface plasmon polariton condenser for optical far/near-field conversion, *Applied Physics Letters*, 86(18), 1108, 2005.
36. L. Yin, V. K. Vlasko-Vlasov, J. Pearson, J. M. Hiller, J. Hua, U. Welp, D. E. Brown, and C. W. Kimball, Subwavelength focusing and guiding of surface plasmons, *Nano Letters*, 5, 1399, 2005.
37. R. Soref and B. Bennett, Electrooptical effects in silicon, *Journal of Quantum Electronics*, 23, 123–129, 1987.
38. V. M. Shalaev, Optical negative-index metamaterials, *Nature Photonics*, 1, 41–48, 2006.
39. S. Linden, C. Enkrich, G. Dolling, M. W. Klein, J. Zhou, T. Koschny, C. M. Soukoulis, S. Burger, F. Schmidt, and M. Wegener, Photonic metamaterials: Magnetism at optical frequencies, *IEEE Journal of Selected Topics in Quantum Electronics*, 12, 1097–1105, 2006.
40. N. Engheta, A. Salandrino, and A. Alù, Circuit elements at optical frequencies: Nanoinductors, nanocapacitors, and nanoresistors, *Physical Review Letters*, 95, 095504, 2005.
41. E. H. Lee and VLSI photonics, Tutorial Lecture, *Optoelectronic and Optical Communication Conference (OECC)*, Hong Kong, China, July 13–28, 2009.
42. E. H. Lee, Optical printed circuit board (O-PCB) and VLSI photonics, The IEEE Photonics Society, Distinguished Lecture Series, 2007–2009.

II

Scientific and Engineering Issues

1

Optoelectronic VLSI

David V. Plant
Joshua D. Schwartz

1.1 Introduction

In this chapter, we will address the subject of optoelectronic very-large-scale integration (OE-VLSI) by describing some relevant designs, fabrication steps, and systems based on a number of successful demonstrations. The significance of this topic cannot be overstated: electrical interconnections, whether inter- or intra-chip, on a system bus at the circuit board level, or even the copper cable that runs into a household, are rapidly becoming critical bottlenecks in the pursuit of higher-speed and higher-bandwidth communication. The Semiconductor Industry Association (SIA) predicts in its technology roadmap [1] that the near future will see the exhaustion of the capabilities of electrical interconnects to move data at the rates required. It is anticipated that optically enabled electronics capable of optical input and output (I/O) can help bridge this gap. The advantage of fiber-optic cable in long-distance interconnections is, of course, already very well known and forms the backbone of modern networks, but optical technology is well situated to challenge copper at shorter and shorter distances. An optically interconnected system can be a practical solution at several levels of the communication hierarchy—between backplanes, boards, chips, and, perhaps someday even at the lowest level, to route data across a single chip. At the moment, several methods have been proposed for evaluating quantitatively the benefits of such optical interconnections compared to copper at distances typically ranging from millimeters (between chips) to meters (between backplanes).

One of the principal connectivity advantages that the optical approach brings to bear is believed to be in the deployment of massively parallel interconnects. High bandwidth links can be established between high-performance electronics through the deployment of parallel optical interconnects. Moreover, compared to copper, optical connections use lighter components at potentially lower costs, have low power consumption, are effectively guarded against electromagnetic interference, and have superior performance by way of reduced signal attenuation and distortion. The many material advantages have been thoroughly documented and can be explored in numerous journals [2–6]; some basic point-to-point parallel optical data links (usually with a dozen channels at 3.25 Gb/channel) can already be obtained commercially.

Generally speaking, the objective of OE-VLSI research is to enable optical I/O operating in unison with traditional electrical I/O systems by directly sourcing and terminating optical signals on silicon [7]. Recent developments have generated key enabling technologies toward this end. One critical development was that of the vertical cavity surface-emitting laser (VCSEL) as an optical source that is oriented in a manner favoring wafer-like production and assembly, making it easy to integrate large laser arrays with familiar silicon technologies. The successful integration of the VCSEL and corresponding PIN photodiodes (PDs) with application-specific integrated circuits (ASICs) [8–10] formed the basis for coining the term "OE-VLSI."

There are a great number of issues to be addressed in an OE-VLSI system. These include: materials processing, integration and compatibility with CMOS, transceiver designs, design of the physical optical link-up, and digital functionality. The complication of packaging for OE-VLSI offers a world of challenges unto itself, forcing the designer to deal with both electrical issues (the usual suspects: signal integrity and loss) and various optical issues (e.g., beam-steering and focusing in a relay system, optical clock detection and recovery, etc.). We will touch upon a few of these issues in this chapter, as we highlight recent work on certain aspects of OE-VLSI technology, emphasizing the development of nanoscale optoelectronic components, and based on successful experimental demonstrations of optically enabled ASICs. The research involved in these demonstrations involves a wide range of areas, encompassing some electronic circuit techniques, design and packaging, and optical device design. We will look at demonstrations for both short-range applications and also for all-optical networks with the goal of bringing fiber-optics directly into the home.

The chapter is organized as follows. Section 1.2 will provide an overview of the key design space criteria of OE-VLSI ASIC design. Section 1.3 will explore heterogeneous integration techniques, including fabrication and material parameters for realizing an OE-VLSI ASIC. Section 1.4 will outline OE-VLSI architectural options. Section 1.5 will explore the underlying electronic transceiver circuits, as demonstrated in CMOS technology, for the purposes of driving the arrays of optical sources and detectors. Section 1.6 will turn our attention to the discussion of burst-mode receivers capable of fast locking, which are expected to form an important part of the deployment of all-optical networking and fiber-to-the-home efforts. Finally, Section 1.7 will provide concluding remarks about the technology.

1.2 Targets of OE-VLSI Design

We begin by identifying the design space in which short-reach OE-VLSI technology is relevant and attractive as a solution in off-chip communications between chips or boards. The International Technology Road Map was consulted. Figure 1.1 condenses data from the relevant projections [1] regarding on- and off-chip clock-speed for VLSI, as well as the number of off-chip clock lines and overall I/O capacity for high-performance electronics. By 2014, off-chip clocks are expected to operate at 1.8 GHz, with an off-chip bus width of 3000 lines and a total capacity for off-chip I/O approaching 5 Tb/s. The on-chip clock is projected to be 13.5 Gb/s. This suggests a target design space that is illustrated in Figure 1.2, in which the data rate per channel to achieve an aggregate 5 Tb/s off-chip bus is plotted against the number of parallel channels. We are bounded by the anticipated on-chip clock rate of 13.5 Gb/s, which sets the maximum per-channel data rate without having to rely on some form of data serialization/de-serialization procedure. At this rate, the aggregate 5 Tb/s can be obtained from 370 parallel optical interconnects. If, on the other hand, we use the projected off-chip clock rate of 1.8 Gb/s, then 2700 optical I/Os become necessary to achieve the aggregate rate. These two numbers can act as rough bounds on the scale of the optical interconnect array in our target system. Two recently reported demonstrations of OE-VLSI ASICs, having 256 and 1080 optical I/O channels ([9] and [10], respectively), are close to the target range, while larger matrix-addressed VCSEL arrays have been demonstrated with 4096 outputs [11].

Having established a target scope, we underline a few basic premises popularly invoked in the design of OE-VLSI technology. First, the optical source is generally assumed to be a VCSEL operating at 850 or 960 nm with a corresponding P-N or PIN photodetector at the receiving end, although designs could

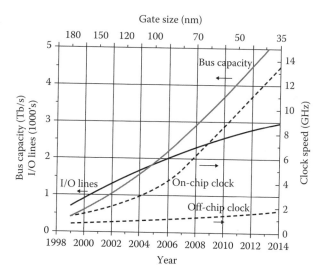

FIGURE 1.1 Projected evolution of on-chip and off-chip clock speeds (dashed lines, right axis), number of high speed I/O lines and total bus capacity (solid lines, left axis) as a function of time and transistor size for high-performance systems. (From Kirk, A. et al., *IEEE J. Sel. Top. Quantum Electron.*, 9(2), 531, 2003, Figure 2. With permission. © 2003 IEEE.)

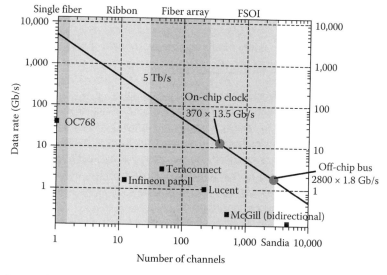

FIGURE 1.2 Projected off-chip I/O requirements for 2014. (From Kirk, A. et al., *IEEE J. Sel. Top. Quantum Electron.*, 9(2), 531, 2003, Figure 3. With permission. © 2003 IEEE.)

easily be adapted for other wavelengths. We anticipate that heterogeneous integration of optoelectronic devices with traditional CMOS circuits will be achieved through flip-chip bonding techniques with substrate removal.

Regarding the optical path itself, which may be either free-space or in optical fiber, we will be concerned with a point-to-point interconnect with a distance of less than 150 mm to represent a typical chip-to-chip or board-to-board distance. Some misalignment tolerance is expected and must be realistic.

An example of a point-to-point optical interconnect is shown in Figure 1.3a [12]. The system was experimentally realized as a free-space interconnect in which two optically enabled ASICs situated 3 in. apart formed a bidirectional link of 256 optical channels at a density of 28 channels per mm². This

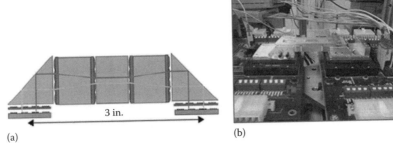

(a)   (b)

FIGURE 1.3   (a) Representative free-space optical interconnect. (From Plant, D. et al., *IEEE J. Lightw. Technol.*, 19(8), 1093, 2001, Figure 4(a). With permission. © 2001 IEEE.) (b) Photograph of an experimentally realized 512 channel inter-chip interconnect system. (From Kirk, A. et al., *IEEE J. Sel. Top. Quantum Electron.*, 9(2), 531, 2003, Figure 4(a). With permission. © 2003 IEEE.)

remains the highest reported density for a free-space optical interconnect at the time of this publication. A bookshelf configuration can be seen in the figure (photographed in Figure 1.3b); two prisms were used to relay the optical signals around the bends, and clustered diffractive lenses were used to perform proper imaging. The system is illustrative of the general hierarchy of components in an OE-VLSI system—some ASICs, a PCB or multi-chip-module substrate, an optical system for transport (free-space or fiber), and the packaging required for the link-up.

1.3 Heterogeneous Integration Techniques

A key enabler in OE-VLSI is the close integration of optoelectronic devices with silicon substrates. In this section, we describe methods for realizing OE-VLSI chips and summarize a heterogeneous integration strategy using flip-chip bonding.

To achieve the greatest possible density of optical I/O, techniques for heterogeneous integration with ASICs must first ensure the compatibility of the metals used in flip-chip bump bonding with the ASIC metals. Care must be taken as well to achieve a proper alignment between the connection points and the optoelectronic contacts on III–V material. To support the kind of highly dense optical interconnect shown in Figure 1.3, some 2-D arrays of VCSELs and corresponding PDs were fabricated on separate substrates before being integrated in an interleaved fashion onto the ASIC die. In Section 1.3.1, we describe the design and operating principles of the VCSELs and PDs and their heterogeneous integration strategies, as well as their layout configurations.

1.3.1 Design of the VCSEL and PD Devices

We articulate the process of assembling a system such as the one shown in Figure 1.3. The optical and electrical properties of suitable VCSEL and PD devices are summarized in Table 1.1, gleaned either directly from manufacturer data sheets (in the case of exact numbers) or otherwise directly measured values. Eight hundred and fifty nanometer (850 nm) VCSELs were popularly used in demonstration and feature typical threshold currents ranging between 1.0 and 4.5 mA with slope efficiencies of 0.25–0.35 mW/mA. VCSELs are often made backside-emitting to comply with the fact that they were to be inverted onto the corresponding ASIC driver circuits during the flip-chip process. This necessitates the removal of the GaAs substrate to allow outward transmission with a minimum of loss. To make this easier, VCSELs can be fabricated having n- and p-contacts on the top surface of the wafer in preparation to be flipped and bonded onto the die. Note that the n-contacted distributed grating reflector (DBR) becomes the top emitting surface once bonded to the ASIC. In Figure 1.4, we present a VCSEL layout geometry with an indicated direction of laser emission depicting the device after die-bonding and substrate removal. The n-contact can be brought to the same surface level as

TABLE 1.1 PD and VCSEL Optical and Electrical Properties

PD parameter		
Responsivity	R	0.317 A/W ± 108 mA/W
Junction capacitance	C_J	500 fF
Dark current	I_{dark}	6 nA
VCSEL parameter		
Threshold current (@25°C)	$I_{TH}|_{25°C}$	1.400 mA ± 75 µA
Slope efficiency (@25°C)	$\eta|_{25°C}$	0.430 W/A ± 10 mW/A
I_{TH} temperature dependence	$\partial I_{TH}/\partial T$	15.1 µA/°C
η temperature dependence	$\partial \eta/\partial T$	−2.2 mW/A°C
Differential resistance (2–5 mA)	$\partial R|_{2-5mA}$	71.7 Ω ± 12.1Ω
Threshold voltage (@2 mA)	$V_{TH}|_{2mA}$	1.580 V ± 17 mV
Junction capacitance	C_J	500

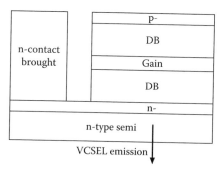

FIGURE 1.4 Schematic of the VCSEL geometry indicating the emission direction, post substrate removal. (From Plant, D. et al., *IEEE J. Lightw. Technol.*, 19(8), 1093, 2001, Figure 1. With permission. © 2001 IEEE.)

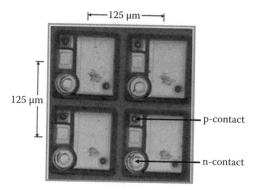

FIGURE 1.5 Photomicrograph of a 2×2 VCSEL sub-array (a) before and (b) after flip-chip bump bonding to a CMOS ASIC. The p- and n-contacts are indicated. The p- and n-contacts and the VCSEL active area are on a 125 µm horizontal and vertical pitch. (From Plant, D. et al., *IEEE J. Lightw. Technol.*, 19(8), 1093, 2001, Figure 2. With permission. © 2001 IEEE.)

the p-contact using mesa-isolation and ion implantation techniques. A microphotograph of such a structure can be seen in Figure 1.5 for a 2×2 array of VCSELs prior to being flipped and bonded onto a CMOS transmitter circuit. The pitch of the contacts and VCSELs in the figure is 125 µm, horizontally and vertically. In a similar manner, PDs were fabricated having the same pitch and square active areas of about 50 µm on a side, with 15×15 µm p- and n-contact pads. Figure 1.6 contains a

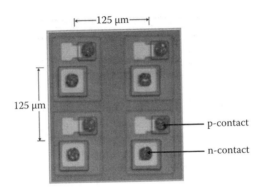

FIGURE 1.6   Photograph of a 2×2 PD sub-array (a) before and (b) after flip-chip bump bonding to a CMOS ASIC. The p- and n-contacts and a unit cell are indicated. (From Plant, D. et al., *IEEE J. Lightw. Technol.*, 19(8), 1093, 2001, Figure 3. With permission. © 2001 IEEE.)

microphotograph of a 2×2 PD sub-array prior to the flip-chip process. An identical contact area for both the VCSEL and PD contacts must be established on the CMOS die for proper performance after the flip-chip procedure.

1.3.2 Heterogeneous Integration

The processes involved in heterogeneous integration are fairly conventional photolithography-based operations for depositing and lifting off contact metals on the wafers, along with an appropriate precision alignment tool to complete the assembly process. The photolithography steps, performed first individually for the CMOS and the optoelectronic I/O devices, involve spinning photoresist polymer onto the wafer printed with the contact metal, followed by development. When indium is evaporated onto the wafer and the photoresist is removed, the contact pad metal remains in place. Individual VCSEL/PD devices are then mechanically diced, and an alignment hybridization tool used to match them up properly with the CMOS die. Following the example of Figure 1.4, the VCSEL is attached first to the CMOS die, and the substrate is then removed by dry etching, and the process gets repeated for the PDs in an interleaved fashion. Force and temperature control make possible a proper bond for the indium contacts of each die.

   This process yielded good electrical isolation between the individual VCSEL/PD elements on the die. Migrating these concepts to work at the wafer level is fairly straightforward, lending itself readily to large-scale arrays and high parallel channel counts that can meet the requirements laid out in Section 1.2.

1.4 OE-VLSI Architectures

Finding the right architecture for an OE-VLSI design is a matter of considering how their strengths and weaknesses play to the needs of the application at hand. The efficiency of the interconnect performance depends on the complexity of the processing to be performed on the data, as well as the amount of data to be handled. The major architectures to choose from are: pixelized, modular, or area-distributed optical padframe [13]. Each interacts differently with the automatic place and route software that would be used in designing an ASIC. We will explore here the pixelized and modular architectures.

1.4.1 Pixelized Architecture

The pixelized or "smart pixel" architecture essentially creates a single unit cell capable of both transmission and reception of data, and uses that as the basis for an array. Each unit cell contains a source, a detector, and the underlying transceiver electronics to operate them both, as well as whatever processing electronics are required to perform local operations on the data. This unit cell does not necessarily need

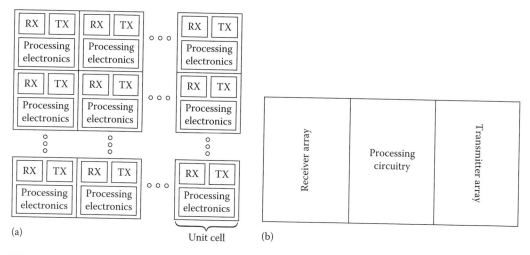

FIGURE 1.7 (a) The unit cell is tiled to form a two dimensional array of processing elements. (b) A modular layout methodology is shown in which the core processing circuitry and the optical transmitter and receiver arrays are physically separate modules.

to communicate with others in the grid. An illustration of this arrangement is depicted in Figure 1.7a. The pitch of the optoelectronics, and the space taken up by their transceiver, puts a limit on the complexity of the processing electronics that can be fitted into each unit cell.

This form of architecture is best suited to those operations that require only very basic data processing. Examples might include multi-chip interconnection fabrics [14], control and routing functions [9,15], and optical backplanes [16]. Other potential employers of this architecture include applications such as image processing [17,18] and binary neural associative memory [19], which handle large data sets but perform only simple instructions on the data. The pixelized architecture is *not* a suitable approach wherever high regularity row/column structures are employed for timing synchronicity or routing traffic, such as in a microprocessor core or in random access memory (RAM). The reason for this is primarily that such sensitive and complex circuit design would be difficult to adapt in design and layout to a coincident pixel matrix [13].

The point-to-point 256-channel interconnect illustrated in Figure 1.3 was built using a pixelized architecture. Several unit cells can be seen in Figure 1.8 containing interleaved VCSELs and PDs in a

FIGURE 1.8 Four clusters after flip-chip bump bonding and substrate removal. The VCSEL device was 110 μm × 112.5 μm and had a 20 μm in diameter active region. The PIN was 125 μm by 90 μm and had a 50 μm × 50 μm active region. (From Plant, D. et al., *IEEE J. Lightw. Technol.*, 19(8), 1093, 2001, Figure 6. With permission. © 2001 IEEE.)

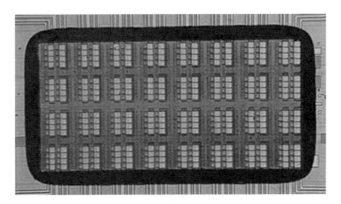

FIGURE 1.9   Photograph of the complete 256 element VCSEL and PD arrays integrated onto a CMOS ASIC. (From Plant, D. et al., *IEEE J. Lightw. Technol.*, 19(8), 1093, 2001, Figure 7. With permission. © 2001 IEEE.)

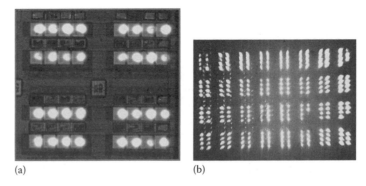

(a)   (b)

FIGURE 1.10   (a) Four clusters with 32 VCSELs forward biased (below threshold). (b) The entire VCSEL array biased above threshold. The distortion seen is caused by the optical system used to image the 3 mm × 6 mm VCSEL array. (From Plant, D. et al., *IEEE J. Lightw. Technol.*, 19(8), 1093, 2001, Figure 8. With permission. © 2001 IEEE.)

clustered configuration. Four rows of eight VCSEL and PD elements feature a pitch of 125 μm in the horizontal and 250 μm in the vertical direction between each successive cell. Each cluster of 8 I/O devices were pitched at 750 μm horizontally and vertically. There were 32 clusters comprising 4 columns and 8 rows for a total of 256 active optical channels on each ASIC. Figure 1.9 is a photograph of the completed ASIC—a 10 mm × 10 mm chip fabricated using a 0.35 μm process having only three metal layers. In Figure 1.10a, four clusters are shown in forward bias (subthreshold), while the entire 3 mm × 6 mm VCSEL array is presented in Figure 1.10b biased above threshold, with some slight distortion evident from the imaging system.

1.4.2 Modular Architecture

In contrast to the pixelized architecture, the core processing circuit, transmitter, and receiver arrays (including the optoelectronic devices and their corresponding transceiver circuits) can all be separated from each other in what is known as the modular architectural approach (Figure 1.7b) [10,13]. The pitch of the VCSELs/PDs determines how much complexity can be included in the transceiver circuits, but is now distinct from the digital core processor. The modular approach necessitates that the modules be themselves interconnected by wide and parallel electrical lines. This imposes a limit on the circuit performance—if the array is especially large, these electrical lines can be of significant length, incurring the very problem that optical interconnects are designed to circumvent; this is particularly bad news for the receivers, which must drive these long lines. To compensate, additional buffering circuitry can be

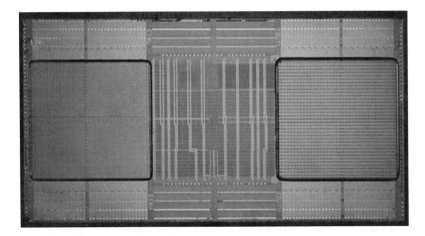

FIGURE 1.11 Photograph of the chip after flip-chip bump bonding of optoelectronic devices to an underlying CMOS device. The principal sections are shown: receiver array, digital section, and transmitter array. (From Venditti, M., *IEEE J. Sel. Top. Quantum Electron.*, 9(2), 61, 2003, Figure 1. With permission. © 2003 IEEE.)

used that will lower the latency involved in driving the electrical lines, or pipelining strategies can be deployed [20].

The advantage of this architecture is that there is no limit on the complexity of the data processing unit, located apart from the array. This architecture has been used in the development of a number of systems: fully interconnected switches [16,21], a crossbar switch [22], an static RAM chip and a fast Fourier transformer [23], a page-buffer [13], and a chip featuring pseudorandom bit sequence (PRBS) generation, forward error correction (FEC), and first-in first-out buffering [10]. As an example, this last chip was 14.6 mm × 7.5 mm and was built in a 0.25 μm process with single-poly and five metal layers; it is illustrated in Figure 1.11, with the key modules (transmitter, processor, and receiver) easily apparent. The optical element pitch was a 250 μm grid of 34 rows and 35 columns, comprising 1190 total devices (1080 of which were actually in use—the rest were not connected to circuitry).

1.5 Analog Circuits

The design of transceivers to drive the optoelectronic components of OE-VLSI is another area where significant research has been undertaken to explore the various topologies. Our intent in this section is not to propose general design procedures for OE-VLSI circuits, but rather to highlight a few specific designs that have been demonstrated to produce results. Many single-ended and differential designs have been demonstrated, but only a few will be explored in this section. We subdivide our discussion to consider laser drivers and receivers separately.

1.5.1 Laser Drivers

In this section, we will report on an optical transmitter circuit built in a 180 nm CMOS technology and designed to operate an 850 nm VCSEL at 5 Gb/s. Although this technology (and data rate) are not cutting-edge, this low-cost design exhibited a total power consumption of 18.35 mW, the lowest power consumption per bandwidth reported (3.67 mW/Gb/s) [24]. This low power consumption of the design helps to mitigate power dissipation issues present in large arrays of parallel optical interconnects. The circuit is compact and has very low susceptibility to power supply switching noise due to the architecture decision to use a dual power rail.

Using this example to motivate discussion, we will summarize the overall implementation of the transmitter—from architectural decisions to device modeling, simulation, and finally test and measurement.

1.5.1.1 Architecture

The choice of using CMOS technology is particularly attractive for transceiver design. There are many reasons, not the least of which is an effort to keep manufacturing and operating costs low. Designing for CMOS takes advantage of the fact that this is the traditional platform used for digital signal processing functions, and having the transceivers integrated on the same die is a significant cost-saving measure. Transceiver area should be kept minimized for the sake of parallel array-based operation in applications that use VCSEL pitches from 125 to 250 μm [25]. As a result, large on-chip resistors or inductors are to be generally avoided.

The specific design under discussion was built in a 180 nm CMOS process to help minimize manufacturing cost. In addition, an emphasis was placed on using NMOS-only circuits wherever possible for the best performance. This is because PMOS transistors are traditionally slower, and so their use was primarily restricted to being active loads. Detailed analysis of the transition frequency (f_T) for the technology was undertaken to identify the best performance point (even though such analysis is typically carried out for small-signal regime, a large-signal can see optimum performance operating near this frequency).

Noise performance is dramatically improved by choosing a differential topology to handle electrical signaling. Current-steering designs (which rely on redirecting in part the flow of one fixed current value rather than adjusting the value directly at the source) provide an additional advantage in transmitter design, as this generates reduced switching noise [25].

1.5.1.2 VCSEL Modeling

It is essential to properly model the behavior of the optoelectronic devices, both to get accurate simulation results and to properly establish a "dummy" load that accurately reflects the VCSEL during the current steering operation. In the specific research demonstration we are following, S-parameter measurements were undertaken on several Emcore Gigalase™ VCSEL samples at varying bias current levels using a high-frequency probe. Various S11 (reflection) measurements for three different dies and four different bias current levels are presented in Figure 1.12. *V–I* curves were used to calculate the differential resistances of each device for each bias point. It was decided upon to work with the 2 mA current model for further simulation in this case.

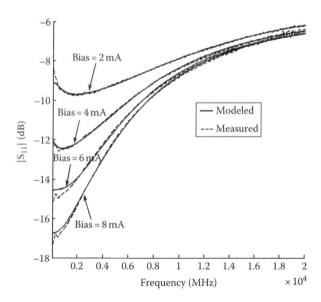

FIGURE 1.12 Experimental and modeled reflection parameter curves at four bias current levels.

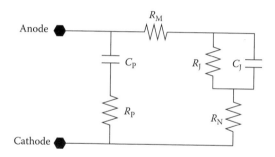

FIGURE 1.13   VCSEL electrical model topology.

TABLE 1.2   Experimental and Modeled VCSEL Differential
Resistance at Four Bias Current Levels

Bias Current (mA)	δR_{DC} (Experimental) (Ω)	δR_{DC} (Modeled) (Ω)
2	103.4	104.0
4	83.3	83.0
6	73.1	73.0
8	67.0	67.0

Figure 1.13 shows the electrical model of the VCSEL, whose topology is derived from common approaches [26,27]. The model features elements representing the relevant mirror (R_N and R_M), pad (R_P and C_P), and junction (R_J and C_J) resistances and capacitances.

The model parameters were fitted to the measured S11 and differential resistance data for each bias current, and were optimized using quasi-Newton and random search on the least-squares error function. The differential resistance of the model was then evaluated against the measured data as an accuracy check (as presented in Table 1.2), showing excellent agreement.

1.5.1.3 Transmitter Design

Our example transmitter circuit now had an accurate VCSEL model to drive and a general architecture (differential current-steering). The circuit itself was built in three-stages: a differential pre-amplifier circuit, a pulse-shaper, and a driver, as illustrated in Figure 1.14 [28].

The pre-amplifier seen here is implemented in a common-gate configuration with approximately a 100 Ω differential input resistance for matching. A differential input voltage of 0.4 V in peak amplitude was the driving input to the preamplifier, which featured an 11 dB gain (and a −3 dB bandwidth of about 6 GHz), driving the output signal peak to be near the 1.8 V power rail.

Following that, a common-source configuration with variable bias was used as a pulse-shaping circuit. As shown in Figure 1.14, the bias of this stage is adjusted in order to change the low-voltage seen by the output current driving stage. This has the effect of reducing the overall ringing at the output if a voltage is selected that is slightly higher than the voltage at node "X" (V_X in Figure 1.14) [29]. Therefore, it becomes possible to use a variety of different modulating currents without taking a significant performance hit.

The third stage is essentially a current driver having an adjustable range of currents between 0 and 4 mA. Since the optical output is single-ended, a dummy load (diode-connected PMOS) is employed in the differential stage to balance the operation.

The VCSEL itself requires a forward voltage of 2 V, which is already higher than the VDD (1.8 V) of the circuit. For this reason, a dual power-level was established by adding another power ring around the circuit at 3.3 V. The traditional approach of using only one supply voltage level for the entire range of transmission [30] has a detrimental effect on both the performance (since thick-oxide transistors are required) and power consumption. It is noteworthy, however, that having differing power supplies for

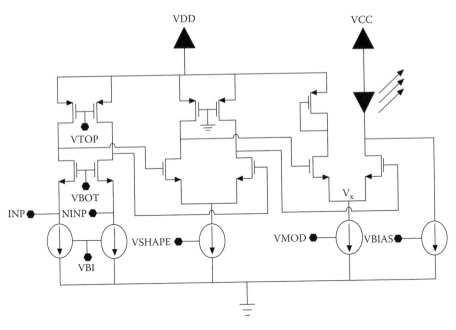

FIGURE 1.14   Three-stage optical transmitter schematic.

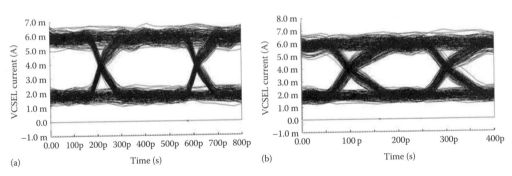

FIGURE 1.15   Simulated eye diagram at (a) 2.5 Gb/s and (b) 5 Gb/s.

two branches in the current-steering stage induces some unwanted power-rail switching noise, which was minimized by adding an on-chip coupling capacitor.

The optical transmitter extracted layout, with added models for bondwires and other aspects of the packaging, as well as the VCSEL model, was simulated with SpectreS at 55°C, since the temperature was expected to rise during operation and no cooling provisions were implemented. The test board itself was unsuitable for operation above 2.5 Gb/s, and so the decision was made to apply the 0.4 V low-voltage differential signal input directly to the transmitter through on-chip probing with 50 Ω high-speed probes.

Eye diagrams at 2.5 and 5 Gb/s were generated using a $2^{10}-1$ PRBS sequence and are presented in Figure 1.15. A 4 mA modulation current was applied around a central bias current of 2 mA (above the 1 mA threshold of the VCSEL). Rise and fall times were 120 and 110 ps, respectively, at 5 Gb/s. Total power consumption was expected to be 21.75 mW (16.8 mW was consumed by the VCSEL modulation and bias current).

1.5.1.4 Experimental Measurements

The final die was fabricated through the Canadian Microelectronics Corporation (CMC), working with the Taiwan Semiconductor Manufacturing Corporation (TSMC). Emcore VCSELs were packaged with it

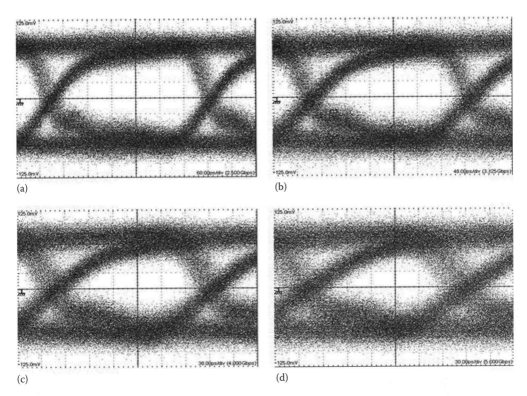

(a)   (b)

(c)   (d)

FIGURE 1.16 Measured eye diagram (250 mV span is displayed) after transmission and conversion at (a) 2.5 Gb/s (60 ps/div), (b) 3.125 Gb/s (48 ps/div), (c) 4 Gb/s (38 ps/div), and (d) 5 Gb/s (30 ps/div).

into a ceramic package mounted on a board to supply power and control voltages, while the chip signals were probed directly with 50 Ω coplanar microprobes. This demonstration chip was *not* flip-chip assembled, but used short (200 μm) bondwires to connect the VCSEL to the transmitter die. Many parallel wirebonds were used around the chip to connect to the voltage supply to keep inductance and resistance low.

The optical output was collected by butt-coupling a multimode optical fiber directly to the VCSEL aperture—an optical loss of about 1.76 dB, or a 66% efficiency, was established. The optical signal was picked up by a 12 GHz photoreceiver and sent to a communications signal analyzer. Measured eye diagrams are presented in Figure 1.16 at four different data rates with a $2^{31}-1$ length PRBS input sequence. With a signal swing of 150 mV and a 3.5 V modulation current swing, an average optical output power of about 2 mW was observed. The bit-error rate (BER) was measured for various data rates and PRBS lengths, as summarized in Table 1.3. Error-free operation at 3 Gb/s was established,

TABLE 1.3 Measured BER at Various Data Rates and PRBS Lengths

Data Rate (Gb/s)	PRBS Length	Bit Error Rate
1.250	$2^{31}-1$	**Error Free**
2.500	$2^{31}-1$	**Error Free**
3.125	$2^{31}-1$	**Error Free**
4.000	$2^{31}-1$	$< 1 \times 10^{-10}$
4.000	$2^{15}-1$	$< 1 \times 10^{-11}$
4.000	$2^{7}-1$	$< 1 \times 10^{-12}$
5.000	$2^{15}-1$	$< 1 \times 10^{-7}$
5.000	$2^{7}-1$	$< 1 \times 10^{-10}$

TABLE 1.4 Comparison of Optical Transmitter Characteristics with Previously Published Designs

	Kucharski [30]	Kromer [31]	Annen [32]	Thibodeau [24]
Technology	CMOS 130 nm	CMOS 80 nm	SiGe 350 nm	CMOS 180 nm
Supply voltage (V)	2.5	1.1	3.3	1.8
Max. data rate (Gb/s)	20	10	10	5
Power consumed (mW)	120	25	132	18.35
Power consumed per bandwidth (mW/Gb/s)	5.99	2.5	13.16	3.67

with the BER slipping a bit to 1×10^{-10} for the longest PRBS at 4 Gb/s; however, performance rapidly degraded by 5 Gb/s.

Overall power consumption, calculated by measuring the currents from each power supply, could be brought as low as 18.35 mW (independent of the data rate) by adjusting the bias current to the minimum.

1.5.1.5 Discussion

The values reported in the example described above compare favorably to other demonstrations with VCSEL-based transmitters, as evidenced in Table 1.4, which draws from previous works [30–32]. Power consumption is clearly the primary advantage of this design architecture. Only [31] had better power consumption per unit bandwidth—and that was in a far more aggressive 80 nm technology node with 1.1 V supply and low-threshold VCSELs. Although better data rates have been reported, this is primarily a reflection on their choices of more aggressive technology nodes in CMOS design, and the design presented here would scale readily to the 80 nm node. The choice of 180 nm technology was primarily motivated to keep a low cost overhead, which makes it attractive for manufacture.

One notable feature of the design highlighted is the fact that it contained multiple controllable parameters that could be adjusted manually—bias current, modulation current, and pulse-shaping was all externally controllable. This opens the way to feedback-controlled channel management by having the receiver dictate control of the transmitter to optimize channel integrity. This could compensate for a number of issues, from slight changes in optical path length or board misalignments developing over time. These adjustments would be low-frequency and would make it unnecessary to implement a more complicated automatic gain control circuit in the transmitter.

One issue of note: CMOS technologies are advantageous for cost and integration, but display poorer noise characteristics than their SiGe or silicon-on-insulator counterparts [33]. Chip-on-board packaging with high-quality printed circuit boards would obviously yield better performance for both noise and data rates, as compared to the use of ceramic packaging on an FR-4 board that cannot sustain beyond 2.5 Gb/s.

1.5.2 Receivers

We now turn our attention to the design of optical receivers, where we will explore one of the primary difficulties associated with operating an array of parallel detectors at high-speed: the problem of inter-channel skew. Skew exists when the channels in a receiver array exhibit different and unpredictable delays relative to each other, making synchronous operation much more difficult. Because a photodetector in an optical channel generates a DC-coupled input photocurrent (since the corresponding VCSEL is modulated around some "on" bias point), there is an obvious advantage in choosing a differential receiver topology to extract the modulated signal. Optically single-ended receivers controlled by some common bias in a large array would inevitably drift apart in their response due to variations in optical received power across the array.

1.5.2.1 Architectures

We will explore the designs of two fully differential preamplifier stages for an optical receiver that make use of common-mode feedback (CMFB) [34] in order to stabilize their operating point during variations of input DC photocurrent. One design is based on a common-gate (CGA) configuration with an active load. The other is a traditional transimpedance amplifier topology (TIA) with a resistor in the feedback branch. Schematics for both designs are presented in Figure 1.17.

Each of these designs shares a common feature—each includes a means to perform and validate performance before integration with the photodiode during assembly. This is achieved through the transistor MP (Figure 1.17), which acts in parallel with the photodiode during final assembly but serves as a source of simulated "photocurrent" even without the actual photodiode in place (as shown in the dashed lines of the figure). Active-low control inputs nV_{tl} and nV_{tr} serve to inject approximately $60\,\mu A$ of differential current into the preamplifier inputs when testing the circuit before integrating the PDs.

The TIA preamplifier contains an adjustable feedback mechanism (represented by resistors RF in Figure 1.17b) that is implemented on-chip using active devices, allowing it to be tuned through some digital control inputs (see Figure 1.18). A reference NMOS transistor (MR) is placed in parallel with four others (M0–M3) having varying width/length ratios. The reference is always "on," establishing a $16\,k\Omega$ resistance branch. The overall feedback resistance can then be reduced from this level by turning on (through digital control signals) some or all of M0–M3, each of which represents a resistance of 16, 8, 4, and $2\,k\Omega$. Sixteen different effective feedback resistances can be established, down to a minimum of $1\,k\Omega$.

In Figure 1.19, we present a block diagram of the receiver design (TIA and CGA), each of which essentially is comprised of four stages: the preamplifier, two post-amplifiers, a Schmitt trigger, and a line driver. The two post-amps are folded cascode differential amplifiers that amplify the received signal to

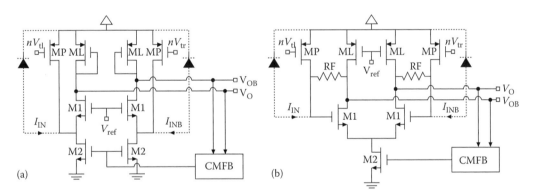

FIGURE 1.17 Transistor-level schematics for the (a) CGA and (b) TIA preamplifier circuits illustrating CMFB, test mode circuitry, and control inputs. The locations of the PDs after heterogeneous OED integration are indicated by dashed lines. (From Venditti, M., *IEEE J. Sel. Top. Quantum Electron.*, 9(2), 61, 2003, Figure 8. With permission. © 2003 IEEE.)

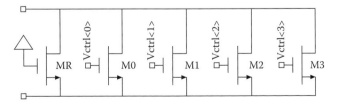

FIGURE 1.18 Transistor level schematic of TIA preamplifier feedback resistance RF. (From Venditti, M., *IEEE J. Sel. Top. Quantum Electron.*, 9(2), 61, 2003, Figure 9. With permission. © 2003 IEEE.)

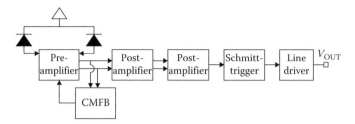

FIGURE 1.19   Receiver block diagram, applicable to either preamplifier design. (From Venditti, M., *IEEE J. Sel. Top. Quantum Electron.*, 9(2), 61, 2003, Figure 10. With permission. © 2003 IEEE.)

nearing the level of the power rails, as well as to convert the signal to a single-ended output. These stages use feedback to maintain a stable bias point [35]. The use of a Schmitt trigger is to provide better immunity to noise generated from switching on the power supply rail [9,36]. Finally, the output line driver is simply composed of cascaded inverters to drive the signal from the receiver to the nearest buffer, which was up to 2 mm away at the target data rate, which for this design was 250 Mb/s. A power dissipation of about 9 mW in each receiver was recorded.

1.5.2.2 Reducing Skew

As mentioned above, one of the challenges of managing parallel links is providing tolerances to such factors as misalignment or process variations in either the optoelectronic devices or the transceivers themselves across a large array of channels. One of the most significant problems is that DC photocurrent levels can vary from element to element in the receiver array, translating into a different latency in each channel. When latency varies unpredictably between elements, synchronous operation becomes extremely different—hence the introduction of a controlled feedback mechanism [37] described in the previous section to stabilize the bias current (and therefore delay) of each element, even as the DC photocurrent may vary.

An interesting challenge that arose in the investigation of this issue was "how do we accurately measure receiver latency and skew?" It would be difficult, under ordinary circumstances, to isolate and identify small changes in latency due to capacitive loading effects of invasive measurements. Instead, a novel approach was used that had the additional benefit of testing the entire signal path of an optical link, from transmitter to receiver. The approach was dubbed an optically enabled ring-oscillator (OERO), modeled after conventional electronic oscillators consisting of an odd number of inverters [38]. The idea is straightforward: two optically enabled chips (with VCSELs and PDs) are aligned "face-to-face," establishing a communication link, and they volley a signal back and forth between them with an inverter in the electrical signal path (Figure 1.20). An imaging system (not shown in the figure) consisting of a few choice lenses and beam splitters can image the optical power onto the appropriate photodetectors. The resulting system will oscillate at some natural frequency determined by the time it takes the signal to make a round trip.

To compare different receiver strategies, the CGA and TIA designs were implemented on a single chip, both with and without the feedback mechanism discussed above for DC photocurrent rejection (DCPR). Each corner of the chip featured one of four receiver designs, which were then wire-bonded to an adjacent bar of VCSELs and PDs, making each chip fully enabled for I/O. Once the system was powered up and aligned, it began to oscillate at a natural frequency. By measuring the frequency of oscillation of the ring, the *relative* changes in latency could be evaluated when different levels of optical power were transmitted (by inserting a simple attenuator in the imaging optics). The results are presented in Figure 1.21 and clearly show that, not only is overall latency improved with the DCPR technique, but also skew is reduced in the case of changing received power levels.

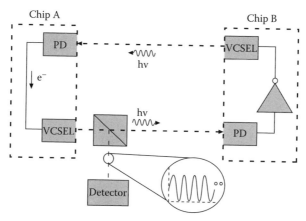

FIGURE 1.20 Diagram of an OERO system. Optical path (dashed line) between the two chips is tapped by a beam-splitter to observe the signal. Imaging system is not shown. (From Schwartz, J.D. et al., *Appl. Opt.*, 43(12), 2456, 2003, Figure 1(b). With permission.)

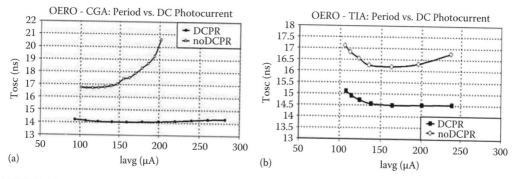

FIGURE 1.21 Demonstration of improved delay behavior (in absolute value and in stability) by examining the ring oscillation period T_{osc} vs. input photocurrent for systems using the DCPR technique in both CGA and TIA receiver architectures. (From Schwartz, J.D. et al., *Appl. Opt.*, 43(12), 2456, 2003, Figure 8. With permission.)

1.6 Burst-Mode Receivers

We will now move from the area of short-range interconnects to a much broader perspective, looking at how OE-VLSI can help shape the future of large-area networking. There is no question that to increase the speed and efficiency people experience in their Internet or local connections, fiber-optics will be required closer and closer to the end user. The ultimate goal is to deploy a network consisting solely of fast, low-cost optical components that can handle all the responsibilities of switching and routing traffic without the need for a cumbersome and bottlenecking conversion of the signal to the electronic domain at some central hub. For this reason, passive optical networks (PON) are seen as the future of the fiber-to-the-home initiative. This kind of uplink, which for each end user is a point-to-multipoint network, means that the kind of traffic on a PON, coming from any number of optical network units (ONUs), will be bursty in nature, having phase and amplitude variations. In order to handle this, the optical line terminal (OLT) located in the center hub will have to employ a burst-mode receiver (BMRx) [39,40]. The receiver must maintain a low packet loss ratio (PLR) to preserve the integrity of the channel. Here, we will explore the PLR performance of a BMRx as a function of the phase-step between consecutive packets, signal power, BER, and mode-partition noise penalty. There are distinct trade-offs between power penalty, resistance to pattern error, and the length of the signal "preamble" used to alert the receiver to incoming data.

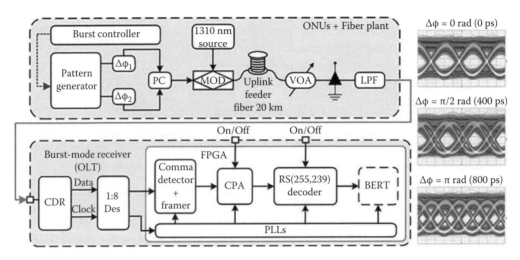

FIGURE 1.22   Block diagram of the 622 Mb/s GPON uplink experimental setup with the BMRx (Des, deserializer; PLL, phase-locked loop; BERT, BER tester). (From Shastri, B.J. et al., *IEEE Photon. Technol. Lett.*, 20(5), 363, 2008, Figure 1. With permission. © 2008 IEEE.)

1.6.1 BMRx Demonstration

A Gigabit-PON (GPON) experimental setup is presented in Figure 1.22. Bursty traffic is generated by a burst controller that inserts random phase steps and a number of consecutive identical digits (CID) between alternating packets coming from ports on a pattern generator. These signals are concatenated with a power combiner (PC) and drive an optical modulator (MOD) [40]. Each packet is composed as follows [41]:

- 16 guard bits
- 0–28 preamble bits
- 20 delimiter bits
- A $2^{15}-1$ bit PRBS payload
- 48 comma bits

In this demonstration, a 1310 nm laser is modulated with the generated data and sent down a 20 km spool of fiber. A variable optical attenuator (VOA) manages the received power level at the photodetector. The photocurrent output is then low-pass filtered (a fourth-order Bessel–Thomson filter, 3 dB frequency at 467 MHz). The BMRx itself includes multi-rate clock data recovery (CDR) supporting key frequencies of interest, a clock-phase aligner (CPA), a 1:8 deserializer, and forward error correction (FEC) implemented on an FPGA using a Reed–Solomon (255,239) (RS) method.

The CDR in the design accommodates 622 Mb/s, 666.43 Mb/s (with FEC for the 15/14 overhead introduced by the RS-codes), and 1.25 Gb/s. Lower-rate parallel data goes to the FPGA, where the payload is automatically detected through a framer and comma-detector. The CPA uses a phase-picking algorithm [40] and is turned on during PLR measurement that uses phase acquisition (otherwise, the CPA is turned off). Realigned data is sent to the RS decoder with FEC and a BER measurement taken.

1.6.2 Results and Discussion

The PLR performance of the link as a function of the phase-step inserted between bursty packets is presented in Figure 1.23 (first in back-to-back configuration, without the fiber or the CPA). The bell-shape

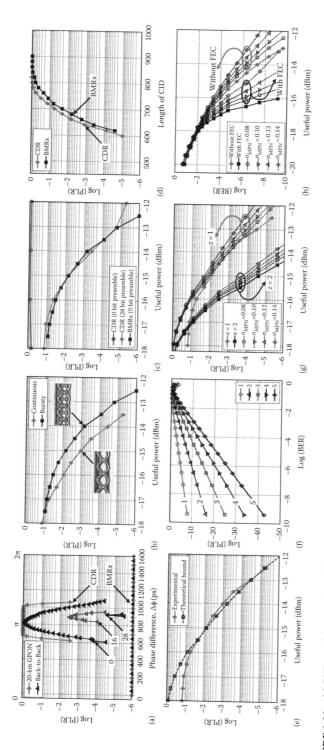

FIGURE 1.23 (a) PLR vs. phase step for different preamble lengths. (b) Burst-mode penalty. (c) Preamble length penalty. (d) PLR vs. CID immunity. (e) Simulated vs. measured PLR. (f) PLR vs. BER performance as a function of pattern correlator error resistance. (g) Effect of MPN on PLR. (h) Effect of MPN on BER. (From Shastri, B.J. et al., *IEEE Photon. Technol. Lett.*, 20(5), 363, 2008, Figure 2. With permission. © 2008 IEEE.)

of the curve around 800 ps is explained because that represents the worst-case phase-step (π radians) for a half-bit period, which causes the CDR to sample precisely at the edge of the eye diagram. An alternating sequence of preamble bits can help the CDR acquire lock; however, this reduces the overall effective throughput and means greater delay. Error-free operation (defined by a PLR < 10^{-6} and a BER = 10^{-10}) occurs for any phase step with 32 preamble bits, but this does not satisfy the established G.984.2 standard, which specifies a 28 bit maximum.

Once the 20 km fiber is introduced, the PLR performance degrades further, as seen in Figure 1.23a.

When the CPA (and therefore burst-mode functionality) is switched on, however, error-free operation is obtained (with or without the fiber spool) for any phase step and zero preamble bits, indicating an instantaneous phase acquisition. We do note a penalty to sensitivity because of the quick extraction of the decision threshold and clock phase from using a very short preamble [42]. By reducing the length of the CPA field, there is more room in bits to perform some amplitude recovery, mitigating the penalty.

In Figure 1.23b, PLR is plotted against received signal power. With a 0 bit preamble, the best-case scenario (continuous data at the bit rate with no phase steps) is compared to the worst case (bursty data having the worst-case phase difference). There is a noted 1 dB power penalty. In the worst-case phase difference between packets, the CDR will not be able to recover any of the packets (irrespective of the signal level), as shown in Figure 1.23c—this is the worst-case PLR of 1. If, however, the allotted 28 bit of preamble are used, the PLR behavior of the CDR compares well to the PLR of the BMRx with 0 preamble bits. There is a natural trade-off between the power penalty with the BMRx oversampling (when there is no phase-step to track) versus the number of preamble bits required when there is some non-zero phase step between packets. Since phase-steps are inevitable, it would seem to suggest that this power penalty had best be paid to avoid packet loss.

In Figure 1.23d, we can see that there is excellent behavior for CID cases (a consecutively repeating data value). The BMRx supports 600 CID instances without error, which far exceeds the minimum standard of 72 set by G.984.2.

The delimiters of each packet must be properly detected, otherwise that packet will be lost. The length of the delimiter, as well as the choice of implementation of the pattern correlator, determines how error-proof recognition will be. If the pattern correlate for a d-bit delimiter has an error resistance of z bits, then the PLR with some given BER p_e is estimated as

$$\text{PLR} \leq \sum_{j=z+1}^{d} P(j); \quad P(x) = \binom{d}{x} p_e^x \left(1 - p_e\right)^{d-x} \tag{1.1}$$

and $P(j) \approx P(z+1)$ for $p_e \ll 1$.

Simulation and measurement results for the PLR are presented in Figure 1.23e and shown to be in agreement. Figure 1.23f illustrates how PLR varies for different BERs (by adjusting the error resistance of the delimiter). Even a simple pattern correlator (1 bit error resistance) yields error-free operation for BER of 10^{-10}. If the correlator is increased to $z = 2$ bits, the PLR improves by eight orders of magnitude. Figure 1.21g shows that coding gain improves by 2 dB at a PLR of 10^{-6}, which more than compensates for the 1 dB power penalty of the BMRx.

FEC is recommended for PONs in part because they tend to use low-cost Fabry–Perot lasers at each terminal, and these suffer from mode partition noise (MPN) leading to a fair bit of signal phase jitter. BER and PLR performance will suffer as a result; in Figure 1.23h, we plot BER as a function of signal power, both with and without FEC. A coding gain of 3 dB is evident at a BER of 10^{-10}. The exact power penalty caused by MPN has been characterized for a 1330 nm system [43]. If the laser spectrum is Gaussian, the mean square variance of the MPN can be described as

$$\sigma_{MPN} = \frac{k}{\sqrt{2}}\left(1 - e^{-\beta^2}\right); \quad \beta = \pi BDL\sigma_{\lambda} \tag{1.2}$$

where

k is the mode partition coefficient

B is the bit rate

D is the fiber delay dispersion per unit length per wavelength

L is the fiber length

σ_{λ} is the laser spectral width

The power penalty resulting from MPN is then

$$\delta_{MPN} = -5\log\left(1 - Q^2\sigma_{MPN}^2\right) \tag{1.3}$$

where Q is the signal-to-noise ratio determined by the BER $p_e = 0.5 \, \text{erfc}(Q/\sqrt{2})$. PLR and BER performance with respect to σ_{MPN} are plotted in Figure 1.23g and h, respectively. It is evident that increased σ_{MPN} impacts performance negatively. Implementing FEC and increasing the pattern correlator error resistance can mitigate the effect of MPN on BER and PLR, respectively. The effective coding gain can help in any number of ways, allowing the reduction of transmitter power, extend the range of the link, or enable more lines on a PON tree.

In summary, we have presented a burst-mode receiver in a 20 km GPON uplink and performed PLR measurements, evaluating how it depends on a number of factors: phase-steps between packets, BER and MPN penalties, signal power, and CID immunity. We have explored the trade-offs between power penalty, preamble length, and pattern correlator error resistance. The receiver has instantaneous phase acquisition for any phase step with only a 1 dB power penalty.

1.7 Discussion

Optically enabled chips, boards, and backplanes are expected, along with all-optical networks, to overcome the limitations of traditional copper interconnects. In this chapter, we have explored several aspects of OE-VLSI design by looking at specific demonstration systems for short-range parallel optical interconnections in large arrays, as well as burst-mode receivers for their role in future PON that bring fiber into the home. We have explored some design rules, fabrication steps, and performance parameters for these systems, which ultimately must bridge the worlds of electrical and optical I/O to ensure that high-bandwidth, high-speed interconnects can keep the pace with the growing demands of multimedia applications and increasing processor speeds.

A common design assumption is that optics provide increased connectivity through parallelism. Specifically, parallel optical interconnects are capable of providing high bandwidth communication links both within and between high-performance electronic systems. The concept of the direct sourcing and termination of optical signals on silicon has been proposed as a method to relieve the off-chip communication bottleneck. The objective of the approach is to create optical I/Os that operate in conjunction with traditional electrical I/O. We have shown examples where this partitioning is effective and have proven that significant progress has been made in the enabling technologies required to achieve this objective. Much of the reported progress is as a result of a series of successful OE-VLSI ASIC demonstrations. Through the realization of these devices, research in a wide range of areas has been accomplished, including optoelectronic device design, electronic circuit design, and optical design and packaging.

Acknowledgments

This work was partially supported by the Natural Sciences and Engineering Research Council of Canada (NSERC) and industrial partners through the Agile All-Photonic Networks (AAPN) research network, the Canadian Microelectronics Corporation (CMC) and the *Fonds Québécois de la Recherche sur la Nature et les Techologies* (FQRNT). The authors gratefully acknowledge contributions from current and former members of the Photonic Systems Group at McGill University.

References

1. Semiconductor Industry Association. 1999. *International Technology Roadmap for Semiconductors*: 1999 ed. Austin, TX: International SEMATECH.
2. N.M. Jokerst and A.L. Lentine, Eds. 1996. *IEEE J. Sel. Top. Quantum Electron.*, Special issue on smart pixels, 2(1):3–148 and references therein.
3. N.M. Jokerst, R. Beats, and K. Kobayashi, Eds. 1999. *IEEE J. Sel. Top. Quantum Electron.*, Special issue on smart photonic components, interconnects & processing, 5(2):143–398 and references therein.
4. Y. Li, E. Towe and M.W. Haney, Eds. 2000. *Proc. IEEE*, Special issue on optical interconnects for digital systems, 88(6):721–876 and references therein.
5. H. Thienpont, Ed. 2003. *IEEE J. Sel. Top. Quantum Electron.*, Special issue on optical interconnects, 9(2), 347–676 and references therein.
6. L.A. Buckman Windover, K.J. Ebeling, J.N. Lee, J. Meindl, and D.A.B. Miller, Eds. 2004. *J. Lightwave Technol.*, Special issue on optical interconnects, 22(9):2017–2222 and references therein.
7. J.W. Goodman, F.J. Leonberger, S.Y. Kung, and R.A. Athale. 1984. Optical interconnects for VLSI systems, *Proc. IEEE*, 72(7):850–866.
8. A.V. Krishnamoorthy, L.M.F. Chirovsky, W.S. Hobson et al. 1999. Vertical cavity surface emitting lasers flip-chip bonded to gigabit/s CMOS circuits, *IEEE Photonics Technol. Lett.*, 11:128–130.
9. D.V. Plant, M.B. Venditti, E. Laprise et al. 2001. A 256 channel bi-directional optical interconnect using VCSELs and photodiodes on CMOS, *J. Lightw. Technol.*, 19(8):1093–1103.
10. M.B. Venditti, E. Laprise, J. Faucher, P.-O. Laprise, J.E. Lugo, and D.V. Plant. 2003. Design and test of an optoelectronic-VLSI (OE-VLSI) chip with 540 element receiver/transmitter arrays using differential optical signaling, *IEEE J. Sel. Top. Quantum Electron.*, 9:361–379.
11. K.M. Geib, K.D. Choquette, D.K. Serkland, A.A. Allerman, and T.W. Hargett. 2002. Fabrication and performance of two-dimensional matrix addressable arrays of integrated vertical-cavity lasers and resonant cavity photodetectors, *IEEE J. Sel. Top. Quantum Electron.*, 8:943–947.
12. M. Châteauneuf, A.G. Kirk, D.V. Plant, T. Yamamoto, J.D. Ahearn, and W. Luo. 2002. 512-channel vertical-cavity surface-emitting laser based free-space optical link, *Appl. Opt.*, 41:5552–5561.
13. F.E. Kiamilev and R.G. Rozier. 1997. Smart pixel IC layout methodology and its application to photonic page buffers, *Int. J. Optoelectron.*, 11:199–215.
14. M.W. Haney, M.P. Christensen, P. Milojkovic et al. 2000. Description and evaluation of the *FAST-Net* smart pixel-based optical interconnection prototype, *Proc. IEEE*, 88:819–828.
15. D.R. Rolston, D.V. Plant, T.H. Szymanski et al. 1996. A hybrid-SEED smart pixel array for a four-stage intelligent optical backplane demonstrator, *IEEE J. Sel. Top. Quantum Electron.*, 2:97–105.
16. A.L. Lentine, K.W. Goossen, J.A. Walker et al. 1997. Optoelectronic VLSI switching chip with greater than 1 Tbit/s potential optical I/O bandwidth, *Electron. Lett.*, 33:894–895.
17. N. McArdle, M. Naruse, and M. Ishikawa. 1999. Optoelectronic parallel computing using optically interconnected pipelined processing arrays, *IEEE J. Sel. Top. Quantum Electron.*, 5:250–260.
18. C.B. Kuznia, J.-M. Wu, C.-H. Chen et al. 1999. Two-dimensional parallel pipeline smart pixel array cellular logic (SPARCL) processors-chip design and system implementation, *IEEE J. Sel. Top. Quantum Electron.*, 5:376–386.

19. D. Fey, W. Erhard, M. Gruber et al. 2000. Optical interconnects for neural and reconfigurable VLSI architectures, *Proc. IEEE*, 88:838–848.

20. M.B. Venditti. 2003. Receiver, transmitter, and ASIC design for Optoelectronic-VLSI applications, PhD dissertation, McGill University, Montreal, Canada.

21. A.L. Lentine, K.W. Goossen, J.A. Walker et al. 1996. High-speed optoelectronic VLSI switching chip with >4000 optical I/O based on flip-chip bonding of MQW modulators and detectors to silicon CMOS, *IEEE J. Sel. Top. Quantum Electron.*, 2:77–84.

22. A.V. Krishnamoorthy, J.E. Ford, F.E. Kiamilev et al. 1999. The AMOEBA switch: An optoelectronic switch for multiprocessor networking using dense-WDM, *IEEE J. Sel. Top. Quantum Electron.*, 5:261–275.

23. R.G. Rozier, F.E. Kiamilev, and A.V. Krishnamoorthy. 1997. Design and evaluation of a photonic FFT processor, *J. Parallel Distrib. Comput.*, 41:131–136.

24. J.-P. Thibodeau. 2005. An ultra low-power 5 Gb/s vertical cavity surface emitting laser (VCSEL) based optical transmitter for optical interconnect applications and access networks, PhD dissertation, McGill University, Montreal, Canada.

25. M.B. Venditti and D.V. Plant. 2003. On the design of large transceiver arrays for OE-VLSI applications, *J. Lightw. Technol.*, 21:3406–3416.

26. P.R. Brusenback, S. Swirhun, T.K. Uchida, M. Kim, and C. Parsons. 1993. Equivalent circuit for vertical cavity top surface emitting lasers, *Electron. Lett.*, 29:2037–2038.

27. D. Wiedenmann, R. King, C. Jung et al. 1999. Design and analysis of single-mode oxidized VCSEL's for high-speed optical interconnects, *IEEE J. Sel. Top. Quantum Electron.*, 5:503–511.

28. K.R. Shastri, K.A. Yanushefski, J.L. Hokanson, and M.J. Yanushefski. 1989. 4.0 Gb/s NMOS laser driver, in *Proceedings of the Custom Integrated Circuit Conference*, San Diego, CA, pp. 14.7.1–14.7.3.

29. K.R. Shastri, K.N. Wong, and K.A. Yanushefski. 1988. 1.7 Gb/s NMOS laser driver, in *Proceedings of the Custom Integrated Circuit Conference*, Rochester, NY, pp. 5.1.1–5.1.14.

30. D. Kucharski, Y. Kwark, D. Kuchta et al. 2005. A 20 Gb/s VCSEL driver with pre-emphasis and regulated output impedance in 0.13 μm CMOS, *Dig. Tech. Pap. IEEE Int. Solid State Circuits Conf.*, 48:172–173+586.

31. C. Kromer, G. Sialm, C. Berger et al. 2005. A 100 mW 4×10 Gb/s transceiver in 80 nm CMOS for high-density optical interconnects, *IEEE J. Solid State Circuits*, 40(12):2667–2677.

32. R. Annen and S. Eitel. 2004. Low power and low noise VCSEL driver chip for 10 Gbps applications, *Conf. Proc. Laser and Electro-Optics Society. Annual Meet CLEO*, 1:312–313.

33. B. Jalali, S. Yegnanarayanan, T. Yoon, T. Yoshimoto, I. Rendina, and F. Coppinger. 1998. Advances in silicon-on-insulator optoelectronics, *IEEE J. Sel. Top. Quantum Electron.*, 4:938–947.

34. P.M. VanPetegham and J.F. Duque-Carillo. 1990. A general description of common-mode feedback in fully-differential amplifiers, *Proc. IEEE Int. Symp. Circuits Syst.*, 4:3209–3212.

35. M. Bazes. 1991. Two novel fully complementary self-biased CMOS differential amplifiers, *IEEE J. Solid State Circuits*, 26:165–168.

36. A.V. Krishnamoorthy, R.G. Rozier, T.K. Woodward et al. 2000. Triggered receivers for optoelectronic VLSI, *Electron. Lett.*, 36:249–250.

37. M.B. Venditti, J.D. Schwartz, and D.V. Plant. 2004. Skew reduction for synchronous OE-VLSI receiver applications, *IEEE Photon. Technol. Lett.*, 16:1552–1554.

38. J.D. Schwartz, M.B. Venditti, and D.V. Plant. 2004. Experimental techniques using optically enabled ring oscillators, *Appl. Opt.*, 43(12):2456–2461.

39. B.J. Shastri, Z.A. El-Sahn, M. Zeng, N. Kheder, L.A. Rusch, and D.V. Plant. 2008. A standalone burst-mode receiver with clock and data recovery, clock phase alignement, and RS(255,239) codes for SAC-OCDMA applications, *IEEE Photon. Technol. Lett.*, 20(5):363–365.

40. J. Faucher, M.Y. Mukadam, A. Li, and D.V. Plant. 2006. 622/1244 Mb/s burst-mode CDR for GPONs, *Conf. Proc. Lasers Electro-Optics Society Annual Meet CLEO*, Paper TuDD3.

41. ITU-T Recommendation G.984.2, 2003.
42. P. Ossieur, X.-Z. Qiu, J. Bauwelinck, and J. Vandewege. 2003. Sensitivity penalty calculation for burst-mode receivers using avalanche photodiodes, *J. Lightw. Technol.*, 21(11):2565–2575.
43. G.P. Agrawal, P.J. Anthony, and T.-M. Shen. 1988. Dispersion penalty for 1.3 μm lightwave systems with multimode semiconductor lasers, *J. Lightw. Technol.*, 6(5):620–625.
44. A. Kirk, D.V. Plant, M.H. Ayliffe et al. 2003. Design rules for highly parallel free-space optical interconnects, *IEEE J. Sel. Top. Quantum Electron.*, 9(2):531–547.

2

Integrated Optical Waveguides for VLSI Applications

Henning Schröder
Elmar Griese

2.1 Introduction

The transmittable bandwidth on electrical circuits is proportional to the circuit diameter and inversely proportional to the square of the length of the circuit trace [1]. Current channel data rates for copper board-backplane systems reach from 3.125 to 6.25 Gbps and must increase to between 10 and 25 Gbps to meet the requirements of future systems [2]. While 10 Gbps × m over the backplane has already been achieved in an optimized design [3,4], challenges include integrating the additional driver and filter components, as well as identifying or producing the high-frequency-compatible base materials required for high signal integration. Furthermore, solutions must be found for the limited channel density imposed by the necessities of electromagnetic shielding, above all in connectors.

Optical interconnection technology is a promising opportunity for transmitting extremely high bandwidth in data systems and is already standard by means of optical waveguide technology using glass (GOF) or polymer (POF) optical fibers in intersystem applications and signal transmission over long distances (telecommunication, machine systems, and railway). However, these assemblies also have disadvantages, particularly the fact that passive optical structures, such as splitters and combiners, cannot be used, leaving splicing technique as the only option. Additionally, such systems cannot be

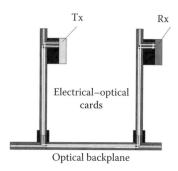

FIGURE 2.1 Subsystem scheme with electrical-optical cards plugged into an optical backplane with a two-layer optical waveguide array and coupling elements with beam crossings.

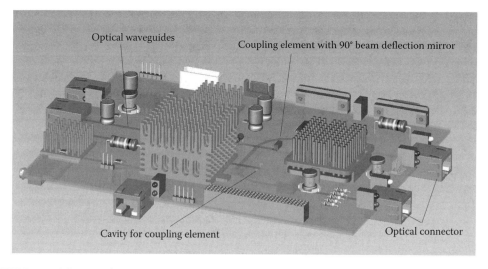

FIGURE 2.2 Schematic drawing of a hybrid electrical-optical daughter card with vertical pluggable module coupling to integrated planar optical waveguides in optical foils.

assembled using standard SMD-processing and additional, costly steps are necessary. Hybrid electro-optically integrated interconnection technology presents a means of overcoming these disadvantages [5,6]. Figure 2.1 presents a schematic example of such a system. An additional optical layer with one or *n* stacked optical-strip waveguide layers (cladding/core/cladding) is employed for the optical links. The layer is fabricated as a foil and processed using lamination technology standards in PCB fabrication processes, with the waveguides incorporated within the circuit board and backplane. Figure 2.2 provides a schematic overview of the electronic-optical circuit board (EOCB) arrangement with electrical/optical carrier, coupling elements, connectors, optoelectronic devices, and drivers. The most important interface issue is the assembly of the EOCB, and, here, pluggable systems seem to be the more promising approach compared to surface mount technology. Different approaches for the optical coupling of optoelectronic devices and waveguides exist, including direct-butt coupling and integration of deflection coupling elements, as shown in Figures 2.1 and 2.3. The latter is based on reflecting facets of an optical pin in front of the waveguide ends. Here, the optical axes of the optoelectronic components are mounted orthogonal to the waveguide axes (Figure 2.2). In the former approach, the devices are direct-butt coupled to the waveguides to minimize the axial offset between the device and waveguide axis, which occurs as a result of distance between the two.

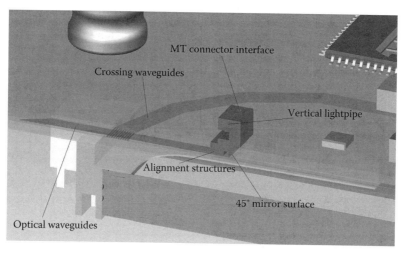

FIGURE 2.3   Optical waveguides and coupling element with 90° light deflection mirror and MT ferrule for plugging of the electrical-optical module. (Part of Figure 2.2, cross section.)

2.2 Waveguide Types and Packaging Concepts

2.2.1 Polymer Step-Index Waveguides and Ion-Exchanged Graded-Index Waveguides

Due to the relatively high tolerances in PCB manufacturing, board-integrated waveguides today require a core diameter of some tens of microns, generally resulting in multimode propagation at a wavelength of 850 nm. This wavelength is becoming standard due to the availability of appropriate transmitter (VCSEL) and receiver components and the high transparency of all suited polymer materials. Two refractive index profiles familiar from fiber technology, step- and graded-index stripe waveguides, can be employed. The two profiles exhibit different wave propagation properties, with the graded-index profile causing less mode dispersion as shown in Figure 2.4.

Each profile is the direct result of the respective structuring technology used. Polymer waveguides can be created by embossing or photolithography (see Figure 2.25) resulting in a sharp material interface between high index core and lower index cladding, and are therefore categorized as step index. Depending on various process parameters, such as temperatures for curing processes and solvent content, a smart graded-index zone is possible but difficult to control. On the other hand, graded-index

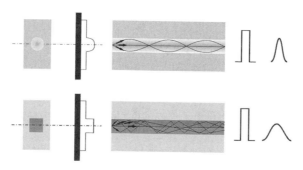

FIGURE 2.4   Refractive index profiles of (a) step-index and (b) graded-index multimode stripe waveguides, wave propagation and signal dispersion.

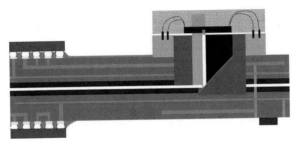

FIGURE 2.5   Scheme of EOCB with inlay integrated waveguide foils (bright core layer) with 90° optical deflection element.

waveguides, integrated in thin glass foils, are produced by an isotropic diffusion process instead of mechanical structuring.

2.2.2 Overlay and Inlay Integration

Optical integration into PCB has existed for several years, and initially simply comprised polymer waveguide foil lamination onto the PCB surface as an overlay. This seemed to be easier than inlay integration because of reduced temperature and pressure requirements. Assembling this layer after soldering of the components was attempted but no fabrication processes were found that matched the alignment requirements. Furthermore, the reliability problems caused by mechanical damage and photodegradation of the polymer foils, as well and surface area required, rendered the approach increasingly unattractive. Today, the symmetrical integration of the waveguide foils into the PCB layer stack has become the most common build-up (Figure 2.5). Apart from overcoming the drawbacks of the overlay approach, inlay integration, as used in the hybrid EOCB, also features improved thermal-mechanical behavior due to the symmetrical layer stacks, which circumvent the bow otherwise caused by the coefficient of thermal expansion (CTE). The challenge in inlay integration is maintaining the lamination parameters of time, pressure, and temperature as close as possible to the standard conditions in PCB manufacturing, to ensure good lamination without damaging the prestructured waveguide foils. In the case of polymer waveguides, the plastic deformation of the waveguides and thermal degradation are the most important problems. Thin glass foils, on the other hand, have to be handled with care to prevent cracks but do not exhibit any problems regarding thermal degradation or changes in shape or refractive index profile.

2.2.3 Fiber Integration

Integrating GOF into or onto the surface of PCBs is a very straightforward approach for optical integration and is used for some applications. Different build-ups have been developed, some of which are commercially available. Indeed, for optical backplanes, flexible polymeric foils with optical fibers in between and standardized connectors at the edges of the foil layers are now standard technology (Figure 2.6). Special routing machines are necessary to lay the fibers onto the bottom foil, particularly to accommodate the drawback of relatively large bending radii, a result of reliability and attenuation considerations. Standard multimode fiber of 250/125/50 μm (coating/cladding/core) has a safe bend radius of approximately 30 mm. Decreasing the bend radius to less than 20 mm causes a considerable increase in optical loss and mechanical stress, thereby reducing the long-term stability. Recently, new fibers with smaller diameters (125/80/50 μm) and increased refractive index differences have begun to emerge on the market, which have a bend radius of about 2 mm [7]. Despite this progress, direct fiber integration is still restricted to a number of pure optical backplane solutions. For hybrid-integrated electrical-optical circuit boards (EOCB) the assembly and interconnection remains too complicated for cost-effective manufacturing technologies.

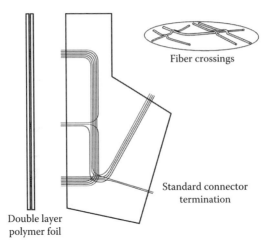

Fiber crossings

Standard connector
termination

Double layer
polymer foil

FIGURE 2.6 Schematic drawing of GOF flexible foil for specific backplane applications.

2.3 Modeling of Optical Multimode Interconnects

2.3.1 General Aspects in the Design of Integrated Optical Interconnects

The industrial design of electronic components and systems is today entirely based on the use of software tools and design rules. Especially for conventional-printed circuit boards, many, very efficient EDA tools are available that streamline and assist the entire design and development process. Instead of developing a completely new process for designing electrical- and optical-printed circuit board technology, research has focused on extending existing printed circuit board design process, to thereby maintain compatibility between the two approaches. The result of these investigations is the design process depicted in Figure 2.7. To ensure that the routing and wiring of optical onboard waveguides has the same degree of freedom standard in the design of electrical-printed circuit boards, the "specification" and "schematic design" steps of the hybrid design process should not be limited to the consideration of the optical interconnects. As the behavior of optical guided wave propagation is completely different from electrical signal propagation, the first task was developing additional design rules for

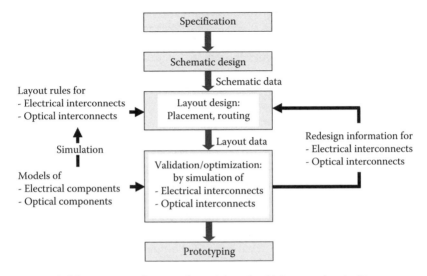

FIGURE 2.7 Extended design process for printed circuit boards with integrated optical interconnects.

optical interconnects applied during the placement and routing process. Examples for such rules are determining the minimum curve radius of a waveguide with respect to the overall optical power budget and determining the geometry of a power splitter to optimize efficiency. Developing these design rules requires simulation algorithms and tools able to analyze and compute the wave propagation within multimode waveguide structures very accurately and quickly.

The second essential task is validating and optimizing already existing fiber optic interconnect standards in terms of timing and signal integrity. This also has to be based on appropriate simulation algorithms and tools, and requires very accurate models of the multimode waveguides, coupling components, and optical transmitters and receivers. This is a very involved task as today's electronic equipment is extremely complex.

While simulation models and algorithms to predict the timing behavior and signal integrity of electrical interconnects have been established for many years [11], comparable models and algorithms allowing time-domain analysis of highly multimode waveguides are not yet commercially available and the following must be developed:

- Simulation models for optical (board-integrated) multimode waveguides
- Simulation models for optical transmitters and receivers
- Simulation models for optical fibers and connectors
- Simulation algorithms for time-domain analysis of optical interconnects and transmission systems
- Design rules for electrical-optical interconnects and transmission systems

To obtain a homogeneous and efficient design process for hybrid EOCBs, these new models and algorithms have to be integrated into the existing design, simulation, and analysis tools for conventional electrical components and systems. This means that the development of the models must prioritize compatibility [8,9].

2.3.2 Modeling of Board-Integrated Multimode Waveguides

Modeling of board-integrated optical multimode waveguides is a very challenging task if the waveguide's cross section dimension is significantly larger than the optical wavelength, in which case hundreds of modes can be guided by the waveguide. Different methods for modeling such highly multimode waveguides are presented below.

In recent years, various optical technology approaches were developed, culminating in today's step-index waveguides and the even more recent graded-index waveguides, which are based on an ion exchange process. Both waveguide types are multimodal, but due to their different index profiles and the manufacturing process–based properties, the required modeling strategies are completely different. Step-index waveguides possess a rough interface between the core and the surrounding cladding, which has a significant impact on the overall propagation characteristics. The profile of graded-index waveguides is shaped by the manufacturing process used and cannot be described analytically. Moreover, in the ion exchange process, the "larger" implanted silver ions can result in additional losses.

2.3.2.1 Modeling of Board-Integrated Step-Index Multimode Waveguides

The main difficulty in modeling board-integrated step-index waveguides is the roughness of the core/cladding interface, the characteristics of which are determined by the manufacturing process. Although the roughness depth is generally much smaller than the optical wavelength, it has a huge influence on the propagation characteristics. The most important effects include increased attenuation and mode coupling. Recently two different approaches were investigated to model these effects. One method, described in [9,13,14], is based on geometrical optics. Following this approach, the wave propagation is modeled by ray tracing. An advantage of this approach is that only a geometrical modeling of the wave guiding structures is required, which means that arbitrary structures such as beam splitters and couplers can be analyzed. However, of course, high numerical effort is necessary. The roughness of the core/cladding interface can be taken into account by a Monte Carlo process, which determines the new direction of the rays, as well as their transported optical power. The required probability functions are derived from the analysis of plane wave scattering by a rough interface dividing two half spaces with different dielectric properties [8,9].

Another approach, which, in principle, yields very fast analysis results, is based on the description of wave propagation by modes. The propagation characteristics of an optical multimode waveguide are described by the guided modes of the corresponding ideal waveguide. This is a very difficult task because the cross sections of board-integrated waveguides are predominantly rectangular and an analytical solution is not available for determining the waveguide modes. This means that appropriate numerical or approximation methods are required. Due to the large number of guided modes, the application of conventional numerical methods is very difficult because the required high grade of discretization leads to very high computation times and requires enormous computing memory. Therefore, semi-analytical methods, in which the electromagnetic field strengths are represented by reasonable approximations, are very effective. One such follows an analytical approximation published in [16], and is described in more detail here. The precondition for this approach, a sufficiently small numerical aperture, must be met by all technologies. Near-field measurements have indicated that the optical power is concentrated mainly in the waveguide's core and only very little energy is carried within the waveguide's cladding. Thus, the field decays very quickly outside of the core, which can be shown by the equation

$$\beta_z^2 - \beta_1^2 \ll \beta_z^2 - \beta_i^2 \quad (i = 2,3,4,5)$$

(2.1)

where

β_z is the propagation constant of the modes propagating in z-direction
β_i is the wave number of the *i*th region

The conclusion of these considerations is that the electromagnetic field, as well as the energy, is mainly represented by the modes far from cutoff. This means that the field in the shaded regions in Figure 2.8 can be neglected.

Due to the sufficiently small numerical aperture, it is necessary to distinguish between the two types of modes. One type is polarized mainly in the x-direction (E_{nm}^x modes) and the other is primarily polarized in the y-direction (E_{nm}^y modes) [12]. The corresponding propagation constants $\beta_{z_{nm}}^{(E^x)}$ and $\beta_{z_{nm}}^{(E^y)}$ are given by

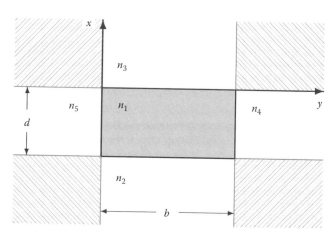

FIGURE 2.8 Cross section of a board-integrated multimode waveguide and its partitioning into different regions. The field in the shaded regions is neglected.

$$\beta_{z_{nm}}^{(E^x)} = \sqrt{\beta_1^2 - \beta_{x_n}^{(E^x)^2} - \beta_{y_m}^{(E^x)^2}}, \tag{2.2}$$

$$\beta_{z_{nm}}^{(E^y)} = \sqrt{\beta_1^2 - \beta_{x_n}^{(E^y)^2} - \beta_{y_m}^{(E^y)^2}}. \tag{2.3}$$

The phase constants $\beta_{x_n}^{(E^x)}$, $\beta_{y_m}^{(E^x)}$, $\beta_{x_n}^{(E^y)}$, and $\beta_{y_m}^{(E^y)}$ are given by the eigenvalue equations

$$\tan\beta_{x_n}^{(E^x)}d = jn_1^2\beta_{x_n}^{(E^x)} \frac{n_2^2\beta_{x_3} + n_3^2\beta_{x_2}}{n_2^2n_3^2\beta_{x_n}^{(E^x)^2} + n_1^2\beta_{x_2}\beta_{x_3}}, \tag{2.4}$$

$$\tan\beta_{y_m}^{(E^x)}b = j\beta_{y_m}^{(E^x)} \frac{\beta_{y_4} + \beta_{y_5}}{\beta_{y_m}^{(E^x)^2} + \beta_{y_4}\beta_{y_5}}, \tag{2.5}$$

$$\tan\beta_{x_n}^{(E^y)}d = j\beta_{x_n}^{(E^y)} \frac{\beta_{x_2} + \beta_{x_3}}{\beta_{x_n}^{(E^y)^2} + \beta_{x_2}\beta_{x_3}}, \tag{2.6}$$

$$\tan\beta_{y_m}^{(E^y)}b = jn_1^2\beta_{y_m}^{(E^y)} \frac{n_4^2\beta_{y_5} + n_5^2\beta_{y_4}}{n_4^2n_5^2\beta_{y_m}^{(E^y)^2} + n_1^2\beta_{y_4}\beta_{y_5}}, \tag{2.7}$$

where

$$\beta_{x_2} = -j\sqrt{\beta_1^2 - \beta_2^2 - \beta_x^2}, \tag{2.8}$$

$$\beta_{x_3} = -j\sqrt{\beta_1^2 - \beta_3^2 - \beta_x^2}, \tag{2.9}$$

$$\beta_{y_4} = -j\sqrt{\beta_1^2 - \beta_4^2 - \beta_y^2}, \tag{2.10}$$

$$\beta_{y_5} = -j\sqrt{\beta_1^2 - \beta_5^2 - \beta_y^2}. \tag{2.11}$$

A detailed derivation of these interrelations is included in [12].

The following illustrates the computation of the guided modes of a multimode waveguide with a quadratic cross section ($50 \times 50\,\mu m^2$) as example. The refractive indices of core and cladding are $n_{core} = 1.595$ and $n_{cladding} = 1.585$, which results in a numerical aperture of $N_A = 0.18$. This fulfills the precondition mentioned above of weakly guiding waveguides. Figures 2.9 and 2.10 depict the graphical representations of the eigenvalue equations of the E_{nm}^x modes. The numerical computation of the intersection points (marked by circles) of the corresponding functions is reduced to the search of zeros, which can be performed very quickly and easily. Due to the equidistant branches of the tangent function for each value of β_{x_n} and β_{y_m}, a finite interval can be given (see Figures 2.9 and 2.10), which contains the solution looked for.

Figure 2.11 shows the normalized propagation constant $\beta_{z_{nm}}^{(E^x)}$ for $n = m$ versus the structure parameter V with

$$V = \frac{2\pi d}{\lambda} \sqrt{n_1^2 - n_2^2} = \beta_0 \cdot d \cdot N_A, \tag{2.12}$$

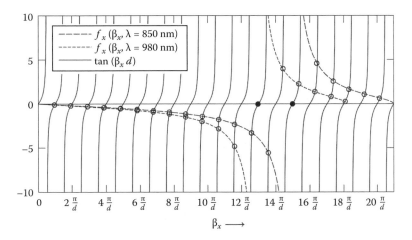

FIGURE 2.9 Graphical representation of Equation 2.4. The function $f_x(\beta_x, \lambda)$ is the right-hand side of Equation 2.4, its pole is indicated by a dot (•).

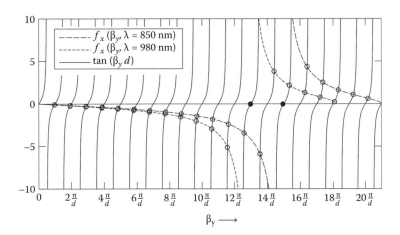

FIGURE 2.10 Graphical representation of Equation 2.5. The function $f_x(\beta_x, \lambda)$ is the right-hand side of Equation 2.4, its pole is indicated by a dot (•).

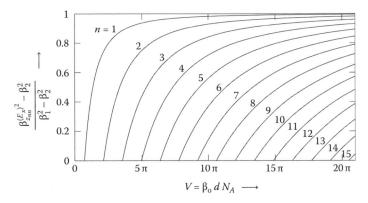

FIGURE 2.11 Normalized propagation constant $\beta_{z\,nn}^{(E^x)}$ of the E_{nn}^x modes versus the structure parameter V.

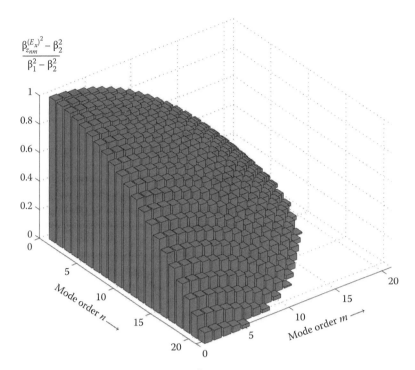

FIGURE 2.12   Normalized propagation constant $\beta_{z\,nm}^{(E^x)}$ of the E_{nm}^x modes versus the mode orders n and m.

where

  β_0 is the free space wave number
  d is the edge length of the waveguide's quadratic cross section

For an approximate overview of the high number of modes and their density, all the possible propagation constants of the E_{nm}^x modes were computed and are depicted in Figure 2.12. A rough estimation shows that the analyzed waveguide is able to guide more than 300 E_x modes. Additionally, exactly the same number of E_y modes has to be considered, leading to a total number of 650–700 guided modes.

As already mentioned, a rough interface between the core and cladding in step-index waveguides results in mode coupling between the guided modes, and between guided modes and the continuous spectrum of radiation modes, which is the cause of additional losses. This coupling mechanism can be analyzed using coupled mode theory. Once the results of such analysis have been attained, coupled power theory [17] can be applied to describe the propagation of the optical power of each guided mode as a function of the initial power distribution on all modes and the propagated distance. These equations can be explained by stating that the change of the amount of power of mode μ after the distance dz is the sum of the loss and the total number of all guided modes.

Figure 2.13 shows an integrated waveguide with the geometric dimensions b and d and the refractive indices n_1 and n_2. The direction of propagation is given by the positive z-direction. The waveguide is not ideal, meaning that the surfaces, the sidewalls ① to ④, are rough. For simplification, the roughness of the waveguide's surfaces is only shown at sidewall ②, but note that every sidewall has the same level of roughness. This roughness is represented by a one dimensional function $f(z)$ with a Gaussian autocorrelation function. In case of $f(z) = 0$, the real waveguide becomes identical with the ideal waveguide. The functions describing the roughness at each sidewall are not correlated.

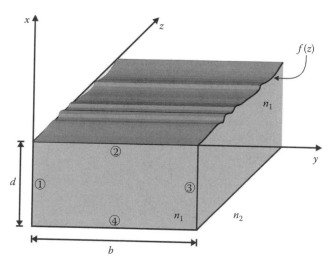

FIGURE 2.13   Multimode waveguide with distortion function $f(z)$.

Application of the coupled power theory [17] results in the so-called coupled power equations

$$\frac{dP_\mu}{dz} = -2\alpha_\mu P_\mu + \sum_{v=1}^{N} h_{\mu v}\left(P_v - P_\mu\right) \tag{2.13}$$

where
  P_μ is the power of mode μ
  $2\alpha_\mu$ is the power-loss coefficient of mode μ
  $h_{\mu v}$ is the coefficient describing the coupling between the modes μ and v
  N is the coupling loss or gain between mode μ and every other individual guided mode

Note that this system of differential equations does not consider changes of the signal in time, only describing continuous wave operations. The coupled power equations can be easily solved by the well-known exponential approach

$$P_\mu(z) = A_\mu \cdot e^{-\sigma \cdot z}. \tag{2.14}$$

The result is a homogenous system of linear equations of the order N:

$$\sum_{v=1}^{N} \left[h_{\mu v} + (\sigma - 2\alpha_v - \sum_{\eta=1}^{N} h_{v\eta}) \cdot \delta_{\mu v} \right] \cdot A_v = 0 \tag{2.15}$$

where $\delta_{\mu v}$ represents the Kronecker symbol. The solution is given by the computation of the N eigenvalues $\sigma^{(n)}$ and eigenvectors $A^{(n)}$ and finally the optical power mode μ can be written as a function of the propagation distance z

$$P_\mu(z) = \sum_{n=1}^{N} c_n \cdot A_\mu^{(n)} \cdot e^{-\sigma^{(n)} \cdot z}, \tag{2.16}$$

with

$$c_n = \sum_{\nu=1}^{N} A_\nu^{(n)} \cdot P_\nu(z=0) \tag{2.17}$$

where $P_\nu(z=0)$ is the initial power of mode ν.

Following the calculations, it is obvious that all eigenvalues $\sigma^{(n)}$ will be positive. Furthermore, it can be stated that the exponential term with the smallest eigenvalue σ_{min} will overwhelm all other terms of exponential functions for sufficiently large values of the propagation distance z. Thus, the exponential expressions with eigenvalues $\sigma^{(n)} > \sigma_{min}$ can be neglected for this sufficiently large z. In this so-called steady state, σ_{min} describes the power-loss coefficient by

$$\alpha_{dB/cm} = 10 \cdot \log_{10}\left(e^{\sigma_{min} \cdot 1cm}\right). \tag{2.18}$$

Before the steady state is reached, the overall power-loss rate is significantly larger. Moreover, here, the question arises as to the distance z_t or at which point the steady state is reached. This transition length z_t can be defined with the aid of the second smallest eigenvalue $\sigma_{min,2}$. Of course a distance z_t exists with

$$\frac{e^{-\sigma_{min} \cdot z_t}}{e^{-\sigma_{min,2} \cdot z_t}} = K \tag{2.19}$$

where K is an arbitrarily chosen value. The equation only states that both exponential functions have a ratio K. Thus, z_t can be given as a function of a given K as in

$$z_t = \frac{\ln K}{\sigma_{min,2} - \sigma_{min}}. \tag{2.20}$$

Another important result of the investigations is the representation of the coupling coefficient $h_{\mu\nu}$ describing the coupling between mode μ and mode ν. The application of the coupled power theory leads to

$$h_{\mu\nu t} = \left|\hat{K}_{\mu\nu}\right|^2 \cdot \left\langle\left|F(\beta_{z\mu} - \beta_{z\nu})\right|^2\right\rangle \tag{2.21}$$

with

$$\left\langle\left|F(\beta_{z\mu} - \beta_{z\nu})\right|^2\right\rangle = \int_{-\infty}^{\infty} R(u) \cdot e^{-j(\beta_{z\mu} - \beta_{z\nu})u} du. \tag{2.22}$$

The coupling coefficients $\hat{K}_{\mu\nu}$ are obtained using coupled mode theory [15] and $R(u)$ is the autocorrelation function of the roughness $f(z)$. This is a very important factor because the function $f(z)$ is no longer necessary but is instead the Fourier transform of its autocorrelation function. The autocorrelation function of realistic roughness can be accurately approximated by a decaying exponential or a Gaussian function [18] and the Fourier transform of both model functions can be obtained analytically. The calculation of the power loss requires knowing the radiation modes, which can be represented by plane waves. Here too, the coupling of the guided modes to the radiation modes is modeled by applying the coupled mode theory.

The following describes the results of an analysis that again uses a rectangular waveguide ($50 \times 50\,\mu m^2$) with the refractive indices $n_{core} = 1.595$ and $n_{cladding} = 1.585$ at an optical wavelength of $\lambda = 850\,nm$. The rms deviation and the correlation length of the roughness between core and cladding are assumed to be

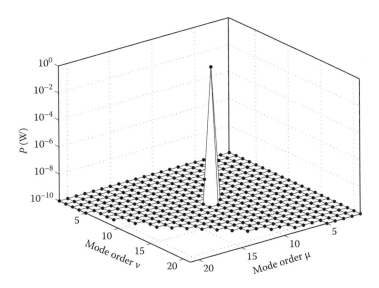

FIGURE 2.14 Excitation of an optical multimode waveguide by the $E^x_{10,10}$ mode.

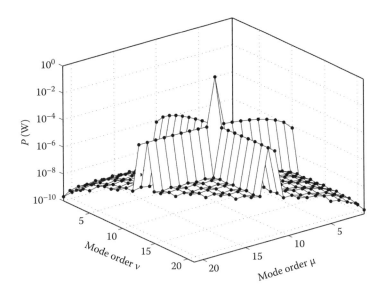

FIGURE 2.15 Power of $E^x_{\mu,\nu}$ modes after a propagation length of 1 cm.

$\bar{\sigma} = 70$ nm and $D = 2\,\mu$m. This waveguide is excited solely with the $E^x_{10,10}$ mode as depicted in Figure 2.14. The coupling mechanism between the propagating modes is illustrated in Figures 2.15 and 2.16. These figures show the optical power versus the mode orders μ and ν. Although the waveguide is excited in the beginning solely with the $E^x_{10,10}$ mode, after a propagation length of only 1 cm, various E^x and E^y modes are excited due to the mode coupling caused by the roughness.

Figures 2.15 and 2.16 illustrate that the coupling is largest between modes of the same mode order μ or ν, that is, the initial excited mode. Moreover, coupling to modes with same polarization is significantly larger than coupling to modes with the opposite polarization. Two effects can be observed as the propagation distance grows. First, the difference of the optical power led by the modes decreases and, second, the coupling with the radiation modes increases with the mode order. This coupling and attenuation mechanism can be summarized as follows: with increasing propagation distance,

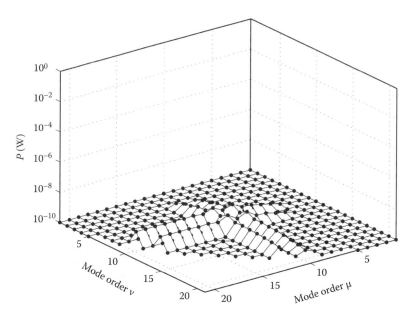

FIGURE 2.16   Power of $E_{\mu,\nu}^{y}$ modes after a propagation length of 1 cm.

increasing power is coupled to the high ordered modes, from where it is coupled to the radiation modes.

Figure 2.17 illustrates the power loss versus the propagation distance z for different excitations of the multimode waveguide. Obviously the power loss is highly dependent on the initial power distribution. The steady state is reached after a propagation length of $z_t = 1.5$ m. After this length, the lines are parallel

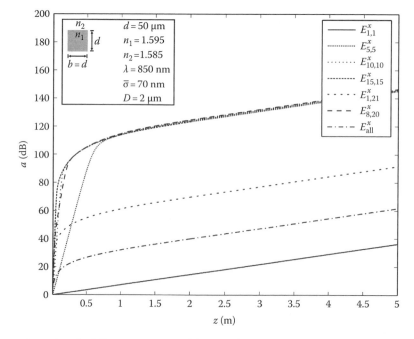

FIGURE 2.17   Power loss for different excitations of the optical multimode waveguide.

and the differential loss of each mode also remains constant. The calculation using Equation 2.20 yields the transition length $z_t = 1.85$ m, using $k = 100$.

Figure 2.18 depicts the power loss in dB cm^{-1} for steady state versus the rms deviation $\bar{\sigma}$ and the correlation length D. The increase in power loss with growing $\bar{\sigma}$ and D follows logically from the theoretical investigations [15]. In Figure 2.19, the transition length for decreasing $\bar{\sigma}$ and D is shown to

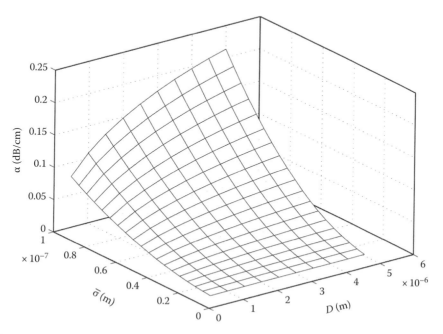

FIGURE 2.18 Power loss in steady state.

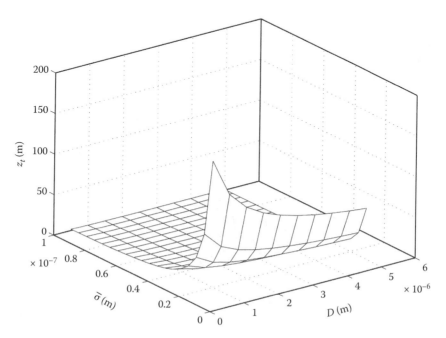

FIGURE 2.19 Transition length z_t versus rms deviation $\bar{\sigma}$ and correlation length D.

be growing because, as the rms deviation and correlation length decreases, the mode coupling must also decrease.

Along with very fast simulation algorithms, the model is also able to explain the coupling and attenuation mechanism of highly multimode waveguides. Another important result of this approach is the fact that the propagation losses depend significantly on the excitation. This means that power losses can be minimized by optimizing the beam properties of the optical source.

2.3.2.2 Modeling of Board-Integrated Graded-Index Multimode Waveguides

A particularly interesting technique for producing optical layers with graded-index waveguides is applying ion exchange processes [34,35] to thin glass sheets to produce optical layers. The promise of this technology for the electrical-optical interconnection technology on PCB lies in the extremely suitable thermal properties of glass.

Traditionally, wave propagation within graded-index structures could not be analyzed by means of wave optical analysis methods. Two recently developed approaches have now overcome this problem. Ray tracing, based on geometrical optics, can be applied if the optical wavelength is small compared to all diffracting and refracting objects. Ray tracing can also be applied in inhomogeneous media if the refractive index variance is sufficiently slight within the wavelength range, which allows accurate description of the field strengths using local plane waves. The key to this approach is solving the so-called eikonal equation. However, efficient numerical ray tracing methods instead solve the ray equation as explained in [19], in which it was also shown that for a given relative error parameter, the rays, which are the phase trajectories, can be accurately computed using a Runge–Kutta method with an automatic step size adaptation.

The second method for modeling wave propagation within graded-index waveguide structures employs conventional numerical methods such as the finite element method (FEM) [21]. One prerequisite for using such methods is avoiding non-physical solutions by selecting approaches that focus on edge-based elements. An additional advantage of the FEM is that the modes of graded-index waveguides with arbitrary index profiles can be computed. Figure 2.20 shows the index profile of a multimode waveguide manufactured by an ion exchange process [34,35]. The normalized power distribution of the fundamental mode of this graded-index waveguide at an optical wavelength of $\lambda = 850 \, nm$ is depicted in Figure 2.21. It can be computed by the longitudinal component of the Poynting vector, and thereby provides important information on the excitation necessary for minimizing the coupling losses.

Measurements with graded-index waveguides manufactured using an ion exchange process, as in Figure 2.20, have shown attenuation significantly larger than the attenuation of the glass substrate. This means that the overall attenuation is given by a basic material attenuation, in general from the transmission characteristics of the substrate, and by an additional attenuation caused by the diffusion process. To model this attenuation, investigating the ion exchange process in more detail is necessary. The ion exchange process is based on the diffusion of relatively large silver ions into the glass. As these silver ions are several times smaller than the optical wavelength, Raleigh scattering is the main cause of the additional losses and has to be modeled together with the diffusion process. An initial attempt at this type of analysis, in which a one-dimensional diffusion process was modeled by Fick's law for time-variant diffusion, is described in [21]. This approach has proved promising and will continue to be developed.

2.4 Polymer Optical Waveguides

2.4.1 Polymer-Integrated Waveguide Core Material

Polymers for use as EOCB-waveguide core materials must fulfill a number of optical, thermo-mechanical, and processing criteria, the key properties of which include

- Low optical loss
- Low birefringence
- Refractive index control

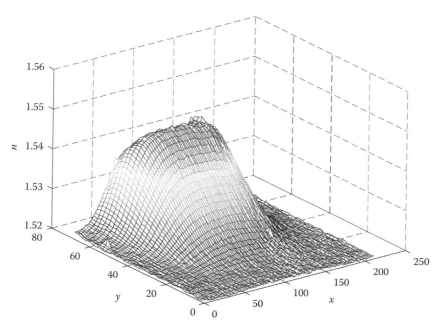

FIGURE 2.20 Index profile of a multimode waveguide manufactured by ion exchange technology. (From Schröder, H. et al., Planar glass waveguides for high performance electrical-optical circuit boards (EOCB)—The glass-layer-concept, *Proceedings of the 53rd Electronic Components and Technology Conference*, New Orleans, LA, 2003, pp. 1053–1059; Schröder, H. et al., High performance electrical-optical circuit boards (EOCB) using planar optical interconnects integrated in thin glass foils, *Proceedings of the EOS Conference on Trends in Optoelectronics*, Munich, Germany, June 2007.)

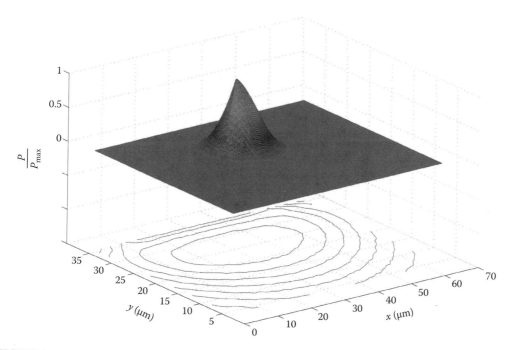

FIGURE 2.21 Normalized power distribution of the fundamental mode of the graded-index waveguide depicted in Figure 2.20 at an optical wavelength of 850 nm.

- High thermal and humidity stability
- Good dimensional stability
- Good adhesion
- High-quality waveguide formation

Naturally, all these characteristics must be stable and reproducible during processing and long-term performance.

Low optical loss at the working wavelength is the first and most important selection factor because optical loss limits the size and design options of the optical structures. Using the wavelength range of 850 nm is an exception to this problem, as a number of polymers are highly transparent (material losses less than 0.1 dB cm^{-1}) at this wavelength, including commercially available poly(meth)acrylates and epoxies. Therefore, development of 850 nm applications mainly focuses on refractive index tuning, processing characteristics and thermal stability.

For other ranges, much effort has been made worldwide to synthesize polymers with low inherent loss at 1.3 and 1.55 μm. Theoretical calculations show that CH-, OH-, and NH-overtones are the main structural elements causing inherent losses of organic polymers at these wavelengths [22]. Therefore, successful synthesis strategies have focused on deuterated and/or halogenated, especially fluorinated, polymers, such as polyacrylates [23–26], polyimides [27,28], polyarylene ethers [29], polysiloxanes [30], polyperfluorcyclobutanes [31], polycyanurates [32], and others. To date, these high-performance and high-cost materials have generally been used for single-mode applications, such as thermo-optic switches or arrayed waveguide gratings.

Apart from the optical properties, the thermo-mechanical and humidity stability of the polymers is the second criterion for their successful use as waveguide material. The materials must be able to withstand thermal treatment at up to 180°C for 1 h during lamination and soldering, as well as elevated temperatures and humidity during long-term use, without significant alteration in characteristics. For this reason, most poly(meth)acrylates can only be used in EOCBs to a limited extent or not at all, regardless of their outstanding optical properties. On the other hand, many epoxies (only for wavelengths <1 μm) and other high-performance thermosets listed above meet both optical and thermostability requirements.

Today the most frequently employed processes for production of planar polymer waveguides are hot embossing/micromolding and photolithography. Alternative techniques, such as reactive ion etching, UV laser ablation, E-beam writing and others, are often too expensive or have not been fully developed. Photolithography and maskless laser direct writing is an attractive alternative for long-waveguide structures, for which the polymer used for forming the waveguides must be UV structured at high quality and at a thickness of up to 100 μm.

Other requirements that must be met by the polymers include special rheological properties, processing without use of solvents, low shrinkage and stress-free cure, good adhesion to substrate but also good deforming from the mold when used for replication technologies such as hot embossing and micromolding. Low-loss UV-curable epoxies have been successfully used for the manufacture of EOCBs at the 850 nm range. Several effects determining attenuation are shown in Figure 2.22.

2.4.2 Optical Substrate Materials

The choice of a suitable optical cladding substrate for the optical EOCB layer is another important factor in material selection. Here, good adhesion to the waveguide core material and the other layers of the PCB, as well as thermo-mechanical and humidity stability, are the key properties. Furthermore, if replication technologies such as hot embossing for polymer waveguides are applied, the substrate also acts as the optical cladding of the waveguides and thus the optical properties of the substrate, especially high transparency, low birefringence, and adjusted refractive indices cladding/core, become crucially important. Here, special, commercially available polycarbonate- and cycloolefine-copolymer foils have been successfully used (Figure 2.23). On the other hand, the importance of the substrate's optical properties

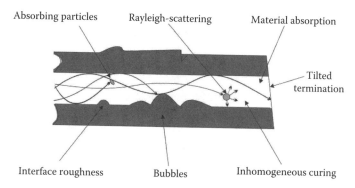

FIGURE 2.22 Effects causing attenuation in a real waveguide.

FIGURE 2.23 Micrograph of hot-embossed polymer waveguides in an OptoFoil before PCB lamination (cross section).

can be minimized by using sandwich-like structures—substrate/cladding/core/cladding—formed by photolithography using a pair of refractive index-adjusted polymers for cladding and core, for which any substrate material suited for PCB-production, such as standard polyimide foils, can be used.

2.5 Polymer Waveguide Structuring

When polymers are chosen for manufacturing the optical structures, care must be taken to ensure their compatibility with the structuring technology (hot embossing, photolithography), as well as with the PCB lamination and soldering processes (see Figure 2.24).

2.5.1 Hot Embossing

The process flow for manufacturing hot-embossed OptoFoils is shown in Figure 2.25. The embossing can be carried by either using an electro-formed Nickel shim or ultra-precise milling. Under increasing temperature and pressure, the tool is imprinted into the thermoplastic optical cladding material. Various thermoplastic materials can be used as foil material in building up the OptoFoil. The properties of COC (cycloolefin-copolymer) and high-temperature polycarbonate meet the required property profile of the EOCB processing very well. In terms of the optical properties, the material has to be highly transparent

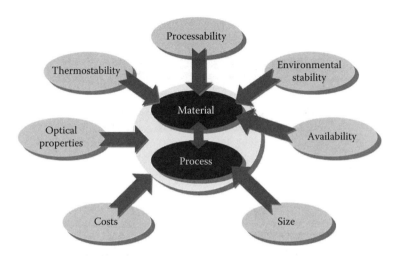

FIGURE 2.24   Determining factors in choosing appropriate materials and processes for EOCB.

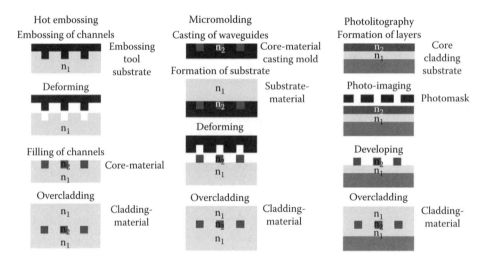

FIGURE 2.25   Planar technologies for polymer optical waveguide structuring.

and have a lower refractive index than the core material. The next fabrication step is filling the embossed channel structures with a core material of higher refractive index compared to the embossed cladding foil. Commercially available optical adhesives are commonly used and can be applied by blade. Good wetting by means of pretreatment of the under-cladding foil is essential, for which UV irradiation and O_2 plasma are well suited (Figure 2.26). Subsequently, the core material is cured and an over cladding seals off the core area as the final step. A cross section photograph of an OptoFoil is shown in Figure 2.23, clearly showing the precisely rectangular shape of the $(80 \times 80)\ \mu m^2$ waveguides. Such replication technologies are very effective, however the structures they can handle with the required precision are limited in size (up to 5 in., approx.) at the current stage of development.

2.5.2 Photolithography

Photopolymers such as SU8 from SOTEC Microsystems SA, the EpoCore/EpoClad-System from Microresist Technologies GmbH [35] or other epoxies can be used for low-loss waveguides at wavelength

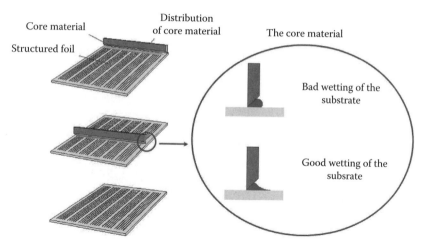

Core material
Structured foil
Distribution of core material
The core material
Bad wetting of the substrate
Good wetting of the subsrate

FIGURE 2.26 Process of channel filling with optical core material in hot-embossed cladding grooves (bed and good wetting).

<1 μm and were successfully applied in EOCB. If a photo mask is used, the mask quality is very important, as well as the process quality with respect to the waveguide core surface roughness, which causes the scattering responsible for the majority of waveguide attenuation. One possible process flow is, firstly, to coat a conventional FR4 substrate with the optical cladding polymer. The solvent is then removed by thermal treatment. While not cross linked, the material has a low molar mass and glass transition temperature, which leads to a high planarization effect. The micro-roughness of uncoated FR4 substrate (Ra = 850 nm) is at Ra < 100 nm after coating with a layer thickness of about 50 μm. After surface UV exposure and subsequent thermal treatment, the layer cannot be dissolved by organic solvents and is not softened by a second coating. Subsequently an optical core layer of the same thickness as the required waveguide height has to be applied by spin, blade, or spray coating. After thermally removing the solvent from the layer, the UV exposure is carried out via a mask with the corresponding structural layout of the waveguide cores.

The lithographic structuring of the waveguides can be carried out by aid of exposure devices designed for wafers. Here, 12-inch technology can be used for the discussed waveguides. After thermal treatment, the developed structures are stable. Figure 2.27 shows that the structures of the waveguide cores have almost vertical card profiles. The waveguide cores are clad by doctor blading of the core material over the exposed waveguide cores, for which a layer thickness of 100 μm is optimal. The width of the blade gap determines the layer thickness. After again removing the solvent with a thermal treatment, subsequent UV surface exposure, and thermal post-processing, the lamination process can proceed. All photolithography processes are restricted in size by the available mask and equipment dimensions and the mask-edge quality.

2.5.3 Laser Direct Writing

The advantage of laser direct writing is design freedom without the requirement of large and expensive masks. Furthermore, a smoother waveguide side wall is possible because of UV laser spot tuning capabilities and maskless irradiation. Apart from the straightforward approach of using UV lasers to cure the core material, followed by a chemical development process and additional over cladding, two photon processing is also possible, for which femtosecond lasers are used. A process for inducing the refractive index increase by two photon process directly at the focus point of the laser without any development process has been developed but is not yet commercially available. The main hurdle is fine-tuning the

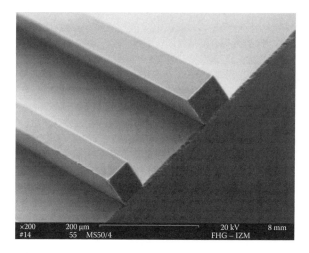

FIGURE 2.27   Photopatterned optical waveguides after core structuring and before overcladding with EpoCore. (Courtesy of Microresist Technology GmbH.)

mechanical system to ensure an extremely smooth waveguide trace, as the x–y stages of the systems tend to cause staircase-like steps or insufficient point-to-point accuracy.

2.6 Thin Glass Graded-Index Waveguides

2.6.1 Thin Glass Sheets and Salt Melts

Thin glass foil can be structured by diffusing multimode waveguides of arbitrary layouts into the glass matrix, for which ion exchange from salt melt, as used in integrated single mode optics, can be used. Due to the increasing requirements for thermal and mechanical stability for printed circuit boards, laminating thin glass foils with low CTE (e.g., $<10 \cdot 10^{-6}\,K^{-1}$) between conventionally used polymer substrate layers is also a promising concept. This results in a decrease of the board CTE and therefore increased system reliability.

Borosilicate glass is used in this approach, which is produced by melting very pure raw materials (Table 2.1), ensuring it is very resistant to chemical attack.

It features

- Low optical attenuation
- Very smooth surface
- Low impurities
- High sodium and potassium content
- High temperature stability
- Mechanically flexibility
- Acid resistance
- Salt melt resistance

TABLE 2.1   Chemical Composition of D 263 T (Compounds in wt.%)

SiO_2	B_2O_3	Al_2O_3	Na_2O	K_2O	ZnO	TiO_2	Sb_2O_3
64.1	8.4	4.2	6.4	6.9	5.9	4.0	0.1

Apart from the physical properties listed in Table 2.2, the most relevant optical values are the refractive index of 1.517 at 850 nm and very good transmission of approximately 92% over a wide wavelength range.

TABLE 2.2 Key Physical Properties of D 263 T

Tg	$\alpha_{(20°C-300°C)}$	ρ
557°C	$7.2 \cdot 10^{-6}\,K^{-1}$	$2.51\,g \cdot cm^{-3}$

2.6.2 Ion Exchange Processing

The ion exchange process flow can be described as follows (Figure 2.28):

1. Determination of diffusion process parameters in the display glass for modifying the refractive index and waveguide structuring.
2. Design of waveguides for optimum coupling of optical sources and detectors without signal distortion (size, numerical aperture).
3. Processing of designed waveguides by ion exchange in hot salt melt. Adjusting of optical and mechanical properties after ion exchange is possible with multi-component salt melts. The maximum steady thermal diffusion speed at 350°C is approximately 0.5 µm per hour. Batch processing is possible. An electrical field can be applied to bury the waveguide.
4. Optical and mechanical characterization requires measuring diffusion parameters with inverse WKB method by mode line spectroscopy and RNF method. The maximum numerical aperture without glass damage after silver ion exchange is 0.39. For the realized waveguides a salt melt is processed to achieve a numerical aperture of 0.22.

The determination of diffusion parameters can be done by fitting the refractive index profiles with complementary error functions as seen in Figure 2.29a. The post diffusion process (Figure 2.29b) is necessary to turn the surface type GI waveguide caused by the silver ion exchange into a GI waveguide with lower refractive index at the surface. Using the derived diffusion parameters, it is possible to design the waveguide shape and the optical properties, that is, to define the appropriate mask dimensions and process parameters like temperature and time. A cross section of diffused double-sided waveguides in a 300 µm thick thin glass sheet is shown in Figure 2.30. Ensuring compatibility between materials and processes is key to establishing this new EOCB technology. One key step here is using sputtering and photolithography to apply a metal diffusion mask for the optical layer, to include the waveguide structures and the alignment structures in the same process. The multimode waveguide structures are

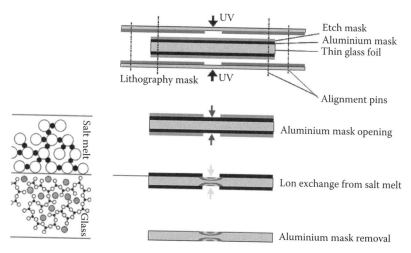

FIGURE 2.28 Double-sided ion exchange process on thin glass foil for double waveguide density design.

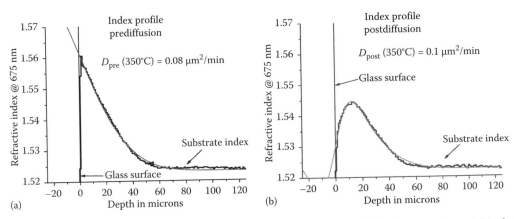

(a)

(b)

FIGURE 2.29 (a) Refractive index profile after silver ion exchange and fitted diffusion parameter and (b) after the post-diffusion step.

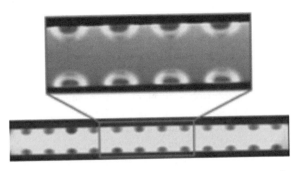

FIGURE 2.30 Double-sided optically functionalized thin glass foil with GI waveguide cross sections of approximately 53 μm × 110 μm, fabricated by silver-ion exchange using a salt melt.

produced using diffused dopands for the core material by thermal silver ion exchange. The metal mask must be able to withstand the very aggressive hot salt melt.

2.7 PCB Lamination

For hybrid electrical-optical circuit boards, the optically structured polymeric or thin glass foils have to be laminated onto commercial polymer-based printed circuit board materials such as FR4. The fact that the optical data transmission allows high bandwidth without the need for high-frequency base materials, such as Rogers or other Teflon-based ones, is an example of the advantages of hybrid electrical-optical technology. However, the lamination is quite challenging and requires care. In Figure 2.31, polymeric waveguide lamination faults are depicted. Temperature and the pressure have to be low enough for waveguide survival but high enough to avoid delamination. Such problems are more relevant in the case of polymer waveguides than for thin glass. On the other hand, for glass, the CTE mismatch and cracks caused by internal stresses in the waveguide region are risky. However, careful parameter adjustment can ensure lamination without any cracks or other damage, as shown in Figure 2.32. The pressing cycle—in other words, the combination of pressure and temperature applied—is very important. Test series have shown that the pressure must be applied within a tight time window. If pressure is applied too soon, the glass is pressed against unmelted resin and shatters. Additionally, pressure should not be maintained for too long and must be eased as soon as the resin gels and hardens to prevent breakage of the glass film. Providing that the right materials and suitable process parameters are selected, manufacturing laminated base materials with an integrated layer of glass is possible.

FIGURE 2.31 Optical micrograph of damaged polymer waveguide after lamination with cracks and delamination of the core and cladding layers.

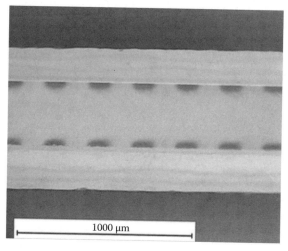

FIGURE 2.32 Cross section of laminated thin glass foil with double-sided waveguides in between layers of FR4 base material. (Courtesy of Würth Electronic GmbH.)

2.8 Summary

Optical interconnection technology using planar optical waveguide foils has been introduced as a very promising approach to fulfill the requirements of future high bandwidth PCB systems. Here, two main approaches, step refractive index waveguides made of polymer materials and graded-index waveguides realized in thin glass foils, were presented and the materials and processes involved discussed. For polymer step-index waveguides several materials and technologies already exist with both specific advantages and drawbacks in terms of optical attenuation due to interface roughness, reliability, and cost. Graded-index waveguides can be manufactured by ion exchange in hot salt melts and hold advantages in terms of wavelength range, reliability, and performance. To optimize the design of such waveguides, the beam propagation has to be modeled and simulated carefully. The PCB lamination technology is challenging and not mature for both types of optical waveguide foils, but very good results have already been achieved. Assembly and optical interconnection of the e/o modules and optical connectors was another crucial point only briefly discussed in the chapter. The two different coupling schemes are direct-butt coupling and surface mounting with 90° optical beam deflection. The main problem here is the very tight mechanical tolerances for efficient optical coupling. To sum up the state of the art, various promising planar optical waveguide technologies exist but standards terms of materials and interfaces have not yet been set. While it is still not clear which approach will carry the day, it appears that the thin glass approach to graded-index waveguides is currently the most advantageous.

References

1. D.A.B. Miller, Physical reasons for optical interconnects, *J. Optoelectron.*, 1997, 11, 155–168.
2. M. Cartier, J. Chan, T. Cohen, B. Kirk, Advanced design techniques to support next generation Backplane links beyond 10 Gbps, *Proceedings of DesignCon 2007*, Santa Clara, CA, 2007.
3. J. Chen, High speed signaling design: From 1 Gbps to 10 Gbps, *Proceedings of DesignCon East 2004*, Boxborough, MA, 2004.
4. G. Havermann, M. Witte, Artificial card edge interfaces for 10 Gbps module cards: How the high speed propagation characteristics are affected by exchanging the PCB-edge with a connector, *Proceedings of DesignCon 2007*, Santa Clara, CA, 2007.
5. J.W. Goodman, Optical interconnections for VLSI systems, *Proc. IEEE*, July 1984, 72(7), 850–867.
6. D. Krabe, F. Ebling, N. Arndt-Staufenbiel, G. Lang, W. Scheel, New technology for electrical/optical systems on module and board level: The EOCB approach, *Proceedings of the 50th ECTC*, Las Vegas, NV, May 21–24, 2000.
7. A. Suzuki, T. Ishikawa, Y. Wakazono, 10-Gb/s × 12-ch downsized optical modules with electrical conductive film connector, *Proceedings of the 58th ECTC*, Lake Buena Vista, FL, May 27–30, 2008, pp. 250–255.
8. E. Griese, Reducing EMC problems through an electrical/optical interconnection technology, *IEEE Trans. Electromagn. Compat.*, November 1999, 41(4), 502–509.
9. E. Griese, Modeling of highly multimode waveguides for time domain simulation, *IEEE J. Sel. Top. Quantum Electron.*, March/April 2003, 9(2), 433–442.
10. T. Bierhoff, A. Wallrabenstein, A. Himmler, E. Griese, G. Mrozynski, Ray tracing technique and its verification for the analysis of highly multimode optical waveguides with rough surfaces, *IEEE Trans. Magn.*, 2001, 37(5, Part 1), 3307–3310.
11. M. Ramme, E. Griese, M. Kurten, Fast simulation method for a transient analysis of lossy coupled transmission lines using a semi-analytical recursive convolution procedure, *Proceeding of the 1997 IEEE International Symposium on EMC*, Austin, TX, 1997, pp. 277–282.
12. K. Halbe, E. Griese, An approach to model board-integrated multimode waveguides, *Proceedings of the 9th IEEE Workshop of Signal Propagation on Interconnects (SPI 2005)*, Bordeaux, France, 2005, pp. 79–82.
13. E. Griese, Entwurf und Simulation elektrisch-optischer Leiterplatten, in W. Scheel (ed.), *Optische Aufbau- und Verbindungstechnik in der elektronischen Baugruppenfertigung—Einführung*, pp. 220–283, Verlag Dr. Markus Detert, Templin, Germany, 2002.
14. Th. Bierhoff, Strahlenoptische Analyse der Wellenausbreitung und Modenkopplung in optisch hoch multimodalen Wellenleitern, Shaker Verlag, Aachen, Germany, 2006.
15. K. Halbe, E. Griese, A modal approach to model integrated optical waveguides with rough core-cladding-interfaces, *Proceedings of the 10th IEEE Workshop on Signal Propagation on Interconnects (SPI 2006)*, Berlin, Germany, 2006, pp. 133–136.
16. E.A.J. Marcatili, Dielectric rectangular waveguide and directional coupler for integrated optics, *Bell Syst. Tech. J.*, 1969, 48, 2071–2102.
17. D. Marcuse, *Theory of Dielectric Optical Waveguides*, Academic Press, New York, 1991.
18. E. Griese, D. Krabe, E. Strake, Electrical-optical printed circuit boards: Technology-design-modeling, in H. Grabinski (ed.), *Interconnects in VLSI Design*, pp. 221–236, Kluwer Publisher, Boston, MA, 2000.

19. Y. Soenmez, A. Himmler, E. Griese, G. Mrozynski, A ray tracing approach to model wave propagation in highly multimode graded index waveguides, *Proceedings of XI International Symposium on Theoretical Electrical Engineering (ISTET 2001)*, Linz, Austria, 2001, pp. 277–282.

20. Th. Kühler, E. Griese, Modeling propagation characteristics of multimode graded-index waveguides with finite elements using edge-based elements, *Proceedings of the 11th IEEE Workshop on Signal Propagation on Interconnects (SPI 2007)*, Genova, Italy, 2007, pp. 173–176.

21. Th. Kühler, E. Griese, An approach to model propagation characteristics of multimode graded-index waveguides manufactured by ion-exchange technology, *Proceedings of the 12th IEEE Workshop on Signal Propagation on Interconnects (SPI 2008)*, Avignon, France, 2008.

22. N. Tanio, Y. Koike, *POF Conference '97*, Kauai, HI, 1997, Session B, TuB-5, 33.

23. L. Eldada, L.W. Shacklette, Advances in polymer integrated optics, *IEEE J. Sel. Top. Quantum Electron.*, 6, 54, 2000.

24. T. Knoche, L. Müller, R. Klein, A. Neyer, Low loss polymer waveguides at 1300 nm and 1550 nm using halogenated acrylates, *Electron. Lett.*, 32, 1284, 1996.

25. R. Yoshimura, M. Hikita, S. Tomaru, S. Imamura, *J. Lightw. Technol.*, 1998, 14, 2338.

26. N. Keil, C. Weinert, W. Wirges, H.H. Yao, S. Yilmaz, C. Zawadzki, J. Schneider, J. Bauer, M. Bauer, *Electron. Lett.*, 2000, 36, 430.

27. J. Kobayashi, T. Matsuura, Y. Hida, T. Maruno, *J. Lightw. Technol.*, 1998, 16, 1024.

28. K. Han, K.H. You, J.H. Kim, E.J. Kim, W.H. Jang, T.H. Rhee, *KJF '99*, 1999, P-I-22, 52.

29. H.J. Lee, E.M. Lee, M.H. Lee, M.C. Oh, J.H. Ahn, S.G. Han, H.G. Kim, *J. Polym. Sci. Part A Polym. Chem.*, 1998, 36, 2881.

30. M. Usui, S. Imamura, S. Sugawara, S. Hayashida, H. Sato, M. Hikiti, T. Izawa, *Electron. Lett.*, 1994, 30, 958.

31. G. Fischbeck, R. Moosburger, C. Kostrzewa, A. Achen, K. Petermann, *Electron. Lett.*, 1997, 33, 518.

32. J. Bauer, C. Dreyer, M. Bauer, C. Zawadzki, S. Yilmaz, W. Wirges, H.H. Yao, N. Keil, *Proc. ACS, Div. PMSE*, 1999, 40, 1307.

33. H. Schröder, J. Bauer, F. Ebling, M. Franke, A. Beier, P. Demmer, W. Süllau, J. Kostelnik, R. Mödinger, K. Pfeiffer, U. Ostrzinski, E. Griese, *Invited talk* Waveguide and packaging technology for optical backplanes and hybrid electrical-optical circuit boards, *Proceedings of the Photonics West 2006*, January 21–26, 2006, San Jose, CA, SPIE 6115–6136.

34. H. Schröder, N. Arndt-Staufenbiel, M. Cygon, W. Scheel, Planar glass waveguides for high performance electrical optical circuit boards (EOCB)—the glass-layer-concept, *Proceedings of the 53rd Electronic Components and Technology Conference*, New Orleans, LA, 2003, pp. 1053–1059.

35. H. Schröder, A. Beier, R. Mödinger, S. Intemann, J. Kostelnik, E. Griese, I. Schlosser, High performance electrical optical circuit boards (EOCB) using planar optical interconnects integrated in thin glass foils, *Proceedings of the EOS Conference on Trends in Optoelectronics*, Munich, Germany, June 2007.

III

Integrated Photonic Technologies: Micro-Ring Structures

3

Silicon Microspheres for VLSI Photonics

Ali Serpengüzel
Yiğit Ozan Yılmaz
Ulaş Kemal Ayaz
Adnan Kurt

3.1 Introduction

Silicon-based microelectronic very large-scale integration (VLSI) is continuously developing in agreement with "Moore's law." This development makes integrated circuits (ICs) physically bigger and electronic signal paths longer. With the current developments in optoelectronics, metal interconnections are no longer the limiting factor for the performance of electronic systems. The replacement of the metal interconnections with optical interconnections could provide low-power dissipation, low latencies, and high bandwidths.

Silicon-based electrophotonic ICs (EPICs) are a possible solution to the synchronous clock distribution within these large ICs. EPICs, or, alternatively, optoelectronic ICs (OEICs) [1] are the natural evolution of the microelectronic IC with the added benefit of photonic capabilities. Future micro-EPIC technology requires the integration of microphotonic circuit elements with the already well-integrated microelectronic circuit elements.

Two-dimensional (2D) silicon planar lightwave circuits (PLCs) are currently being manufactured with microrings acting as active and passive microphotonic circuit elements. Light coupling is either performed from microcavity to microcavity or through the intermediation of optical waveguides. A three-dimensional (3D) extension of these PLCs can be envisioned as volumetric lightwave circuits (VLCs). Spherical silicon microshells or microspheres, which can be integrated in three dimensions, are the natural extensions of silicon microrings and microdisks. Silicon microspheres, with their high-quality-factor morphology-dependent resonances (MDRs), can be used in wavelength division multiplexing (WDM) applications, such as resonant filters, detectors, modulators, switches, light sources, wavelength converters, and variable attenuators. In this chapter, we will discuss the potential of silicon microspheres for VLSI photonics.

3.2 Silicon Photonics

Silicon is the most widely available semiconducting material, since it is the second most abundant material on the earth's crust after oxygen. Silicon has been the material of choice for the microelectronics industry for more than half a century [2]. Silicon is a relatively inexpensive and well-understood material for producing complementary metal-oxide semiconductor (CMOS)-based microelectronic devices [3].

Additionally, the need for low-cost photonic devices [4] and the need for high-speed intrachip communication [5,6] have stimulated a significant amount of research in silicon photonics [7]. Although silicon photonics [8,9] is less well developed as compared to the direct-bandgap III–V semiconductor photonics, silicon is poised to make a serious impact on optical communications [10]. Traditionally, the microelectronic VLSI industry has been based on group IV silicon, whereas the microphotonic industry on group III–V semiconductors. However, silicon-based photonic microdevices have been making strands in "siliconizing" photonics.

Figure 3.1 shows the transmission spectrum of crystalline silicon. Crystalline silicon, with a near-infrared (near-IR) bandgap of 1.1 eV, is transparent in the standard near-IR communication bands with wavelengths greater than 1.1 μm, and is a suitable high-refractive-index optoelectronic group IV material for WDM applications. The refractive index of the silicon is 3.5, which increases the optical path-length and makes the use of smaller silicon devices possible, thus offering an additional miniaturization advantage over devices fabricated from lower-refractive-index dielectric materials.

Recent progress in silicon photonics integration is being heralded by the observation of first the Raman gain [11,12] and then stimulated Raman scattering (SRS) [13] in crystalline silicon waveguides, SRS lasing first in pulse-modulated [14–16] and later in continuous-wave (cw) [17] silicon Raman lasers, and finally the hybrid silicon Raman laser [18]. Additionally, silicon modulators [19,20] have been developed first using a metal-oxide-semiconductor (MOS) capacitor [21], a Mach–Zehnder [22] configuration, SRS [23], and a micro-ring [24] configuration. Recently, a silicon micro-ring–based wavelength converter has been realized [25].

With well-established CMOS-processing techniques, it will be possible to integrate lasers, waveguides, modulators, switches, wavelength converters, and resonant cavity enhanced (RCE) photodetectors into silicon motherboards [26] for long-haul and short-haul WDM [27], and for rack-to-rack, backplane, interchip, and intrachip communications.

Passive and active silicon racetrack resonators [28], micro-ring resonators [29], and silicon waveguides with feedback [30,31] are some of the geometries pursued for 2D silicon PLCs.

Silicon microspheres, which are the natural extensions of silicon microracetracks, microdisks, and microrings, can therefore be integrated in three dimensions to fabricate future VLCs. Figure 3.2 shows a scanning electron micrograph (SEM) of a silicon microsphere.

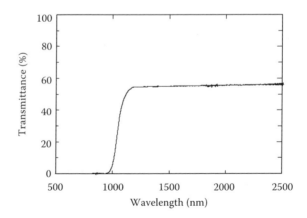

FIGURE 3.1   Transmission spectrum of crystalline silicon from the visible to the near-IR.

FIGURE 3.2   SEM of the 1 mm diameter silicon microsphere.

Table 3.1 summarizes various optical microcavity geometries with their optical modes, their respective quantization numbers, and degeneracies. The photonic crystal (PC)-based "defect" microcavities are not included in Table 3.1.

3.3 Modes of Various Geometry Optical Microresonators

In all the possible geometries, if the optical round-trip path is equal to an integral number of wavelengths, an optical resonance is formed. The resonance wavelengths (λ_n) satisfy the resonance condition:

$$\lambda_n = \frac{Lm}{n},$$

TABLE 3.1   Various Geometry Optical Microresonators, Relevant Dimensions, Mode Numbers, and Degeneracies

Resonator Geometry	Resonator Schematic	D	Mode Quantization	Mode Degeneracy
Fabry–Perot		1	Longitudinal (n)	Lateral Transverse
Micro-ring Microracetrack Microtoroid		1	Longitudinal (n)	Lateral Transverse
Microdisk		2	Radial (l) Azimuthal (n)	2 fold ($\pm n$)
Microspiral		2	Radial (l) Azimuthal (n)	None
Micropolygon		2	Radial (l) Azimuthal (n)	2 fold ($\pm n$)
Microsphere		3	Radial (l) Polar (n) Azimuthal (m)	($2n+1$)-fold
Microspheroid		3	Radial (l) Polar (n) Azimuthal (m)	2 fold ($\pm m$)

where

  n is the mode number
  L is the round-trip distance
  m is the refractive index of the microcavity with respect to the surroundings

3.3.1 1D Fabry–Perot Microresonators

In a planar microcavity, light is confined in the longitudinal dimension using two parallel mirrors. There are degeneracies in the transverse and the lateral dimensions. These degeneracies are lifted by the transverse and the lateral dimensions of the mirrors that form the planar microcavity.

3.3.2 2D Microresonators

In a circular microcavity, light is confined in the radial and azimuthal angular dimensions [32]. The microcirculator resonances are labeled by the radial mode number (l) and the azimuthal mode number (n). Additionally, there is a twofold (clockwise [CW] and counterclockwise [CCW]) degeneracy in the azimuthal dimension.

Among the 2D whispering gallery mode (WGM) microresonators are single [33] and coupled 1D chains of [34] microtoroids, single [35] and coupled 1D chains of [36] microspirals, single [37] micropolygons, single [38] and coupled 1D chains [39] and 2D arrays of [40] microrings, and single [41] and coupled 1D chains [42] and 2D arrays of [43] microdisks. Microspirals are novel structures, which break the CW and CCW symmetry of the other WGM microresonators, and have therefore directional optical output. These tiny optical cavities, whose diameters vary from a few to several hundred micrometers, have MDRs with reported Q-factors as large as 10^{10} [44].

3.3.3 3D Microspheres

The 3D analogue of the well-known 2D microracetrack, micro-ring, microtoroid, or microdisk resonators are the microsphere resonators, which are ideal 3D microphotonic building blocks due to their small volumes and high quality factors [45].

In a spherical microcavity, light is confined in the radial, polar, and azimuthal dimensions. The microsphere resonances are described by a set of vector-spherical harmonics [46] labeled by a polar mode number (n), a radial mode order (l), and an azimuthal mode number (m). The azimuthal mode number (m) takes values from $\pm n$, $\pm(n-1)$, ..., ± 1, to 0. For a microsphere, the azimuthal resonances for a given n are spectrally ($2n + 1$) degenerate [47], and MDRs, differing only in the azimuthal mode number, have identical resonance frequencies [32]. This azimuthal degeneracy is lifted in the ellipsoidal geometry. However, there is a twofold (CW and CCW) degeneracy in the azimuthal dimension even in ellipsoids.

A physical interpretation of MDRs is based on the propagation of rays around the inside surface of the microsphere [48], confined by an almost total internal reflection (TIR) [49]. The rays approach the internal surface at an angle beyond the critical angle and are totally internally reflected each time. After propagating around the microsphere, the rays return to their respective entrance points exactly in phase and then follow the same path all over again without being attenuated by destructive interference [50]. It takes longer for the energy of an MDR to leak out of the microsphere, and extremely large energy densities can accumulate in the MDR.

3.3.4 Relevant Resonance Parameters of Microspheres

MDR wavelengths depend on the size, the shape, and the refractive index of the microspheres [51]. MDRs satisfy resonance conditions for specific values of the size parameter

$$x = \frac{2\pi a}{\lambda},$$

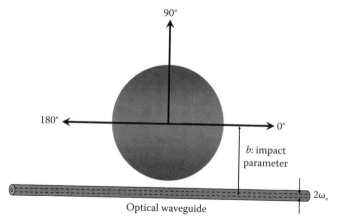

FIGURE 3.3 Optical waveguide (Gaussian beam) excitation geometry of a microsphere.

where

 a is the radius of the sphere
 λ is the vacuum wavelength of light [52]

The MDRs of a microsphere are analyzed by the localization principle [53] and the generalized Lorenz–Mie theory (GLMT) [54,55]. Figure 3.3 shows an example of GLMT elastic light-scattering geometry for a microsphere excited by an optical waveguide (Gaussian beam). The waveguide is placed at an impact parameter (b) from the microsphere, and the Gaussian beam has a full-width half-maximum (FWHM) of $2\omega_0$.

The linewidth ($\delta\lambda$) of the resonance is related to the quality factor (Q) by the following relation:

$$Q = \frac{\lambda}{\delta\lambda},$$

and to the finesse (F) by the following relation:

$$F = \frac{\Delta\lambda}{\delta\lambda},$$

where $\Delta\lambda$ is the spectral wavelength separation (mode spacing) between the adjacent MDR peaks with the same mode order (l) and consecutive mode numbers (n). The mode spacing ($\Delta\lambda$) is approximately given by the following relation:

$$\Delta\lambda = \frac{\lambda^2 tg^{-1}((m^2-1)^{1/2})}{(2\pi a(m^2-1)^{1/2})},$$

where m is the refractive index ratio of the microsphere to the surroundings [56].

Figure 3.4 shows the schematic of a typical elastic light-scattering spectrum indicating the mode spacing ($\Delta\lambda$) and the mode linewidth ($\delta\lambda$) of resonances with mode numbers around n.

Coupled microspheres have been used to construct 1D chains [57], 2D arrays [58], and 3D lattices [59]. Dielectric microspheres have found various applications in photonics [60] and optoelectronics [61]. Numerous potential applications have been proposed by using microsphere MDRs: microlasers [62], narrow filters [63], optical switching [64], ultrafine sensing [65], displacement measurement [66], detection of rotation [67], high-resolution spectroscopy [60], and Raman sources [61]. In addition to the microtoroids, the dielectric [68] and the silicon [69] microsphere is the only 3D

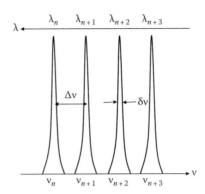

FIGURE 3.4   Schematic of the elastic scattering spectrum indicating the mode spacing ($\Delta\lambda$) and linewidth ($\delta\lambda$).

microresonator with a high quality factor. Additionally, microsphere resonators [63] are uniquely applicable for future 3D OEICs [70].

An impediment to the use of microsphere resonators in practical devices has been the difficulty of efficiently coupling light into and out of the microspheres. Light can be coupled into the microsphere using different types of coupling devices: side-polished optical fibers [71], prisms [72], and tapered optical fibers [73]. The principle of these devices is based on providing efficient energy transfer to the microsphere MDR through the evanescent field of the optical fiber. Efficient coupling can be expected if two main conditions are satisfied: phase synchronism and significant overlap of the two waves, the MDR mode and the optical fiber mode [70].

Fiber-optic add-drop filters based on a silica microsphere system on a taper-resonator-taper coupler have been developed [74]. Optical resonances from silicon microspheres in the near-IR telecommunication bands have also been observed [69].

3.4 Photonic Properties of Silicon Microspheres

The photonic properties of the silicon microsphere are measured using the following procedure. The silicon microsphere is coupled to an optical-fiber half coupler (OFHC). A tunable distributed feedback (DFB) laser is used to excite the MDRs of the silicon microsphere. Wavelength tuning is achieved by tuning the temperature of the DFB laser with a laser diode controller (LDC). The pigtailed DFB laser is coupled directly to the OFHC. The optical fiber used in the OFHC is a standard single-mode optical fiber (SMOF). The silicon microsphere used in the experiment is placed on the interaction region of the OFHC. A red positioning laser is used to indicate the coupling region on the OFHC. The elastically scattered light from the microsphere at 90° is collected by a microscope lens through a Glan polarizer and detected by an InGaAs photodiode (PD), as shown in Figure 3.5. The transmitted power through the optical fiber is detected by an optical multimeter (OMM) with InGaAs power/wave head (PWH). The InGaAs PD signal is sent to the digital storage oscilloscope (DSO) for signal monitoring and data acquisition. A near-IR viewer is used to observe the elastically scattered light from the microsphere. Data acquisition and control are performed with an IEEE-488 GPIB interface.

Figure 3.6 shows the elastic scattering spectrum (lower curve) at a scattering angle of 90° from the silicon microsphere and the power transmission spectrum (upper curve). In the elastic scattering spectrum, there is a background due to OFHC surface imperfections. The DFB laser is temperature-tuned in the o-band from 1302.2 to 1302.7 nm. The peaks in the elastic scattering correlate well with the dips in the transmission. The dips in the transmission are due to the coupling of the microsphere MDRs, which results in incident power losses. The mode spacing ($\Delta\lambda$) is in good agreement with the size and the refractive index of the silicon microsphere. The linewidths ($\delta\lambda$) of the MDRs are measured to be approximately 0.01 nm, which correspond to a quality factor on the order of 10^5.

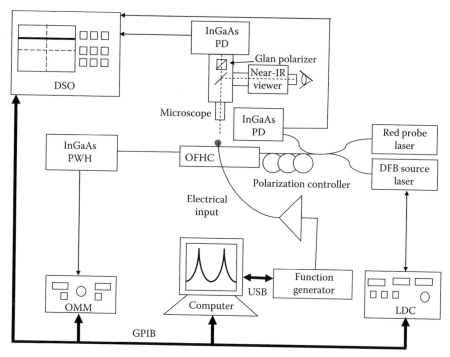

FIGURE 3.5 Schematic of the experimental setup for silicon microsphere measurements.

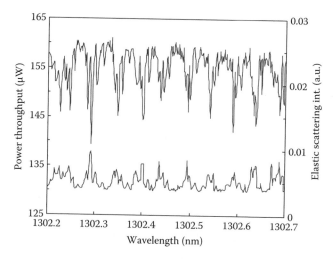

FIGURE 3.6 Elastic scattering (lower curve) and transmission (upper curve) spectra from the silicon sphere.

3.5 Electronic Properties of Silicon Microspheres

The silicon microsphere is electrically contacted by two metallic (M) electrodes, which double as positioners for the microsphere. The simple optoelectronic device has a metal-silicon-metal (MSM) geometry. Figure 3.7 shows a silicon microsphere contacted by two metal electrodes.

Figure 3.8 shows the measured current–voltage (IV) curve of the MSM silicon microsphere structure. The IV curve shows that the MSM structure acts as two back-to-back Schottky contacts. The asymmetry of the graph is a result of the asymmetry of the electrical contacts.

FIGURE 3.7 Image of the 1 mm diameter silicon (S) microsphere between two metal (M) electrodes.

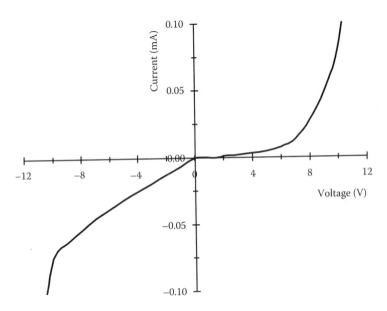

FIGURE 3.8 IV curve of the silicon microsphere MSM structure.

This electronic response together with the photonic response heralds the silicon microsphere as a potential optoelectronic device platform for RCE photodetectors, modulators, switches, sources, wavelength converters, and variable attenuators for future WDM telecommunications and VLSI photonics.

3.6 Conclusions

We have introduced silicon photonics as well as the mode properties of various geometry optical microresonators as a prelude to the electronic and photonic properties of silicon microspheres. The photonic response of the microsphere shows MDR peaks in the elastic scattering spectra and associated dips in the transmission spectra. The electronic response of the silicon microsphere is of a metal-silicon-metal (MSM) device. With the appropriate electrophotonic integration geometry, silicon microsphere–based optoelectronic devices show promise as future building blocks of VLSI photonics and potential VLCs.

Acknowledgments

We would like to acknowledge the partial support extended toward this research by TÜBİTAK Grant No. EEEAG-106E215 and EC Grant Nos. FP6-IST-003887 NEMO and FP6-IST-511616 PHOREMOST.

References

1. G. Guillot and L. Pavesi, *Optical Interconnects: The Silicon Approach*, Springer Verlag, Berlin (2006).
2. L. Pavesi, Will silicon be the photonic material of the next millennium? *J. Phys. Condens. Matter* 15, 1169–1196 (2003).
3. C. Gunn, CMOS photonics for high-speed interconnects, *IEEE Micro* 26, 58–66 (2006).
4. M. J. Paniccia, M. Morse, and M. Salib, Integrated photonics, in L. Pavesi and D. J. Lockwood (Eds.), *Silicon Photonics*, pp. 51–85, Springer Verlag, Berlin, Germany (2004).
5. G. T. Reed, G. Z. Mashanovich, W. R. Headley, S. P. Chan, B. D. Timotijevic, and F. Y. Gardes, Silicon photonics: Are smaller devices always better? *Jpn. J. Appl. Phys.* 45, 6609–6615 (2006).
6. L. C. Kimerling, L. Dal Negro, S. Saini, Y. Yi, D. Ahn, S. Akiyama, D. Cannon, J. Liu, J. G. Sandland, D. Sparacin, J. Michel, K. Wada, and M. R. Watts, Monolithic silicon microphotonics, in L. Pavesi and D. J. Lockwood (Eds.), *Silicon Photonics*, pp. 89–119, Springer Verlag, Berlin, Germany (2004).
7. L. Pavesi and D. J. Lockwood, *Silicon Photonics*, Springer-Verlag, Berlin, Germany (2004).
8. G. T. Reed and A. P. Knights, *Silicon Photonics: An Introduction*, John Wiley & Sons, New York (2004).
9. R. Soref, The past, present, and future of silicon photonics, *IEEE J. Sel. Top. Quantum Electron.* 12, 1678–1687 (2006).
10. H. Zimmermann, *Integrated Silicon Optoelectronics*, Springer Verlag, Berlin, Germany (2000).
11. A. Liu, H. Rong, M. J. Paniccia, O. Cohen, and D. Hak, Net optical gain in a low loss silicon-on-insulator waveguide by stimulated Raman scattering, *Opt. Express* 12, 4261–4268 (2004).
12. H. Rong, A. Liu, R. Nicolaescu, and M. J. Paniccia, Raman gain and nonlinear optical absorption measurements in a low-loss silicon waveguide, *Appl. Phys. Lett.* 85, 2196–2198 (2004).
13. R. Jones, H. Rong, A. Liu, A. Fang, and M. J. Paniccia, Net continuous wave optical gain in a low loss silicon-on-insulator waveguide by stimulated Raman scattering, *Opt. Express* 13, 519–525 (2005).
14. O. Boyraz and B. Jalali, Demonstration of a silicon Raman laser, *Opt. Express* 12, 5269–5273 (2004).
15. H. Rong, A. Liu, R. Jones, O. Cohen, D. Hak, R. Nicolaescu, A. Fang, and M. J. Paniccia, An all-silicon Raman laser, *Nature* 433, 292–294 (2005).
16. O. Boyraz and B. Jalali, Demonstration of directly modulated silicon Raman laser, *Opt. Express* 13, 796–800 (2005).
17. H. Rong, R. Jones, A. Liu, O. Cohen, D. Hak, A. Fang, and M. J. Paniccia, A continuous-wave Raman silicon laser, *Nature* 433, 725–728 (2005).
18. A. W. Fang, H. Park, O. Cohen, R. Jones, M. J. Paniccia, and J. E. Bowers, Electrically pumped hybrid AlGaInAs-silicon evanescent laser, *Opt. Express* 14, 9203–9210 (2006).
19. A. Liu, D. Samara-Rubio, L. Liao, and M. J. Paniccia, Scaling the modulation bandwidth and phase efficiency of a silicon optical modulator, *IEEE J. Sel. Top. Quantum Electron.* 11, 367–372 (2005).
20. L. Liao, A. Liu, R. Jones, D. Rubin, D. Samara-Rubio, O. Cohen, M. Salib, and M. J. Paniccia, Phase modulation efficiency and transmission loss of silicon optical phase shifters, *IEEE J. Quantum Electron.* 41, 250–257 (2005).
21. A. Liu, R. Jones, L. Liao, D. Samara-Rubio, D. Rubin, O. Cohen, R. Nicolaescu, and M. J. Paniccia, A high-speed silicon optical modulator based on a metal-oxide-semiconductor capacitor, *Nature* 427, 615–618 (2004).

22. L. Liao, D. Samara-Rubio, M. Morse, A. Liu, D. Hodge, D. Rubin, U. D. Keil, and T. Franck, High speed silicon Mach-Zehnder modulator, *Opt. Express* 13, 3129–3135 (2005).

23. R. Jones, A. Liu, H. Rong, M. J. Paniccia, O. Cohen, and D. Hak, Lossless optical modulation in a silicon waveguide using stimulated Raman scattering, *Opt. Express* 13, 1716–1723 (2005).

24. Q. Xu, S. Manipatruni, B. Schmidt, J. Shakya, and M. Lipson, 12.5 Gbit/s carrier-injection-based silicon microring silicon modulators, *Opt. Express* 15, 431–436 (2007).

25. S. F. Preble, Q. Xu, and M. Lipson, Changing the colour of light in a silicon resonator, *Nat. Photon.* 1, 293–296 (2007).

26. B. Jalali, S. Yegnanarayanan, T. Yoon, T. Yoshimoto, I. Rendina, and F. Coppinger, Advances in silicon on insulator optoelectronics, *IEEE J. Sel. Top. Quantum Electron.* 4, 938–947 (1998).

27. B. J. Offrein, R. Germann, F. Horst, H. W. M. Salemink, R. Beyerl, and G. L. Bona, Resonant coupler-based tuneable add-after-drop filter in silicon–oxynitride technology for WDM networks, *IEEE J. Sel. Top. Quantum Electron.* 5, 1400–1406 (1999).

28. B. D. Timotijevic, F. Y. Gardes, W. R. Headley, G. T. Reed, M. J. Paniccia, O. Cohen, D. Hak, and G. Z. Masanovic, Multi-stage racetrack resonator filters in silicon-on-insulator, *J. Opt. A* 8, S473–S476 (2006).

29. Q. F. Xu and M. Lipson, All-optical logic based on silicon micro-ring resonators, *Opt. Express* 15, 924–929 (2007).

30. B. Jalali, Teaching silicon new tricks, *Nat. Photon.* 1, 193–195 (2007).

31. A. Alduino and M. J. Paniccia, Interconnects—Wiring electronics with light, *Nat. Photon.* 1, 153–155 (2007).

32. P. W. Barber and R. K. Chang (Eds.), *Optical Effects Associated with Small Particles*, World Scientific, Singapore (1988).

33. S. M. Spillane, T. J. Kippenberg, K. J. Vahala, K. W. Goh, E. Wilcut, and H. J. Kimble, Ultrahigh-Q toroidal microresonators for cavity quantum electrodynamics, *Phys. Rev. A* 71, 013817 (2005).

34. M. Hossein-Zadeh and K. J. Vahala, Free ultra-high-Q microtoroid: A tool for designing photonic devices, *Opt. Express* 15, 166–175 (2007).

35. M. Kneissl, M. Teepe, N. Miyashita, N. M. Johnson, G. D. Chern, and R. K. Chang, Current-injection spiral-shaped microcavity disk laser diodes with unidirectional emission, *Appl. Phys. Lett.* 84, 2485–2487 (2004).

36. G. D. Chern, G. E. Fernandes, R. K. Chang, Q. Song, L. Xu, M. Kneissl, and N. M. Johnson, High-Q-preserving coupling between a spiral and a semicircle μ-cavity, *Opt. Lett.* 32, 1093–1095 (2007).

37. C. Li, N. Ma, and A. W. Poon, Waveguide-coupled octagonal microdisk channel add-drop filters, *Opt. Lett.* 29, 471–473 (2004).

38. J. Niehusmann, A. Vörckel, P. H. Bolivar, T. Wahlbrink, W. Henschel, and H. Kurz, Ultrahigh-quality-factor silicon-on-insulator microring resonator, *Opt. Lett.* 29, 2861–2863 (2004).

39. B. E. Little, S. T. Chu, P. P. Absil, J. V. Hryniewicz, F. G. Johnson, F. Seiferth, D. Gill, V. Van, O. King, and M. Trakalo, Very high-order microring resonator filters for WDM applications, *IEEE Photon. Technol. Lett.* 16, 2263–2265 (2004).

40. B. E. Little, S. T. Chu, W. Pan, and Y. Kokubun, Microring resonator arrays for VLSI photonics, *IEEE Photon. Technol. Lett.* 12, 323–325 (2000).

41. T. J. Johnson, M. Borselli, and O. Painter, Self-induced optical modulation of the transmission through a high-Q silicon microdisk resonator, *Opt. Express* 14, 817–831 (2006).

42. A. Nakagawa, S. Ishii, and T. Baba, Photonic molecule laser composed of GaInAsP microdisks, *Appl. Phys. Lett.* 86, 041112 (2005).

43. C. P. Michael, M. Borselli, T. J. Johnson, C. Chrystal, and O. Painter, An optical fiber-taper probe for wafer-scale microphotonic device characterization, *Opt. Express* 15, 4745–4752 (2007).

44. A. A. Savchenkov, V. S. Ilchenko, A. B. Matsko, and L. Maleki, Kilohertz optical resonances in dielectric crystal cavities, *Phys. Rev. A* 70, 051804(R) (2004).

45. K. J. Vahala (Ed.), *Optical Microcavities*, World Scientific, Singapore (2004).

46. C. F. Bohren and D. R. Huffman, *Absorption and Scattering of Light by Small Particles*, John Wiley & Sons, New York (1983).

47. H.-M. Lai, P. T. Leung, K. Young, P. W. Barber, and S. C. Hill, Time-independent perturbation for leaking electromagnetic modes in open systems with application to resonances in microdroplets, *Phys. Rev. A* 41, 5187–5198 (1990).

48. M. L. Gorodetsky, A. A. Savchenkov, and V. S. Ilchenko, Ultimate Q of optical microsphere resonators, *Opt. Lett.* 21, 453–455 (1996).

49. S. Arnold, Microspheres, photonic atoms and the physics of nothing, *Am. Sci.* 89, 414–421 (2001).

50. P. W. Barber and S. C. Hill, *Light Scattering by Particles: Computational Methods*, World Scientific, Singapore (1990).

51. P. R. Conwell, P. W. Barber, and C. K. Rushforth, Resonant spectra of dielectric spheres, *J. Opt. Soc. Am. A* 1, 62–66 (1984).

52. M. Pelton and Y. Yamamato, Ultralow threshold laser using a single quantum dot and a microsphere cavity, *Phys. Rev. A* 59, 2418–2421 (1999).

53. H. M. Nussenzveig, *Diffraction Effects in Semiclassical Scattering*, Cambridge University Press, Cambridge, U.K. (1992).

54. J. A. Lock and G. Gouesbet, Rigorous justification of the localized approximation to the beam shape coefficients in generalized Lorenz–Mie theory. I. On-axis beams, *J. Opt. Soc. Am. A* 11, 2503–2515 (1994).

55. G. Gouesbet and J. A. Lock, Rigorous justification of the localized approximation to the beam shape coefficients in generalized Lorenz–Mie theory. II. Off-axis beams, *J. Opt. Soc. Am. A* 11, 2516–2525 (1994).

56. R. Jia, D. Jiang, P. Tan, B. Sun, J. Zhang, and Y. Lin, Photoluminescence study of $CdSe_xS_{1-x}$ quantum dots in a glass spherical microcavity, *Chin. Phys. Lett.* 18, 1350–1352 (2001).

57. B. M. Möller, U. Woggon, and M. V. Artemyev, Band formation in coupled-resonator slow-wave structures, *Opt. Express* 15, 17362–17370 (2007).

58. V. N. Astratov, J. P. Franchak, and S. P. Ashili, Optical coupling and transport phenomena in chains of spherical dielectric microresonators with size disorder, *Appl. Phys. Lett.* 85, 5508–5510 (2004).

59. V. N. Astratov and S. P. Ashili, Percolation of light through whispering gallery modes in 3D lattices of coupled microspheres, *Opt. Express* 15, 17351–17361 (2007).

60. R. K. Chang and A. J. Campillo (Eds.), *Optical Processes in Microcavities*, World Scientific, Singapore (1996).

61. S. M. Spillane, J. T. Kippenberg, and K. J. Vahala, Ultralow threshold Raman laser using a spherical dielectric microcavity, *Nature* 415, 621–623 (2002).

62. M. Cai, O. Painter, and K. J. Vahala, Fiber-coupled microsphere laser, *Opt. Lett.* 25, 1430–1432 (2000).

63. T. Bilici, S. Isci, A. Kurt, and A. Serpengüzel, Microsphere-based channel dropping filter with an integrated photodetector, *IEEE Photon. Technol. Lett.* 16, 476–478 (2004).

64. H. C. Tapalian, J. P. Laine, and P. A. Lane, Thermooptical switches using coated microsphere resonators, *IEEE Photon. Technol. Lett.* 14, 1118–1120 (2002).

65. I. Teraoka, S. Arnold, and F. Vollmer, Perturbation approach to resonance shifts of whispering-gallery modes in a dielectric microsphere as a probe of a surrounding medium, *J. Opt. Soc. Am. B* 20, 1937–1946 (2003).

66. J. P. Laine, H. C. Tapalian, B. E. Little, and H. A. Haus, Acceleration sensor based on high-Q optical microsphere resonator and pedestal antiresonant reflecting waveguide coupler, *Sens. Actuat. A* 93, 1–7 (2001).

67. A. B. Matsko, A. A. Savchenkov, V. S. Ilchenko, and L. Maleki, Optical gyroscope with whispering gallery mode optical cavities, *Opt. Commun.* 233, 107–112 (2004).

68. M. L. Gorodetsky, A. D. Pryamikov, and V. S. Ilchenko, Rayleigh scattering in high-Q microspheres, *J. Opt. Soc. Am. B* 17, 1051–1057 (2000).

69. Y. O. Yilmaz, A. Demir, A. Kurt, and A. Serpengüzel, Optical channel dropping with a silicon micro-sphere, *IEEE Photon. Technol. Lett.* 17, 1662–1664 (2005).

70. A. Serpengüzel, S. Arnold, G. Griffel, and J. A. Lock, Enhanced coupling to microsphere resonances with optical fibers, *J. Opt. Soc. Am. B* 14, 790–795 (1997).

71. A. Serpengüzel, S. Arnold, and G. Griffel, Excitation of resonances of microspheres on an optical fiber, *Opt. Lett.* 20, 654–656 (1995).

72. M. L. Gorodetsky and V. S. Ilchenko, High-Q optical whispering gallery microresonators: Precession approach for spherical mode analysis and emission patterns, *Opt. Commun.* 113, 133–143 (1994).

73. J. C. Knight, G. Cheung, F. Jacques, and T. A. Birks, Phased matched excitation of whispering gallery modes by a fiber taper, *Opt. Lett.* 22, 1129–1131 (1997).

74. M. Cai, G. Hunziker, and K. J. Vahala, Fibre optic add-drop device based on a silica microsphere-whispering gallery mode system, *IEEE Photon. Technol. Lett.* 11, 686–687 (1999).

Silicon Micro-Ring Resonator Structures: Characteristics and Applications

Vilson Rosa
Almeida
Roberto Ricardo
Panepucci

4.1 Introduction

Integrated silicon photonics has its roots in the groundbreaking work of Soref et al. [1], which outlined the opportunities in transmission, modulation, and possible paths to detection and light generation. Today, silicon photonics is seen by many experts in the microelectronics industry as a strong candidate to answer the increasing demand for chip-to-chip and intra-chip communications in the decades to come [2–4]. In fact, the need arising from the microelectronics industry provides a "killer-application" for VLSI micro/nanophotonics [5]. The high-volumes commanded in the microelectronics industry are driving research into practical high integration solutions for on-chip photonics. Silicon has offered a unique platform in which to develop microelectronics, and now it appears poised to dominate the drive into large-scale photonic integration, partly due to the fabrication technology and knowledge base developed during the microelectronics era.

The initial interest in silicon for photonics arose from a material platform and processing perspective, and followed the design guidelines borrowed from III–V materials, such as InP and GaAs, with similar indices of refraction [1]. The approach in III–Vs relies on relatively easy-to-manufacture ridge waveguides, or low-index contrast claddings, with relatively low lateral confinement. Several important results using silicon ridge waveguides indicated a viable path to commercial success based on the needs of the fiber-optic communications market, and substantial advantages in integration over competing technologies, such as silica-on-silicon [6]. However, this approach began to lose ground over designs that took advantage of the high index of refraction contrast between silicon and silica, which led to strong optical confinement. High-contrast silicon photonic designs offered significantly smaller devices, and also offered compatibility with intriguing applications based on photonic crystals [7]. The development and deployment of fiber-optic amplifiers with wide wavelength bandwidth shifted the scaling strategy in fiber-optic systems toward dense-wavelength-division multiplexing (DWDM) solutions. With channel counts expected to reach over 300 channels, it became clear that high levels of on-chip integration would be required for fiber-optic communication systems to handle future scaling. The demonstration of wavelength filters using ultrasmall micro-ring resonators and high-contrast silicon waveguides by Little et al. [8] offered a very enticing motivation for this emerging technology, which was followed by the important demonstrations of all-optical switching [9], and viable compact electro-optical modulation devices [10,11], such that in recent years silicon nanowire waveguides have become the platform of choice for silicon photonics.

In this chapter, silicon micro-ring resonator structures in high-contrast silicon photonics and their role in the photonic VLSI effort are reviewed. In the next sections, we will cover the basics of the silicon micro-ring resonator—(1) the theory of operation, including its application as add-drop filters; (2) the fabrication methods and challenges; and (3) the tuning of micro-ring resonator filters. This introduction is followed by a description of the *slot-waveguide* ring-resonator structures and applications derived from the high optical power guided in the low-index slot surrounded by silicon. Relevant nonlinear optical effects arising in micro-ring devices are investigated with applications to VLSI photonics and, finally, the subject of ultrahigh-speed all-optical and electro-optical modulation schemes using silicon micro-ring resonators is described.

4.1.1 The Role of Polarization

Silicon—while not intrinsically birefringent—is used in rectangular cross-section devices that lead to significant waveguide birefringence. In addition, fabrication methods that result in differential roughness for the top/bottom versus the lateral confining boundaries of the waveguides lead to further differences, particularly in losses, for quasi-TE and quasi-TM polarizations. With different losses, the performance of the micro-ring resonator devices is markedly different for each polarization of the propagating field. This has been addressed by different approaches: (1) single polarization systems, (2) carefully designed non-birefringent devices [12–15], and (3) split-polarization followed by subsequent recombination [16]. Polarization dispersion is often encountered in fiber-optic systems, however, in chip-to-chip and intrachip communications one can afford to use a single polarization. This chapter focuses on devices that favor a single polarization, the quasi-TE mode. The quasi-TE and quasi-TM nomenclature refers to the silicon slab waveguide from which the rectangular cross-sections are fabricated. In quasi-TE modes the electric field is mostly parallel to the substrate interface, whereas in quasi-TM modes the magnetic field is mostly parallel to the substrate interface. A polarization-maintaining fiber must be used to ensure the predictive launching of the desired mode.

4.1.2 Silicon Photonic Nanowires as a High-Bandwidth Platform

The development of silicon as a practical photonic platform has been systematically addressed by the IBM group, which demonstrated very low losses for single mode waveguides of 3.6 dB/cm for TE-polarization at 1550 nm. Bending losses of 0.086 and 0.013 dB were measured for 90° bends of 1 and 2 μm, respectively [17]. Furthermore, they have measured group index and group velocity dispersion values that can be three orders of magnitude higher than those in single mode fibers [18]. In subsequent work they have shown the ability to reduce this dispersion through the addition of a thin over layer of silicon nitride [19]. Nonlinear effects leading to cross-channel modulation present limitations to DWDM systems that were explored through the simultaneous transmission of 32 wavelength channels carrying data across a single 50 mm silicon photonic waveguide wire with throughput of 1.28 Tb/s data (40 Gb/s per wavelength channel) [20].

4.2 Single Ring Resonator

Ring resonators, as well as disk resonators belong to the traveling-wave category of resonators, in contrast to the well-known standing-wave resonators, such as the Fabry–Perot cavity. In stand-alone disk or ring cavities, light circulates in a closed path and is coupled out by perturbations in the geometry of the structure. When coupled to one or more waveguides, ring resonators are a versatile multifunctional photonic component for optical circuits. The theoretical analysis of the ring resonator coupled to a single waveguide predicts that it is possible to extract all the optical power from the resonator at the so-called critical coupling condition [21,22]; under such a condition, the rate of optical radiation that leaves the ring resonator equals the rate of light being scattered from the ring resonator due to surface roughness, provided that absorption inside the ring resonator can be considered negligible. As mentioned earlier, these parameters are strongly polarization dependent.

The schematic shown in Figure 4.1 depicts a ring resonator coupled to a single waveguide, whose geometric parameters are the radius of curvature (r), the gap distance between waveguide and ring (g), the waveguide width (w), and the height (h). The complex coupling coefficient (κ) represents the fraction of the optical field that is coupled between waveguide and ring resonator, whereas the real through-transmission coefficient (τ) represents the optical field component that is not coupled. Owing to energy conservation, the following relation holds:

$$\tau = \sqrt{1 - |\kappa|^2}.$$

$$(4.1)$$

The optical field attenuation coefficient (α) accounts for all mechanisms of attenuation that the optical field undergoes while propagating inside the ring resonator, given by

$$\alpha = \frac{2\pi}{\lambda_0} \mathrm{Im}[n_{\mathrm{eff}}],$$

$$(4.2)$$

where

n_{eff} is the complex effective index of the optical eigenmode guided inside the resonator
λ_0 is the free-space wavelength of light
$\mathrm{Im}[\cdot]$ denotes the imaginary part of a complex quantity

For simplicity, we assume that the absorption inside the ring is negligible—what makes the attenuation become related exclusively with light scattering processes.

Resonances in the ring/disk resonator occur when the following condition is satisfied:

$$\frac{L}{\lambda_0} \mathrm{Re}[n_{\mathrm{eff}}] = m, \quad m \in Z,$$

$$(4.3)$$

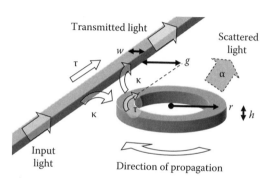

FIGURE 4.1 Schematic of an in-plane single-coupled ring resonator. The cross-section ($w \times h$) is assumed to be the same for both the waveguide and the ring resonator.

where

$L = 2\pi r$ is the ring resonator circumference.

Re[·] denotes the real part of a complex quantity.

Neglecting any optical power back-reflected into the input side waveguide, the optical power transmittance (T_p) under resonance condition becomes simply

$$T_p = \left(\frac{e^{-\alpha L} - \tau}{1 - \tau e^{-\alpha L}} \right)^2.$$ (4.4)

The *critical coupling* condition ($T_p = 0$) is attained when the following relationship is satisfied:

$$\tau = e^{-\alpha L}.$$ (4.5)

The quality factor (Q) of a resonance of a single-coupled ring resonator [23] is approximately given by

$$Q \cong \frac{\lambda_0}{\Delta\lambda_{\mathrm{FWHM}}} \cong \frac{\pi \sqrt{\tau e^{-\alpha L}} L n_g}{\left(1 - \tau e^{-\alpha L}\right) \lambda_0},$$ (4.6)

where

$\Delta\lambda_{\mathrm{FWHM}}$ is the full-width-at-half-maximum (FWHM) resonance bandwidth

n_g is the group index of light guided in the ring resonator

Under *critical coupling* condition from Equation 4.6, it is reduced to

$$Q \cong \frac{\lambda_0}{\Delta\lambda_{\mathrm{FWHM}}} \cong \frac{\pi e^{-\alpha L} L n_g}{\left(1 - e^{-2\alpha L}\right) \lambda_0}.$$ (4.7)

Although not explicitly shown above, the parameters: α, κ, τ, n_{eff}, and n_g are actually wavelength dependent, making numerical methods a requirement for accurate calculations.

It is straightforward to obtain an approximate estimate for n_g at the resonance wavelength of interest by making use of the average free spectral range (FSR) of a resonator, which is defined as the wavelength interval between two adjacent resonances. This approximation is given by

$$n_g \cong \frac{\lambda_0^2}{\left(\dfrac{\mathrm{FSR_L} + \mathrm{FSR_R}}{2} \right) L},$$ (4.8)

where $\mathrm{FSR_L}$ and $\mathrm{FSR_R}$ are the FSR to shorter and longer adjacent resonance wavelengths, respectively, with respect to the resonance of interest.

T_p, Q, FSR, and $\Delta\lambda_{\mathrm{FWHM}}$ are observable quantities that are measured by simple transmission spectra. It is possible to extract values for the internal parameters of the ring from these expressions. For more details the reader is directed to [24].

If $\alpha \cdot L \ll 1$, we obtain from Equations 4.7 and 4.8, an expression for the optical field propagation losses inside a ring resonator, in the vicinity of a resonance under critical coupling condition given by

$$\alpha \cong \frac{\pi \Delta\lambda_{\mathrm{FWHM}}}{L\left(\mathrm{FSR_L} + \mathrm{FSR_R}\right)} = \frac{\Delta\lambda_{\mathrm{FWHM}}}{2r\left(\mathrm{FSR_L} + \mathrm{FSR_R}\right)}.$$ (4.9)

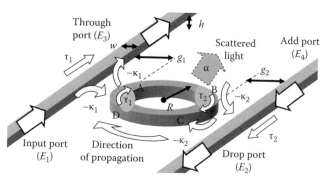

FIGURE 4.2   Schematic of an in-plane double-coupled (add-drop configuration) ring resonator. The cross-section ($w \times h$) is assumed to be the same for waveguides and ring resonator.

It is noteworthy to say that it is more widely accepted to express optical losses in terms of the optical power propagation losses α_p ($\alpha_p = 2\alpha$), which is interchangeably represented simply by the symbol α.

This simple single-coupled configuration has found applications as filters [8,25], as single-mode laser stabilizers [26], as resonant photodetectors [27], and as optical modulators, as described later in this chapter.

Ring and disk resonator cavities symmetrically coupled to two waveguides (double-coupled) correspond to the most widely used geometry to attain add-drop wavelength filters [28,29]. For ring resonators in the add-drop configuration shown in Figure 4.2, the expressions take on a more elaborate form and their derivation can be found elsewhere [21]. These analytical expressions can yield valuable approximations to the problem; however, as mentioned earlier the wavelength dependence of α, κ, τ, n_{eff}, and n_g must be taken into account to adequately describe the wavelength response. Simple back-of-the-envelope calculations will yield incorrect FSR, as will numerical simulations that do not include the material wavelength dependence.

4.3 Fabrication of Silicon Micro-Ring Resonators

The fabrication requirements for the critical dimension in silicon nanophotonics described here are currently met by the CMOS requirements as set forth in the ITRS roadmap for the 65 nm node and beyond [30]. The line edge roughness (LER) defined in the roadmap indicates RMS values below 3.5 nm; however, no specifications exist for the spatial frequency distribution of this roughness. Smith et al. [31] warn that relying on values for resolution and LER requirements is a misleading assumption as the requirements for photonics are based on unique phase fidelity properties of propagating waves instead of the statistics of electron transit times. These requirements, while addressing both the pattern generation and the etching of the structures, are typically challenging to the former. Challenges to the photolithographic definition of the photonic structures exist and need to be addressed. This is especially true when precise wavelength values must be matched, as the high sensitivity of micro-ring filter designs on the dimensions of micro-ring resonators leads to a 20 GHz resonator frequency shift for a 1 nm dimensional change in a radius of 5000 nm [32]. Stepper photolithography has been successfully used for silicon micro-ring resonator fabrication [33], however, access to high-performance systems providing the needed resolution is difficult, and mask costs can be substantial. Dumon et al. [33] describe a systematic characterization of deep UV for the fabrication of waveguides and micro-ring resonators. Their findings for waveguide loss versus waveguide width are consistent with the detailed analysis by Popovic [34], and measured Q-factors of 8000 were reported with losses per round trip for 500 nm wide waveguides of 0.2 dB, corresponding to 5 dB/mm on a 5 μm radius race-track resonator.

In the current scientific literature, most of the reported silicon photonic micro-ring resonator devices are based on electron beam lithography pattern generation. This is due to the high resolution available and the ease with which patterns can be modified. At universities, many such systems are based on adapted scanning electron microscopes (SEMs), which suffer from small fields around 300 μm and low writing speeds. Stitching write-fields on such systems is difficult and requires specialized techniques such as spatial phase locking [35,36]. In order to avoid stitching problems, several groups have devised approaches based on fabricating devices in single fields [37]. Commercial high-resolution electron beam systems have enabled significant developments in the field [9,38–40], however losses on wave-guides are still high [17]. Certain challenges are unique to this technique, such as stitching boundaries, square exposure grid, and electron proximity effect, all of which affect the resulting pattern. This has led researchers to a design-for-manufacturing approach, by exploring the design space for micro-ring resonators in search for reduced sensitivity in fabrication requirements [34,37].

Nanoimprint lithography has been demonstrated in the direct fabrication of silicon photonic structures [41] as the technique is capable of very high fidelity in the pattern transfer [42]. Currently there are several systems in the market that offer this capability at affordable prices with limited alignment capabilities. This technique requires a master mold with 1:1 dimensions, typically fabricated by electron beam lithography. Plachetka et al. [43] report on a soft UV-nanoimprint fabrication scheme used to fabricate micro-ring resonators with quality factors $Q = 47,700$. A waveguide loss of 3.5 dB/cm was extracted in contrast with a 1.9 dB/cm loss from a comparable e-beam patterned sample. A soft template, fabricated by replicating the desired pattern onto polydimethylsiloxane (PDMS) material, was used. The template patterned an UV-curable resist material (Amonil, AMO GmbH) over a 100 nm SiO_2 hardmask used to etch the silicon device layer. The ring diameter was 20 μm with coupling distance of 265 and 443 nm waveguide width. An alternative approach was initially reported by Scheerlinck [44], where the authors used a mold fabricated by deep UV lithography in silicon, and treated with anti-stiction layer, prior to thermal nanoimprinting into a PMMA layer. This pattern was then etched into a silicon on insulator (SOI) wafer with SF6 plasma. PMMA has shown very high resolution in thermal nanoimprint; however, dimensional control needs to be demonstrated.

Using careful calibration of exposures requiring repetitive experiments has allowed researchers to achieve 1 GHz matching in third-order filters with average ring-waveguide widths matched to within 26 pm dimensions [36]. Nonetheless, it will be important to find approaches that relief the fabrication requirements for VLSI photonic systems. From Equation 4.5 for the critical coupling, it is possible to increase roundtrip losses to achieve this condition, for example using ion-implantation [45,46], and more generally to tune device characteristics, such as done by Chu et al. [47] with a polysilane polymer whose refractive index is adjusted by UV exposure. Due to silicon's low electro-optic effect, the simplest approach to tuning such devices is through the use of thermo-optic effect. Therefore design for manufacture together with post-fabrication adjustments and tuning will be required in future applications.

4.4 Tuning of Micro-Ring Resonators

The thermo-optic coefficient of silicon ($\partial n/\partial T \cong 1.86 \times 10^{-4} K^{-1}$) [48] can be used to tune the resonance wavelength in ultrasmall micro-ring resonators at the rate of approximately 0.1 nm/K [49]. While thermo-optic tuning has moderate speed compared with electro-optic [11] and all-optic tuning [9], with silicon's high thermo-optic coefficient, a much wider wavelength-tunable range can be realized. The large tunable range and large FSR of silicon micro-ring resonators has enabled the development of wavelength-reconfigurable multi-channel matrix switch [50,51]. Baehr-Jones et al. [52] developed a semi-analytic approach to evaluate the tuning sensitivity of micro-ring resonators of different dimensions as a function of temperature. They found good agreement between their model and measured data for varying cladding index shift. In [53], the authors demonstrate an optimized heater design for Si-rich SiN micro-rings, with efficient on-chip thermal tuning of a second-order silicon-rich silicon nitride resonator. The tuning achieved required $80\,\mu W/GHz$ with closed loop control maintaining the target frequency within $0.3\,GHz$, achieved with a resonant wavelength tuning of $30\,pm/K$.

4.5 Slot-Waveguide Resonators

The optical slot-waveguide concept was introduced by Almeida et al. [54] and first demonstrated by Xu et al. [55], having been originated as an unexpected outcome of theoretical studies on metal-oxide-semiconductor (MOS) electro-optic modulation in high-confinement silicon waveguides (in http://en.wikipedia.org/wiki/Slot_waveguide, as seen on October 7, 2008). These references have shown theoretically and experimentally, respectively, that the electric component of the optical field profile can be efficiently enhanced and concentrated inside a low-refractive-index region, the so-called slot, with light guided by total internal reflection (TIR) principle, instead of by interference phenomena as in other approaches [54]. Figure 4.3a schematically shows a 3D slot-waveguide structure, where n_C and n_S are the low refractive indexes (e.g., silica and/or air) for the cladding and slot, respectively, and n_H is the high refractive index in the core (e.g., silicon). Figure 4.3b shows the transverse electric field (E-field) profile for the quasi-TE eigenmode of a 3D slot-waveguide based on SOI platform; notice that inside the slot region the E-field amplitude is relatively high and that the E-lines are oriented almost perpendicular to the interfaces.

The reason why the major E-field distribution concentrates preferentially inside the slot region for a specific polarization state relies naturally on Maxwell's equations and constitutive relations. Considering a high-index-contrast interface, Maxwell's equations state that in order to satisfy the continuity of the component of electric flux density D perpendicular to the interface, the correspondent E-field must undergo a discontinuity proportional to n_H^2/n_S^2, with higher amplitude in the low-refractive-index side; the ratio $n_H^2/n_S^2 = 6$ for a Si–SiO$_2$ interface, and $n_H^2/n_S^2 = 12$ for a Si–air interface. This principle of operation can be readily extended to horizontal and/or multiple slots [56]. The fortunate fact that the E-field becomes stronger inside the slot region is of foremost importance for applications such as optical amplification, lasing, and nonlinear optics [57], where the square of the E-field amplitude is the relevant quantity to be taken into account. Although the slot-waveguide can be fabricated on virtually any material platform, the silicon-silica/air (i.e., SOI) approach is straightforward and provides the high-index contrast required to cause high electric field concentration inside the slot region.

Even though the E-field is strongly confined in low-refractive-index material, the slot waveguides stay compatible with highly integrated photonics, allowing for the fabrication of micrometer-sized photonic structures and devices [55,58] such as silicon micro-ring resonators [55,59], as seen in Figure 4.4. The silicon micro-ring resonators based on slot-waveguides have direct applications in optical filtering, high-sensitivity refractive-index-change and gas sensors [60], as well as ultrafast optical modulation and switching [59]; the latter application relies on the fact that a slot waveguide presents much shorter

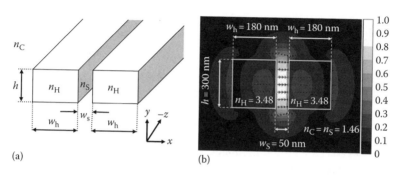

FIGURE 4.3 3D slot-waveguide. (a) Schematic of a rectangular cross-section channel structure. (b) Transverse E-field profile of a typical quasi-TE mode, showing the contour of the total E-field amplitude and lines. (Reproduced and adapted from Almeida, V. R. et al., *Opt. Lett.*, 29(11), 1209, 2004. With permission. © 2004 Optical Society of America, Inc.)

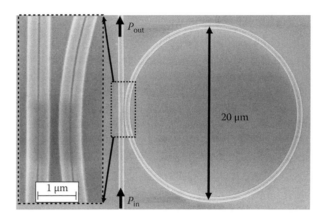

FIGURE 4.4 Scanning electron micrograph of a micro-ring resonator based on slot waveguide structure. The coupling region is enlarged in the inset. (Reproduced from Xu, Q. et al., *Opt. Lett.*, 29(14), 1626, 2004. With permission. © 2004 Optical Society of America, Inc.)

free-carrier lifetime than that of a conventional silicon channel waveguide, as a direct consequence of presenting a larger area of non-passivated sidewalls, allowing for a faster recombination lifetime of free-carriers used for silicon modulation via free-carrier optical dispersion effect.

Although propagation losses in silicon micro-ring resonators based on slot-waveguides strongly depend on the surface roughness of the waveguide walls, particularly those located in the slot region, experimental results have shown it is possible to attain high resonance quality-factor levels even for a micro-ring with a radius of only $10\,\mu m$ [61].

The slot waveguide concept, whether or not implemented in a micro-ring structure, has been proposed and demonstrated for several distinct applications, devices, and material platforms, where its high confinement optical properties have been proved to be very advantageous over other approaches, such as in: nano-opto-electro-mechanical devices and sensors (NOEMS) [62], electro-optic modulators based on polymers [63], light emission enhancement in erbium-doped silicon nanoclusters [64], ultrasmall cavity mode volumes [65], optical gain and low-threshold lasing [57], biosensing [66], temperature dependence control [67], photonic crystal planar waveguides [68,69] and photonic crystal fibers [70]. Surprisingly, the slot waveguide structure appears as a natural solution for high confinement devices, spontaneously emerging from simulations employing evolutionary algorithm simulations [71].

4.6 Nonlinear Optical Effects

The initial challenge in using crystalline silicon for optical high-speed active devices, such as optical modulators and routers, arises from the fact that crystalline silicon does not intrinsically present a strong and fast nonlinear optical mechanism: the optical Pockels effect caused by the second-order optical nonlinearity ($\chi^{(2)}$) is absent due to the symmetry properties of silicon crystals, and the optical Kerr effect, caused by the real part of the third-order optical susceptibility ($\chi^{(3)}$), is usually relatively small in bulk and large-cross-section waveguides [72]; however, it may become very relevant for highly integrated silicon photonics. The Raman effect is a phonon-assisted optical nonlinear mechanism that has enabled silicon highly integrated photonics as an active test bed for optical amplification, wavelength conversion and lasing [73–76].

Silicon waveguides and structures with submicron dimensions, including micro-ring resonators, can be engineered in order to provide appropriate linear optical properties, such as the chromatic dispersion coefficients, thus enabling efficient use or prevention of silicon optical nonlinearities. The appropriate tailoring of silicon waveguides and micro-structure geometries and dimensions enables the combination of chromatic dispersion and nonlinear optical properties on innumerous ways, allowing for

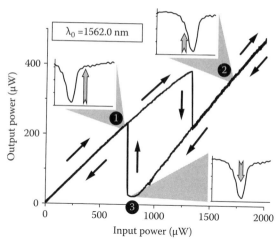

FIGURE 4.5 Nonlinear thermo-optical hysteresis curve for a quasi-TE mode resonance of a SOI-based micro-ring resonator. The markers define points of operation in the hysteresis curve associated with the associated transmission spectra shown in the insets: ❶ non-shifted spectrum on the linear region before hysteresis loop; ❷ shifted spectrum due to high input optical power, beyond the hysteresis loop; ❸ resonance shifted to the optical pump wavelength during decreasing input power while returning from beyond the hysteresis loop. (Reproduced from Almeida, V. R. and Lipson, M., *Opt. Lett.*, 29(20), 2387, 2004. With permission. © 2004 Optical Society of America, Inc.)

functional and compact on-chip nonlinear optical devices with suitable parameters and functionalities; indeed, the strong optical confinement also modifies the effective nonlinear optical coefficients [77]. The most relevant silicon optical nonlinear effects and parametric processes are: spontaneous/stimulated Raman scattering/emission (both Stokes and anti-Stokes), optical Kerr effect, four-wave mixing (FWM), self-phase modulation (SPM), cross-phase modulation (XPM), and two-photon absorption (TPA); the latter being related to the imaginary part of $\chi^{(3)}$ [77]. The third-order optical nonlinear effects (optical Kerr and TPA) and parametric processes (such as FWM, SPM, and XPM) in silicon are related to its nonlinear electronic polarizability with a response time of tens of femtoseconds, whereas silicon Raman effect presents a frequency shift of 15.6 THz and a spectral width of 105 GHz, providing a response time of approximately 10 ps.

The thermo-optical effect [78], which is strong though relatively slow, on the order of microseconds, originates from silicon's large thermo-optic coefficient, $\partial n/\partial T = 1.86 \times 10^{-4}\,\mathrm{K}^{-1}$, and has also become a promising alternative mechanism for some applications employing micro-ring resonators [79,80] as mentioned earlier. Figure 4.5 shows a typical hysteresis curve obtained from thermo-optical effect in a silicon micron-ring resonator, thus enabling the achievement of all-optical memory and switching capabilities via thermo-optical bistability [79].

On the other hand, the most promising nonlinear electro-optical mechanism in silicon—the free-carrier optical dispersion effect, by means of which the presence of free-carriers alters the optical properties of silicon—shows only a weak dependence for both the refractive index and the absorption coefficients on the free-carrier concentration [81], which are quantitatively expressed, for a wavelength of $\lambda_0 = 1.55\,\mu\mathrm{m}$, by

$$\Delta n = \Delta n_e + \Delta n_h = -[8.8 \times 10^{-22}\,\Delta N + 8.5 \times 10^{-18}\,(\Delta P)]^{0.8}, \tag{I}$$

and

$$\Delta \alpha = \Delta \alpha_e + \Delta \alpha_h = 8.5 \times 10^{-18}\,\Delta N + 6.0 \times 10^{-18}\,\Delta P, \tag{II}$$

respectively, where Δn_e and Δn_h are the refractive index variations due to electron and hole concentrations change, respectively; ΔN and ΔP are the electron and hole concentrations change, respectively, in cm^{-3}; and $\Delta\alpha_e$ and $\Delta\alpha_h$ (in cm^{-1}) are the free-carrier absorption (FCA) coefficient variations due to ΔN and ΔP, respectively. For submicron cross-sectional waveguides and micro-ring resonators, time response for the free-carrier optical dispersion effect may range from tens of picoseconds [82], through the usual hundreds of picoseconds [9], eventually reaching the nanosecond scale [73]. The free-carrier dispersion effect influences the efficiency of the Raman effect, via an intricate combination of TPA and FCA, as well as by free-carrier-based XPM [74]; it also allows for attaining fast optical bistability in a silicon micro-ring resonator [83].

Micro-ring silicon resonators, being small-footprint traveling-wave optical cavities, present a natural and straightforward approach for enhancing the overall optical intensity inside silicon on resonances, thereby reducing the total optical power input demanded for unleashing its nonlinear optical, thermo-optical, and electro-optical properties. Furthermore, the proper choice of geometrical parameters and dimensions (cross-section and ring) of micro-ring silicon resonators also makes it possible to engineer their optical linear properties, such as resonance wavelengths, bending losses, and chromatic dispersion. The proper combination of its linear and nonlinear optical properties has turned micro-ring silicon resonators into efficient and useful devices, as follows.

4.7 All-Optical and Electro-Optical Modulation

Silicon-based optical active devices present optical properties that can be modified as desired by means of an external physical agent, which may be of electrical, thermal, or optical nature [59]. In the case that only optical means are used to change its optical properties, the device is regarded as an all-optical one. Silicon micro-ring resonators have been widely exploited as both all-optical and electro-optical modulators, by means of the free-carrier dispersion effect [9,11,83–85]. For all-optical modulators based on silicon micro-ring resonators, both linear and nonlinear absorption mechanisms have been used for allowing the photo-generation of free-carriers by ultrafast pump pulses [9,84]. On the other hand, for the electro-optical counterpart, the structure of the device consists of a micro-ring resonator with a p-i-n junction embedded inside it [85], eventually encompassing the coupling waveguide for achieving fast (12.5 Gb/s) and efficient free-carrier injection and extraction [11]; further improvements in its structure have been proposed toward creating a p-i-n-i-p junction in order to push the speed limit well above, perhaps reaching the 100 Gb/s landmark [86]. Figure 4.6 shows a SEM of a conventional silicon single-coupled micro-ring resonator with a radius of only 5 μm.

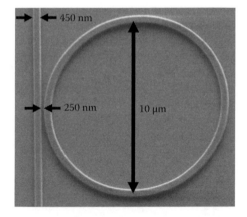

FIGURE 4.6   SEM of a SOI-based single-coupled micro-ring resonator. The height of all structures is $h = 250$ nm. (Reproduced from Almeida V. R. et al., *Opt. Lett.*, 29(24), 2867, 2004. With permission. © 2004 Optical Society of America, Inc.)

The speed performance of both all-optical and electro-optical silicon micro-ring modulators based on free-carrier dispersion effect is primarily limited by the free-carrier dynamics during injection and extraction, which may be controlled inside the optical device by a combination of several distinct mechanisms, including: photo-excitation by means of linear or nonlinear optical absorption (such as TPA) [9,84], fast electrical free-carrier injection and extraction [11], free-carrier recombination inside the doped core and/or on the unpassivated structure sidewalls [59], etc. The intrinsic highly integrated silicon free-carrier lifetime, which is primarily due to fast recombination mechanisms on the unpassivated sidewalls of the fabricated structures, is usually much shorter than that in bulk silicon; nonetheless, it does not represent a fundamental limit for the device speed, because it can be further shortened by manipulating the degree of surface passivation and/or by using ion implantation [87].

Besides direct optical modulation, silicon micro-ring resonators have been used for taking its advantageous linear and nonlinear optical properties toward harnessing photons as demanded, such as: by indirect modulation, for low-power wavelength converter based on the free-carrier dispersion effect in silicon [88]; and by direct parametric wavelength conversion, via FWM using ultralow peak pump powers [89]. Furthermore, not only direct or indirect amplitude modulation due to the spectral shift of a single micro-ring resonator has been investigated, but also quality-factor (Q) electro-optic modulation of a coupled double-ring microcavity has allowed for advanced functionalities [90].

An initial source of criticism regarding the use of silicon micro-ring resonators as optical modulators arose from the temperature dependence of the resonant spectrum, caused by silicon large thermo-optical effect. In order to mitigate such temperature effects on the device performance, strain in the silicon devices could be employed [91], introduced in the fabrication process by controlling the overcladding deposition conditions [92]; such a strain introduced on the silicon device induces a decrease of the refractive index with temperature, which counterbalances the thermo-optical effect in silicon [91]. Additionally, micro-ring resonators presenting intrinsic free-carrier recombination times of about 450 ps may alternatively be used as 20 GHz modulators by adopting a scheme that employs a reduced modulation depth of only 10% [9,84], reducing the device modulation time scale to about 50 ps; the dissipated pump pulse energy needed for this modulation depth may then reach a mere 17 fJ, much lower than that demanded for near 100% modulation depth (150 fJ). In such a case, taking into account the thermal conductivity of Si and SiO_2, the average dissipated power under 20 GHz modulation scheme induces a temperature increase around the micro-ring resonator of approximately 3 K, which is expected to have a minimum effect on the performance of the ring resonators since the typical time scale of the thermo-optical effect is on the order of a few microseconds [79].

References

1. R. A. Soref and J. P. Lorenzo, All-silicon active and passive guided-wave components for $\lambda = 1.3$ and 1.6 μm, *IEEE Journal of Quantum Electronics*, 22, 873–879 (1986).
2. Y. Vlasov et al., High-throughput silicon nanophotonic wavelength-insensitive switch for on-chip optical networks, *Nature Photonics*, 2, 242–246, April 2008.
3. A. Barkai et al., Integrated silicon photonics for optical networks [Invited], *Journal of Optical Networking*, 6, 25–47 (2007).
4. D. Vantrease et al., Corona: System implications of emerging nanophotonic technology, *Proceedings of the 35th International Symposium on Computer Architecture* (ISCA-35), Beijing, China (2008).
5. R. A. Soref, The past, present, and future of silicon photonics, *IEEE Journal of Selected Topics in Quantum Electronics*, 12, 1678–1687 (2006).
6. G. T. Reed and A. P. Knights, *Silicon Photonics: An Introduction.* Chichester, U.K.; Hoboken, NJ: John Wiley & Sons, 2004.
7. L. Pavesi and D. J. Lockwood, *Silicon Photonics.* Berlin; New York: Springer, 2004.
8. B. E. Little et al., Ultra-compact Si-SiO₂ micro-ring resonator optical channel dropping filters, *IEEE Photonics Technology Letters*, 10, 549–551 (1998).

9. V. R. Almeida, C. A. Barrios, R. Panepucci, and M. Lipson, All-optical control of light on a silicon chip, *Nature*, 431, 1081–1084 (2004).

10. C. A. Barrios et al., Electrooptic modulation of silicon-on-insulator submicrometer-size waveguide devices, *Journal of Lightwave Technology*, 21, 2332–2339 (2003).

11. Q. Xu et al., 12.5 Gbit/s carrier-injection-based silicon micro-ring silicon modulators, *Optics Express*, 15(2), 430–436 (2007).

12. W. R. Headley et al., Polarization-independent optical racetrack resonators using rib waveguides on silicon-on-insulator, *Applied Physics Letters*, 85, 5523–5525 (2004).

13. G. T. Reed et al., Issues associated with polarization independence in silicon photonics, *IEEE Journal of Selected Topics in Quantum Electronics*, 12, 1335–1344 (2006).

14. M. K. Chin, Polarization dependence in waveguide-coupled micro-resonators, *Optics Express*, 11, 1724–1730 (2003).

15. Z. C. Wang et al., Polarization-insensitive ultrasmall micro-ring resonator design based on optimized Si sandwich nanowires, *IEEE Photonics Technology Letters*, 19, 1580–1582 (2007).

16. T. Barwicz et al., Polarization-transparent microphotonic devices in the strong confinement limit, *Nature Photonics*, 1, 57–60 (2007).

17. Y. A. Vlasov and S. J. McNab, Losses in single-mode silicon-on-insulator strip waveguides and bends, *Optics Express*, 12, 1622–1631 (2004).

18. E. Dulkeith et al., Group index and group velocity dispersion in silicon-on-insulator photonic wires, *Optics Express*, 14, 3853–3863 (2006).

19. J. I. Dadap et al., Nonlinear-optical phase modification in dispersion-engineered Si photonic wires, *Optics Express*, 16, 1280–1299 (2008).

20. B. G. Lee et al., Ultrahigh-bandwidth silicon photonic nanowire waveguides for on-chip networks, *IEEE Photonics Technology Letters*, 20, 398–400 (2008).

21. A. Yariv, Universal relations for coupling of optical power between microresonators and dielectric waveguides, *Electronics Letters*, 36, 321–322 (2000).

22. Y. Xu et al., Scattering-theory analysis of waveguide-resonator coupling, *Physical Review E*, 62, 7389–7404 (2000).

23. P. Rabiei et al., Polymer micro-ring filters and modulators, *Journal of Lightwave Technology*, 20, 1968–1975 (2002).

24. S. J. Xiao et al., Modeling and measurement of losses in silicon-on-insulator resonators and bends, *Optics Express*, 15, 10553–10561 (2007).

25. P. P. Absil et al., Compact micro-ring notch filters, *IEEE Photonics Technology Letters*, 12, 398–400 (2000).

26. L. Bach et al., Wavelength stabilized single-mode lasers by coupled micro-square resonators, *IEEE Photonics Technology Letters*, 15, 377–379 (2003).

27. L. Chen et al., High performance germanium photodetectors integrated on submicron silicon waveguides by low temperature wafer bonding, *Optics Express*, 16, 11513–11518 (2008).

28. D. Rafizadeh et al., Waveguide-coupled AlGaAs/GaAs microcavity ring and disk resonators with high finesse and Z1.6-nm free spectral range, *Optics Letters*, 22, 1244–1246 (1997).

29. B. E. Little et al., Micro-ring resonator channel dropping filters, *Journal of Lightwave Technology*, 15, 998–1005 (1997).

30. W. Conley et al., in *International Technology Roadmap for Semiconductors* (ITRS), 2007 edition, http://www.itrs.net/Links/2007ITRS/2007_Chapters/2007_Lithography.pdf (2007).

31. H. I. Smith et al., Strategies for fabricating strong-confinement micro-ring filters and circuits, *Optical Fiber Communication and the National Fiber Optic Engineers Conference, 2007* (OFC/NFOEC 2007), San Diego, CA, pp. 1–3 (2007).

32. C. W. Holzwarth et al., Accurate resonant frequency spacing of micro-ring filters without postfabrication trimming, *Journal of Vacuum Science and Technology B*, 24, 3244–3247 (2006).

33. P. Dumon et al., Low-loss SOI photonic wires and ring resonators fabricated with deep UV lithography, *IEEE Photonics Technology Letters*, 16, 1328–1330 (2004).

34. M. Popovic, Theory and design of high-index-contrast microphotonic circuits, PhD thesis, Department of Electrical Engineering and Computer Science, Massachusetts Institute of Technology, Cambridge, MA (2008).

35. J. Sun et al., Accurate frequency alignment in fabrication of high-order micro-ring-resonator filters, *Optics Express*, 16, 15958–15963 (2008).

36. T. Barwicz et al., Fabrication of add-drop filters based on frequency-matched micro-ring resonators, *Journal of Lightwave Technology*, 24, 2207–2218 (2006).

37. Q. F. Xu et al., Silicon micro-ring resonators with 1.5-µm radius, *Optics Express*, 16, 4309–4315 (2008).

38. T. Baehr-Jones et al., High-Q ring resonators in thin silicon-on-insulator, *Applied Physics Letters*, 85, 3346–3347 (2004).

39. F. N. Xia et al., Ultra-compact high order ring resonator filters using submicron silicon photonic wires for on-chip optical interconnects, *Optics Express*, 15, 11934–11941 (2007).

40. S. J. Xiao et al., Silicon-on-insulator micro-ring add-drop filters with free spectral ranges over 30 nm, *Journal of Lightwave Technology*, 26, 228–236 (2008).

41. C. Y. Chao and L. J. Guo, Polymer micro-ring resonators fabricated by nanoimprint technique, *Journal of Vacuum Science and Technology B*, 20, 2862–2866 (2002).

42. S. Y. Chou et al., Nanoimprint lithography, *Journal of Vacuum Science and Technology B*, 14, 4129–4133 (1996).

43. U. Plachetka et al., Fabrication of photonic ring resonator device in silicon waveguide technology using soft UV-nanoimprint lithography, *IEEE Photonics Technology Letters*, 20, 490–492 (2008).

44. S. Scheerlinck et al., Vertical fiber-to-waveguide coupling using adapted fibers with an angled facet fabricated by a simple molding technique, *Applied Optics*, 47, 3241–3245 (2008).

45. B. E. Little and S. T. Chu, Theory of loss and gain trimming of resonator-type filters, *IEEE Photonics Technology Letters*, 12, 636–638 (2000).

46. N. M. Wright et al., Free carrier lifetime modification for silicon waveguide based devices, *Optics Express*, 16, 19779–19784 (2008).

47. S. T. Chu et al., Wavelength trimming of a micro-ring resonator filter by means of a UV sensitive polymer overlay, *IEEE Photonics Technology Letters*, 11, 688–690 (1999).

48. G. Cocorullo, F. G. Della Corte, and I. Rendina, Temperature dependence of the thermo-optic coefficient in crystalline silicon between room temperature and 550 K at the wavelength of 1523 nm, *Applied Physics Letters*, 74(22), 3338–3340 (1999).

49. M. S. Nawrocka et al., Tunable silicon micro-ring resonator with wide free spectral range, *Applied Physics Letters*, 89, 071110–071113 (2006).

50. N. Sherwood-Droz et al., Optical 4×4 hitless silicon router for optical Networks-on-Chip (NoC), *Optics Express*, 16, 15915–15922 (2008).

51. H. Y. Ng et al., 4×4 wavelength-reconfigurable photonic switch based on thermally tuned silicon micro-ring resonators, *Optical Engineering*, 47, 044601 (2008).

52. T. Baehr-Jones et al., Analysis of the tuning sensitivity of silicon-on-insulator optical ring resonators, *Journal of Lightwave Technology*, 23, 4215–4221 (2005).

53. R. Amatya et al., Precision tunable silicon compatible micro-ring filters, *IEEE Photonics Technology Letters*, 20, 1739–1741 (2008).

54. V. R. Almeida, Q. Xu, C. A. Barrios, and M. Lipson, Guiding and confining light in void nanostructure, *Optics Letters*, 29(11), 1209–1211 (2004).

55. Q. Xu, V. R. Almeida, R. R. Panepucci, and M. Lipson, Experimental demonstration of guiding and confining light in nanometer-size low-refractive-index material, *Optics Letters*, 29(14), 1626–1628 (2004).

56. F. Ning-Ning, J. Michel, and L. C. Kimerling, Optical field concentration in low-index waveguides, *IEEE Journal of Quantum Electronics*, 42, 885–890 (2006).

57. J. T. Robinson, K. Preston, O. Painter, and M. Lipson, First-principle derivation of gain in high-indexcontrast waveguides, *Optics Express*, 16(21), 16659–16669 (2008).

58. P. A. Anderson, B. S. Schmidt, and M. Lipson, High confinement in silicon slot waveguides with sharp bends, *Optics Express*, 14(20), 9197–9202 (2006).

59. V. R. Almeida, Building blocks for silicon nanophotonics, PhD dissertation, Cornell University, Ithaca, NY (2005).

60. J. T. Robinson, L. Chen, and M. Lipson, On-chip gas detection in silicon optical microcavities, *Optics Express*, 166, 4296–4301 (2008).

61. S. Xiao, M. H. Khan, H. Shen, and M. Qi, Compact silicon micro-ring resonators with ultra-low propagation loss in the C band, *Optics Express*, 15(22), 14467–14475 (2007).

62. V. R. Almeida and R. R. Panepucci, NOEMS devices based on slot-waveguides, *CLEO-QELS'2007*, *JThD104*, Baltimore, MD (2007).

63. G. Wanga, T. Baehr-Jones, M. Hochberg, and A. Scherer, Design and fabrication of segmented, slotted waveguides for electro-optic modulation, *Applied Physics Letters*, 91, 143109 (2007).

64. M. Galli et al., Direct evidence of light confinement and emission enhancement in active silicon-on-insulator slot waveguides, *Applied Physics Letters*, 89, 241114 (2006).

65. J. T. Robinson, C. Manolatou, L. Chen, and M. Lipson, Ultrasmall mode volumes in dielectric optical microcavities, *Physical Review Letters*, 95, 143901 (2005).

66. C. A. Barrios et al., Label-free optical biosensing with slot-waveguides, *Optics Letters*, 33(7), 708–710 (2008).

67. J.-M. Lee et al., Controlling temperature dependence of silicon waveguide using slot structure, *Optics Express*, 16(3), 1645–1652 (2008).

68. F. Riboli, P. Bettotti, and L. Pavesi, Band gap characterization and slow light effects in one dimensional photonic crystals based on silicon slot-waveguides, *Optics Express*, 15(19), 11769–11775 (2007).

69. T. Liu and R. Panepucci, Confined waveguide modes in slot photonic crystal slab, *Optics Express*, 15(7), 4304–4309 (2007).

70. G. S. Wiederhecker et al., Field enhancement within an optical fibre with a subwavelength air core, *Nature Photonics*, 1, 115–118 (2007).

71. A. Gondarenko et al., Spontaneous emergence of periodic patterns in a biologically inspired simulation of photonic structures, *Physical Review Letters*, 96, 143904 (2006).

72. R. A. Soref, Silicon-based optoelectronics, *Proceedings of the IEEE*, 81(12), 1687–1706 (1993).

73. Q. Xu, V. R. Almeida, and M. Lipson, Demonstration of high Raman gain in a submicrometer-size silicon-on-insulator waveguide, *Optics Letters*, 30(1), 35–37 (2005).

74. Q. Xu, V. R. Almeida, and M. Lipson, Time-resolved study of Raman gain in highly confined silicon-on-insulator waveguides, *Optics Express*, 12(19), 4437–4442 (2004).

75. B. Jalali, R. Claps, D. Dimitropoulos, and V. Raghunathan, Light generation, amplification, and wavelength conversion via stimulated Raman scattering in silicon microstructures, *Topics in Applied Physics*, 94, 199–238 (2004).

76. H. Rong et al., A continuous-wave Raman silicon laser, *Nature*, 433, 725–728 (2005).

77. R. M. Osgood Jr. et al., Engineering nonlinearities in nanoscale optical systems: Physics and applications in dispersion-engineered silicon nanophotonic wires, *Advances in Optics and Photonics*, 1(1), 162–235 (2009).

78. H. M. Gibbs, *Optical Bistability: Controlling Light with Light*, Academic Press, Inc., New York (1985).

79. V. R. Almeida and M. Lipson, Optical bistability on a silicon chip, *Optics Letters*, 29(20), 2387–2389 (2004).

80. G. Priem et al., Optical bistability and pulsating behaviour in silicon-on-insulator ring resonator structures, *Optics Express*, 13(23), 9623–9628 (2005).

81. R. A. Soref and B. R. Bennett, Kramers–Kronig analysis of electro-optical switching in silicon, *Proceeding of SPIE*, 704, 32–37 (1987).
82. S. F. Preble, Q. Xu, B. S. Schmidt, and M. Lipson, Ultrafast all-optical modulation on a silicon chip, *Optics Letters*, 30(21), 2891–2893 (2005).
83. Q. Xu and M. Lipson, Carrier-induced optical bistability in silicon ring resonators, *Optics Letters*, 31(3), 341–343 (2006).
84. V. R. Almeida et al., All-optical switching on a silicon chip, *Optics Letters*, 29(24), 2867–2869 (2004).
85. M. Lipson, Compact electro-optic modulators on a silicon chip, *IEEE Journal of Selected Topics in Quantum Electronics*, 12(6), 1520–1526 (2006).
86. S. Manipatruni, Q. Xu, and M. Lipson, PINIP based high-speed high-extinction ratio micron-size silicon electro-optic modulator, *Optics Express*, 15(20), 13035–13042 (2007).
87. A. Chin, K. Y. Lee, B. C. Lin, and S. Horng, Picosecond photoresponse of carriers in Si ion-implanted Si, *Applied Physics Letters*, 69(5), 653–655 (1996).
88. Q. Xu, V. R. Almeida, and M. Lipson, Micrometer-scale all-optical wavelength converter on silicon, *Optics Letters*, 30(20), 2733–2735 (2005).
89. A. C. Turner, M. A. Foster, A. L. Gaeta, and M. Lipson, Ultra-low power parametric frequency conversion in a silicon micro-ring resonator, *Optics Express*, 16(7), 4881–4887 (2008).
90. S. Manipatruni, C. B. Poitras, Q. Xu, and M. Lipson, High-speed electro-optic control of the optical quality factor of a silicon microcavity, *Optics Letters*, 33(15), 1644–1646 (2008).
91. S. M. Weiss, M. Molinari, and P. M. Fauchet, Temperature stability for silicon-based photonic bandgap structures, *Applied Physics Letters*, 83(10), 1980–1982 (2003).
92. P. Cheben, D.-X. Xu, S. Janz, and A. Delâge, Scaling down photonic waveguide devices on the SOI platform, *Proceedings SPIE—The International Society for Optical Engineering*, Maspalomas, Spain, 2003, vol. 5117, pp. 147–156.

5

Nanophotonics with Microsphere Resonators

Oliver Benson
S. Götzinger
A. Mazzei
G. Zumofen
V. Sandoghdar
L. de S. Menezes

Optical microresonators are efficient traps for light [Vah03]. The long storage time and strong confinement of the electromagnetic field lead to a pronounced enhancement of the interaction between light and matter. Complementarily, the narrow resonances associated with specific modes in the resonator are attractive for filtering [IYM99, CV00] or for detecting changes in the electromagnetic environment, an ideal prerequisite for ultrasensitive sensing applications [VA08]. Microresonators have been used to enhance nonlinear effects [MKB06, SKV02, SQC85] to detect small particles or biomolecules [VAK08] and to collect and guide light with high efficiency [GMM06, MWA05]. From the viewpoint of fundamental physics, microresonators have been extensively studied to investigate quantum electrodynamic effects [MGM07, PCW06]. In this chapter, we highlight several of these aspects and provide an overview of recent experiments. We concentrate on a special class of microresonators, that is, microsphere resonators [BGI89, CLB93, MK94, MI06a, MI06b]. Even more specifically, we consider silica resonators and do not discuss other materials such as polymers [KJD95, AWW01, GRD07] or crystals [SIM04]. The chapter is organized as follows: Section 5.1 introduces the concept of light storage in spherical resonators, and is followed by a description of their basic properties in Section 5.2. Section 5.3 is devoted to nonlinear effects, that is, Raman lasing, whereas Sections 5.4 and 5.5 introduce experiments with scanning probes as controllable local scatterers and fluorescent nanoprobes. Section 5.6 points out the similarity of resonantly enhanced scattering to cavity quantum electrodynamic effects. Section 5.7 finally concludes with a brief summary and an outlook.

5.1 Introduction

The confinement of light is possible with a large variety of microresonators. In this section, we provide basic parameters to quantify their properties. We then focus on the description of optical resonances in dielectric spheres and introduce whispering-gallery modes (WGMs).

5.1.1 Optical Microresonators

An important parameter of a microresonator that quantifies its ability to store light is its so-called *quality factor* or *Q factor*. It is a measure for energy losses and is defined as the time-averaged energy in the cavity divided by the energy loss per cycle [Jac98].

$$Q = \omega_0 \frac{\text{Stored energy}}{\text{Power loss}} \tag{5.1}$$

Here, ω_0 denotes the angular resonance frequency, assuming no losses.

The Q factor can also be expressed as

$$Q = \frac{\omega_0}{\delta\omega} = \omega_0 \tau \tag{5.2}$$

Another parameter to characterize microresonators is the mode volume, V_{mode}, which can be defined as

$$V_{\text{mode}} \approx \frac{\int_{V_Q} \varepsilon(\vec{r})|\vec{E}|^2 \, d^3 r}{\max \varepsilon(\vec{r})|\vec{E}|^2} \tag{5.3}$$

where
  $\varepsilon(\vec{r})$ is the dielectric constant of the material
  $\vec{E}$ is the cavity field
  V_Q is the integration volume

The ratio $Q/\sqrt{V_{\text{mode}}}$ is proportional to the electric field per photon and of paramount importance to observe quantum electrodynamic effects as will be discussed in Section 5.6. A suitable parameter taking both the photon storage time, that is, the Q factor, and the resonator size, that is, the mode volume into account, is the *finesse*. It relates the free spectral range (FSR) to the resonance linewidth by [Sie86]

$$F = 2\pi \frac{\text{FSR}}{\delta\omega} = 2\pi Q \frac{\text{FSR}}{\omega_0} \tag{5.4}$$

The FSR represents the spacing between two longitudinal modes and is given by

$$\text{FSR} = \frac{c}{2LN} \tag{5.5}$$

where

c is the speed of light
L is the resonator length
$N = \sqrt{\varepsilon}$ the refractive index of the medium inside the resonator

The smaller the resonator, the larger the FSR. The relation between the finesse and the Q factor then follows as

$$F = \pi c \frac{Q}{LN\omega_0} \tag{5.6}$$

5.1.2 Whispering-Gallery Modes

Spherical resonators support a specific class of high-Q optical modes, the so-called WGMs, sometimes also called morphology-dependent resonances. These resonances were first studied by Lord Rayleigh for acoustic waves propagating along close curved walls, for example, along the inside of the dome of St. Paul's Cathedral, London. The optical analogy is light confined inside a dielectric sphere due to repeated total internal reflection (see Figure 5.1). If the surface of the sphere is very smooth and internal losses are minimized, light is trapped very efficiently. For glass microspheres, Q factors of up to 8×10^9 have been reported [GSI96]. To get an intuitive idea of the extremely high Q factor of silica microspheres, one can consider a mechanical analog, for example, a tuning fork with a resonance frequency of $440\,\text{Hz}$ (standard pitch). If such a tuning fork had a Q as high as a microsphere resonator with $Q = 8 \times 10^9$, it would oscillate for 33 days.

WGMs belong to the wider family of Mie resonances, already described theoretically in 1908 by Gustav Mie in his pioneering work on the scattering of colloidal metal solutions [Mie08]. Since then, scattering of light by small spheres has been studied extensively both experimentally and theoretically [Str41, BC88, BH98].

Starting from Maxwell's equations and applying appropriate boundary conditions (continuity of the tangential and vertical field components of $\vec{E}$ and $\vec{H}$, respectively), one can derive the fields inside a dielectric sphere analytically. Resonances are found as roots of a transcendental equation (here as an example for TE-polarized modes):

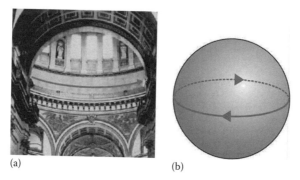

(a) (b)

FIGURE 5.1 Analogy between acoustical and optical whispering-gallery modes: (a) the whispering-gallery in the dome of St. Pauls Cathedral. (Reproduced from Wallraff, A. and Ustinov, A.V., *Phys. Unserer Zeit*, 33(4), 184, 2002. With permission. Copyright Wiley-VCH Verlag GmbH & Co. LGaA.) (b) Sketch of a microsphere: light is guided inside the sphere by total internal reflection.

$$\frac{\left[N\rho j_l\left(N\rho\right)\right]'}{\mu_1 j_l\left(N\rho\right)} = \frac{\left[\rho h_l^1\left(\rho\right)\right]'}{\mu_2 h_l^1\left(\rho\right)} \tag{5.7}$$

where the prime denotes the derivative. Here, j_l and h_l^1 denote the spherical Bessel- and Hankel-functions of first kind, respectively. Resonant modes in dielectric spheres are labeled by three mode numbers n, l, and m, similar to atomic bound states. The index l is related to the mode's angular momentum and the mode number m to its projection on the z-axis. N is the index of refraction of the sphere and $\rho = ka$ with sphere diameter a, and wave vector (outside the sphere) k. The roots correspond to eigenmodes that are labeled by the radial mode number $n = 1, 2, 3, \ldots$, where $n = 1$ is associated with the lowest frequency. The solutions are degenerate in m for a perfect sphere. In a similar way, the transcendental equation for the TM modes can be derived. Of all the modes, a special subclass with small n and large l is of particular interest. These modes are the WGMs mentioned above. They provide the highest Q factor together with smallest mode volume.

The radial extent ΔR of the mode with small n and large l can be approximated by [Sch93]

$$\frac{\Delta R}{a} \approx 2.2 \left[\frac{\left(n - 1/4\right)}{Nx_{l,n}}\right]^{2/3} \tag{5.8}$$

where $x_{l,n} = (2\pi/\lambda)a$ is the sphere's *size parameter* that relates the dimensions of the microresonator (its radius a) to the wavelength. n counts the number of intensity maxima inside the sphere or, $n - 1$ is the number of zeros of the Bessel function $j_l(kR)$ inside the resonator. The higher the n, the more the field leaks out of the sphere (see Figure 5.2).

An approximate formula can be derived for the intensity distribution of the modes along the θ- and ϕ directions. For large l and $|m| \approx l$ the intensity distribution on the entire sphere surface can be approximated by [KDS95]

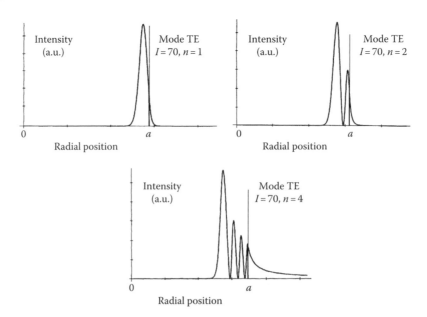

FIGURE 5.2 Radial intensity distribution for three whispering-gallery modes differing in n. (From Collot, L., *Etude théorique et expérimentale des résonances de galerie de microsphère de silice: pièges à photons pour des expériences d'électrodynamique en cavité*, PhD thesis, Laboratoir Kastler Brossel de l'Ecole Normale Supérieur, Paris, France, 1994.)

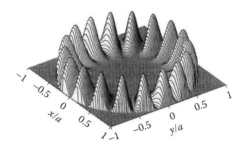

FIGURE 5.3 Intensity distribution in the equatorial plane for a TE whispering-gallery mode with $n=1$ and $m=l=9$. (From Chang, R.K. and Campillo, A.J., *Optical Processes in Microcavities*, World Scientific, Singapore, 1996. With permission.)

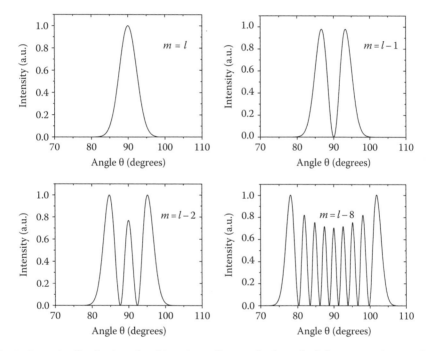

FIGURE 5.4 Intensity distribution of a whispering-gallery mode along the θ direction with $n = 1$, $l = 350$ and different m values.

$$I_{l,m}(\theta,\phi) \propto \left| H_{l-|m|}\left(l^{1/2}\cos\theta\right)\sin^{m}(\theta)\exp(im\phi)\right|^{2} \tag{5.9}$$

where $H_{l-|m|}$ represents a Hermite polynomial [GR00].

In Figure 5.3 the intensity distribution in the equatorial plane (along ϕ) is shown for a TE WGM with $n=1$ and $m=l=9$. The radius of the sphere is normalized to 1.

Figure 5.4 shows plots of $I_{l,m}(\theta)$ for four different m values and $l=350$. A mode characterized by n, l, and m has $l-|m|+1$ lobes in the θ direction, centered around $\pi/2$, the equatorial plane of the sphere. The most confined mode with $l=m$ is called a *fundamental whispering-gallery* mode of the microsphere. From Figure 5.4 one can also see that with increasing difference between m and l, the modes become more extended, while at the same time the distance h between adjacent intensity maxima becomes smaller.

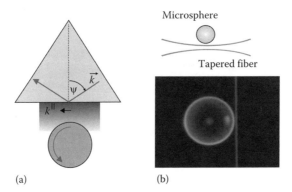

FIGURE 5.5 Schematics of efficient coupling to whispering-gallery modes with external couplers. (a) Coupling via frustrated total internal reflection at the surface of a prism. (b) Coupling via the evanescent field of an adiabatically tapered fiber.

5.1.3 Efficient Coupling to Whispering-Gallery Modes

Light in WGMs is tightly confined. However, if no light is coupled out of the resonator, there is no trivial way to couple light into it. A solution for efficient coupling is offered by the evanescent field outside the sphere. The wave vector k of a field in the mode can be decomposed in a complex perpendicular component $k_\perp$ and a real parallel component $k_\parallel$. To couple external light efficiently into the sphere, its vector $k_\parallel$ has to be matched to the vector $k_\parallel$ of the light inside the WGMs. There are different approaches to such an evanescent coupling: prism couplers [BGI89], fiber tapers [KCJB97], eroded monomode fibers [DKL95], angle polished single mode fibers [IYM99], and pedestal antiresonant reflecting waveguides [LLL00]. Two examples are shown in Figure 5.5.

In a WGM, light travels almost at glancing angle ($k_\parallel \cong k$) along the inner surface of a sphere. In case of a prism coupler with an index of refraction of N_{prism}, the phase matching condition reads (see Figure 5.5)

$$kN_{prism} \sin \Psi \cong kN_{sphere} \tag{5.10}$$

or

$$\sin \Psi \cong \frac{N_{sphere}}{N_{prism}} \tag{5.11}$$

where N_{sphere} is the index of refraction of the sphere.

For this condition to be satisfied, the refractive index of the prism has to be larger than the refractive index of the sphere.

5.2 General Properties of Microsphere Resonators

In this section we introduce silica (SiO_2) microspheres as optical resonators. A crucial point for experiments with these resonators is the efficient coupling to the tightly confined modes. We describe how scanning probes can be used not only to characterize the modes precisely, but also to optimize the coupling to external couplers.

5.2.1 Fabrication of Silica Microspheres

For the fabrication of silica microspheres with a very high Q factor, ultrapure material is required. In the ultraviolet range, metallic impurities reduce the Q, while in the infrared range water contamination

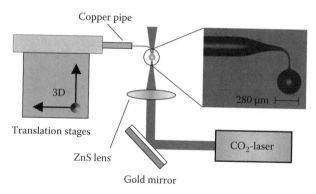

FIGURE 5.6   Sketch of the setup used to fabricate silica microspheres. Silica fibers are pulled and mounted in copper tubes. The tip of the fiber is melted in the focus of a CO_2 laser. The inset shows a microscope image of a larger sphere of 280 μm in diameter.

can spoil the Q substantially. In our experiments, high quality spheres were produced from small (few millimeters) Suprasil 300 glass rods with an extremely low ion contamination (e.g., $OH^-\leq 1$ ppm). The cleaned rods were then pulled in the flame of an oxy-hydrogen torch to thin fibers and cut into small pieces, which were glued into small copper pipes. These copper pipes could be fixed on a three-dimensional translation stage to position the fiber in the focus of a CO_2 laser beam (see Figure 5.6) where surface tension during the melting process finally forms the sphere at the end of the fiber. With this procedure, silica microspheres with diameters between 20 and 300 μm can be produced routinely. The inset in Figure 5.6 shows a picture of a larger sphere. An atomic force microscope measurement confirmed a surface roughness of less than 1 nm.

5.2.2 Mode Mapping of Whispering-Gallery Modes

For many experiments, identification of the fundamental WGM is crucial. As pointed out, this mode not only has a high Q factor, but also the smallest mode volume, providing the strongest field per photon. Since the field outside the microsphere is evanescent, a scanning near-field optical microscope (SNOM) [DPR86] is ideally suited to probe and identify the modes.

Figure 5.7 shows an experimental setup to perform mode mapping with a SNOM probe. An optical fiber tip is actively stabilized via the shear-force method [BFW92] at a distance of approximately 10 nm from the sphere's surface. It scatters photons out of the evanescent field into an optical fiber that is directly coupled to a photomultiplier tube (PMT). By moving the tip, one can probe the intensity on the microsphere's surface locally. When the tip is moved in one dimension perpendicular to the sphere equator and a tunable diode laser is scanned over several modes, a spatio-spectral mode map is obtained. Figure 5.8 shows a part of such a map for a sphere with a diameter of 200 μm. Modes with a different number of lobes corresponding to different values of $l-|m+1|$ are clearly visible. Lifting of the degeneracy of the modes is caused by a deviation from a perfect spherical symmetry. The spacing of 3.7 GHz between the fundamental mode ($l-|m+1|=1$) and the mode with two lobes ($l-|m+1|=2$) corresponds to a 1% oblate deformation of the sphere. This value is derived by fitting to an approximate formula for the mode splitting given in reference [GI94].

Due to the sensitive detection of the scattered light via the near-field probe, it is also possible to detect much fainter resonances that can be attributed to modes with higher radial mode number n. These modes are less effectively excited, because the incident angle of the laser was optimized for coupling to lowest-order n modes. In order to visualize these modes, a strongly nonlinear color scale was used in Figure 5.8.

The method described above allows a fast screening of a wide spectral range and unequivocal identification of a particular mode. However, for a complete study of a WGM and especially to determine the

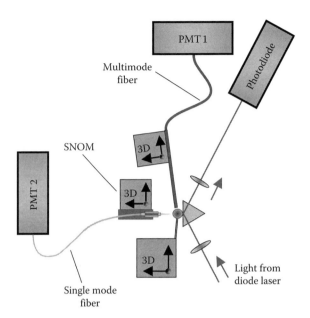

FIGURE 5.7   Sketch of a typical experimental setup for experiments with a microsphere resonators and near-field probe. Light from a tunable (diode) laser is coupled into a microsphere via frustrated total internal reflection at a prism. The microsphere is mounted and temperature stabilized on a translation stage (3D). Resonances are detected as dips in the transmitted signal at a photodiode or as peaks in the light scattered into a multi-mode fiber sensor by a photomultiplier (PMT1). The near-field probe can be moved at a constant height of a few nanometers above the sphere's surface. Photons from evanescent fields on the microsphere surface are collected and detected at the end of the fiber with at photomultiplier (PMT2).

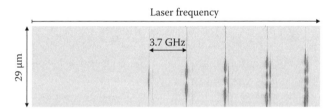

FIGURE 5.8   Spatio-spectral mode map of a microsphere with diameter 200 μm. The fainter modes correspond to higher-order radial mode numbers n. The contrast in the plot is adjusted such that even modes which are less efficiently excited (by a factor of about 100) than the strong low-order n modes are visible. (From Götzinger, S. et al., *Appl. Phys. B*, 73, 825, 2001. With kind permission from Springer Science + Business Media.)

relative position of the equatorial plane with respect to the horizontal, a two-dimensional mapping is necessary. Then it is possible to adjust and optimize the coupling [MG05]. In order to map the intensity distribution of an individual mode of a microsphere, the diode laser was locked to a single WGM, using a side-band modulation technique similar to the one described in reference [DHK83]. The SNOM is used to collect photons from the evanescent field while it is scanned across the microsphere's surface. In Figure 5.9a the intensity scattered into the SNOM fiber is recorded while the tip is scanned, leading to an image of the intensity distribution on the microsphere's surface. The scanned area is approximately centered vertically to the equator of the sphere. Several features are clearly distinguished. The various stripes of high intensity are inclined to the horizontal by an angle of about 13°. These rings, parallel to the microsphere's "optical equator" (the plane of the fundamental mode), are a signature of the spatial intensity distribution of a WGM with $l-|m|+1 \approx 50$. The modulation along these stripes is an

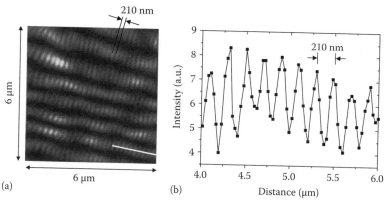

FIGURE 5.9 (a) Two-dimensional mapping of the intensity distribution of a whispering-gallery mode with the SNOM. During the scan, the laser was locked to a particular mode. The scanned area is centered vertical around the sphere's equator. (b) Cross section of (a) marked by the white line. Details are discussed in the text.

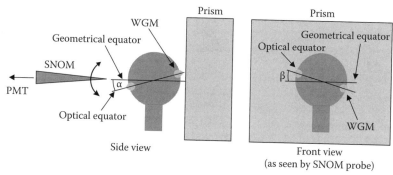

FIGURE 5.10 Schematic side and front view of the microsphere-prism system. The fundamental mode is inclined with respect to the horizontal plane due to the fabrication process and therefore does not couple efficiently to the prism.

interference pattern caused by the clockwise and counterclockwise propagating fields. Figure 5.9b shows a cross section. Measuring this standing wave pattern allows an accurate estimation of the radial mode number n [KDS96] and thus gives full control over the WGM in terms of n and $l - |m|$.

A further application of mode mapping is the optimization of coupling to the fundamental WGM with highest Q and lowest mode volume [MG05]. This can be accomplished by aligning the "geometric equator" given by the coupler's reference frame to the optical equator given by the propagating WGM (see Figure 5.10). Both can be measured by scanning the SNOM probe in the direction perpendicular to the geometric equator along the sphere surface. In the following experiment, at each pixel the laser frequency was tuned across the resonance of the fundamental WGM and both the optical and topography signals were recorded. The resulting topographical-spectral and spatial-spectral maps are shown in Figure 5.11b and d. The topographical signal does not depend on the frequency of the diode laser as expected, but an optical signal from the fundamental WGM is only detected in resonance. The single maximum of the measured intensity distribution of the WGM proves that the fundamental mode has been detected. White lines are plotted to highlight the geometric equator (topographical map) and the optical equator (spatial map). From the shift of the two lines with respect to each other, it is possible to directly determine the angle α (see Figure 5.10). Then, the microsphere is tilted with the rotation stage to annul the shift.

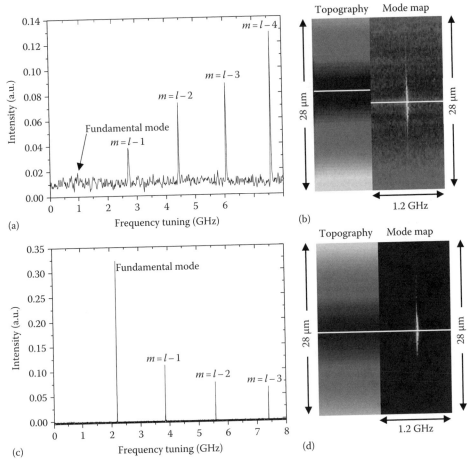

FIGURE 5.11   (a) Recorded spectrum of modes with $m \geq l$ before optimization. (b) Corresponding topographical-spectral and a spatial-spectral mode map of the fundamental WGM. (c) Recorded spectrum of modes after optimization. The best coupling is now observed for the fundamental mode. (d) Corresponding topographical-spectral and a spatial-spectral mode map of the fundamental WGM. (Reprinted from Mazzei, A. et al., *Opt. Commun.*, 250, 428, 2005, Figures 3a and 5a. With permission.)

The alignment of the second rotational degree of freedom (angle β in Figure 5.10) can be done with the mode mapping technique. In order to do this, a two-dimensional mode map similar to that described above is performed. Then, it is straightforward to derive and correct the tilting angle β with respect to the horizontal axis.

Figure 5.11 shows a spectrum before (a) and after (c) the optimization procedure, that is, the correction of the two tilting angles α and β. Figure 5.11b and d shows a topographical-spectral and a spatial-spectral map. After optimization, both white lines, which indicate the geometric and optical equator, respectively, are collinear. The microsphere resonator is now perfectly aligned with respect to the prism coupler. Accordingly, the spectrum shows that the coupling is now optimized for the fundamental WGM, which is by far stronger than modes with $l \neq m$.

With prism couplers, we were able to obtain coupling efficiencies to high-Q WGMs exceeding 70%. A reliable procedure is crucial, as spherical resonators inevitably require external couplers, which then have to be aligned and optimized. This is of paramount importance for experiments aiming at the observation of nonlinear effects at extremely low pump powers as described in the following section.

5.3 Nonlinear Optics in Microspheres

In this section, we investigate the Raman effect as an example for enhanced nonlinear effects in microspheres. C.V. Raman observed in 1928 [RK28] that light scattered from a dielectric material contained frequencies different from those of the excitation source [BH98]. The physical origin is inelastic scattering involving the generation (Stokes emission) or absorption (anti-Stokes emission) of phonons [Boy92].

5.3.1 Ultralow Threshold Raman Laser

We study Raman scattering with a similar setup as shown in Figure 5.7, but with the diode laser replaced with a tunable Ti:sapphire laser (linewidth 500 kHz) that was continuously scanned across a single WGM. By scanning a near-field probe with a typical tip diameter of $\approx$100 nm along the sphere surface, information about the local intensity distribution of the mode under investigation was obtained with subwavelength precision.

In a first experiment we demonstrated Raman lasing. Raman lasing builds up from spontaneously scattered Stokes photons. These photons may be stored in high-Q modes of a microsphere resonator and stimulate coherent emission of further Stokes light. If loss is overcome, Raman lasing occurs. As pointed out, a remarkable feature of microsphere resonators is that even though the Raman gain in silica is very low, lasing is still possible. The Raman gain spectrum of silica glass has two maxima, shifted by 13.9 and 14.3 THz from the pump frequency, and a full width half maximum (FWHM) of about 10 THz, which is much broader than the free spectral range (FSR) of a microsphere resonator used in our experiments. This guarantees that the Stokes emission overlaps with high-Q WGMs. By pumping resonantly with the Ti:sapphire laser at a wavelength of $\lambda_p = 785$ nm, the wavelength of the Stokes emission is expected at $\lambda_s = 814$ nm. The Stokes emission is coupled out of the microsphere by the prism coupler. Together with the transmitted pump beam it is sent to a spectrometer. Additional filters can be used to suppress the pump light. Figure 5.12 shows the Stokes emission as a function of the absorbed pump power. A threshold behavior is clearly visible. For a Q factor exceeding 10^9 we observed thresholds as low as 4.3 µW.

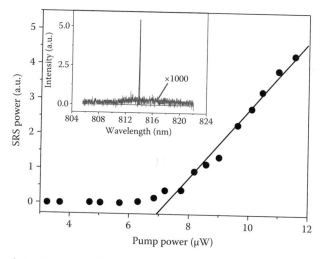

FIGURE 5.12 Raman lasing in a microsphere with a diameter of 70 µm and a Q factor of 3×10^8. The measured threshold is $P_T = 7.5$ mW of absorbed pump power. The inset shows the spectrum above ($20 \times P_T$) and below threshold. (Reprinted from Mazzei, A. et al., *Appl. Phys. Lett.*, 89, 101105, 2006, Figures 1, 2, and 3. With permission. © 2006 American Institute of Physics.)

5.3.2 Mode Mapping of Lasing Modes in a Raman Laser

The onset of Raman lasing can also be observed in the spectrum of the Stokes emission coupled out of the microsphere (inset of Figure 5.12). If the pump power is above threshold, a distinct peak, shifted by 13.4 THz from the pump frequency, appeared. With the resolution of the spectrometer of 0.02 nm, it is not possible to resolve modes with same l but different m mode numbers that differ by typically 0.003 nm. In order to determine the number of modes that participate in lasing, we performed mode mapping using the SNOM probe. We continuously scanned the laser across a single WGM, which was verified by detecting the resonance dip in the absorbed pump beam. Then, we simultaneously scanned the SNOM probe along the polar direction and performed a mode mapping as described in the previous section. By using appropriate spectral filters we could detect the mode profile of the pump mode and the lasing mode. Figure 5.13a and c shows the mapping of the pump mode for the fundamental WGM with $l - m = 0$ and a WGM with $l - m = 1$, respectively. The pump power was reduced below threshold for this measurement. Figure 5.13b and d show the according lasing modes for a pump power of 20 times and 130 times the threshold pump power $P_T = 7.5\,\mu$W, respectively. We carefully checked that all pump light was filtered out in this measurement. Under these conditions single mode Raman lasing occurs. Obviously, the mode profile of the lasing mode is the same as that of the pump mode that is expected by considering maximum mode overlap.

There is a systematic deviation of the detected mode profile of the lasing mode from the detected pump mode close to the maxima in polar direction. We further investigated this effect by reducing the pump power during mode map. Mode maps with a pump power of five times the threshold pump power $P_T = 7.5\,\mu$W are shown in Figure 5.14. In this case, lasing is completely quenched for certain positions of the scanning probe. Qualitatively, this behavior is explained by a Q factor degradation by photon scattering through the scanning probe [GBS02], which raises the laser threshold above the pump power.

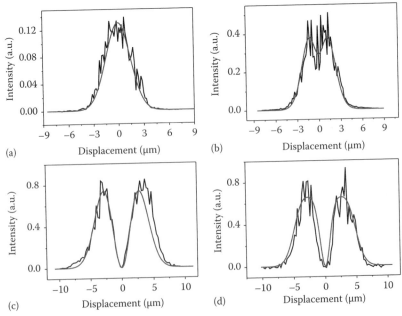

FIGURE 5.13   Mode map of the pump (a and c) and lasing (b and d) modes for WGMs with $l - m = 0$ (a and b) and $l - m = 1$ (c and d). The pump power was 20 and 130 times $P_T = 7.5\,\mu$W, respectively. Gray solid lines are theoretical fits (see text). (Reprinted from Mazzei, A. et al., *Appl. Phys. Lett.*, 89, 101105, 2006, Figures 1, 2, and 3. With permission. © 2006 American Institute of Physics.)

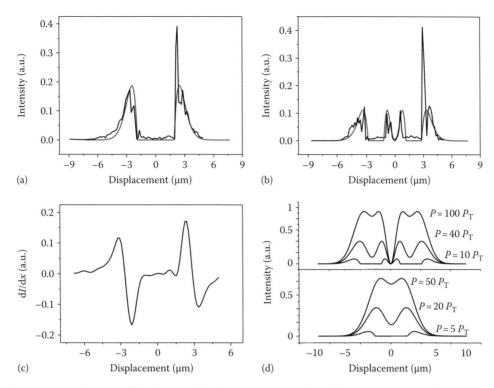

FIGURE 5.14 Mode maps of the lasing modes for a pump power of 5 and 10 times $P_T = 7.5$ µW for WGMs with $l - m = 0$ (a) and $l - m = 1$ (b). Gray solid lines are theoretical fits. In (c) the intensity gradient as derived from the smoothed experimental data in (a) is plotted. (d) Shows the theoretical curves for different pump powers and the two different modes with $l - m = 1$ (top) and $l - m = 0$ (bottom). (Reprinted from Mazzei, A. et al., *Appl. Phys. Lett.*, 89, 101105, 2006, Figures 1, 2, and 3. With permission. © 2006 American Institute of Physics.)

In a theoretical model, we consider the scanning probe tip as a scatterer in the evanescent field of the mode, which introduces an additional loss channel. This degradation of the Q factor thus depends on the position of the probe. The total quality factor Q_{tot} of a mode will have three contributions:

$$\frac{1}{Q_{tot}} = \frac{1}{Q_0} + \frac{1}{Q_K} + \frac{1}{Q_s} \tag{5.12}$$

where

Q_0 is the intrinsic (unloaded) Q factor of the sphere
Q_K is the coupling-limited Q factor (given by the distance between sphere and the prism coupler)
$Q_s(\theta)$ is due to scattering by the probe

$Q_s(\theta)$ depends on the position of the probe in the following way:

$$Q_s(\theta) = \overline{Q}_s / \left[H\left(l - m, \sqrt{l}\cos\theta\right)\sin(\theta)^l \right] \tag{5.13}$$

with $H(n, x)$ being the normalized Hermite polynomial of order $l - m$ (which describes the intensity profile of the WGM along the polar direction [KDS95]), and $\overline{Q}_s$ is the maximum reduction of the Q factor. This means that the tip will influence the Q factor mostly when it is positioned in the maximum of the mode, and will produce no effect if it is not in the evanescent field of the mode ($Q_s = \infty$).

The gray solid curves in Figures 5.13b and d and 5.14a and b show a fit to the experimental data according to the theoretical model [MKB06]. From the measurement we can also derive the exact radius of the tip that can be modeled as a small sphere [GBS02]). A radius of 70 nm is in very good agreement with values measured for such tips using scanning electron microscopy.

The measurements nicely demonstrate that scanning probes are versatile tools to study microlasers. Furthermore, the tip of a scanning probes can act as a well controllable nanoscopic scatter. Here we studied the intricate interplay between the probe's position and quenching of laser emission by additional scattering. In the next chapters, we will give more examples for controlled experiments with microsphere resonators and scanning probes as local probes or scatterers.

5.4 Controlled Coupling of Single Nanoscopic Emitters to High-Q Optical Modes

Scanning probes are not only useful to optimize spectroscopy of high-Q modes in microsphere resonators. They also represent a versatile tool to study the evanescent interaction of nanoscopic emitters with WGMs in highly controlled experiments.

5.4.1 Preparation of a Scanning Probe as Controllable Nanoscopic Emitter

In this section we discuss the realization of an on-demand coupling between a dye-doped nanoparticle and the WGMs of a microsphere. As depicted in Figure 5.15a, our strategy is to attach the nanoparticle to the end of a fiber tip (see inset) and to use the SNOM stage to manipulate the emitter in the vicinity of the microsphere surface. The recipe for the production of such probes is discussed in [KHS01] in more detail. As described above, the Q factor of the sphere could be measured by direct spectroscopy using a narrow-band diode laser at $\lambda = 670$ nm. The fluorescent nanoparticle at the tip was excited with a cw frequency-doubled Nd:vanadate laser through the fiber, and a prism was used to extract part of the particle emission that coupled to the WGMs [GMB04]. Alternatively, a microscope objective (NA = 0.75, not shown in Figure 5.15a allowed collection of both the free space components of the nanoparticle fluorescence and the scattering from the sphere [GMB04].

Figure 5.15b shows part of the fluorescence spectrum recorded via the prism when a bead (red fluorescent, Molecular Probes, Inc.) of 500 nm in diameter was placed within a few nanometers of the microsphere. The spectrum shows a FSR of 0.85 nm, which is expected for the 90 μm sphere used in this measurement. By using polarized detection, we identified the two dominant peaks in each FSR as TM and TE modes. Since the small numerical aperture of our detection path via the prism was optimized for coupling to low n modes [GI94], we attribute these resonances to $n = 1$ and the weaker ones to higher n modes.

5.4.2 On-Demand Coupling of a Nanoemitter to High-Q Modes of a Microsphere Resonator

The SNOM now allows for examination of the position dependence of the bead's coupling to the WGMs. To do this, we tuned the emitter's angular coordinate θ (see insets in Figure 5.16b) to the sphere equator [MG05] and then varied its radial separation to the sphere surface using the SNOM distance stabilization and scanning machinery. The fluorescence emitted into the WGMs was collected through the prism coupler and detected by a PMT. Figure 5.16a shows a characteristic decrease expected for the evanescent coupling between the bead and the WGMs of the microsphere. Next, we fixed the particle-sphere separation to 5–10 nm and scanned the bead about the equator along θ. The symbols in Figure 5.16b show that the fluorescence signal detected through the prism drops quickly within 5°, or equivalently 4 μm, about the equator. We note that at room temperature the broad spectrum of the dye molecules couples

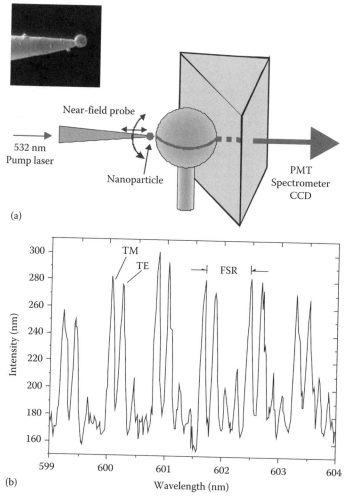

FIGURE 5.15 (a) The schematics of the experimental setup. The inset shows an SEM image of a single 500 nm bead attached to a glass tip. Fluorescence can be detected via the prism or via the microscope objective. (b) Spectrum recorded via the prism when the bead was close to the sphere's surface.

to many modes of different m. However, because WGMs with higher $l - |m|$ values have larger mode volumes and therefore lower electric fields at the equator, their coupling efficiencies to both the bead fluorescence and to the prism is reduced. The solid curve in Figure 5.16b displays a fit to the data, accounting for the contributions of the first 10 WGMs of different m. The profiles of the first five are plotted under the experimental data where the blue curve represents the fundamental mode, and the profile heights reflect the respective weighting factors in the fit procedure. The data in Figure 5.16 demonstrate the local and controlled coupling of a single nanoemitter to the WGMs of a microsphere.

5.5 Efficient Energy Transfer via Shared Modes in Microsphere Resonators

In this section we describe an experiment where a scanning probe is utilized to realize bipartite interaction of two nanoscopic emitters. We discuss possible ways of photon transfer and show how high-Q modes can boost the efficiency for this process.

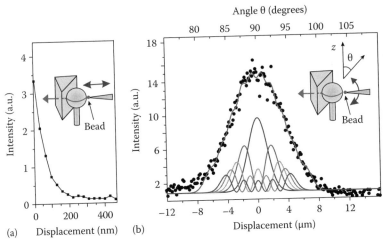

FIGURE 5.16 Total fluorescence intensity detected through the prism coupler as a function of the sphere-bead distance (a) and the particle's lateral position (b). The thick solid line in (b) is a fit using the first 10 WGMs. The weighted intensity distributions of the first 5 WGMs are plotted for clarity. (Reprinted from Götzinger, S. et al., *Nano Lett.*, 6, 1151, 2006, Figures 2 and 3. With permission. © 2006 American Chemical Society.)

5.5.1 Principles of Photon Transfer

The most elementary system for studying optical interactions consists of two coupled dipole emitters. If these are separated by a distance r much less than a wavelength λ, and if their resonance frequencies are very close to each other, they can undergo strong coherent dipole–dipole coupling, leading to sub- and super-radiance [Dic54, HSZ02]. If their transitions are broadened, dipole–dipole coupling becomes incoherent as in the case of fluorescence resonant energy transfer (FRET), where the energy from a "donor" is transferred to an "acceptor," provided there is sufficient overlap between the former's emission spectrum and the latter's absorption line. The efficiency of FRET is proportional to $(1 + (r/r_0)^6)^{-1}$ and drops to 50% over a Förster radius r_0, which is typically smaller than 10 nm [Foe65].

If $r > \lambda$, optical communication between the two emitters takes place via propagating photons. The direct free space emission and absorption process scales as $1/r^2$, amounting to an efficiency of about 3×10^{-13} at a distance of 50 µm, considering a typical room temperature absorption cross section of $\sigma_A \approx 10^{-16}$ cm^2 [Sch90]. In order to improve this, one could funnel the energy from one emitter to the other by using optical elements such as lenses and waveguides, or even surface plasmons in metal structures [Ab04]. However, the coupling remains limited because each photon flies by the atom only once. Resonant structures such as cavities provide much longer effective interaction times and can also influence radiative processes by modifying the density of states [Ber94, CC96, Vah03]. In addition, if the cavity is made very small, the field per photon becomes large, leading to a stronger coupling between the emitter and the resonator mode. First applications of microcavities to enhance the energy transfer rate have been performed on ensembles of molecules in microdroplets [FAD85, AHD96] and more recently in polymer microcavities [HGD99]. Figure 5.17 provides a cartoon picture of three possible ways of long-distance photon transfer.

5.5.2 Cavity-Mediated Long-Distance Photon Transfer between a Single Emitter/Absorber Pair

Here, we discuss energy transfer between a donor and an acceptor nanoparticle via the WGMs in a microsphere resonator. In this experiment a sphere of $2R = 35$ µm was dip-coated with a solution of acceptor beads (Crimson, Molecular Probes, Inc.) of 200 nm in diameter. After coating there were a total of 5–10 particles on the surface of the microsphere and the Q factor of the fundamental mode was

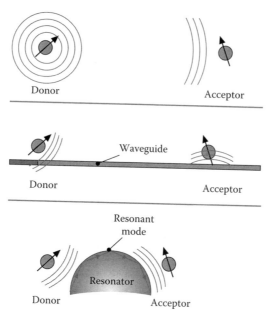

FIGURE 5.17 Possible configurations for photon transfer between a donor dipole and an acceptor dipole (from top to bottom): free-space emission, transfer via a dielectric waveguide, and transfer mediated by optical modes in a microsphere resonator.

measured to be 3×10^7. Then, we retracted the prism to avoid losses due to output coupling and imaged the location of the acceptor beads on the microsphere using a CCD camera [GMB04]. We located a single nanoparticle close to the sphere's equator and centered it in the confocal detection path of a spectrometer (see Figure 5.18a). A single donor bead (Red Fluorescent, Molecular Probes, Inc.) of 500 nm in diameter was attached to a SNOM tip and was excited through the fiber with a laser power of $\approx 20\,\mu W$. The curves with a maximum at around 619 and 650 nm in Figure 5.18b) show the fluorescence spectra of the donor and acceptor beads, respectively. Finally, we approached the donor to the sphere and recorded the spectrum of the single acceptor bead through the microscope objective.

The strongly modulated uppermost curve in Figure 5.18b plots the spectrum obtained from the location of the acceptor. The fast spectral modulations again provide a direct evidence of coupling to high-Q modes. Comparison of this spectrum with the spectra of the donor and acceptor reveals the coexistence of contributions from the donor and the acceptor fluorescence. We point out that although our confocal detection efficiently discriminates against light emitted at the donor location, it is possible for this emission to couple to the WGMs and get scattered into our collection path by the acceptor bead. To take this into account, we subtracted the donor fluorescence spectrum from the recorded (strongly modulated) spectrum after normalizing their short wavelength parts. Furthermore, to rule out the possibility of direct excitation of the acceptor by the laser light, we retracted the donor from the sphere, photobleached it with an intense illumination of the excitation light and approached it again to the sphere. The signal at the acceptor position was then collected under exactly the same conditions and is shown in Figure 5.18b. This contribution is clearly negligible compared to the total emission, verifying that the acceptor fluorescence has been almost entirely pumped by the donor emission. After subtracting this small contribution, we arrive at the modulated curve in Figure 5.18c, which coincides very well with the fluorescence spectrum of the acceptor (also shown for convenience in Figure 5.18c). These measurements clearly demonstrate photon exchange between two well-defined nanoemitters via shared high-Q modes of a microresonator.

It is possible to provide an estimation for the photon transfer efficiency η_i, defined as the probability β_i, of the donor emitting a photon into the ith WGM and subsequent absorption by the acceptor.

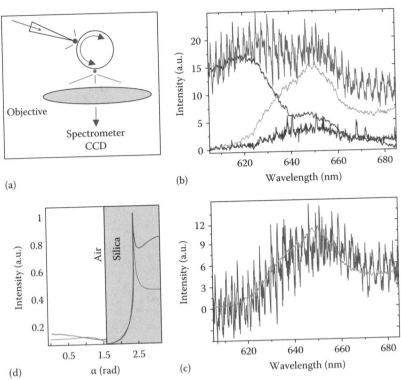

FIGURE 5.18 (a) Scheme of the cavity-mediated energy transfer measurement. (b) The curves with a maximum at around 619 and 650 nm display the fluorescence spectra of the naked donor and the acceptor beads, respectively. The strongly modulated uppermost curve shows the recorded spectrum when a donor is brought close to the sphere's surface. The bottommost curve shows the same measurement after bleaching of the donor. (c) The modulated curve plots the net emission spectrum of the acceptor as a result of energy transfer (see text for details). The smoother curve shows the normalized spectrum of a free acceptor (see (b)) for comparison. (d) The light gray and dark gray curves show the angular dependence of the emission intensity of dipoles parallel and perpendicular to an air-silica interface. (Reprinted from Götzinger, S. et al., *Nano Lett.*, 6, 1151, 2006, Figures 2 and 3. With permission. © 2006 American Chemical Society.)

Then, the total efficiency η for transferring a photon between the donor and the acceptor can be written as

$$\eta = \sum_i \beta_i \frac{\sigma_{A,abs}}{\sigma_{A,abs} + \sigma_{D,sca} + \sigma_{D,abs} + \sigma_{i,Q}} \tag{5.14}$$

A similar expression is derived in [LY88]. The parameter $\sigma_{A,abs}$ denotes the absorption cross section of the acceptor whereas cross sections $\sigma_{D,sca}$ and $\sigma_{D,abs}$ quantify losses out of the mode due to scattering and absorption of a photon by the donor. Finally, $\sigma_{i,Q}$ is a cross section signifying all losses associated with the measured Q of the microsphere, including those caused by scattering from the acceptor.

The absorption cross section of the acceptor particle can be taken as $\sigma_{A,abs} \approx 3 \times 10^{-11}$ cm^2, assuming $\sigma_{A,abs} \approx 10^{-16}$ cm^2 per molecule [Sch90] and 10^5 molecules per particle. Since due to the Stokes shift of molecular fluorescence the donor does not absorb its own emission very efficiently, we can neglect $\sigma_{D,abs}$. In addition, we obtain $\sigma_Q \approx 2 \times 10^{-12}$ cm^2, for a fundamental mode based on $Q = 3 \times 10^7$. Since in our experiment the measured Q remained unchanged as the tip approached the microsphere, we conclude that $\sigma_{D,sca}$ was negligible compared to σ_Q [GBS02] in this experiment. We find therefore, that the quotient in the equation for η given above is of the order of 1 for a fundamental mode.

A quantitative calculation of β_i for an arbitrary WGM has to consider all available modes (including the free space) and requires extensive numerical calculations. In Ref. [GMM06] we derived an estimation for β_i taking into account the enhanced emission into high-Q WGMs (see Figure 5.18d) as well as the homogeneous and inhomogeneous broadening of the emitters. We found $\beta_i \approx 10^{-8}$ and therefore $\eta_i \approx 10^{-8}$ for a fundamental WGM. Counting all WGMs that contribute to the energy transfer we finally obtain $\eta_i \approx 10^{-9}$ for transfer of donor photons to a single molecule in the acceptor particle, which is about 4 orders of magnitude larger than the free space rate for the absorption of a photon emitted by a donor placed at a comparable distance of about 50 μm.

It is important to note that in this experiment the effect of a high Q factor is to circulate each photon a large number of times, increasing its chance of interaction with the acceptor. All experiments described in this chapter are performed at room temperature. At cryogenic temperatures, however, the emission from the molecules is narrow band, and there is the promise to achieve controlled coupling of the emitters to single WGMs. This could enhance the achievable coupling efficiency to $\beta > 50\%$ and would thus allow for far enhanced photon transfer rates. Even coherent coupling between single donor and acceptor molecules over large distances (several 10 μm) may be envisioned.

5.6 Cavity (Quantum) Electrodynamics

In this section we introduce another experiment with a scanning probe. Here, the tip of the SNOM probe acts as a single subwavelength scatterer and is used to examine and control the backscattering induced coupling between counterpropagating high-Q modes of a microsphere resonator. It is interesting that the observed phenomena of the underlying purely classical physics resemble effects well known in strongly coupled systems of quantum electrodynamics.

5.6.1 Cavity Quantum Electrodynamics

The radiative properties of atoms can be strongly modified by coupling them to resonators [Ber94]. A historical cornerstone of this field of research, known as cavity quantum electrodynamics (CQED), was set in 1946 by E. M. Purcell who proposed that the radiation rate of an oscillating dipole at wavelength λ can be enhanced by a factor $F = 3Q\lambda^3/4\pi^2 V_m$ in a resonant cavity of quality factor Q and mode volume V_m [Ber94]. This so-called *Purcell effect* holds in the dissipative *weak* coupling regime where the cavity finesse is small so that the atomic radiation remains dominated by its coupling to the bath of the electromagnetic modes. In the *strong* coupling regime, coherent exchange of energy between the atom and the resonator causes the atomic resonance to lose its identity and to become replaced by a doublet. These phenomena have been studied for more than three decades [DGR00, KLN06, ADW06, HBW07], although the *in situ* manipulation of a single emitter in a single mode of a high-Q microresonator remains a challenge [ADW06, HBW07].

5.6.2 Mode-Splitting in Microsphere Resonators

About 10 years ago, it was discovered that the high-Q resonances of microsphere resonators are often composed of doublets [WSH95]. Such a mode splitting has since been discussed in conjunction with various WGM resonators [GPI00, KSV02, BJP05, SP07]. It turns out that mode splitting has been observed in other ring resonators and has been explained as the result of the coupling between the electric fields E_c and E_{cc} of the degenerate clockwise (c) and counterclockwise (cc) modes via backscattering. The new superposition states (+) and (−) are described by

$$E_+ = aE_c + bE_{cc}; \quad E_- = aE_c - bE_{cc} \tag{5.15}$$

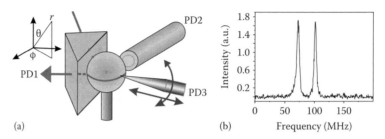

FIGURE 5.19   (a) Experimental setup to study mode splitting: Whispering-gallery modes of a microsphere are excited via a prism. A glass fiber tip can be positioned in (r, θ, ϕ) close to the sphere surface. The resonator spectrum can be recorded on PD1 in transmission, on PD2 from global scattering out of the sphere via a multimode fiber, and on PD3 through the fiber tip. (b) An example of the doublet spectrum recorded on PD2. (Reprinted from Mazzei, A. et al., *Phys. Rev. Lett.*, 99, 173603, 2007, Figures 1, 2, and 4. With permission. © 2007 American Physical Society.)

Here a and b are complex coefficients. In the simplest case, the coupling between E_c and E_{cc} can be caused by a reflector [SW91, VKG93]. In the case of WGM resonators, however, it has been suggested that backscattering from a distribution of residual subwavelength inhomogeneities in the glass matrix or on its surface is the source of this coupling [WSH95, KSV02, GPI00]. The orders of magnitude of the doublet splitting can be correctly estimated from classical electrodynamic considerations following this hypothesis [WSH95, GPI00, KSV02, BJP05, BG05, SP07]. Nevertheless, the direct link between the spectral features of a doublet and the nanoscopic details of the backscattering sources has not been demonstrated experimentally, and a proper treatment of the losses inflicted by the scatterers is missing in the literature. In fact, an intuitively perplexing and interesting question arises in this context: given that the radiation of a subwavelength scatterer is nearly isotropic and that the angle subtended by a typical cavity mode is merely about 10^{-4} rad [GMM06], how could the rate of scattering back into a cavity mode dominate the rate of scattering out of the resonator to ensure the population of E_+ and E_-?

The schematics of our experimental arrangement are similar to those in the previous sections. For convenience we provide a cartoon of the basic experimental tools again in Figure 5.19a. Figure 5.19b displays a typical doublet with a splitting of 29 MHz and $Q = 8 \times 10^7$ recorded on the photodiode PD2. The peaks represent the intensities $|E_+|^2$ and $|E_-|^2$ of the symmetric and antisymmetric eigenmodes. Following the procedure described in Section 5.2, we applied SNOM techniques to map the spatial intensity distribution of the WGMs on photodiode PD3 and to identify the fundamental mode of the resonator, which exhibits a single intensity maximum in the r and θ directions.

Equation 5.15 implies that the interference between the c and cc running modes should give rise to sine and cosine standing waves along the equator. The locations of the nodes and antinodes of E_+ and E_- are automatically established by the random distribution of a large number of inhomogeneities in the silica sphere [WSH95, GPI00]. To visualize this effect, we have scanned a sharp fiber tip along the equator (i.e., in the ϕ direction) of an already split fundamental WGM and have recorded spectra at each point (note that the radial coordinate of the tip is kept constant using a shear-force feedback [MG05]). Figure 5.20a shows that the intensities of the two peaks of a doublet undergo out of phase periodic modulations as a function of the tip location. A slight slope of the middle base line is attributed to a drift in the shear-force tip-sphere stabilization. At location (i), the tip is positioned in the node of the symmetric mode and the antinode of the antisymmetric mode. Thus, it induces loss in the E_+ mode while it leaves E_- nearly unaffected. Position (iii) shows the opposite counterpart of (i), whereas at position (ii) both modes are affected equally strongly.

As shown in Figure 5.20b, the three spectra reveal that in addition to a change in the intensity balance of the doublets, their splittings are also modified. Figure 5.20d displays the modulation of the splitting about its initial value of 24 MHz shown by the dotted line. Interestingly, we find that the tip can not only

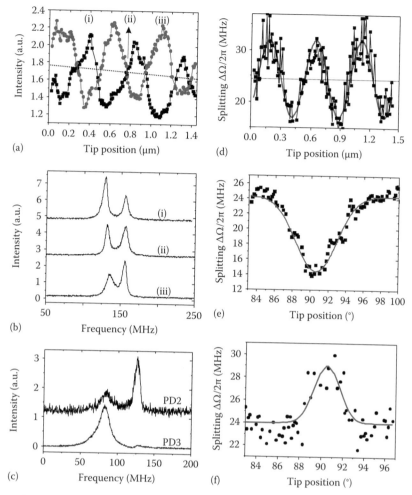

FIGURE 5.20 (a) The intensities of the two peaks in a doublet as a function of the tip position along the equator. The dotted line marks the slight intensity drift in the detection. (b) Spectra recorded at positions (i–iii) in Figure 5.20a. The spectra are displaced vertically for clarity. (c) Simultaneously measured spectra on PD2 and PD3. (d) The recorded splitting corresponding to the data in Figure 5.20a. The solid curve shows a sinusoidal fit. (e and f) The variations of the mode splitting as a function of the tip in the θ direction for positions (i) and (iii), respectively. The solid curves display fits according to the spatial mode function of the fundamental WGM. (Reprinted from Mazzei, A. et al., *Phys. Rev. Lett.*, 99, 173603, 2007, Figures 1, 2, and 4. With permission. © 2007 American Physical Society.)

increase the mode splitting, but it can also decrease it. This is due to a destructive interference between the field scattered by the tip and the field scattered by the inhomogeneities in the microsphere that gave rise to the initial splitting. Figure 5.20e and f provides further data on the increase and decrease of the mode splitting as the tip was scanned in the θ direction for two different φ positions spaced by half of the interference period along the equator. Finally, Figure 5.20c plots the resonance spectra recorded simultaneously on the photodiodes PD2 and PD3, that is, via global scattering from the sphere and via the fiber tip. The different lineshapes on the two channels might seem unexpected at first. However, this effect shows that if the tip is placed in an antinode of E_- or E_+ it efficiently extracts photons out of that mode, leading to a larger signal in the fiber tip and thus a lower intensity in the cavity mode. On the contrary, the mode that is less perturbed is stronger in the resonator and is nearly uncoupled to the fiber tip. It is evident that the mode that is coupled to the tip has experienced an additional broadening.

5.6.3 The Classical Mode-Splitting Problem in a Quantum Optical Light

In this section we show that the radiation properties of a subwavelength object such as its scattering rate are modified much in the same manner as those of the spontaneous emission of an atom. Our guiding thought is that many central features of CQED, including the modification of the mode density in a resonator, can be traced to the spatial character of the modes and should be thus shared by classical cavity electrodynamics. We first present a simple treatment of the *free-space* Rayleigh scattering using a semi-quantum electrodynamic (semi-QED) approach, where the material scatterer is treated classically while the field is quantized. Then, we will discuss the modification of the scattering rate when the scatterer is coupled to a resonator.

Let us assume that a freely propagating photon is incident on a subwavelength spherical scatterer of radius a and refractive index N. We take the photon to be in a mode $\hat{E}_k$ with volume V_k, frequency ω_k, and a linear polarization along the unit vector ε_k such that $\hat{E}_k = E_k^{(0)}\varepsilon_k(\hat{a}_k^\dagger + \hat{a}_k)$ where $\hat{a}_k^\dagger$ and $\hat{a}_k$ are the usual creation and annihilation operators and $E_k^{(0)} = \sqrt{\hbar\omega_k/2\varepsilon_0 V_k}$. In the limit where the scatterer is considerably smaller than l, it can be described by a dipolar polarizability α [Jac98] so that the induced dipole moment operator reads $\hat{p}_k = \varepsilon_0\alpha E_k^{(0)}\left(\hat{a}_j^\dagger + \hat{a}_k\right)\varepsilon_k$. Thus the interaction energy between this dipole moment and another mode $\hat{E}_j$ becomes

$$\hat{V}_{k,j} = -\hat{p}_k \cdot \hat{E}_j = \hbar g_{kj}\left(\hat{a}_k^\dagger\hat{a}_j + \hat{a}_j^\dagger\hat{a}_k\right) \tag{5.16}$$

if we neglect the terms that do not conserve photon numbers. Here we have taken $g_{kj} = -\alpha\sqrt{\omega_k\omega_j}\left|\varepsilon_j^* \cdot \varepsilon_k\right|/2\sqrt{V_kV_{vac}}$ and have set $V_j = V_{vac}$ for all vacuum modes j into which the incident beam is scattered. The system Hamiltonian becomes [MGM07]

$$\hat{H} = \hbar\omega_k\hat{a}_k^\dagger\hat{a}_k + \sum_j \hbar\omega_j\hat{a}_j^\dagger\hat{a}_j + \sum_j \hbar g_{kj}\left(\hat{a}_k^\dagger\hat{a}_j + \hat{a}_j^\dagger\hat{a}_k\right) \tag{5.17}$$

leading to the Heisenberg equation of motion

$$i\frac{d}{dt}\hat{a}_k = \omega_k\hat{a}_k + \sum_j g_{kj}\hat{a}_j \tag{5.18}$$

The last term in Equation 5.18 signifies the scattering of the incident field into all vacuum modes. Following the Weisskopf–Wigner formalism [Mil94], we find the rate

$$\Gamma_R = \frac{2\omega_k^2 V_{vac}}{3\pi c^3}g_R^2 = \frac{\alpha^2\omega_k^2}{6\pi c^3 V_k} \tag{5.19}$$

for this scattering event [MGM07]. Here we have restricted ourselves $\omega_k = \omega_j$ for elastic scattering and have used the notation $g_R = -\alpha\omega_k/2\sqrt{V_kV_{vac}}$. Now we can calculate the Rayleigh scattering cross section σ_R [Jac98], by considering the total power radiated by the scatterer according to $I_{inc}\sigma_R = \hbar\omega_k\Gamma_R$. Given that $I_{inc} = \hbar\omega_k c/V_k$ and $\alpha = 4\pi a^3\left|\frac{N^2-1}{N^2+2}\right|$, one obtains the well-known relation

$$\sigma_R = \frac{8\pi k^4 a^6}{3}\left|\frac{N^2-1}{N^2+2}\right|^2 \tag{5.20}$$

We note that a rigorous quantum optical treatment of scattering is not frequently discussed in the literature [DBK99]. However, the fact that σ_R can be derived via the Weisskopf–Wigner formalism using quantized fields provides a robust support for the intuitive expectation that a modification of the mode density, for example in a resonator or in front of a mirror, could also lead to a change in the Rayleigh scattering rate. The corresponding Purcell effect offers a physical explanation for the question posed earlier. The rate with which energy is transferred from E_c to E_{cc} is enhanced by the Purcell factor F and is given by ηF where η is the geometric factor determining the fraction of the solid angle subtended by the mode. Therefore, a fundamental WGM with $V_m \cong 130\,\mu m^3$ (sphere diameter 30 μm) and $Q = 10^8$ yields $F \sim 10^4$, compensating for the very small geometric acceptance of the order of 10^{-4}. The influence of the Purcell factor F also explains why reducing the cavity Q results in the disappearance of light in the counterpropagating mode as reported previously [WSH95].

Having shown that the density of states plays a central role in the description of Rayleigh scattering, we consider next the coupling of two counterpropagating *cavity* modes E_c and E_{cc} via a single Rayleigh scatterer. We skip all the details here and refer to Ref. [MGM07] instead. Returning to a classical notation, we find

$$\frac{d}{dt}\tilde{E}_- = \left(-i\Delta - \gamma_0\right)\tilde{E}_- + \kappa_0$$

$$\frac{d}{dt}\tilde{E}_+ = \left(-i\Delta + 2ig - \gamma_0 - \Gamma\right)\tilde{E}_+ + \kappa_0 \tag{5.21}$$

Here we have defined $\tilde{E} = e^{i\omega t}E$, $\Delta = \omega - \omega_c$ shows the detuning of the laser frequency ω, $\kappa = \kappa_0 e^{i\omega t}$ is the mode excitation rate, and $2\gamma_0$ denotes the unperturbed cavity linewidth. When dealing with Rayleigh scattering out of a cavity, we have to take into account the spatial variation $f(r)$ of E_c and E_{cc} in the resonator mode. Going back to the definition of g_{kj} and noting that $V_j = V_k$ is the WGM volume V_m of the two modes, we thus obtain [MGM07]

$$2g = -\alpha f^2(r)\omega_c/V_m$$

$$\Gamma = \alpha^2 f^2(r)\omega_c^4/6\pi c^3 V_m \tag{5.22}$$

for the mode splitting and broadening, respectively.

A close scrutiny of Equation 5.21 shows that if $|2g|$ is sufficiently large to overcome γ_0 and if $2\left|\frac{N^2-1}{N^2+2}\right| < \left(\frac{\lambda}{2\pi a}\right)^3$ to assure that $|2g| > \Gamma$, a mode splitting is resolved. This is similar to the case of the strong coupling in CQED where the coherent exchange of energy between the cavity mode and an atom leads to a mode splitting if the coupling coefficient $|g|$ becomes larger than the cavity linewidth. As in CQED, the mode splitting $|2g|$ grows with decreasing V_m and increasing ω, but the atomic dipole strength is replaced here by α, and the splitting is asymmetric. The antisymmetric mode remains unperturbed (see Equation 5.21) whereas the symmetric mode undergoes an additional broadening given by Γ. This is a consequence of the fact that the phases of the new eigenmodes are self-adjusted so that the local scatterer is placed in a node (antinode) of the antisymmetric (symmetric) mode.

To realize an ideal scenario for studying the interaction of a single well-defined scatterer with the fundamental mode of a microsphere, we have searched for spheres in which no mode splitting was observable in the beginning and it was created only when the tip was introduced. Figure 5.21 presents four spectra of a resonance as the tip was scanned in the θ-direction (see Figure 5.19) from outside the mode to its maximum at the equator at constant distance from the sphere's surface. The E_+ mode is shifted in frequency by 13 MHz and broadened by about 6 MHz. Given that $|2g|/\Gamma = 3\lambda^3/4\pi^2\alpha$ according to Equations 5.22, the observed ratio of the splitting to broadening implies a radius of $a \sim 140$ nm for a spherical Rayleigh scatterer. This is in very good quantitative agreement with the experimental parameters, considering a typical value of 50–100 nm for the radii of curvature of SNOM tips and accounting for the overlap of the conical tip taper with the evanescent part of the mode.

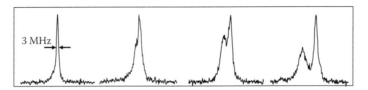

FIGURE 5.21   WGM spectra recorded at different θ values of the tip location. The leftmost spectrum corresponds to a position where the tip is nearly outside the spatial profile of the mode whereas for the rightmost spectrum it is in the mode maximum, i.e., the sphere equator. (Reprinted from Mazzei, A. et al., *Phys. Rev. Lett.*, 99, 173603, 2007, Figures 1, 2, and 4. With permission. © 2007 American Physical Society.)

The results presented in this section demonstrate that although the introduction of a scatterer into a high-finesse resonator might be commonly thought to introduce losses, it can mediate a coherent coupling of the resonator modes and cause their consequent normal mode splitting. Interestingly, the phenomenon of Rayleigh scattering in the presence of a resonator can be described in the framework of a semi-QED approach. The modification in classical scattering results from the modification of the density of states in analogy to Purcell enhancement in CQED.

5.7 Discussion and Outlook

Microsphere resonators are ideal systems to study the evanescent interaction of nano emitters with confined modes of the electromagnetic field. At room temperature the observed phenomena are governed by classical physics. However, the framework developed in quantum optics can be applied and sometimes is more intuitive, as was discussed in the last section. In order to reach the true quantum regime, narrow-band emitters have to be studied. However, controlled CQED experiments with microsphere resonators face several problems:

- Coupling of light into and out of tightly confined modes requires external couplers that have to be mode-matched to obtain high coupling efficiencies.
- The mode volume of microspheres with sizes of several tens of micrometers in diameter is still too large for experiments in the strong-coupling regime of CQED.
- It is difficult to implement systems where several emitters (more than two) are located at well-defined positions within the cavity mode where they have strongest interaction with the field. We have introduced scanning probes that may act as single emitters or scatterers, but scaling up is very challenging.
- There are only very few emitters, even at cryogenic temperatures, which are photostable and show no spectral diffusion or blinking.

Figure 5.22 shows a system that can possibly solve some of these problems. It consists of another type of resonator, a *microtoroid* [AKS03] that also supports WGM. A typical resonator made from SiO_2 by optical lithography, selective etching, and subsequent partial melting by a CO_2 laser is shown in Figure 5.22a. The optical equator in a toroid is obviously always well defined in contrast to a microsphere (compare with discussion in Section 5.2). This facilitates coupling to WGMs via external couplers, such as tapered optical fibers. Figure 5.22b shows an artist's view of how to establish coupling to few quantum emitters. We suggest utilizing nano manipulation by a scanning probe, for example, an atomic force microscope, to push individual nanodiamonds containing single color centers into the evanescent field of high-Q modes. It has been shown that single color or defect centers are photostable even at room temperature and have a lifetime-limited zero phonon line at low temperature [JW06]. Also tuning via Stark effect is possible [TGR06]. First results concerning the coupling of these nanodiamonds to small microsphere resonators on-demand were successful [SSB08]. Together with the possible efficient coupling via

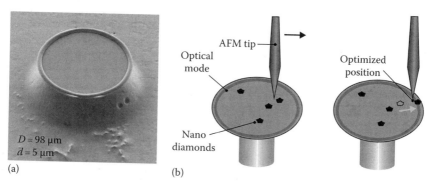

FIGURE 5.22 (a) Image of a toroidal microresonator where D and d indicated the diameter of the toroid and its melted rim, respectively. Coupling to evanescent modes can be obtained via a tapered optical fiber. (b) Cartoon of an approach to realize a CQED system with several quantum emitters. An atomic force microscope (AFM) is used to push previously deposited nanodiamonds containing single color centers into the evanescent field of WGMs in a toroidal microresonator.

tapered optical fibers (see Figure 5.5), such a platform may be very promising to scale up experiments to more complex configurations.

Acknowledgments

We acknowledge funding from the DFG (SFB 448) and from the Alexander von Humboldt foundation. We thank M. Gregor and R. Henze for help with editing.

References

[Ab04] P. Andrew and W. L. Barnes. Energy transfer across a metal film mediated by surface plasmon polaritons. *Science* 306, 1002–1005, 2004.

[ADW06] T. Aoki, B. Dayan, E. Wilcut, W. P. Bowen, A. S. Parkins, T. J. Kippenberg, K. J. Vahala, and H. J. Kimble. Observation of strong coupling between one atom and a monolithic microresonator. *Nature* 443, 671–674, 2006.

[AHD96] S. Arnold, S. Holler, and S. D. Druger. Imaging enhanced energy transfer in a levitated aerosol particle. *J. Chem. Phys.* 104, 7741–7748, 1996.

[AKS03] D. K. Armani, T. J. Kippenberg, S. M. Spillane, and K. J. Vahala. Ultra-high-Q toroid microcavity on a chip. *Nature* 421, 925–928, 2003.

[AWW01] M. V. Artemyev, U. Woggon, R. Wannemacher, H. Jaschinski, and W. Langbein. Light trapped in a photonic dot: Microspheres act as a cavity for quantum dot emission. *Nano Lett.* 1, 309–314, 2001.

[Ber94] P. Berman, Ed., *Cavity Quantum Electrodynamics*. Academic Press, London, U.K., 1994.

[BC88] P. W. Barber and R. K. Chang. *Optical Effects Associated with Small Particles*. World Scientific, Singapore, 1988.

[BFW92] E. Betzig, L. Finn, and L. S. Weiner. Combined shear-force and nearfield scanning optical microscopy. *Appl. Phys. Lett.* 60, 2484–2486, 1992.

[BG05] P.-Y. Bourgeois and V. Giordano. Simple model for the mode-splitting effect in whispering-gallery-mode resonators. *IEEE Trans. Microw. Theory Tech.* 53, 3185–3190, 2005.

[BGI89] V. B. Braginsky, M. L. Gorodetsky, and V. S. Ilchenko. Quality-factor and nonlinear properties of optical whispering-gallery modes. *Phys. Lett. A* 137, 393–397, 1989.

[BH98] C. F. Bohren and D. R. Huffman. *Absorption and Scattering of Light by Small Particles*. John Wiley & Sons, New York, 1998.

[BJP05] M. Borselli, T. J. Johnson, and O. Painter. Beyond the Rayleigh scattering limit in high-Q silicon microdisks: Theory and experiment. *Opt. Express* 13, 1515–1530, 2005.

[Boy92] R. W. Boyd. *Nonlinear Optics.* Academic Press Inc., San Diego, CA, 1992.

[BS91] I. N. Bronstein and K. A. Semendjajew. *Taschenbuch der Mathematik*, 25th edn. Teubner Verlagsgesellschaft, Stuttgart, Germany, 1991.

[CC96] R. K. Chang and A. J. Campillo. *Optical Processes in Microcavities.* World Scientific, Singapore, 1996.

[CLB93] L. Collot, V. Lefevre-Seguin, M. Brune, J. M. Raimond, and S. Haroche. Very high-Q whispering-gallery mode resonances observed in fused-silica microspheres. *Eur. Phys. Lett.* 23, 327–334, 1993.

[CV00] M. Cai and K. Vahala. Highly efficient optical power transfer to whispering-gallery modes by use of a symmetrical dual-coupling configuration. *Opt. Lett.* 25, 260–262, 2000.

[DBK99] B. J. Dalton, S. M. Barnette, and P. L. Knight. A quantum scattering theory approach to quantum-optical measurements. *J. Mod. Opt.* 46, 1107–1121, 1999.

[DGR00] P. Domokos, M. Gangl, and H. Ritsch. Single-atom detection in high-Q multimode cavities. *Opt. Commun.* 185, 115–123, 2000.

[DHK83] R. W. P. Drewer, J. L. Hall, F. V. Kowalski, J. Hough, G. M. Ford, A. J. Munley, and H. Ward. Laser phase and frequency stabilization using an optical resonator. *Appl. Phys. B* 31, 97–105, 1983.

[Dic54] R. H. Dicke. Coherence in spontaneous radiation processes. *Phys. Rev.* 93, 99–110, 1954.

[DKL95] N. Dubreuil, J. C. Knight, D. K. Leventhal, V. Sandoghdar, J. Hare, and V. Lefèvre. Eroded mono-mode optical fiber for whispering-gallery mode excitation in fused-silica microspheres. *Opt. Lett.* 20, 813–815, 1995.

[DPR86] U. Dürig, D. W. Pohl, and F. Rohner. Near-field optical-scanning microscopy. *J. Appl. Phys.* 59, 33133317, 1986.

[FAD85] L. Folan, S. Arnold, and S. Druger. Enhanced energy-transfer within a microparticle. *Chem. Phys. Lett.* 118, 322–327, 1985.

[Foe65] T. Förster. *Modern Quantum Chemistry*, Vol. III, O. Sinanoglu (Ed.), pp. 93–137, Academic Press, New York, 1965.

[GBS01] S. Götzinger, O. Benson, and V. Sandoghdar. Towards controlled coupling between a high-Q whisperinggallery mode and a single nanoparticle. *Appl. Phys. B*, 73, 825–828, 2001.

[GBS02] S. Götzinger, O. Benson, and V. Sandoghdar. Influence of a sharp fiber tip on high-Q modes of a microsphere resonator. *Opt. Lett.* 27, 80–82, 2002.

[GI94] M. L. Gorodetsky and V. S. Ilchenko. High-Q optical whispering-gallery microresonators: Precession approach for spherical mode analysis and emission patterns with prism couplers. *Opt. Comm.* 113, 133–143, 1994.

[GMB04] S. Götzinger, L. de S. Menezes, O. Benson, D. V. Talapin, N. Gaponik, H. Weller, A. L. Rogach, and V. Sandoghdar. Confocal microscopy of nanocrystals on a high-Q microsphere resonator. *J. Opt. B* 6, 154–158, 2004.

[GMM06] S. Götzinger, L. de S. Menezes, A. Mazzei, O. Benson, S. Kühn, and V. Sandoghdar. Controlled energy transfer between two individual nanoemitters via shared high-Q modes of a microsphere resonator. *Nano Lett.* 6, 1151–1154, 2006.

[GPI00] M. L. Gorodetsky, A. D. Pryamikov, and V. S. Ilchenko. Rayleigh scattering in high-Q microspheres. *J. Opt. Soc. Am. B* 17, 1051–1057, 2000.

[GR00] I. S. Gradshteyn and I. M. Ryzhik. *Table of Integrals, Series, and Products*, 6th edn. Academic Press, San Diego, CA, 2000.

[GRD07] M. Gerlach, Y. P. Rakovich, and J. F. Donegan. Radiation-pressure-induced mode splitting in a spherical microcavity with an elastic shell. *Opt. Express* 15, 3597–3606, 2007.

[GSI96] M. L. Gorodetsky, A. A. Savchenkov, and V. S. Ilchenko. Ultimate Q of optical microsphere resonators. *Opt. Lett.* 21, 453–455, 1996.

[HBW07] K. Hennessy, A. Badolato, M. Winger, D. Gerace, M. Atature, S. Gulde, S. Falt, E. L. Hu, and A. Imamoglu. Quantum nature of a strongly coupled single quantum dot-cavity system. *Nature* 445, 896–899, 2007.

[**HGD99**] M. Hopmeier, W. Guss, M. Deussen, E. O. Gobel, and R. F. Mahrt. Enhanced dipole-dipole interaction in a polymer microcavity. *Phys. Rev. Lett.* 82, 4118–4121, 1999.

[**HSZ02**] C. Hettich, C. Schmitt, J. Zitzmann, S. Kühn, I. Gerhardt, and V. Sandoghdar. Nanometer resolution and coherent optical dipole coupling of two individual molecules. *Science* 298, 385–389, 2002.

[**IYM99**] V. S. Ilchenko, X. S. Yao, and L. Maleki. Pigtailing the high-Q microsphere cavity: a simple fiber coupler for optical whispering-gallery modes. *Opt. Lett.* 24, 723–725, 1999.

[**Jac98**] J. D. Jackson. *Classical Electrodynamics*, 3rd edn. John Wiley & Sons, New York, 1998.

[**JW06**] F. Jelezko and J. Wrachtrup. Single defect centres in diamond: A review. *Phys. Stat. Sol. A* 203, 3207–3225, 2006.

[**KCJB97**] J. C. Knight, G. Cheung, F. Jacques, and T. A. Birks. Phase-matched excition of whispering-gallery-mode resonances by a fiber taper. *Opt. Lett.* 22, 1129–1131, 1997.

[**KDS95**] J. C. Knight, N. Dubreuil, V. Sandoghdar, J. Hare, V. Lefèvre-Seguin, J. M. Raimond, and S. Haroche. Mapping whispering-gallery modes in microspheres with a near-field probe. *Opt. Lett.* 20, 1515–1517, 1995.

[**KDS96**] J. C. Knight, N. Dubreuil, V. Sandoghdar, J. Hare, V. Lefèvre-Seguin, J. M. Raimond, and S. Haroche. Characterizing whispering-gallery modes in microspheres by direct observation of the optical standing-wave pattern in the near-field. *Opt. Lett.* 21, 698–700, 1996.

[**KHS01**] S. Kühn, C. Hettich, C. Schmitt, J.-Ph. Poizat, and V. Sandoghdar. Diamond colour centres as a nanoscopic light source for scanning near-field optical microscopy. *J. Microsc.* 202, 2–6, 2001.

[**KJD95**] M. Kuwata-Gonokami, R. H. Jordan, A. Dodabalapur, H. E. Katz, M. L. Schilling, R. E. Slusher, and S. Ozawa. Polymer microdisk and microring lasers. *Opt. Lett.* 20, 2093–2095, 1995.

[**KLN06**] J. Klinner, M. Lindholdt, B. Nagorny, and A. Hemmerich. Normal mode splitting and mechanical effects of an optical lattice in a ring cavity. *Phys. Rev. Lett.* 96, 023002, 2006.

[**KS99**] F. K. Kneubühl and M. W. Sigrist. *Laser*, Vol. 5. Teubner, Stuttgart, Leipzig, 1999.

[**KSV02**] T. J. Kippenberg, S. M. Spillane, and K. J. Vahala. Modal coupling in traveling-wave resonators. *Opt. Lett.* 27, 1669–1671, 2002.

[**LY88**] P. T. Leung and K. Young. Theory of enhanced energy transfer in an aerosol-particle. *J. Chem. Phys.* 89, 2894–2899, 1988.

[**LLL00**] B. E. Little, J.-P. Laine, D. R. Lim, H. A. Haus, L. C. Kimerling, and S. T. Chu. Pedestal antiresonant reflecting waveguides for robust coupling to microsphere resonators and for microphotonic circuits. *Opt. Lett.* 25, 73–75, 2000.

[**MG05**] A. Mazzei, S. Götzinger, L. de S. Menezes, V. Sandoghdar, and O. Benson. Optimization of prism coupling to high-Q modes in a microsphere resonator using a near-field probe. *Opt. Commun.* 250, 428–433, 2005.

[**MGM07**] A. Mazzei, S. Götzinger, L. de S. Menezes, G. Zumofen, O. Benson, and V. Sandoghdar. Controlled coupling of counterpropagating whispering-gallery modes by a single Rayleigh scatterer: A classical problem in a quantum optical light. *Phys. Rev. Lett.* 99, 173603, 2007.

[**MI06a**] A. B. Matsko and V. S. Ilchenko. Optical resonators with whispering-gallery modes—Part I: Basics. *IEEE J. Sel. Top. Quantum Electron.* 12, 3–14, 2006.

[**MI06b**] V. S. Ilchenko and A. B. Matsko. Optical resonators with whispering-gallery Modes—Part I: Applications. *IEEE J. Sel. Top. Quantum Electron.* 12, 15–32, 2006.

[**Mie08**] G. Mie. Beiträge zur Optik trüber Medien, speziell kolloidaler Metallösungen. *Ann. Phys.* 25(4):377–445, 1908.

[**Mil94**] P. W. Milonni. *The Quantum Vacuum.* Academic Press, New York, 1994.

[**MK94**] H. Mabuchi and H. J. Kimble. Atom galleries for whispering atoms—Binding atoms in stable orbits around an optical microresonator. *Opt. Lett.* 19, 749–751, 1994.

[**MKB06**] A. Mazzei, H. Krauter, O. Benson, and S. Gtzinger. Influence of a controllable scatterer on the lasing properties of an ultra-low threshold Raman microlaser. *Appl. Phys. Lett.* 89, 101105, 2006.

[**MWA05**] B. M. Möller, U. Woggon, and M. W. Artemyev. Coupled-resonator optical waveguides doped with nanocrystals. *Opt. Lett.* 30, 2116–2118, 2005.

[**PCW06**] Y. S. Park, A. K. Cook, and H. L. Wang. Cavity QED with diamond nanocrystals and silica microspheres. *Nano Lett.* 6, 2075–2079, 2006.

[**RK28**] C. V. Raman and K. S. Krishnan. Temporal behavior of radiation-pressure-induced vibrations of an optical microcavity phonon mode. *Nature* 121, 711, 1928.

[**Sch90**] F. P. Schäfer. *Dye Lasers.* Springer Verlag, Berlin, 1990.

[**Sch93**] S. Schiller. Asymptotic expansion of morphological resonance frequencies in Mie scattering. *Appl. Opt.* 32, 2181–2185, 1993.

[**Sie86**] A. E. Siegman. *Lasers.* University Science Books, Sausalito, CA, 1986.

[**SIM04**] A. A. Savchenkov, V. S. Ilchenko, A. B. Matsko, and L. Maleki. Kilo-hertz optical resonances in dielectric crystal cavities. *Phys. Rev. A* 70, 051-804, 2004.

[**SKV02**] S. M. Spillane, J. J. Kippenberg, and K. J. Vahala. Ultralow-threshold Raman laser using a spherical dielectric microcavity. *Nature* 415, 621–623, 2002.

[**SP07**] K. Srinivasan and O. Painter. Mode coupling and cavity-quantum-dot interactions in a fiber-coupled microdisk cavity. *Phys. Rev. A* 55, 023814, 2007.

[**SQC85**] J. B. Snow, S.-X. Qian, and R. K. Chang. Stimulated Raman-scattering from individual water and ethanol droplets at morphology-dependent resonances. *Opt. Lett.* 10, 37–39, 1985.

[**SSB08**] S. Schietinger, T. Schröder, and O. Benson. One-by-one coupling of single defect centers in nano-diamonds to high-Q modes of an optical microresonator. *Nano Lett.* 8, 3911–3915, 2008.

[**Str41**] J. A. Stratton. *Electromagnetic Theory.* McGraw-Hill Book Company, New York, 1941.

[**Str99**] E. Streed. Spectroscopy of high-Q whispering-gallery modes in silica microspheres. Senior thesis experimental, California Institute of Technology, Pasadena, CA, 1999.

[**SW91**] R. J. C. Spreeuw and J. P. Woerdman. The driven optical ring resonator as a model system for quantum optics. *Phys. Rev. B* 75, 96–110, 1991.

[**TGR06**] Ph. Tamarat, T. Gaebel, J. R. Rabeau, M. Khan, A. D. Greentree, H. Wilson, L. C. L. Hollenberg, S. Prawer, P. Hemmer, F. Jelezko, and J. Wrachtrup. Stark shift control of single optical centers in diamond. *Phys. Rev. Lett.* 97, 083002, 2006.

[**VA08**] F. Vollmer and S. Arnold. Whispering-gallery-mode biosensing: Label-free detection down to single molecules. *Nat. Methods* 5, 591–596, 2008.

[**VAK08**] F. Vollmer, S. Arnold, and D. Keng. Single virus detection from the reactive shift of a whispering-gallery mode. *PNAS* 105, 20701–20704, 2008.

[**Vah03**] K. Vahala. Optical microcavities. *Nature* 424, 839–846, 2003.

[**VKG93**] A. Venugopalan, D. Kumar, and R. Ghosh. Optical mode-coupling in a ring due to a back-scatterer. *Pramana J. Phys.* 40, 107–112, 1993.

[**Wat22**] G. N. Watson. *A treatise on the Theory of Bessel Functions.* Cambridge University Press, Cambridge, U.K., 1922.

[**WSH95**] D. S. Weiss, V. Sandoghdar, J. Hare, V. Lefevre-Seguin, J. M. Raimond, and S. Haroche. Splitting of high-Q Mie modes induced by light scattering in silics microspheres. *Opt. Lett.* 20, 1835–1837, 1995.

[**WU02**] A. Wallraff and A. V. Ustinov. Flüsternde Flussquanten. *Phys. Unserer Zeit*, 33(4), 184, 2002.

IV

Integrated Photonic Technologies: Photonic Crystals and Integrated Circuits

6

Controlling Dispersion and Nonlinearities in Mesoscopic Silicon Photonic Crystals

C. W. Wong
X. Yang
J. F. McMillan
R. Chatterjee
S. Kocaman

6.1 Controlling Dispersion in Silicon Photonic Crystals

6.1.1 Bound Surface States in Negative Refraction Photonic Crystals for Subdiffraction Imaging

6.1.1.1 Negative Refraction Metamaterials

The prediction that in specific circumstances a certain class of metamaterials, known as left-handed metamaterials (LHMs), have a negative index of refraction has recently generated increasing scientific interest. First introduced as homogeneous electromagnetic media characterized by simultaneous negative permittivity and permeability,[1] LHMs were predicted to possess remarkable optical properties, such as refraction governed by an inverted Snell law, negative light pressure, or reversal of Vavilov–Cherenkov radiation and Doppler shift.[1] Although these intriguing phenomena have implications in fundamental physics, the intense current interest in these materials is sparked by the prediction that a lossless LHM slab will have the characteristics of a "perfect lens."[2] In particular, it has been shown theoretically[2] that an LHM lens amplifies evanescent waves, and therefore information contained in both propagating and evanescent fields can be reconstructed into a perfect image. This property of "superlensing," or, more rigorously, imaging beyond the diffraction limit, has recently been demonstrated in a series of experiments involving different classes of LHMs.[3–6] In most of these experiments, the LHMs consist of metallo-dielectric periodic microstructures[3–10] that operate at microwave frequencies. Metal-based LHMs, however, currently show large absorption losses especially when scaled to visible or near-infrared wavelengths[11]; this loss can possibly act as a practical limit on the resolution of subdiffraction limit imaging at optical frequencies,[12] although recent remarkable efforts have examined important pathways for improvements to these losses.[13–15] A promising alternative is to use dielectric-based photonic crystal (PhC) LHMs,[16–23] whose optical losses could be considerably smaller. Here, we report the first observation of imaging beyond the diffraction limit in negative refraction PhCs, using near-field techniques at near-infrared wavelengths. Our experimental results indicate an effective negative-index figure of merit (FOM) [$-\text{Re}(n)/\text{Im}(n)$] of at least 380, approximately 125× improvement over recent experimental efforts,[13] with an operating bandwidth of 0.4 THz. The observed resolution is 0.47λ, clearly smaller than the diffraction limit of 0.61λ, and is supported by detailed theoretical analyses and comprehensive numerical simulations. Importantly, we clearly show for the first time that this subdiffraction limit imaging is due to the resonant excitation of surface slab modes, permitting amplification of evanescent waves.

Metal-based LHMs containing resonant metallic microstructures operating at optical frequencies have recently been remarkably demonstrated to possess negative refraction[13,24] or negative magnetic permeability.[25] For certain artificially engineered metal-based LHMs, the resonant response of metallic inclusions leads to changes in the effective parameters of the composite structure, thus forming a metamaterial with both an effective electric permittivity, ε, and an effective magnetic permeability, μ, that are negative in a frequency band close to the resonant wavelength.[7] Since for such metamaterials, the resonant wavelength is much larger than the characteristic length of the primary unit cell of the bulk material, near this wavelength the material acts as a homogeneous optical medium. Other approaches, such as materials with indefinite permittivity and permeability tensors,[26] domain twin structures,[27] isotropic nonmagnetic two-component molecular media,[28] and dielectric metamaterials in the mid-infrared[29,30] and near-infrared,[15] also suggest interesting potential for negative refraction. An alternative approach is to use PhCs, whose dispersion properties can be engineered so that at specific frequencies they possess a negative refraction. While theoretical studies have predicted subwavelength imaging in PhC-based LHMs at optical and near-infrared frequencies,[16–21] which is the frequency region of most interest for technological applications, an experimental demonstration of this phenomenon at these frequencies is still critically lacking, primarily due to the significant challenges in device nanofabrication and measurements. Here, by using a two-dimensional (2D) PhC slab, which emulates a negative refractive index over a specific frequency range in the near-infrared spectrum, we achieve the first near-field observation of subdiffraction limit imaging at near-infrared frequencies with a large effective negative-index FOM of at least 380. This FOM is about 125× improvement over recent experimental efforts[13] and more than 15× better than the best theoretical efforts.[14] Furthermore and importantly, we also clearly show, for the first time, that this subdiffraction limit imaging is due to the resonant excitation of surface slab modes, permitting the amplification of evanescent waves.

6.1.1.2 Understanding Negative Refraction and Subdiffraction Imaging in PhCs

The possibility of left-handed behavior of wave propagation in a PhC is intimately related to the existence of photonic modes with a negative phase index, i.e., modes for which the Bloch wavevector, k, in the first Brillouin zone and the group velocity, v_g, defined as the gradient of the mode frequency $\omega(k)$, $v_g = \nabla_k \omega(k)$, are antiparallel, i.e., $k \cdot v_g < 0$.[16,17] In other words, in a PhC, photonic bands are folded back into the first Brillouin zone, and it is therefore possible to find regions in the k-space where the mode frequency decreases with the mode wavevector. If this condition is satisfied, the wavevector k and the Poynting vector S, which in an infinite PhC is parallel to v_g,[31] have opposite directions. In the corresponding frequency range, $k \cdot v_g < 0$, i.e., the PhC emulates certain optical properties of a homogeneous negative-index medium.[1] In particular, negative refraction and subdiffraction limit imaging in such PhCs have been predicted theoretically[16–21] and demonstrated at microwave frequencies[4,6,10] and in the near-infrared without subdiffraction limit imaging.[22,23] Although these phenomena would ideally be studied in a PhC whose phase index is isotropic, as is usually the case near the Γ symmetry point, they can be observed even when the phase index shows a certain degree of optical anisotropy.[4] This ability to tolerate PhC anisotropy[4] is particularly important for the 2D PhC slab geometries used in our study, since in this case modes with the wavevector k close to the Γ symmetry point are leaky modes characterized by large optical losses, and thus it is not possible to operate in this near-Γ region of the k-space.

In the case of a PhC slab, Luo et al.[18] have clearly shown that there are two mechanisms that can lead to the amplification of evanescent waves by the slab, and therefore to the possibility of imaging beyond the diffraction limit. In the first case, the amplification of the evanescent waves is due to the resonant excitation of bound surface states at the air–PhC interfaces, which results in a *single-interface resonance* divergence of the transmission through the interface; in the second case, resonant amplification is caused by the excitation of bound transverse guiding states of the PhC slab, with the effect of an *overall resonance* divergence of the transmission through the slab. In addition, in the latter case, the bound transverse guiding states can be either *slab modes*, similar to those supported by a *homogeneous* dielectric slab waveguide, or *surface slab modes*, which are formed by the linear superposition of surface states excited at the air–PhC interface (such surface slab modes decay exponentially from the air–PhC interface, both into the slab and into the air). In our case, we have designed our PhC slab such that for our wavelengths, the symmetry properties of the slab modes allow only the excitation of *surface slab modes*, which, as explained below, allows us to achieve the subdiffraction limit resolution.

6.1.1.3 Photonic Crystal Design

The PhC slab is designed by employing full three-dimensional (3D) frequency- and time-domain computations. To determine dispersive properties of the photonic bands and the spatial distribution of the electromagnetic field, a mode solver based on a plane wave decomposition algorithm and the finite-difference time-domain (FDTD) method, respectively, was used. The photonic band structure of the PhC slab was computed using the supercell method, with a basis consisting of $32 \times 128 \times 32$ plane waves. In our FDTD simulations, we utilized a rectangular computational domain, $L_x \times L_y \times L_z$, of dimensions $13.5 \times 0.87 \times 11\,\mu m^3$, surrounded by a $0.25\,\mu m$ thick perfectly matched layer (PML). The reflectivity of the PML was set to 10^{-8}, so that the wave reflections at the boundaries are negligible. A $15 \times 15 \times 15\,nm^3$ uniform grid covered both the computational domain and the PML. The time-averaged spatial distribution of the electromagnetic field was determined by averaging the spatial field distributions, which were recorded at 100 time points equally distributed within one wave period. The severe computational requirements of full 3D simulations were satisfied using a Linux cluster consisting of 18 2.8 GHz Pentium 4 processors.

The PhC-based materials investigated in this study were designed to consist of a slab with a hexagonal lattice of holes (*p6m* symmetry group) etched into a silicon film of thickness $t = 320\,nm$, on a SiO_2 insulator substrate (we consider the median plane of the slab to be the x-z plane, with the y-axis normal to this plane). The normal to the input facet is oriented along the Γ-M symmetry axis of the crystal. The designed PhC slab has a lattice constant $a = 430\,nm$ and a radius $r = 120\,nm$ ($r = 0.279a$). Vertical optical confinement is obtained through total internal reflection, with air cladding on the top and buried silicon oxide (1 μm thick) at the bottom. Our subwavelength source is a tapered $450 \times 320\,nm^2$ waveguide (tapered from 450 to 765 nm) placed 250 nm away from the PhC slab and fabricated onto the same silicon-on-insulator (SOI) platform. At the wavelengths used in our measurements, the SOI waveguide supports only two modes, $E_{11}^x (E_{11}^y)$, whose electric (magnetic) field is predominantly oriented along the x-axis.

The photonic band structure of the guided modes is presented in Figure 6.1b; the electromagnetic field of these modes is strongly confined to the slab, with the energy flow parallel to the plane of the slab. By contrast, the electromagnetic field of photonic modes above the light-cone can be shown to have a real propagation vector component normal to the slab. These leaky modes or guided resonances have large optical losses, which make them unsuitable for observing subdiffraction limit imaging. In designing our slab, we also computed the transverse-propagating guiding modes (surface and slab modes) of the complete slab (several such bands are shown in Figure 6.1b, the bottom-right panel), whose **k** wavevector is along the Γ-K crystal axis. We classify the photonic bands in Figure 6.1b according to how the modes transform upon applying the reflection operators Σ_i ($i = x, y, z$), and for this we calculated the parameters σ_i, which quantify the parity of the modes: $\sigma_i = \int_{BZ} E_k(r) \cdot \Sigma_i E_k(r) dr$,

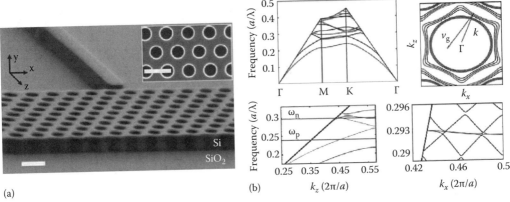

FIGURE 6.1 Designed and fabricated negative refraction PhC metamaterial. (a) SEM of fabricated PhC slab. Inset: Higher-resolution SEM of low-disorder PhC lattice fabricated. Disorder variation is statistically quantified to less than $0.004a$. Scale bar: 500 nm. (b) Photonic band structure of an air-hole hexagonal PhC slab with $r = 0.279a$; $t = 0744a$ with $a = 430$ nm (top-left panel). The TE-like (TM-like) photonic bands are depicted in grey (black) and correspond to effective positive (negative) values of the parity σ_y. Only guided modes, situated below the light-cone, are shown. The bottom-left panel shows a zoom-in of the spectral domain in which the experiments are performed. The photonic bands between ~0.22 to ~0.30 correspond to effective positive (negative) values of σ_x. The horizontal lines correspond to two frequencies at which the effective index of refraction is negative (ω_n) or positive (ω_p). The top-right panel shows the EFCs of the photonic band with negative refraction, computed for $\omega = 0.282$ (outermost), $\omega = 0.285$ (middle), and $\omega = 0.288$ (innermost). The circles represent sections through the light-cones. The bottom-right panel shows a zoom-in of the photonic band structure of the transverse guiding modes of the PhC slab. The low dispersion (zigzagged) photonic bands correspond to effective positive (negative) values of σ_x. The substrate light-line is also shown. (Reprinted from Chatterjee, R. et al., *Phys. Rev. Lett.*, 100, 187401, 2008. With permission.)

the integration being performed over one unit cell. For a slab that is invariant to the reflection operator Σ_y, the modes are either even, $\sigma_y = 1$ (TE modes), or odd, $\sigma_y = -1$ (TM modes). In our case, the PhC slab is not invariant to the reflection operator Σ_y; however, the computed values of the parity parameter σ_y ($\sigma_y \sim \pm 0.98$ to ± 0.99) show that the modes can be meaningfully classified as TE like and TM like.

These symmetry properties of the PhC slab modes are crucial to understanding the optical response of the photonic structure and, in particular, how the slab transverse evanescent modes may be excited. Since the excitation wavelengths fall in the bandgap of TE-like modes, only the TM-like modes are excited by any of our input wavelengths (see Figure 6.1b). In addition, because of the y-axis symmetry of these TM-like modes, they can only be excited by the E_{11}^y mode of the SOI waveguide for this frequency range. Moreover, of these two TM-like bands, only one has the same (x-axis) symmetry properties upon Σ_x reflection as the E_{11}^y mode does, namely, $\sigma_x > 0$; this band spans approximately between 0.27 to 0.30 in Figure 6.1b, bottom-left panel. Furthermore, we have designed and our numerical computations demonstrate that this band emulates a negative index of refraction: equi-frequency curves (EFCs) that correspond to increasing frequency shrink toward the Γ symmetry point (see top-right panel in Figure 6.1b). For the wavelength domain considered in our experiments, the effective refractive index varies between $n = -1.64$ at $\lambda = 1480$ nm and $n = -1.95$ at $\lambda = 1580$ nm, determined from the relation $|k| = n\,\omega/c$. Note, though, that while the shape of the EFC presented in Figure 6.1b shows that our PhC slab does not have negative refraction[18] for all angles, the refraction of incident light at the input facet of our PhC slab is negative for a large domain of incident wavevectors k, so that the PhC slab does possess one of the main ingredients necessary to observe subdiffraction limit imaging.

As discussed above, in order to have subdiffraction limit imaging by the PhC, it is equally important to properly design the PhC slab so that it excites the needed transverse guided modes. With this in

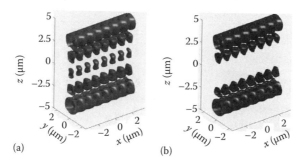

(a) (b)

FIGURE 6.2 Isosurface plots of the surface slab modes. (a) The symmetric and (b) the antisymmetric surface slab modes, computed at the K crystal symmetry point. For clarity, seven periods along the transverse direction of the PhC slab (x-axis) are presented.

mind, we have designed our slab so that it contains two such modes with the same symmetry as the E_{11}^{y} mode of the SOI waveguide; both of these modes are *surface slab modes* (see Figure 6.2). They extend from ~1481 to ~1484 nm (see bottom-right panel in Figure 6.1b with $a = 433.5$ nm), and their symmetry properties with respect to Σ_z reflections are such that these two modes are a symmetric and antisymmetric superposition of surface modes excited at each air–PhC interface. All the other modes are guiding modes of the entire PhC slab and do not have the required symmetry with respect to Σ_z reflections, and thus cannot be excited. Thus, to summarize, our design is such that it meets all criteria for achieving imaging beyond the diffraction limit, namely, amplification of evanescent waves (see below), excitation of surface slab modes, and emulating negative-index behavior.

6.1.1.4 Photonic Crystal Fabrication and Quality Characterization

A number of these designed PhC devices along with the input waveguides were fabricated on an SOI platform in a standard CMOS foundry at the Institute of Microelectronics, Singapore. Deep UV lithography with 180 nm critical dimension resolution was used to define the patterns, and reactive ion etching (RIE) was used to etch these patterns on the silicon. A scanning electron micrograph (SEM) of the PhC slab is shown in Figure 6.1a. The fabrication disorder in the PhC slab (shown in Figure 6.1a) was statistically parameterized through edge-detection algorithms and fit to the probability density distribution.[32] For the rectangular slab, the resulting statistical parameters have a radius of 119.36 ± 2.29 nm, a lattice period of 433.50 ± 1.53 nm, an ellipticity of 6.29 ± 0.51 nm, and a random edge roughness correlation length of 27 nm, suggesting a remarkably small random disorder in the fabricated PhC slab against the designed lattice. The positional variation of holes is therefore $\sim 0.004a$, and the size, ellipticity, and roughness variations are therefore $\sim 0.003a$, with the SEM resolution limit $\sim 0.01a$. These actual variations are satisfactorily below the $\sim 0.05a$ disorder theoretical target investigated for distortions of the image position, intensity, and side peaks.[33]

6.1.1.5 Near-Field Measurements

The near-field optical measurement is done using an Aurora-3 near-field scanning optical microscope (NSOM) from Veeco Instruments, and the measurement setup is shown in Figure 6.3a. The input light is from 1480 to1580 nm (0.2744–0.2929 in normalized units of $\omega a/2\pi c$) tunable laser source and modulated by an internal chopper. A polarization controller is used for TM propagation in both the waveguide and the PhC. Coupling of light into the waveguide is achieved by a UV-cured adhesive bonding of an SMF-28 fiber to the input waveguide, both on a glass base. A secondary wafer is installed to support the fiber and also to provide the correct height for the coupling (see Figure 6.3b).

The NSOM probe tip is pulled from an SMF-28 fiber using a laser-based micropipette puller (Sutter Instruments P-2000). The fiber tip was then mounted on a quartz tuning fork. Scattered evanescent waves on the PhC are collected by the fiber tip and then registered by a high-gain pigtailed femto-Watt

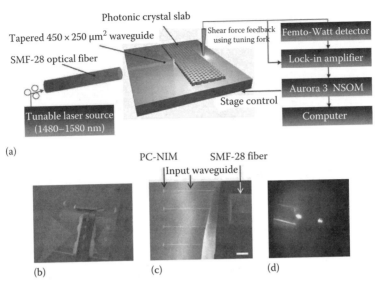

FIGURE 6.3   Near-field measurements on the negative refraction PhC chip. (a) Schematic of the experiment setup, illustrating NSOM measurement and setup, and input fiber and waveguide coupling to the PhC-NIM. (b) Photograph of the UV adhesive bonded chip coupled to the SMF-28 optical fiber. (c) SEM image of UV-cured adhesive bonded optical fiber to Si chip with input waveguides and the PhC-NIM devices. (d) Collected far-field infrared light scattered from the fiber output, waveguide input, and PhC-NIM, to confirm successful coupling between the fiber and the waveguides.

InGaAs/PIN photodetector. The sampled signal goes through a lock-in amplifier for noise removal and then arrives at the IO board on the Aurora-3. The scanning speed of the NSOM is set slow enough for efficient lock-in time averaging at each point of the sample. The proportional-integral-derivative (PID) controller of the tuning fork feedback is carefully chosen so that the fiber tip constantly stays about 20 nm above the sample. The all-fiber setup effectively isolates exterior noise in the experimental measurements.

A $20 \times 20\,\mu m^2$ area is scanned by the NSOM tip covering the whole surface (top and facet surfaces) of the PhC slab, including the regions where light enters and exits the crystal. The cross-section profile of the NSOM measurements is deconvolved with the simulated waveguide output profile to obtain the non-fiber-perturbed characteristics. Each near-field image measurement is averaged over 27 scans to reduce white noise.

6.1.1.6 Experimental Observations and Comparison to Simulations

Figure 6.4a through d shows a compilation of the near-field measurements. This figure illustrates that while the full-width half-maximum (FWHM) of the input-source waist remains approximately constant at ~1.0 μm (1.5 μm before deconvolution) across all wavelengths, a minimum FHWM waist of the image, of 0.7 μm, i.e., 0.47λ (1.29 μm before deconvolution), is observed at 1489 nm. This image size is clearly below the ideal-lens Abbe diffraction limit of 0.91 μm (0.61λ) that corresponds to λ = 1489 nm. As a comparison, the FHWM waist is 1.9 μm (1.23λ) at 1550 nm. Uncertainty in the FHWM waist is estimated as 0.05 μm. Further investigation of the experimental data indicates that the subdiffraction limit imaging bandwidth for this sample is ~3 nm (or 0.4 THz), as given by the 3 dB width of Lorentzian fit for the FWHM wavelength dependences (see Figure 6.4d). The additional scattering at the top-right side of the source in Figure 6.4a and b is due to fabrication defects, which are not seen in Figure 6.1a. Although these defects increased somewhat the light scattering at the output of the SOI waveguide, the large-angle nature of the scattering caused it to not affect the quality of the images. To further confirm the subdiffraction limit imaging in our PhC, we measured the intensity of the optical near-field along

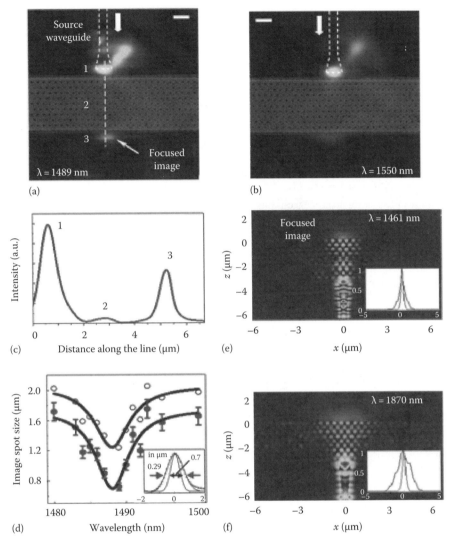

FIGURE 6.4 Near-field scanning optical microscope confirmation of near-infrared negative refraction imaging beyond the diffraction limit in PhC metamaterials. (a, b) Near-field measurements of the PhC LHM with superimposed SEM image. The high-index-contrast silicon waveguide serves as a subwavelength source (filled arrow indicates incident direction) to the PhC LHM with band structure calculated in Figure 6.1a. At $\lambda = 1489$ nm, a subdiffraction limit image with an FWHM waist of $0.7 \pm 0.05\,\mu m$ is observed, illustrative of effective negative-index phenomena at the second band of our PhC lattice. The subdiffraction limit image has a Gaussian spatial profile. At $\lambda = 1550$ nm, the operation does not fall into the anomalous transmission regime and an image waist of $1.9\,\mu m$ is observed. Scale bar: $1\,\mu m$. (c) Plot of the intensity profile of the optical near-field along the Γ-M symmetry axis of the PhC, at $\lambda = 1489$ nm. An intensity peak (peak 2) is clearly observed and corresponds to a focused spot inside the PhC slab. (d) Dependence of the FWHM waist on the wavelength, before and after deconvolution (round circles and dots, respectively; the curves are Lorentzian fit of the experimental results), illustrating a minimum waist at $\lambda = 1489$ nm and a 3 dB bandwidth of 0.4 THz. The inset shows the intensity cross section across the focused image at 1489 nm, before (open circles) and after (filled dots) deconvolution, parallel to the PhC surface. (e, f) FDTD simulations of subwavelength imaging in the designed PhC LHM at different wavelengths (1461 nm and 1870 nm, respectively). A magnetic field is plotted to best illustrate the image formation. The wavelengths 1461 and 1870 nm correspond to the frequencies ω_n and ω_p in Figure 6.1b, respectively. The inset shows the corresponding intensity cross section across the input spots (inner curve) and the focused image (outer curve). (Reprinted from Chatterjee, R. et al., *Phys. Rev. Lett.*, 100, 187401, 2008. With permission.)

the Γ-M symmetry axis of the crystal, at λ = 1489 nm (see Figure 6.4c). At this wavelength, an intensity peak is observed inside the PhC slab; such a mid-slab image peak is a well-known characteristic of a material with an effective negative index.[1,2] The spatial FWHM (along the *x*-direction) of this second focus spot, inside the PhC slab, was analyzed to be ~0.96 ± 0.02 μm.

These experimental features can also be seen clearly via rigorous 3D FDTD simulation, as summarized in Figure 6.4e and f. This figure shows this numerical simulation of the field propagating through the SOI waveguide and the PhC slab. Negative refraction subdiffraction limit imaging is optimized at a wavelength of λ = 1461 nm. For comparison, we also show the field profile that corresponds to λ = 1870 nm, a wavelength that falls in the first TM-like band; this band has a positive index of refraction. Notice that the image in this case is degraded and larger than that at 1461 nm. Returning now to the wavelength in the negative index (NIM) band, at λ = 1461 nm, our numerical computations suggest a subdiffraction limit imaging resolution of 0.67 μm (0.46λ), which is very close to our NSOM measurements (free-space propagation over the same distance would lead to an image size of ~15.8 μm). Finally, note that the difference in the wavelength for the minimum imaging resolution between our simulations and our NSOM experiment is consistent with a 2% fabrication numerical difference. Small fabrication imperfections such as this would shift the measured band structure compared to that of the numerically computed one, and thus change the effective refractive index of the PhC slab. In addition, the difference between simulation and measurement is also expected on the basis of the finite imaging impulse response of the NSOM fiber tip. A fiber probe with an aperture <100 nm would provide improved measurement resolution.

In our experiments, the wavelength (1489 nm) for the minimum image size is closely related to the resonant excitation of a surface slab mode near this wavelength. Thus, for wavelengths shorter than this value, the modes become leaky and the increased losses impair the imaging efficiency, whereas for longer wavelengths, the frequency separation from the surface slab modes increases and therefore the amplification of the evanescent waves decreases; this behavior has also been reported in a recent theoretical study.[18] In addition, the anisotropy of the phase refractive index increases with the wavelength, again reducing the imaging efficiency. Note also that the anisotropy of the phase refractive index increases this wavelength, again reducing the imaging efficiency.

6.1.1.7 Experimental Verification of Evanescent Wave Amplification

A crucial aspect of our design is the presence of evanescent wave amplification. Figure 6.5 presents experimental evidence that at λ = 1489 nm, the optimum for subdiffraction limit imaging resolution, evanescent waves are amplified by means of excitation of a surface slab mode. Thus, Figure 6.5a shows the wavelength dependence of the field intensity measured at three locations on the facet of the PhC slab. Note that in order to eliminate the effects of local near-field structure, we have averaged each of the signals over spatial domains with a length of ~3*a* along the crystal facet. At *any* location along the slab facet, in addition to the three shown, the averaged transmitted signal has a maxima at 1489 nm. For wavelengths shorter than 1489 nm, the modes become leaky, and the increased losses impair the imaging efficiency; for longer wavelengths, the frequency separation from the surface slab modes increases, and thus recovery of the evanescent wave decreases. This behavior has also been suggested in a recent theoretical study.[18] Anisotropy of the phase refractive index also increases with the wavelength, again reducing imaging efficiency.

Moreover, due to refocusing of the non-propagating waves, we note that local field intensity along the focus axis can be enhanced at on-resonance (λ = 1489 nm), compared to off-resonance, albeit only at locations close to the focus axis. This is shown in Figure 6.5c. At all other wavelengths (such as at the 1480 nm shown), there is less refocusing. Moreover, in comparison, a homogenous slab does not show strong frequency dependence or refocusing of the evanescent waves. Moreover, our FDTD simulations likewise verify this local refocusing of the evanescent wave by the excited surface slab modes. We note that the local field (not energy flux, which is conserved) is the quantity measured by the NSOM.

Since both propagating and evanescent waves contribute to the image formation, the image created by the PhC slab is fundamentally different from an image formed by a conventional lens. Thus, the field

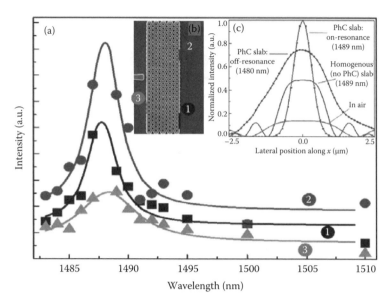

FIGURE 6.5 Observation of bound surface states excitation at a wavelength of highest subdiffraction limit imaging resolution. (a) Wavelength dependence of the field intensity at the input and the output facet of the PhC slab, averaged over three spatial domains, each with a length of ~3a, marked in the inset (b) locations of examined field intensities. (c) Refocusing of the near-field intensity, comparing measured local intensity at the output facet of the PhC slab (on-resonance: 1489 nm; off-resonance: 1480 nm) versus calculated homogenous (no PhC) slab (1489), and in air. (Reprinted from Chatterjee, R. et al., *Phys. Rev. Lett.*, 100, 187401, 2008. With permission.)

distribution in the *x-z* plane, at the location of the image, displays a local maximum, which suggests that the PhC lens provides a moderate amplification of the evanescent waves, i.e., it operates in a moderate subwavelength regime.[18] By contrast, in an extreme subwavelength regime, the strength of the evanescent waves would dominate over the propagating waves (this regime is absent in conventional lenses) and no local maximum of the near-field could be observed. Another factor that influences the quality of the image is the optical losses in the PhC slab. To estimate these losses, we integrated the optical intensity in the transverse direction at the two slab positions in the PhC; an optical loss of 0.59 ± 0.003 dB is found at the wavelength $\lambda = 1500$ nm (near-resonance but not at 1489 nm, so as not to include the surface state enhancements). This gives an estimated $\text{Im}(n)$ to be $\sim 4.4 \times 10^{-3}$ and correspondingly an FOM of ~ 380. Coupling losses from the waveguide into the PhC account for ~ 2.2 dB, mainly due to impedance and mode mismatch. This negative refraction FOM is about 125× improvement over recent experimental efforts and more than 15× better than the best current theoretical efforts,[14] and the negative refraction PhC is scalable to optical frequencies using wider bandgap materials and smaller lattice constants in the PhC.

6.1.1.8 Negative Refraction in PhC: Wedge-Shaped Structures

As an alternative verification that the PhC slab emulates a negative index of refraction at the operating frequency, we also fabricated both a wedge-shaped slab with a similar PhC structure, namely, a lattice constant $a = 443$ nm, a hole radius $r = 135$ nm $(r = 0.3047a)$, and a thickness $t = 320$ nm, and a *homogeneous* Si slab with identical geometry and dimensions. We chose this shape since a PhC slab with this particular geometry would form an image on the opposite side of the normal,[9] as compared to the position of the image predicted by a positive refraction medium. Although not excited by a perfect plane wave, the predominant direction of wavevectors' excitation (along the longitudinal axis of the waveguide) could provide another indication of negative refraction. We performed a large series of near-field measurements for these two samples, and two results are illustrated in Figure 6.6. The right inset of

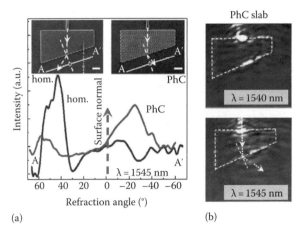

(a)   (b)

FIGURE 6.6 Observation of negative refraction and subdiffraction limit imaging in wedge-like PhC metamaterials. (a) Spatial distribution of the field intensity measured along straight lines parallel to the output facets of a homogeneous Si slab and a PhC slab, whose SEMs are shown in the top panels. The dark grey and vertical dashed arrow curves correspond to the PhC and homogenous slab, respectively. The light corresponds to the location of the normal at the slab output facets (also marked in top panels). Note that in the two cases, the maximum field intensity is observed at opposite sides with respect to the normal, a clear indication that the PhC slab emulates negative refraction phenomena. Scale bar for insets: 2 μm. (b) NSOM measurements of the PhC slab at two wavelengths. For the PhC slab, a focused image is clearly observed at $\lambda = 1545$ nm. An FWHM waist of 2.2 ± 0.15 μm is observed. Scale bar: 5 μm. The data presented in (a) and (b) allows us to estimate an analogous refractive index of -1.29 at $\lambda = 1545$ nm for the PhC media. (Reprinted from Chatterjee, R. et al., *Phys. Rev. Lett.*, 100, 187401, 2008. With permission.)

Figure 6.6a clearly shows that the PhC slab forms an image at the "negative" side of the surface normal at the PhC–air interface[34] at $\lambda = 1545$ nm, in contrast to the homogeneous slab (left inset of Figure 6.4a), which shows the image to be on the *opposite* side of the normal. The measurement of the angle of the image with respect to the surface normal enables us to calculate an effective negative index of ~ -1.29 at $\lambda = 1545$ nm. The minimum image spatial FWHM occurs at 1545 nm and is estimated as 2.2 ± 0.15 μm. Although the image by this wedge-shaped slab is not beyond the diffraction limit, the minimum spatial FWHM in the wedge-shaped slab occurs at the same normalized frequency (a/λ) that corresponds with the earlier and different rectangular PhC slab (Figures 6.4 and 6.5). The difference (1545 nm in the wedge versus 1489 nm in the rectangular slab) is due to different geometrical a and r parameters of the two PhC structures nanofabricated and designed. Moreover, because the wedge-shaped slab has different surface termination at the input and output facets, it does not support surface modes at the same frequency, and hence does not show subdiffraction limit imaging. In addition, we note that our PhC structures are not designed to support super-collimation recently observed,[35] as shown in the band structure of Figure 6.1b and in the larger intensity at the output facet than the input facet, which points strikingly to the excitation of surface slab modes. The comparison of the position of the image with respect to the surface normal for both the homogeneous and PhC slabs (see Figure 6.6a) therefore provides another clear indication of negative refraction in our PhC material.

The details of our instrumentation show that our imaging measurements overestimate the image size. In particular, when our NSOM fiber probe scans the sample surface, it remains ~ 20 nm above the surface, due to sensing by a high-Q piezoelectric tuning fork attached on the fiber probe. As the fiber probe approaches but is not yet above the PhC, the tip of the NSOM probe will rise above the PhC sample surface due to the finite diameter of the fiber tip. Thus, in our NSOM measurements, the collected images at PhC input and output are taken with the fiber probe above the PhC surface, but not through the median plane of the PhC slab; this geometry causes our measurement to record a larger image size.

6.1.1.9 Conclusion and Future Outlook

This first experimental observation of subdiffraction limit imaging through bound surface states in negative refraction PhCs in the near-infrared suggests unprecedented opportunities for scaling to visible and shorter wavelengths for imaging, detection, and nanolithography applications, where the commonly perceived diffraction limit is no longer a barrier. Further fundamental investigations include extension to higher-fold symmetry quasicrystals that can extend the subdiffraction imaging phenomena to non-near-field focusing regimes,[36,37] zero-effective-index superlattice metamaterials with optical bandgaps that are independent of incident wavevectors,[38,39] as well as transformation optics[40,41] for manipulating the flow of light.[42]

6.1.2 Zero-$\bar{n}$ Photonic Crystal Superlattices

6.1.2.1 Introduction

LHMs are artificial composites having both negative permittivity and permeability[43,44]; conditions that allow LHMs to have negative refraction indices. Their uncommon physical properties have sparked intense interest in these materials, and many groups have studied LHMs.[45,46] It has also been demonstrated that a periodic superlattice consisting of alternating layers of LHMs and regular dielectric materials with positive permittivity and permeability (RHMs, right-handed metamaterials) has an omnidirectional bandgap that remains unaltered to the different wave polarizations, angles of incidence, or periods of the structure.[47] This gap is termed as a zero-n gap, since it occurs when the volume average index is equal to zero. Here, we report an experimental verification of zero-n materials in the near-infrared region. As suggested in a previous work,[48] for the LHM, one can use dielectric-based 2D photonic crystals (PC), which we recently fabricated,[49] whose optical losses are much smaller than metal-based LHMs.

6.1.2.2 Design and FDTD Simulations

The photonic superlattice shown in Figure 6.7 is a periodic structure with a period of $\Lambda = d_1 + d_2$, where d_1 and d_2 are the thicknesses of the PC and the RHM in the primary unit cell, consecutively. Here, we only consider the structures that have a zero spatial average of the refractive index and, as a result, the zero-n gap occurs when the corresponding frequency is insensitive to the periodicity. We consider the RHM as a regular dielectric slab, and the PC slab here has a hexagonal lattice with air holes etched into a dielectric background with a thickness of $t = 320$ nm. The optical characteristics of the PC slab depend on the period a, the ratio of the hole radius, the period r/a, and the background refractive index n. In our design, $a = 430$ nm, $r/a = 0.279$, and $n = 3.46$. TM-polarized waves are used because TM polarization provides a larger frequency range of the isotropic negative refractive index, which can be seen in Figure 6.8, where the photonic band structure of the guided modes is shown.

Furthermore, the orientation of the normal to the input facet has been chosen along the Γ-M symmetry axis (z-axis) of the PC slab. In order to demonstrate the existence of the zero-n gap, we have

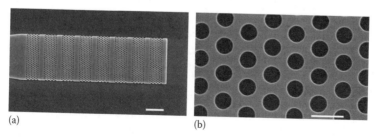

(a)   (b)

FIGURE 6.7 Designed and fabricated negative index photonic superlattice. (a) SEM of the fabricated samples with eight superperiods whose PC slabs have a thickness $dl = 3.5\sqrt{3}$. Scale bar: 5 μm. (b) SEM of the PC lattice with a higher resolution. Scale bar: 500 nm. (Panel (a) reprinted from S. Kocaman et al., *Phys. Rev. Lett.*, 102, 203905, 2009. With permission.)

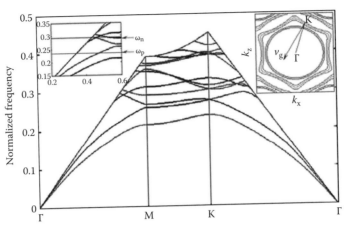

FIGURE 6.8   Photonic band structure of an air-hole hexagonal PC slab in the sample with $r = 0.279a$; $t = 0.744a$ with $a = 430$ nm. The TE-like (TM-like) photonic bands are depicted in grey (black). The inset on the left shows a zoom-in of the spectral domain in which the experiments are performed. The inset on the right shows the EFCs of the photonic band with negative index of refraction, computed for $\omega = 0.282$ (outermost), $\omega = 0.285$ (middle), and $\omega = 0.288$ (innermost). The circles represent sections through the light-cones. (Reprinted from Chatterjee, R. et al., *Phys. Rev. Lett.*, 100, 187401, 2008. With permission.)

designed three different superlattices by altering the period. They have 7, 11, and 15 unit cells along the z-axis so that their PC thicknesses are $d_1 = 3.5\sqrt{3}a$, $d_1 = 5.5\sqrt{3}a$, and $d_1 = 7.5\sqrt{3}a$, respectively. The corresponding RHM slab thicknesses have been calculated to get a zero average refractive index according to the formula $n = (n_1 d_1 + n_2 d_2)/\Lambda = 0$[48] without changing the ratio between the PC and the RHM. We have calculated the effective index of the RHM slab as $n_2 = 2.64$ and obtained the negative effective index value from the second band shown in Figure 6.8 for a normalized frequency $\omega = 0.273$. In Figure 6.9a, the results of the 3D FDTD simulations for all three superlattices have been illustrated. There are several photonic gaps, and except for that located near the normalized frequency $\omega = 0.276$, their mid-gap locations are shifted by the changing period of the superlattice. Varying mid-gap frequency is the expected behavior for the regular Bragg gaps; the lack of variation shows that this band is the zero-n gap. In addition, we changed the ratio and the gap locations also varied (zero-n gap $\omega = 0.272$) at as seen in Figure 6.9b which also show that the gap cannot come from the PC. In addition, in order to determine the type of each gap, we have also calculated $k_o \Lambda/\pi$.[50] We used the corresponding negative index value from Figure 6.8 and an approximate effective index value for the RHM slab, and our calculations also showed that the gaps, which are located at $\omega = 0.276$ in Figure 6.9a and $\omega = 0.272$ in Figure 6.9b are a zero-order gap. Table 6.1 shows the calculated orders for the gaps in Figure 6.9a,b. Furthermore, we have also studied the dependence of the gap locations to the superperiod number. The results of these calculations are present in Figure 6.9c, and it shows that the zero-n gap location has not changed and gaps have become clearer. Finally, we have checked the possibility that the unvarying gap may come from the PC itself. Figure 6.9d shows the transmission spectrum with increasing number of unit cells which also supports the zero-n gaps are not due to PC itself.

6.1.2.3 Experimental Verification and Discussions

Based on the simulation results, we fabricated samples with three superperiods whose PC slabs have a thickness of $d_1 = 3.5\sqrt{3}a$ using a silicon film, on a SiO_2 insulator substrate. The fabricated PC slab has a lattice constant $a = 422$ nm and a radius $r = 122$ nm ($r = 0.29a$). The wavelength of the light coupled into the chip has been tuned between 1300 and 1650 nm (0.255 to 0.324 in normalized units of $\omega a/2\pi c$). Figure 6.10a,b shows the experimental result for the fabricated sample and the simulation data from Figure 6.9a,b. Finally, Figure 6.10c shows the zero-n gap evolution with increasing number of superperiods.

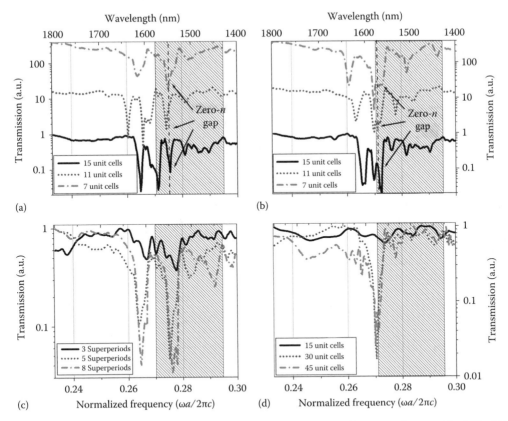

FIGURE 6.9 (a) Transmission highlighting the invariant zero-n gaps for a superlattice with $d_2/d_1 = 0.746$ (design 1), containing 7, 11, and 15 unit cells in each PhC slab. Three superperiods in the PhC superlattice are used for the case of PhC layers with 11 and 15 unit cells, and 5 superperiods for the case of PhC layers with 7 unit cells. (b) Same as in (a) for $d_2/d_1 = 0.794$ (design 2). The shifting gaps are the conventional Bragg gaps. The transmission plots are offset vertically for viewing clarity. (c) Transmission for a superlattice with $d_2/d_1 = 0.794$, containing 3, 5, and 8 superperiods. Each PhC layer contains 7 unit cells. (d) Transmission through a regular PhC slab (nonsuperlattice) containing 15, 30, and 45 unit cells, illustrating a regular Bragg gap. All designs are obtained through full 3D FDTD numerical simulations. Shaded region in all plots illustrate the negative-index range, in order to observe the zero-n gap. (Reprinted from S. Kocaman et al., *Phys. Rev. Lett.* 102, 203905, 2009.)

TABLE 6.1

Figure	Unit cells of PhC	Gap frequency	d_2/d_1	Effective index PhC	Effective index Slab	Average index, $\bar{n}$	$k_o\Lambda/\pi$	Gap order
Figure 2a (design 1)	7	0.276	0.746	−1.988	2.648	−0.007	−0.044	0
Figure 2a (design 1)	11	0.283	0.746	−1.844	2.684	0.091	0.874	1
Figure 2b (design 2)	7	0.272	0.794	−2.080	2.622	0.001	0.007	0
Figure 2b (design 2)	7	0.283	0.794	−1.856	2.683	0.153	0.962	1

6.1.2.4 Conclusion

Here, we demonstrated a zero-n gap, at near-infrared wavelengths with a photonic superlattice consisting of periodic layers of 2D PCs as LHMs and regular RHMs. Our experiments show good correspondences with our 3D FDTD simulations.

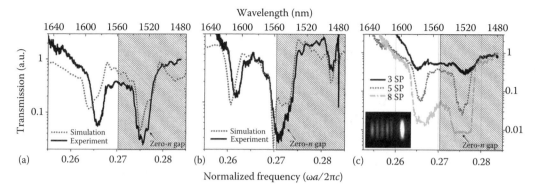

FIGURE 6.10 (a) Measured transmission for a superlattice with $d_2/d_1 = 0.746$, with 7 unit cells in the PhC layers and 5 superperiods; for comparison, results of numerical simulations are also shown. (b) The same as in (a), but for a superlattice with 0.794. (c) Measured transmission for a superlattice with $d_2/d_1 = 0.746$, with 3, 5, and 8 superperiods and 7 unit cells in the PhC layers. Both gaps become deeper as the number of superperiods increases. Inset: Example of near-infrared top view image of 3 superperiods, under transmission measurement at 1550 nm. In all plots, the shaded region illustrates the negative-index region.

6.2 Controlling Nonlinearities in Silicon Photonic Crystals

6.2.1 Femto-Joule Optical Bistability in Fano-Type Nanocavities

6.2.1.1 Introduction

Two-dimensional PhC slabs confine light by the Bragg reflection in-plane and total internal reflection in the third dimension. The introduction of point and line defects into 2D PhC slabs creates localized resonant cavities and PhC waveguides, respectively, with ab initio arbitrary dispersion control. Such defect cavities in high-index-contrast materials possess strong confinement with subwavelength modal volumes (V_m) at $\sim (\lambda/n)^3$, corresponding to high field intensities per photon for increased nonlinear interaction. Moreover, cavities with remarkable high quality factors (Q)[51,52] have been achieved recently, now permitting nanosecond photon lifetimes for enhanced light–matter interactions. The strong localization and long photon lifetimes in these high-Q/V_m PhC nanocavities point to enhanced nonlinear optical physics, such as Lorentzian-cavity-based bistability[53–56] in silicon photonics.

The interference of a discrete energy state with a continuum can give rise to sharp and asymmetric lineshapes, referred to as Fano resonances.[57,58] Compared to a Lorentzian resonance, these lineshapes arise from temporal pathways that involve direct and indirect (e.g., a resonance) transmission, with a reduced frequency shift required for nonlinear switching due to its sharper lineshape. If the indirect pathway can be further strongly localized (such as in a high-Q/V_m 3D cavity instead of a 1D Fabry–Perot cavity), the nonlinear characteristic switching thresholds can be further reduced. Optical bistability involving Fano resonances due to Kerr effect in PhC cavities has been theoretically studied based on Green's function solution of Maxwell's equations.[59] Fano resonances have also been studied by the transfer matrix technique[58,60] and coupled-mode equations.[61] We present our measurements on Fano-based optical bistability as well as a temporal nonlinear coupled-mode framework for numerical analysis.

6.2.1.2 Coupled-Mode Theory

Figure 6.11a shows a schematic of the theoretical model. A waveguide with two partially reflecting elements is side-coupled to a cavity. a is the amplitude of the cavity mode, which is normalized to represent the energy of the cavity mode $U = |a|^2$, and s is the amplitude of the waveguide mode, which is normalized to represent the power of the waveguide mode $P = |s|^2$. With the coupled-mode formalism,[62] the dynamic equation for the amplitude $a(t)$ of the resonance mode is[63]

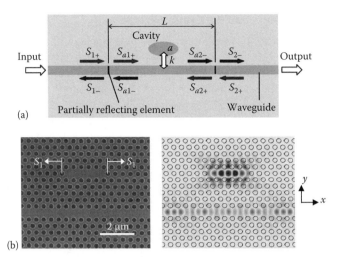

(a)

(b)

FIGURE 6.11 Designed and fabricated Fano-type nanocavity. (a) Schematic of an optical system including a waveguide side-coupled to a cavity. Two partially reflecting elements are placed in the waveguide. (b) SEM of PhC L5 point-defect cavity side-coupled to line-defect waveguide. The input and output facets of the high-index-contrast waveguide form the partially reflecting elements. (c) E_y field of the resonance mode mid-slab from 3D FDTD simulations. (Reprinted from Yang, X. et al., *Appl. Phys. Lett.*, 91, 051113, 2007. With permission. Copyright 2007, American Institute of Physics.)

$$\frac{da}{dt} = \left(-\frac{1}{2\tau_{\text{total}}} + i(\omega_0 + \Delta\omega - \omega_{\text{wg}}) \right) a + \kappa s_{a1+} + \kappa s_{a2+} \tag{6.1}$$

As shown in Figure 6.11a, $s_{a1-} = \exp(-i\phi)s_{a2+} + \kappa a$ and $s_{a2-} = \exp(-i\phi)s_{a1+} + \kappa a$. $\phi = \omega_{\text{wg}}n_{\text{eff}}L/c$ is the phase shift. κ is the coupling coefficient between the waveguide modes $s(t)$ and $a(t)$, and $\kappa = i\exp(-i\phi/2)/\sqrt{2\tau_{\text{in}}}$.[63] For a lossy partially reflecting element with amplitude reflectivity r and transmissivity t, $(r^2 + t^2 \leq 1)$,

$$\begin{pmatrix} s_{aj+} \\ s_{aj-} \end{pmatrix} = \frac{1}{it} \begin{pmatrix} -(r^2 + t^2) & -r \\ r & 1 \end{pmatrix} \begin{pmatrix} s_{j+} \\ s_{j-} \end{pmatrix}, \quad j = 1,2 \tag{6.2}$$

In Equation 6.1, the total loss rate for the resonance mode $1/\tau_{\text{total}}$ is

$$\frac{1}{\tau_{\text{total}}} = \frac{1}{\tau_{\text{in}}} + \frac{1}{\tau_{\text{v}}} + \frac{1}{\tau_{\text{lin}}} + \frac{1}{\tau_{\text{TPA}}} + \frac{1}{\tau_{\text{FCA}}} \tag{6.3}$$

where $1/\tau_{\text{in}}$ and $1/\tau_{\text{v}}$ are the loss rates into waveguide (in-plane) and into free space (vertical), and $1/\tau_{\text{in/v}} = \omega/Q_{\text{in/v}}$, $1/\tau = 1/\tau_{\text{in}} + 1/\tau_{\text{v}}$. The linear material absorption, $1/\tau_{\text{lin}}$, is assumed small since the operation is within the bandgap of the silicon material. $1/\tau_{\text{TPA}}$ and $1/\tau_{\text{FCA}}$ are the loss rates due to two-photon absorption (TPA) and free-carrier absorption (FCA), respectively. The $\Delta\omega$ detuning of the cavity resonance from ω_0 is modeled due to the Kerr effect, free-carrier dispersion (FCD), and thermal dispersion effects under first-order perturbation. $(\omega_0 + \Delta\omega)$ is the shifted resonant frequency of the cavity and ω_{wg} is the input light frequency in the waveguide. With the modeled TPA-generated carrier dynamics and thermal transients due to the total absorbed optical power (Equations 6.25 and 6.35), the coupled nonlinear dynamical behavior of the Fano optical system is numerically integrated.

6.2.1.3 Experimental Results and Analysis

The optical system consists of a PhC waveguide side-coupled to a high-Q/V_m nanocavity with five linearly aligned missing air holes ($L5$) in an air-bridge triangular-lattice PhC slab with a thickness of $0.6a$ and air holes radius of $0.29a$, where the lattice period $a = 420$ nm, as shown in Figure 6.11b. The shift S_1 of two air holes at the cavity edge is $0.02a$ to tune the radiation mode pattern for increasing the Q factors. The waveguide-to-cavity separation is five layers of holes. The index contrast at the waveguide input and output facets acts as partially reflecting elements with a distance L of around 1.9 mm to form a Fabry–Perot resonator and perturb the phase of the waveguide mode. Figure 6.11c shows the E_y field of the resonance mode mid-slab from 3D FDTD simulations.

The devices were fabricated with the standard integrated circuit techniques in an SOI substrate. A polarization controller and a lensed fiber are used to couple transverse-electric polarization light from a tunable laser source into the waveguide. A second lensed fiber collects the transmission from the waveguide output that is sent to the photodetector and lock-in amplifier. The input power coupled to the waveguide is estimated from the transmitted power through the input lensed fiber, the waveguide, and the output lensed fiber.[55] The total transmission loss of the whole system is around 24.8 dB at a wavelength of 1555 nm. At a low input power of 20 μW, the measured resonant wavelength, λ_0, is 1556.805 nm. To measure the Q factor, the vertical radiation from the top of only $L5$ nanocavity is collected by a 40× objective lens and a 4× telescope through an iris (spatial filter), which will isolate the cavity region only, so that there is no other influence other than the radiation of the cavity mode. The estimated Q, based on the full width at half maximum (FWHM) $\Delta\lambda$ of 52 pm, is around 30,000. From the 3D FDTD method, the vertical Q factor, Q_v, is around 100,000 and the in-plane Q factor, Q_{in}, is around 45,000 so that the total Q factor $Q_{tot} = 1/(1/Q_v + 1/Q_{in}) = \sim31,000$.

Figure 6.12a shows the measured transmission spectrum of the waveguide with different input powers. Each transmission shown is repeated over multiple scans. Sharp and asymmetric Fano lineshapes are observed. The spectral lineshapes depend on the position of the cavity resonance in a Fabry–Perot background, highlighting Fano interference pathways. Here, the spectra show ascending Fano resonances. The Fabry–Perot fringe spacing, $d\lambda$, is around 230 pm, which corresponds to the distance between two waveguide facets, $d = 1.902$ mm ($d = \lambda^2/(2 * d\lambda * n_{eff})$) and an effective index of 2.77 from FDTD simulations. As the input power increases, the Fano lineshapes were redshifted due to TPA-induced thermo-optic nonlinearities in silicon.[53-55] Figure 6.12b shows the calculated transmission spectrum from a nonlinear coupled-mode model with the input powers used in the experiment. All parameters used in calculation are from either reference papers or FDTD results.[64] When the input power is 1 μW or less, the cavity response is in the linear regime. As the input power increased, the Fano lineshapes were redshifted.

Figure 6.13a shows the observed hysteresis loop of Fano resonance at a red detuning δ of 22 pm ($\delta/\Delta\lambda = 0.423$). For comparison, the inset of Figure 6.13a shows the measured hysteresis loop for Lorentzian resonance at the detuning of 25 pm for an $L5$ nanocavity, where the resonance wavelength is $\lambda_0 = 1535.95$ nm, $Q \sim 80,000$, and the detuning $\delta/\Delta\lambda = 2.6 > \sqrt{3}/2$. The bistable loops of ascending Fano lineshapes are very distinct from Lorentzian lineshapes. Firstly, one suggestive indication is the asymmetry in the hysteresis loop, with a sharp increase (gentle decrease) in increasing (decreasing) power for the lower (upper) branch, resulting from the asymmetric Fano lineshape. Secondly, for ascending Fano resonances, an important indication is the upward slope (*increase* in transmission) for *increasing* input power for a side-coupled cavity. For a symmetric Lorentzian in a side-coupled drop cavity, a downward slope (or decrease in transmission) should be expected for increasing input power.[65-67] Thirdly, the dip in the transmission (as indicted by the light grey circled region in Figure 6.13a) is another signature of the Fano resonance. This feature is not observable with a symmetric Lorentzian and in fact is an aggregate result of the three self-consistent solutions of the nonlinear Fano system, such as that predicted using Green's function method in Ref. [59]. Our nonlinear coupled-mode theory framework cannot trace out the individual solutions[59] but show the aggregate behavior, and is in remarkable agreement with our experimental measurements and Green's function predictions.

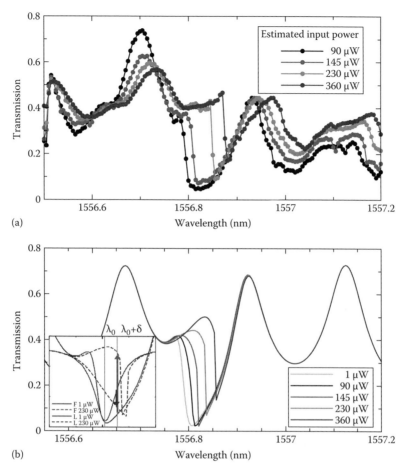

FIGURE 6.12 Transmission spectrum at different input powers, illustrating the asymmetric lineshapes. (a) Measured and (b) CMT-calculated results. The side-coupled *L5* cavity has a total Q of ~30,000. The inset of (b) plots the calculated transmission spectra of Fano lineshapes (lighter grey) and Lorentzian lineshapes (darker grey) with input power of 1 μW (solid line) and 230 μW (dashed line), respectively. (Reprinted from Yang, X. et al., *Appl. Phys. Lett.*, 91, 051113, 2007. With permission. Copyright 2007, American Institute of Physics.)

The Fano bistable "off" power (p_{off}) is estimated at 147 μW and the "on" power (p_{on}) at 189 μW for a 22 pm detuning, as shown in Figure 6.13a. These threshold powers are determined experimentally from half the total system transmission losses. From the 189 μW (147 μW) p_{on} (p_{off}) thresholds, this corresponds to an estimated internally stored cavity energy[53] of 4.5 fJ (1.5 fJ) based on a numerical estimate of a waveguide-to-cavity coupling coefficient (κ^2) of 13.3 GHz. The consumed energy, in terms of the definition used in Ref. [66], is ~540 fJ (60 fJ) based on the numerically estimated thermal relaxation time of 25 ns and 11.4% (1.6%) of input power absorbed by the TPA process for the "on" ("off") state, although this could be much lower with minimum detuning to observe bistability. The femto-joule level switching in the stored cavity energy is due to the lowered threshold from the sharp Fano interference lineshape, the small mode volume, and high-Q PhC cavities. For the 22 pm detuning, the switching intensity contrast ratio is estimated at 8.5 dB (from the regions with sharp discrete bistable "jumps") with a p_{on}/p_{off} ratio of 1.286. Figure 6.13b shows the calculated Fano bistable hysteresis at the detuning of 22 pm from the nonlinear coupled-mode theory. The calculated p_{off} and p_{on} thresholds are 151 μW (with the stored cavity energy of 1.5 fJ) and 186 μW (4.5 fJ), respectively, with a switching contrast of 9.3 dB and p_{on}/p_{off} ratio of 1.232, in excellent agreement with the experimental results.

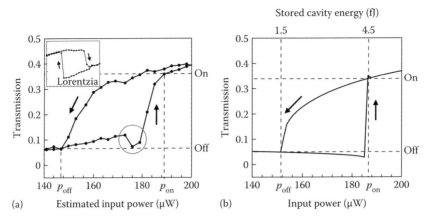

FIGURE 6.13 Asymmetric hysteresis loops for Fano resonance at a detuning of 22 pm. (a) Measured and (b) CMT-calculated results. The light grey circled region in panel (a) highlights a dip in transmission with increasing input power, a signature not present in Lorentzian-type resonances, and indicative of nonlinear Fano-type solutions. The inset of (a) shows the measured hysteresis loop for Lorentzian resonance. The arrows depict ascending and descending input powers to the Fano system. (Reprinted from Yang, X. et al., *Appl. Phys. Lett.*, 91, 051113, 2007. With permission. Copyright 2007, American Institute of Physics.)

Now we examine parametrically the dependence of the Fano-type bistability against achievable device characteristics, with our developed nonlinear model. Figure 6.14 summarizes the extensive numerically calculated effects of normalized detuning ($\delta/\Delta\lambda$), mirror reflectivity r, cavity Q, the position of cavity resonance on the characteristic threshold power p_{on}, and switching contrast. A baseline Q of 30,000, an r of 0.5 with 11% mirror loss, a λ_0 of 1556.805 nm, and a detuning of $\delta/\Delta\lambda = 22\,\text{pm}/52\,\text{pm} = 0.423$ is used, which correspond to the current experimental parameters and are represented as the light grey filled symbols in Figure 6.14. In Figure 6.14a, for both Fano bistability (solid lines) and Lorentzian bistability ($r = 0$, dashed lines), the threshold power increases for increasing normalized detuning (further normalized shift of incident laser frequency from the cavity resonance) due to the larger shift in resonance needed for bistable switching. The minimum detuning required for Lorentzian bistability is $\delta/\Delta\lambda$ ~ 0.7, which agrees well with the theoretical threshold detuning $\delta/\Delta\lambda = \sqrt{3}/2$.[68] However, the threshold detuning for Fano bistability to appear is $\delta/\Delta\lambda \sim 0.3$, which is much smaller than in the Lorentzian case. The threshold power is similar for both cases (around 140 μW) at the threshold detuning. For both cases, the switching contrast *decreases* with increasing detuning due to the reduced contrast in the transmission at the higher input powers needed for a bistable operation. Compared to Lorentzian bistability, Fano bistability has a higher switching contrast (9.87 dB at threshold detuning), and it decreases more slowly with increased wavelength detuning. The low threshold detuning and high switching contrast for Fano bistability is due to the sharp and asymmetric Fano lineshapes. The inset of Figure 6.12b plots the transmission spectra of Fano lineshapes and Lorentzian lineshapes with an input power of 1 and 230 μW, respectively. The sharp transition right before the cavity resonance makes Fano bistability occur with much lower wavelength detuning, while Lorentzian bistability needs a larger detuning to reach the multivalue transmission regime. For the detuning shown in the figure, the switching contrast of Fano bistability (marked as a light grey upward pointing arrow) is higher than Lorentzian bistability (marked as a dark grey downward pointing arrow). There is significant difference in terms of wavelength detuning required for bistability to occur and the bistable switching contrast between Fano bistability and Lorentzian bistability. This switching contrast can significantly increase when the mirror reflectivity, r, increases from 0.35 to 0.8 (at a detuning of $\delta/\Delta\lambda = 0.423$) at the expense of increasing p_{on} (Figure 6.14b). The increase in p_{on} is due to a higher mirror reflectivity, resulting in lower power coupled into the Fano system. A limit of 0.35 is used, because for smaller r, a combination of both

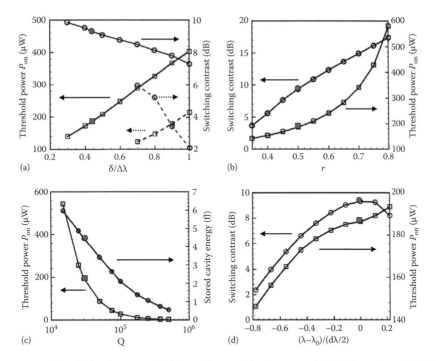

FIGURE 6.14 CMT-calculated parametrical dependence of the Fano-type bistability against achievable device characteristics. (a) Effects of the wavelength detuning, $\delta/\Delta\lambda$. Dashed lines (–) represent Lorentzian bistability that, compared to this particular Fano-type bistability, requires higher detuning to observe bistability and results in a lower switching contrast. (b) Effects of mirror reflectivity, r. (c) Effects of cavity Q factor. (d) Effects of the position of cavity resonance on the switching threshold power, p_{on}, and switching contrast. (The ligh grey filled symbols correspond to the experimental parameters.) (Reprinted from Yang, X. et al., *Appl. Phys. Lett.*, 91, 051113, 2007. With permission. Copyright 2007, American Institute of Physics.)

Lorentzian and ascending Fano resonance starts to appear. For r greater than 0.8, the threshold power will go up even higher.

Figure 6.14c plots the threshold power and the stored cavity energy with different cavity Q factors at a detuning of $\delta/\Delta\lambda = 0.423$ and $r = 0.5$. Note that p_{on} shows a $(1/Q^{1.569})$ dependence, while the stored cavity energy needed for bistability shows a $(1/Q^{0.689})$ dependence. For a cavity Q factor of half a million, the Fano threshold power is estimated at 2.4 µW, which corresponds to the stored cavity energy of 0.55 fJ. This stored cavity energy is much lower than a Fano resonance with a cavity Q of 30,000 (4.5 fJ). The comparison between Fano bistability and Lorentzian bistability at threshold detuning shows that the Fano system has a similar threshold power and stored cavity energy as the Lorentzian system. We also note that direct comparisons between an ascending Fano-type bistability and a Lorentzian-type bistability in terms of threshold power are difficult, because the Fano system has a different threshold detuning from the Lorentzian system and the Fano system also depends on additional parameters such as mirror reflectivity, r.

Figure 6.14d illustrates the influence of different positions of the cavity resonance, λ, relative to experimental λ_0 within the half period of Fabry–Perot background $d\lambda/2$ with $Q = 30,000$, $r = 0.5$, and $\delta/\Delta\lambda = 0.423$, where the limits of $(\lambda - \lambda_0)/(d\lambda/2)$ are from −0.8 to 0.2 for ascending Fano resonances. These limits are chosen because they cover the region where the ascending Fano resonances are dominant over the Lorentzian or the descending Fano resonances. For ascending Fano resonances at a detuning of $\delta/\Delta\lambda = 0.423$, both the threshold power and the switching contrast increase as the cavity resonance, λ, shifts from the Fabry–Perot background maximum to its minimum. The switching contrast has a maximum at a region close to the minimum Fabry–Perot background, illustrating an interesting trade-off when selecting an optimum set of Fano-type bistable operating parameters. The Fano resonance has little control in the current configuration, which depends on the Fabry–Perot cavity formed by two wave-guide facets. For different devices, either an ascending or a descending Fano resonance can be obtained, depending on the position of the cavity resonance in the Fabry–Perot background. In future, integrated partially reflecting mirrors embedded inside a PhC waveguide can be adopted. By tuning the distance between the air holes, the Fabry–Perot background can be tuned to achieve the designed Fano lineshapes.

6.2.1.4 Conclusion

We have demonstrated experimentally all-optical bistability arising from sharp Fano resonances in high-Q/V_m silicon PhC nanocavities. Using the TPA-induced thermo-optic nonlinearity, an "on"-state threshold of 189 μW and stored cavity energy of 4.5 fJ are observed, and are in good agreement with the nonlinear coupled-mode formalism. Although the thermo-optic nonlinear mechanism is slow (on the order of μs), other nonlinear mechanisms, such as TPA-induced FCD, can remarkably achieve ~50 ps switching in silicon. The threshold power can be further reduced to the μW level (or sub-fJ of stored cavity energy) with higher-Q/V_m nanocavities or further optimization of the detuning for reduced threshold and a large contrast ratio. Our observations of Fano-type bistability highlight the feasibility of an ultralow energy and high-contrast switching mechanism in monolithic silicon benefiting from the sharp Fano lineshapes, for scalable functionalities such as all-optical switching, memory, and logic for information processing.

6.2.2 Raman Scattering in Photonic Crystals

6.2.2.1 Photonic Crystal Nanocavities

6.2.2.1.1 Introduction

Raman scattering in silica-based high-Q microcavities, such as microspheres,[69] microdisks, and micro-toroids,[70] has shown remarkable ultralow lasing thresholds. In addition, Raman lasing in silicon wave-guides has also been observed[71-79] where the bulk Raman gain coefficient, g_R, is 10^3–10^4 times larger in silicon than in silica. To achieve significant amplification and ultimately lasing, the gain medium should be placed in a cavity with a sufficiently high Q and ultrasmall modal volumes, V_m. The enhanced stimulated Raman amplification and ultralow threshold Raman lasing in high-Q/V_m photonic bandgap nanocavities were suggested.[80] Stimulated Raman scattering (SRS) in periodic crystals with a slow group velocity was theoretically studied with a semiclassical model,[81] and an enhancement in line-defect pho-tonic crystal slow-light waveguides was also proposed.[82]

The coupled-mode equations for SRS in SOI waveguides,[74,83-85] fiber Bragg grating,[86] and silica micro-sphere[87,88] have been studied. Here, we present the coupled-mode framework to describe the dynamics of coupling between the pump cavity mode and the Stokes cavity mode. The loss rates of cavity modes due to radiation, linear material absorption, TPA, and FCA are included in the model. The refractive index shift from the Kerr effect, FCD, and thermal dispersion are also considered in the coupled-mode equations. These equations are numerically integrated to describe the dynamical behavior of the system for the designed $L5$ photonic bandgap nanocavities. Specific examples, such as lasing threshold and pulsed Raman frequency conversion, are investigated.

6.2.2.1.2 Design Concept and Coupled-Mode Theory

SRS is an inelastic two-photon process, where an incident photon interacts with an excited state of the material (the LO and TO phonons of single-crystal silicon). The strongest Stokes peak arises from single first-order Raman phonon (threefold degenerate) at the Brillouin zone center. We have proposed a pho-tonic bandgap cavity with five linearly aligned missing air holes ($L5$) in an air-bridged triangular-lattice photonic crystal slab with a thickness of $0.6a$ and air holes radius of $0.29a$, where the lattice period

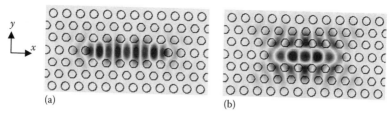

FIGURE 6.15 The electric field profile (E_y) of cavity modes for *L5* cavity. (a) Pump mode and (b) Stokes mode. (Reprinted from Yang, X. and Wong, C.W., *Opt. Express*, 15, 4763, 2007. With permission. Copyright 2007, Optical Society of America.)

$a = 420$ nm.[80] The designed cavity supports two even modes, the pump mode and the Stokes mode, with a spacing 15.6 THz, corresponding to the optical phonon frequency in monolithic silicon. Figure 6.15a and b shows the electric field profile (E_y) at the middle of the slab for the pump mode and the Stokes mode calculated from the 3D FDTD method.

Coupling between the pump mode and the Stokes mode in SRS can be understood classically with nonlinear polarizations $\mathbf{P}_{NL}^{(3)}$. The dynamics of SRS is governed through a set of time-dependent coupled nonlinear equations (in MKS)[81]:

$$\nabla \times \nabla \times \mathbf{E}_p + \frac{\varepsilon_p}{\varepsilon_0 c^2} \frac{\partial^2 \mathbf{E}_p}{\partial t^2} = -\frac{1}{\varepsilon_0 c^2} \frac{\partial^2 \mathbf{P}_{NL}^{(3)}(\omega_p)}{\partial t^2} \tag{6.4}$$

$$\nabla \times \nabla \times \mathbf{E}_S + \frac{\varepsilon_S}{\varepsilon_0 c^2} \frac{\partial^2 \mathbf{E}_S}{\partial t^2} = -\frac{1}{\varepsilon_0 c^2} \frac{\partial^2 \mathbf{P}_{NL}^{(3)}(\omega_S)}{\partial t^2} \tag{6.5}$$

where

$\mathbf{E}_p$ and $\mathbf{E}_S$ are the electric fields of the pump mode and the Stokes mode, respectively
$\mathbf{P}_{NL}^{(3)}(\omega_p)$ and $\mathbf{P}_{NL}^{(3)}(\omega_S)$ are the third-order nonlinear polarizations
ε_p and ε_S are the dielectric constants

The third-order nonlinear polarizations are given by

$$\mathbf{P}_{NL}^{(3)}(\omega_p) = 6\varepsilon_0 \chi_{ijkl}^{(3)}(\omega_p) \mathbf{E}_S \mathbf{E}_S^* \mathbf{E}_p \tag{6.6}$$

$$\mathbf{P}_{NL}^{(3)}(\omega_S) = 6\varepsilon_0 \chi_{ijkl}^{(3)}(\omega_S) \mathbf{E}_p \mathbf{E}_p^* \mathbf{E}_S \tag{6.7}$$

where $\chi_{ijkl}^{(3)}$ is the third-order nonlinear electric susceptibility.

Assume that the electric fields of the pump mode and the Stokes mode are

$$\mathbf{E}_p(\mathbf{r}, t) = a_p(t) \frac{\mathbf{A}_p(\mathbf{r})}{\sqrt{N_p}} e^{-i\omega_p t} \tag{6.8}$$

$$\mathbf{E}_S(\mathbf{r}, t) = a_S(t) \frac{\mathbf{A}_S(\mathbf{r})}{\sqrt{N_S}} e^{-i\omega_S t} \tag{6.9}$$

where

$\mathbf{A}_p(\mathbf{r})$ and $\mathbf{A}_S(\mathbf{r})$ are the spatial parts of the modes
$a_p(t)$ and $a_S(t)$ are slowly varying envelopes of the pump and Stokes modes, respectively

The amplitude is normalized by the spatial part $N_i = \frac{1}{2}\int \varepsilon_i(\mathbf{r})|\mathbf{A}_i(\mathbf{r})|^2 d^3r$ $(i = p, S)$ to represent the energy

of the mode (in units of Joules) $U_i = |a_i|^2 = \frac{1}{2}\int \varepsilon_i(\mathbf{r})|\mathbf{E}_i(\mathbf{r},t)|^2 d^3r$.

This results in the coupled-mode rate equations, relating the pump and Stokes evolutions without any loss terms currently:

$$\frac{da_p}{dt} = \left(\frac{\omega_p}{\omega_S}\right) g_S^c |a_S|^2 a_p \tag{6.10}$$

$$\frac{da_S}{dt} = g_S^c |a_p|^2 a_S \tag{6.11}$$

where the Raman gain coefficient in the photonic bandgap nanocavities, g_S^c $(\mathrm{J^{-1}\,s^{-1}})$, is

$$g_S^c = \left(\frac{c^2}{2n_p n_S V_R}\right) g_R^B \tag{6.12}$$

and g_R^B (m/W) is the bulk gain coefficient; $n_{p,S}$ are the refractive indices at the pump and Stokes wavelengths λ_p and λ_S, respectively; and $n_{p,S}^2 = \varepsilon_{p,S}/\varepsilon_0$. The Raman gain is assumed constant, without saturation or parametric instability. The effective modal volume, V_R, for Raman scattering indicates the spatial overlap between the pump mode and the Stokes mode:

$$V_R = \frac{\int n_p^2(\mathbf{r})|\mathbf{A}_p(\mathbf{r})|^2 d^3r \cdot \int n_S^2(\mathbf{r})|\mathbf{A}_S(\mathbf{r})|^2 d^3r}{\int_{Si} n_p^2(\mathbf{r})|\mathbf{A}_p(\mathbf{r})|^2 \cdot n_S^2(\mathbf{r})|\mathbf{A}_S(\mathbf{r})|^2 d^3r} \tag{6.13}$$

Note that in this classical formulation, the Raman gain coefficient in the photonic bandgap nanocavities, g_S^c, is still equivalent to the bulk Raman gain coefficient, g_R^B, since possible cavity quantum electrodynamics enhancements are not yet considered.

The electric fields of the input pump wave and the input Stokes wave in the waveguide are

$$\mathbf{E}_{i,\text{in}}(\mathbf{r},t) = s_i(t)\frac{\mathbf{S}_i(\mathbf{r})}{\sqrt{N_{i,\text{in}}}} e^{-i\omega_i t} \tag{6.14}$$

where the field amplitude is normalized by $N_{i,\text{in}} = \frac{1}{2}\int n_{i,\text{in}}(\mathbf{r})\varepsilon_0 c|\mathbf{S}_i(\mathbf{r})|^2 d^2r$ $(i = p, S)$, to represent the

input power $P_{i,\text{in}} = |s_i|^2 = \frac{1}{2}\int n_{i,\text{in}}(\mathbf{r})\varepsilon_0 c|\mathbf{E}_{i,\text{in}}(\mathbf{r},t)|^2 d^2r$. Now, considering the in-plane waveguide coupling

loss, $1/\tau_{i,\text{in}}$, and the vertical radiation loss, $1/\tau_{i,v}$, the coupled-mode rate equations are

$$\frac{da_p}{dt} = -\frac{1}{2\tau_p}a_p - \left(\frac{\omega_p}{\omega_S}\right)g_S^c|a_S|^2 a_p + \kappa_p s_p \tag{6.15}$$

$$\frac{da_S}{dt} = -\frac{1}{2\tau_S}a_S + g_S^c|a_p|^2 a_S + \kappa_S s_S \tag{6.16}$$

where $1/\tau_i = 1/\tau_{i,\text{in}} + 1/\tau_{i,v}$, and $1/\tau_{i,\text{in/v}} = \omega_i/Q_{i,\text{in/v}}$, $1/\tau_i = \omega_i/Q_i$, and $(i = p, S)$. $1/\tau_{i,\text{in}}$ and $1/\tau_{i,v}$ are the loss rates into waveguide (in-plane) and into free space (vertical). κ_i is the coupling coefficient of the input pump wave, $s_p(t)$, or the Stokes wave, $s_S(t)$, coupled to the pump mode, $a_p(t)$, or the Stokes mode, $a_S(t)$, of the cavity; and $\kappa_i = \sqrt{1/\tau_{i,\text{in}}}$. The threshold pump power with the cavity radiation loss for stimulated Raman lasing is obtained from Equations 6.15 and 6.16:

$$P_{\text{in,th}} = \frac{\pi^2 n_p n_S}{\lambda_p \lambda_S} \frac{V_R}{g_R^B} \left(\frac{Q_{p,\text{in}}}{Q_p^2 Q_S} \right) \tag{6.17}$$

The lasing threshold scales with $V_R/Q_p Q_S$, as illustrated in Equation 6.17. This, therefore, suggests the motivation for small V_R cavities with high Q factors.

Now, considering the total loss rate, $1/\tau_{i,\text{total}}$, and the shifted resonant frequency, $\Delta\omega_i$, of the pump mode and the Stokes mode, the coupled-mode rate equations are therefore

$$\frac{da_p}{dt} = \left(-\frac{1}{2\tau_{p,\text{total}}} + i\Delta\omega_p \right) a_p - \left(\frac{\omega_p}{\omega_S} \right) g_S^c |a_S|^2 a_p + \kappa_p s_p \tag{6.18}$$

$$\frac{da_S}{dt} = \left(-\frac{1}{2\tau_{S,\text{total}}} + i\Delta\omega_S \right) a_S + g_S^c |a_p|^2 a_S + \kappa_S s_S \tag{6.19}$$

This framework has been reported earlier in Johnson et al.[89] and Uesugi et al.[55] for a single frequency in cavities. We further advance these investigations for the pump–Stokes interactions, as well as study the lasing thresholds and dynamics under various conditions. The total loss rate for each cavity mode is

$$\frac{1}{\tau_{i,\text{total}}} = \frac{1}{\tau_{i,\text{in}}} + \frac{1}{\tau_{i,v}} + \frac{1}{\tau_{i,\text{lin}}} + \frac{1}{\tau_{i,\text{TPA}}} + \frac{1}{\tau_{i,\text{FCA}}} \tag{6.20}$$

The linear material absorption, $1/\tau_{\text{lin}}$, is assumed small since the operation is within the bandgap of the silicon material. $1/\tau_{\text{TPA}}$ and $1/\tau_{\text{FCA}}$ are the loss rates due to TPA and FCA, respectively. The mode-averaged TPA loss rates are[53]

$$\frac{1}{\tau_{p,\text{TPA}}} = \frac{\beta_{\text{Si}} c^2}{n_p^2 V_{p,\text{TPA}}} |a_p|^2 + \frac{\beta_{\text{Si}} c^2}{n_p^2 V_{o,\text{TPA}}} 2 |a_S|^2 \tag{6.21}$$

$$\frac{1}{\tau_{S,\text{TPA}}} = \frac{\beta_{\text{Si}} c^2}{n_S^2 V_{S,\text{TPA}}} |a_S|^2 + \frac{\beta_{\text{Si}} c^2}{n_S^2 V_{o,\text{TPA}}} 2 |a_p|^2 \tag{6.22}$$

where the first terms represent the TPA due to two pump photons or two Stokes photons. The second terms represent that one pump photon and one Stokes photon are absorbed simultaneously. β_{Si} is the TPA coefficient of bulk silicon. The effective mode volume for TPA, $V_{i,\text{TPA}}$, is

$$V_{i,\text{TPA}} = \frac{\left(\int n_i^2(\mathbf{r}) |\mathbf{A}_i(\mathbf{r})|^2 dr^3 \right)^2}{\int_{\text{Si}} n_i^4(\mathbf{r}) |\mathbf{A}_i(\mathbf{r})|^4 dr^3} \tag{6.23}$$

$V_{o,TPA}$ indicates the spatial overlap between the pump mode and the Stokes mode, and $V_{o,TPA} = V_R$. We note that the $(\beta_{Si}/V_{i,TPA})$ TPA term is the effective silicon-air contribution, summing over the cavity modal distributions within the solid and negligible contribution from air, and neglecting crystal anisotropy. The bulk TPA coefficient is also assumed to be frequency independent,[90] and without surface modification for simplicity. The mode-averaged FCA loss rates are

$$\frac{1}{\tau_{i,FCA}} = \frac{c}{n_i}\alpha_{i,FCA} = \frac{c}{n_i}(\sigma_{i,e} + \sigma_{i,h})N(t) \tag{6.24}$$

where $\sigma_{i,e/h}$ are the absorption cross sections for electrons and holes from the Drude model.[91]

The mode-averaged free-carrier density (electron–hole pairs) generated by TPA is $N(t)$, which is governed by the rate equation[89]

$$\frac{dN}{dt} = -\frac{N}{\tau_{fc}} + G \tag{6.25}$$

The mode-averaged generation rate of free carriers, G, can be calculated from the mode-averaged TPA loss rate:

$$G = \frac{\beta_{Si}c^2}{2\hbar\omega_p n_p^2 V_{p,FCA}^2}|a_p|^4 + \frac{\beta_{Si}c^2}{2\hbar\omega_S n_S^2 V_{S,FCA}^2}|a_S|^4$$

$$+ \frac{\beta_{Si}c^2}{\hbar(\omega_p + \omega_S)n_p^2 V_{Sp,FCA}^2}2|a_S|^2|a_p|^2 + \frac{\beta_{Si}c^2}{\hbar(\omega_p + \omega_S)n_S^2 V_{pS,FCA}^2}2|a_p|^2|a_S|^2 \tag{6.26}$$

The expressions of the effective mode volume for FCA, V_{FCA}, are

$$V_{i,FCA}^2 = \frac{\left(\int n_i^2(\mathbf{r})|\mathbf{A}_i(\mathbf{r})|^2 dr^3\right)^3}{\int_{Si} n_i^6(\mathbf{r})|\mathbf{A}_i(\mathbf{r})|^6 dr^3} \tag{6.27}$$

$$V_{Sp,FCA}^2 = \frac{\left(\int n_p^2(\mathbf{r})|\mathbf{A}_p(\mathbf{r})|^2 dr^3\right)^2 \left(\int n_S^2(\mathbf{r})|\mathbf{A}_S(\mathbf{r})|^2 dr^3\right)}{\int_{Si} n_p^4(\mathbf{r})|\mathbf{A}_p(\mathbf{r})|^4 \cdot n_S^2(\mathbf{r})|\mathbf{A}_S(\mathbf{r})|^2 dr^3} \tag{6.28}$$

$$V_{pS,FCA}^2 = \frac{\left(\int n_S^2(\mathbf{r})|\mathbf{A}_S(\mathbf{r})|^2 dr^3\right)^2 \left(\int n_p^2(\mathbf{r})|\mathbf{A}_p(\mathbf{r})|^2 dr^3\right)}{\int_{Si} n_S^4(\mathbf{r})|\mathbf{A}_S(\mathbf{r})|^4 \cdot n_p^2(\mathbf{r})|\mathbf{A}_p(\mathbf{r})|^2 dr^3} \tag{6.29}$$

τ_{fc} is the effective free-carrier lifetime accounting for both recombination and diffusion. Time constants of radiative and Auger recombination, as well as from bulk defects and impurities, are assumed to be significantly slower than the free-carrier recombination and diffusion lifetime.[90] We note that while the free-carrier lifetime can vary with carrier density, and carrier density can vary spatially with intensity in the cavity, an effective lifetime is used here for simplicity. A quiescent carrier density of $N_0 = 10^{22}\,\text{m}^{-3}$ is used in the initial condition for silicon.

In Equations 6.18 and 6.19, $\Delta\omega_i$ is the detuning of the resonance frequency of the cavity from the input light frequency due to the Kerr effect, FCD, and thermal dispersion. $\Delta\omega_i = \omega_i' - \omega_i$, ω_i' is the shifted resonant frequency of the cavity and ω_i is the input light frequency in the waveguide. Under first-order perturbation, the detuning of the resonance frequency can be expressed as[53]

$$\frac{\Delta\omega_i}{\omega_i} = -\frac{\Delta n_i}{n_i} = -\left(\frac{\Delta n_{i,\text{Kerr}}}{n_i} + \frac{\Delta n_{i,\text{FCD}}}{n_i} + \frac{\Delta n_{i,\text{th}}}{n_i}\right) \tag{6.30}$$

The detuning due to Kerr effect is

$$\frac{\Delta n_{p,\text{Kerr}}}{n_p} = \frac{cn_2}{n_p^2 V_{p,\text{Kerr}}}|a_p|^2 + \frac{cn_2}{n_p^2 V_{o,\text{Kerr}}}2|a_S|^2 \tag{6.31}$$

$$\frac{\Delta n_{S,\text{Kerr}}}{n_S} = \frac{cn_2}{n_S^2 V_{S,\text{Kerr}}}|a_S|^2 + \frac{cn_2}{n_S^2 V_{o,\text{Kerr}}}2|a_p|^2 \tag{6.32}$$

where the effective modal volumes for the Kerr effect are $V_{i,\text{Kerr}} = V_{i,\text{TPA}}$ and $V_{o,\text{Kerr}} = V_{o,\text{TPA}}$. The first terms represent self-phase modulation and the second terms represent cross-phase modulation. The detuning due to FCD is related by

$$\frac{\Delta n_{i,\text{FCD}}}{n_i} = -\frac{1}{n_i}\left(\zeta_{i,e} + \zeta_{i,h}\right)N(t) \tag{6.33}$$

where $\zeta_{i,e/h}$ are the material parameters with units of volume from the Drude model.[91] The detuning due to thermal effect is

$$\frac{\Delta n_{i,\text{th}}}{n_i} = \frac{1}{n_i}\frac{dn_i}{dT}\Delta T \tag{6.34}$$

The mode-averaged temperature difference between the photonic crystal cavity and its environment, ΔT, is governed by[89]

$$\frac{d\Delta T}{dt} = -\frac{\Delta T}{\tau_{\text{th}}} + \frac{P_{\text{abs}}}{\rho_{\text{Si}}c_{p,\text{Si}}V_{\text{cavity}}} \tag{6.35}$$

where $\rho_{\text{Si}}, c_{p,\text{Si}}$, and V_{cavity} are the density of silicon, the constant-pressure specific heat capacity of silicon, and the volume of the cavity, respectively. The temperature decay lifetime, τ_{th}, is determined by the thermal resistance, R, of the air-bridged silicon photonic crystal cavities:

$$\tau_{\text{th}} = \rho_{\text{Si}}c_{p,\text{Si}}V_{\text{cavity}}R \tag{6.36}$$

The total absorbed power is given by

$$P_{\text{abs}} = P_{p,\text{abs}} + P_{S,\text{abs}} + P_{R,\text{abs}} \tag{6.37}$$

$$P_{i,\text{abs}} = \left(\frac{1}{\tau_{i,\text{lin}}} + \frac{1}{\tau_{i,\text{TPA}}} + \frac{1}{\tau_{i,\text{FCA}}}\right)|a_i|^2 \tag{6.38}$$

The absorbed power due to Raman scattering–generated optical phonon, $P_{R,abs}$, is

$$P_{R,abs} = 2\left(\omega_p/\omega_S - 1\right)g_S^c\left|a_p\right|^2\left|a_S\right|^2 \tag{6.39}$$

Equations 6.18 through 6.20, 6.25, 6.30, and 6.35, therefore, describe the dynamic behavior of SRS in photonic crystal nanocavities and are numerically integrated in our work to describe the dynamical behavior of pump–Stokes interactions in our *L5* photonic crystal cavity system that supports the desired two-mode frequencies at the appropriate LO/TO phonon spacing.

6.2.2.1.3 Numerical Analysis

Around the lasing threshold, the Stokes gain equals the losses, and the Stokes mode energy, $|a_S|^2$, is much smaller than the pump mode energy, $|a_p|^2$; thus, Equation 6.19 is simplified to

$$g_S^c\left|a_p\right|_{th}^2 = \frac{1}{2\tau_{S,total}} \tag{6.40}$$

The loss rate due to the TPA of the pump mode is

$$\frac{1}{\tau_{S,TPA}} = \frac{\beta_{Si}c^2}{n_S^2 V_{o,TPA}} 2\left|a_p\right|_{th}^2 \tag{6.41}$$

The free carrier generated by the TPA of the pump mode is

$$N = \tau_{fc}G \tag{6.42}$$

$$G = \frac{\beta_{Si}c^2}{2\hbar\omega_p n_p^2 V_{p,FCA}^2}\left|a_p\right|_{th}^4 \tag{6.43}$$

Figure 6.16 shows the threshold pump mode energy, $\left|a_p\right|_{th}^2$, as a function of Q_S for different free-carrier lifetimes, τ_{fc}. All the parameters used in the calculation are presented in Table 6.2. By comparing the curve in the absence of TPA and FCA ($\beta_{Si} = 0$) and the curve with TPA but without FCA ($\tau_{fc} = 0$), it is observed that TPA increases the threshold pump mode energy but the effect of TPA is relatively weak. The effect of TPA-induced FCA is much more dramatic as shown for different free-carrier lifetimes, τ_{fc}. The lasing threshold increases when τ_{fc} is larger. There is a minimum Stokes Q_S required for lasing, as seen in the solutions plotted in Figure 6.16. If Q_S is lower than a critical value for a certain τ_{fc}, there is no solution numerically, and physically this translates into an absence of a lasing threshold regardless of the pump intensity. For increasing τ_{fc}, the critical value of Q_S increases monotonically, as seen in Figure 6.16. The solid and dotted curves show the lasing and shutdown thresholds, respectively.[84] The shutdown threshold is the pump power in which the lasing output power returns to zero due to increasing TPA and FCA. For the *L5* cavity studied in this present article, $Q_S = 21,000$, the maximum τ_{fc} is around 0.175 ns, and the threshold pump mode energy is 29 fJ. For air-bridged silicon photonic bandgap nanocavities, $\tau_{fc} = 0.5$ ns,[55] which is much higher than the maximum τ_{fc}. In order to get lasing for this cavity, instead of using a continuous wave (CW) pump signal, a pulse pump signal can be used to reduce the TPA-induced FCA for loss reduction and an increase in the net gain.

We now consider the dynamics of the Raman lasing interactions. Equations 6.18 through 6.20, 6.25, 6.30, and 6.35 are numerically integrated with a variable-order Adams–Bashforth–Moulton predictor–corrector. All the parameters used in the calculation are presented in Table 6.2. Figure 6.17 shows the dynamics of Raman amplification with a 60 mW CW pump wave and a 10 µW CW Stokes seed signal; free-carrier lifetime is 0.5 ns. In the beginning when the free-carrier density is low, the Raman gain is

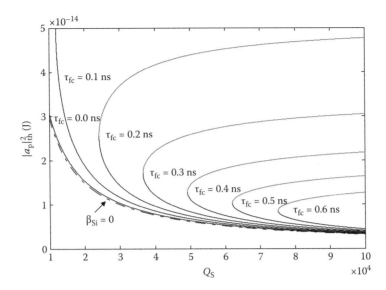

FIGURE 6.16 Threshold pump mode energy versus Q_S of $L5$ cavity for different values of free-carrier lifetimes, τ_{fc}. The solid curve and dotted curve show the lasing and shutdown thresholds, respectively. (Reprinted from Yang, X. and Wong, C.W., *Opt. Express*, 15, 4763, 2007. With permission. Copyright 2007, Optical Society of America.)

greater than the loss and there is amplification. When the free-carrier density increases, FCA dominates the loss and the Stokes signal is suppressed. The temperature difference then increases significantly. The cavity resonance is redshifted, and the pump mode energy goes down. From the numerical results, the Kerr effect is predominantly weak. The FCD effect dominates at first when the temperature difference is low, with a resulting blueshift. Eventually, the thermal effect dominates, with a resulting redshift. Figure 6.18 shows the dynamics of Raman amplification with a pump pulse of 60 mW peak power, a pulse width of $T_{FWHM} = 50$ ps, and a 10 μW CW Stokes signal. The free-carrier density and the temperature difference are significantly reduced, and a strong Stokes pulse is generated by the pump pulse. The resonance frequency shift is also significantly reduced by the pump pulse operation.

6.2.2.1.4 Conclusion

We have derived and presented the coupled-mode equations for SRS in high-Q/V_m silicon photonic bandgap nanocavities toward optically pumped silicon lasing. Both the lasing threshold and the lasing dynamics are numerically studied in the presence of cavity radiation losses, linear material absorption, TPA, and FCA, together with the refractive index shift from the Kerr effect, FCD, and thermal dispersion. The results show that compact Raman amplifiers and lasers based on high-Q/V_m silicon photonic bandgap nanocavities are feasible.

6.2.2.2 Slow-Light Photonic Crystal Waveguides

6.2.2.2.1 Introduction

The SOI technology is one of the leading platforms for integrated photonic devices due to its low absorption at telecommunication wavelengths and high optical confinement. The characteristic of high modal confinement means that devices can be made incredibly compact; this feature along with silicon as the material of choice in integrated circuit fabrication makes SOI an ideal choice for cheap integrated optical devices. However, the inability of silicon to amplify light with band-to-band recombination due to its indirect bandgap severely limits its capabilities.

For this reason, the subject of Raman scattering in silicon-integrated photonic devices has received much interest in the last 6 years. Due to silicon having an intrinsically large Raman gain coefficient (being

TABLE 6.2   Parameters Used in Coupled-Mode Theory

Parameter	Symbol	Value	Source
Refractive index of silicon	n_i	3.485	Ref. [92]
Wavelength of pump mode	λ_p	1496.7 nm	FDTD
Wavelength of Stokes mode	λ_S	1623.1 nm	FDTD
Pump mode in-plane Q	$Q_{p,in}$	960	FDTD
Pump mode vertical Q	$Q_{p,v}$	960	FDTD
Stokes mode in-plane Q	$Q_{S,in}$	42,000	FDTD
Stokes mode vertical Q	$Q_{S,v}$	42,000	FDTD
Linear material absorption loss	$1/\tau_{lin}$	0.86 GHz	Ref. [89]
Raman mode volume	V_R	$0.544868 \times 10^{-18}\,m^3$	FDTD
TPA mode volume of pump mode	$V_{p,TPA}$	$0.258387 \times 10^{-18}\,m^3$	FDTD
TPA mode volume of Stokes mode	$V_{S,TPA}$	$0.396806 \times 10^{-18}\,m^3$	FDTD
FCA mode volume of pump mode	$V_{p,FCA}$	$0.202601 \times 10^{-18}\,m^3$	FDTD
FCA mode volume of Stokes mode	$V_{S,FCA}$	$0.299289 \times 10^{-18}\,m^3$	FDTD
FCA mode volume	$V_{p,FCA}$	$0.337575 \times 10^{-18}\,m^3$	FDTD
FCA mode volume	$V_{S,FCA}$	$0.368572 \times 10^{-18}\,m^3$	FDTD
Bulk Raman gain coefficient	g_R^B	$2.9 \times 10^{-10}\,m/W$	Ref. [72]
TPA coefficient	β_{Si}	$4.4 \times 10^{-12}\,m/W$	Ref. [71]
Kerr coefficient	n_2	$4.4 \times 10^{-18}\,m^2/W$	Ref. [93]
Free-carrier lifetime	τ_{fc}	0.5 ns	Ref. [55]
Absorption cross sections for electrons	$\sigma_{i,e}$	$8.5 \times 10^{-22}\,m^2$	Refs. [91,94]
Absorption cross sections for holes	$\sigma_{i,h}$	$6.0 \times 10^{-22}\,m^2$	Refs. [91,94]
FCD parameter for electrons	$\zeta_{i,e}$	$8.8 \times 10^{-28}\,m^3$	Refs. [91,94]
FCD parameter for holes	$\zeta_{i,h}$	$4.6 \times 10^{-28}\,m^3$	Refs. [91,94]
Density of silicon	ρ_{Si}	$2.33 \times 10^3\,kg/m^3$	Ref. [95]
Constant-pressure specific heat capacity of silicon	$c_{p,Si}$	$0.7 \times 10^3\,J/kg/K$	Ref. [95]
Volume of cavity	V_{cavity}	$0.462 \times 10^{-18}\,m^3$	$\sim L \times W \times H$
Thermal resistance	R	50 K/mW	Ref. [55]
Temperature dependence of refractive index	dn_i/dT	$1.85 \times 10^{-4}\,K^{-1}$	Ref. [96]

Source: Reprinted from Yang, X. and Wong, C.W., *Opt. Express*, 15, 4763, 2007. With permission. Copyright 2007, Optical Society of America.

10^3–10^4 times greater than that for silica), it is possible to compact on-chip gain media at desired telecommunication frequencies. The first observations of spontaneous Raman scattering in large-mode-area silicon waveguides[97] were soon followed by pulsed[73,75] and CW[71] amplification in silicon waveguides. In addition, the investigation of Raman scattering in small mode area nanowire waveguides, with both enhancement of spontaneous emission[98] and stimulated amplification[72] due to the increased optical intensity in these devices, was performed.

Not long after these initial results in scattering and amplification were the first reports of silicon behaving as a gain medium in a laser reported. The first demonstration of a silicon Raman laser utilized a fiber loop as a cavity.[76] The first all-silicon pulsed laser utilized a Fabry–Perot cavity made from a silicon rib waveguide.[77] By overcoming high nonlinear optical losses due to FCA via an integrated p-i-n reverse-biased diode, it was possible to create the first continuous-wave all-silicon Raman laser.[77] With the integration of a ring cavity,[99] a further reduction in the threshold powers of silicon continuous-wave Raman lasers was observed.

In order to further enhance Raman scattering and also reduce the size of the device, optical nanostructures exhibiting slow group velocities or high quality factors are currently being explored. Raman

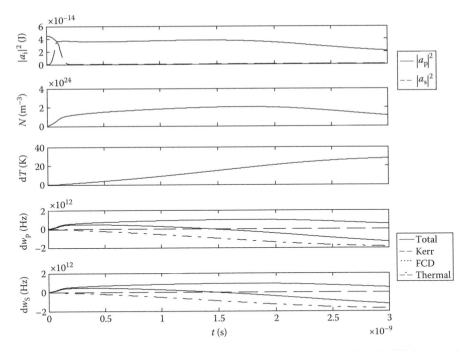

FIGURE 6.17   Dynamics of Raman amplification with 60 mW CW pump wave and 10 μW CW Stokes seed signal; τ_{fc} is 0.5 ns. (Reprinted from Yang, X. and Wong, C.W., *Opt. Express*, 15, 4763, 2007. With permission. Copyright 2007, Optical Society of America.)

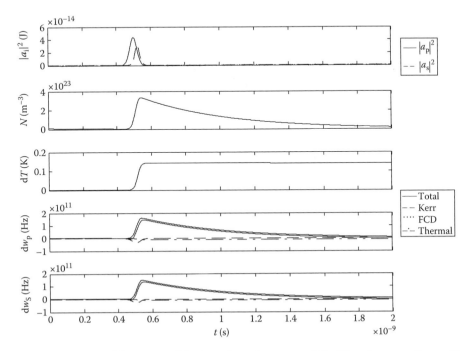

FIGURE 6.18   Dynamics of Raman amplification with pulse pump of 60 mW peak power; $T_{FWHM} = 50$ ps and 10 μW CW Stokes seed signal; τ_{fc} is 0.5 ns. (Reprinted from Yang, X. and Wong, C.W., *Opt. Express*, 15, 4763, 2007. With permission. Copyright 2007, Optical Society of America.)

scattering in GaAs Bragg stacks has been experimentally shown to enhance emission by 12,000 times.[100] The enhancement of the photonic density of states that PhCs possess has been theoretically shown to enhance the Raman scattering rate.[101] Enhanced Raman scattering has been observed in bulk hollow-core, slow-light guided-wave structures[102] and has also recently been suggested for PhC defect nanocavities.[64,80] In addition, a semi-classical model of Raman scattering in bulk PhCs has been introduced.[81] Observations of spontaneous Raman scattering have been made in GaAs PhC waveguides.[103]

6.2.2.2.2 Photonic Crystal Waveguides

Another nanostructure exhibiting low group velocities is the PhC waveguide. The measured and theoretical group velocities of the fundamental mode of a silicon PhC waveguide can be seen in Figure 6.19. They have been successfully fabricated in a number of materials, including silicon, and exhibit only slightly higher losses than regular channel waveguides.[104] The group velocity of these devices has been successfully measured by the authors and others[105] to be as low as 0.01*c*. Here, we will explore how the unique dispersion properties of a silicon photonic crystal waveguide (PhCWG) can be used to enhance the scattering over regular channel waveguide structures. Using a Bloch–Floquet formalism, we show what effect the unique modes of a silicon PhCWG have on the Raman susceptibility and explicitly the enhancement one can expect from the low group velocities that these modes exhibit.

A single-line defect in a hexagonal silicon PhC slab (air above and below), the so-called W1 waveguide, can be seen in Figure 6.19, and examples of its dispersion relation, calculated using the plane-wave expansion method,[106] can be seen in Figure 6.20. This structure supports two tightly confined modes with TE-like polarization that exhibit small group velocities as it approaches the Brillouin zone edge and the mode onset frequency. This slow group velocity can be visualized as multiple photon scattering processes off the bulk PhC lattice, and so the total forward movement of the photon is slow. The group velocity of these modes can be determined from the flatness of the dispersion curve ($v_g \equiv d\omega/dk$). The theoretically derived group velocity can be seen in Figure 6.19. In addition to these TE-like modes, which are confined in-plane by the PhC bandgap, there also exists a TM-like mode, which is a totally index-guided mode. The dispersion of this mode can be seen in Figure 6.20c. Even though a photonic bandgap does not exist for the TM-like mode, it is not completely unaffected by periodic lattice. Due to the periodic modulation of the effective index in the direction of propagation caused by the holes of the PhC, this mode exhibits a stopgap at the Brillouin zone edge. The edges of this stopband themselves exhibit low group velocities, but not with an equal bandwidth seen with the TE-like modes. The

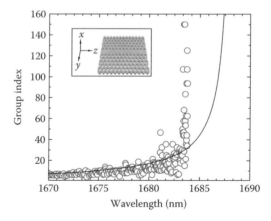

FIGURE 6.19 The group velocity of the fundamental mode of a silicon PhC waveguide, measured data (open circles), and theoretically derived curve (solid line) from differentiating the band structure. Inset: W1 PhC waveguide. The group velocity was measured using the Fabry–Perot spacing method (From Notomi, M. et al., *Phys. Rev. Lett.*, 87, 253902, 2001.).

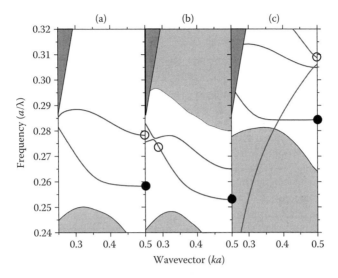

FIGURE 6.20 Projected band structure of three proposed Raman scattering pumping schemes (open circles, pump frequency; closed circles, Stokes frequency). (a) Scheme 1: $(r/r) = 0.29$. (b) Scheme 2: $(r/a) = 0.22$. (c) Scheme 3: $(r/a) = 0.34$. $h/a = 0.6$ for all cases.

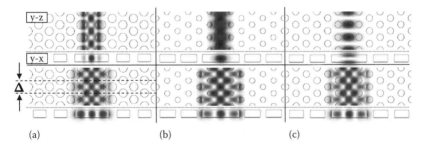

FIGURE 6.21 Calculated modes of the W1 PhC waveguide for the three different Raman pumping schemes. Top row: pump mode, bottom row: Stokes mode. (a) Scheme 1. (b) Scheme 2. (c) Scheme 3. (Top panels reprinted from MaMillan, J.F. et al., *Opt. Lett.*, 31, 1235, 2006. With permission. Copyright 2006, Optical Society of America.)

computed mode distributions for the corresponding mode in Figure 6.20 can be seen in Figure 6.21. The strong subwavelength modal confinement of the high-index-contrast PhCWG leads to increased field intensities in the silicon gain media, permitting increased nonlinear interactions.

6.2.2.2.3 Theory

In SRS for silicon, an incident photon interacts with the LO and TO phonons; the strongest Stokes peak arises from the single first-order Raman phonon at the center of the electron Brillouin zone. The generation of the Stokes photons can be understood classically as a third-order nonlinear effect, and this formalism has been used to model SRS in SOI waveguides, both in CW[107] and pulsed[85] operation. It can be modeled in bulk materials as a degenerate four-wave-mixing problem involving the pump and Stokes beams. The important material parameter is the third-order nonlinear Raman susceptibility, χ^R. For silicon, at resonance, χ^R is defined by the components $\chi^R_{ijij} = -i\chi^R = -i11.2 \times 10^{-18}\,m^2 \cdot V^{-2}$ $(i, j = 1, 2, 3)$[107]. An additional symmetry, imposed by the crystal point group (*m3m* for Si), is $\chi^R_{iiji} = 0.5\chi^R_{iiji}$.[108] These components, and their permutations as defined by the crystal point group, define the SRS in a silicon crystal. For our purpose, we shall consider scattering in silicon along the $[1\,\bar{1}\,0]$ direction, since practical devices are fabricated along this direction due to the favorable cleaving of silicon along this direction.

For bulk silicon, the evolution of the Stokes beam is defined by the following equation[109]:

$$\frac{dI_S}{dz} = -\frac{3\omega_S Im(\chi_{eff}^R)}{\varepsilon_0 c^2 n_p n_S} I_p I_S \tag{6.44}$$

where $\chi_{eff}^R = \sum_{ijkl} \chi_{ijkl}^R \hat{\alpha}_i^* \hat{\beta}_j \hat{\beta}_k \hat{\alpha}_l$. Here, $\hat{\alpha}$ and $\hat{\beta}$ are unit vectors along the polarization directions of the pump and Stokes beams, respectively. Equation 6.44 describes the gain of the Stokes intensity, I_S. It shows an intrinsic dependence on the polarization and the phonon selection rules through χ^R and the intensity of the pump beam by I_p. The bulk solution also describes SRS in dielectric waveguides, where χ_{eff}^R is averaged over the waveguide mode field distribution.

A PhCWG presents a very different field distribution than the bulk or dielectric waveguide case. As shown in the computed modal profiles of Figure 6.21, the mode differs from that of a conventional channel waveguide in that it exhibits a periodic variation in the direction of propagation. We introduce the modal distribution of the pump and Stokes modes in a Bloch–Floquet formalism:

$$\mathbf{E}_{n,k_n}(\mathbf{r},\omega_n) = \mathbf{E}_{n,k_n}(\mathbf{r},\omega_n)\exp[i\mathbf{k}(\omega_n)\cdot\mathbf{r}] \tag{6.45}$$

where

n is a mode index ($n = p, s$)

$\mathbf{k}_n = \mathbf{k}(\omega_n)$ is the mode wave-vector

$\mathbf{E}_{n,k_n}(\mathbf{r},\omega_n)$ is the modal distribution within a unit cell of the PhC, defined in Figure 6.20, and obeys the Bloch boundary condition $\mathbf{E}_{n,k_n}(\mathbf{r}+\Delta,\omega_n) = \mathbf{E}_{n,k_n}(\mathbf{r},\omega_n)$

Here, Δ defines the length of the unit cell in the direction of propagation (see Figure 6.21), and for a W1 waveguide, this equals the PhC lattice constant a. To develop an equation that relates the evolution of the Stokes mode to the pump mode, we employ the Lorentz reciprocity theorem[85,109]:

$$\frac{\partial}{\partial z}\int_A [\mathbf{E}_{n,k_n}^* \times \tilde{\mathbf{H}} + \tilde{\mathbf{E}} \times \mathbf{H}_{n,k_n}^*]\cdot\check{\mathbf{e}}_z \, dA = i\omega\int_A \mathbf{P}^R \cdot \mathbf{E}_{n,k_n} \, dA \tag{6.46}$$

This relates the unperturbed linear PhCWG modes of the pump or Stokes wavelengths, $\{\mathbf{E}_{n,k_n},\mathbf{H}_{n,k_n}\}$, to those of the nonlinearly induced fields, $\{\tilde{\mathbf{E}},\tilde{\mathbf{H}}\}$. The envelopes of the fields are defined as

$$\tilde{\mathbf{E}}(\mathbf{r}) = u_S(z)\mathbf{E}_{S,k_S}(\mathbf{r},\omega_S) + u_p(z)\mathbf{E}_{p,k_p}(\mathbf{r},\omega_p) \tag{6.47}$$

$$\tilde{\mathbf{H}}(\mathbf{r}) = u_S(z)\mathbf{H}_{S,k_S}(\mathbf{r},\omega_S) + u_p(z)\mathbf{H}_{p,k_p}(\mathbf{r},\omega_p) \tag{6.48}$$

with the assumption that the change in the pump and Stokes field amplitudes, $u_p(z)$ and $u_S(z)$, respectively, over the length of the unit cell of the waveguide is very small ($\Delta\frac{du_{p,S}}{dz} \ll 1$). Taking the fields as defined in Equation 6.46 and substituting in Equation 6.46, we derive the dependence of the Stokes amplitude on the longitudinal distance, z:

$$\frac{du_S(z)}{dz} = \frac{i\omega_S}{4P_S\Delta}\int_{V_0} \mathbf{P}^R(\mathbf{r},\omega_S)\cdot\mathbf{E}_{S,k_S}(\mathbf{r},\omega_S)dV \tag{6.49}$$

where P_S is the mode power and $\mathbf{P}^R(\mathbf{r}, \omega_S) = 6\varepsilon_0 \hat{\chi}^R : \mathbf{E}^*_{p,k_p}(\mathbf{r})\mathbf{E}_{p,k_p}(\mathbf{r})\mathbf{E}_{S,k_S}(\mathbf{r})|u_p|^2 u_S$. The integral in Equation 6.49 is taken over the volume (V_0) of the unit cell of the PhCWG mode. Furthermore, the group velocity of the modes can be expressed by the following equation[109]:

$$v_g^{p,S} = \frac{P_{p,S}\Delta}{\frac{1}{2}\varepsilon_0 \int_{V_0} \varepsilon(\mathbf{r})\,|\,\mathbf{E}_{p,S}(\mathbf{r},\omega_{p,S})|^2\,dV} \tag{6.50}$$

With Equations 6.46 and 6.50, and by rewriting Equation 6.49 in terms of the mode's intensity, an equation for the intensity of the Stokes mode inside the PhCWG is obtained:

$$\frac{dI_S}{dz} = -\frac{3\omega_S}{\varepsilon_0 v_g^p v_g^S}\kappa I_p I_S, \tag{6.51}$$

where

$$\kappa = \frac{\Delta A_{\text{eff}}\,\text{Im}\left(\displaystyle\int_{V_0}\mathbf{E}^*(\omega_S)\cdot\hat{\chi}^R:\mathbf{E}^*(\omega_p)\mathbf{E}(\omega_p)\mathbf{E}(\omega_S)dV\right)}{\left(\dfrac{1}{2}\displaystyle\int_{V_0}\varepsilon(\mathbf{r})\,|\,\mathbf{E}(\omega_p)|^2\,dV\right)\left(\dfrac{1}{2}\displaystyle\int_{V_0}\varepsilon(\mathbf{r})\,|\,\mathbf{E}(\omega_S)|^2\,dV\right)} \tag{6.52}$$

is the effective susceptibility. Here, the effective area, A_{eff}, is defined as the average modal area across the volume, V_0:

$$A_{\text{eff}}^2 = \frac{\left(\displaystyle\int_{V_0}x^2\,|\,\mathbf{E}(\omega_S)|^2\,dV\right)\left(\displaystyle\int_{V_0}y^2\,|\,\mathbf{E}(\omega_S)|^2\,dV\right)}{\left(\displaystyle\int_{V_0}|\,\mathbf{E}(\omega_S)|^2\,dV\right)^2} \tag{6.53}$$

The final equation, Equation 6.51, shows the explicit inverse dependence the Stokes mode amplification has on the group velocities of the pump and Stokes modes. When compared to Equation 6.44, which shows an inverse dependence on c^2, it can be seen that equivalent Raman gains at lower pump powers (I_p) can be achieved in a PhCWG at frequencies with low group velocities.

6.2.2.2.4 Analysis

In our analysis, we consider three different schemes, in all of which the generated Stokes is situated at the frequency corresponding to the onset of the fundamental TE-like mode. For each scheme, the placement of the pump mode differs (Figure 6.20): (1) the pump is situated on the mode onset frequency of the odd TE-like mode; (2) the pump resides on the fundamental TE-like mode, away from the onset; and (3) the pump is situated on the upper TM-like stopgap edge. Each scheme has its advantages and disadvantages both theoretically and experimentally. For Scheme 1, both modes have the same polarization and are confined by the PhC at their respective mode onset frequencies, ensuring high modal confinement and low group velocities. However, experimentally, Scheme 1 is at a disadvantage due to the difficulty in efficiently pumping the higher-order mode. For Scheme 2, the experimental disadvantage of Scheme 1 is overcome by placing both the pump and the Stokes on the same fundamental TE-like mode. However, this comes with the disadvantage that the pump is no longer experiencing any enhancement

TABLE 6.3 Effective Area, Group Velocities, and Effective Susceptibility in the Three PhCWG Raman Schemes

Scheme	v_g^p	v_g^s	$\kappa (\times 10^{-19}\,\text{m}^2/\text{V})$
1	$0.0067c$	$0.0067c$	0.55
2	$0.24c$	$0.0067c$	2.02
3	$0.017c$	$0.0067c$	0.63

from having a low group velocity, since it is placed at a position on the dispersion curve where the mode exhibits a group velocity similar to that of a regular channel waveguide. For Scheme 3, both the disadvantages of Schemes 1 and 2 are overcome by placing the pump mode on the TM-like band; now both the Stokes and pump travel at experimentally easy-to-excite modes and are experiencing high effective indices. There are disadvantages to Scheme 3 also: TM stopgap does not have as strong a dispersion as the TE bandgap, and so the bandwidth of slow light is less and the maximum group index is lower than in the TE gap. In addition, experimental reports[110] indicate that the TM-like mode can experience significantly more loss than the TE-like modes.

To determine the effect the modes of the PhCWG have on Equation 6.51, the effective susceptibility of each pumping scheme can be calculated using Equation 6.52 by using the simulated electric field of the PhCWG modes. The modes (see Figure 6.21) are simulated using the plane-wave expansion method.[106] With the susceptibility and by selecting values for the group velocities that correspond to those observed experimentally, we can see the expected enhancement from Equation 6.51 that we would expect when compared to bulk silicon. The summary of the results can be seen in Table 6.3.

The results in Table 6.3 explicitly show that the Raman gain, which is proportional to $\kappa / v_g^s v_g^p$, is enhanced by up to four orders of magnitude, as in the case of Scheme 1 (Scheme 1: 12,000×, Scheme 2: 1,300×, Scheme 3: 5,500×). The calculated effective susceptibility for the PhCWG is in the same order of magnitude as those calculated for a regular silicon channel waveguide.[85] However, there is a reduction in the susceptibility when comparing Scheme 2 with Schemes 1 and 3. This can be attributed to the large modal overlap that occurs between the pump and Stokes frequencies of Scheme 2, since they both reside on the fundamental mode of the PhCWG, while the other schemes involve the pump frequency situated on either a mode of a different parity (Scheme 1) or a different polarization (Scheme 2), reducing the modal overlap in these two cases. However, the advantage Scheme 1 holds in the effective susceptibility is outweighed by the increased group velocity its pump mode has due to it residing at a low group index frequency.

6.2.2.2.5 Conclusion

The results presented here show the advantage slow-light structures can exhibit when it comes to nonlinear optical phenomenon; however, no analysis of how the other phenomenon scales with the group velocity has been explored. Theoretical analysis[111] and experiments[104,112] have shown that loss in PhC waveguides scales with the group index. The same increase in the interaction length gained from the slow group velocity will not only increase the light–matter interaction, but also increase the chance for the photon to be scattered by the fabrication disorder or couple to a backscattered mode, thus increasing the optical loss. State-of-the-art electron beam fabrication has been able to reduce the fabrication disorder to the point where propagation losses in PhCWGs approach regular channel waveguides,[104] as theory predicts it would.[113] In addition to linear loss, recent results have shown experimentally that two-photon absorption also scales with the group velocity.[114] Such effects could be reduced with the integration of a reverse-biased p-i-n diode in order to sweep photo-generated carriers out of the optical mode region.[77] In conclusion, we have shown how, through the unique dispersion properties and high modal confinement of silicon, PhCWGs can be used theoretically to enhance SRS in order to build compact, low-power optical amplifiers and lasers.

References

1. V. G. Veselago, Electrodynamics of substances with simultaneously negative values of ε and μ, *Sov. Phys. Usp.* **10**, 509 (1968).
2. J. B. Pendry, Negative refraction makes a perfect lens, *Phys. Rev. Lett.* **85**, 3966 (2000).
3. N. Fang, H. Lee, C. Sun, and X. Zhang, Subdiffraction-limited optical imaging with a silver superlens, *Science* **308**, 534 (2005).
4. E. Cubukcu, K. Aydin, E. Ozbay, S. Foteinopolou, and C. M. Soukoulis, Subwavelength resolution in a two-dimensional photonic-crystal-based superlens, *Phys. Rev. Lett.* **91**, 207401 (2003).
5. A. Grbic and G. V. Eleftheriades, Overcoming the diffraction limit with a planar left-handed transmission-line lens, *Phys. Rev. Lett.* **92**, 117403 (2004).
6. Z. Lu et al., Three-dimensional subwavelength imaging by a photonic-crystal flat lens using negative refraction at microwave frequencies, *Phys. Rev. Lett.* **95**, 153901 (2005).
7. R. A. Shelby, D. R. Smith, and S. Schultz, Experimental verification of a negative index of refraction, *Science* **292**, 77 (2001).
8. A. Houck, J. B. Brock, and I. L. Chuang, Experimental observations of a left-handed material that obeys Snell's law, *Phys. Rev. Lett.* **90**, 137401 (2003).
9. G. Parazzoli, R. B. Greegor, K. Li, B. E. C. Koltenbah, and M. Taniellian, Experimental verification and simulation of negative index of refraction using Snell's law, *Phys. Rev. Lett.* **90**, 107401 (2003).
10. P. V. Parimi, W. T. Lu, P. Vodo, and S. Sridhar, Imaging by flat lens using negative refraction, *Nature* **426**, 404 (2003).
11. N. C. Panoiu and R. M. Osgood, Influence of the dispersive properties of metals on the transmission characteristics of left-handed materials, *Phys. Rev. E* **68**, 016611 (2003).
12. N. Fang and X. Zhang, Imaging properties of a metamaterial superlens, *Appl. Phys. Lett.* **82**, 161 (2003).
13. V. M. Shalaev, Optical negative-index metamaterials, *Nat. Photon.* **1**, 41 (2007).
14. S. Zhang et al., Optical negative-index bulk metamaterials consisting of 2D perforated metal-dielectric stacks, *Opt. Express* **14**, 6778 (2006).
15. G. Dolling et al., Low-loss negative-index metamaterial at telecommunication wavelengths, *Opt. Lett.* **31**, 1800 (2006).
16. M. Notomi, Theory of light propagation in strongly modulated photonic crystals: Refraction-like behavior in the vicinity of the photonic band gap, *Phys. Rev. B* **62**, 10696 (2000).
17. S. Foteinopoulou and C. M. Soukoulis, Electromagnetic wave propagation in two-dimensional photonic crystals: A study of anomalous refractive effects, *Phys. Rev. B* **72**, 165112 (2005).
18. C. Luo, S. G. Johnson, J. D. Joannopoulos, and J. B. Pendry, Subwavelength imaging in photonic crystals, *Phys. Rev. B* **68**, 045115 (2003).
19. S. Xiao, M. Qiu, Z. Ruan, and S. He, Influence of the surface termination to the point imaging by a photonic crystal slab with negative refraction, *Appl. Phys. Lett.* **85**, 4269 (2004).
20. T. Decoopman et al., Photonic crystal lens: From negative refraction and negative index to negative permittivity and permeability, *Phys. Rev. Lett.* **97**, 073905 (2006).
21. Y. Jin and S. He, Canalization for subwavelength focusing by a slab of dielectric photonic crystal, *Phys. Rev. B* **75**, 195126 (2007).
22. A. Berrier et al., Negative refraction at infrared wavelengths in a two-dimensional photonic crystal, *Phys. Rev. Lett.* **93**, 073902 (2004).
23. E. Schonbrun, T. Yamashita, W. Park, and C. J. Summers, Negative-index imaging by an index-matched photonic crystal slab, *Phys. Rev. B* **73**, 195117 (2006); T. Matsumoto, K.-S. Eom, and T. Baba, Focusing of light by negative refraction in a photonic crystal slab superlens on silicon-on-insulator substrate, *Opt. Lett.* **31**, 2786 (2006).
24. S. Zhang et al., Experimental demonstration of near-infrared negative-index metamaterials, *Phys. Rev. Lett.* **95**, 137404 (2005).

25. A. N. Grigorenko et al., Nanofabricated media with negative permeability at visible frequencies, *Nature* **438**, 335 (2005).

26. D. R. Smith and D. Schurig, Electromagnetic wave propagation in media with indefinite permittivity and permeability tensors, *Phys. Rev. Lett.* **90**, 077405 (2003).

27. Y. Zhang, B. Fluegel, and A. Mascarenhas, Total negative refraction in real crystals for ballistic electrons and light, *Phys. Rev. Lett.* **91**, 157404 (2003).

28. Y.-F. Chen, P. Fischer, and F. W. Wise, Negative refraction at optical frequencies in nonmagnetic two-component molecular media, *Phys. Rev. Lett.* **95**, 067402 (2005).

29. T. Taubner et al., Near-field microscopy through a SiC superlens, *Science* **313**, 1595 (2006).

30. A. J. Hoffman et al., Negative refraction in semiconductor metamaterials, *Nat. Mater.* **6**, 946 (2007).

31. K. Sakoda, *Optical Properties of Photonic Crystals*, Springer, New York, 2004.

32. M. Skorobogatiy, G. Bégin, and A. Talneau, Statistical analysis of geometrical imperfections from the images of 2D photonic crystals, *Opt. Express* **13**, 2487 (2005).

33. X. Wang and K. Kempa, Effects of disorder on subwavelength lensing in two-dimensional photonic crystal slabs, *Phys. Rev. B* **71**, 085101 (2005).

34. J. Pacheco Jr., T. M. Grzegorczyk, B.-I. Wu, Y. Zhang, and J. A. Kong, Power propagation in homogeneous isotropic frequency-dispersive left-handed media, *Phys. Rev. Lett.* **89**, 257401 (2002).

35. P. T. Rakich et al., Achieving centimeter-scale supercollimation in a large-area two-dimensional photonic crystal, *Nat. Mater.* **5**, 93 (2006).

36. W. Man, M. Megens, P. J. Steinhardt, and P. M. Chaikin, Experimental demonstration of the photonic properties of icosahedral quasicrystals, *Nature* **436**, 993 (2005).

37. Z. Feng et al., Negative refraction and imaging using 12-fold-symmetry quasicrystals, *Phys. Rev. Lett.* **94**, 247402 (2005).

38. J. Li, L. Zhou, C. T. Chan, and P. Sheng, Photonic band gap from a stack of positive and negative index materials, *Phys. Rev. Lett.* **90**, 083901 (2003).

39. N. C. Panoiu, R. M. Osgood, S. Zhang, and S. R. J. Brueck, Zero-n bandgap in photonic crystal superlattices, *J. Opt. Soc. Am. B* **23**, 506 (2006).

40. D. Schurig et al., Metamaterial electromagnetic cloak at microwave frequencies, *Science* **314**, 977 (2006).

41. U. Leonhardt, Optical conformal mapping, *Science* **312**, 1777 (2006).

42. K. L. Tsakmakidis, A. D. Boardman, and O. Hess, 'Trapped rainbow' storage of light in metamaterials, *Nature* **450**, 397 (2007).

43. D. R. Smith et al., Composite medium with simultaneously negative permeability and permittivity, *Phys. Rev. Lett.* **84**, 4184–4187 (2000).

44. R. A. Shelby et al., Experimental verification of a negative index of refraction, *Science* **292**, 77–79 (2001).

45. L. Ran et al., Microwave solid-state left-handed material with a broad bandwidth and an ultralow loss, *Phys. Rev. B* **70**, 073102 (2004).

46. H. Chen et al., Negative refraction of a combined double S-shaped metamaterial, *Appl. Phys. Lett.* **86**, 151909 (2005).

47. H. Jiang et al., Omnidirectional gap and defect mode of one-dimensional photonic crystals containing negative-index materials, *Appl. Phys. Lett.* **83**, 5386–5388 (2003).

48. N. C. Panoiu et al., Zero-n bandgap in photonic crystal superlattices, *J. Opt. Soc. Am. B* **23**, 506–513 (2006).

49. R. Chatterjee et al., Near-field observation of negative refraction superlensing at the near-infrared, *CLEO 2006*, Long Beach, CA, paper CPDA4.

50. Y. Yuan et al., Experimental verification of zero order bandgap in a layered stack of left-handed and right-handed materials, *Opt. Express* **14**, 2220 (2006).

51. B. S. Song, S. Noda, T. Asano, and Y. Akahane, Ultra-high-Q photonic double-heterostructure nanocavity, *Nat. Mater.* **4**, 207 (2005).

52. E. Kuramochi, M. Notomi, S. Mitsugi, A. Shinya, T. Tanabe, and T. Watanabe, Ultrahigh-Q photonic crystal nanocavities realized by the local width modulation of a line defect, *Appl. Phys. Lett.* **88**, 041112 (2006).

53. P. E. Barclay, K. Srinivasan, and O. Painter, Nonlinear response of silicon photonic crystal microresonators excited via an integrated waveguide and fiber taper, *Opt. Express* **13**, 801 (2005).

54. T. Tanabe, M. Notomi, S. Mitsugi, A. Shinya, and E. Kuramochi, All-optical switches on a silicon chip realized using photonic crystal nanocavities, *Appl. Phys. Lett.* **87**, 151112 (2005).

55. T. Uesugi, B. Song, T. Asano, and S. Noda, Investigation of optical nonlinearities in an ultra-high-Q Si nanocavity in a two-dimensional photonic crystal slab, *Opt. Express* **14**, 377 (2006).

56. V. R. Almeida, C. A. Barrios, R. R. Panepucci, and M. Lipson, All-optical control of light on a silicon chip, *Nature* **431**, 1081 (2004).

57. U. Fano, Effects of configuration interaction on intensities and phase shifts, *Phys. Rev.* **124**, 1866 (1961).

58. S. Fan, Sharp asymmetric line shapes in side-coupled waveguide-cavity systems, *Appl. Phys. Lett.* **80**, 908 (2002).

59. A. R. Cowan and J. F. Young, Optical bistability involving photonic crystal microcavities and Fano line shapes, *Phys. Rev. E* **68**, 046606 (2003).

60. V. Lousse and J. P. Vigneron, Use of Fano resonances for bistable optical transfer through photonic crystal films, *Phys. Rev. B* **69**, 155106 (2004).

61. S. Fan, W. Suh, and J. D. Joannopoulos, Temporal coupled-mode theory for the Fano resonance in optical resonators, *J. Opt. Soc. Am. A* **20**, 569 (2003).

62. H. A. Haus, *Waves and Fields in Optoelectronics*, Prentice-Hall, Englewood Cliffs, NJ, 1984, Chap. 7, p. 197.

63. See for example: B. Maes, P. Bienstman, and R. Baets, Switching in coupled nonlinear photonic-crystal resonators, *J. Opt. Soc. Am. B* **22**, 1778 (2005).

64. X. Yang and C. W. Wong, Coupled-mode theory for stimulated Raman scattering in high-Q/V_m silicon photonic band gap defect cavity lasers, *Opt. Express* **15**, 4763 (2007).

65. M. F. Yanik, S. Fan, and M. Soljačić, High-contrast all-optical bistable switching in photonic crystal microcavities, *Appl. Phys. Lett.* **83**, 2739 (2003).

66. M. Notomi, A. Shinya, S. Mitsugi, G. Kira, E. Kuramochi, and T. Tanabe, Optical bistable switching action of Si high-Q photonic-crystal nanocavities, *Opt. Express* **13**, 2678 (2005).

67. M.-K. Kim, I.-K. Hwang, S.-H. Kim, H.-J. Chang, and Y.-H. Lee, All-optical bistable switching in curved microfiber-coupled photonic crystal resonators, *Appl. Phys. Lett.* **90**, 161118 (2007).

68. M. Soljačić, M. Ibanescu, S. G. Johnson, Y. Fink, and J. D. Joannopoulos, Optimal bistable switching in nonlinear photonic crystals, *Phys. Rev. E* **66**, 055601 (2002).

69. M. Spillane, T. J. Kippenberg, and K. J. Vahala, Ultralow-threshold Raman laser using a spherical dielectric microcavity, *Nature* **415**, 621–623 (2002).

70. T. J. Kippenberg, S. M. Spillane, D. K. Armani, and K. J. Vahala, Ultralow-threshold microcavity Raman laser on a microelectronic chip, *Opt. Lett.* **29**, 1224–1226 (2004).

71. R. Claps, D. Dimitropoulos, V. Raghunathan, Y. Han, and B. Jalali, Observation of stimulated Raman amplification in silicon waveguides, *Opt. Express* **11**, 1731–1739 (2003).

72. R. L. Espinola, J. I. Dadap, R. M. Osgood Jr., S. J. McNab, and Y. A. Vlasov, Raman amplification in ultrasmall silicon-on-insulator wire waveguides, *Opt. Express* **12**, 3713–3718 (2004).

73. T. K. Liang and H. K. Tsang, Efficient Raman amplification in silicon-on-insulator waveguides, *Appl. Phys. Lett.* **85**, 3343–3345 (2004).

74. A. Liu, H. Rong, M. Paniccia, O. Cohen, and D. Hak, Net optical gain in a low loss silicon-on-insulator waveguide by stimulated Raman scattering, *Opt. Express* **12**, 4261–4268 (2004).

75. Q. Xu, V. R. Almeida, and M. Lipson, Time-resolved study of Raman gain in highly confined silicon-on-insulator waveguides, *Opt. Express* **12**, 4437–4442 (2004).

76. O. Boyraz and B. Jalali, Demonstration of a silicon Raman laser, *Opt. Express* **12**, 5269–5273 (2004).

77. R. Jones, H. Rong, A. Liu, A. W. Fang, M. J. Paniccia, D. Hak, and O. Cohen, Net continuous wave optical gain in a low loss silicon-on-insulator waveguide by stimulated Raman scattering, *Opt. Express* **13**, 519–525 (2005); H. Rong, A. Liu, R. Jones, O. Cohen, D. Hak, R. Nicolaescu, A. Fang, and M. Paniccia, An all-silicon Raman laser, *Nature* **433**, 292–294 (2005); H. Rong, R. Jones, A. Liu, O. Cohen, D. Hak, A. Fang, and M. Paniccia, A continuous-wave Raman silicon laser, *Nature* **433**, 725–728 (2005).

78. O. Boyraz and B. Jalali, Demonstration of directly modulated silicon Raman laser, *Opt. Express* **13**, 796–800 (2005).

79. R. Jones, A. Liu, H. Rong, M. Paniccia, O. Cohen, and D. Hak, Lossless optical modulation in a silicon waveguide using stimulated Raman scattering, *Opt. Express* **13**, 1716–1723 (2005).

80. X. Yang and C. W. Wong, Design of photonic band gap nanocavities for stimulated Raman amplification and lasing in monolithic silicon, *Opt. Express* **13**, 4723–4730 (2005).

81. L. Florescu and X. Zhang, Semiclassical model of stimulated Raman scattering in photonic crystals, *Phys. Rev. E* **72**, 016611 (2005).

82. J. F. MaMillan, X. Yang, N. C. Paniou, R. M. Osgood, and C. W. Wong, Enhanced stimulated Raman scattering in slow-light photonic crystal waveguides, *Opt. Lett.* **31**, 1235–1237 (2006).

83. D. Dimitropoulos, B. Houshmand, R. Claps, and B. Jalali, Coupled-mode theory of the Raman effect in silicon-on-insulator waveguides, *Opt. Lett.* **28**, 1954–1956 (2003).

84. M. Krause, H. Renner, and E. Brinkmeyer, Analysis of Raman lasing characteristics in silicon-on-insulator waveguides, *Opt. Express* **12**, 5703–5710 (2004).

85. X. Chen, N. C. Panoiu, and R. M. Osgood, Theory of Raman-mediated pulsed amplification in silicon-wire waveguides, *IEEE J. Quantum Electron.* **42**, 160–170 (2006).

86. V. E. Perlin and H. G. Winful, Stimulated Raman scattering in nonlinear periodic structures, *Phys. Rev. A* **64**, 043804 (2001).

87. B. Min, T. J. Kippenberg, and K. J. Vahala, Compact, fiber-compatible, cascaded Raman laser, *Opt. Lett.* **28**, 1507–1509 (2003).

88. D. Braunstein, A. M. Khazanov, G. A. Koganov, and R. Shuker, Lowering of threshold conditions for nonlinear effects in a microsphere, *Phys. Rev. A* **53**, 3565–3572 (1996).

89. T. J. Johnson, M. Borselli, and O. Painter, Self-induced optical modulation of the transmission through a high-Q silicon microdisk resonator, *Opt. Express* **14**, 817–831 (2006).

90. H. W. Tan, H. M. van Driel, S. L. Schweizer, and R. B. Wehrspohn, Influence of eigenmode characteristics on optical tuning of a two-dimensional silicon photonic crystal, *Phys. Rev. B* **72**, 165115 (2005).

91. R. A. Soref and B. R. Bennett, Electrooptical effects in silicon, *IEEE J. Quantum Electron.* **23**, 123–129 (1987).

92. E. Palick, ed., *Handbook of Optical Constants of Solids*, Academic Press, Boston, MA, 1985.

93. M. Dinu, F. Quochi, and H. Garcia, Third-order nonlinearities in silicon at telecom wavelengths, *Appl. Phys. Lett.* **82**, 2954–2956 (2003).

94. A. Cutolo, M. Iodice, P. Spirito, and L. Zeni, Silicon electro-optic modulator based on a three terminal device integrated in a low-loss single-mode SOI waveguide, *J. Lightw. Technol.* **15**, 505–518 (1997).

95. S. Sze, *Physics of Semiconductor Devices*, 2nd edn. John Wiley & Sons, New York, 1981.

96. G. Cocorullo, F. G. Della Corte, and I. Rendina, Temperature dependence of the thermo-optic coefficient in crystalline silicon between room temperature and 550 K at the wavelength of 1523 nm, *Appl. Phys. Lett.* **74**, 3338–3340 (1999).

97. R. Claps, D. Dimitropoulos, Y. Han, and B. Jalali, Observation of Raman emission in silicon waveguides at 1.54 μm, *Opt. Express* **10**, 1305–1313 (2002).

98. J. I. Dadap, R. L. Espinola, R. M. Osgood, S. J. McNab, and Y. A. Vlasov, Spontaneous Raman scattering in ultrasmall silicon waveguides, *Opt. Lett.* **29**, 2755–2757 (2004).

99. H. Rong, Y.-H. Kuo, S. Xu, A. Liu, R. Jones, M. Paniccia, O. Cohen, and O. Raday, Monolithic integrated Raman silicon laser, *Opt. Express* **14**, 6705–6712 (2006).

100. A. Fainstein, B. Jusserand, and V. Thierry-Mieg, Raman scattering enhancement by optical confinement in a semiconductor planar microcavity, *Phys. Rev. Lett.* **75**, 3764–3767 (1995).

101. S. V. Gaponenko, Effects of photon density of states on Raman scattering in mesoscopic structures, *Phys. Rev. B* **65**, 140303 (2002).

102. S. O. Konorov, D. A. Akimov, A. N. Naumov, A. B. Fedotov, R. B. Miles, J. W. Haus, and A. M. Zheltikov, Coherent anti-stokes Raman scattering of slow light in a hollow planar periodically corrugated waveguide, *JETP Lett.* **75**, 66 (2002).

103. H. Oda, K. Inoue, N. Ikeda, Y. Sugimoto, and K. Asakawa, Observation of Raman scattering in GaAs photonic-crystal slab waveguides, *Opt. Express* **14**, 6659–6667 (2006).

104. E. Kuramochi, M. Notomi, S. Hughes, A. Shinya, T. Watanabe, and L. Ramunno, Disorder-induced scattering loss of line-defect waveguides in photonic crystal slabs, *Phys. Rev. B* **72**, 161318 (2005).

105. M. Notomi, K. Yamada, A. Shinya, J. Takahashi, C. Takahashi, and I. Yokohama, Extremely large group-velocity dispersion of line-defect waveguides in photonic crystal slabs, *Phys. Rev. Lett.* **87**, 253902 (2001).

106. S. Johnson and J. Joannopoulos, Block-iterative frequency-domain methods for Maxwell's equations in a planewave basis, *Opt. Express* **8**, 173–190 (2001).

107. B. Jalali, R. Claps, D. Dimitropoulos, and V. Raghunathan, Light generation, amplification, and wavelength conversion via stimulated Raman scattering in silicon microstructures, *Top. Appl. Phys.* **94**, 199 (2004).

108. R. Loudon, The Raman effect in crystals, *Adv. Phys.* **13**, 423 (1964).

109. A. W. Snyder and J. D. Love, *Optical Waveguide Theory*, Chapman-Hall, London, U.K., 1983.

110. E. Dulkeith, S. J. McNab, and Y. A. Vlasov, Mapping the optical properties of slab-type two-dimensional photonic crystal waveguides, *Phys. Rev. B* **72**, 115102 (2005).

111. S. Hughes, L. Ramunno, J. F. Young, and J. E. Sipe, Extrinsic optical scattering loss in photonic crystal waveguides: Role of fabrication disorder and photon group velocity, *Phys. Rev. Lett.* **94**, 033903 (2005).

112. L. O'Faolain, T. P. White, D. O'Brien, X. Yuan, M. D. Settle, and T. F. Krauss, Dependence of extrinsic loss on group velocity in photonic crystal waveguides, *Opt. Express* **15**, 13129–13138 (2007).

113. M. L. Povinelli, S. G. Johnson, E. Lidorikis, J. D. Joannopoulos, and M. Soljačić, Effect of a photonic band gap on scattering from waveguide disorder, *Appl. Phys. Lett.* **84**, 3639 (2004).

114. Y. Hamachi, S. Kubo, and T. Baba, Low dispersion slow light and nonlinearity enhancement in lattice-shifted photonic crystal waveguide, QTuC1, *CLEO/QELS 2008*, San Jose, CA, May 2008.

115. R. Chatterjee, N. C. Panoiu, K. Liu et al., Achieving subdiffraction imaging through bound surface states in negative refraction photonic crystals in the near-infrared range, *Phys. Rev. Lett.* **100**, 187401 (2008).

116. X. Yang et al., Observation of femtojoule optical bistability involving Fano resonances in high-Q/V_m silicon photonic crystal nanocavities, *Appl. Phys. Lett.* **91**, 051113 (2007).

117. S. Kocaman, R. Chatterjee, N. C. Panoiu, J. F. McMillan, M. B. Yu, R. M. Osgood, D. L. Kwong, and C. W. Wong, Observations of zero-order bandgaps in negative-index photonic crystal superlattices at the near-infrared, *Phys. Rev. Lett.* **102**, 203905 (2009).

7

Functional Devices in Photonic Crystals for Future Photonic Integrated Circuits

A. Shinya
T. Tanabe
E. Kuramochi
H. Taniyama
S. Kawanishi
M. Notomi

7.1 Introduction

Photonic crystal (PhC) is a promising candidate as a platform on which to construct devices with dimensions of several wavelengths for future photonic integrated circuits [1] (Figure 7.1). A PhC with a photonic band gap (PBG) functions as a light insulator in the PBG. A line-defect and a point-defect cavity in the PhC form a waveguide (WG) and a resonator that does not leak light, respectively. In this chapter, we discuss PhC-WGs and resonators and a system in which they are coupled. This unique system is one in which single-mode WGs are effectively coupled with an ultrasmall cavity with a high-Q factor. The applications of this system include ultrasmall optical passive devices, such as optical lines, resonators, and filters. Moreover, since the photon density in the cavity is extremely high owing to its smallness and effective coupling to PhC-WGs, which results in a large optical nonlinearity, ultrasmall optical active devices such as optical switches and memories can be realized. Moreover, it is easy to integrate these elements on one-chip PhC platform.

We describe a low-loss WG and ultrasmall resonators with a high-Q factor in Sections 7.2 and 7.3, respectively. In Section 7.4, we consider a way of coupling WGs effectively with resonators, and we design an ultrasmall multi-port channel drop filter as a signal multiplexer. In Section 7.5, we discuss the

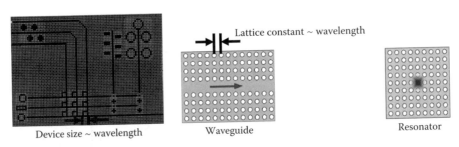

FIGURE 7.1   Photonic-crystal-based integrated circuit.

use of a coupled system as an optical logic gate and an optical memory. Their combination enables us to realize all-optical digital processing, which we describe in Section 7.6.

7.2 Photonic Crystal Waveguide

Our optical circuit is based on a two-dimensional PhC with a triangular air-hole lattice. Figure 7.2a is a scanning electron microscope (SEM) image of a PhC fabricated on a silicon-on-insulator (SOI) substrate by a combination of electron-beam lithography and dry etching. This crystal has a frequency band called a PBG where light cannot exist in the PhC, as shown in Figure 7.2b. This means a line defect in the PhC can function as a WG that does not leak light in the in-plane direction. However, since there is no PBG in the thickness direction, it is necessary to consider out-plane radiation loss.

7.2.1 Dispersion of Photonic Crystal Waveguide

Figure 7.3a shows a dispersion curve of a PhC-WG. The dotted lines show the waveguiding mode, and the solid line is the light line of the cladding layer. The increment in the light line is inversely proportional to the refractive index of the cladding. The modes above the light line leak toward the cladding layer, and the modes outside the PBG leak toward the PhC cladding when the WG is bent. In short, only the modes below the light line and in the PBG function as WGs. Here, it is important to use a low refractive index material as a cladding layer. Normally, an air-bridge structure, whose upper and lower cladding layers is air, is used to suppress light leakage toward the cladding. Figure 7.3b is a SEM image

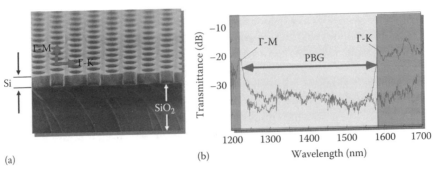

FIGURE 7.2   Two-dimensional PhC. (a) SEM of PhC on SOI substrate. (b) PGB of triangular air-hole lattice PhC.

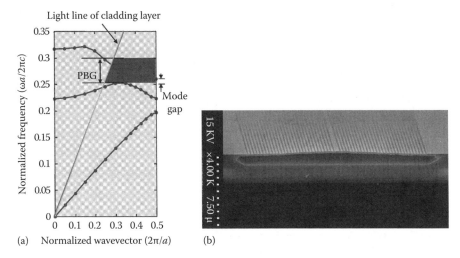

(a)   Normalized wavevector ($2\pi/a$)   (b)

FIGURE 7.3   PhC-WG. (a) Dispersion curve of PhC-WG. The dotted lines are the waveguiding-mode, and the solid line is the light line of the cladding layer. The mode below the light line and in the PGB functions as a WG. (b) SEM image of air-bridge PhC-WG. The air-hole size fluctuation is 2% or less and its propagation loss is less than 2 dB/cm.

of a PhC-WG. The air hole size fluctuation is 2% or less. We can achieve a WG with a very low loss of less than 2 dB/cm [2].

7.2.2 Slow Light

Promising applications of the PhC-WG are optical delay lines [3,4], all-optical buffers [5,6], and optical storage devices [7]. As shown in Figure 7.3a, a PhC-WG has a very large dispersion. As a consequence, it has an extremely large group refractive index in the frequency region where the dispersion curve is flat, and this results in very slow propagation in the PhC-WG at speeds of less than $c/100$. But the large dispersion results in the deformation of the signal pulse propagating in the WG. Some coupled PhC-WGs have been proposed that overcome this problem [3,8]. Figure 7.4 shows an example of a coupled PhC-WG for slow light and an experimental result with respect to its group velocity. This device can reduce the light speed to about $0.017c$ [8].

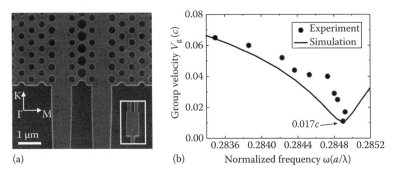

(a)   (b)   Normalized frequency $\omega(a/\lambda)$

FIGURE 7.4   Coupled-PhC-WG for slow light. (a) SEM image of coupled PhC-WG. (b) Experimental group velocity result. The light speed is about $0.017c$. (From Huang, S.-C. et al., *Opt. Express*, 15, 3543, 2007. With permission. Copyright 2007, Optical Society of America.)

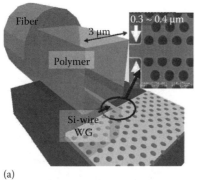
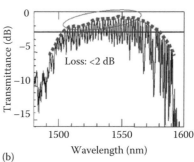

(a)   (b)

FIGURE 7.5 Spot size converter for highly efficient coupling between PhC and fiber. (a) Schematic diagram of an adiabatic coupling system consisting of a polymer WG and a tapered silicon-wire WG. The spot size conversion loss from the fiber to the silicon-wire is about 0.8 dB. (b) Transmission spectrum of the coupling between the silicon-wire WG and the PhC-WG. The coupling loss is less than 2 dB.

7.2.3 Highly Efficient Coupling between Photonic Crystal-Waveguide and Fiber

We must achieve highly efficient coupling between a fiber and a PhC-WG because we want to introduce high-power light into the PhC. Figure 7.5a is a schematic diagram of an adiabatic coupling system consisting of a polymer WG and a tapered silicon wire WG. Figure 7.5b shows a transmission spectrum of the coupling between the silicon wire WG and the PhC-WG. The coupling loss is less than 2 dB. The spot size conversion loss from the fiber to the silicon wire is about 0.8 dB [9], so we can input light from a fiber into the PhC-WG with a coupling loss of less than 3 dB [10].

7.3 Photonic Crystal Resonators

We must confine light in a very small space if we are to realize high-density photonic integrated circuits. A PhC cavity offers the potential for realizing ultrasmall photonic devices. The quality of the light confinement is estimated in terms of the Q-factor of the resonator. With conventional photonic resonators it is very difficult to minimize the volume while keeping the Q-factor high. This is because the Q-factor decreases as the volume decreases, thus limiting the use of light confinement by total internal reflection, as used in conventional photonic devices. By contrast, there is no such limitation with a PhC cavity because the light can be completely confined by the PBG. However, since our PhC is a two-dimensional structure, we must also consider the radiation loss in the cladding layer direction as with a PhC-WG.

7.3.1 Point-Defect Resonator

Figure 7.6a is a single-point-defect cavity that suppresses the radiation loss [11,12]. One air hole is removed and the six surrounding holes shrink and are pushed away from the center. This cavity has a hexagonally symmetrical mode profile called the hexapole mode. Since the phases of adjacent lobes are opposite, destructive interference suppresses the vertically radiating field [13].

The vertical Q_V and V_{eff} are plotted in Figure 7.6b. Here, Q_V is the Q-factor of an isolated cavity in a two-dimensional PhC, and is inversely proportional to the radiation loss, and V_{eff} is the effective mode volume. When the resonant frequency is near the PBG edge, the Q_V has a maximum value of about 2,000,000, and the V_{eff} of 1.4 is even smaller than that of a conventional resonator, which is about 5. The Q-factor of a fabricated PhC cavity is about 320,000 [14].

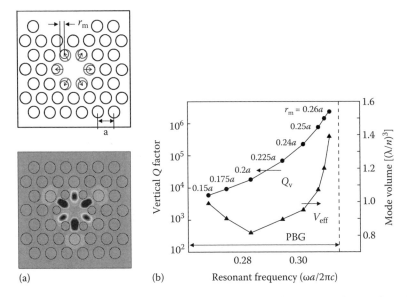

(a)   (b)

FIGURE 7.6 Single-point-defect cavity. (a) Structure of single-point-defect cavity and its mode profile. This cavity has a hexagonally symmetrical mode profile. Since the phases of adjacent lobes are opposite, the destructive interference suppresses the vertically radiating field. (b) Q-Factor and mode volume. When the resonant frequency is near the PBG edge, the Q_v has a maximum value of about 2,000,000, and the V-effective of 1.4 is still smaller than that of a conventional resonator (about 5). The Q-factor of a fabricated cavity is about 320,000.

To understand the high-Q mechanism near the gap edge, we analyzed the field distributions in a Fourier space and the results are shown in Figure 7.7. The white circle indicates the light cone boundary. In this circle, light in the cavity radiates toward the cladding layer. As can be seen, the Fourier component in the light cone is greatly reduced when $r_m = 0.26a$. When the resonant frequency is near the gap edge, the light confinement becomes weak, and the mode profile becomes delocalized. As a result, the contrast of the mode distribution in k-space is enhanced, and the Fourier component near the light cone boundary is greatly reduced. The above mechanisms can be employed for multi-point–defect cavities.

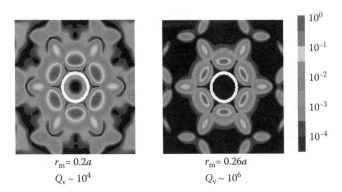

$r_m = 0.2a$ $r_m = 0.26a$
$Q_v \sim 10^4$ $Q_v \sim 10^6$

FIGURE 7.7 Fourier-space representation of field profile of single-point-defect cavity. The white circle indicates the light cone boundary. In this circle, light in the cavity radiates toward the cladding layer. The Fourier component near the light cone boundary is greatly reduced and the Q-factor becomes large when the resonant frequency is near the gap edge.

7.3.2 Resonator Using Mode Gap Confinement

It is known that there are two schemes for confining light in a PhC cavity. One is to use a PBG as described above; the other is to use the mode gap of a PhC-WG [15–19]. As shown in Figure 7.3a, a line-defect WG has a large dispersion and a mode gap near the PBG edge of the longer wavelength side. The position of the mode gap can be shifted to a shorter wavelength by narrowing the WG or reducing the size of the PhC [10,20,21]. That is, we can form potential barriers in the WG by changing the geometrical parameters and can confine light between the barriers. The advantage of the mode gap confinement is that there is no structural scattering at the edge of the cavity because the cavity is not terminated by the air holes. This is quite different from the situation with point defect cavities. As a result, the Fourier component in the light cone is greatly reduced and the cavity has a very large Q-factor.

Figure 7.8a is a cavity structure realized by using mode gap confinement [18]. The air holes surrounding the cavity (A, B, and C) are shifted away from the center of the WG by distances of x, $2x/3$, and $x/3$, respectively. Figure 7.8b shows the calculated magnetic field of the resonant mode. It reveals that the resonant mode was strongly confined near the center of the cavity. Figure 7.8c shows the resonant wavelength, Q_V, and modal volume V_{eff} calculated by FDTD as a function of hole shift x. The maximum Q_V value was 7×10^7 obtained at $x = 9$ nm, which is a very high theoretical Q_V value for a PhC cavity, and the minimum V_{eff} of 1.4 is almost the same as that of the point-defect cavity. The theoretical Q-factor is much higher than that of the point-defect cavity, and we can actually realize a Q-factor of over 1,000,000 [19].

7.3.3 Trapping Light

The light input into the cavity is output after it has remained in the cavity for a time. Since this time is inversely proportional to the Q-factor, a cavity with a high-Q factor can slow the light speed. Figure 7.9 shows a pulse delay result obtained by using a mode gap confinement cavity. The black and gray curves are the outputs from samples without and with a cavity, respectively. We can clearly observe a time delay between the gray and black curves. The delay time is about 1.45 ns, which is one of the highest values yet obtained for slow-light systems in PhC. The estimated transit speed from the input to output WGs,

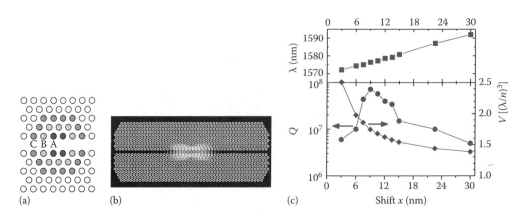

FIGURE 7.8 PhC cavity using a mode-gap confinement effect. (a) Cavity structure using mode gap confinement and its mode profile. The air holes surrounding the cavity, A, B and C, shift away from the center of the WG by distances of x, $2x/3$, $x/3$, respectively. (b) Field profile of the resonant mode. (c) Resonant wavelength, Q, and modal volume V calculated by FDTD as a function of hole shift x. The maximum Q value was 7×10^7 obtained at $x = 9$ nm. The Q-factor of the fabricated cavity is over 1,000,000.

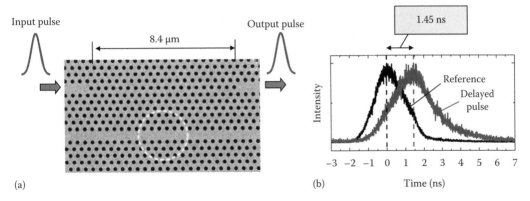

(a)   (b)   Time (ns)

FIGURE 7.9 Experimental light trapping result. (a) SEM image of high-Q resonator using mode-gap confinement. A cavity is arranged in the dotted circle. Two PhC-WGs are connected to the cavity for the input/output WGs. The distance between them is 8.4 μm, (b) Experimental pulse delay result. The black and gray curves are the outputs from the sample without and with a cavity, respectively. The delay time is about 1.45 ns and the estimated transit speed between two WGs is 5800 m/s, which means that the pulse speed is reduced to $2 \times 10^{-5}c$.

which is a distance of 8.4 μm, is 5800 m/s. This means that the pulse speed is reduced to $2 \times 10^{-5}c$ [19]. This is the slowest value reported for any dielectric slow-light medium.

7.4 Coupled Resonator–Waveguide System

Next, we discuss a resonator–WG coupled system. The combination of WGs and cavities is a key feature of a photonic integrated circuit. One promising device is a resonant tunneling filter, which can form a basic element of a PhC-based wavelength division multiplexing (WDM) system. This filter can also be part of an optical switch and a memory, as described later. In this section, we describe a way of coupling PhC-WGs effectively with PhC cavities.

7.4.1 Two-Port Resonant Tunneling Filter

In a resonant tunneling system, the transmittance (T) and the light power in the cavity (P) depend on the balance between T, Q_V, and Q_H. Here, Q_H is a horizontal Q-factor that is inversely proportional to the coupling efficiency between the cavity and the WGs.

The total balance of the system is represented as $1/Q_T = 1/Q_V + 1/Q_H$, $T = (Q_T/Q_H)^2$, and $P = Q_T^2/Q_H$. Here, Q_T is the total Q-factor of the system. The cavity enhances the interaction between the light and the material when the Q_T value is large. However, T decreases as Q_T increases. On the other hand, a high T is an important factor when the systems are cascaded. Since Q_V is mainly determined by the cavity structure, the total balance of the system is determined by Q_H. In other words, it is very important to control Q_H.

Since the coupling between the cavity and the WG modes is determined by the superposition of their fields, we must consider the field profiles of the cavity and the WGs if we are to control the Q_H. For example, the fields of a four-point-defect cavity and a terminated WG expand in ΓM directions. The fields are shown simply by a circle and arrows and a rectangle and arrows, respectively in Figure 7.10. In Figure 7.10b, the WGs and cavity are arranged on the same ΓK axis. In Figure 7.10c, the cavity and WGs are not on the same axis. Although all the structures can constitute a resonant tunneling filter, it is easy to control the Q_H by using the structures shown in Figure 7.10c because their expanding fields overlap on the same ΓM axis [22]. We use the ΓM coupling structure in the following sections.

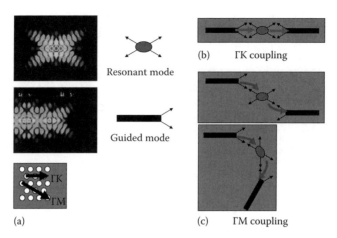

FIGURE 7.10 Two-port resonant tunneling filter. (a) Field profiles of four-point-defect PhC cavity and PhC-WG. They are shown simply by a circle and arrows and a rectangle and arrows, respectively. (b) ΓK coupling structure. The WGs and cavity are arranged on the same ΓK axis. (c) ΓM coupling structure. The cavity and WG are not on the same axis. These structures easily control the Q_H because their expanding fields overlap on the same ΓM axis.

7.4.2 Three-Port Resonant-Tunneling Filter

A two-port system is a very useful filter. However, it cannot be used to realize a multi-port filter because any light that does not resonate with the cavity is reflected back to the input port. So we first develop a three-port resonant tunneling filter that has a through WG. The simplest way to construct such a filter is to arrange one cavity between parallel bus and drop WGs. The input light from the bus WG is output to a drop WG when the light resonates with the cavity, and is output to a through WG when it does not resonate. However, the dropping efficiency is only 25% because the light is divided into the four ends of the bus and the drop WGs when the light is output from the cavity. Some schemes have been reported that overcome this problem [23–29]. In this section, we introduce our simple solution [27,29].

Figure 7.11a is a schematic showing the structure of a three-port resonant tunneling filter [27]. The through WGs (P2) and additional WGs (P4) are attached to a two-port system. The combination of P1 and P2 is the same as that of P3 and P4, to keep the structure symmetric. The resonant frequency is tuned in a frequency band where there is a waveguiding mode in P1 and P3 but not in P2 and P4, as shown in Figure 7.11b. When the frequency is resonant, the equivalent filters can be regarded as a two-port system, and the light is output solely from P3, as shown in Figure 7.11c. When the frequency is in the common band where there are waveguiding modes in both P1 and P2, the light is output from only P2, because there is no resonating mode and the light cannot reach P3, as shown in Figure 7.11d.

7.4.3 Multi-Port Resonant Tunneling Filter

Next, we construct a multi-port filter. The bus WGs of three-port filters are connected in series. Figure 7.12a shows the fabricated five-port channel drop filter on an SOI substrate [29]. We used six different PhCs to control the propagation band of the WG. Figure 7.12b shows the transmission spectra at each drop port. Five signals with different frequencies are output from each drop port. All the transmittance values are about 80%. The device functions in the L-band and the device size is only 18 μm.

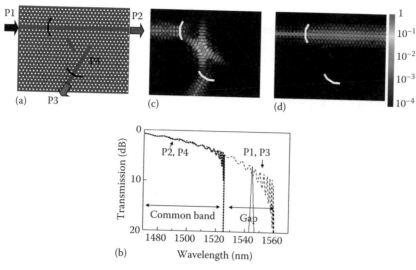

FIGURE 7.11 Three-port resonant tunneling filter. (a) Schematic structure of a three-port resonant tunneling filter. (b) Transmission spectra of the WG and the resonator. The resonant frequency is tuned in a frequency band where there is a waveguiding mode in P1 and P3 but not in P2 and P4. (c) When the frequency is resonant, the equivalent filters can be regarded as a two-port system, and the light is output from only P3. (d) When the frequency is in the common band where there are waveguiding modes in both P1 and P2, the light is output from only P2. (From Shinya, A. et al., *Opt. Express*, 13, 4202–4209, 2005. With permission. Copyright 2005, Optical Society of America.)

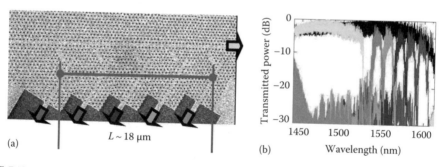

FIGURE 7.12 Five-port channel drop filter. (a) SEM image of a fabricated five-port channel drop filter on an SOI substrate. The device size is only 18 μm. (b) Transmission spectra at each drop port. Five signals with different frequencies are output from each drop port. All the transmittance values are about 80% in the L-band. (From Shinya, A. et al., *Opt. Express*, 14, 12394, 2006. With permission. Copyright 2006, Optical Society of America.)

7.5 Nonlinear Optical Switch and Memory

Our coupled resonator–WG system can effectively confine light in an ultrasmall and high-Q cavity. The very high photon density in the cavity effectively induces optical nonlinearity, which enables us to realize optical switching and memory functions. The intense light in the cavity can control the transmittance of the system because it changes the refractive index of the cavity, and shifts the resonant frequency. The shifted value of the resonant frequency is proportional to Q/V_{eff}. The value required to switch the state from ON (high T) to OFF (low T) is determined by the width of the resonant mode, and is inversely proportional to Q. Therefore, the switching power is proportional to V/Q^2. As described, this PhC cavity value is much smaller than that of a conventional resonator. That is, our coupled resonator–WG system can function as a nonlinear switch with lower power than a conventional optical switch.

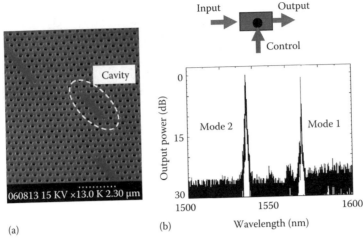

FIGURE 7.13   Two-port resonant tunneling filter for nonlinear operation. (a) SEM image of a fabricated two-port resonant tunneling filter on an SOI substrate. A four-point-defect cavity is coupled with two WGs. (b) Transmission spectrum of the filter. There are two resonating modes. We modulate the signal light resonating with mode 2 by using the control light resonating with mode 1.

Figure 7.13 shows a fabricated two-port resonant tunneling filter on an SOI substrate and its measured transmission spectrum. A four-point-defect cavity is coupled with two WGs in the ΓM direction. As can be seen, there are two resonating modes. So the intense light resonating with mode 1 changes the refractive index of the cavity, and shifts the resonant frequency of mode 2. That is, the control light resonating with mode 1 can modulate the signal light resonating with mode 2 [30]. This section describes refractive index modulation by using the thermo-optic (TO) effect and carrier plasma effect to achieve switching and memory functions, respectively.

7.5.1 Thermo-Optic Effect

We can use the TO effect to control the cavity resonance. This effect induces a red shift in the cavity resonance. Figure 7.14a shows the input power dependence of the resonant frequency. As the input power is increased, the resonant frequency shifts toward a longer wavelength, because of the TO effect caused by the two-photon absorption in silicon. Since the input light is scanned from a short to a long wavelength, the measured spectra are deformed toward longer wavelengths [30].

We can also observe the bistable performance of the two-port resonant tunneling filters. Figure 7.14b shows bistable transmission spectra for several detuning values of the operation wavelengths. The minimum limit of detuning for bistable operation is about 80 pm, and the minimum power is 40 μW. This shows that we can use optical memory with very low power [30].

We can also operate the resonant tunneling filter as an optical transistor, as shown in Figure 7.15. A control light is input as a bias light with a power slightly below the threshold of the hysteresis curve. And an AC signal light with a low power of 0.7 dB is input to modulate the bias light. The AC signal is copied to the bias light and amplified to the power of the 3 dB AC signal [30].

7.5.2 Carrier Plasma Effect

We can utilize the carrier-plasma effect to control the cavity resonance. This effect induces a blue shift in the cavity resonance [31]. Its response time is faster than the TO effect. We used two continuous wave signals with two detuning values. To modulate these signals, we used a 6.4 ps pulsed light as a control

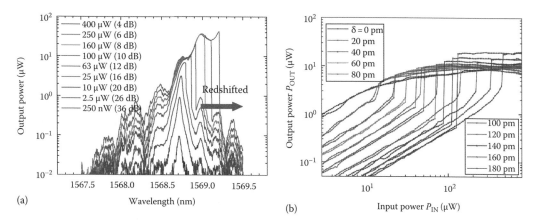

FIGURE 7.14 TO effect. (a) Input power dependence of resonant frequency. As the input power is increased, the resonant frequency shifts toward a longer wavelength due to two-photon absorption in silicon. (b) Bistable transmission spectra for several de-tuning operations. The minimum limit of de-tuning for bistable operation is about 80 pm, and the minimum power is 0.4 mW. (From Notomi, M. et al., *Opt. Express*, 13, 2678, 2005. With permission. Copyright 2005, Optical Society of America.)

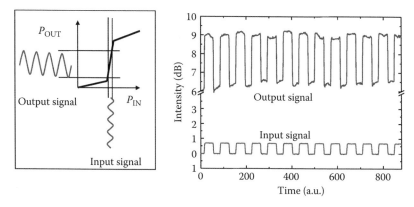

FIGURE 7.15 Transistor operation. AC signal light with small power of 0.7 dB is inputted. The AC signal is amplified to 3 dB. (From Notomi, M. et al., *Opt. Express*, 13, 2678, 2005. With permission. Copyright 2005, Optical Society of America.)

light. Figure 7.16a shows an experimental result. The output signal indicated by the solid line was in the ON state after the control pulse had been applied when the signal detuning was $\delta = -0.45$ nm. The signal indicated by the dotted line is the reverse of $\delta = 0.01$ nm. It show that two-port resonant tunneling filter can function as AND and NOT gates with a switching energy of about 100 fJ and a switching window with a width of about 100 ps. The switch-on time is limited by the Q-factor of the cavity, where the switching recovery time is determined by the effective carrier relaxation time. According to our numerical calculation, the effective carrier relaxation time is 80 ps, which is surprisingly short. This results from the small device size because diffusion plays a dominant role.

Figure 7.16b is an experimental result related to the memory operation [32]. We use the input CW light as a bias light and set the power slightly below the bistable threshold power of 0.4 mW. And we use the control light as a set pulse. If there is no set pulse, the output signal level remains low. When a set pulse is applied, the output signal turns ON. This system stayed in the ON state. When the bias light is reset, the output signal returns to a low level. The measured switching time was ~100 ps with a set pulse energy of 74 fJ.

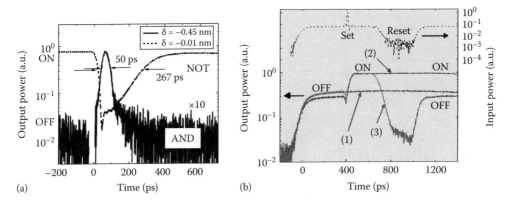

FIGURE 7.16 Carrier-plasma effect. (a) AND and NOT operation. The detuning values are set at −0.45 nm for AND operation and 0.01 nm for NOT operation. The switching energy is about 100 fJ and the switching window is about 100 ps wide. (b) Memory operation. The bias power is set at slightly below the bistable threshold power of 0.4 mW. The measured switching time was ~100 ps with a set pulse energy of 74 fJ. (a) Used with permission. Copyright American Institute of Physics. (b) From Tanabe, T. et al., *Opt. Letters*, 30, 2575, 2005. With permission. Copyright 2005, Optical Society of America.

7.6 Optical Digital Circuit

As described above, our resonator–WG coupled system can function as logic gates and optical memories. Their combination enables us to achieve simple optical digital processing. In this section, we design optical flip-flop (FF) circuits. Electrical FF circuits are indispensable in electrical circuits for signal processing. By contrast, optical FFs are not yet practical, because most are composed of optical devices combined with optical fibers. These fibers constitute feedback loops designed to achieve a memory function. However, a long loop results in a slow response. Since the PhC ultimately shortens the loop, it can be used to minimize the optical circuit and to increase the speed of optical devices.

One of the functions we want to achieve in all optical devices is that of a D(delayed)-FF composed of three SR(set/reset)-FFs. Since it can synchronize input digital data with the system's clock, optical D-FF is expected to be an indispensable device in an OTDM system [33], and a combination of D-FFs can be widely used as a shift register and a serial parallel converter. In this section, we design an optical SR-FF and a sequential circuit that has a D-FF function.

7.6.1 Set-Reset Flip-Flop Circuit

A SR-FF is composed of two switches (SWs) and their ON/OFF states are switched by SR pulses. When the set pulse is applied, one SW turns ON and the SR-FF outputs a signal (Q). This ON state is maintained after the set pulse has been cut. When the reset pulse is applied, the ON-SW turns OFF and another SW turns ON, and the SR-FF outputs a signal ($\bar{Q}$) instead of Q.

Figure 7.17a is a schematic diagram of an SR-FF on a PhC [34]. It contains three WGs, WG_B, WG_{SR}, and WG_{out} for the bias light, SR pulse, and output signals, respectively, and two different cavities (C1 and C2). C1 and C2 have one identical resonant frequency ($\lambda 1$) and two different resonant frequencies ($\lambda 2$ for C1 and $\lambda 3$ for C2), and exhibit bistability for the bias light. The WGs are designed so that $\lambda 1$ light can propagate in WG_B and WG_{Out}, and $\lambda 2$ and $\lambda 3$ can propagate in WG_B and WG_{SR}. We use $\lambda 1$ resonant modes with a symmetrical mode profile for C1 and an asymmetrical profile for C2 so that the direct coupling between them is very weak. WG_B is connected to C1 and C2, and the bias light power can maintain a bistable ON state for one cavity, but not for two. The bias light is output from the WG_{out}, which is connected to the ON state cavity.

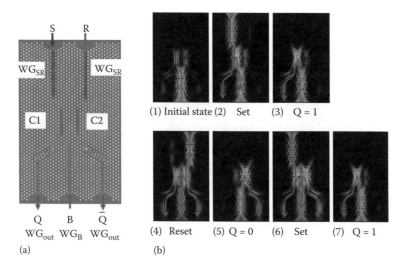

FIGURE 7.17   SR-FF circuit. (a) Schematic diagram of SR-FF based on 2D-PhC. It contains two different cavities (C1 and C2) and three kinds of WGs (WG$_B$, WG$_{SR}$ and WG$_{out}$) for the bias light, SR pulse and output signal, respectively. (b) Calculated result obtained with 2D-FDTD. (1) Initial state. (2) Set pulse is applied. The signal (Q) is output from the left WG$_{out}$. (3) Set pulse is cut. Q is output. (4) Reset pulse is applied. Q is cut and $\bar{Q}$ is output from the right WG$_{out}$. (5) Reset pulse is cut. $\bar{Q}$ is output. (6) Set pulse is applied. $\bar{Q}$ is cut and Q is output from the left WG$_{out}$. (7) Set pulse is cut. The situation is the same as (3). The output ports for Q and $\bar{Q}$ are switched by the SR pulses. This switching operates at 44 GHz. (From Notomi, M. et al., *Opt. Express*, 15, 17458–17481, 2007. With permission. Copyright 2007, Optical Society of America.)

Figure 7.17b is a simulated result obtained with 2D-FDTD. (1) Initial state. Bias light (λ1) is input from WG$_B$. C1 and C2 are in the OFF state because the bias light is insufficient to turn them ON. (2) Set pulse (λ2) is applied. C1 turns ON and the signal (Q) is output from the left WG$_{out}$. (3) Set pulse is cut. Since the bias light is concentrated in C1, it is sufficient to keep C1 ON. (4) Reset pulse (λ3) is applied. C2 turns ON and the bias light is divided into C1 and C2. Since the bias light in C1 is insufficient to keep the C1 ON, C1 gradually turns OFF. The signal Q is cut and $\bar{Q}$ is output from the right WG$_{out}$. (5) Reset pulse is cut. Since the bias light is concentrated in C2, its power is sufficient to keep C2 ON. (6) Set pulse is applied. C1 turns ON and the bias light is divided into C1 and C2. Since the bias light in C2 is insufficient to keep C2 ON, C2 gradually turns OFF. (7) Set pulse is cut. The situation is the same as (3). As a result, the output ports for Q and $\bar{Q}$ are switched by the SR pulses. This switching operation requires 44 GHz.

7.6.2 Sequential Circuit

An FF is a kind of sequential circuit whose output signals depend on the input signal sequence. In this section, we introduce our sequential circuit, which has a similar function to a D-FF, to synchronize the input data with a system clock.

Figure 7.18a shows the structure of a new sequential circuit based on 2D-PhC [35]. We simulated the system using the 2D-FDTD method. This circuit contains two different cavities (C1 and C2) and two kinds of WGs (WG1 and WG2). The two cavities have one common resonant frequency (λ1) and one that is different (λ2 for C1 and λ3 for C2). The cavities can function as two logic gates. The widths of the WGs and input ports are determined so that the λ1 and λ2 lights can propagate only in WG1, and the λ3 light can propagate only in WG2.

To make it easy to understand the mechanism of the sequential circuit, we consider the equivalent circuit shown in Figure 7.18b. This circuit is composed of two logic gates (G1 and G2). The gate is initially in the closed state. When signal λ1 is input, G1 remains closed. When λ1 and λ2 are input, G1 opens.

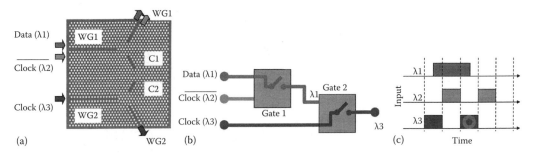

FIGURE 7.18 Sequential circuit for synchronizing input data with a system clock. (a) Structure of sequential circuit based on 2D-PhC. This circuit contains two different cavities and two kinds of WGs. (b) Equivalent circuit. This circuit is composed of two logic gates (G1 and G2). (c) Input light sequence. (1) The $\lambda 3$ light is input. Both gates close. (2) The $\lambda 1$ light is input. But both gates remain closed. (3) The $\lambda 3$ light is cut and $\lambda 1$ and $\lambda 2$ lights are input into G1. G1 opens. (4) The $\lambda 2$ light is cut but G1 remains open. G2 opens and the $\lambda 3$ light is output. (5) The $\lambda 1$ light is cut. G1 closes but G2 remains open and the $\lambda 3$ light is output. (6) The $\lambda 3$ light is cut and G2 closes. It should be noted that the first $\lambda 3$ input pulse is not output and the second $\lambda 3$ input pulse is output sustaining its width. (From Shinya, A. et al., *Opt. Express*, 14, 1230–1235, 2006. With permission. Copyright 2006, Optical Society of America.)

Once G1 has opened, it remains open while either or both signals are being input. When both signals are cut, G1 closes. G2 functions in the same way for $\lambda 1$ and $\lambda 3$. Here, we consider the input light sequence, as shown in Figure 7.18c. (1) The $\lambda 3$ light is input. Both gates close. (2) The $\lambda 1$ light is input. But both gates remain closed. (3) The $\lambda 3$ light is cut and the $\lambda 1$ and $\lambda 2$ lights are input into G1. G1 opens. (4) The $\lambda 2$ light is cut but G1 remains open. So the $\lambda 1$ and $\lambda 3$ lights are input into G2. G2 opens and the $\lambda 3$ light is output. (5) The $\lambda 1$ light is cut. G1 closes but G2 remains open and the $\lambda 3$ light is output. (6) The $\lambda 3$ light is cut and G2 closes. It should be noted that the first $\lambda 3$ input pulse is not output and the second $\lambda 3$ input pulse sustains its width when it is output. The mechanism of our sequential circuit means that its output condition depends on the input signal sequence.

Figure 7.19 is a time chart calculated with the 2D-FDTD method. The input NRZ-format DATA ($\lambda 1$) deviates slightly from an ideal signal synchronized with the clock shown by the dotted line. As a result, the sequential circuit can regenerate the ideal DATA with a return-to-zero (RZ) format ($\lambda 3$) synchronized

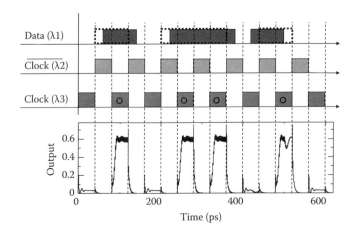

FIGURE 7.19 Time chart of the sequential circuit calculated by the 2D-FDTD method. The input NRZ-format DATA ($\lambda 1$) deviates slightly from the ideal signal synchronized with the clock shown by the dotted line. This circuit can regenerate the ideal DATA with a RZ format ($\lambda 3$) synchronized with the clock. The input light power is 61 mW and the response time is about 10 ps. (From Shinya, A. et al., *Opt. Express*, 14, 1230–1235, 2006. With permission. Copyright 2006, Optical Society of America.)

with the clock. We set the input light power at 61 mW. This figure shows that the system can function as a sequential circuit with a response time of about 10 ps, which is limited by the Q-factor of the system. The response time can be improved by using a low-Q system with highly nonlinear material.

7.7 Conclusion

We investigated a WG, resonator, and a coupled system based on a two-dimensional PhC with a PBG, in only the in-plane direction. In this study, we showed how to design WGs and resonators to overcome radiation loss in the thickness direction. We fabricated a very low loss WG, and designed an ultrasmall and high-Q resonator. Moreover, we achieved a very low connection loss between a fiber and a PhC-WG. We then effectively coupled the PhC-WGs to the PhC resonators, and devised a multi-port channel drop filter based on a two-port resonant tunneling filter.

Next, we showed that a resonant tunneling filter can act as an AND gate or a NOT gate with an operating power of 100 fJ and as an optical memory with a bias power of 0.4 mW. The operation of these devices is based on the carrier-plasma effect and their switching speed is about 100 ps. Finally, we described an all-optical sequential circuit that can synchronize input NRZ optical data with its clock and regenerate input data with an RZ format. The response time is about 10 ps and the operating power is 61 mW. These ultrasmall passive and active devices have the potential to provide various signal processing functions in PhC-based optical circuits.

References

1. J. D. Joannopoulos, P. R. Villeneuve, and S. Fan, Photonic crystals: Putting a new twist on light, *Nature* 386, 143 (1997).
2. E. Kuramochi, S. Hughes, T. Watanabe, L. Ramunno, A. Shinya, and M. Notomi, Low loss photonic crystal slab waveguides: Fabrication, experiment, and theory, *the 17th Annual Meeting of the IEEE Lasers and Electro-Optics Society* (LEOS2004), Rio Grande, Puerto Rico, November, WF2, 2004.
3. D. Mori and T. Baba, Dispersion-controlled optical group delay device by chirped photonic crystal waveguides, *Appl. Phys. Lett.* 85, 1101 (2004).
4. M. L. Povinelli, S. G. Johnson, and J. D. Joannopoulos, Slow-light, band-edge waveguides for tunable time delays, *Opt. Express* 13, 7145 (2005).
5. A. Yariv, Y. Xu, R. K. Lee, and A. Sherer, Coupled-resonator optical waveguide: A proposal and analysis, *Opt. Lett.* 24, 711 (1999).
6. T. J. Karle, D. H. Brown, R. Wilson, M. Steer, and T. F. Krauss, Planar photonic crystal coupled cavity waveguides, *IEEE J. Sel. Top. Quantum Electron.* 8, 909 (2002).
7. M. F. Yanik, W. Suh, Z. Wang, and S. Fan, Stopping light in a waveguide with an all-optical analog of electromagnetically induced transparency, *Phys. Rev. Lett.* 93, 233903 (2004).
8. S.-C. Huang, M. Kato, E. Kuramochi, C.-P. Lee, and M. Notomi, Time-domain and spectral-domain investigation of inflection-point slow-light modes in photonic crystal coupled waveguides, *Opt. Express* 15, 3543 (2007).
9. T. Shoji, T. Tsuchizawa, T. Watanabe, K. Yamada, and H. Morita, Low loss mode size converter from 0.3 μm square Si wire waveguide to single mode fibers, *Electron. Lett.* 38, 1669–1670 (2002).
10. A. Shinya, M. Notomi, E. Kuramochi, T. Shoji, T. Watanabe, T. Tsuchizawa, K. Yamada, and H. Morita, Functional components in SOI photonic crystal slabs, *Proc. SPIE* 5000, 104–117 (2003).
11. H.-Y. Ryu, M. Notomi, and Y.-H. Lee, High-quality-factor and small-mode-volume hexapole modes in photonic-crystal-slab nanocavities, *Appl. Phys. Lett.* 83, 4294–4296 (2003).
12. G.-H. Kim, Y.-H. Lee, A. Shinya, and M. Notomi, Coupling of small, low-loss hexapole mode with photonic crystal slab waveguide mode, *Opt. Express* 26, 6624–6631 (2004).
13. S. G. Johnson, S. Fan, A. Mekis, and J. D. Joannopoulos, Multipole-cancellation mechanism for high-Q cavities in the absence of a complete photonic band gap, *Appl. Phys. Lett.* 78, 3388–3390 (2001).

14. T. Tanabe, A. Shinya, E. Kuramochi, S. Kondo, H. Taniyama, and M. Notomi, Single point defect photonic crystal nanocavity with ultrahigh quality factor achieved by using hexapole mode, *Appl. Phys. Lett.* 91, 021110 (2007).

15. K. Inoshita and T. Baba, Lasing at bend, branch and intersection of photonic crystals, *Electron. Lett.* 39, 844 (2003).

16. M. Notomi, A. Shinya, S. Mitsugi, E. Kuramochi, and H.-Y. Ryu, Waveguides, resonators and their coupled elements in photonic crystal slabs, *Opt. Express* 12, 1551 (2004).

17. B. S. Song, S. Noda, T. Asano, and Y. Akahane, Ultra-high-Q photonic double-heterostructure nanocavity, *Nat. Mater.* 4, 207–210 (2005).

18. E. Kuramochi, M. Notomi, S. Mitsugi, A. Shinya, and T. Tanabe, Ultrahigh-Q photonic crystal nanocavities realized by the local width modulation of a line defect, *Appl. Phys. Lett.* 88, 041112 (2006).

19. T. Tanabe, M. Notomi, E. Kuramochi, A. Shinya, and H. Taniyama, Trapping and delaying photons for one nanosecond in an ultrasmall high-Q photonic-crystal nanocavity, *Nat. Photonics* 1, 49 (2007).

20. M. Notomi, K. Yamada, A. Shinya, J. Takahashi, C. Takahashi, and I. Yokohama, Extremely large group velocity dispersion of line-defect waveguides in photonic crystal slabs, *Phys. Rev. Lett.* 87, 253902 (2001).

21. M. Notomi, A. Shinya, K. Yamada, J. Takahashi, C. Takahashi, and I. Yokohama, Structural tuning of guiding modes of line-defect waveguides of silicon-on-insulator photonic crystal slabs, *IEEE J. Quantum Electron.* 38, 736 (2002).

22. S. Mitsugi, A. Shinya, G. Kira, E. Kuramochi, and M. Notomi, QH control and transmittance measurement of M-direction couple resonant tunneling filters on 2D photonic slabs, *65th Autumn Meeting of The Japan Society of Applied Physics*, Sendai, Japan, p. 932, 2004.

23. S. Fan, P. R. Villeneuve, and J. D. Joannopoulos, Channel drop tunneling through localized states, *Phys. Rev. Lett.* 80, 960–963 (1998).

24. C. Manolatou, M. J. Khan, S. Fan, P. R. Villeneuve, H. A. Haus, and D. J. Joannopoulos, Coupling of modes analysis of resonant channel add-drop filters, *IEEE Quantum Electron.*, 35, 1322–1331, 1999.

25. B. S. Song, S. Noda, and T. Asano, Photonic devices based on in-plane hetero photonic crystals, *Science* 300, 1537 (2003).

26. H. Takano, Y. Akahane, T. Asano, and S. Noda, In-plane-type channel drop filter in a two-dimensional photonic crystal slab, *Appl. Phys. Lett.* 84, 2226–2228 (2004).

27. A. Shinya, S. Mitsugi, E. Kuramochi, and M. Notomi, Ultrasmall multi-channel resonant-tunneling filter using mode gap of width-tuned photonic-crystal waveguide, *Opt. Express* 13, 4202–4209 (2005).

28. H. Takano, B.-S. Song, T. Asano, and S. Noda, Highly efficient multi-channel drop filter in a two-dimensional hetero photonic crystal, *Opt. Express* 14, 3491–3496 (2006).

29. A. Shinya, S. Mitsugi, E. Kuramochi, and M. Notomi, Ultrasmall multi-port channel drop filter in two-dimensional photonic crystal on silicon-on-insulator substrate, *Opt. Express* 14, 12394 (2006).

30. M. Notomi, A. Shinya, S. Mitsugi, G. Kira, E. Kuramochi, and T. Tanabe, Optical bistable switching action of Si high-Q photonic-crystal nanocavities, *Opt. Express* 13, 2678 (2005).

31. T. Tanabe, M. Notomi, S. Mitsugi, A. Shinya, and E. Kuramochi, All-optical switches on a silicon chip realized using photonic crystal nanocavities, *Appl. Phys. Lett.* 87, 151112 (2005).

32. T. Tanabe, M. Notomi, S. Mitsugi, A. Shinya, and E. Kuramochi, Fast bistable all-optical switch and memory on a silicon photonic crystal on-chip, *Opt. Lett.* 30, 2575 (2005).

33. Kawanishi et al., *the 2nd International Symposium on Ultrafast Photonic Technologies*, University of St. Andrews, U.K., August 2, 2005.

34. S. Mitsugi, A. Shinya, T. Tanabe, M. Notomi, and I. Yokohama, JSAP Spring meeting 30p-YV-11 (2005).

35. A. Shinya, S. Mitsugi, T. Tanabe, M. Notomi, I. Yokohama, H. Takara, and S. Kawanishi, All-optical flip-flop circuit composed of coupled two-port resonant tunneling filter in two-dimensional photonic crystal slab, *Opt. Express* 14, 1230–1235 (2006).

V

Integrated Photonic Technologies: Plasmonics and Integration

8

Surface Plasmon-Polariton Waveguides and Components

Pierre Berini

8.1 Introduction

A surface plasmon-polariton (SPP) waveguide is a metallo-dielectric structure along which SPP modes are guided. This is an important class of optical waveguide having interesting and potentially useful attributes. SPP waveguides exhibit an energy asymptote in their dispersion curve, high surface and bulk sensitivities, and sub-wavelength confinement in some designs (or for operation near the energy asymptote). Furthermore, having a metal in the optical path allows electric fields or heat to be applied, potentially with good overlap to the SPP mode, and the chemical properties of a metal may be useful for certain studies or for applications such as (bio)chemical sensing. Unfortunately, SPP waveguides are also generally characterized by a high attenuation, especially near the energy asymptote. However, a subset of structures can be designed to support long-range SPPs (LRSPPs) having a substantially lower attenuation, especially at long wavelengths (away from the energy asymptote). This chapter reviews and discusses the performance of various SPP waveguides and components, initially as one-dimensional (1D) structures, then as two-dimensional (2D) variants.

8.1.1 Notation

Throughout this chapter, an $e^{+j\omega t}$ time dependence is assumed with modes propagating in the $+z$ direction according to $e^{-\gamma z}$. The complex propagation constant, γ, in m^{-1} expands as $\gamma = \alpha + j\beta$ where

α and β are the attenuation and phase constants, respectively. The normalized propagation constant is $\gamma_{eff} = \gamma/\beta_0 = \alpha/\beta_0 + j\beta/\beta_0 = k_{eff} + jn_{eff}$ where $\beta_0 = 2\pi/\lambda_0 = \omega/c_0$ is the phase constant of plane waves in free space, λ_0 is the wavelength in free space, and c_0 is the speed of light in free space. The complex effective index of a mode N_{eff} is then given by $N_{eff} = -j\gamma/\beta_0 = \beta/\beta_0 - j\alpha/\beta_0 = n_{eff} - jk_{eff}$. The mode power attenuation (MPA) in dB/m is given by $MPA = \alpha\, 20 \log_{10} e$, and the $1/e$ mode propagation length by $L_e = 1/(2\alpha)$. The relative permittivity is obtained from the optical parameters n, k in the usual way $\varepsilon_r = (n - jk)^2$, and the relative permittivity of a metal is written in terms of real and imaginary parts as $\varepsilon_{r,m} = -\varepsilon_R - j\varepsilon_I$.

8.2 One-Dimensional SPP Waveguide Structures

8.2.1 Structures

Three 1D SPP waveguides of general interest are sketched in Figure 8.1. These waveguides provide field confinement along the y direction, with propagation occurring at any angle in the x–z plane, say, specifically along the $+z$ direction (out of the page) for simplicity. The relative permittivities of the metal and dielectric regions in these structures are denoted $\varepsilon_{r,m}$ and $\varepsilon_{r,1}$, respectively. Figure 8.1a shows the conventional single-interface metal–dielectric waveguide (e.g., [1,2]). Figure 8.1b shows a symmetric dielectric-cladded metal film of thickness t, henceforth referred to as the IMI (insulator–metal–insulator) waveguide (e.g., [3–5]). Figure 8.1c shows a symmetric metal-cladded dielectric film

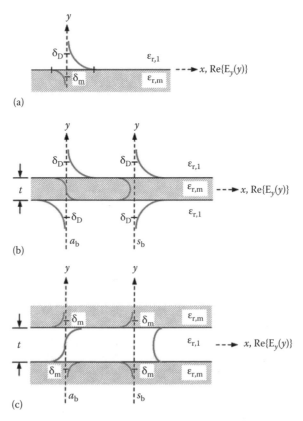

FIGURE 8.1 1D SPP waveguides: (a) single-interface, (b) symmetric dielectric-cladded metal slab—IMI, and (c) symmetric metal-cladded dielectric slab—MIM. The relative permittivity of the metal and dielectric regions are denoted $\varepsilon_{r,m}$ and $\varepsilon_{r,1}$, respectively. The distribution of the main transverse electric field component of the bound modes supported by each waveguide is sketched as the thick gray curves.

of thickness t, henceforth referred to as the MIM (metal–insulator–metal) waveguide (e.g., [4–7]). The distribution of the main transverse electric field component (E_y) of the non-radiative purely bound mode(s) supported by each structure is also sketched. In the single-interface (Figure 8.1a) it is the usual SPP mode, and in the IMI and MIM the modes are the asymmetric and symmetric coupled modes, denoted herein as a_b and s_b, respectively, following the distribution of E_y. These modes are all transverse magnetic (TM).

8.2.2 Materials

Highly conductive metals and good transparent dielectrics are generally used to implement SPP waveguides. Ag is often preferred given its low loss, but it is reactive, so care must be taken during fabrication and use in order to avoid degradation that can occur if Ag is exposed to air or water (for example). Alternatively, Au is a good choice given its chemical stability. SiO_2 or polymers are often used as the dielectric.

The operating wavelength and the materials are generally selected such that interband transitions causing increased absorption are avoided. For dielectrics and semiconductors, this means operating below the bandgap energy and for metals, this means operating in or near the Drude region. The Drude model represents the permittivity of metals via the following relationship [8]:

$$\varepsilon_{r,m} = -\varepsilon_R - j\varepsilon_I = 1 - \frac{\omega_p^2}{\omega^2 + 1/\tau_D^2} - j\frac{\omega_p^2/\tau_D}{\omega(\omega^2 + 1/\tau_D^2)} \tag{8.1}$$

where
ω_p is the plasma frequency
τ_D is the relaxation time

The Drude region corresponds to that portion of the electromagnetic spectrum where this equation holds. For many metals, this region spans the range from long visible wavelengths into the infrared. The metal approaches a perfect electric conductor (PEC) as the wavelength increases through the infrared and beyond.

The measured optical parameters (n, k) of Ag and SiO_2 [9–12] were used in the computations discussed in the next subsections. The n and k values were splined and interpolated at the desired wavelengths, and then used to compute the relative permittivities, which are plotted in Figure 8.2 for both materials. The plasma frequency (ω_p) and relaxation time (τ_D) of Ag were obtained by fitting Equation 8.1 to this data (following [13]), yielding $\omega_p = 1.26 \times 10^{16}$ rad/s and $\tau_D = 8.40 \times 10^{-15}$ s. The Drude region was taken as $\lambda_0 \geq 725$ nm. From Figure 8.2 it is noted that Re{$\varepsilon_{r,Ag}$} is negative and has a large magnitude over this wavelength range.

8.2.3 Single-Interface SPP

One purely bound (non-radiative) SPP mode is supported by the single interface (Figure 8.1a), having fields (E_y, E_z, and H_x) that peak at the interface and decay exponentially away into both media. The fields in the dielectric are coupled to a charge density wave in the metal so the SPP propagates as a coupled excitation (plasmon-polariton).

The normalized propagation constant of the single-interface SPP is [1]

$$\gamma_{eff} = \left(\frac{\varepsilon_{r,m}\varepsilon_{r,l}}{\varepsilon_{r,m} + \varepsilon_{r,l}} \right)^{1/2} \tag{8.2}$$

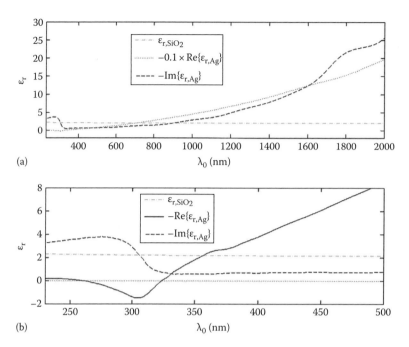

FIGURE 8.2 Relative permittivity of Ag and SiO$_2$: (a) over the range $230 \leq \lambda_0 \leq 2000$ nm and (b) over the range $230 \leq \lambda_0 \leq 500$ nm.

For a lossless dielectric cladding ($\text{Im}\{\varepsilon_{r,1}\} = 0$), the above simplifies to the following approximate expressions for the effective index and normalized attenuation [1]:

$$n_{\text{eff}} \cong \left(\frac{\varepsilon_R \varepsilon_{r,1}}{\varepsilon_R - \varepsilon_{r,1}} \right)^{1/2} \quad \text{and} \quad k_{\text{eff}} \cong \frac{\varepsilon_I}{2\varepsilon_R^2} \left(\frac{\varepsilon_{r,1}\varepsilon_R}{\varepsilon_R - \varepsilon_{r,1}} \right)^{3/2} = \frac{\varepsilon_I}{2\varepsilon_R^2} n_{\text{eff}}^3 \tag{8.3}$$

Ag and SiO$_2$ have Re$\{\varepsilon_r\}$ of opposite sign over a large wavelength range (Figure 8.2), so the interface between these materials supports a purely bound SPP. As the operating wavelength increases, the metal approaches a PEC and the confinement of the SPP decreases (SPP's are not supported at the interface between a smooth PEC and a semi-infinite dielectric).

If ε_I of Ag is neglected, then from Equation 8.3 and Figure 8.2b, n_{eff} would diverge as λ_0 decreases from the infrared, and thus as $\varepsilon_R \to \varepsilon_{r,1}$, becoming infinite when $\varepsilon_R = \varepsilon_{r,1}$ at $\lambda_0 \sim 360$ nm. In reality though, $\varepsilon_I \neq 0$, so n_{eff} never becomes infinite although it may still be large. From Equation 8.3 and Figure 8.2, it is apparent that k_{eff} also increases as λ_0 decreases, and evidently, at a much greater rate. Figure 8.3a and b plots n_{eff} and the MPA of the single-interface SPP computed using Equation 8.2 for the Ag/SiO$_2$ interface, highlighting these trends, especially the sharp increase in both at $\lambda_0 \sim 360$ nm.

Figure 8.3c plots the dispersion curve of the single-interface SPP ($E = \hbar\omega$ in eV versus β, $\hbar$ is Planck's reduced constant) along with the light line in SiO$_2$. The SPP energy asymptote is readily observed near $E \sim 3.4$ eV ($\lambda_0 \sim 360$ nm). The group velocity decreases and the optical density of states increases as the energy asymptote is approached. The SPP bend-back [14] is also observed for wavelengths shorter than the asymptote ($\lambda_0 < 360$ nm, $E > 3.4$ eV) and links the non-radiative SPP to the radiative one on the left side of the light line.

Figure 8.3d plots the 1/e mode field magnitude width, δ_w [5], of the single-interface SPP along with the diffraction-limited width in SiO$_2$, $\lambda_0/(2n_{\text{SiO}_2})$ [15], which corresponds approximately to the smallest

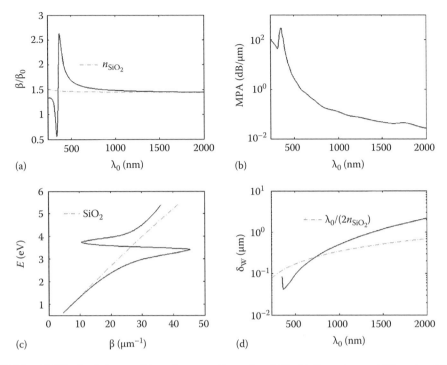

FIGURE 8.3 (a) Effective index n_{eff}, (b) mode power attenuation MPA, (c) dispersion, and (d) mode width, δ_w, of the single-interface SPP along an Ag/SiO$_2$ interface as a function of λ_0.

mode width achievable in a slab waveguide having SiO$_2$ as the core. It is noted from Figure 8.3d that the SPP mode width, δ_w, becomes smaller than this measure for $\lambda_0 < 750$ nm, indicating that sub-wavelength confinement is provided in this range as the energy asymptote is approached.

Figure 8.3 also highlights an important trade-off known to exist for SPP waveguides, namely, that confinement and attenuation increase or decrease together [5,16]. This is especially evident from Figure 8.3a, b, and d as the energy asymptote is approached.

Figure 8.4 plots the normalized group velocity, v_g/c_0, computed from

$$v_g^{-1} = \frac{\partial \beta}{\partial \omega} = \frac{1}{c_0}\left(n_{eff} - \lambda_0 \frac{\partial n_{eff}}{\partial \lambda_0}\right) \tag{8.4}$$

the lifetime, τ, computed from

$$\tau = \frac{1}{v_g 2\alpha} \tag{8.5}$$

and the quality factor, Q, given by

$$Q = \omega\tau \tag{8.6}$$

of the single-interface SPP along the Ag/SiO$_2$ interface as a function of λ_0 [5]. Asymptotic forms for these three quantities are also plotted over the Drude region. From Figure 8.4a, it is noted that the group velocity decreases by a factor of about 6.5 as the energy asymptote is approached. However, its lifetime and quality factor decrease by about two orders of magnitude over the same range.

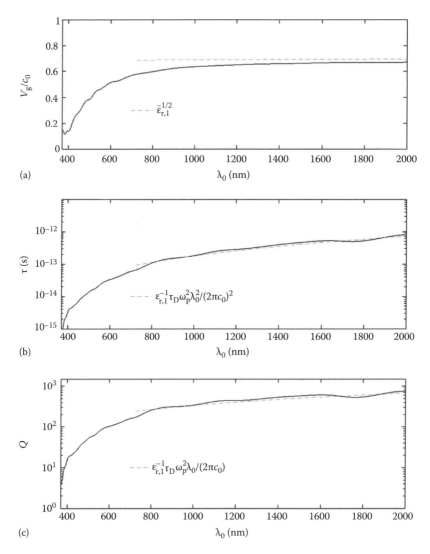

FIGURE 8.4 (a) Normalized group velocity v_g/c_0, (b) lifetime τ, and (c) quality factor Q of the single-interface SPP along an Ag/SiO$_2$ interface as a function of λ_0.

8.2.4 Insulator–Metal–Insulator

Two purely bound (non-radiative) SPP modes are supported by the symmetric IMI (metal film cladded by identical dielectrics—Figure 8.1b). The modes are formed via coupling through the metal film of the individual SPPs supported by the top and bottom interfaces, so they are termed "coupled modes" or "supermodes." The supermodes have transverse field components E_y and H_x that exhibit asymmetry or symmetry along the y axis and so are identified as the a_b and s_b modes, respectively [3]. The distribution of the E_y field component of both modes is sketched in Figure 8.1b. These modes are not supported if the metal film becomes a PEC (say by operating at very long wavelengths).

The propagation characteristics of the a_b and s_b modes depend on the thickness, t, of the metal film. Figure 8.5a and b gives the MPA and n_{eff} of the a_b and s_b modes along a SiO$_2$/Ag/SiO$_2$ IMI as a function of t at $\lambda_0 = 1550$ nm. These results were computed using the method of lines (MoL) following [17,18],

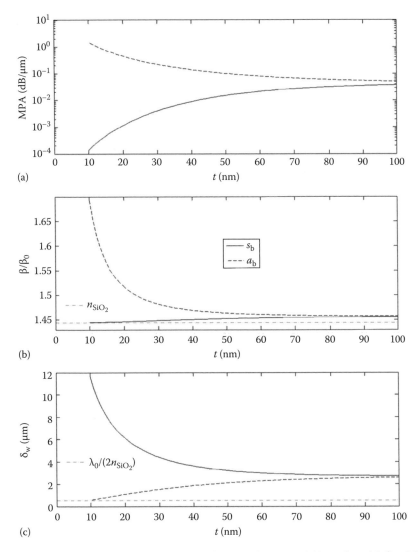

FIGURE 8.5 (a) Mode power attenuation MPA, (b) effective index n_{eff}, and (c) mode width δ_{w} of the a_{b} and s_{b} modes along an $SiO_2/Ag/SiO_2$ IMI as a function of t at $\lambda_0 = 1550$ nm.

but alternatively, the modes can be found by solving the transcendental equation of this three-layer structure [3].

From Figure 8.5a and b, it is observed that the modes become degenerate with increasing t. As the separation between the top and bottom interfaces increases, the a_{b} and s_{b} modes split into a pair of uncoupled SPP modes each localized at a metal–dielectric interface. The propagation constant of the a_{b} and s_{b} modes thus tend toward that of an SPP supported by the interface between semi-infinite Ag and SiO_2 regions (Equation 8.2).

As t decreases, the MPA and n_{eff} of the a_{b} mode increase, becoming very large for very thin films. This is due to the fact that the fields of this mode penetrate progressively deeper into the metal as t is reduced. In the case of the s_{b} mode, a decreasing t causes the opposite effect, that is, the fields penetrate progressively more into the top and bottom claddings and less into the metal. The MPA and n_{eff} of this mode thus tend asymptotically toward those of a transverse electromagnetic (TEM) wave propagating in an infinite homogeneous medium having the same permittivity as the claddings, i.e., MPA $\to$ 0 as observed in Figure 8.5a (since absorption is negligible in SiO_2) and $n_{eff} \to n_{SiO_2}$ as observed in Figure 8.5b. The a_b and s_b modes have no cutoff thickness.

Figure 8.5c plots the 1/e mode field magnitude width δ_w [5] of the a_b and s_b modes, along with the diffraction-limited width in SiO_2, $\lambda_0/(2n_{SiO_2})$ [15]. The mode width of the s_b mode increases with decreasing t as the mode evolves into the TEM wave of the background, whereas the mode width of the a_b mode decreases with decreasing t. The diffraction-limited width in SiO_2 is attained by the a_b mode for $t = 10$ nm at this operating wavelength.

8.2.5 Metal–Insulator–Metal

The MIM (dielectric film cladded by identical metals—Figure 8.1c) differs from the previous two structures in that it supports guided modes as the metals become PECs (say, due to operation at long wavelengths). Thus, the optical modes of the MIM [6,7] evolve from the respective TE and TM modes of the parallel-plate waveguide [19,20]; as the PECs become real metals, the modes develop a plasmonic character and their fields penetrate increasingly into the metals [19]. So the first two optical modes of the MIM originate from the TEM and TM_1 modes of the parallel-plate waveguide [19], and are denoted herein as the plasmonic s_b and a_b modes respectively, highlighting the symmetric and asymmetric distribution of their E_y field component, as sketched in Figure 8.1c.

The propagation characteristics of the a_b and s_b modes depend on the thickness of the dielectric film t. Figure 8.6a and b give the MPA and n_{eff} of the a_b and s_b modes along a Ag/SiO$_2$/Ag MIM as a function of t at $\lambda_0 = 1550$ nm. These results were computed using the MoL, but alternatively, the modes can be found by solving the transcendental equation of this three-layer structure.

As t decreases, the MPA of the a_b and s_b modes increases, becoming very large for very small t. This is due to the fact that the mode fields penetrate progressively deeper into the metal as t is reduced. n_{eff} of the s_b mode also increases with decreasing t, but n_{eff} of the a_b mode decreases, cutting off for $t \sim 500$ nm.

Figure 8.6c plots the 1/e mode field magnitude width δ_w [5] of the a_b and s_b modes, along with the diffraction-limited width in SiO_2, $\lambda_0/(2n_{SiO_2})$ [15]. The mode width of both modes decreases with decreasing t, bearing in mind that the a_b mode cuts off for $t \sim 500$ nm. Mode widths well below the diffraction-limited width in SiO_2 are readily attainable by the s_b mode.

8.2.6 End-Fire Coupling

The symmetric (s_b) mode of the IMI and MIM has transverse field components E_y and H_x that exhibit symmetry along the y axis (Figure 8.1) and so can be excited using TM-polarized beams through end-fire (butt) coupling at the waveguide end facets. This makes operating in these modes attractive given the ease and maturity of this technique.

Figure 8.7 depicts an end-fire coupling arrangement using single-mode polarization-maintaining optical fibers butt-coupled to the end facets of an IMI and aligned to excite the s_b mode therein. Efficient coupling to optical fibers can be achieved for IMIs having a small t, because the s_b mode fields extend over dimensions comparable to those of the fundamental optical fiber mode. On the other hand, efficient coupling to high-confinement dielectric waveguides can be achieved for the s_b mode in the MIM if the latter has a thickness t comparable to that of the core of the dielectric waveguide.

8.2.7 Confinement: Attenuation Trade-Off

From Figures 8.5 and 8.6 it is noted that the s_b mode in the IMI and MIM exhibit complementary behavior as t decreases, in that they capture different confinement-attenuation trade-offs [5]. The single-interface is, nominally, in the "middle" of the trade-off, whereas the MIM and IMI are at

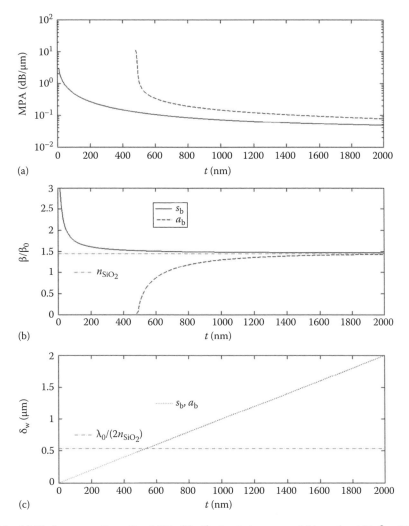

FIGURE 8.6 (a) Mode power attenuation MPA, (b) effective index n_{eff}, and (c) mode width δ_w of the a_b and s_b modes along an Ag/SiO$_2$/Ag MIM as a function of t at $\lambda_0 = 1550$ nm.

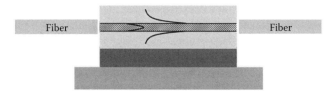

FIGURE 8.7 End-fire coupling scheme using single-mode polarization-maintaining optical fibers butt-coupled to the end facets of an IMI and aligned to excite the s_b mode.

opposite ends for small t. For small t, the s_b mode of the MIM is strongly confined but also strongly attenuated and thus often termed an SRSPP (short-range SPP), whereas the s_b mode in the IMI is weakly confined but also weakly attenuated and thus is often termed an LRSPP (long-range SPP). This confinement-attenuation trade-off applies to all known SPP waveguides in that generally attenuation increases with confinement.

8.2.8 Bulk and Surface Sensitivities

The propagation characteristics and the surface and bulk sensitivities of the SPP in the single-interface, the s_b mode in the MIM, and the s_b mode in IMI, are strongly dependent on the structure design (t) and the operating wavelength, as revealed by computations performed as a function of t for wavelengths spanning the range $600 \leq \lambda_0 \leq 1600$ nm assuming Au as the metal ($\varepsilon_{r,m}$), H_2O as the dielectric ($\varepsilon_{r,1}$), and thin adlayers representative of biochemical matter deposited along the metal/H_2O interfaces [21]. The s_b mode surface sensitivity in the thin MIM is about 100× larger than that of the SPP in the single interface, whereas the s_b mode surface sensitivity in the thin IMI is about 5× smaller. The s_b mode bulk sensitivity in the thin MIM is about 3× larger than that of the SPP in the single interface, whereas the s_b mode bulk sensitivity in the thin IMI is slightly smaller. The mode sensitivities increase with confinement and attenuation.

A generic single-output Mach–Zehnder interferometer (MZI) implemented with attenuating waveguides has an optimal sensing length equal to the propagation length (L_e) of the mode used [21]. Furthermore, the overall surface and bulk sensitivities of the MZI are proportional to the ratio of the corresponding mode sensitivity to its normalized attenuation. Maximizing the ratio maximizes the overall MZI sensitivity and minimizes its detection limit leading to preferred waveguide designs and operating wavelengths. The ratio of s_b mode surface sensitivity to normalized attenuation in the thin MIM is about 3× larger than the ratio for the SPP in the single interface, whereas the ratio of s_b mode surface sensitivity to normalized attenuation in the IMI is about 10× larger. The ratio of s_b mode bulk sensitivity to normalized attenuation in the thin MIM is about 10× smaller than the ratio for the SPP in the single interface, whereas the ratio of s_b mode bulk sensitivity to normalized attenuation in the thin IMI is about 10× larger.

Thus, for transducers exploiting an optical interaction length such as a single-output MZI, it is the ratio of mode sensitivity to normalized attenuation that matters. From this perspective, the IMI and the MIM both offer an improvement for surface sensing over the single-interface SPP, despite being at opposite ends of the confinement-attenuation trade-off. With regards to bulk sensing, only the IMI offers an improvement over the single-interface SPP. From the same perspective, preferred wavelengths maximizing the ratio of mode surface sensitivity to normalized attenuation occur near the short-wavelength edge of the Drude region [21].

8.3 Two-Dimensional SPP Waveguide Structures

1D SPP waveguides are limits of similar 2D structures having a large width and operating in their main mode. For example, the metal stripe in an asymmetric dielectric environment [22–27] supports modes localized along either the top or bottom metal–dielectric interfaces having characteristics (effective index, attenuation, perpendicular confinement, etc.) comparable to the corresponding single-interface SPP. The metal stripe bounded by symmetric dielectrics [17,18] supports a mode similar to the s_b mode of the IMI. The gap, wedge, and channel waveguides [28–33] supports a mode that resembles the s_b mode of the MIM. Some of the trends uncovered for the 1D waveguides should therefore carry over to their respective 2D variants.

8.3.1 Metal Stripe Waveguide

Among the many 2D SPP waveguides investigated [17,18,22–33], the metal stripe depicted in front cross-sectional view in Figure 8.8 is interesting due to ease of fabrication, the ability of the structure to provide confinement in the plane transverse to the direction of propagation, and its ability to support a LRSPP under appropriate conditions.

Similarly to the IMI, a mode of the metal stripe is a supermode created from the coupling of "corner" and "edge" modes supported by isolated corners and finite-width edges [18,23]. This leads to a very rich

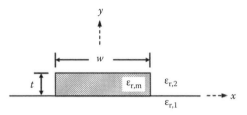

FIGURE 8.8   Metal stripe waveguide; the relative permittivity of the metal and dielectric regions are denoted $\varepsilon_{r,m}$ and $\varepsilon_{r,1}$, $\varepsilon_{r,2}$ respectively.

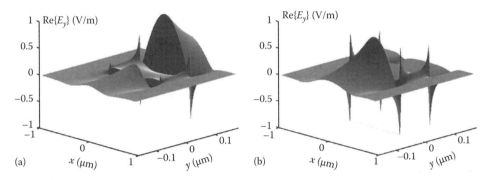

FIGURE 8.9   (a) ss_b^1 and (b) sa_b^1 modes supported by an asymmetric metal stripe (Figure 8.8a). In both cases, $t = 100\,\mathrm{nm}$, $w = 1\,\mu\mathrm{m}$, $\varepsilon_{r,m} = -19 - j0.53$, $\varepsilon_{r,1} = 4$, $\varepsilon_{r,2} = 3.61$, and $\lambda_0 = 633\,\mathrm{nm}$. The field distributions are normalized such that $\max|\mathrm{Re}\{E_y\}| = 1$.

set of supermodes that depend nontrivially on structural and material parameters and on the operating wavelength. The supermodes are essentially TM polarized if $w \gg t$.

The asymmetric metal stripe ($\varepsilon_{r,1} \neq \varepsilon_{r,2}$) [22–27] supports many such supermodes, including low order ones that have one E_y field maximum along the stripe width, localized at the center of either the top or bottom metal–dielectric interface. Figure 8.9 shows the E_y field distribution for two such modes, the ss_b^1 and sa_b^1 supermodes in parts (a) and (b), respectively, supported by an asymmetric metal stripe having $t = 100\,\mu\mathrm{m}$, $w = 1\,\mu\mathrm{m}$, $\varepsilon_{r,m} = -19 - j0.53$ (Ag), $\varepsilon_{r,1} = 4$, and $\varepsilon_{r,2} = 3.61$ at $\lambda_0 = 633\,\mathrm{nm}$ [23]. From Figure 8.9, it is observed that the top and bottom edge modes participating in the ss_b^1 and sa_b^1 supermodes are different from each other. In Figure 8.9a, the bottom edge mode has three extrema and is of higher order than the top edge mode which has one extremum. In Figure 8.9b, the bottom edge mode has one extremum while the top one has none. In this structure, $\varepsilon_{r,1} > \varepsilon_{r,2}$ so the effective index of a bottom edge mode will be higher than that of the same edge mode at the top interface. Since a supermode is created through coupling of edge modes and corner modes having similar effective indices, it is expected that in an asymmetric structure different edge modes may couple to create a supermode. The ss_b^1 supermode of Figure 8.9a is localized along the $\varepsilon_{r,m} - \varepsilon_{r,2}$ interface and so has an effective index and MPA comparable to those of the corresponding single-interface SPP. Likewise, the sa_b^1 supermode of Figure 8.9b is localized along the $\varepsilon_{r,1} - \varepsilon_{r,m}$ interface and so has an effective index and MPA comparable to those of the corresponding single-interface SPP.

Some of the supermodes of the asymmetric metal stripe cut off with diminishing metal width or thickness (w or t), and in so doing, exhibit vanishing attenuation and confinement [23]. Beyond cut-off such modes are radiative, into the substrate for $\varepsilon_{r,1} > \varepsilon_{r,2}$. Operating a mode near cut-off is possible but requires tight tolerances on fabrication and material parameters. Well-confined and localized supermodes, such as those shown in Figure 8.9, can be excited using prism coupling. Alternatively, end-fire

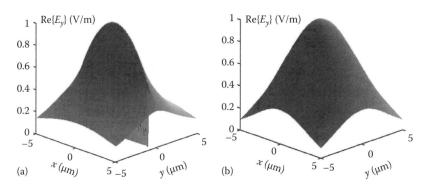

FIGURE 8.10   (a) ss_b^0 mode supported by a symmetric metal stripe (Figure 8.8a) for $t = 20$ nm, $w = 8$ μm, $\varepsilon_{r,m} = -131.95 - j12.65$, $\varepsilon_{r,1} = \varepsilon_{r,2} = (1.444)^2$, and $\lambda_0 = 1550$ nm. (b) Field distribution of a standard single-mode fiber (SMF-28, core diameter = 8.2 μm, numerical aperture = 0.14). The field distributions are normalized such that $\max|\mathrm{Re}\{E_y\}| = 1$.

coupling can be used if the input beam is small enough and aligned with the main lobe of the mode to be excited such that reasonable coupling efficiency is achieved.

The symmetric metal stripe ($\varepsilon_{r,1} = \varepsilon_{r,2}$) [17,18] also supports many supermodes, and in this structure trends are readily identifiable. For example, the evolution of the supermodes with decreasing t resembles the evolution of the s_b and a_b modes of the IMI in that all supermodes eventually become partitioned into either lower loss or higher loss modes, as determined by the symmetry or asymmetry of E_y along y: the ss_b^m and as_b^m modes are s_b-like, and the aa_b^m and sa_b^m modes are a_b-like.

The ss_b^0 supermode evolves smoothly and predictably as the metal film vanishes into the TEM wave supported by the background. As this happens, the mode's E_y and H_x fields evolve from being highly localized near the metal corners to being distributed smoothly over the waveguide cross section. This evolution is accompanied by a reduction in MPA of a few orders of magnitude due to reduced field penetration into the metal. The resulting field distribution can be well-matched to polarization-aligned modes of dielectric waveguides such as single-mode fiber, leading to efficient end-fire coupling in the scheme of Figure 8.7. The ss_b^0 supermode therefore evolves into the LRSPP in the symmetric metal stripe.

Figure 8.10a shows the E_y field distribution of the ss_b^0 supermode (LRSPP) supported by a symmetric metal stripe for $t = 20$ nm, $w = 8$ μm, $\varepsilon_{r,m} = -131.95 - j12.65$ (Au), $\varepsilon_{r,1} = \varepsilon_{r,2} = (1.444)^2$ (SiO$_2$), and $\lambda_0 = 1550$ nm. The MPA in this case is 0.93 dB/mm, which is at least one order of magnitude smaller than the MPA of the corresponding single-interface SPP. For comparison, Figure 8.10b shows the field distribution of a single-mode fiber having a core diameter of 8.2 μm and a numerical aperture of 0.14 (SMF-28); the coupling loss in the scheme of Figure 8.7 computed between the modes of Figure 8.10a and b is 0.27 dB per facet.

The as_b^0 supermode evolves in a similar manner as the metal vanishes except that its E_y field develops two extrema along the stripe width and the mode cuts off for certain stripe dimensions. This mode evolves with decreasing t into the first long-range higher-order mode. Long-range supermodes of order higher than the as_b^0 supermode also exist, originating from the families ss_b^m ($m > 0$, odd) and as_b^m ($m > 0$, even). They have cut-off dimensions that increase with mode order m. All supermodes of the aa_b^m and sa_b^m families exhibit increasing loss with decreasing t and do not butt-couple efficiently with Gaussian-like beams. This is due to the asymmetry in the distribution of E_y along y for all of these supermodes.

So the symmetric metal stripe can be dimensioned such that the ss_b^0 mode propagates with low loss, good horizontal and vertical confinement, and butt-couples efficiently to the TM mode of a dielectric waveguide. Furthermore, all long-range higher-order modes can be cut off and any remaining modes are unexcited or excited with very low efficiency and rapidly absorbed. These attributes render the symmetric metal stripe useful for application to integrated optical structures and devices.

8.3.2 Passive Components Based on the LRSPP in the Metal Stripe

Excitation of the LRSPP in the symmetric metal stripe is readily achieved experimentally by butt-coupling polarization-maintaining single-mode fibers [34–36] to the end facets of the waveguides. The LRSPP was first excited in this manner along an 8 μm wide 20 nm thick Au stripe buried in SiO_2 at $\lambda_0 = 1550$ nm [34].

The MPA and coupling efficiency were measured at $\lambda_0 = 1550$ nm in straight waveguides constructed from metal stripes on 15 μm of SiO_2 on Si and covered with index-matched polymer [35]. Au stripes 31, 25, and 20 nm thick, and Ag stripes 20 nm thick, were used. The lowest MPA measured among these sets were 0.42 and 0.32 dB/mm ($L_e \sim 1$ and 1.4 cm), respectively. These MPAs are 93× and 138× lower than those of the corresponding Au and Ag single-interface SPPs, respectively. The largest coupling efficiency measured among the set of Au structures was 98%, corresponding to a coupling loss of 0.09 dB per facet. Theoretical results were obtained for all of the structures characterized, and theory and experiment agreed to within about 5% for the 31 and 25 nm thick Au structures.

Insertion loss measurements were reported at $\lambda_0 = 1550$ nm for the LRSPP in Au stripes of various lengths, 3–12 μm wide and 10 nm thick, on 15 μm of benzocyclobutene (BCB) spin-coated and cured on Si, and covered by 15 μm of BCB [36]. MPAs in the range of about 0.6–0.8 dB/mm and coupling losses of about 0.5 dB were estimated.

The design of curved metal stripes is delicate because a vanishing stripe supporting the LRSPP does so with low loss but also with low confinement compromising the ability of the LRSPP to round bends due to excessive radiation. The MoL in cylindrical coordinates was used to model the LRSPP along curved symmetric metal stripes of various dimensions [37]. The 90° insertion and radiation losses, the effective index, mode contours and profiles, and the transition loss to a straight section were computed as a function of the radius of curvature. An optimal radius of curvature minimizing the insertion loss exists for each design, with the loss on one side of the optimal radius dominated by propagation loss and the loss on the other side dominated by radiation loss.

Straight and curved waveguides [34–38] are the main building blocks from which more complex integrated structures can be constructed, such as S-bends, four-port couplers, Y-junctions and Mach–Zehnder interferometers [39–41]. Figure 8.11 shows LRSPP mode outputs measured at $\lambda_0 = 1550$ nm from various passive elements implemented with the symmetric metal stripe [39]. In these structures, $t = 25$ nm thick Au was used for the stripes, 15 μm of SiO_2 was used as the lower cladding, and thick index-matched polymer was used as the upper cladding. Figure 8.11a shows the output of straight waveguides of the same length but of different widths, $w = 2, 4, 6,$ and 8 μm (top to bottom); the LRSPP is observed to become more tightly bound and more attenuated as the width of the stripe increases. Figure 8.11b shows the output of S-bend waveguides implemented with $w = 8$ μm wide stripes and different radii of curvature, $R = \infty, 20, 15, 10,$ and 5 mm (top to bottom); a reduction in output intensity is noted for $R = 10$ and 5 mm compared to the larger radii. Figure 8.11c shows the output of a Y-junction implemented with $w = 8$ μm wide stripes and $R = 12.53$ mm; an even split is noted in the output intensities. Figure 8.11d shows the output of a series of couplers implemented with $w = 8$ μm wide stripes, a coupling length CL = 1.5 mm, and edge-to-edge stripe separations $S = 8, 7, 6, 5, 4, 3,$ and 2 μm (top to bottom), with light launched into the left input waveguide in each case; an increasingly large fraction of the input power is noted to emerge from the coupled port as S decreases, with 3 dB coupling and total coupling occurring for $S \sim 5$ and 3 μm, respectively.

8.3.3 Bragg Gratings Based on the LRSPP in the Metal Stripe

Bragg gratings operating in the LRSPP can be implemented in the symmetric metal stripe as a step-in-width [42–44] or a step-in-thickness [45,46] periodic perturbation applied to the stripe. In such gratings the reflected wave originates from the multiplicity of small perturbations in effective index associated with the steps. A simple model for the gratings was proposed consisting of cascaded dielectric slabs,

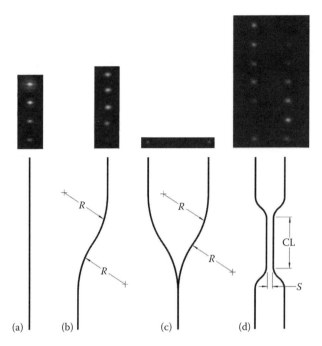

FIGURE 8.11 ss_b^0 mode output measured at $\lambda_0 = 1550\,\text{nm}$ for various passive elements implemented with the symmetric metal stripe (Figure 8.8a) using Au as the stripe, SiO_2 as the lower cladding and index-matched polymer as the upper cladding. (a) Output of straight waveguides of width $w = 2$, 4, 6, and 8 μm (top to bottom). (b) Output of S-bend waveguides, $w = 8\,\mu\text{m}$, and radius of curvature $R = \infty$, 20, 15, 10, and 5 mm (top to bottom). (c) Output of a Y-junction, $w = 8\,\mu\text{m}$ and $R = 12.53\,\text{mm}$. (d) Output of couplers, $w = 8\,\mu\text{m}$, coupling length CL = 1.5 mm, and separation $S = 8$, 7, 6, 5, 4, 3, and 2 μm (top to bottom); light is launched in the left input waveguide.

each slab taking on the complex effective index of the LRSPP supported by the associated waveguide segment as determined using a mode solver for uniform waveguides [43,44].

Figure 8.12a shows in isometric view a uniform periodic step-in-width Bragg grating of length L_g and Figure 8.12b shows in top view a unit cell of the grating [42]. Figure 8.12c compares the measured and computed reflectance responses of a third-order uniform grating implemented as a 3 mm long $t = 22\,\text{nm}$ thick $w_1 = 8\,\mu\text{m}$ wide Au stripe periodically stepped in width to $w_2 = 2\,\mu\text{m}$, deposited on SiO_2 on Si with an index-matched polymer used as the top cladding, and characterized by butt-coupling to polarization-maintaining single-mode fibers [43]. The Au stripe was patterned using contact lithography. The maximum reflectance of this design is about 40% at the center (Bragg) wavelength of $\lambda_B = 1544.22\,\text{nm}$ and its bandwidth is about 0.3 nm full width at half maximum. The inset located at the bottom center of Figure 8.12c shows the measured LRSPP output from the grating. A very low level of background radiation was observed from images captured over a broad wavelength range including on resonance indicating a very low level of scattering. A maximum reflectance close to 97% is predicted for first-order gratings [44].

8.3.4 Active Components Based on the LRSPP in the Metal Stripe

Thermo-optic components operating in the LRSPP have been implemented in the symmetric metal stripe [47–51]. In such components, the stripe can also be used as an Ohmic heating element driven by the passage of current, significantly improving the efficiency due to the excellent overlap that is achieved between the heated region and the LRSPP. Optically noninvasive electrical contacts to the metal stripe are achieved through patterning such that a few contact arms and pads extend from the stripe along the

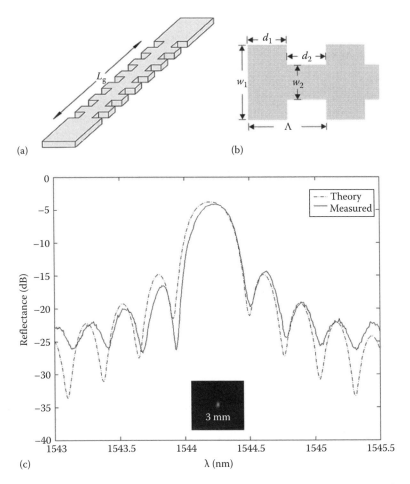

(a)   (b)

(c)

FIGURE 8.12 (a) Isometric view of a step-in-width Bragg grating of length L_g. (b) Top view of a unit cell of a step-in-width Bragg grating. (c) Measured and theoretical reflection responses of a third-order step-in-width Bragg grating. The inset shows a measured mode output from this 3 mm long grating.

perpendicular direction. Knowing the electromigration current density limit of the metal and operating well away from this limit is essential to avoid burnout.

Thermo-optic variable optical attenuators (VOAs) based on mode cut-off induced by index asymmetry in a straight waveguide section were demonstrated using Au on SiO_2 with an index-matched polymer uppercladding [47,48]. The Au stripe in the devices of [48] was used to drive the VOA through Ohmic heating by passing current through the stripe, and as a temperature monitor by simultaneously measuring its resistivity. Similarly, a thermo-optic switch implemented as a four-port coupler [49], a thermo-optic VOA implemented as a Mach–Zehnder interferometer [49], and thermo-optic VOAs implemented as a straight section [50,51] were demonstrated using Au stripes in polymer.

Electro-optic components operating in the LRSPP have also been implemented in the symmetric metal stripe [52]. The design of straight and curved waveguides, Bragg gratings, and modulators was considered. Structures consisting of metal stripes buried in $LiNbO_3$ were fabricated by wafer-bonding, and intensity modulation of light by the structure was demonstrated.

(Bio)chemical sensors operating in the LRSPP have also been implemented in the symmetric metal stripe [21,53,54]. In order to propagate the LRSPP along the stripe in any liquid or gaseous medium irrespective of its refractive index, the stripe was supported by an optically not-too-invasive thin free-standing dielectric membrane, and the sensing medium was allowed to surround the structure thus essentially ensuring index symmetry ($\varepsilon_{r,1} = \varepsilon_{r,2}$) [53]. Operation of such structures in air and various liquids was demonstrated [53,54].

In order to overcome the residual attenuation of the LRSPP, the incorporation of an optically-pumped gain medium near the metal stripe was investigated theoretically [55]. A model was proposed taking into account the nonuniformity of the gain medium close to the metal surface due to position-dependent dipole lifetime and pump irradiance, and then used to model the LRSPP propagating along a physically realizable structure comprising R6G dye molecules. The results indicate that net amplification is possible at visible wavelengths using reasonable pump power and molecular concentration.

8.4 Concluding Remarks

Research on SPP waveguides and components is proceeding at a vigorous pace motivated by the interesting and potentially useful properties of SPP waves, such as an energy asymptote in their dispersion curve, high surface and bulk sensitivities, and sub-wavelength confinement. SPP waveguides are well-suited to applications where such features are exploited or where significant advantage can be derived from having a metal(s) collocated with the optical path, such as using them as heating elements in thermo-optic devices, as electrodes in electro-optic devices, as contacts in charge-carrier devices, or as a surface onto which receptor chemistries are attached for (bio)chemical sensor applications.

Materials effects for device applications are generally weak, and surface mass loading in (bio)chemical sensors usually occurs as a (sub-)monolayer (nm's), so significant phase changes (for example) need to be accumulated over long propagation lengths, suggesting architectures using long-range waves. Also, using a long-range wave is advantageous if net amplification and lasing is to be achieved. These perspectives motivate research on LRSPP-based components, and the symmetric metal stripe is a suitable structure for propagating them. All of the fundamental passive elements required to enable an integrated technology based on the symmetric metal stripe and operating in the LRSPP have been demonstrated and robust theoretical models have been constructed and validated experimentally. Most of the implementation reported to date has occurred using Au and low index claddings targeting operation near $\lambda_0 = 1310$ and $1550\,\mathrm{nm}$ but other wavelengths and materials can be used. Promising applications include thermo-optic devices and (bio)chemical sensors addressing a variety of needs.

Acknowledgments

The author is grateful to past and present coworkers, Robert Charbonneau, Stéphanie Jetté-Charbonneau, Ian Breukelaar, Christine Scales, Junjie Lu, Guy Gagnon, Nancy Lahoud, Greg Mattiussi, Robin Buckley, and Ewa Lisicka-Skrzek.

References

1. H. Raether, *Surface Plasmons on Smooth and Rough Surfaces and on Gratings*, Springer, Berlin, 1988.
2. W. L. Barnes, Surface plasmon-polariton length scales: A route to sub-wavelength optics, *J. Opt. A Pure Appl. Opt.*, 8, S87–S93, 2006.
3. J. J. Burke, G. I. Stegeman, and T. Tamir, Surface-polariton-like waves guided by thin, lossy metal films, *Phys. Rev. B*, 33, 5186–5201, 1986.
4. R. Zia, M. D. Selker, P. B. Catrysse, and M. L. Brongersma, Geometries and materials for subwavelength surface plasmon modes, *J. Opt. Soc. Am. A*, 21, 2442–2446, 2004.
5. P. Berini, Figures of merit for surface plasmon waveguides, *Opt. Express*, 14, 13030–13042, 2006.
6. E. N. Economou, Surface plasmons in thin films, *Phys. Rev.*, 182, 539–554, 1969.
7. J. A. Dionne, L. A. Sweatlock, H. A. Atwater, and A. Polman, Plasmon slot waveguides: Towards chip-scale propagation with subwavelength-scale localization, *Phys. Rev. B*, 73, 035407, 2006.

8. N. W. Ashcroft and N. D. Mermin, *Solid State Physics*, Saunders College, Philadelphia, PA, 1976.

9. E. D. Palik (Ed.), *Handbook of Optical Constants of Solids*, Academic Press, Orlando, FL, 1985.

10. P. Winsemius, F. F. van Kampen, H. P. Lengkeek, and C. G. van Went, Temperature dependence of the optical properties of Au, Ag and Cu, *J. Phys. F Metal Phys.*, 6, 1583–1606, 1976.

11. G. Leveque, C. G. Olson, and D. W. Lynch, Reflectance spectra and dielectric functions for Ag in the region of interband transitions, *Phys. Rev. B*, 27, 4654–4660, 1983.

12. B. Brixner, Refractive-index interpolation for fused silica, *J. Opt. Soc. Am.*, 57, 674–676, 1967.

13. D. J. Nash and J. R. Sambles, Surface plasmon-polariton study of the optical dielectric function of silver, *J. Mod. Opt.*, 43, 81–91, 1996.

14. E. T. Arakawa, M. W. Williams, R. N. Hamm, and R. H. Ritchie, Effect of damping on surface plasmon dispersion, *Phys. Rev. Lett.*, 31, 1127–1129, 1973.

15. J. Takahara, S. Yamagishi, H. Taki, A. Morimoto, and T. Kobayashi, Guiding of a one-dimensional optical beam with nanometer diameter. *Opt. Lett.*, 22, 475–477, 1997.

16. R. Buckley and P. Berini, Figures of merit for 2D surface plasmon waveguides and application to metal stripes, *Opt. Express*, 15, 12174–12182, 2007.

17. P. Berini, Plasmon-polariton modes guided by a metal film of finite width, *Opt. Lett.*, 24, 1011–1013, 1999.

18. P. Berini, Plasmon-polariton waves guided by thin lossy metal films of finite width: Bound modes of symmetric structures, *Phys. Rev. B*, 61, 10484–10503, 2000.

19. R. Buckley and P. Berini, Radiation suppressing metallo-dielectric optical waveguides, *J. Lightw. Technol.*, 27, 2800–2808, 2009.

20. D. K. Cheng, *Field and Wave Electromagnetics*, 2nd ed., Addison-Wesley, Reading, MA, 1989.

21. P. Berini, Bulk and surface sensitivities of surface plasmon waveguides, *New J. Phys*, 10, 105010, 2008.

22. P. Berini, Plasmon-polariton modes guided by a metal film of finite width bounded by different dielectrics, *Opt. Express*, 7, 329–335, 2000.

23. P. Berini, Plasmon-polariton waves guided by thin lossy metal films of finite width: Bound modes of asymmetric structures, *Phys. Rev. B*, 63, 125417, 2001.

24. J.-C. Weeber, A. Dereux, C. Girard, J. R. Krenn, and J.-P. Goudonnet, Plasmon polaritons of metallic nanowires for controlling submicron propagation of light, *Phys. Rev. B*, 60, 9061–9068, 1999.

25. B. Lamprecht, J. R. Krenn, G. Schider, H. Ditlbacher, M. Salerno, N. Felidj, A. Leitner, and F. R. Aussenegg, Surface plasmon propagation in microscale metal stripes, *Appl. Phys. Lett.*, 79, 51–53, 2001.

26. J.-C. Weeber, J. R. Krenn, A. Dereux, B. Lamprecht, Y. Lacroute, and J.-P. Goudonnet, Near-field observation of surface plasmon polariton propagation on thin metal stripes, *Phys. Rev. B*, 64, 045411, 2001.

27. R. Zia, M. D. Selker, and M. L. Brongersma, Near-field characterization of guided polariton propagation and cutoff in surface plasmon waveguides, *Phys. Rev. B*, 74, 165415, 2006.

28. I. V. Novikov and A.A. Maradudin, Channel polaritons, *Phys. Rev. B*, 66, 035403, 2002.

29. D. F. P. Pile and D. K. Gramotnev, Channel plasmon-polariton in a triangular groove on a metal surface, *Opt. Lett.*, 29, 1069–1071, 2004.

30. D. F. P. Pile, T. Ogawa, D. K. Gramotnev, Y. Matsuzaki, K. C. Vernon, K. Yamaguchi, T. Okamoto, M. Haraguchi, and M. Fukui, Two-dimensionally localized modes of a nanoscale gap plasmon waveguide, *Appl. Phys. Lett.*, 87, 261114, 2005.

31. G. Veronis and S. Fan, Guided subwavelength plasmonic mode supported by a slot in a thin metal film, *Opt. Lett.*, 30, 3359–3361, 2005.

32. S. I. Bozhevolnyi, V. S. Volkov, E. Devaux, and T. W. Ebbesen, Channel plasmon-polariton guiding by subwavelength metal grooves, *Phys. Rev. Lett.*, 95, 046802, 2005.

33. L. Liu, Z. Han, and S. He, Novel surface plasmon waveguide for high integration, *Opt. Express*, 13, 6645–6650, 2005.

34. R. Charbonneau, P. Berini, E. Berolo, and E. Lisicka-Skrzek, Experimental observation of plasmon-polariton waves supported by a thin metal film of finite width, *Opt. Lett.*, 25, 844–846, 2000.

35. P. Berini, R. Charbonneau, N. Lahoud, and G. Mattiussi, Characterisation of long-range surface plasmon-polariton waveguides, *J. Appl. Phys.*, 98, 043109, 2005.

36. R. Nikolajsen, K. Leosson, I. Salakhutdinov, and S. I. Bozhevolnyi, Polymer-based surface-plasmon-polariton stripe waveguides at telecommunication wavelengths, *Appl. Phys. Lett.*, 82, 668–670, 2003.

37. P. Berini and J. Lu, Curved long-range surface plasmon-polariton waveguides, *Opt. Express*, 14, 2365–2371, 2006.

38. A. Degiron and D. Smith, Numerical simulations of long-range plasmons, *Opt. Express*, 14, 1611–1625, 2006.

39. R. Charbonneau, N. Lahoud, G. Mattiussi, and P. Berini, Demonstration of integrated optics elements based on long-ranging surface plasmon polaritons, *Opt. Express*, 13, 977–984, 2005.

40. A. Boltasseva, T. Nikolajsen, K. Leosson, K. Kjaer, M. S. Larsen, and S. I. Bozhevolnyi, Integrated optical components utilizing long-range surface plasmon polaritons, *J. Lightw. Technol.*, 23, 413–422, 2005.

41. R. Charbonneau, C. Scales, I. Breukelaar, S. Fafard, N. Lahoud, G. Mattiussi, and P. Berini, Passive integrated optics elements based on long-ranging surface plasmon polaritons, *J. Lightw. Technol.*, 24, 477–494, 2006.

42. S. Jetté-Charbonneau, R. Charbonneau, N. Lahoud, G. Mattiussi, and P. Berini, Demonstration of Bragg gratings based on long-ranging surface plasmon polariton waveguides, *Opt. Express*, 13, 4674–4682, 2005.

43. S. Jetté-Charbonneau, R. Charbonneau, N. Lahoud, G. Mattiussi, and P. Berini, Bragg gratings based on long-range surface plasmon-polariton waveguides: Comparison of theory and experiment, *IEEE J. Quantum Electron.*, 41, 1480–1491, 2005.

44. S. Jetté-Charbonneau and P. Berini, Theoretical performance of Bragg gratings based on long-range surface plasmon-polariton waveguides, *J. Opt. Soc. Am. A*, 23, 1757–1767, 2006.

45. S. I. Bozhevolnyi, A. Boltasseva, T. Søndergaard, T. Nikolajsen, and K. Leosson, Photonic bandgap structures for long-range surface plasmon polaritons, *Opt. Commun.*, 250, 328–333, 2005.

46. A. Boltasseva, S. I. Bozhevolnyi, T. Nikolajsen, and K. Leosson, Compact Bragg gratings for long-range surface plasmon polaritons, *J. Lightw. Technol.*, 24, 912–918, 2006.

47. I. Breukelaar, R. Charbonneau, and P. Berini, Long-range surface plasmon-polariton mode cutoff and radiation in embedded strip waveguides, *J. Appl. Phys.*, 100, 043104, 2006.

48. G. Gagnon, N. Lahoud, G. Mattiussi, and P. Berini, Thermally activated variable attenuation of long-range surface plasmon-polariton waves, *J. Lightw. Technol.*, 24, 4391–4402, 2006.

49. T. Nikolajsen, K. Leosson, and S. I. Bozhevolnyi, Surface plasmon polariton based modulators and switches operating at telecom wavelengths, *Appl. Phys. Lett.*, 85, 5833–5835, 2004.

50. T. Nikolajsen, K. Leosson, and S. I. Bozhevolnyi, In-line extinction modulator based on long-range surface plasmon polaritons, *Opt. Commun.*, 244, 455–459, 2005.

51. S. Park and S. H. Song, Polymeric variable optical attenuator based on long range surface plasmon polaritons, *Electron. Lett.*, 42, 20060236, 2006.

52. P. Berini, R. Charbonneau, S. Jetté-Charbonneau, N. Lahoud, and G. Mattiussi, Long-range surface plasmon-polariton waveguides and devices in lithium niobate, *J. Appl. Phys.*, 101, 113114, 2007.

53. P. Berini, R. Charbonneau, and N. Lahoud, Long-range surface plasmons on ultrathin membranes, *Nano Lett.*, 7, 1376–1380, 2007.

54. P. Berini, R. Charbonneau, and N. Lahoud, Long-range surface plasmons along membrane-supported metal stripes, *IEEE J. Sel. Top. Quantum Electron.*, 14, 1479–1495, 2008.

55. I. De Leon and P. Berini, Theory of surface plasmon-polariton amplification in planar structures incorporating dipolar gain media, *Phys. Rev. B*, 78, 161401(R), 2008.

VI

Integrated Photonic Technologies: Quantum Devices and Integration

9

Quantum Dot Lasers: Theory and Experiment

Peter M. Smowton
Peter Blood

9.1 Introduction

The MASER and later the LASER were conceived as devices that exploited optical transitions between electron states of atoms or molecules,[*] though Bernard and Duraffourg (1961) pointed out that laser action could also be achieved between the continua of states forming the valence and conduction bands of semiconductors. In this realization, electrons are thermally distributed across the continuum of energy states and, while laser action is provided by transitions between a small number of states separated by the laser transition energy, the drive current has to supply carriers lost by recombination between states across the full thermal distribution. The availability of a continuum of energy states also enables the carrier distribution to spread out as the temperature is increased, leading to an increase in threshold with increasing temperature. Nevertheless, there are significant advantages to be derived using semiconductor diodes as lasers due to direct electrical pumping by a p-n junction, leading to the development of compact, low-cost, battery-powered, portable sources of coherent light that can also be integrated into optoelectronic systems. The use of quantum well semiconductor structures has enabled these advantages to be retained while using confinement to restrict the number of states available for

---

[*] Two general books give excellent accounts of the historical development of the laser: *The History of the Laser*, Mario Bertollotti (IoP Publishing, Philadelphia, PA, 2005), and *How the Laser Happened*, Charles H. Townes (Oxford University Press, New York, 1999).

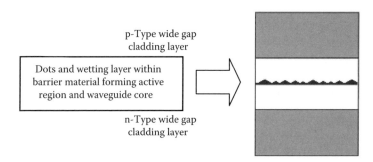

FIGURE 9.1 Schematic diagram showing a layer of dots within the core of a slab waveguide with n- and p-type cladding layers, forming the active region of a diode laser.

transitions outside the laser spectrum, thereby reducing the threshold current and its temperature dependence. Quantum dots restrict electron motion in all three directions, leading to well-defined atomic-like states, reminiscent of the original laser concept, with recombination only occurring at the photon energies of the laser emission. Furthermore, if these states are separated by more than a few integer multiples of (k_BT), the carriers cannot be redistributed in energy as the temperature is increased, leading to a low, temperature insensitive, threshold current, as pointed out originally by Arakawa and Sakaki (1982) and treated in greater detail by Asada et al. (1986).

The quantum dot laser offered the prospect of combining the merits of discrete, atomic-like states with those of direct electrical injection from a p-n junction. While these merits have not been fully realized, due chiefly to the distribution of dot sizes in an ensemble leading to an inhomogeneous distribution of state energies, quantum dot lasers have led to improvements in device performance as described in Section 9.5.1. Furthermore, device structures containing dots and incorporating the waveguide and p-n junction necessary for a laser diode (Figure 9.1) can be made by conventional epitaxial crystal growth processes currently used for quantum well structures. In the longer term it may prove possible to utilize colloidal nano-crystals to produce large area emitters by relatively simple technologies,* though probably at the cost of forgoing the convenience of direct electrical injection.

The purpose of this chapter is to provide an introduction to the basic concepts of quantum dot lasers, covering their formation, electronic and optical properties, and their recombination and optical gain characteristics. This provides the basis for a survey, in the later sections of the chapter, of the current status of device performance with particular attention to threshold current and its temperature dependence and concludes with remarks on the use of quantum dots in integrated systems. We assume a basic knowledge of semiconductor physics and the principles of laser action in semiconductor structures such as covered in texts by Coldren and Corzine (1995) and Chuang (1995). Furthermore, we focus attention on the distinctive characteristics of the quantum dot gain medium and do not provide an account of waveguide characteristics and device structures, both of which are similar to quantum well lasers. Two early books provide valuable background reading on quantum dot lasers that add detail to the account presented here: the monograph by Bimberg et al. (1999) (abbreviated as BGL, 1999) and the multi-author text edited by Sugawara (1999).

9.2 Formation of Dots

Quantum dots are regions where electrons and holes are spatially localized in all three directions within dimensions comparable with their de Broglie wavelength or excitonic Bohr radius. The confining potential is established by a semiconductor embedded in a matrix of wider-gap "barrier" material. Quantum dots may be formed within a waveguide by a number of processes, though realization of devices with useful performance only came with the development of growth of self-assembled dots by

---

* http://www.engineering.utoronto.ca/news/news_4192006.htm

the Stranski–Krastanow process in the mid-1990s—the explosion of interest in quantum dot lasers dates from this time (see BGL, 1999 for a detailed account). Conventional Frank–van der Merwe or layer-by-layer epitaxial growth utilizes lattice-matched materials although dislocation-free mismatched layers with a strain of a few percent, with thickness limit set approximately by the Matthews–Blakeslee critical thickness criterion can also be produced. For greater mismatch, growth proceeds by the Stranski–Krastanow mode, where after the deposition of about 1.5–2 monolayers of material forming a wetting layer, clusters begin to nucleate with the eventual formation of islands. The driver for island formation is the balance between the total energy of a strained film and that of the same number of atoms in a strain-free island. Dots may be grown by molecular beam epitaxy and metal organic vapor deposition. In the specific case of deposition of InAs on GaAs where the mismatch is about 7%, dislocation-free, coherent islands of InAs are formed, together with a continuous wetting layer, which have the shape of flat pyramids lying in the plane of the growth surface with base width of 10s of nm, height of a few nm, and typical densities of 10^{10}–10^{11} dots cm^{-2} in a single layer. The wetting layer is sufficiently thin that it forms a quantum well and some dot structures are grown in such a way as to place the dots within a purpose-grown quantum well to form a "Dot in a well" or DWELL structure (Lester et al. 1999, Liu et al. 2000).

The total dot density may be increased by stacking several dot layers above each other, and most laser diodes employ such structures to obtain sufficient optical gain. In closely stacked structures, there is evidence that the dots are aligned vertically, probably by the residual strain field of the underlying dot layers, and the electronic properties may be modified further by quantum mechanical coupling.

The shape and size of the dots is determined by the growth conditions, particularly temperature and the amount of material deposited, and is also dependent upon the index of the surface on which the material is deposited. In practice, the dots are rarely pure InAs, but rather InGaAs alloy of uncertain composition; likewise the wetting layer composition may not be precisely known. Since the nucleation of dots is a random process, in any ensemble there is a distribution of dot sizes and possibly composition that causes an inhomogeneous broadening of the optical spectra. It is possible to determine the size distribution of a sample of dots using atomic force microscopy (AFM) on an uncapped layer of dots and to examine the effect of modifications in the growth process (an example is shown in Figure 9.2), but the results may not be typical of the dots in the final device structure because capping and thermal annealing during the growth of subsequent layers may change the dot properties. Typical size distributions have a width at half height of a few nm about a mean value of 6–12 nm.

9.3 Energy States

In principle, electrons and holes are localized in a dot by the conduction and valence band offsets between dot material and the wider-gap wetting layer with the allowed energy states obtained by solution of Schrödinger's equation in all three directions. These solutions are determined by the dimensions of the sample and for a single *quantum well* the dimensions are very large (~1 mm) in the plane of the layer so the solutions are a continuum of energies and electron motion is permitted throughout the plane of the sample. Perpendicular to the plane, the layer thickness is very small and the states are widely spaced in energy and localized to the quantum well. This continuum of states enables an equilibrium carrier population to be established throughout the plane of a single well, though multiple wells remain quantum-mechanically isolated. In a *quantum dot*, the size of a single dot, in the growth plane as well as the growth direction, is such that the carrier is localized at the dot in all three directions and the allowed states form a series of discrete energy levels made up of contributions from the quantization condition in each direction. The concept can be illustrated by considering an infinitely deep, three-dimensional rectangular quantum "box" with orthogonal dimensions L_x, L_y, and L_z. The energy states are given by

$$E = E_{nx} + E_{ny} + E_{nz} = \frac{\hbar^2}{2m^\star}\left\{\left(\frac{n_x\pi}{L_x}\right)^2 + \left(\frac{n_y\pi}{L_y}\right)^2 + \left(\frac{n_z\pi}{L_z}\right)^2\right\} \tag{9.1}$$

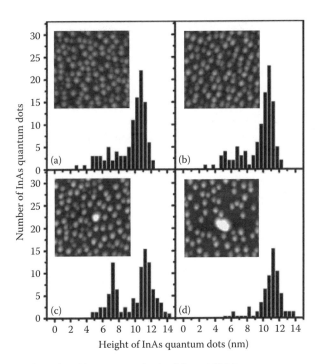

FIGURE 9.2 AFM images (insets) and histograms obtained from AFM data on separate structures where growth is stopped after a particular layer. Growth is stopped after the first (a) and fifth (b) layers for structures grown with the so-called HGTSL approach (see Section 9.5.2) with the intention of reducing the number of large defective dots that are evident in the second (c) and fifth (d) layers of structures grown without such an approach. The AFM image size is 500×500 nm, and by using the histograms, a dot density of 3×10^{-10} cm^{-2} can be measured for (a) and (b). (Reprinted from Liu, H.Y. et al., *J. Appl. Phys.*, 96, 1988, 2004. With permission. Copyright 2004, American Institute of Physics.)

and are defined by combinations of the positive integers (n_x, n_y, n_z). The wave function of each state is made up of three orthogonal components:

$$F(\mathbf{r}) = F_{nx}(\mathbf{x}) F_{ny}(\mathbf{y}) F_{nz}(\mathbf{z}) \tag{9.2}$$

and the amplitudes of these components are normalized such that the probability of finding an electron in a state is unity. The ground state is given by the state numbers (1, 1, 1) and a ladder of higher states, termed excited states, is formed by integer steps of these numbers (1, 1, 2), etc., the sequence in energy depending upon the relative dimensions.

In a potential well of finite depth, the wave functions penetrate some distance into the adjacent barrier material, though it is usually assumed that the dot spacing is such that the wave functions of adjacent dots do not overlap. The energy spacing of transitions between successive pairs of states is often observed to be approximately the same, so it has been suggested (Deppe et al. 1999, Park et al. 2000) that the carriers are confined by a harmonic potential $V = (C/2)(r_i)^2$, where r_i is distance in one of the orthogonal directions. The reason for the potential being of this form are unclear; it may arise from composition variation in the dot (Deppe et al. 1999). For a harmonic potential in one dimension

$$E_n = \left(n + \frac{1}{2}\right) h\nu \tag{9.3}$$

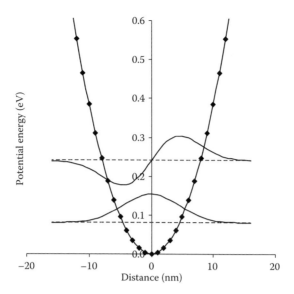

FIGURE 9.3 Ground and first excited state energy levels and wave functions for a harmonic potential.

and the natural frequency is given by

$$\nu = \frac{1}{2\pi}\sqrt{\frac{C}{m^*}} \tag{9.4}$$

where m^* is the effective mass of the carrier. The size of the potential well determines the constant C. Figure 9.3 shows a parabolic potential in the z-direction (growth direction) and the wave functions for ground and first excited state. For flat pyramidal dots, if the dimensions L_x, L_y are equal, then the states (2, 1, 1) and (1, 2, 1) have the same energy and the first excited state is doubly degenerate (Kim et al. 2003). A number of detailed calculations of dot states are described in BGL (1999, section 5.2).

While the precise form of the potential is usually uncertain and the composition and size of the dot are not well known, a number of general features emerge. Over the range of sizes that give confinement effects, the number of states is small and does not increase in a continuous manner with the volume of the dot, but in a stepwise fashion as further states are bound; consequently, it is not meaningful to define a number of dot states per unit volume. Since the hole effective mass is usually greater than the electron effective mass, the hole states are much closer in energy, and although the valence band offsets are smaller than for the conduction band, this is likely to result in a larger number of confined hole states.

The quantum states in an individual dot possess two important characteristics that distinguish them from quantum wells and bulk materials:

1. The energies are not closely spaced to form a continuum or band but are sufficiently separated that distinct transitions can be observed. Concepts of semiconductor physics developed for extended state systems (wells and bulk materials) such as bands, plane-wave states, and densities of states are not immediately transferable to dots.
2. The wave functions are localized at a dot in all three directions. Electrons can only move from one dot to another by quantum mechanical tunneling or by excitation into the continuum states of the wetting layer where they are free to move to be subsequently captured at another dot. The latter is the dominant inter-dot transport mechanism. The consequences of localization are that (a) the distribution of carriers among an ensemble of dots depends upon exchange of carriers with the wetting layer and thermal equilibrium among the dots in an ensemble may not always be achieved and (b) recombination can only occur between electrons and holes in the same dot.

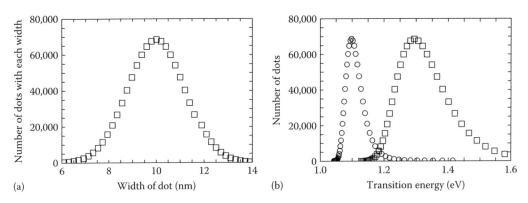

FIGURE 9.4   Figure (a) shows a Gaussian distribution in dot size in the z-direction (standard deviation 1.2 nm) and (b) shows the dot number distribution as a function of transition energy for ground (circles) and excited (squares) states computed from this size distribution using a harmonic potential. (From Blood, P. and Pask, H.J., PhD, Cardiff University, Wales, U.K., unpublished.)

We noted in Section 9.2 that the growth process results in dots of a range of sizes and it follows that this results in a range of energy states for dots in an ensemble. If the randomness occurs in the *size* of the dots, then we can represent the distribution of dot size L by a Gaussian function centered on a mean size L_{mean} and standard deviation σ_L. The probability of a dot having a size in the range $L \pm (\Delta L/2)$ is

$$P(L)\Delta L = \frac{1}{\sigma_L\sqrt{2\pi}}\exp\left\{-\frac{\left(L - L_{mean}\right)^2}{2\sigma_L^2}\right\}\Delta L \tag{9.5}$$

The distribution of energy states arising from a Gaussian size distribution using a harmonic potential is not a Gaussian but has a high energy tail. This inhomogeneity is manifest as a distribution in transition energy, observable in absorption and emission spectra, given by the sum of the confined state energies for electrons (E_{1e}) and holes (E_{1h}) in the dot, plus the band gap of the dot material (E_{gdot}):

$$E_{trans} = E_{1e} + E_{1h} + E_{gdot} \tag{9.6}$$

A calculated transition energy distribution arising from a Gaussian size distribution using an harmonic potential is shown in Figure 9.4. It may be that a harmonic potential is not realized in practice; nevertheless, whatever the potential, we do not expect a Gaussian size distribution to result in a Gaussian distribution in energy. If, however, there is a Gaussian variation in composition from dot to dot, and the band gap of the dot is linearly related to composition, then we expect a Gaussian distribution in transition energy through Equation 9.6. In the analysis of spectra and calculation of gain characteristics, it is usually assumed that the distribution of transition energies is Gaussian. In experimental data it is almost impossible to distinguish a Gaussian in energy from the tailed energy distribution of a Gaussian size distribution (Figure 9.4), because inhomogeneous ground and excited state distributions usually overlap, so the high energy tail of the ground state distribution is obscured by excited state transitions.

Figure 9.5 illustrates schematically the interactions between an inhomogeneous distribution of dot states and the continuum states in the wetting layer. In a laser, electrons are injected from the p-n junction into the wetting layer (WL) where they are mobile in the plane and are captured into empty dot states. The capture process depends upon (1) the probability that an electron in the WL is in the vicinity of a dot, which is associated with the random motion of carriers in the WL, and (2) the ability of the carrier to lose energy and occupy one of the localized dot states. This energy loss between WL and confined states and between the confined states themselves is achieved by phonon emission or by an Auger-type

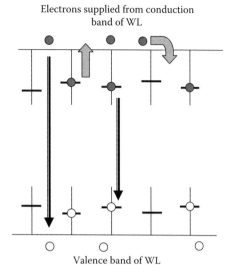

FIGURE 9.5 Schematic illustration of the interaction of carriers in the wetting layer with an inhomogeneous distribution of dots, showing processes of capture and emission of carriers at the dots and recombination between dot states containing an electron and a hole and between the conduction and valence bands of the wetting layer.

process involving excitation of an electron into a high energy continuum state in the WL (Uskov et al. 1998). The phonon emission process can only occur when the energy spacing of confined states is resonant with a multiple of phonon energies; however, due to the Auger process it is unlikely that there is a significant bottleneck to capture when this condition is not met. Electrons may also be thermally re-emitted from dots, and this is more likely for dot states closer to the wetting layer, consequently a thermal distribution of electrons among the localized dot states may be established through thermal coupling to the carrier reservoir in the wetting layer where the carriers are in equilibrium. There is good experimental evidence to show that at room temperature, a Fermi–Dirac function is a good representation of the distribution of electrons on the localized dot states, and this probably holds in many cases to temperatures below 200 K; however, below about 100 K the dots are probably populated randomly with carriers, irrespective of the energy of the states (Summers et al. 2001). The random population model has been developed by Grundmann and Bimberg (1997) to describe the behavior of dots in this regime. This coupled system of dot and wetting layer states forms the basis of a number of detailed rate equation analyses of quantum dot laser operation, particularly with regard to dynamical characteristics (e.g., Fiore and Markus 2007) and multiple state lasing (e.g., Markus et al. 2006). Huang and Deppe (2001) have described a non-equilibrium rate equation model in which the dots are coupled to a thermal distribution in the wetting layer and Summers and Rees (2007) have described how the Fermi–Dirac function may be modified for nonthermal conditions. A number of very detailed calculations of the electronic properties of dots have been published that are specific to the particular structure assumed. Examples include papers by Wojs et al. (1996), Stier et al. (1999), Williamson et al. (2000), and Andreev and O'Reilly (2005).

9.4 Absorption, Gain, and Recombination

9.4.1 Optical Transitions in Dots

Optical transitions can only occur between states within an individual dot, so to determine the behavior of an ensemble the probability of occupation of each individual dot by electrons and holes must be considered. There are two possible mechanisms:

- Each dot may be populated with electrons and holes independently.
- Due to Coulomb attraction a dot is populated simultaneously by an electron and a hole, that is, the dot is populated with excitons. This mechanism is assumed in the random population model (Grundmann and Bimberg 1997).

The measured inversion factor of carriers in an inhomogeneous ensemble of dots at room temperature can be represented by a Fermi–Dirac function using the measured lattice temperature (Summers et al. 2001) and numerical modeling shows that this would not be the case for dots populated simultaneously with electron and holes (Pask 2006). We conclude that at room temperature, dots are populated independently with occupation probability for each carrier type determined by a Fermi function dependent upon the energy of the state, and hence the size of the dot. On this basis, we consider absorption, gain, and recombination in the following sections.

9.4.2 Absorption Cross Section

We begin by considering the interaction of electromagnetic radiation of angular frequency ϖ with a single two-level dot system where the levels are separated by the energy E_i (Figure 9.6); we take the lower state to be occupied with an electron and the upper state to be empty. For an optical flux of Φ_0 photons per second incident upon unit area normal to the light beam and containing a *single* dot per unit area (1 dot per unit area), the fraction of light absorbed by a dot comprising a semiconductor of index n and having an interband transition matrix element M is

$$\frac{\Delta\Phi(\hbar\varpi)}{\Phi_0(\hbar\varpi)} = 1 \times \left[\frac{2 \times 4\pi\hbar}{cn\varepsilon_0\hbar\varpi} \left(\frac{e}{2m_0} \right)^2 M^2 \left\{ \int_{\text{dot}} F_1(\mathbf{r}) F_u(\mathbf{r}) d\mathbf{r} \right\}^2 \left\{ \frac{1}{\pi} \frac{\Delta E_{\text{hom}}}{(\hbar\varpi - E_i)^2 + (\Delta E_{\text{hom}})^2} \right\} \right] \quad (9.7)$$

where $F_u(\mathbf{r})$ and $F_1(\mathbf{r})$ are the envelope wave functions in three dimensions for the upper (conduction) and lower (valence) states, respectively (Equation 9.2); the integral represents their overlap.[*] The factor 2 takes account of the two transitions between two pairs of states of the same spins, one pair spin up and one pair spin down.

Equation 9.7 is derived by integrating the solution of the time-dependent Schrödinger equation over a period of time for which there is a coherent interaction between the electron and the optical field before the phase relation is destroyed. This period is called the "dephasing time" (usually called T_2) and it determines the homogeneous linewidth $\Delta E_{\text{hom}} = (\hbar/T_2)$ in Equation 9.7. (There may be further contributions to the homogeneous linewidth due to changes in the population of carriers occupying the states (T_1 time), in addition to dephasing.[†])

The final {} term in Equation 9.7 is the normalized Lorentzian homogeneous lineshape function $L(E_i, \hbar\varpi)$, which is plotted in Figure 9.7 as a function of photon energy relative to the state separation for a linewidth of 1 meV. A direct consequence of this term is that it is possible for a photon to interact with the two-level system even when its energy does not exactly equal the energy separation of the states,

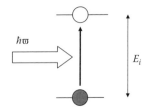

FIGURE 9.6 Diagram showing interaction of a light beam of photons of energy $\hbar\varpi$ with the two-level system of a quantum dot with state separation E_i.

---

[*] The derivation of this equation follows from textbook treatments of the interaction of light with a two-level system (e.g., Dicke and Wittke 1960, section 15-6), modified to take account of the fact that the levels are in a semiconductor matrix with spatially varying envelope functions. A general account of the light–matter interaction is given by Loudon 1983, see equation 2.5.2 and associated text.

[†] See Siegman (1986, p. 210), for a discussion of this notation, as well as a helpful introduction to the concepts of homogeneous and inhomogeneous broadening.

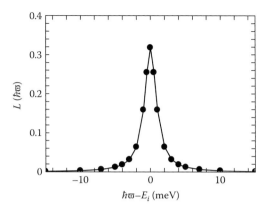

FIGURE 9.7 Normalized Lorentzian function calculated for a homogeneous broadening ΔE_{hom} of 1 meV.

and $[L(E_i, \hbar\varpi)]\Delta(\hbar\varpi)$ is the probability that an interaction occurs for a photon in the energy range $\hbar\varpi \pm 1/2(\Delta\hbar\varpi)$. This causes broadening of the absorption line for a single dot that follows the Lorentzian in Figure 9.7. Experimental measurements on an ensemble of dots in an optical amplifier by Borri et al. (2002) show that the homogeneous linewidth due to dephasing increases with drive current and at room temperature ranges between 5 and 18 meV. At low temperature the linewidth is reduced to around 5 meV at 150 K and to 2 meV and less at 26 K.[*]

The [] term in Equation 9.7 has dimensions $[L^2]$ and therefore defines the absorption cross section of a single dot, which we denote by $\sigma(E_i, \hbar\varpi)$; its photon energy dependence arises almost entirely from the Lorentzian lineshape. For light incident normal to a layer of N *identical* dots per unit area, each with the lower state occupied and upper state empty, the fraction of light absorbed is given by summing Equation 9.7 over all the dots, which in this case is equivalent to multiplying by the number of identical dots:

$$\left[\frac{\Delta\Phi(\hbar\varpi)}{\Phi_0}\right] = N\left[\sigma(E_i, \hbar\varpi)\right] = N\left[\sigma_0(E_i)L(E_i, \hbar\varpi)\right] \tag{9.8}$$

where $\sigma_0(E_i)$ is the area under the absorption cross section spectrum of a single dot (the Lorentzian is normalized), which from Equation 9.7 is given by

$$\sigma_0 = \left[\frac{2 \times 4\pi\hbar}{cn\varepsilon_0\hbar\varpi}\left(\frac{e}{2m_0}\right)^2 M^2 \left\{\int_{\text{dot}} F_l(\mathbf{r})F_u(\mathbf{r})d\mathbf{r}\right\}^2\right] \tag{9.9}$$

This fundamental quantity determines the strength of the interaction of the optical field and a pair of energy levels in a dot. Using text book values for InAs and an overlap integral of unity, the integrated cross section spectrum is $\sigma_0 = 1.4 \times 10^{-15}$ cm^2 eV and the peak value for the cross section is 4.5×10^{-14} cm^2 for an homogeneous linewidth of 10 meV.

The optical cross section does not depend directly upon the physical size of the dot (itself much smaller than the wavelength of light) but is determined primarily by the overlap integral and the transition

---

[*] Lineshape broadening of the transitions also occurs in quantum well lasers where it is also incorporated in calculations of the gain (Coldren and Corzine 1995, section 4.3.2). Because the density of states function is slowly varying on the energy scale of the homogeneous linewidth ($\approx$13 meV), except near the sub-band energies the effect of broadening on the final spectra is not pronounced, other than to smear out changes in gain on a scale less than the linewidth, for example, at the sub-band edges. This contrasts with quantum dots where the energy states are widely spaced compared with the linewidth and the effect of broadening is significant.

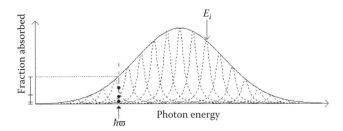

FIGURE 9.8 Illustration of the fraction of light absorbed by an inhomogeneous system of dots as a function of photon energy, showing how the spectrum is made up of contributions from dots of different size with state separation energies, E_i, each with homogeneously broadened transitions. The absorption at any specific photon energy $\hbar\omega$ is obtained by summing over all different size dots (E_i) that contribute to the absorption at $\hbar\omega$.

matrix element of the constituent material. The structure of the dot determines the energy states (and hence transition wavelength) and the wave functions. Internal strain fields and composition variations that destroy the symmetry of the potential reduce the overlap integral. Andreev and O'Reilly (2005), for example, have made detailed calculations of the optical matrix element for InAs dots.

In a real ensemble of dots, there is an inhomogeneous distribution of dot transitions energies, E_i, specified by a probability function, $P(E_i)$, often taken to be a Gaussian. As illustrated in Figure 9.8, the absorption at any photon energy ($\hbar\omega$) is the sum over all different dots with homogeneously broadened absorption lines that contribute to absorption at ($\hbar\omega$).

9.4.3 Modal Absorption and Gain

In a conventional laser structure, light is usually guided along the plane of dots in a slab waveguide as depicted in Figure 9.9. The optical field across the waveguide is described by $\xi(z)$ and the optimum width of the waveguide ($\approx$300 nm) that maximizes the field at the dot layer (ξ_{dot}) is significantly greater than the thickness of the dot layer consequently, provided the number of dot layers is not too large, the field is the same for all the dots. The modal gain (or absorption) for light propagating along a layer of N *identical* dots per unit area in the waveguide with *states in all dots inverted* is given by

$$G(\hbar\omega) = N\sigma(\eta\omega, \hbar\omega)\left[\frac{\xi^2_{dot}}{\int \xi^2(z)\,dz}\right]_{E_i} \tag{9.10}$$

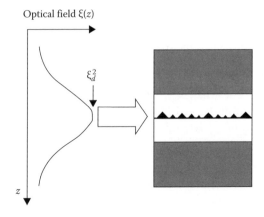

FIGURE 9.9 Illustration of a layer of dots within a waveguide showing the optical field profile, $\xi(z)$.

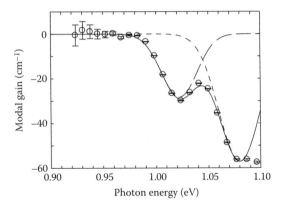

FIGURE 9.10 Measured modal absorption (shown as negative gain) spectrum for a three layer dot system, with the ground and excited state peaks fitted with Gaussian energy distributions. (Reproduced from Osborne, S.W. et al., *J. Phys. Condens. Matter*, 16, 3749, 2004a. With permission. Institute of Physics Publishing.)

If the optical cross section does not depend on the occupancy of the dot (i.e., Coulomb interactions are neglected), we may equate this cross section with that which characterizes the absorption, though the cross section may depend on the relative orientation of the electric field vector and the dot layer. Even if this assumption does not hold, the form of Equation 9.10 remains valid. The term in [] describes the coupling of the dots to the optical field and defines an effective mode width (which can be calculated using an optical mode solver) as

$$w_{\text{eff}} = \frac{1}{\xi_{\text{dot}}^2} \int \xi^2(z)\,dz \tag{9.11}$$

The gain due to an inhomogeneous distribution of dots is given by

$$G(\hbar\omega) = N \frac{1}{w_{\text{eff}}} \int \sigma_0(E_i) L(\hbar\omega - E_i) P(E_i) \left\{ \sum_{lm} Q_{lm}(E_i, E_{fu}, E_{fl}) \right\}_i dE_i \tag{9.12}$$

where σ_0 is the spectrally integrated optical cross section (Equation 9.9). The nature of an optical transition depends upon the occupation of each individual dot specified as l, m = number of electrons in the upper state and lower state. For example, ground states with occupation numbers (0, 2) and (0, 1) provide two and one absorbing transitions, respectively, whereas occupation numbers (2, 0) and (1, 0) provide two and one stimulated transitions. The number of dots of a particular size having a specific configuration (l, m) is given by the probability that the dot contains l and m electrons in the upper (conduction) and lower (valence) states and the individual occupation probabilities are determined by the energy of the state and the corresponding quasi-Fermi energy (E_{fu}, and E_{fl}) of the upper and lower states. The transition "number" and probability of a specific configuration are expressed by the term Q_{lm} in Equation 9.12, which is then summed over all dots of a particular size with transition energy E_i.[*] The gain is then obtained by integrating over dots with different E_i with homogeneous lines that contribute at the photon energy in question. The interplay between homogeneous and inhomogeneous broadening is important for the collective behavior of the dot ensemble, and temperature dependence of the homogeneous broadening determines the evolution of the laser spectra (Sugawara et al. 2000).

Figure 9.10 shows experimental data for the modal absorption spectrum of a three-layer system of InAs dots with Gaussians in energy fitted to the ground and excited state transitions (Osborne et al.

---

[*] This is the basis of the calculations by Pask (2006) used to obtain the data in Figures 9.12 and 9.13.

2004a). The area under the ground state Gaussian corresponds to an integrated optical cross section per dot of $\sigma_0 = 4.3 \times 10^{-16} \, \text{cm}^2$ eV and the standard deviation of the Gaussian is 16 meV. The value for σ_0 depends on the value used for the dot density (in this case (three layers) $\times 2.5 \times 10^{10} \, \text{cm}^{-2}$). The optical cross section at the photon energy of the absorption peak is $1 \times 10^{-14} \, \text{cm}^2$ and the measured peak modal absorption from the ground state is 30 cm^{-1}, or about 10 cm^{-1} per dot layer. The measured value for σ_0 is about three times smaller than estimated above for InAs (Section 9.4.2) and the difference could be due to incomplete overlap of the wave functions, inclusion of some Ga in the dot and error in the dot density. Given the simplicity of Equation 9.7 this numerical agreement is satisfactory. Eliseev et al. (2001) have obtained a value for the gain cross section of $7 \times 10^{-15} \, \text{cm}^2$ from the saturation gain and the calculated optical confinement factor, which is similar to the measured peak absorption cross section above.

The area under the excited state Gaussian is twice that of the ground state. If the dimensions of the dot are similar in the two directions in the plane of the layer ($L_x \sim L_y$), there are two states at the energy of the first excited states that we can represent by numbers (2, 1, 1) and (1, 2, 1). Figure 9.3 shows that the wave function of the $n = 1$ and $n = 2$ states are symmetric and anti-symmetric, respectively, so the overlap integral between them is zero. Consequently, transitions are only possible between excited states with the same set of quantum numbers and of the same spin, resulting in a total of four possible transitions, twice that of the ground state.

Gain characteristics and the role of the gain in device performance are further discussed in detail in Section 9.5.4; we show in Figure 9.11 some typical spectra for the same sample as Figure 9.10, to introduce some general points. The peak ground state gain increases with current, begins to saturate at about 10 cm^{-1} (measured net modal gain 5 cm^{-1} minus the measured waveguide loss, i.e., "gain" of (-5 cm^{-1})) followed by an increase in gain from the first excited state at higher current. The gain peak shifts to lower energy relative to the absorption peak due to Coulomb interactions in the presence of injected carriers, particularly the large number of carriers in the wetting layer (Schneider et al. 2001). The ground state gain appears to saturate at a smaller peak gain than indicated by the absorption of 30 cm^{-1}. This may be due in part to Coulomb interactions; however, it is observed that at lower temperatures the saturation gain value increases with decreasing temperature and saturation does not occur at 100 K. This is because as the Fermi levels move toward the high density of wetting layer states at high current, most injected carriers populate the wetting layer rather than the dot states (Matthews et al. 2002). Due to this rapid increase in recombination current from the wetting layer, the ground state gain does not increase significantly with increasing current above about 100 mA, and saturation of the gain–total current plot does not necessarily indicate that the ground states of all the dots are fully inverted (see also Section 9.4.5).

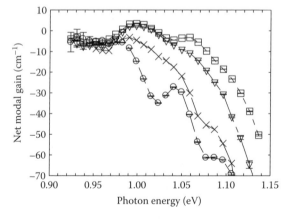

FIGURE 9.11 Modal gain spectra measured for the same sample as Figure 9.10, for drive currents of 20 (crosses), 100 (triangles) and 200 (squares) mA, together with the absorption spectrum (circles). (Reproduced from Osborne, S.W. et al., *IEEE J. Quantum Electron.*, 40, 1639, 2004b. With permission. © 2004 IEEE.)

TABLE 9.1 Summary of Optical Loss and Gain Values

		Maximum Peak Modal Value, cm^{-1}	
	Integrated Optical Cross Section, σ_0 cm^2 eV	Single Layer 2.5 × 10^{10} InAs Dots cm^{-2} Mode Width 0.28 μm	Single InAs Quantum Well Mode Width 0.28 μm
"Theoretical"	1.39 × 10^{-15}	31	210
Experimental absorption	(0.43 ± 0.1) × 10^{-15}	Absorption = 10	
Experimental modal gain		Gain = 3	

The values for absorption and gain for InAs dots are summarized in Table 9.1. From the calculated integrated absorption cross section, the modal absorption in a waveguide with effective mode width of 0.28 μm (corresponding to the experimental data) is 31 cm^{-1}. Using the same value for the transition matrix element and effective masses, the modal absorption due to $n = 1$ sub-band transitions of an InAs *quantum well* (Blood 2000) is 210 cm^{-1}, considerably larger than for the layer of dots, though the ratio depends on the dot density. The measured modal absorption is about 10 cm^{-1} per dot layer. Under carrier injection, the measured maximum peak modal gain from the ground state is about 3.3 cm^{-1} per layer, due chiefly to suppression of gain by the high density of wetting layer states. Since the optical loss of a typical diode laser with uncoated facets and cavity length 250 μm is about 50 cm^{-1}, to achieve laser action using dots it is necessary to reduce the optical loss (e.g., by increasing the cavity length) and increase the dot density (e.g., by using multiple layers of dots, Herrmann et al. (2001)).

The key factors that control the gain are therefore as follows. Equation 9.7 shows that the absorption (and gain) cross section is controlled by the transition matrix element and the overlap integral; these also determine the radiative recombination rate that contributes to the threshold current. Within an ensemble of dots, the peak modal gain is then determined by the dot density, overlap of the optical mode with the dot system, the inhomogeneous broadening, and the proximity in energy of wetting layer states. It is clearly desirable to minimize the broadening (which increases the peak gain but not the total recombination current) and maximize the energy separation of dots and wetting layer states.

9.4.4 Recombination

Knowledge of the radiative and non-radiative recombination processes is important because these processes control the threshold current for a given degree of inversion specified by the quasi-Fermi level separation. Radiative recombination within a single dot is given by an equation similar to Equation 9.7, together with the occupation of the upper and lower states. By applying the Einstein relations to a single dot, the spontaneous lifetime τ_{spon} can be calculated from the integrated cross section σ_0. The recombination rate of an ensemble is given by an equation similar to Equation 9.12 with the factor Q_{lm} now representing the probability of a particular configuration (l, m) and the number of downward spontaneous transitions associated with it. Since the number of dots that are occupied by an electron and a hole where recombination occurs is less than the total number of electrons, n_d, in the dot system, the effective carrier lifetime of the ensemble determined from the total recombination rate R_{spon}

$$\tau_{eff} = \frac{n_d}{R_{spon}} \tag{9.13}$$

is significantly longer than the lifetime τ_{spon} in a single dot.

Localization affects the dependence of the ensemble recombination rate on the total number of carriers populating the ensemble for each process: radiative, deep state (Shockley–Read–Hall recombination), and Auger recombination (Blood et al. 2007). Auger recombination of carriers within dots, involving excitation of an electron to a continuum state, is thought to be an important process in quantum dot lasers, identified by Marko et al. (2003) by pressure experiments. The dependence of these processes on the population of electrons in the dots cannot be observed because the electron population cannot be measured directly; however, this behavior does affect how the relative magnitudes of radiative and non-radiative processes vary with total drive current. For example, localized defect recombination and radiative recombination at dots vary in the same way with electron population (Pask et al. 2005), unlike extended state systems where radiative recombination comes to dominate at high injection. Experimental results that support this statement are reported in Section 9.5.4.

An estimate of the maximum radiative recombination rate when all the dots are populated can be made using the measured integrated absorption cross section in Table 9.1, assuming that the carrier lifetime when the dots are occupied is the same as that deduced from the absorption when the dots are empty. The spontaneous lifetime for a single transition is 0.96 ns, so for a layer of 2.5×10^{10} dots cm^{-2}, the ground state recombination current density is 8.3 A cm^{-2}, and for the excited states the recombination current is 16.6 A cm^{-2} when all upper states are fully occupied. Examples of recombination spectra are shown in Section 9.5.4.

9.4.5 Gain–Current Relations and Threshold Current

Here we draw together the concepts introduced in the previous sections to identify the physics that determines the threshold current and its temperature dependence to provide the basis for the discussion of the performance currently achieved by quantum dot lasers in Section 9.5.4.

The threshold current density of a particular device is determined by the optical loss of the cavity and the relation between recombination current and peak modal gain of the active region. Theoretically, this relation is obtained from calculations of gain and spontaneous emission spectra at increasing quasi-Fermi level separation with the intrinsic current determined as the spectrally integrated spontaneous emission. Theoretical curves do not usually include non-radiative contributions, although in some cases Auger recombination, an intrinsic process, has been included. The threshold gain is determined by those dots of a size that contribute to the gain at the peak, whereas recombination occurs for all populated dots in the inhomogeneous distribution; consequently, the inhomogeneous broadening has a major influence on the relation between peak gain and total recombination current.

Figure 9.12 shows calculations using a model for localized recombination and gain (Pask 2006) of ground state peak modal gain for a single layer of 3×10^{10} InAs dots cm^{-2} with Lorentzian line width of 10 meV, inhomogeneous broadening giving a full width at half maximum of about 80 meV, wetting layer band gap 0.45 eV greater than that of the dot material and wetting layer radiative recombination rate specified by a recombination coefficient of $B = 3.5 \times 10^{-7}$s^{-1}m^2. (This is an illustrative example and does not correspond to any particular device structure discussed in Section 9.5.) The dots are populated independently by electrons and holes according to Fermi–Dirac statistics. The quasi-Fermi levels also determine the carrier populations in the wetting layer and are specified such that the total numbers of electrons and holes are equal. The points show calculations for the ground state peak gain as a function of (1) the ground state radiative recombination current, (2) the total radiative recombination current (ground and excited states and wetting layer), and (3) the total recombination current including non-radiative recombination in the wetting layer (non-radiative lifetime of 0.3 ns). The ground state peak gain increases initially linearly with ground state recombination current since both processes depend upon the number of states that contain both an electron and a hole and the curvature is due to an increase in recombination current due to states being populated across the inhomogeneous distribution that do not contribute gain at the wavelength of the peak. The maximum peak modal gain is 12.4 cm^{-1} and is inversely proportional to the inhomogeneous linewidth. In real structures, there are contributions to the current from excited states and the wetting layer, plots (2) and (3), which "stretch" the current axis and the gain appears to saturate at a lower value than the maximum possible with all dots

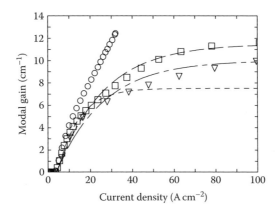

FIGURE 9.12 Calculated ground state peak modal gain as a function of current density (points) using localized recombination model for a single layer of 3×10^{10} dots cm^{-2} with a Lorentzian linewidth of 10 meV and inhomogeneous width at half maximum of 80 meV. (From Blood, P. and Pask, H.J., unpublished.) The current axis represents (1) the ground state radiative current (circles), (2) total radiative recombination current (ground and excited states and wetting layer) (squares), and (3) total recombination current including that due to a non-radiative lifetime of 0.3 ns in the wetting layer (triangles). The lines are fits of Equation 9.14 to the calculated points over the full current range (dashed lines) or a reduced current range to 40 A cm^{-2} (dotted line). (From Blood, P. and Pask, H.J., unpublished.)

inverted. In practice, the observed saturation level represents the maximum gain that is accessible in a practical device (see Figure 9.11). For a cavity loss of 8 cm^{-1} these calculations suggest the threshold current density should lie between 20 and 40 A cm^{-2}.

It is common to use an empirical relation introduced by Zhukov et al. (1999) to describe experimental data and to estimate the saturation peak gain:

$$G_{\text{pk}} = G_0 \left\{ 1 - \exp\left[\frac{-\gamma(J - J_0)}{J_0} \right] \right\} \tag{9.14}$$

where

G_0 is the saturation peak gain
J_0 is the transparency current (where $G_{\text{pk}} = 0$)

At low inversion when $J \approx J_0$ Equation 9.14 becomes linear in J and this corresponds to the ideally linear relation between gain and current curve for the ground state alone (as shown in Figure 9.12 at low gain) and for an ideal dot system the parameter γ is unity (Zhukov et al. 1999). If values of γ less than unity are required to fit experimental data, it may indicate the presence of other contributions to the current, such as excited state or wetting layer recombination.

Applying this relation to plots (2) and (3) yields G_0 of 11.5 cm^{-1} ($\gamma \approx 0.17$) and 10 cm^{-1}, respectively, lower than the maximum peak gain available from the ground state of 12.4 cm^{-1} due to the rapidly increasing population of the wetting layer. The transparency current is only increased from $J_0 = 4.0$ A cm^{-2} to 4.0 and 4.5 A cm^{-2} for plots (2) and (3), respectively. Experimental data may only be available over a limited current range and as an example of the consequences of fitting such data, the figure shows that points to about 40 A cm^{-2} give an extrapolated saturated gain of only 7.5 cm^{-1} ($\gamma \approx 0.45$). Note that values of γ significantly less than 1 are required to obtain these fits as a consequence of current contributions in addition to those from the ground state. This illustrates that for data based on the total device current, the saturation peak gain value should not necessarily be taken as the maximum gain available at full inversion. Furthermore, when looking at experimental data, remember that the maximum peak gain available is inversely proportional to the inhomogeneous linewidth, which may vary from one structure to another and, while the linewidth does not change the total radiative recombination rate

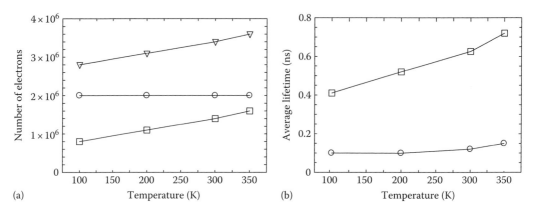

FIGURE 9.13 Calculations using a localized recombination model for (a) the number of electrons (ground state, circles; excited state, squares; and total, triangles) as a function of temperature required to provide a ground state peak modal gain of $4\,\mathrm{cm}^{-1}$ and (b) calculated temperature dependence of the total effective carrier lifetime for a fixed electron density for average populations of 2 (squares) and 4 (circles) electrons per dot (the ground and excited states of a dot can accommodate 2 and 4 electrons, respectively). (Reproduced from Blood, P. et al., *IEEE J. Quantum Electron.*, 43(12), 1140, 2007. With permission. © 2007 IEEE.)

for a given inversion, the inhomogeneous linewidth does influence the relation between gain and current through its effect on the peak gain.

The temperature dependence of threshold current arises from two effects: (1) the temperature dependence of the number of electrons to achieve the required peak modal gain (n_{th}) and (2) the temperature dependence of the recombination processes themselves. These effects occur as follows:

1. In an ideal system with all dots of the same size and with excited states separated from the ground states by many ($k_B T$), the number of electrons required to achieve a specific gain does not depend on temperature; however, in the presence of inhomogeneous broadening, increasing the temperature spreads the electrons over a wider range of energy states and hence a larger number of dots, reducing the probability of an electron and hole in the same dot. This reduces the gain at a given wavelength so it is necessary to inject more electrons to achieve the required peak gain, represented by an increase in the quasi-Fermi level separation. In many structures, the ground and excited state inhomogeneous distributions overlap in energy, so population of excited states also contributes to $n_{th}(T)$. The computed behavior of $n_{th}(T)$ for ground state modal gain of $4\,\mathrm{cm}^{-1}$ is shown in Figure 9.13, where the chief contribution to the temperature dependence comes from the excited state (Blood et al. 2007). The electron population in the wetting layer also increases as the quasi-Fermi level separation increases and this is a strong effect due to the high density of states.

2. The effect of distributing a given number of electrons across more dots is also to reduce the probability of recombination, so for a temperature-independent spontaneous lifetime in each dot, the *effective recombination lifetime* (per number of dot electrons) (Equation 9.13) decreases with increasing temperature and this partly mitigates the increase in recombination current that arises from the increase in n_{th}. Calculated temperature dependence of the effective lifetime $\tau_{spon}(T)$ for 2 electrons per dot (ground state almost full) and 4 electrons per dot (excited states also populated) is shown in Figure 9.13 (Blood et al. 2007). The radiative recombination coefficient of the wetting layer also decreases with increasing temperature (as T^{-1}, assuming it to be a quantum well).

It follows that the temperature dependence of threshold of a dot system is particularly sensitive to the inhomogeneous broadening and is also dependent upon the gain required.

9.5 Threshold Current Density and its Variation with Temperature

9.5.1 Summary of Present Position

Since the initial pioneering report of a self-assembled QD laser (Kirstaedter et al. 1994) with a threshold current density of ~1 k A cm^{-2} at 300 K, significant progress has been made, with many groups world-wide achieving threshold current densities in the InAs on GaAs QD system below 45 A cm^{-2} (which is the lowest threshold current density for a QW laser—Chand et al. (1991)), with the lowest (CW) threshold current density at room temperature to date being 11.7 A cm^{-2} (Freisem et al. 2008) for a 2 cm long laser with uncoated facets. Equally impressive improvements have been made to the temperature stability at 300 K with lasers being reported that have little (Shchekin and Deppe 2002), or no obvious temperature dependence of the threshold current density (Fathpour et al. 2004) ($T_0 = \infty$ up to just above room temperature[*]). These two achievements were the original reasons envisaged for pursuing QD lasers (Arakawa and Sakaki 1982, Asada et al. 1986) and are of considerable practical importance for the application of semiconductor lasers in integrated circuits and systems. In such applications, low current and low voltage are required for compatibility with existing electronics, high efficiency is required to minimize any additional heat load, and low temperature dependence of threshold current is required because of the variation in temperatures that can exist in such systems. However, even though the results to date are impressive, further progress is still required. A low threshold current does not necessarily follow the achievement of low threshold current density and the temperature dependence of threshold current deteriorates, even in the best performing structures, at elevated temperatures. The purpose of this section is to evaluate the factors that make up the threshold current density, to establish how far the current state-of-the-art material is from the intrinsic or fundamental limits (which may be assessed from the material in Section 9.4) and to establish how to achieve the required performance.

9.5.2 Evaluation of Device Operation

To illustrate the important physics involved, we will examine the behavior of a set of high performance InGaAs on GaAs DWELL structures emitting in the 1.3 µm wavelength range, which are described in Figure 9.14 and Table 9.2. Currently, lasers fabricated from self-assembled quantum dots in other material systems, for example, InAs on InP (1.5–2 µm emission) and InP on GaAs (650–780 nm emission) are less developed than InGaAs on GaAs structures, but we expect the fundamental issues described in Sections 9.3 and 9.4 to be similar subject to, for example, different matrix elements and dot densities. The DWELL approach has to date produced some of the lowest threshold current densities (13 A cm^{-2}, Eliseev et al. 2001 and 17 A cm^{-2}, Sellers et al. 2004) and produces good performance even for lasers with a relatively high threshold gain requirement. The design principles of these DWELLs are that the dots are grown on an InGaAs surface, which increases the dot density compared to growth on GaAs (Lester et al. 1999, Park et al. 2000a); that the dots are capped by InGaAs, which reduces the strain at the apex of the dot resulting in a taller/less intermixed dot in the capped structure (Nishi et al. 1999), and that the InGaAs quantum well captures carriers effectively and keeps them close to the dots (Liu et al. 2000). The structures of Figure 9.14 were grown by solid source molecular beam epitaxy (SS-MBE) and the DWELL was formed from 3.0 monolayers of InAs grown on 2 nm of In$_{0.15}$Ga$_{0.85}$As and capped by 6 nm of In$_{0.15}$Ga$_{0.85}$As. The dots and the In$_{0.15}$Ga$_{0.85}$As well, formed from the 6 nm layer above and the 2 nm layer below the dots, were grown at a temperature of 510°C. Such conditions result in a dot density of ~3 × 10^{10} cm^{-2} per self-assembled dot layer,

---

[*] Although not based on a physical law, T_0 has been used for many years as an empirical means to characterize the temperature (T) dependence of threshold current density (J_{th}) and relies on fitting experimental data with the equation $J_{th} = J_0$ exp(T/T_0). When rewritten as $\dfrac{1}{T_0} = \dfrac{1}{J_{th}} \dfrac{dJ_{th}}{dT}$ it is clear that T_0 should be used with caution because T_0 can be increased (made to appear better) by an artificially high threshold.

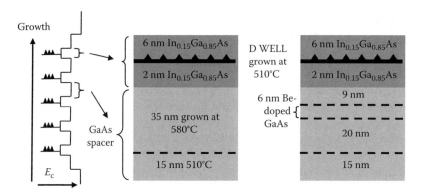

FIGURE 9.14   A schematic diagram representing, on the left, the band gap of the deposited materials, with the direction of growth indicated, for a five layer DWELL (dots in well) structure. In the middle is a detail of one cycle of the DWELL for a standard sample, indicating the temperature of the substrate during growth. On the right is the detail for one cycle for the samples containing p-modulation doping.

TABLE 9.2   Summary of Variations in Growth of DWELL Structure of Figure 9.14

Sample	Number of Layers	Spacer Thickness/nm	Growth Temperature of Spacer/°C[a]	Modulation p-Doping/ Dopant Atoms per Dot
1	5	50	580	0
2	5	50	580	0
3	5	50	580	0
4	3	50	580	0
5	7	50	580	0
6	10	50	580	0
7	5	35	580	0
8	5	35	620	0
9	5	50	580 (510 last 15 nm)	15
10	7	50	580 (510 last 15 nm)	0
11	7	50	580 (510 last 15 nm)	18

[a] The first 15 nm of the spacer is always grown at 510°C.

as deduced from AFM on uncapped test structures (see Figure 9.2). Different approaches are described in the literature to minimize the effect of the first grown layer of dots on subsequent layers (e.g., Ledentsov 2003). The samples described in Figure 9.14 incorporate the so-called High Growth Temperature Spacer Layer (HGTSL) (Liu et al. 2004). The initial 15 nm of the GaAs spacer is grown at 510°C with the rest at a higher temperature, which allows the surface to replanarize before the growth of subsequent layers of dots (Liu et al. 2004). The temperature is reduced back to 510°C for the growth of the next DWELL. This results in the reduction of large defective dots seen in the images of Figure 9.2c and d.

Measurements on sample 1 (Table 9.2), when fabricated into 20 μm wide, 5 mm long ridge waveguide (RW) lasers, are reported in Liu et al. (2004). Devices with uncoated facets had a CW threshold current of 37 mA at room temperature corresponding to a threshold current density of 37 A cm^{-2}. The lasing wavelength was 1306 nm at 300 K and the devices lased on the ground state up to 100°C.

9.5.3 Characterization

One of the key issues with improving the performance of quantum dot lasers is assessing what is present following a particular growth. Even for QW lasers this issue can be important, but for QDs we have the

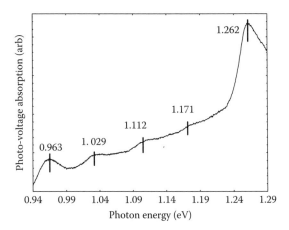

FIGURE 9.15 Photovoltage absorption spectrum measured at 300 K, where light is coupled into the waveguide of laser chip at the facet using light polarized TE (the electric field vector in the same plane as the epitaxial layers). The photovoltage generated at the device contacts has peaks at energies that correspond to the inhomogeneously broadened quantum dot transitions and the wetting layer absorption edge and these are indicated by the solid lines. (Reproduced from Sandall, I.C. et al., *IEEE Photon. Technol. Lett.*, 18, 965, 2006b. With permission. © 2006 IEEE.)

added complexity of variations in dot size, shape, and compositional variations within each dot. The dot formation process can also be seriously affected by the initial growth surface, which can be modified by the growth of underlying dot layers. To fully understand laser performance, it is necessary to perform characterization that separates the effects due to the electronic states present in any dot structure and those due to the population of those states.

Measurements of absorption spectra, such as photoluminescence excitation (PLE), are useful to characterize the available transitions in a structure. In Figure 9.15, the results of an edge photovoltage absorption measurement (Smowton et al. 1996) are plotted, but note that the measurement is in arbitrary units and that the shape of the spectrum is affected by the variation in distance over which absorption occurs as the absorption coefficient changes with wavelength. The spectrum shown, which is for light polarized with the electric field parallel to the growth plane (TE), can be used to identify four low energy features, corresponding to transitions within the quantum dots, with centre energies of approximately 0.963, 1.029, 1.112, and 1.171 eV, and a more pronounced feature at 1.262 eV that corresponds to absorption by the wetting layer. As described in Section 9.3, the dot transitions are approximately equally spaced in energy. The larger number of excited state transitions and the smaller energy separation between dot ground state and wetting layer observed here compared to the illustrative calculation used in Section 9.4.5 is expected to increase the effect of the higher energy states on the gain and current, beyond that seen in Figures 9.12 and 9.13. TM polarized light produces negligible absorption at energies corresponding to the dot states, which is a common observation for dots where the vertical dimension is small relative to the lateral dimensions.*

9.5.4 Gain and Recombination

Net modal gain spectra $(G - \alpha_i)$ measured at a temperature of 300 K are shown in Figure 9.16 as a function of the drive current density (Walker et al. 2005). The modal loss spectrum $(\alpha + \alpha_i)$, which is equivalent to a gain spectrum for zero current density, is also plotted. At long wavelength, where both

---

* Some work has been done, growing dots with a large vertical dimension with the so-called columnar dots having more equal absorption and gain in the TE and TM polarizations (Kita et al. 2002) and are of particular interest for semiconductor optical amplifiers where polarization insensitivity is required (Akiyama et al. 2007).

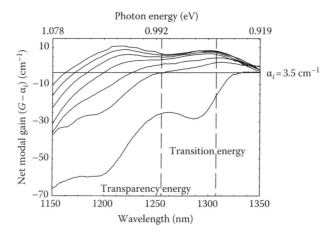

FIGURE 9.16 Measured net modal gain ($G - \alpha_i$) spectra as a function of current density (76–457 A cm^{-2}) and modal loss ($\alpha + \alpha_i$) data at 300 K. Features corresponding to the inhomogeneously broadened quantum dot ground state and first excited state are apparent in both gain and loss spectra. The dashed lines indicate the energy of transparency (for the lowest current density gain spectrum) and the dot state transition energy. (Reproduced from Walker, C.L. et al., *IEEE Photon. Technol. Lett.*, 17, 2011, 2005. With permission. © 2005 IEEE.)

the gain (G) and the absorption (α) tend to zero, the spectra tend to the value of α_i of 3 ± 2 cm^{-1}. Over the wavelength range plotted, two of the features observed in Figure 9.15 are seen in these spectra and correspond to the inhomogeneously broadened ground and excited state transitions. As discussed in Section 9.3, the degree of inhomogeneous broadening can be characterized by fitting Gaussians in energy to the transitions observed in the absorption spectrum. A value of standard deviation of 16 meV for the ground state is obtained for the spectrum of Figure 9.16.

For the gain spectra taken at higher current density in Figure 9.16, the gain at the excited state exceeds that at the ground state meaning that lasers with a large threshold gain requirement would lase at a wavelength corresponding to the excited state (~1.2 μm). The ground state does provide sufficient gain ($G = \alpha_i + 1/L \ln(1/R) = 5.5$ cm^{-1}) for lasing in a 5 mm long device with uncoated facets even for temperatures significantly above room temperature, consistent with the observed ground state laser operation above 100°C.

An unamplified spontaneous emission rate spectrum recorded at a current density of 350 A cm^{-2}, which is approximately 9 times the threshold current density of the 5 mm long laser, is shown in Figure 9.17. Spontaneous recombination is observed at energies corresponding to all the transitions identified in the photovoltage absorption spectrum of Figure 9.15 including the wetting layer, albeit at a low level. This suggests that, at high carrier density, carriers are present in all the dot states and the two-dimensional wetting layer states and, since non-radiative recombination can be at a significantly higher level than the spontaneous radiative recombination—a point to which we will return, the presence of carriers in the higher energy states, even at a relatively low level, may be significant. At lower drive current density, as illustrated in Figure 9.18, the spontaneous emission is principally from the inhomogeneously broadened QD ground state transition centered at ~0.959 eV. At a drive current density of 40 A cm^{-2} (the upper spectrum of Figure 9.18), corresponding to that required for laser action in the 5 mm long uncoated device, the ground state related spontaneous recombination current makes up approximately two thirds of the total spontaneous recombination current and 80% of the total at 22 A cm^{-2} (the lower spectrum of Figure 9.18). This demonstrates that further progress in threshold current density should be possible if the excited state transition can be separated from the ground state—perhaps by reducing the dot size and increasing the quantum confinement. However, an even more significant factor becomes apparent when we consider the level of non-radiative recombination.

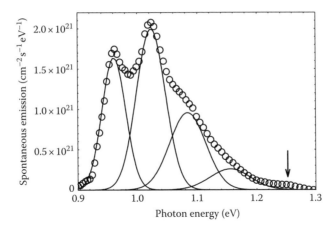

FIGURE 9.17 Measured spontaneous emission spectrum at a drive current density of 350 A cm^{-2} at 300 K and four Gaussians fitted to the experimental spectrum to represent each of the four inhomogeneously broadened quantum dot transitions. The arrow indicates the emission at energies corresponding to the wetting layer. (Reproduced from Sandall, I.C. et al., *IEEE Photon. Technol. Lett.*, 18, 965, 2006b. With permission. © 2006 IEEE.)

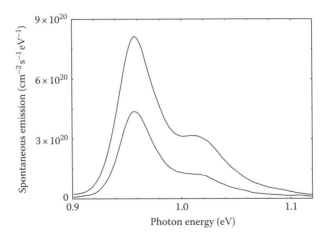

FIGURE 9.18 Measured spontaneous emission spectra at drive current densities of 22 and 40 A cm^{-2} at 300 K.

By integrating the spontaneous emission rate spectra and multiplying by the electronic charge, we can calculate the current density associated with spontaneous recombination events (here denoted as radiative current density) and, by subtraction from the total drive current density determine the non-radiative current density.[*] The non-radiative current density is plotted in Figure 9.19 as a function of the radiative current density for different heat-sink temperatures. At all temperatures, the non-radiative current density is much larger than the corresponding radiative current density, indicating that this material that can produce very low threshold current density lasers, is extremely inefficient. At a current density of 40 A cm^{-2}, non-radiative recombination makes up approximately 80% of the total. Examining the form of the data of Figure 9.19, we note that all five sets of data exhibit an approximately linear increase of the non-radiative current density with increasing radiative current density at low current density, followed by a steeper increase at higher current density. When characterizing QW lasers, the form of the

---

[*] This assumes that there are no parallel current leakage paths and that the stimulated recombination rate is negligible in the single-pass multi-section stripe length experiment used to obtain this data, which can be shown to be the case using the approach described by Blood et al. (2003).

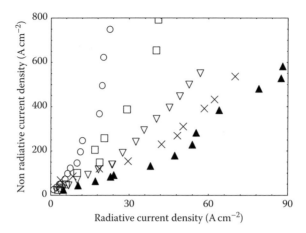

FIGURE 9.19   Non-radiative recombination current density (difference of total drive current density and sponta-neous recombination current density) versus spontaneous current density measured at temperatures of 200 K (full triangles), 250 K (crosses), 300 K (triangles), 340 K (squares), and 360 K (circles). (Reproduced from Sandall, I.C. et al., *IEEE Photon. Technol. Lett.*, 18, 965, 2006b. With permission. © 2006 IEEE.)

data of Figure 9.19 would often be used to differentiate between the various recombination processes. As discussed in Section 9.4.4, the relations between recombination processes and overall carrier density are not as easy to define in quantum dot structures. However, at low current density, the data does follow the linear increase of non-radiative recombination with radiative recombination as expected from the description of Section 9.4.4. The faster increase of non-radiative current density at high radiative current density could be due to either Auger- or Shockley–Read–Hall-recombination in the higher energy quantum dot states, as these higher degeneracy states become populated as observed in Figure 9.17. Shockley–Read–Hall recombination could be exacerbated because the wave functions of the higher-lying states have a larger overlap with defects on the dot surface and Auger recombination could be more important at high current density because the higher degeneracy of the higher energy states favors the Auger process. It could be due to the thermally activated loss of carriers to the wetting layer, which as already seen in Figure 9.17 is populated at high current density. Non-radiative recombination in the wetting layer may be stronger (inefficient for device operation) because the continuum states of the wetting layer allow free movement of carriers to any defects. Finally, thermally activated loss of carriers to the GaAs bulk layers and recombination there may be important (Hasbullah et al. 2008). To avoid modifying the dots, the GaAs immediately above them tends to be grown at a temperature below that normally considered optimal for the growth of GaAs, and may result in the presence of significant numbers of point defects.

9.5.5 Radiative Efficiency

In Figure 9.20 the radiative efficiency, which is defined as the ratio of the radiative current density to the drive current density, is plotted as a function of the (transparency energy – transition energy) to represent the degree of inversion* (which we denote by Δ in the following text), for heat-sink temperatures of 200, 300, and 360 K for sample 1. Similar data for nominally identical structures grown at subsequent times in the same MBE growth reactor is also included. These three samples are separated by approximately 200 growth runs, between samples 1 and 2, and a further 200 growth runs between samples 2 and 3 and give some idea of the range of data that can be expected for similar structures. The efficiency is

---

* Previously introduced to provide a relative measure of the injection level in different samples or at different temperatures, whether or not the carrier distribution can be described by Fermi–Dirac statistics (Sandall et al. 2006). The transparency and transition energies are defined in Figure 9.16.

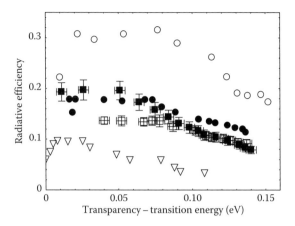

FIGURE 9.20 Efficiency (ratio of spontaneous emission recombination current density to total current density) versus (transparency – transition energy) for sample 1 at temperatures of 200 K (open circles), 300 K (filled circles), and 360 K (triangles); and for sample 2 (filled squares) and sample 3 (open squares) at 300 K.

fairly constant for low values of transparency – transition energy and then decreases at higher values for all the sets of data. At 300 K and low Δ, sample 1 has a constant efficiency of ~18%, sample 2 a constant efficiency of ~19%, and sample 3 a constant efficiency of ~14%. These low values of efficiency signal the magnitude of improvement (reduction in threshold current density from an already low value) that may be accessible if the process responsible can be addressed. All three samples exhibit a similar reduction in efficiency above a Δ value of 0.07 ± 0.01 eV. The form of the efficiency data is consistent with the non-radiative plot in Figure 9.19 and suggests that one non-radiative process is present at low Δ (resulting in the less than 100% efficiency) with a second process becoming important at higher Δ.

The absolute value of efficiency at low Δ varies with temperature, the constant value being approximately 30% at 200 K, 25% at 250 K, and ~9% at 360 K. These values indicate that the non-radiative current density also has the significant effect on the temperature dependence of threshold current density. The value of Δ at which the efficiency starts to decrease reduces with increasing temperature as would be expected if the non-radiative process is exacerbated by the population of higher energy dot, wetting layer, or GaAs spacer states. In Figure 9.21 the efficiency at 300 K is plotted as a function of Δ for a series

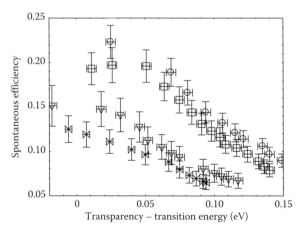

FIGURE 9.21 Efficiency at 300 K as a function of the (transparency – transition energy) for samples 4 (3 layer, circles), 2 (5 layer, squares), 5 (7 layer, triangles), and 6 (10 layer, crosses). (Reprinted from Sandall, I.C. et al., *IEEE J. Quantum Electron.*, 43, 698, 2007. With permission. © 2007 IEEE.)

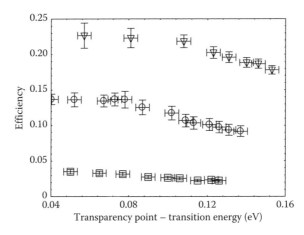

FIGURE 9.22   Efficiency at 300 K as a function of the (transparency − transition energy) for samples 3 (circles), 7 (squares), and 8 (triangles).

of samples with different layer number, the five layer sample being sample 2 of Figure 9.20. The multi-layer samples marked as multilayer 3, 5, 7, and 10 are for samples that were grown in a pseudorandom order sequence and differ only in the number of dot and spacer layers. Results for these samples are taken from Sandall et al. (2007), where plots of the radiative and non-radiative current density are also included. These show that the reduction in the constant value of radiative efficiency seen for higher layer number in Figure 9.21 is due to increases in the non-radiative current density per layer. An increased non-radiative recombination per layer at a particular value of Δ is consistent with an increased number of defects in large layer number samples, resulting in increased Shockley–Read–Hall recombination, whereas both Auger recombination and loss of carriers to higher energy states would be expected to be unchanged. In Figure 9.22, the efficiency is plotted as a function of Δ for three samples grown with different spacer layers. Sample 3 is the data already presented in Figure 9.20. Sample 4 is for a sample where the GaAs spacer layer is reduced from 50 to 35 nm with the first 15 nm grown at the same temperature of 510°C as samples 1–3 and the last 20 nm grown at the same temperature of 580°C as samples 1–3. Sample 5 has a 35 nm spacer with the first 15 nm grown at 510°C and the last 20 nm grown at 620°C. This latter sample exhibits a larger radiative efficiency than all the previous samples, whereas the sample with the thinner 35 nm spacer grown at the same temperature exhibits the lowest efficiency seen to date. One of the key advantages of growing the spacer layer at higher temperature (580°C) is believed to be the reduction in defects (Liu et al. 2004), and associated Shockley–Read–Hall recombination, and it appears that this has been further reduced by increasing the growth temperature of a section of the spacer to 620°C. Together, these results indicate that Shockley–Read–Hall recombination is still a significant factor affecting the efficiency in the region where low loss lasers operate. We will return to a discussion of the non-radiative processes after considering p-doped samples.

9.5.6 p-Doping and Carrier Distribution

P-modulation doping is known to increase the gain available from the quantum dot ground state (Sandall et al. 2006c), and this arises because p-modulation doping partially compensates the large asymmetry in the movement of the quasi-Fermi levels with increasing injection, and leads to an increased gain at a fixed quasi-Fermi level separation (Smowton et al. 2007b) (or the same gain for a reduced injection level) compared to an undoped sample. Of most practical significance is that p-modulation doped lasers have demonstrated a temperature insensitive threshold current around room temperature (Shchekin and Deppe 2002, Fathpour et al. 2004).

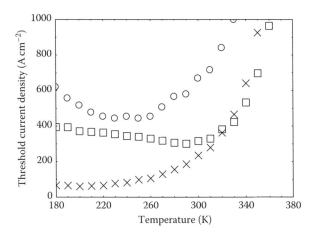

FIGURE 9.23 Pulsed (300 ns pulse, 1 ms period) threshold current density measured for 50 μm wide, 2000 μm long, oxide isolated stripe lasers with uncoated facets fabricated from the undoped samples 2 (crosses), and the doped sample 9 (square) and a 1000 μm long doped sample 9 (circles). (Reprinted from Sandall, I.C. et al., *Appl. Phys. Lett.*, 89, 151118, 2006a. With permission. © 2006 American Institute of Physics.)

Immediately following sample 2, which was previously described, sample 9, incorporating modulation doping was grown and is also represented schematically in Figure 9.14. Be atoms were grown within 6 nm of the GaAs spacer, placed 9 nm from the next grown DWELL, at a level of 7.5×10^{17} cm^{-3}, which, based on AFM measurements of dot density, corresponds to approximately 15 atoms per dot. Figure 9.23 shows the pulsed threshold current density measured for 50 μm wide oxide isolated stripe lasers with uncoated facets fabricated from these samples.

The undoped lasers exhibit a monotonically increasing threshold current from low to high temperatures. The p-doped lasers exhibit more unusual behavior. The threshold current density decreases as the temperature increases from 200 K, reaching a minimum at 280 K and then increasing at higher temperatures. Different length p-doped devices exhibit the minimum at different temperatures. One thousand micrometer long p-doped lasers exhibit an even more pronounced minimum at a lower temperature of ~240 K, whereas 3000 μm long p-doped lasers exhibit a minimum around 300 K. This type of behavior has been widely reported in the literature and is largely responsible for the very large T_0s that have been achieved in p-doped quantum dot lasers.

In undoped quantum dot lasers such an effect is observed at lower temperatures (around 200 K and below) and at the temperatures where the threshold has a minimum there is also a minimum in the energies over which the carriers are distributed, which manifests itself as a minimum in the spectral width of photoluminescence data (Zhukov et al. 1997) or in the width of laser emission spectra (Patane et al. 1999). The minimum in the carrier distribution width is caused by a transition from a nonthermal distribution of carriers among the available states at low temperature to a thermal distribution as the temperature is increased (Zhukov et al. 1997) (see also discussion in Section 9.3). In p-doped quantum dot lasers, the effect occurs around room temperature (Deppe et al. 2005, Marko et al. 2005, Sandall et al. 2006a) with a clear demonstration of a change from a nonthermal distribution of carriers to a thermal distribution of carriers (Smowton et al. 2007a) shown in Figure 9.24. At higher temperatures, where a thermal distribution is achieved, it has been shown that p-doping leads to a shift in the quasi-Fermi levels towards the valence states for a given quasi-Fermi level separation (Smowton et al. 2007b), and this makes the escape of an electron from a dot, into which it was initially captured, more difficult since escape is controlled by the quasi-Fermi level to barrier energy step. Since unimpeded carrier capture and escape are required for thermalization, this supports the idea that it is the electrons in the conduction states that move from a nonthermal to a thermal distribution causing the minimum in the threshold current density characteristic. Further control over the temperature at which thermalization occurs may be possible by modification

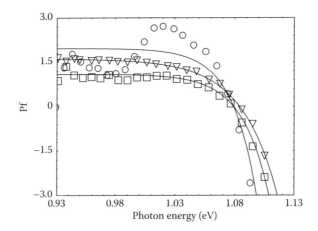

FIGURE 9.24 Experimentally determined Pf (carrier distribution factor) at 250 K (circles), 280 K (triangles), and 350 K (squares), and those calculated assuming a thermal carrier distribution (lines). (Modified from Smowton, P.M. et al., *IEEE J. Sel. Top. Quantum Electron.*, 13, 1261, 2007a. With permission. © 2007 IEEE.)

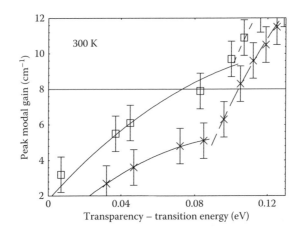

FIGURE 9.25 Peak modal gain versus transparency – transition energy for the doped structure 9 (squares) and undoped structure 2 (crosses). The solid line indicates the threshold gain requirement of a 2 mm long laser with uncoated facets. (Modified from Smowton, P.M. et al., *IEEE J. Sel. Top. Quantum Electron.*, 13, 1261, 2007a. With permission. © 2007 IEEE.)

of the energy barrier for electron loss. However, in structures where the carriers are not thermally distributed, the maximum modulation speed that can be achieved may be reduced (Deppe et al. 2005).

The detail of how the distribution of carriers controls the threshold current density via its effect on the peak modal gain can be seen in further measurements, remembering that the peak gain occurs at a specific wavelength on the gain spectrum, whereas the whole recombination spectrum contributes to the current. In Figure 9.25, the peak modal gain is plotted versus the (transparency energy – transition energy) = Δ at 300 K derived from data like that shown in Figure 9.16 for samples 2 and 9. At any given value of Δ, the gain available is larger for the p-doped material. The internal optical loss was found to be the same for the two samples at $2 \pm 2 \, \text{cm}^{-1}$. Also marked on Figure 9.25 is the threshold gain requirement, the sum of the measured internal optical mode loss and the calculated mirror loss for a 2000 μm long laser with uncoated facets. The value of Δ to achieve the modal gain requirement corresponding to a 2000 μm long laser is 0.085 eV for sample 9 (doped) and 0.103 eV for sample 2 (undoped). In Figure 9.26, the value of Δ necessary to achieve the modal gain requirement corresponding to a 2000 μm long

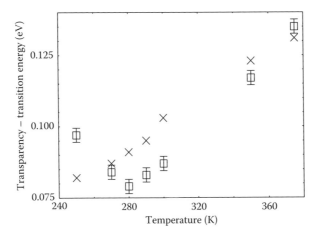

FIGURE 9.26 Transparency – transition energy to achieve a fixed value of modal gain of 8 cm^{-1} for the doped structure 9 (squares) and undoped structure 2 (crosses) samples as a function of temperature. (Modified from Smowton, P.M. et al., *IEEE J. Sel. Top. Quantum Electron.*, 13, 1261, 2007a. With permission. © 2007 IEEE.)

laser is plotted as a function of temperature. The value of Δ required for the fixed gain for the p-doped devices has a minimum at a temperature of 280 K, the same temperature as the minima in the threshold current density versus temperature characteristics in Figure 9.23, whereas there is no minimum for the undoped structure. For a nonthermal distribution, where the inhomogeneous broadening is greater than the width of a thermal distribution at that temperature, the carriers are widely spread in energy and a smaller proportion of the carriers occupy the states taking part in the lasing process, so a higher total number of carriers are required to produce the same gain. While the transparency energy does not reflect the quasi-Fermi level separation in this case, Δ is still an indicator of the degree of population inversion and will follow the number of carriers in the system. As the temperature is increased, the carrier distribution becomes more thermal, a larger proportion of the carriers occupy the states involved in the lasing process and the total carrier population and Δ reduces. At still higher temperatures, the carriers increasingly spread throughout the available states in a thermal distribution (as described in Section 9.4.5) and so again there is a smaller proportion of carriers in the states involved in the lasing process, the total number of carriers has to be increased to achieve the threshold gain requirement and Δ increases.

9.5.7 Temperature Dependence

The radiative and non-radiative recombination rates decrease and increase with increasing temperature following the behavior of the Δ. However, in addition, the non-radiative recombination itself is temperature dependent. In Figure 9.27 the non-radiative recombination is plotted as a function of temperature for a fixed gain of 8 cm^{-1} and for a fixed Δ. The non-radiative current density at fixed gain decreases with increasing temperature reaching a minimum at about 280 K and then increases at higher temperature. The non-radiative recombination at fixed Δ is constant within the experimental uncertainty at low temperatures and then increases with temperature above 300 K. These results suggest that the threshold current that decreases with increasing temperature in Figure 9.23 is due entirely to the decrease in the Δ necessary to achieve the required modal gain, but that at elevated temperatures the increasing threshold current is a combination of the increase in the Δ necessary to achieve the required modal gain and an increase in the non-radiative recombination at fixed injection level. The constant level of non-radiative recombination below 300 K for the fixed injection case is approximately 300 A cm^{-2}. Measuring the increase of non-radiative recombination above this level at 350 K and at 375 K for the two cases of fixed Δ and fixed gain requirement, it can be seen that the increase in non-radiative recombination at fixed Δ, which is the change arising from the non-radiative process itself, is

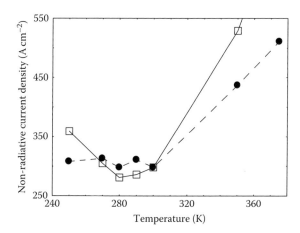

FIGURE 9.27   Non-radiative current density as a function of temperature for the doped structure 9 at a fixed modal gain of 8 cm⁻¹ (solid lines, squares) and at fixed transparency energy – transition energy (dashed lines, circles) of 0.085 eV. (Reprinted from Sandall, I.C. et al., *Appl. Phys. Lett.*, 89, 151118, 2006a. With permission. © 2006 American Institute of Physics.)

55% of the total at 350 K and 47% at 375 K. The remainder is the change in non-radiative recombination resulting from the change in Δ.

At higher temperatures, the non-radiative recombination process itself increases in both p-doped and undoped structures. In Figure 9.17 and in similar data for p-doped structures (Smowton et al. 2008), the spontaneous emission at energies corresponding to the wetting layer indicates the increasing presence of carriers in the wetting layer as the temperature is increased. For the p-doped structures, the increase in wetting layer recombination occurs over the same temperature range as the increasing non-radiative recombination at fixed Δ seen in Figure 9.27. The most likely origin of the increasing part of the non-radiative recombination with temperature is the population and subsequent extra recombination in the higher lying dot, wetting layer, or GaAs spacer states. The temperature-independent part of the non-radiative recombination may be due to defect-related Shockley–Read–Hall process, or an Auger process, or a combination of the two in the lower energy dot states.

9.5.8 Increasing Energy State Separation

The population of higher-lying energy states should be minimized and partly to this end, several groups have reported that they use a composite InAlAs–InGaAs capping layer for the dots (Chen et al. 2006, Liu et al. 2005, Chang et al. 2005, Wei et al. 2002). The use of such capping layers has resulted in a record ground- to excited-state separation of 108 meV (Wei et al. 2002). However, conflicting reports have been published showing either a red or a blue shift in the emission wavelength of the ground state transition (Chen et al. 2006, Liu et al. 2005, Wei et al. 2002). It is believed that there are several competing factors that result in this shift; the first is the change in the confinement potential, which will give rise to a blue shift of the spectra. The second effect is due to less material migrating from the quantum dots to the wetting layer during the capping process when an InAlAs cap is used, resulting in larger dots and a longer wavelength. Therefore the combination of these two effects will give rise to the overall effect seen. The results of capacitance–voltage measurements on InAlAs capped structures have been interpreted as showing an increased emission time for electrons between ground and excited states (Chen et al. 2006), due to the increased barriers around the dots. However, there has as yet been no direct measurement of improved laser performance from the use of an InAlAs cap. In fact, one study showed that a longer cavity length was required to achieve ground state lasing from an InAlAs capped device (Sellers et al. 2003), suggesting either a reduced gain and/or a higher internal loss.

9.5.9 Summary

We have based this section on a discussion of the properties of a set of specific laser structures to enable us to give a consistent account of the key factors that determine their performance. From this account we can draw some conclusions that apply in some degree (depending on the details of the particular device) to quantum dot lasers in general:

1. Even in high performance, low threshold quantum dot lasers non-radiative recombination makes a significant contribution to the threshold current, more than 80% in samples we have studied.

2. While Auger recombination is intrinsic, in our samples the non-radiative recombination is at least in part made up from defect-related recombination and therefore there is scope for a further improvement in performance following further growth optimization.

3. Of the radiative recombination current (about 20% of the threshold current density of a 5 mm long, low loss laser), the majority (80% in our samples) is made up from ground state recombination.

4. At higher values of inversion, as measured by the quantity (transparency energy − transition energy), which are required for driving higher threshold loss structures, or at higher temperatures, in addition to increased spontaneous recombination, a faster increase in non-radiative recombination occurs, which is believed to be related to the increasing population of higher energy states. Although initial efforts to reduce threshold current density by shifting the energy of the dot states away from the higher energy states have not been conclusive, this appears to be necessary to extend excellent performance to high modal gain structures for operation at higher temperatures.

Finally, from Section 9.4, we can add further general conclusions:

5. The recombination processes in quantum dot structures cannot necessarily be distinguished by their dependence on the total electron population on the dots, as has been done for quantum wells.

6. The observed saturation value of gain observed at high drive current may not correspond to the maximum gain available at full inversion due to the effect of the wetting layer on both Fermi level position and current.

7. The peak modal gain and hence the gain–current relation depend upon the inhomogeneous broadening, and may therefore vary from one structure to another.

9.6 Dot Lasers for Integration

Integration of optoelectronic functions requires that the performance of individual components is not significantly degraded by the presence of the other functionality. Here we introduce some of the properties of quantum dot lasers that favor their use for integration, leaving a more complete description to Chapter 5.

For reasons introduced in Section 9.3, quantum dots can spatially localize electrons and holes within the active region (Kim et al. 1999, 2000) and hence only current* and optical confinement need be provided by the device structure. In addition, the carrier localization can result in carriers not "seeing" etched structures and exposed surfaces, allowing these to be placed closer to the active region than in QW devices. This permits the use of, for example, deep etched ridges that pass completely through the active region (Ouyang et al. 2003). Figure 9.28 shows the relative insensitivity of ~1 μm wavelength QD lasers to the width of the deep etched ridge in comparison to a QW laser, where the assumption

---

* We separate the control of, what some would term, current spreading into current confinement (which we think of as the definition of the current path using a contact geometry) and/or controlling current flow in layers other than the dots and associated layers (e.g., DWELL structure) and carrier confinement (which we define to mean current spreading within the dots and associated layers) in a manner consistent with Kim et al. (2000).

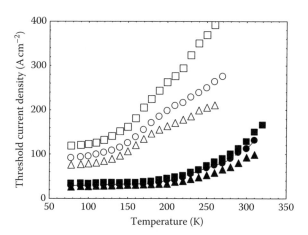

FIGURE 9.28 The temperature dependence of J_{th} as a function of ridge width (10 μm, triangles; 15 μm, circles; 20 μm, squares) for both QD (solid symbols) and QW (open symbols) lasers where the ridge is etched from the p-side through into the n-type cladding layer. Note the significant decreased sensitivity of the threshold current density of QD lasers to the ridge width.

is that in the case of the QW laser, carriers can migrate to the surface defects. Reduced in-plane diffusion would be beneficial in integrated devices where the cleaved mirrors of edge emitting devices must be replaced by deep etched structures. QD devices have been shown to exhibit reduced sensitivity to the processing defects, which result from dry etching (Schoenfeld et al. 1998), which may be a result of reduced diffusion and/or reduced surface recombination velocity for dots. Several groups have reported measurements of carrier diffusion (quantified as the diffusion length) in quantum dot material. Cathodoluminescence studies showed a reduced diffusion length of 0.5 μm in quantum dot material at 300 K compared to 1.43 μm in quantum well material at 10 K (Kim et al. 1999). Others have found small diffusion lengths only at low temperature (Shaw et al. 2003). A spectrally and spatially resolved study using confocal photoluminescence demonstrated no carrier migration when carriers were generated within the InGaAs quantum well of a DWELL structure at low density but displacement characterized by a diffusion length of 1.95 μm for carriers initially generated within the GaAs layers within the structure (Popescu et al. 2003). This highlights that to take advantage of the low carrier diffusion in the quantum dot material, carriers must not be allowed to spread within the other layers within the laser structure and that the threshold-gain requirement must be satisfied at carrier densities such that carriers are primarily contained within the quantum dots. The carrier localization provided by QDs, when coupled with oxide layers within the device structure to minimize diffusion in other layers, has allowed the fabrication of small volume, low threshold current (and current density) devices (Park et al. 2000b). Further comparative measurements on LED structures, where carrier spreading in layers other than the active material was carefully controlled, have been interpreted to yield diffusion lengths varying from 2.7 μm in InGaAs quantum wells to <100 nm in quantum dots (Fiore et al. 2004). There is also some evidence of reduced surface sidewall recombination as characterized by a recombination velocity of 5×10^4 cm/s for quantum dot material compared to 5×10^5 cm/s for comparable quantum well lasers (Moore et al. 2006). Together these properties should give greater freedom in the positioning of etched surfaces with respect to the active region of the laser, resulting in major opportunities for increased complexity while at the same time employing relatively simple processing techniques.

Further advantages result from the low threshold current density, which has already been described, and an increased internal differential efficiency, due to the improved carrier localization, which leads to higher external efficiencies and consequently a reduced thermal budget that is essential for close packing of devices.

9.7 Summary

This chapter has provided an introduction to the basic concepts of self-assembled quantum dot lasers covering their electronic and optical properties, their recombination and gain characteristics, and how these depend on quantum dot device structure and growth.

The factors that affect the achievable optical gain and the radiative recombination current (which determines the intrinsic threshold current) in the same sense are the transition matrix element, the wave function overlap integral, and the dot density. The overlap of the optical mode with the dot electronic wave functions, the inhomogeneous broadening, and the proximity of higher-lying dot- and wetting layer-energy states on the other hand can be engineered to reduce the current required to achieve a given peak gain.

Measurements on state-of-the-art structures indicate that a substantial part of the threshold current is due to non-radiative recombination (about 80%), which is at least in part made up of defect-related recombination allowing scope for further improvements in performance.

The population of higher-lying energy states for operation at higher threshold gain requirements or at higher temperatures has a significant impact on both the radiative and particularly the non-radiative recombination currents and further progress in tailoring energy state separations is still required.

Self-assembled quantum dot lasers are good choices as optical sources in integrated systems due to the low threshold current density, small variation of threshold current with temperature, and the degree of spatial localization that can already be achieved.

Acknowledgments

Our research on self-assembled quantum dot lasers has been supported by the UK Engineering and Physical Sciences Research Council. We are indebted to Helen Pask, Ian O'Driscoll, Ian Sandall, Craig Walker, Simon Osborne, and HuiYun Liu of University College London and to the large team of coworkers from the University of Sheffield, including Mark Hopkinson, David Mowbray, and Kris Groom, for their contributions to the work described in this chapter.

References

Akiyama T., Sugawara M., and Arakawa Y., 2007. Quantum dot semiconductor optical amplifiers. *Proceedings of the IEEE* 95, 1757–1766.

Andreev A.D. and O'Reilly E.P., 2005. Optical matrix element in InAs/GaAs quantum dots: Dependence on quantum dot parameters. *Applied Physics Letters* 87, 213106.

Arakawa Y. and Sakaki H., 1982, Multidimensional quantum well laser and temperature dependence of its threshold current. *Applied Physics Letters* 40, 939–941.

Asada M., Miyamoto Y., and Suematsu Y., 1986. Gain and threshold of three-dimensional quantum box lasers. *IEEE Journal of Quantum Electronics* 22, 1915–1921.

Bernard M.G.A. and Duraffourg G., 1961. Lasing conditions in semiconductors. *Physica Status Solidi* 1, 699.

Bimberg D., Grundmann M., and Ledentsov N.N., 1999. *Quantum Dot Heterostructures*, John Wiley & Sons, Chichester, U.K.

Blood P., 2000. On the dimensionality of optical absorption, gain and recombination in quantum confined structures. *IEEE Journal of Quantum Electronics* 36, 354–362.

Blood P., Lewis G.M., Smowton P.M. et al., 2003. Characterisation of semiconductor laser gain media by the segmented contact method. *IEEE Journal of Selected Topics in Quantum Electronics* 9, 1275–1282.

Blood P., Pask H., Summers H.D. et al., 2007. Localized Auger recombination in quantum-dot lasers. *IEEE Journal of Quantum Electronics* 43(12), 1140–1146.

Borri P., Langbein W., Schneider S. et al., 2002. Exciton relaxation and dephasing in quantum-dot amplifiers from room to cryogenic temperature. *Journal of Selected Topics in Quantum Electronics* 8, 984–991.

Chand N., Becker E.E., van der Ziel J.P. et al., 1991. Excellent uniformity and very low (less than 50 A cm^{-2}) threshold current density strained InGaAs quantum well diode lasers on GaAs substrate. *Applied Physics Letters* 58, 1704.

Chang K.P., Yang S.L., Chuu D.S. et al., 2005. Characterisation of self-assembled InAs quantum dots with InAlAs/InGaAs strain reducing layers by photoluminescence spectroscopy. *Journal of Applied Physics* 97, 083511.

Chen J.F., Hsiao R.S., Hsieh F. et al., 2006. Effect of incorporating an InAlAs layer on electron emission in self-assembled InAs quantum dots. *Journal of Applied Physics* 99, 014303.

Chuang S.L., 1995. *Physics of Optoelectronic Devices*, John Wiley & Sons, New York.

Coldren L.A. and Corzine S.W., 1995. *Diode Lasers and Photonic Integrated Circuits*, John Wiley & Sons, New York.

Deppe D.G., Huffaker D.L., Csutak S. et al., 1999. Spontaneous emission and threshold characteristics of 1.3-μm InGaAs-GaAs quantum-dot GaAs-based lasers. *IEEE Journal of Quantum Electronics* 35, 1238–1246.

Deppe D.G., Freisem S., Huang H. et al., 2005. Electron transport due to inhomogeneous broadening and its potential impact on modulation speed in p-doped quantum dot lasers. *Journal of Physics D* 38, 2119–2125.

Dicke R.H. and Wittke J.P., 1960. *Introduction to Quantum Mechanics*, Addison-Wesley, Reading, MA.

Eliseev P.G., Li H., Liu G.T. et al., 2001. Ground state emission and gain in ultralow-threshold InAs-InGaAs quantum dot lasers. *IEEE Journal of Selected Topics on Quantum Electronics* 70, 135–142.

Fathpour S., Mi Z., Bhattacharya P. et al., 2004. The role of Auger recombination in the temperature-dependent output characteristics ($T_0 = \infty$) of p-doped 1.3 μm quantum dot lasers. *Applied Physics Letters* 85, 5164–5166.

Fiore A. and Markus A., 2007. Differential gain and gain compression in quantum-dot lasers. *IEEE Journal of Quantum Electronics* 43, 287–294.

Fiore A., Rossetti M., Alloing B. et al., 2004. Carrier diffusion in low dimensional semiconductors: A comparison of quantum wells, disordered quantum wells and quantum dots. *Physical Review B* 70, 205311.

Freisem S., Ozgur G., Shavritranuruk K. et al., 2008. Very-low-threshold current density continuous-wave quantum-dot laser diode. *Electronics Letters* 44, 679.

Grundmann M. and Bimberg D., 1997. Theory of random population for quantum dots. *Physical Review B* 55, 9740–9745.

Hasbullah N.F., Ng J.S., Liu H.-Y. et al., 2008. Dependence of the electroluminescence on the spacer layer growth temperature of multi-layer quantum dot laser structures. *IEEE Journal of Quantum Electronics* 45(1), 79–85.

Herrmann E., Smowton P.M., Ning Y. et al., 2001. Performance of lasers containing three, five and seven layers of quantum dots. *IEEE Proceedings-Optoelectronics* 148, 238–242.

Huang H. and Deppe D.G., 2001. Rate equation model for nonequilibrium operating conditions in a self-organized quantum-dot laser. *IEEE Journal of Quantum Electronics* 37, 691–698.

Kim J.K., Strand T.A., Naone R.L. et al., 1999. Design parameters for lateral carrier confinement in quantum-dot lasers. *Applied Physics Letters* 74, 2752–2754.

Kim J.K., Naone R.L., and Coldren L.A., 2000. Lateral carrier confinement in miniature lasers using quantum dots. *IEEE Journal of Selected Topics in Quantum Electronics* 6, 504–510.

Kim K., Norris T.B., Ghosh S. et al., 2003. Level degeneracy and temperature-dependent carrier distributions in self-organized quantum dots. *Applied Physics Letters* 82, 1959–1961.

Kirstaedter N., Ledentsov N.N., Grundmann M. et al., 1994. Low threshold, large to injection laser emission from In(Ga)As quantum dots. *Electronics Letters* 30, 1416–1417.

Kita T., Wada O., Ebe H. et al., 2002. Polarization-independent photoluminescence from columnar InAs/GaAs self-assembled quantum dots. *Japanese Journal of Applied Physics Part 2-Letters* 41(10B), L1143–L1145.

Ledentsov N.N., November 25, 2003. Semiconductor device and method of making the same, United States Patent 6,653,166.

Lester L.F., Stintz A., Li H. et al., 1999. Optical characteristics of 1.24 μm InAs quantum dot laser diodes. *IEEE Photonics Technology Letters* 11, 931–933.

Liu G.T., Stintz S., Li H. et al., 2000. The influence of quantum-well composition on the performance of quantum dot lasers using InAs/InGaAs dots-in-a-well (DWELL) structures. *IEEE Journal of Quantum Electronics* 36, 1272–1279.

Liu H.Y., Sellers I.R., Gutierrez M. et al., 2004. Influences of the spacer layer growth temperature on multi-layer InAs/GaAs quantum dot structures. *Journal of Applied Physics* 96, 1988–1992.

Liu H.Y., Tey C.M., Sellers I.R. et al., 2005. Mechanism for improvements in optical properties of 1.3 μm InAs/GaAs quantum dots by a combined InAlAs–InGaAs cap layer. *Journal of Applied Physics* 98, 083516.

Loudon R. 1983. *The Quantum Theory of Light*, Clarendon Press, Oxford, U.K.

Marko I.P., Andreev A.D., Adams A.R. et al., 2003. The role of auger recombination in 1.3 μm quantum dot laser investigated using high hydrostatic pressure. *IEEE Journal of Selected Topics in Quantum Electronics* 9, 1300–1307.

Marko I.P., Masse N.F., Sweeney S.J. et al., 2005. Carrier transport and recombination in p-doped and intrinsic 1.3 μm InAs/GaAs quantum-dot lasers. *Applied Physics Letters* 87, 211114.

Markus A., Rosetti M., Calligari V. et al., 2006. Two-state switching and dynamics in quantum dot two-section lasers. *Journal of Applied Physics* 100, 113104.

Matthews D.R., Summers H.D., Smowton P.M. et al., 2002. Experimental investigation of the effect of wetting-layer states on the gain-current characteristic of quantum-dot lasers. *Applied Physics Letters* 81, 4904–4906.

Moore S.A., O'Faolain L., Cataluna M.A. et al., 2006. Reduced surface sidewall recombination and diffusion in quantum dot lasers. *IEEE Photonics Technology Letters* 18, 1861–1863.

Nishi K., Saito H., Sugou S. et al., 1999. A narrow photoluminescence linewidth of 21 meV at 1.35 μm from strain reduced InAs quantum dots covered by In0.2Ga0.8As grown on GaAs substrates. *Applied Physics Letters* 74, 1111–1113.

Osborne S.W., Blood P., Smowton P.M. et al., 2004a. Optical absorption cross section of quantum dots. *Journal Physics C: Condensed Matter* 16, S3749–S3756.

Osborne S.W., Blood P., Smowton P.M. et al., 2004b. State filling in InAs quantum-dot laser structures. *IEEE Journal of Quantum Electronics* 40, 1639–1645.

Ouyang D., Ledentsov N.N., Bimberg D. et al., 2003. High performance quantum dot lasers with etched waveguide. *Semiconductor Science and Technology* 18, L53–L54.

Park G., Shchekin O.B., and Deppe D., 2000a. Temperature dependence of gain saturation in multilevel quantum dot lasers. *IEEE Journal of Quantum Electronics* 36, 1065–1071.

Park G., Shchekin O.B., Huffaker D.L. et al., 2000b. Low threshold oxide confined 1.3 μm quantum dot laser. *IEEE Photonics Technology Letters* 13, 230–232.

Pask H.J. 2006. Model for localised recombination in quantum dots. PhD thesis, Cardiff University, Cardiff, Wales, U.K.

Pask H.J., Blood P., and Summers H.D., 2005. Light-current characteristics of quantum dots with localized recombination. *Applied Physics Letters* 87, 083109.

Patane A., Polimeni A., Henini H. et al., 1999. Thermal effects in quantum dot lasers. *Journal of Applied Physics* 85, 625–627.

Popescu D.P., Eliseev P.G., Stintz A. et al., 2003. Carrier migration in structures with InAs quantum dots. *Journal of Applied Physics* 94, 2454–2458.

Sandall I.C., Smowton P.M., Thomson J.D. et al., 2006a. Temperature dependence of threshold current in p-doped quantum dot lasers. *Applied Physics Letters* 89, 151118.

Sandall I.C., Smowton P.M., Walker C.L. et al., 2006b. Recombination mechanisms in 1.3 μm InAs quantum-dot lasers. *IEEE Photonics Technology Letters* 18, 965–967.

Sandall I.C., Smowton P.M., Walker C.L. et al., 2006c. The effect of p doping in InAs quantum dot lasers. *Applied Physics Letters* 88, 111113.

Sandall I.C., Smowton P.M., Liu H.Y. et al., 2007. Nonradiative recombination in multiple layer In(Ga)As quantum-dot lasers. *IEEE Journal of Quantum Electronics* 43, 698–703.

Schneider H.C., Chow W.W., and Koch S.W., 2001. Many-body effects in the gain spectra of highly excited quantum-dot lasers. *Physical Review B* 64, 115315.

Schoenfeld W.V., Chen C.-H., Petroff P.M. et al., 1998. Argon ion damage in self-assembled quantum dots structures. *Applied Physics Letters* 73, 2935.

Sellers I.R., Liu H.Y., Hopkinson M. et al., 2003. 1.3 μm lasers with AlInAs-capped self assembled quantum dots. *Applied Physics Letters* 83, 4710.

Sellers I.R., Liu H.Y., Groom K.M. et al., 2004. 1.3 mm InAs/GaAs multilayer quantum-dot laser with extremely low room-temperature threshold current density. *Electronics Letters* 40, 1412–1413.

Shaw A., Folliot H., and Donegan J.F., 2003. Carrier diffusion in InAs/GaAs quantum dot layers and its impact on light emission from etched microstructures. *Nanotechnology* 14, 571–577.

Shchekin O.B. and Deppe D.G., 2002. High temperature performance of self-organised quantum dot laser with stacked p-doped active region. *Applied Physics Letters* 80, 3277–3279.

Siegman A.E., 1986. *Lasers*, University Science Books, Mill Valley, CA.

Smowton P.M., Blood P., Mogensen P.C. et al., 1996. Role of sublinear gain-current relationship in compressive and tensile strained 630 nm GaInP lasers. *International Journal of Optoelectronics* 10, 383–391.

Smowton P.M., Sandall I.C., Mowbray D.J. et al., 2007a. Temperature dependent gain and threshold in p-doped quantum dot lasers. *IEEE Journal of Selected Topics in Quantum Electronics* 13, 1261–1266.

Smowton P.M., Sandall I.C., Liu H.Y. et al., 2007b. Gain in p-doped quantum dot lasers. *Journal of Applied Physics* 101, 013107.

Smowton P.M., George A.A., Sandall I.C. et al., 2008. Origin of temperature-dependent threshold current in p-doped and undoped In(Ga)As quantum dot lasers. *IEEE Journal of Selected Topics in Quantum Electronics* 14, 1162–1170.

Stier O., Grundmann M., and Bimberg D., 1999. Electronic and optical properties of strained quantum dots modeled by 8-band k.p theory. *Physical Review B* 59, 5688–5701.

Sugawara M. (ed.), 1999. *Self Assembled InGaAs/GaAs Quantum Dots*, Vol. 60 (Semiconductors and Semimetals), Academic Press, San Diego, CA.

Sugawara M., Mukai K., Nakata Y. et al., 2000. Effect of homogeneous broadening of optical gain on lasing spectra in self-assembled $In_xGa_{1-x}As$/GaAs quantum dot lasers. *Physical Review B* 61, 7595–7603.

Summers H.D. and Rees P., 2007. Derivation of a modified Fermi–Dirac distribution for quantum dot ensembles under nonthermal conditions. *Journal of Applied Physics* 101, 073106.

Summers H.D., Thomson J.D., Smowton P.M. et al., 2001. Thermodynamic balance in quantum dot lasers. *Semiconductor Science and Technology* 16, 140.

Uskov A.V., Boucher Y., Le Behan J. et al., 1998. Theory of a self-assembled quantum-dot semiconductor laser with Auger carrier capture: Quantum efficiency and nonlinear gain. *Applied Physics Letters* 73, 1499–1501.

Walker C.L., Sandall I.C., Smowton P.M. et al., 2005. The role of high growth temperature GaAs spacer layers in 1.3 μm In(Ga)As quantum-dot lasers. *IEEE Photonics Technology Letters* 17, 2011.

Wei Y.Q., Wang S.M., Ferdos F. et al., 2002. Large ground to first excited state transition energy separation in InAs quantum dots emitting at 1.3 µm. *Applied Physics Letters* 81, 1621.

Williamson A.J., Wang L.W., and Zunger A., 2000. Theoretical interpretation of the experimental electronic structure of lens-shaped self-assembled InAs/GaAs quantum dots. *Physical Review B* 62, 12963.

Wojs A., Hawrylak P., Fafard S. et al., 1996. Electronic structure and magneto-optics of self-assembled dots. *Physical Review B* 54, 5604.

Zhukov A.E., Ustinov V.M., Egorov A. Yu. et al., 1997. Negative characteristic temperature of InGaAs quantum dot injection lasers. *Japanese Journal of Applied Physics* 36 (Part 1), 4216–4218.

Zhukov A.E., Kovsh A.R., Ustinov V.M. et al., 1999. Gain characteristics of quantum dot injection lasers. *Semiconductor Science and Technology* 14, 118–123.

10

Quantum Dot Microcavity Lasers

Tian Yang
Adam Mock
John O'Brien

10.1 Introduction

Microcavity lasers with InAs quantum dot active regions combine the high-quality-factor, small-volume optical modes of microcavities with the discrete and narrow gain profile of quantum dots. These are promising candidates for high-speed, low-power devices in high-density photonic integrated circuits [7,40,60]. In this chapter, we describe room temperature lasing behavior of InAs quantum dot microdisk and photonic crystal nanocavity lasers.

A quantum dot has a three-dimensional confinement of the electrons and holes on the scale of the de Broglie wavelength. Therefore, the quantum dot behaves like a quantum box. In practice, there are quantum dots with different shapes, for example, pyramid and disk [8]. The details regarding carrier states' spatial and spectral profiles depend on the specific size, shape, material composition, and strain profile of the quantum dot.

Quantum well materials have largely replaced bulk materials as the gain medium for semiconductor lasers because of their higher differential gain and lower oscillation threshold. A quantum dot laser technology is expected to improve on the existing quantum well–based laser technology in several key laser specifications. First, quantum dot lasers are expected to exhibit very low threshold currents. The general formula for the threshold of any kind of diode laser is $J_{th} = eV_a n_{th}/\tau_r$, where J_{th} is the threshold current density, e is the charge of an electron, V_a is the volume of active materials that participate in the lasing, n_{th} is the threshold carrier density, and τ_r is the recombination lifetime of carriers below threshold. For lasers with a quantum dot active material, V_a is significantly smaller than for all other types of active materials, because only a small fraction of the material is composed of quantum dots. n_{th} is also significantly reduced compared to other types of active materials when the lasing threshold is not far above transparency, because along with the fact that the quantum dots comprise a small fraction of the material, the narrow quantum dot material gain spectrum means that quantum dots are very efficient in providing gain at the lasing wavelength.

Second, quantum dot lasers are expected to have excellent temperature stability. The threshold current of a semiconductor laser increases with temperature because as the temperature increases, the carrier population spreads out in energy, as required by the Fermi–Dirac statistics. This reduces the available peak gain. Lasers with a quantum dot active material can have enhanced temperature stability compared to other types of active materials, because the quantum dots have a discrete set of allowed energy states.

Third, increased modulation bandwidths might be expected. The relaxation resonance frequency and damping are important factors in determining the available small-signal modulation bandwidth. The relaxation oscillation resonance frequency of a modulated semiconductor laser is proportional to the square root of the differential gain dg/dn, where g is the optical gain per unit length of active material and n is the electronic carrier density [75]. Quantum dot active materials have higher differential gain values than active regions of higher dimensionality. The narrow gain spectrum of a quantum dot results in a larger increase in gain at the lasing wavelength per added carrier than in a material with gain spectra that is spread more broadly in energy about the lasing energy.

Finally, lasers operating with quantum dot active regions have long been expected to operate with less chirp than their quantum well laser counterparts. A chirp is a modulation in the operating wavelength that results from the modulation of the bias point. When semiconductor lasers are modulated, the modulation in electronic carrier density leads to the modulation of the refractive index of the laser cavity. This change in the index of refraction results in a change of the resonant frequency of the laser cavity and the laser output chirps [75]. Chirped signals have a broadened linewidth and suffer more dispersion when being transmitted in fibers. This frequency shift in response to carrier density change is proportional to the linewidth enhancement factor $\alpha = -dn/dn_i$, where n is the real part of the refractive index and n_i is the imaginary part of the refractive index [17]. n and n_i are not independent variables but are dependent on each other via the Kramers–Kronig relationship. Quantum dot active media can be modeled with a gain profile that is symmetric in frequency about the gain peak. As a result of the Kramers–Kronig relation, changes in the gain peak of a symmetric function of frequency result in no change in the index of refraction at the frequency of the gain peak. Bulk, quantum well, and quantum wire material systems have asymmetric gain profiles as a function of frequency as a result of the extended density of states available on these systems.

10.2　Microcavities: Why Are They Interesting?

Small-volume cavities are interesting for many reasons, and which specific reason is the most important depends on the technology goals. For integrated optics, small optical cavities provide small footprint sources. Small-volume-cavity lasers will also exhibit smaller threshold currents than larger-volume-cavity lasers, assuming that the cavity quality factor is more or less constant in both cases. This constancy of the cavity quality factor between a small-volume-diode laser and a larger-volume-diode laser is often practically true because the choice of the cavity quality factor is constrained by the available material gain and the desire to efficiently couple optical power out of the laser. Small-volume optical cavities also modify the available optical mode spectrum overlapping the material gain spectra. This can modify spontaneous emission processes and result in systems in which the material dipole emitters are strongly coupled to the optical field. An important parameter in characterizing the strength of this coupling is the Rabi frequency [3,23]. An expression for the Rabi frequency, Ω, is given in Equation 10.1:

$$\Omega = \frac{\vec{d}_{cv} \cdot \vec{E}}{\hbar} \tag{10.1}$$

where
   $\vec{d}_{cv}$ is the dipole moment for the c to v transition
   $\vec{E}$ is the vector electric field amplitude

The Rabi frequency describes the rate at which an electronic system can coherently exchange energy with an electromagnetic field mode. It is proportional to the electric field intensity. The electric field intensity inside a resonant cavity is proportional to the ratio of cavity quality factor to mode volume, Q/V, at a constant power flow. Microcavities have, by definition, a small volume, and they typically can be made to have very high Q values. Increasing Q/V values is one important way to strengthen the interaction between light and matter.

The behavior of microcavity systems can be classified according to whether the rate at which the energy can be coherently exchanged between the electronic system and the optical field is greater than or less than the decoherence rate of the system, which comes from photons escaping out of the cavity and the decoherence of the material dipole caused by scattering and recombination mechanisms. When the rate at which energy is exchanged between the electronic system and the field mode is higher than the decoherence rate, the quantum dot will emit and absorb a single photon several times before the photon escapes out of the cavity. This regime is called strong coupling and leads to splitting in the spectrum of the quantum dot–photon system [27,78]. The ability to achieve a strong coupling is important for quantum information processing, because it allows the quantum dot–photon system to go through several operations within the coherence time.

Microcavity diode lasers do not operate in the strong coupling regime. The more moderate Q/V regime is arrived at through the constraints of operating the laser in the linear gain regime and not in the gain saturation regime to improve the modulation characteristics of the laser, and the typical desire to efficiently couple the optical power from a semiconductor laser. Efficient output coupling of the laser radiation is important in microcavity lasers used for integrated optical communication systems, because error-free communication at 10 GB/s will require optical powers of the order of 50 μW to be incident on the detector. This requirement of efficient output coupling moderates the desired Q value, because in order to achieve this, the cold cavity Q must be lower, preferably by an order of magnitude or more, than the Q value that is associated with the internal material losses in a laser, such as free-carrier absorption loss. Nevertheless, a microcavity laser operates at much larger Q/V values than more traditional types of semiconductor diode lasers. The increased coupling between the optical field and the radiating dipole of microcavity lasers compared to traditional diode lasers results in modifications of the spontaneous emission rate. In particular, an increase in the spontaneous emission rate at the transition energy corresponding to that of the resonant optical mode is expected and observed in microcavity lasers. This enhanced spontaneous emission rate is known as the Purcell effect [48]. This effect states that the spontaneous emission rate into the cavity mode is the spontaneous emission rate in vacuum multiplied by the Purcell factor, F_P:

$$F_P = \frac{3Q(\lambda/n_{\text{eff}})^3}{4\pi^2 V} \tag{10.2}$$

where
  λ is the wavelength of the cavity mode in vacuum
  n_{eff} is the effective refractive index of the cavity mode

Here, it is assumed that the frequency of the quantum dot transition is equal to the frequency of the cavity mode, the polarization of the optical transition is aligned with the polarization of the cavity mode, and the spatial position of the quantum dot is the same as the peak point of the electric field density of the cavity mode. Purcell factors between 10 and 100 might be typical for a microcavity laser.

10.3 Microcavities: Photonic Crystal and Microdisks

Solving Maxwell's equations in their most general form for the three electric and three magnetic vector field components is often difficult. In materials that are characterized by a linear and isotropic relative permittivity, solutions to Maxwell's equations are defined by the spatial variation of the refractive index. Numerous techniques exist for finding exact or approximate solutions for the electromagnetic fields in several common dielectric geometries. For example, rectangular waveguides, optical fibers, and one-dimensional Bragg reflectors have been studied for several decades, and approximate analytical solutions may be obtained [25,39,75].

In many microcavity laser geometries, an exact analytical solution of Maxwell's equations is not available. Most photonic crystal lasers, for example, are defined by a two-dimensional photonic crystal patterned into a semiconductor slab that is about one-half wavelength thick. This semiconductor slab confines the optical mode in the third direction. In two-dimensional photonic crystal slabs, obtaining the electric and magnetic fields is difficult. With a typical slab containing hundreds of holes, writing down a solution satisfying the boundary conditions everywhere is not feasible. In many cases, assuming a structure that is uniform in one or two dimensions greatly simplifies the analysis. For example, it is possible to write down exact solutions for the one-dimensional slab waveguide problem, but only approximate solutions may be obtained from the rectangular waveguide problem. No such approximations exist for photonic crystals. If one assumes that the structure is uniform along the hole perforation direction, then Maxwell's equations separate into TE and TM polarizations, but the complete solution still requires a numerical solution. Common numerical approaches to analyzing photonic crystals include plane wave expansion [49,51], finite element method [19,29,31], and finite-difference time-domain (FDTD) method [32,33,61,76].

In this section, we discuss the FDTD method due to its simplicity and ease of implementation, its linear scaling with the problem size, and its inherent parallelizability. The method explicitly discretizes Maxwell's curl equations in space and time. The electric and magnetic fields are staggered relative to each other on a grid, and this results in a second-order accuracy in the first-order space and time derivatives. This approach is known as Yee's algorithm [61,76]. When analyzing optical resonators, the electromagnetic quantities of interest include resonance frequency, quality factor, and electric and magnetic field profiles.

Figure 10.1a shows the time evolution of a typical field component at a user-specified location in a photonic crystal double-heterostructure cavity for 10^5 time steps. The top axis of Figure 10.1a indicates how much real time has elapsed. A typical FDTD run progresses a few tens of picoseconds for optical and near-infrared frequencies. The computational domain used to obtain the time sequence of Figure 10.1a included $950 \times 340 \times 200$ spatial discretization points. To resolve the circular features associated with photonic crystal perforations, we use twenty discretization points per lattice constant. Due to the staggered nature of the Yee grid, the electric field samples the geometry twice per grid point in each direction, resulting in effectively 40 electric field samples per lattice constant. The large domain sizes associated with such fine spatial resolution require parallelization on the order of one hundred processors. At this level of parallelization, computing 10^5 time steps requires about 10 h on USC's linux cluster (http://www.usc.edu/hpcc/).

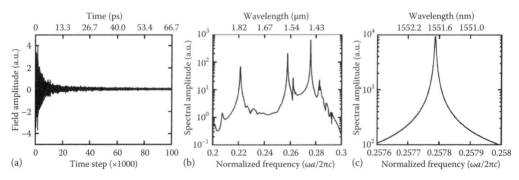

FIGURE 10.1 (a) Amplitude of the z component of **H** as a function of time in a photonic crystal double-heterostructure cavity. (b) Discrete Fourier transform of the time sequence in (a). (c) Pade interpolation of the spectrum in (b).

Once a sufficiently long time sequence is obtained, we use a discrete Fourier transform to find the resonance frequencies [12,14]. The discrete Fourier transform of the time sequence of Figure 10.1a is shown in Figure 10.1b. The bottom axis is in units of normalized frequency, where a is the photonic crystal lattice constant. The top axis corresponds to wavelengths for a lattice constant of 400 nm. Several resonance peaks are apparent in the spectrum. More details about the resonance spectra of specific cavities will be discussed below. Once the resonance frequencies are known, discrete time-filtering techniques may be used to isolate the electric and magnetic spatial field distributions associated with the different resonances.

In addition to the resonance frequency and electromagnetic field distribution, an important quantity in characterizing laser cavity designs is the quality factor, Q. The cavity Q is an indication of how well the cavity stores electromagnetic energy. Laser cavities that dissipate a large percentage of their total stored energy per optical cycle have low Q factors. The cavities discussed below have large Q values and small mode volumes, which make them attractive for low-threshold laser applications. There are several ways to estimate Q using the FDTD method. One method uses the following formula: $Q = \omega_0 U/(-dU/dt)$, in which ω_0 is the resonance frequency and U is the electromagnetic energy stored in the cavity [2]. Alternatively, one may solve the above equation, which indicates that the individual field components have an exponentially decaying envelope function, $\exp(-\omega_0 \cdot t/2Q)$. From a time sequence, one can then pick out Q by analyzing the rate of decay of the oscillating time signal either by directly analyzing the single-mode oscillation or by harmonic inversion techniques [38,67]. A third way is to look at the discrete Fourier transform and calculate Q from the width of the resonance according to $Q = f/\Delta f$. For large Q values, the spectral resolution is often insufficient to estimate Q directly. In this case, we use a Padé interpolation to fill in the space between the discrete Fourier transform frequency samples [18,24,50]. An example interpolation is shown in Figure 10.1c, which corresponds to $Q = 59$ K. From the raw discrete Fourier transform data alone, only Q values as large as a few hundred can be accurately extracted.

As a final remark on using the FDTD method to analyze Q, it is important to terminate the simulation domain with appropriate absorbing boundary conditions (ABCs), so that radiation that leaks out of the cavity is not reflected from the boundaries back into the cavity. In our simulations, we use 15 perfectly matched layers of ABCs [6,61]. This reduces reflections by several orders of magnitude. The efficacy of the perfectly-matched-layer approach is another attractive feature of the FDTD method.

One particularly interesting photonic crystal microcavity is known as the L3 cavity. The L3 cavity consists of removing three neighboring holes along a single direction from a hexagonal photonic crystal lattice, as shown in Figure 10.2a. In-plane confinement of the resonant mode is due to the Bragg reflection provided by the photonic crystal lattice, and out-of-plane confinement is provided by the total internal reflection at the slab–air interface. The first investigation of this cavity geometry in 2003

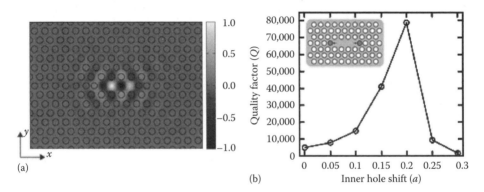

(a)

(b)

FIGURE 10.2 (a) Profile of the z component of **H** at the midplane of the slab for the optimized L3 cavity. (b) Quality factor as a function of hole shift for the L3 cavity.

obtained *Q* values reaching 45,000. This was one of the highest values yet predicted for two-dimensional photonic crystal cavities at the time [1]. The y component of the electric field at the midplane of the slab for the highest *Q* resonance mode of an L3 cavity is shown in Figure 10.2a along with a superimposed diagram of the cavity. The tight field confinement is apparent. In this structure, the hole radius to lattice constant ratio is $r/a = 0.29$ and the slab thickness to lattice constant ratio is $d/a = 0.6$. The index of the slab is $n = 3.4$.

In [1], the authors suggest a method for boosting the *Q* of the L3 cavity by shifting the neighboring holes along the linear defect away from the cavity center by a small fraction of the lattice constant. This is illustrated in the inset of Figure 10.2b. We have performed calculations verifying this effect, where it is seen in Figure 10.2b that the *Q* can increase by almost a factor of 10 for an optimized shift of $0.20a$. This effect is explained by pointing out that moving the two neighboring holes slightly outward smooths the evanescent tail of the optical mode as it penetrates the photonic crystal cladding. To be more precise, this relaxation of the spatial mode confinement removes some of the higher-frequency spatial wavevectors of the mode. This results in a smaller overlap of the k-space distribution of the mode with the light cone, and thus lower out-of-plane radiation. This is an example of a general design approach for two-dimensional photonic crystal cavities that has been discussed by several authors [20,58]. In more recent work, the same group has improved their optimization of the L3 cavity by shifting the second and third neighboring holes. *Q* values as high as 2×10^5 have been achieved [2].

Recently, it was shown that by perturbing the lattice constant of a few periods of an otherwise uniform photonic crystal single-line defect waveguide, a high-*Q* cavity could be formed [56]. This structure is known as a photonic crystal double heterostructure, and a diagram depicting its basic geometry is shown in Figure 10.3a. The lattice constant of the lighter holes in the figure is stretched along the *x* direction by a few percent. Confinement along the *y* direction is due to the Bragg reflection provided by the photonic crystal lattice, and out-of-plane confinement is provided by the total internal reflection at the slab–air interface. Stretching the lattice constant of a few periods, as depicted in Figure 10.3a, shifts the photonic crystal waveguide band down in frequency, causing the band edge of the perturbed region to fall into the mode gap of the neighboring photonic crystal waveguide section. As a result, the field becomes evanescent along the waveguide direction. Because the propagating modes of the perturbed region must fall into the mode gap of the neighboring waveguides in order to be confined, bound states form near local minima in the photonic crystal waveguide dispersion diagram [43]. If the perturbation is made by decreasing the lattice constant of a few periods, then bound states form

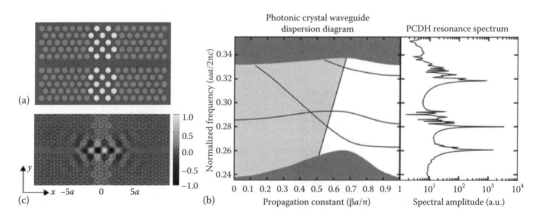

FIGURE 10.3 (a) Schematic diagram of a photonic crystal double-heterostructure cavity. The white holes are stretched along the *x*-direction. (b) Photonic crystal waveguide dispersion diagram. Dark-gray areas are photonic crystal cladding modes. Light-gray area is the light cone. To the right is a photonic crystal double-heterostructure resonance spectrum rotated so that its frequency axis is aligned with the photonic crystal waveguide dispersion diagram. (c) Profile of the z component of **H** at the midplane of the slab for a 5% perturbed double heterostructure.

near local maxima. Figure 10.3b shows a photonic crystal dispersion diagram for a single-line defect photonic crystal waveguide. A discrete Fourier transform resonance spectrum for a photonic crystal double heterostructure formed by stretching the lattice constant by 5% is shown to the right. The resonance spectrum is rotated so that its frequency axis is aligned with that of the dispersion diagram. It is apparent that the resonance peaks are located near the minima of the waveguide bands.

The highest Q photonic crystal bound state has a center frequency $\omega a/2\pi c = 0.261$ and a $Q = 320$ K. The associated mode profile is shown in Figure 10.3c. By a similar argument discussed for the L3 cavity, if the perturbation is weakened by decreasing the amount by which the

FIGURE 10.4 Schematic illustration of a microdisk cavity.

lattice is stretched, then there are fewer high-frequency wave vectors in the mode, and the out-of-plane radiation is reduced. In this way, the Q can be increased, and reports of Q values in excess of 10^6 have been reported for 2.5% perturbations [57,62].

Forming photonic crystal double-heterostructure cavities by stretching the lattice in one dimension is one method of introducing a defect in an otherwise uniform photonic crystal waveguide. However, researchers have explored other ways of perturbing the photonic crystal waveguide. Instead of stretching the lattice constant in only one dimension, the lattice constant in both in-plane directions may be perturbed [43]. Furthermore, one may locally modulate the air hole radius [35] or perturb the photonic crystal waveguide width [34]. Patterning a photorefractive material, such as chalcogenide glass, with a photonic crystal waveguide and illuminating a portion of the waveguide changes the local refractive index and can create a high-Q cavity [64]. Researchers have also explored local air hole infiltration to create a defect in the photonic crystal waveguide [55,65].

Another common microcavity configuration is a microdisk. Figure 10.4 is a schematic illustration of a microdisk cavity. In the plane of the disk, the optical confinement of the high Q modes occurs by light reflecting back into the disk at the edge of the disk along the circumference. Optical confinement in the normal-to-the-disk direction occurs through the total internal reflection at the disk's upper and lower boundaries. There are TE-like and TM-like high Q modes in microdisk cavities. TE-like modes have their dominant electric field components in the plane of the disk, and TM-like modes have their dominant electric field component perpendicular to the plane of the disk. The optical confinement factor of the fundamental TE (TM) mode in a 220 nm thick GaAs disk suspended in air is analytically calculated to be 0.87 (0.84). The optical confinement factor is defined here as the integral of $\vec{E} \times \vec{H}$ over a cross section of the waveguide divided by the same integral over the cross section of the waveguide and the cladding layers.

The Luttinger–Kohn approach to calculating the electronic dispersion shows that the lowest hole states of self-assembled InAs quantum dots are expected to be predominantly of heavy hole character [16,42]. As a result, the TM-like modes in a microdisk cavity with an InAs quantum dot active region will experience less gain than the TE-like mode. Therefore, the discussion will focus on the TE-like modes.

Maxwell equations for monochromatic electromagnetic fields in a uniform medium of refractive index n yield for the magnetic field:

$$\nabla^2 \vec{H} = -k^2 \vec{H} \tag{10.3}$$

where $k = n2\pi/\lambda$ and λ is the wavelength in vacuum. It is useful to decompose **H** into cylindrical components:

$$\vec{H} = \hat{e}_\rho H_\rho + \hat{e}_\varphi H_\varphi + \hat{e}_z H_z \tag{10.4}$$

where

ê$_\rho$, ê$_\varphi$, ê$_z$ are unit vectors for the cylindrical coordinate system
ρ is the radial distance from the center of the disk
φ is the azimuthal angle
z is the distance from the midplane of the disk

Substituting (10.4) into (10.3), it is found that the equation for H_z is independent of the equations for H_ρ and H_φ:

$$\nabla^2 H_z = -k^2 H_z \qquad (10.5)$$

Separating variables in this equation gives

$$H_z = R(\rho)\Phi(\varphi)Z(z) \qquad (10.6)$$

Then the solution to Equation 10.5 for the TE modes inside a microdisk cavity with even symmetry with respect to the midplane of the microdisk is

$$R(\rho) = a \times J_m(k_r\rho), \; Z(z) = b \times \cos(k_z z), \; \Phi(\varphi) = c \times e^{-im\varphi} \qquad (10.7)$$

where

a, b, c are constants
m is an integer that determines the number of periods of oscillation the mode goes through within one round-trip along the circumference, $k_r^2 + k_z^2 = k^2$
J_m is the Bessel function of the first kind of order m

　　The high Q optical modes in microdisk cavities can be found approximately by making the whispering gallery approximation. The whispering gallery approximation treats the vertical guiding with the effective index method, and then assumes that the radial electric field goes to zero at the boundary of the disk. The results of the effective index method give $k_r = 2 \cdot \pi \cdot n_{eff}/\lambda$, where n_{eff} is the effective refractive index.

　　Given **H**, the **E** field of a TE-like mode is obtained via Ampere's law. **E** also contains the form $J_m(2 \cdot \pi \cdot n_{eff} \cdot \rho/\lambda) \exp(im \, \varphi)$. The whispering gallery modes (WGMs) are labeled as TE$_{m,n}$, where the Bessel function reaches the n_{th} zero at the circumference.

　　The electric field profile of the TE$_{17,1}$ WGM at the midplane of a microdisk is plotted in Figure 10.5. It is clear that such modes with a reasonably large m number (and $n = 1, 2$) have their energy

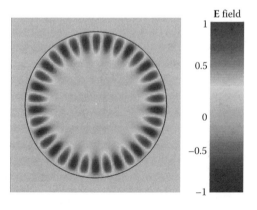

FIGURE 10.5　Electric field profile of the TE$_{17,1}$ WGM in a microdisk.

concentrated near the circumference, and the post beneath the center of the disk has a negligible effect.

10.4 Growth and Nanofabrication of Microcavity Devices Embedded with InAs Quantum Dots

The realization of quantum dot microcavity devices is largely dependent on advanced semiconductor growth and nanofabrication techniques. Semiconductor microcavity lasers containing quantum dot active regions have primarily been demonstrated using self-assembled InAs quantum dots in the active region. These self-assembled InAs quantum dots are grown by molecular beam epitaxy operating in the Stranski–Krastanow (SK) growth mode. This is one of the most well-developed methods to form semiconductor quantum dots [40,42,45,60]. In the SK growth mode, a strained layer of InAs is deposited onto a coherently strained InGaAs surface or onto a GaAs surface. A wetting layer is formed initially, after which the increase in the strain energy with increasing InAs layer thickness forces the formation of three-dimensional coherent islands of InAs, in which the compressive strain is partially relieved [53]. These InAs islands are the quantum dots. After the formation of InAs quantum dots, InGaAs or GaAs is deposited as a covering layer. Typically, quantum dot microcavity lasers are formed on GaAs substrates and the quantum dot active region is embedded in an $Al_xGa_{1-x}As$ waveguide. For the devices used to illustrate the discussion in this chapter, the epitaxial structure for the photonic crystal devices was grown on a GaAs substrate and consisted of a GaAs buffer layer, a 1 μm thick sacrificial or oxidation layer of $Al_{0.94}Ga_{0.06}As$, and a waveguide layer that was 220 nm thick and contained the active region clad with 20 nm thick $Al_{0.25}Ga_{0.75}As$ barriers. This active region contained five layers of self-assembled InAs quantum dots, each in a 5 nm thick $In_{0.15}Ga_{0.85}As$ quantum well. The quantum dot density in each layer was approximately $2 \times 10^{10}/cm^2$. These layers were separated by 30 nm of GaAs. The average diameter of the quantum dots was approximately 25 nm with a height of 3 nm. The photoluminescence (PL) spectrum of the samples at room temperature showed a peak wavelength around 1315 nm with a full-width half-maximum (FWHM) of around 45 nm for the ground-state transition, as in Figure 10.6.

The InAs quantum dot samples for the microdisk devices had a slightly different structure. The sacrificial or oxidation layer was 800 nm of AlAs, and the quantum dot density in each layer was approximately $3 \times 10^{10}/cm^2$. The PL spectrum of the samples showed a peak wavelength at around 1287 nm with an FWHM of around 40 nm for the ground-state transition.

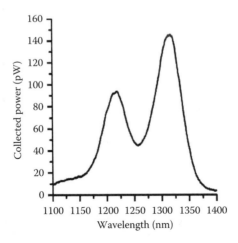

FIGURE 10.6 PL of an InAs quantum dot sample. Pumping conditions were 0.5 mW CW incident pumping power.

Photonic crystal and microdisk microcavity device geometries have been investigated by the authors. These resonant cavities were defined by electron bean lithography in the electron beam resist polymethyl-methacrylate (PMMA). This cavity pattern was subsequently transferred into the semiconductor by dry etching, using the PMMA as a mask. Other researchers have incorporated a silicon nitride or silicon dioxide mask with excellent results [26,46]. After the resist was developed, the PMMA was submitted to a blanket exposure in the electron beam lithography system. This blanket exposure process served to increase the hardness of the PMMA. This was advantageous for the subsequent dry-etching step [44]. It should be noted that this blanket exposure of the PMMA also has a beneficial effect on the smoothness of the sidewalls in some materials that are dry-etched with the PMMA mask.

The dry-etching step transferred the patterns through the semiconductor waveguide membrane and into the sacrificial oxidation AlGaAs layer. To form suspended membrane photonic crystal cavities or microdisks supported by posts, the AlGaAs layer underneath the waveguide layers was completely or partially etched away using a wet etch. A standard method to wet-etch AlGaAs selectively over GaAs, or to wet-etch high-Al-content AlGaAs selectively over low-Al-content AlGaAs is to use diluted HF [68] or buffered oxide etch [30]. This method has been used extensively to remove AlGaAs sacrificial layers to form a suspended GaAs membrane with photonic crystal patterns. Alternatively, a selective wet etching in 1:3 $H_3PO4:H_2O$ can be used. This wet-etch step was followed by an oxidation step. The wet oxidation is necessary to prevent the membranes from cracking. High-aluminum-content AlGaAs cracks in air because of the natural oxidation and the resulting strain. A short oxidation was done to form a thin layer of stable aluminum oxide on the surface of the sacrificial layer to prevent this cracking.

The wet-etching step in the formation of photonic crystal membrane resonant cavities undercut the entire device. The compressive strain from the InAs quantum dot layers can cause significant bowing in the membranes after they have been undercut. To release the strain and eliminate any bowing of the photonic crystal membranes, the devices were made in a cantilever shape. SEM (secondary electron micrograph) images of a finished device are shown in Figure 10.7 [72].

Two different microdisk resonant cavity geometries were formed. Devices that were suspended on a central pillar, as mentioned above, have been investigated [70,71]. Characteristics of a series of devices with diameters ranging from 2.5 to 4 μm will be described in the next section. An SEM of a finished device is shown in Figure 10.8a.

Microdisk cavities without a central post but with a uniform aluminum oxide layer as the lower cladding and a heat dissipation layer were also fabricated in order to achieve continuous-wave (CW) lasing at room temperature [71]. The wet-etching step was eliminated in these devices and, instead, wet oxidation was performed on the AlGaAs layer. A SEM image of a finished device with a diameter of 6.9 μm is shown in Figure 10.8b.

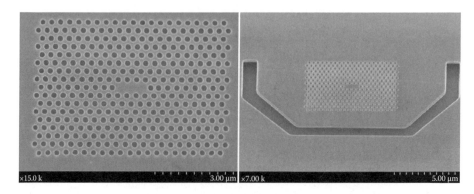

FIGURE 10.7   SEM images of a finished photonic crystal laser cavity in a suspended membrane.

FIGURE 10.8 (a) SEM image of a 3 μm diameter microdisk cavity supported on a post. (b) SEM image of a 6.9 μm diameter microdisk cavity on aluminum oxide.

These high-quality-factor, small-mode-volume microcavities have drawn interest from the research community for a long time because of their ability to provide high spontaneous-emission coupling factors and low-threshold lasing conditions [22,54]. Early work in this area reported lasing demonstrated in microdisk cavities with InAs quantum dot active regions near 980 nm by optically pumping at liquid helium or liquid nitrogen temperatures [10,41]. An electrically pumped demonstration at 920 nm has also been reported, in which the laser operates at liquid helium temperature [79].

10.5 Microcavity Laser Characteristics: Laser Output versus Pump Power and Spectra

10.5.1 Photonic Crystal Microcavity Lasers

This section describes basic laser characteristics for photonic crystal microcavities lasing near 1300 nm at room temperature. The discussion begins with an overview of the typical behavior of quantum dot photonic crystal lasers. The specific cavities used for illustration are the L3 cavity and the photonic crystal heterostructure [46,72–74]. Photonic crystal coupled-cavity lasers with InAs quantum dot active regions have been reported [77]. Efforts on electrical injection devices have also been reported [66].

The L3 cavities were optically pumped with a diode laser at 850 nm. Because these lasers are formed in a suspended membrane, heat dissipation is not very efficient and the lasers only operate under pulsed pumping conditions. Typical pulse conditions are 10–20 ns pulse widths with a 1% duty cycle at room temperature. In our work, the pump beam was focused at normal incidence to a 1–2 μm spot size on the sample using a microscope objective. The emission was collected through the same objective into an optical fiber that was connected to an optical spectrum analyzer. A typical lasing spectrum from the L3 cavity shown in Figure 10.7 is shown in Figure 10.9a. This spectrum was taken at an incident pump power level of 2.5 mW. The spectrum shows single-mode lasing behavior. No other modes were observed within a 100 nm wavelength range. This is consistent with the calculated cavity spectra described earlier. The side-mode suppression ratio (SMSR) was 20 dB. Figure 10.9b shows the output power versus the incident pump power of the same device. The pump power absorbed by the membrane at the threshold was estimated to be approximately 25 μW.

In an effort to characterize the quality factor of these lasers, the linewidth of the devices was measured under a CW operation with an optical spectrum analyzer. Q values of this lasing mode were estimated using devices with larger lattice constants and, therefore, high Q modes at longer wavelengths, where the gain or absorption from the InAs quantum dots was smaller. (These long-wavelength devices were not lasing). This estimation of the Q value was limited by the resolution of our optical spectrum analyzer, but a lower bound of 6000 was obtained. Other researchers working with devices at liquid helium temperatures obtained results for the Q values of L3 cavities containing InAs quantum dots that were not more than 20,000 [59,78]. These Q values are thought to be limited by the surface oxide of GaAs.

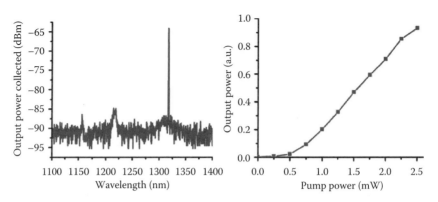

FIGURE 10.9   (a) Lasing spectrum of an L3 cavity at 2.5 mW peak incident pumping power. (b) Output power versus incident pump power characteristic of the same device.

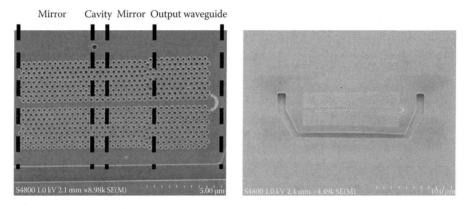

FIGURE 10.10   SEM images of a photonic crystal double-heterostructure nanocavity.

In an effort to work toward microcavity lasers that can be outcoupled efficiently for planar lightwave circuit applications, we have also investigated photonic crystal heterostructure lasers with quantum dot active regions [73,74]. The dot density in the active region is $2 \times 10^{10}/cm^2$ per layer. Figure 10.10a and b shows SEM images of one of these devices [73]. A small amount of bowing, as can be seen in the figure, is typical of devices of this length.

The double-heterostructure cavity design followed that of Song et al. [56], with a photonic crystal lattice constant, a, of 343 nm in the central region and 335 nm in the mirrors. The photonic crystal hole radius was approximately $0.3a$ everywhere. An output waveguide was butt-coupled to the mirror on one side of the cavity. The output waveguide had a lattice constant of 351 nm in order for the lasing wavelength to fall within the high-group-velocity, low-loss wavelength range of the waveguide. No effort was made in this work to optimize the coupling between the laser cavity and the output waveguide.

These heterostructure lasers were optically pumped with a semiconductor laser diode at 850 nm using pump conditions similar to those of the L3 cavities described above. Figure 10.11 shows the lasing spectrum and the output power versus incident pump power characteristic of the device shown in Figure 10.10. The peak pump power absorbed by the cavity at the threshold was estimated to be 12 μW.

The effectiveness of the edge emission and butt coupling to the output waveguides was simply characterized in a qualitative manner. We mounted the sample at the objective plane of a two-lens system and scanned a single-mode fiber in the image plane along the direction of the waveguide to obtain an intensity-versus-position profile for the device shown in Figure 10.10. This was carried out under optical

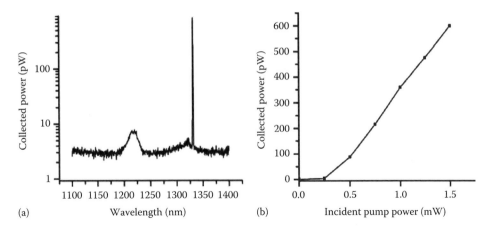

FIGURE 10.11 (a) Lasing spectrum of a photonic crystal double-heterostructure nanocavity. (b) Output power versus incident pumping power.

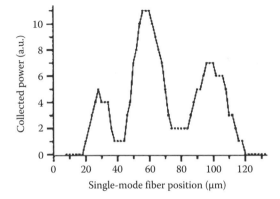

FIGURE 10.12 Laser emission intensity along the direction of the waveguide at the image plane of a two-lens system for the device shown in Figure 10.9.

pumping conditions of 16 ns pulse width, 2% duty cycle, and 2.5 mW peak pump power. The measured profile is plotted in Figure 10.12 [73]. The three peaks in this profile correspond to the light-scattering centers on the sample that are from left to right, the cavity, the mirror-output waveguide junction, and the end of the output waveguide. The magnification of the two-lens system was 7.7× at a wavelength of 780 nm. This surface normal scattering profile indicates that this laser had significant edge emission and part of the edge emission was coupled into the output waveguide. We did not observe any significant scattering from the facet of the mirror on the left side because the mirror on the left was four periods longer than the mirror on the right.

Both types of photonic crystal laser cavities, the L3 and the heterostructure, suffer from absorption due to the quantum dots in regions outside the optical pump spot. A rough estimate of the importance of this absorption loss can be obtained as follows. Non-excited InAs quantum dots are expected to have an absorption rate that is equivalent to a Q value of 900. This value was obtained based on the calculated modal gain for similar quantum dots in [47] and by using the confinement factor of a suspended slab waveguide for the confinement factor of the photonic crystal cavity mode. The field intensity decay rate in the mirror was calculated by Song et al. to be $\exp(-x/1.14a)$, where x is the distance into the mirrors and a is the lattice constant (online supplementary information to [56]). If the pump spot is $5a$ across, which is 1.75 μm for our double-heterostructure devices, and if constant field intensity inside the $2a$-long cavity is assumed, the absorption from the quantum dots outside the pump spot is equivalent to a Q value of 6600. In our experiment, the actual pumped area could be larger than the pump spot due to carrier diffusion, of course.

It is expected that this type of edge-emitting nanocavity structure will develop into a critical component for photonic crystal integrated circuits. It is a good candidate for photon sources where small footprint, high quality factor, high collection efficiency, and simple integration scheme are desired. Such photon sources include lasers for traditional optical communications and single-photon sources for quantum communication and quantum information. In this cavity geometry, we have demonstrated more than 50 µW of peak pulsed output power in a photonic crystal laser with a quantum well active region, and also shown that disordering of the quantum wells in the mirrors reduces the laser threshold and increases the slope efficiency [36,37].

10.5.2 Microdisk Lasers

Microdisk lasers were optically pumped at normal incidence at 850 nm using a semiconductor diode laser. Lasing was observed under a range of pulse conditions at and above room temperature [70,71]. Because these devices are not suspended, they are able to operate under much longer pulse lengths than the suspended membrane photonic crystal lasers described above. The pulse conditions used to collect the data from these devices were 104 ns pulse widths at a 9.4% duty cycle at room temperature. The pump beam was typically focused to a diameter of 2 µm by a microscope objective. The emission from the disk was collected through the same objective and analyzed with an optical spectrum analyzer. A subthreshold spectrum and a lasing spectrum for a 3.1 µm diameter microdisk cavity supported on a thin post are shown in Figure 10.13.

These microdisk cavities were modeled in the WGM approximation. Because the lowest hole states in the self-assembled InAs quantum dots are of predominant heavy hole characteristic [42], the TE-like modes see higher gain values than the TM-like modes, and we assumed that all the strong resonances observed were TE WGMs. For such $TE_{m,n}$ modes, the calculation results showed that the four resonances in Figure 10.13a at the wavelengths of 1254, 1266, 1302, and 1319 nm can be well approximated by $TE_{i+1,1}$, $TE_{i-3,2}$, $TE_{i,1}$, and $TE_{i-4,2}$, $i = 17 \pm 1$.

Figure 10.14 shows the optical power in the lasing mode versus the peak optical pump power for the same device under the same optical pumping conditions. The threshold pump power is 0.6 mW, which corresponds to an incident pump power of 0.38 mW and an estimated absorbed power of 28 µW. Lasing under pulsed conditions with pulses as long as 400 ns and a duty cycle of 29% was also obtained at room temperature in microdisk cavities with slightly larger aluminum oxide post diameters. In these cavities, thermal rollover of the input power versus output power characteristic occurred at about 2.5 times the threshold pump power for 400 ns pulse widths.

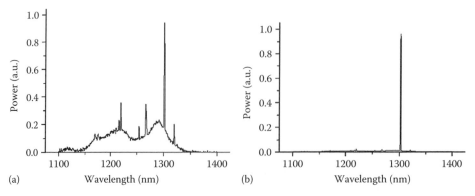

FIGURE 10.13   Subthreshold and lasing spectra of a 3.1 µm diameter microdisk supported on an aluminum oxide post.

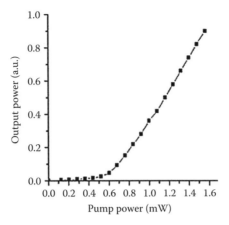

FIGURE 10.14 Optical power in the lasing mode of a 3.1 μm diameter cavity supported on an aluminum oxide post versus peak optic power under pulsed conditions.

For some microdisk cavities, we did not form a pillar underneath but instead formed a uniform aluminum oxide layer as described above [71]. Aluminum oxide provides much better heat sinking than air. At the same time, it lowers the Q factors of the resonant cavities because it has a higher refractive index than air. This lowering of the Q factor was compensated by making the microdisks larger. The fabricated devices were able to operate as CW lasers under optical pumping conditions. The pump beam was focused onto these microdisk cavities with spot diameters ranging from 4 μm to less than 2 μm. Figure 10.15 shows a spectrum of a 6.7 μm diameter microdisk cavity just above the threshold under 2 mW of CW optical pump power. Multimode lasing was observed above the threshold in each of the four modes indicated in Figure 10.15 as A, B, C, D. The particular combination of these four modes and the relative powers in each of these modes depended upon the pump spot size, position, and pumping power.

Figure 10.16 shows, for the same microdisk cavity, the power in mode A versus the CW optical pump power at a constant pump spot size and position. The threshold pump power was 2 mW. This value corresponds to an incident pump power of 1.25 mW and an estimated absorbed power of 125 μW.

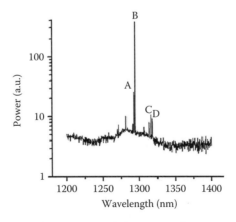

FIGURE 10.15 Spectrum of a 6.7 μm diameter microdisk cavity on aluminum oxide just above threshold under 2 mW of CW optical pumping power.

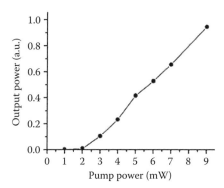

FIGURE 10.16   Optical power in mode A, as in Figure 10.14, versus CW optical pumping power.

10.6 Temperature Dependence of Lasing Behavior

The introduction points to the expectation that quantum dot microcavity lasers will have improved operating characteristics compared to their quantum well counterparts in several ways. Explicitly mentioned were lower threshold currents, improved modulation response and reduced chirp, and better temperature stability. Very low threshold conditions have been demonstrated [46,74], and low threshold operation is described above. There has been little or no investigation of microcavity laser dynamics in devices with quantum dot active regions. There have not been many publications of small-signal laser dynamics measured in a microcavity for any type of active region [4,5]. This section describes the temperature dependence of quantum dot microcavity lasers. Studying the lasing behavior's dependence on multiple factors is important for understanding and engineering the quantum dot microcavity devices in order for them to be integrated into high-density photonic integrated circuits.

Better thermal stability is predicted for quantum dot lasers because of their discrete energy states. A characteristic temperature for the lasing threshold, T_0, as high as 126 K was reported for Fabry–Perot lasers under CW operation with InAs quantum dot active material that is very similar to the active region in our microcavity devices [13]. However, the thermal stability of quantum dot lasers is not necessarily high when electron–hole recombination happens mostly outside the quantum dots. In our InAs quantum dot microdisks and photonic crystal nanocavities, because of the large surface recombination velocity of the material and the large surface-to-volume ratio, surface recombination dominates the carrier recombination at and below the lasing threshold near room temperature. Weak thermal stability of the lasing behavior has been observed, which is attributed to the strong temperature dependence of surface recombination [74]. Figures 10.17 and 10.18 show the output lasing power versus incident pump power characteristic of a microdisk and a photonic crystal L3 cavity with InAs quantum dot active material at various substrate temperatures. The devices were mounted on a copper bar whose temperature was controlled by a thermoelectric cooler. The devices were optically pumped at normal incidence with an 850 nm diode laser under pulsed conditions. The pump beam was focused to a 1–1.5 μm diameter spot. A characteristic temperature for the lasing threshold, $T_0 = 32$ K, and a characteristic temperature for the lasing slope efficiency, $T_\eta = 65$ K, were obtained for the microdisk. $T_0 = 14$ K and $T_\eta = 18$ K were obtained for the photonic crystal L3 cavity. When extracting these characteristic temperatures for the photonic crystal nanocavity, heating of the cavity from the optical pump was considered.

In order to achieve lower threshold, higher slope efficiency, high thermal stability, and CW operation, surface recombination in these devices must be significantly reduced. Passivation of the GaAs surface [21], better fabrication quality [46], p-type modulation doping [52], and incorporating Al-bearing materials around the quantum dots are several possible routes. An experimental characterization of the lasing behavior of InAs quantum dot microdisk lasers from 3 K to room temperature can be found in [28], in which surface recombination is suppressed at low temperatures.

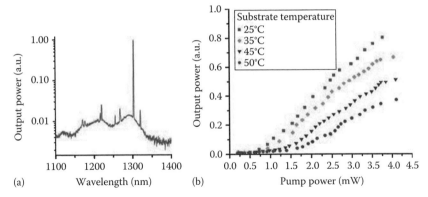

(a) Wavelength (nm) (b) Pump power (mW)

FIGURE 10.17 (a) Lasing spectrum of a 3.1 μm diameter microdisk at room temperature under pumping conditions of a 1.6 mW peak pump power, 104 ns pulse widths, and a 9.4% duty cycle. (b) Output power versus pump power characteristic of a lasing mode at 1.30 μm in a 3.2 μm diameter microdisk under varied substrate temperatures. Pumping conditions are 24 ns pulse widths and a 1% duty cycle.

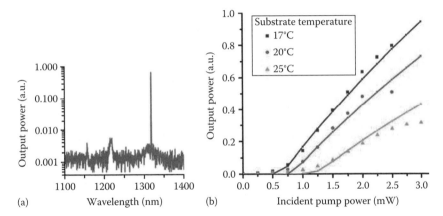

(a) Wavelength (nm) (b) Incident pump power (mW)

FIGURE 10.18 (a) Lasing spectrum of photonic crystal nanocavity at room temperature under 2.5 mW peak incident pump power. (b) The data points are output power versus incident pump power characteristic of a lasing mode at 1.33 μm in another photonic crystal nanocavity at varied substrate temperatures. The curves are fitting results with $T_0 = 14$ K and $T_\eta = 18$ K, the pump heating rate of 0.19 K/pJ included. Pumping conditions for both (a) and (b) are 16 ns pulse widths and a 1% duty cycle.

A phenomenon closely related to the temperature dependence of the lasing behavior is the dependence on the lasing wavelength. In a microcavity, one expects a strong dependence of the threshold current on the lasing wavelength with a minimum threshold condition occurring when the gain peak is aligned to the microcavity resonance. This single-mode cavity behavior has been observed in vertical cavity surface emitting lasers (VCSELs) as well as in photonic crystals (see, for example, [11,15,63]). The discussion that follows applies to the photonic crystal lasers but not to the microdisk lasers, because unlike many microcavities, microdisks often support multiple modes within the gain bandwidth. An experimental characterization of the InAs quantum dot photonic crystal L3 nanocavities for their lasing threshold's wavelength dependence was reported in [74]. In this work, 90 photonic crystal L3 nanocavities were fabricated in an array. The lattice constant, a, varied from 227 to 358 nm; the hole radii had values of 0.30a, 0.31a, and 0.32a; and shift distances of the two holes on the two ends of the L3 cavity were 0.12a and 0.15a. Lasing threshold values were recorded at room temperature. Pumping conditions were 8 ns pulse width and 0.5% duty cycle in order to reduce heating from the pump. The threshold

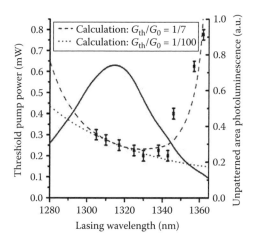

FIGURE 10.19 Data points: Experimental results of threshold incident pump power versus lasing wavelength for photonic crystal nanocavities. Dashed curves: Calculation results of threshold versus lasing wavelength for photonic crystal nanocavities. Solid curve: PL spectrum from an unpatterned area on the same sample.

values that were the lowest among the devices with similar lasing wavelengths are plotted in Figure 10.19. Also plotted in Figure 10.19 is the PL spectrum from an unpatterned region on the same sample. This PL spectrum was taken with the peak pump power large enough, so that the resulting spectrum approximately represents the quantum dot density distribution as a function of the wavelength, but not so large that any spectral shift caused by pump heating is significant.

In Figure 10.19, the threshold pump power decreases with increasing wavelength until the quantum dot density is relatively low. The wavelength corresponding to the lowest threshold is red-shifted with respect to the wavelength corresponding to the maximum quantum dot density. This phenomenon is attributed to the fact that the threshold gain required by the quality factor of the cold cavity is much smaller than the maximum gain available from the quantum dots. The 0.2 mW threshold incident pump power around 1330–1340 nm corresponds to an estimated absorbed threshold pump power of 9 μW. As expected, quantum dot microcavity lasers can have extremely low-threshold pump conditions. At the wavelength corresponding to the maximum quantum dot density, approximately 24 quantum dots on an average are estimated to contribute to the lasing. For the device with the longest lasing wavelength in Figure 10.19, 4–5 quantum dots are estimated to contribute to the lasing on an average. In this estimation, a 45 nm inhomogeneous broadening and a 5 nm room temperature homogeneous broadening of the InAs quantum dot ensemble's gain spectrum was used.

The temperature and wavelength dependences of InAs quantum dot microcavity lasers can be qualitatively explained by a simple numerical model [74]. The fact that both dependences can be explained under the same model indicates that they are closely connected to each other. The model is based on the assumption that surface recombination is the dominant carrier recombination at the lasing threshold. This is true because of the high surface-to-volume ratios of these microcavity devices and the large surface recombination velocities of GaAs and InGaAs near room temperature [69]. This assumption can be confirmed by a simple calculation of radiative and non-radiative recombination rates [72,74].

The model calculates the carrier density outside the quantum dots, N, at the lasing threshold. N is assumed to be approximately proportional to the surface recombination rate and the lasing threshold. The calculation is based on the Fermi–Dirac distribution of carriers, and basically solves for the quasi Fermi levels of electrons and holes, E_{Fe} and E_{Fh}, under the condition that the population inversion of carriers inside the quantum dots just provides the threshold gain required to reach the lasing threshold. Several assumptions were made to simplify the problem without losing the essence of the physics for a qualitative explanation of the temperature and wavelength dependences. By making the assumption that $N = (N_e N_h)^{1/2}$, where N_e and N_h are the electron density and the hole density outside the quantum dots, respectively, the carrier density, N, can be written as a function of temperature, T, and $(E_{Fe} - E_{Fh})$. By further assuming that the occupancy of a quantum dot's conduction band ground state by electrons equals the occupancy of the same quantum dot's valence band ground state by holes, $(E_{Fe} - E_{Fh})$ can be written as a function of temperature T, wavelength λ, and quantum dot density distribution ρ_{QD}, with one fitting parameter G_{th}/G_0. ρ_{QD} is obtained from the PL spectrum in Figure 10.19. G_0 is the gain at the wavelength corresponding to the maximum quantum dot density, λ_0, if all quantum dots are fully inverted in carrier population. G_{th} is the threshold gain. k_B is the Boltzmann constant. h is the Planck constant. c is the speed of light in vacuum.

$$E_{Fe} - E_{Fh} = -2k_B T \ln\left(\frac{2}{\frac{\rho_{QD}(\lambda_0)}{\rho_{QD}(\lambda)} \cdot \frac{G_{th}}{G_0} + 1} - 1 \right) + \frac{hc}{\lambda} \tag{10.8}$$

By calculating $N(T, E_{Fe} - E_{Fh})$, the model predicts $T_0 = 36.0$ K when $T = 40°C$ and $\lambda = 1.30$ μm, and $T_0 = 30.9$ K when $T = 20°C$ and $\lambda = 1.33$ μm. The modeling results compare well with the observation made of microdisk lasers, which was $T_0 = 31$ K at $T = 40°C$ at a lasing wavelength of 1.30 μm. The experimental observation of the photonic crystal laser, which is $T_0 = 14$ K at $T = 20°C$ and a lasing wavelength of 1.33 μm, gives a much smaller T_0 value than the modeling result and the experimental result from the microdisk. The reason for this much smaller T_0 is not clear. This difference may be attributable to the temperature-dependent carrier diffusion behavior. Carriers in the microdisks are confined within the disk, while there are no such confinement boundaries in photonic crystal cavities.

The large difference between the strong temperature dependence of the InAs quantum dot microcavity lasers and the weak temperature dependence of the conventional InAs quantum dot lasers can be understood by looking at the Fermi–Dirac occupancy, f, of a carrier state at energy level E:

$$f = \frac{1}{1 + e^{(E - E_F)/k_B T}} \tag{10.9}$$

For the conventional quantum dot lasers, E in Equation 10.9 predominantly represents the energy levels of the carriers in the quantum dots. For the microcavity lasers, E predominantly represents the energy levels of the surface recombination carriers, which are in the bulk and the quantum wells surrounding the quantum dots. Therefore $(E - E_F)/k_B T$ is a much larger number for the microcavity lasers than for the conventional lasers, which leads to a much stronger temperature dependence of the value of the Fermi function. Assuming that the quantum dots have a carrier capture time of 1 ps [9], it is estimated that the carrier density outside the quantum dots is not clamped above the threshold, and that surface recombination remains an important portion of the total recombination rate above the threshold. This explains the rapid decrease of the slope efficiency with increasing temperature and the rollover of the output power versus pump power characteristic in Figures 10.17 and 10.18. These phenomena are not observed in Fabry–Perot InAs quantum dot lasers [13].

The wavelength dependence can be explained under the same model. Define $G(\lambda) = [\rho_{QD}(\lambda)/\rho_{QD}(\lambda_0)]G_0$. $G(\lambda)$ represents the full population inversion gain at a wavelength λ. When $G_{th} \ll G(\lambda)$, which indicates a high quality factor of the cavities, we see that as λ increases the value of $(E_{Fe} - E_{Fh})$ will predominantly follow the change of the term hc/λ and decrease. Therefore, the surface recombination carrier density, N, and the lasing threshold will decrease. Consequently, in Figure 10.19, the threshold decreases with increasing wavelength up to around 1340 nm. When G_{th} is not much less than $G(\lambda)$, the decrease in $\rho_{QD}(\lambda)$ becomes important. This corresponds to the fast increase in the lasing threshold for wavelengths longer than 1340 nm in Figure 10.19. In this figure, the modeling results with the fitting parameter,

$G_{th}/G_0 = 1/7$, agree well with the experimental results. The modeling also predicts that if we have higher-quality-factor cavities such that $G_{th}/G_0 = 1/100$, the threshold will keep decreasing within a wider wavelength range. A constant threshold value within a broad wavelength range is favorable for photonic integrated circuits.

References

1. Y. Akahane, T. Asano, B.-S. Song, and S. Noda, High-Q photonic nanocavity in a two-dimensional photonic crystal, *Nature*, 425, 944 (2003).
2. Y. Akahane, T. Asano, B.-S. Song, and S. Noda, Fine-tuned high-Q photonic-crystal nanocavity, *Opt. Express*, 13, 1202–1214 (2005).
3. L. Allen and J. H. Eberly, *Optical Resonance and Two-Level Atoms*, John Wiley & Sons, New York, 1975.
4. M. Bagheri, M. H. Shih, Zhi-Jian Wei, S. J. Choi, J. D. O'Brien, P. D. Dapkus, and W. K. Marshall, Linewidth and modulation response of two-dimensional microcavity photonic crystal lattice defect lasers, *IEEE Photon. Technol. Lett.*, 18, 1161–1163, (2006).
5. M. Bagheri, M.-H. Shih, W. K. Marshall, S.-J. Choi, J. D. O'Brien, and P. D. Dapkus, High small-signal modulation bandwidth and narrow linewidth microdisk lasers, *CLEO/QELs*, Paper CTUJJ5, San Jose, CA, 2008.
6. J. P. Berenger, A perfectly matched layer for the absorption of electromagnetic waves, *J. Comput. Phys.* 114, 185 (1994).
7. P. Bhattacharya, S. Ghosh, and A. D. Stiff-Roberts, Quantum dot opto-electronic devices, *Annu. Rev. Mater. Res.*, 34, 1–40 (2004).
8. D. Bimberg, *Semiconductor Nanostructures*, Springer, Berlin, 2008.
9. T. F. Boggessa, L. Zhang, D. G. Deppe, D. L. Huffaker, and C. Cao, Spectral engineering of carrier dynamics in In(Ga)As self-assembled quantum dots, *Appl. Phys. Lett.*, 78, 276–278 (2001).
10. H. Cao, J. Y. Xu, W. H. Xiang, Y. Ma, S.-H. Chang, S. T. Ho, and G. S. Solomon, Optically pumped InAs quantum dot microdisk lasers, *Appl. Phys. Lett.*, 76, 3519–3521 (2000).
11. J. R. Cao, W. Kuang, S.-J. Choi, P.-T. Lee, J. D. O'Brien, and P. D. Dapkus, Threshold dependence on the spectral alignment between the quantum-well gain peak and the cavity resonance in InGaAsP photonic crystal lasers, *Appl. Phys. Lett.*, 83, 4107–4109 (2003).
12. C. T. Chan, Q. L. Yu, and K. M. Ho, Order-N spectral method for electromagnetic waves, *Phys. Rev. B* 51, 16635 (1995).
13. H. Chen, Z. Zou, O. B. Shchekin, and D. G. Deppe, InAs quantum-dot lasers operating near 1.3 m with high characteristic temperature for continuous-wave operation, *Electron. Lett.*, 36, 1703–1704 (2000).
14. D. H. Choi and W. J. R. Hoefer, The finite-difference time-domain method and its application to eigenvalue problems, *IEEE Trans. Microw. Theory Techn.*, MTT-34, 1464 (1986).
15. K. D. Choquette, R. P. Schneider, Jr., and J. A. Lott, Lasing characteristics of visible AlGaInP/AlGaAs vertical-cavity lasers, *Opt. Lett.*, 19, 969–971 (1994).
16. S. L. Chuang, *Physics of Optoelectronic Devices*, John Wiley & Sons, New York, 1995.
17. L. A. Coldren and S. W. Corzine, *Diode Lasers and Photonic Integrated Circuits*, John Wiley & Sons, New York, 1995.
18. S. Dey and R. Mittra, Efficient computation of resonant frequencies and quality factors of cavities via a combination of the finite-difference time-domain technique and Pade approximation, *IEEE Microw. Guided Wave Lett.*, 8, 415–417 (1998).
19. D. C. Dibben and R. Metaxas, Frequency domain vs. time domain finite element methods for calculation of fields in multimode cavities, *IEEE Trans. Magn.*, 33, 1468 (1997).
20. D. England, I. Fushman, and J. Vuckovic, General recipe for designing photonic crystal cavities, *Opt. Express*, 13, 5961–5975 (2005).

21. D. Englund, H. Altug, and J. Vučković, Low-threshold surface-passivated photonic crystal nanocavity laser, *Appl. Phys. Lett.*, 91, 071124 (2007).
22. B. Gayral, J. M. Gerard, A. Lemaitre, C. Dupuis, L. Manin, and J. L. Pelouard, High-Q wet-etched GaAs microdisks containing InAs quantum boxes, *Appl. Phys. Lett.*, 75, 1908–1910 (1999).
23. H. M. Gibbs, Incoherent resonance fluorescence from a Rb atomic beam excited by a short coherent optical pulse, *Phys. Rev. A*, 8, 446–455 (1973).
24. W.-H. Guo, W.-J. Li, and Y.-Z. Huang, Computation of resonant frequencies and quality factors of cavities by FDTD technique and Pade approximation, *IEEE Microw. Wirel. Compon. Lett.*, 11, 223 (2001).
25. H. Haus, *Waves and Fields in Optoelectronics*, Prentice Hall, Inc., Englewood Cliffs, NJ, 1984.
26. K. Hennessy, C. Reese, A. Badolato, C. F. Wang, A. Imamoglu, P. M. Petroff, and E. Hu, Fabrication of high Q square-lattice photonic crystal microcavities, *J. Vac. Sci. Technol. B*, 21, 2918–2921 (2003).
27. K. Hennessy, A. Badolato, M. Winger, D. Gerace, M. Atatüre, S. Gulde, S. Fält, E. L. Hu, and A. Imamoğlu, Quantum nature of a strongly-coupled single quantum dot-cavity system, *Nature*, 445, 896–899 (2007).
28. T. Ide and T. Baba, Lasing characteristics of InAs quantum-dot microdisk from 3 K to room temperature, *Appl. Phys. Lett.*, 85, 1326–1328 (2004).
29. J. Jin, *The Finite Element Method in Electromagnetics*, 2nd edn., John Wiley & Sons, Inc., New York (2002).
30. J.-H. Kim, D. H. Lim, and G. M. Yang, Selective etching of AlGaAs/GaAs structures using the solutions of citric acid/H_2O_2 and de-ionized H_2O/buffered oxide etch, *J. Vac. Sci. Technol. B*, 16, 58–560 (1998).
31. W.-J. Kim and J. D. O'Brien, Optimization of a two-dimensional photonic-crystal waveguide branch by simulated annealing and the finite-element method, *J. Opt. Soc. Am. B*, 21, 289 (2004).
32. W. Kuang, W.-J. Kim, A. Mock, and J. O'Brien, Propagation loss of line-defect photonic crystal slab waveguides, *IEEE J. Sel. Top. Quantum Electron.*, 12, 1183 (2006).
33. W. Kuang, W.-J. Kim, and J. O'Brien, Finite-difference time-domain method for nonorthogonal unit-cell two-dimensional photonic crystals, *J. Lightw. Technol.*, 25, 2612 (2007).
34. E. Kuramochi, M. Notomi, S. Mitsugi, A. Shinya, T. Tanabe, and T. Watanabe, Ultrahigh-Q photonic crystal nanocavities realized by the local width modulation of a line defect, *Appl. Phys. Lett.*, 88, 041112 (2006).
35. S.-H. Kwon, T. Sunner, M. Kamp, and A. Forchel, Ultrahigh-Q photonic crystal cavity created by modulating airhole radius of a waveguide, *Opt. Express*, 16, 4605 (2008).
36. L. Lu, T. Yang, A. Mock, M. H. Shih, E. H. Hwang, M. Bagheri, A. Stapleton, S. Farrell, J. O'Brien, and P. D. Dapkus, 60 microWatts of fiber-coupled peak output power from an edge-emitting photonic crystal heterostructure laser, *Technical Digest Conference on Lasers and Electro-Optics*, Baltimore, MD, CMV3, 2007.
37. L. Lu, E. H. Hwang, J. O'Brien, and P. D. Dapkus, Double-heterostructure photonic crystal lasers with reduced threshold pump power and increased slope efficiency obtained by quantum well intermixing, *Integrated Photonics and Nanophotonics Research and Applications, Topical Meeting*, Boston, MA, ITuB3, 2008.
38. V. A. Mandelshtam and H. S. Taylor, Harmonic inversion of time signals and its applications, *J. Chem. Phys.*, 107 6756 (1997).
39. D. Marcuse, *Theory of Dielectric Optical Waveguides*, 2nd edn., Academic Press Inc., San Diego, CA, 1991.
40. Y. Masumoto and T. Takagahara, *Semiconductor Quantum Dots: Physics, Spectroscopy, and Applications*, Springer, New York, 2002.
41. P. Michler, A. Kiraz, C. Becher, L. Zhang, E. Hu, A. Imamoglu, W. V. Schoenfeld, and P. M. Petroff, Quantum dot lasers using high-Q microdisk cavities, *Phys. Status Solidi B* 224, 797–801 (2001).

42. P. Michler (Ed.), *Single Quantum Dots: Fundamentals, Applications, and New Concepts*, Springer-Verlag, Berlin, 2003.

43. A. Mock, L. Lu and J. O'Brien, Spectral properties of photonic crystal double heterostructures, *Opt. Express*, 16, 9391–9397 (2008).

44. W. M. Moreau, *Semiconductor Lithography: Principles, Practices, and Materials*, Plenum Press, New York, 1988.

45. K. Nishi, Device applications of quantum dots, in *Semiconductor Quantum Dots: Physics, Spectroscopy, and Applications*, Y. Masumoto and T. Takagahara (Eds.), Springer, Berlin, 2002.

46. M. Nomura, S. Iwamoto, K. Watanabe, N. Kumagai, Y. Nakata, S. Ishida, and Y. Arakawa, Room temperature continuous-wave lasing in photonic crystal nanocavity, *Opt. Express*, 14, 6308–6315 (2006).

47. G. Park, O. B. Shchekin, and D. G. Deppe, Temperature dependence of gain saturation in multilevel quantum dot lasers, *IEEE J. Quantum Electron.*, 36, 1065–1071 (2000).

48. E. M. Purcell, Spontaneous emission probabilities at radio frequencies, *Phys. Rev.*, 69, 681 (1946).

49. M. Plihal and A. A. Maradunin, Photonic band structure of two-dimensional systems: The triangular lattice, *Phys. Rev. B*, 44, 8565 (1991).

50. M. Qiu, Micro-cavities in silicon-on-insulator photonic crystal slabs: Determining resonant frequencies and quality factors accurately, *Microw. Opt. Technol. Lett.*, 45, 381 (2005).

51. K. Sakoda, *Optical Properties of Photonic Crystals*, Springer-Verlag, New York, 2001.

52. O. B. Shchekin and D. G. Deppe, Low-threshold high-T_0 1.3-μm InAs quantum-dot lasers due to P-type modulation doping of the active region, *IEEE Photon. Technol. Lett.*, 14, 1231–1233 (2002).

53. O. B. Shchekin, Charge control and energy level engineering in quantum-dot laser active regions, PhD dissertation, University of Texas at Austin, Austin, TX, 2003.

54. R. E. Slusher, A. F. J. Levi, U. Mohideen, S. L. McCall, S. J. Pearton, and R. A. Logan, Threshold characteristics of semiconductor microdisk lasers, *Appl. Phys. Lett.*, 63, 1310–1312 (1993).

55. C. L. C. Smith, D. K. C. Wu, M. W. Lee, C. Monat, S. Tomljenovic-Hanic, C. Grillet, B. J. Eggleton, D. Freeman, Y. Ruan, S. Madden, B. Luther-Davies, H. Giessen, and Y.-H. Lee, Microfluidic photonic crystal double heterostructures, *Appl. Phys. Lett.*, 91, 121103 (2007).

56. B.-S. Song, S. Noda, T. Asano, and Y. Akahane, Ultra-high-Q photonic double-heterostructure, *Nat. Mater.*, 4, 207–210 (2005).

57. B.-S. Song, T. Asano, and S. Noda, Heterostructures in two-dimensional photonic-crystal slabs and their application to nanocavities, *J. Phys. D*, 40, 2629–2634 (2007).

58. K. Srinivasan and O. Painter, Momentum space design of high-Q photonic crystal optical cavities, *Opt. Express*, 10, 670–684 (2002).

59. S. Strauf, K. Hennessy, M. T. Rakher, A. Badolato, P. M. Petroff, E. L. Hu, and D. Bouwmeester, Photonic crystal quantum-dot laser with ultra-low threshold, *Technical Digest Quantum Electronics and Laser Science Conference*, Baltimore, MD, 2005, pp. 404–406.

60. M. Sugawara, Self-assembled InGaAs/GaAs quantum dots, *Semiconductors and Semimetals*, Vol. 60, Academic Press, San Diego, CA, 1999.

61. A. Taflove and S. Hagness, *Computational Electrodynamics*, 3rd edn., Artech House, Norwood, MA, 2005.

62. Y. Takahashi, H. Hagino, Y. Tanaka, B.-S. Song, T. Asano, and S. Noda, High-Q nanocavity with a 2-ns photon lifetime, *Opt. Express*, 15, 17206–17213 (2007).

63. B. Tell, K. F. Brown-Goebeler, R. E. Leibenguth, F. M. Baez, and Y. H. Lee, Temperature dependence of GaAs-AlGaAs vertical cavity surface emitting lasers, *Appl. Phys. Lett.*, 60, 683–685 (1992).

64. S. Tomljenovic-Hanic, M. J. Steel, C. M. de Sterke, and D. J. Moss, High-Q cavities in photosensitive photonic crystals, *Opt. Lett.*, 32, 542–544 (2007).

65. S. Tomljenovic-Hanic, C. M. de Sterke, and M. J. Steel, Design of high-Q cavities in photonic crystal slab heterostructures by air-holes infiltration, *Opt. Express*, 14, 12451–12456 (2006).

66. J. Topol'ancik, S. Chakravarty, P. Bhattacharya, and S. Chakrabarti, Electrically injected quantum-dot photonic crystal microcavity light sources, *Opt. Lett.*, 31, 232–234 (2006).

67. M. R. Wall and D. Neuhauser, Extraction, through filter-diagonalization, of general quantum eigenvalues or classical normal mode frequencies from a small number of residues or a short-time segment of a signal. I. Theory and applications to a quantum-dynamics model, *J. Chem. Phys.*, 102, 8011 (1995).
68. E. Yablonovitch, D. M. Hwang, T. J. Gmitter, L. T. Florez, and J. P. Harbison, Van der Waals bonding of GaAs epitaxial liftoff films onto arbitrary substrates, *Appl. Phys. Lett.*, 56, 2419–2421 (1990).
69. E. Yablonovitch, R. Bhat, C. E. Zah, T. J. Gmitter, and M. A. Koza, Nearly ideal InP/In$_{0.53}$Ga$_{0.47}$As heterojunction regrowth on chemically prepared In$_{0.53}$Ga$_{0.47}$As surfaces, *Appl. Phys. Lett.*, 60, 371–373 (1992).
70. T. Yang, J. Cao, P. Lee, M. Shih, R. Shafiiha, S. Farrell, J. O'Brien, O. Shchekin, and D. Deppe, Microdisks with quantum dot active regions lasing near 1300 nm at room temperature, *Technical Digest Conference on Lasers and Electro-Optics*, Baltimore, MD, CWK3, 2003.
71. T. Yang, O. Shchekin, J. D. O'Brien, and D. G. Deppe, Room temperature, continuous-wave lasing near 1300 nm in microdisks with quantum dot active regions, *Electron. Lett.*, 39, 1657–1658 (2003).
72. T. Yang, S. Lipson, J. D. O'Brien, and D. G. Deppe, InAs quantum dot photonic crystal lasers and their temperature dependence, *IEEE Photon. Tech. Lett.*, 17, 2244–2246 (2005).
73. T. Yang, Samuel Lipson, Adam Mock, J. D. O'Brien, and D. G. Deppe, Edge-emitting photonic crystal double-heterostructure nanocavity lasers with InAs quantum dot active material, *Opt. Lett.*, 32, 1153–1155 (2007).
74. T. Yang, S. Lipson, A. Mock, J. D. O'Brien, and D. G. Deppe, Lasing characteristics of InAs quantum dot microcavity lasers as a function of temperature and wavelength, *Opt. Express*, 15, 7281–7289 (2007).
75. A. Yariv, *Optical Electronics in Modern Communications*, 5th edn., Oxford University Press, U.K., 1997.
76. K. S. Yee, Numerical solution of initial boundary value problems involving Maxwell's equations in isotropic media, *IEEE Trans. Antennas Propag.*, 14, 302 (1966).
77. T. Yoshie, O. B. Shchekin, H. Chen, D. G. Deppe, and A. Scherer, Quantum dot photonic crystal lasers, *Electron. Lett.*, 38, 967–968 (2002).
78. T. Yoshie, A. Scherer, J. Hendrickson, G. Khitrova, H. M. Gibbs, G. Rupper, C. Ell, O. B. Shchekin, and D. G. Deppe, Vacuum Rabi splitting with a single quantum dot in a photonic crystal nanocavity, *Nature*, 432, 200–203 (2004).
79. L. Zhang and E. Hu, Lasing from InGaAs quantum dots in an injection microdisk, *Appl. Phys. Lett.*, 82, 319–321 (2003).

11

Quantum Dot Integrated Optoelectronic Devices

S. Mokkapati

H. H. Tan

C. Jagadish

11.1 Introduction

Chapters 9, 10, 12, and 23 of this book have introduced the importance of quantum dot–based optoelectronic devices and discussed in detail various optoelectronic devices with quantum dot active region. In analogy to the microelectronic device integration that has revolutionized the electronics industry, optoelectronic device integration is expected to reduce the cost of optical subsystems significantly and increase their usage. The concept of optoelectronic device integration was first proposed by Miller et al. in 1969 (Miller 1969, 2059). The monolithic integration of several optoelectronic devices on the same chip reduces the parasitic currents and capacitances inherent to external interconnects and will lead to high-speed devices. Optoelectronic device integration would also reduce the number of components and hence the level of packaging, which would lead to lower cost modules.

A typical optoelectronic integrated circuit consists of several active and passive components that are involved in generation, amplification, modulation, and guiding of light. The realization of several functionalities on a single chip requires the ability to tune the in-plane bandgap energy of epitaxial layers. For instance, the waveguiding regions of the integrated circuit require epitaxial layers with larger bandgap energy than the epitaxial layers in the active region of the lasers. This is essential to minimize the absorption of laser radiation in the waveguides. The in-plane bandgap tuning of quantum dots can be achieved using either regrowth techniques like selective area epitaxy (SAE) or post-growth processing techniques like ion implantation–induced intermixing (Fafard and Allen 1999, 2374; Perret et al. 2000, 5092; Babinski et al. 2001, 2576; Surkova et al. 2001, 6044; Lever et al. 2003, 2053; Salem et al. 2005,

TABLE 11.1   Composition, Doping Level, and Growth Temperature for Each Layer in the Thin p-Clad Laser Structures Studied in This Chapter

Layer	Composition	Thickness (μm)	Doping (cm^{-3})	Temperature (°C)
p-Contact layer	GaAs	0.10	1e19	650
p-Cladding	$Al_{0.45}Ga_{0.55}As$	0.45	5e17	650
Graded index (GRIN) layer	$GaAs \rightarrow Al_{0.3}Ga_{0.7}As$	0.16	Intrinsic	650
Active region	5 QD layers in GaAs barriers	0.18	Intrinsic	550
Graded index (GRIN) layer	$Al_{0.3}Ga_{0.7}As \rightarrow GaAs$	0.16	Intrinsic	750–650
n-Cladding	$Al_{0.3}Ga_{0.7}As$	0.25	3e17	750
n-Cladding	$Al_{0.45}Ga_{0.55}As$	0.3	3e17	750
n-Cladding	$Al_{0.45}Ga_{0.55}As$	1.5	2e18	750
n-Buffer layer	GaAs	0.25	2e18	750

241115) and impurity-free vacancy disordering (IFVD) (Bhattacharya et al. 2000, 4619; Lever 2004, 109–122; Lever et al. 2004b, 7544; Gordeev et al. 2007, 29).

SAE involves epitaxial regrowth on processed surfaces and can be used to simultaneously grow epitaxial layers with different bandgap energies in a single growth step. Ion implantation–induced intermixing and IFVD are post-growth bandgap-tuning techniques that rely on introducing point defects into the epitaxial structures. Variation in the concentration of point defects across the sample causes differential interdiffusion upon annealing and results in epitaxial layers of different bandgap energies in different parts of the substrate. Each of these techniques has its own advantages and disadvantages. This chapter discusses the principle involved in achieving in-plane bandgap tuning of quantum dots, presents results on integrated quantum dot devices fabricated using each of the above-mentioned techniques, and compares and contrasts these techniques.

All the devices discussed in this chapter are 4 μm wide ridge waveguide lasers based on thin p-clad device structures with nominally 50% InGaAs quantum dot active regions. The quantum dots are grown in the Stranski–Krastanow (SK) growth mode in a low pressure, horizontal flow metal-organic chemical vapor deposition (MOCVD) reactor. As quantum dots are highly strained structures, exposure to elevated temperatures causes enhanced interdiffusion in the quantum dots which leads to a blueshift in their emission wavelength. It has also been reported that upon heat treatment the emission efficiency of the quantum dots could be reduced due to the creation of non-radiative recombination centers from the interfacial defects. In order to minimize the effect of elevated temperatures on the electronic structure of the quantum dots in the active region of the devices, the lasers presented in this chapter have been designed with minimal thickness of p-cladding layers. Also the p-cladding layers in these laser structures are designed to have lower Al content. Lower Al content helps in reducing the growth temperature without much compromise in the quality of the AlGaAs layers. C is used as the p-type dopant in the laser structures studied in this chapter, because of its low diffusivity compared to that of Zn. This is especially important for IFVD studies, where impurity-enhanced disordering could also cause intermixing. The structures presented in this chapter have been designed to have a spot size of 0.375 μm. The layer sequence, composition, doping level, and growth temperature for each of the layers in the laser structure is given in Table 11.1. Further details on the design and structure of thin p-clad lasers can be found in Buda et al. (2003, 625). All the devices were tested as cleaved (no facet coatings applied), at room temperature in pulsed mode with a duty cycle of 5% (25 kHz, 2 μs pulses).

11.2 Selective Area Epitaxy

SAE is a growth process in which growth in certain regions of the semiconductor substrate is prohibited. In the experiments described in this chapter, a SiO$_2$ mask is used to inhibit epitaxial growth in certain regions of the GaAs substrates. In a MOCVD reactor, the semiconductor substrate catalyses the pyrolysis

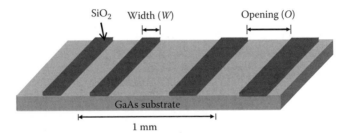

FIGURE 11.1 Schematic of the mask pattern used for illustrating the controlled nucleation and bandgap tuning of quantum dots using selective area epitaxy. (From Mokkapati, S. et al., *J. Phys. D: Appl. Phys.*, 41, 085104, 2008. With permission.)

of group V hydride sources, but does not affect the pyrolysis of group III alkyl molecules. Pyrolysis of group V sources is achieved at relatively lower temperatures (~500°C) in the presence of the semiconductor substrate and requires much higher temperatures in its absence (Coleman 1997, 1715). The pyrolysis of group III alkyl molecules begins above the surface of the semiconductor substrate. But the group V hydride molecules remain relatively intact until they adsorb on the semiconductor surface. Hence masking certain regions of the substrate with SiO_2 inhibits growth in these regions. As a consequence, the deposition in unmasked regions of the substrate is enhanced in accordance with the law of conservation of mass. The adatoms from the masked regions of the substrate migrate toward the unmasked regions through gas-phase diffusion and surface migration. The local growth rate enhancement across the wafer depends on the dimensions of the mask and can be varied by varying the mask dimensions.

Figure 11.1 shows the schematic of the mask pattern used for the experiments studying bandgap tuning of quantum dots, described in this chapter. The mask pattern consists of pairs of stripes. The separation between consecutive pairs of stripes is 1 mm, longer than the diffusion length of the adatoms. The width (W) of the stripes and the openings (O) between the stripes can be varied independently to vary the local growth rate enhancement.

11.2.1 Effect of Coverage and Deposition Rate on Quantum Dots Grown by MOCVD: Brief Review

The characteristics (size, density, bandgap, optical quality) of quantum dots formed by MOCVD are determined by the growth parameters used during their deposition. MOCVD growth of quantum dots depends critically on growth parameters like coverage (total amount of material deposited), deposition rate, growth temperature, and V/III ratio.

It is essential to understand the effect of various growth parameters in order to control the dot properties by controlling the growth parameters. As described earlier in this section, during SAE, migration of adatoms from masked to unmasked regions of the substrate leads to variation in local growth parameters like the coverage and deposition rate across the substrate. Here, a brief review of the effect of coverage and deposition rate on the formation of In(Ga)As quantum dots is presented.

11.2.1.1 Coverage

When In(Ga)As is deposited on GaAs, initial growth takes place in 2D layer-by-layer mode. In(Ga)As adjusts its lattice constant to match with that of the substrate, and hence is under compressive strain. 2D growth continues until the thickness of In(Ga)As exceeds a certain thickness known as critical thickness for island nucleation. As more material is deposited, island nucleation is initiated in order to reduce the strain energy of the system. The island size and density change continuously with further increase in coverage, until the formation of dislocations in the islands, which can penetrate/propagate into subsequent layers (Lever 2004, 33; Sears 2006, 40).

11.2.1.2 Deposition Rate

It is now generally accepted that for a fixed coverage, higher deposition rates lead to the formation of smaller dots with high density and lower deposition rates lead to the formation of larger dots with a low density (Shiramine et al. 2002, 332; El-Emawy et al. 2003, 213). Lower deposition rates allow more time for the adatoms to diffuse across the substrate surface, before they attach to an island surface, leading to the formation of larger, widely spaced islands. On the other hand, higher deposition rates create larger density of adatoms on the surface at any given time, leading to nucleation of more islands. However, reduced AsH_3 flow during the growth of quantum dots at lower deposition rates may also reduce adatom mobility and lead to the formation of smaller islands (Sears 2006, 45).

Thus the total amount of material deposited (coverage) determines whether or not island nucleation takes place, and both coverage and deposition rate play an important role in determining the density and size of the nucleated islands. There exists only a narrow window over which the growth parameters can be varied in order to obtain high density of defect-free dots, necessary for device applications (Lever 2004, 33; Sears 2006, 37).

11.2.2 Controlled Nucleation of Quantum Dots

Controlling the area of nucleation of quantum dots is essential for integration of passive waveguides with quantum dot lasers. The bandgap energy difference between the epitaxial layers in the laser section and the waveguide section of the integrated devices has to be maximized in order to achieve low passive losses. Figure 11.2 shows representative luminescence spectra from quantum dots with different bandgap energies, formed in the SK growth mode, and a quantum well. Due to inherent size distribution of quantum dots formed in the SK growth mode, even though the luminescence peaks of the quantum dots are shifted with respect to each other by ~90 nm, there is considerable overlap in the emission signal from the quantum dots with different bandgap energies, Figure 11.2a. The overlap between the emission spectra of the quantum dots suggests that there could be significant absorption of the lasing radiation in

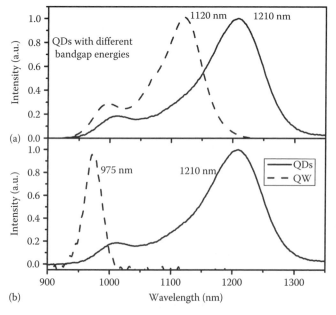

FIGURE 11.2 Luminescence spectra of (a) quantum dots with different bandgap energies and (b) quantum well and quantum dots. The quantum dots are formed via the Stranski–Krastanow growth mode.

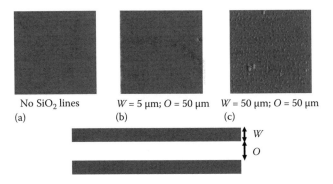

No SiO$_2$ lines W = 5 μm; O = 50 μm W = 50 μm; O = 50 μm

(a)   (b)   (c)

FIGURE 11.3 AFM images (1 μm × 1 μm) of different regions across a patterned semiconductor substrate, after InGaAs deposition for 2.0 s at a rate of ~1.7 ML s^{-1}. The local mask dimensions are also shown with the corresponding images.

the passive section of the integrated devices if quantum dots (with different bandgap energies) are used in both the active and passive sections. Instead of quantum dots, if quantum wells are used to fabricate the passive section of the integrated devices, the differential bandgap shift between the epitaxial layers in the active and passive sections of the integrated devices is maximized. And also, due to the characteristic sharp emission signal from the quantum wells, the overlap between the emission signals from the active and passive sections of the integrated devices is minimized, Figure 11.2b. This section presents results on controlled nucleation of In(Ga)As quantum dots using SAE.

Figure 11.3 shows the AFM images of different regions of a patterned substrate, on which InGaAs has been deposited for 2.0 s at a deposition rate of ~1.7 ML s^{-1}. In regions without dielectric patterning, the thickness of InGaAs deposited (~3.4 ML) is less than the critical thickness for quantum dot nucleation in the InGaAs/GaAs system (~4 ML). As described in Section 11.2.1, the growth in such regions of the substrate proceeds in 2D layer-by-layer mode and the AFM images of these regions show planar surfaces (Figure 11.3a), which are essentially very thin quantum wells (few ML thick). In regions between SiO$_2$ stripes of width (W) 5 μm and opening (O) 50 μm (Figure 11.3b), the growth rate enhancement is minimal and the thickness of InGaAs deposited is still less than the critical thickness for quantum dot nucleation and growth still proceeds in layer-by-layer mode. By increasing the stripe width W to 50 μm, the local growth rate enhancement in the regions between the stripes is sufficient to cause nucleation of quantum dots due to additional adatoms diffusing away from the wider oxide stripes into the openings. The AFM images in these regions of the substrate (Figure 11.3c) show quantum dots with a density ~3 × 10^{10} cm^{-2}.

Figure 11.4 shows the cathodoluminescence spectra from the selectively grown quantum wells and quantum dots shown in Figure 11.3a and c. The differential shift between the peak emission wavelengths of the quantum wells and the quantum dots is ~185 nm. Large differential shift in the peak emission wavelengths and minimal overlap of the emission signal suggest that the selectively grown quantum dots and quantum well can be used to fabricate the active and passive sections, respectively, of a quantum dot laser integrated with a passive waveguide. Section 11.2.5.1 describes in detail the characteristics of quantum dot laser integrated with passive waveguide using the selectively grown quantum dots and quantum wells.

11.2.3 Bandgap Tuning of Quantum Dots Using SAE

As discussed in Section 11.2.2, simultaneous growth of quantum wells and quantum dots is important for fabrication of quantum dot lasers integrated with passive waveguides while bandgap tuning of quantum dots is essential for fabrication of multiple-wavelength quantum dot lasers. Quantum dot growth is very sensitive to the growth parameters and there exists only a narrow range of growth parameters,

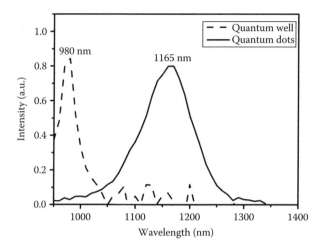

FIGURE 11.4   Room-temperature cathodoluminescence spectra of selectively formed quantum well and quantum dots.

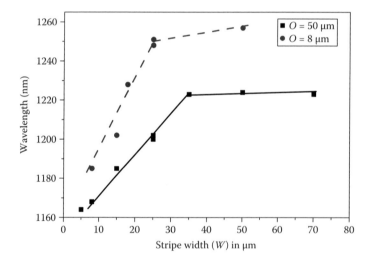

FIGURE 11.5   Room temperature peak emission wavelength from InGaAs quantum dots as a function of SiO$_2$ stripe width (*W*) for fixed openings (*O*) of 8 and 50 μm.

over which defect-free, high-density quantum dots (necessary for devices) can be grown. In this section, results are presented on the bandgap tuning of In(Ga)As quantum dots using SAE.

Figure 11.5 shows the room temperature peak emission wavelength of selectively grown InGaAs quantum dots as a function of the width of the SiO$_2$ stripes (*W*) for fixed openings (*O*) of 8 and 50 μm. The growth parameters (deposition rate: ~1.7 ML s^{-1} and deposition time: 3.0 s) have been optimized so that quantum dots are formed for all stripe widths. For a fixed opening of 50 μm, as the stripe width is initially increased, the peak emission wavelength from the quantum dots initially increases, and then starts to saturate for stripe widths greater than 35 μm. Increase in the growth rate enhancement with increasing stripe width may result in quantum dots with larger mean height, which results in redshift of their emission wavelength. Figure 11.6 shows the variation of quantum dot height and density with increasing stripe width, *W*, for fixed opening, *O* = 50 μm. The average quantum dot height initially

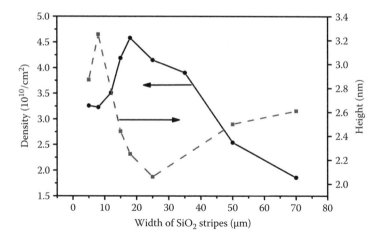

FIGURE 11.6 Variation of InGaAs quantum dot density and height as a function of SiO_2 stripe width, for a fixed opening of 50 μm, as determined by AFM. (Reused from Mokkapati, S., *Appl. Phys. Lett.*, 86, 113102, 2005. With permission. Copyright 2005 American Institute of Physics.)

increases with increasing stripe width and then starts to decrease by further increasing the stripe width. The initial redshift of the quantum dot emission wavelength for smaller stripe widths can thus be correlated to increase in average dot height. By further increasing the stripe width, a decrease in average quantum dot height is still accompanied by an increase in the emission wavelength. These results suggest that the redshift of quantum dot peak emission wavelength with increasing stripe width is not just due to increase in quantum dot height, but also due to increase in the In content in the quantum dots. Increased In incorporation into the quantum dots with increasing stripe width is a consequence of larger diffusion length of In adatoms compared to Ga adatoms.

For stripe widths greater than 35 μm, the peak emission wavelength from the quantum dots starts to saturate because of the formation of defects. An increase in quantum dot height and/or In content increases the strain and promotes formation of defects, which makes the quantum dots optically inactive. The saturation of the emission wavelength from the quantum dots is also accompanied by decrease in the intensity of the emission signal. The intensity of the emission signal is reduced due to decrease in the number of optically active quantum dots. Once dislocations start to form, the relaxed quantum dots act as material sinks, and the average dot density decreases and the average dot height slightly increases (Figure 11.6).

The peak emission wavelength from quantum dots grown in the 8 μm openings between the SiO_2 stripes shows similar behavior with increasing stripe width (Figure 11.5). Quantum dots grown in the 8 μm openings emit at longer wavelengths than the quantum dots grown in the 50 μm openings for a fixed stripe width. This is a consequence of increase in growth rate enhancement by reducing the size of the opening, *O* for a given stripe width, *W*. For the same reason, the peak emission wavelength from the quantum dots grown in the 8 μm openings saturates for smaller stripe width. It is also interesting to note that the emission wavelength from dots grown in the 8 μm openings saturates at a higher value, possibly due to increased In incorporation into the dots.

The quantum dots with different emission energies can be used to fabricate multiple-wavelength quantum dot lasers. The characteristics of such devices are discussed in Section 11.2.5.2.

11.2.4 Growth of Device Structures

A three-step growth process is employed to grow the device structures using SAE for fabrication of integrated devices. The growth process was developed to avoid problems with the regrowth of Al containing

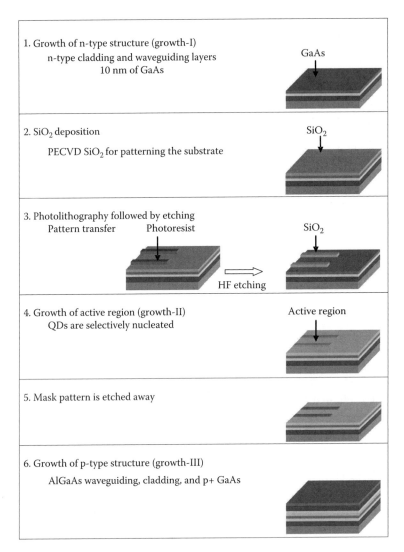

1. Growth of n-type structure (growth-I)
   n-type cladding and waveguiding layers
   10 nm of GaAs

   GaAs

2. SiO$_2$ deposition

   PECVD SiO$_2$ for patterning the substrate

   SiO$_2$

3. Photolithography followed by etching
   Pattern transfer   Photoresist   SiO$_2$

   HF etching

4. Growth of active region (growth-II)
   QDs are selectively nucleated

   Active region

5. Mask pattern is etched away

6. Growth of p-type structure (growth-III)
   AlGaAs waveguiding, cladding, and p+ GaAs

FIGURE 11.7 Flowchart showing the three-step growth process used for growing the device structures using SAE.

compounds (Cockerill et al. 1994, 115). Each growth is terminated with a thin layer of GaAs. The three-step growth process ensures that all the devices have the same structure, except for variations in the active region. Figure 11.7 shows a flowchart of the steps involved in the growth of the device structures and the steps are described here.

- Growth of the n-type structure (growth-I):

The first epitaxial growth step involves growth of the n-buffer layer, n-cladding layers and the graded index (GRIN) layer on a n+ doped (100) GaAs substrate. The layer thickness and composition are the same as indicated in Table 11.1. The growth is then terminated with a 10 nm thick GaAs.

- Patterning for the growth of the active region (steps 2 and 3 in the flowchart):

The sample is taken out of the reactor and is patterned for the growth of the active region. A layer of SiO$_2$ is deposited on the sample using plasma enhanced chemical vapor deposition (PECVD) system. After

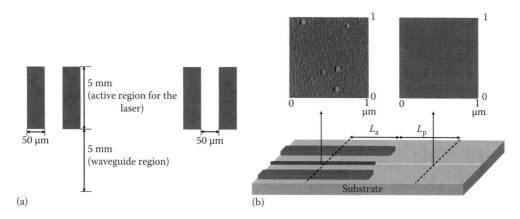

FIGURE 11.8 Schematic of the mask pattern used for growing the active region for a quantum dot laser integrated with a passive waveguide. (a) Top view of the mask pattern and its dimensions (not to scale). (b) Three-dimensional view of the patterned substrates. Regions where quantum dots and quantum wells are selectively formed are indicated. The position of the device mesas with respect to the mask pattern used for growing the active region is shown. L_a and L_p are the lengths of the active and passive sections, respectively, of the cleaved devices.

deposition of the SiO_2 layer, the required mask pattern is transferred onto the sample using standard photolithography and wet chemical etching. The schematic of the mask pattern used for growing the active region to fabricate a laser integrated with a passive waveguide is shown in Figure 11.8. The dimensions of the pattern are indicated on the top view of the mask pattern. The mask pattern used for growing the active region of multiple-wavelength lasers is similar to that shown in Figure 11.1. The SiO_2 stripe width, W was either 5 or 10 or 15 or 20 μm and the size of the openings, O between the SiO_2 stripes was fixed at 50 μm. The sample is then thoroughly cleaned and etched to remove native oxide from the surface. A trim etch is then performed to etch a few monolayers (ML) of GaAs from the sample surface. The trim etch is necessary to eliminate any damage caused due to various processing steps and make the sample surface suitable for epitaxial regrowth.

- Growth of the active region (growth-II):

The patterned sample is then introduced into the reactor for the second epitaxial growth step. This step involves growth of the active region for device fabrication. This step is the most critical of all the three growth steps. For fabrication of a quantum dot laser integrated with a passive waveguide, this growth step involves simultaneous growth of quantum dots, that form the gain region of the active section of the integrated devices and quantum wells, that form the passive section of the integrated devices. It is critical to carefully optimize the growth parameters, so that growth still proceeds in 2D layer-by-layer fashion in the passive section, whereas high-density quantum dots are formed in the active section. A little enhancement of growth rate in the passive section from the optimum value may cause 3D island nucleation; and the presence of quantum dots in the passive section will cause an absorption tail at the lasing wavelength of the active section which increases the passive losses in the devices and is thus undesirable. Growth rate enhancement lower than the optimum value in the active section would cause a low quantum dot density, which may lead to ground-state gain saturation and subsequent lasing from quantum dot excited states or even wetting layer states. On the other hand, growth enhancement higher than the optimum value will promote formation of dislocations that act as non-radiative recombination centers and are not desirable for optimum device performance.

The growth parameters used are listed in Table 11.2. The deposition rate and the composition indicated are the nominal values on unpatterned substrates. Growth on patterned substrates induces local variations in these parameters that depend on the local dimensions of the mask pattern (Sections 11.2.2

TABLE 11.2   Deposition Parameters Used for Growing the
Active Region for Fabrication of Integrated Devices Using SAE

Parameter	Value
No. of layers	5
V/III ratio	20
Deposition time	2.0 s (laser integrated with waveguide)/2.9 s (multicolor lasers)
Deposition rate	1.7 ML s^{-1}
Temperature	550°C
Composition (Ga/Ga + In)	0.46

and 11.2.3). Under the deposition conditions used, the total thickness of InGaAs layers deposited on unpatterned regions is below the critical thickness for 3D island nucleation in the InGaAs/GaAs system. But in the openings between the SiO$_2$ stripes, due to growth rate enhancement, the thickness of InGaAs deposited exceeds the critical thickness for island nucleation. The growth rate enhancement is sufficient to form quantum dots of high density (3×10^{10} cm^{-2}). The 2D layers formed in the unpatterned regions and the 3D islands formed in patterned regions are used for fabricating the waveguide section and the laser section, respectively, of the integrated devices.

For fabrication of multicolor quantum dot lasers, this growth step requires careful optimization of local growth parameters by choosing the right mask dimensions, in order to grow defect-free, high-density quantum dots with varying composition and/or sizes. As already mentioned in Section 11.2.1 the quantum dot growth is very sensitive to the growth parameters and there exists only a narrow range of growth parameters that are suitable for the growth of device quality quantum dots. Thus the growth of active region for fabrication of multiple-wavelength lasers requires very fine control on the local growth parameters. The growth parameters used for the quantum dot deposition are listed in Table 11.2. The deposition rate and the composition indicated are the nominal values on unpatterned substrates. Under the deposition conditions used, the thickness of InGaAs deposited on unpatterned regions of the substrates is just below the critical thickness for quantum dot nucleation, so that no quantum dots are formed in these regions. But due to local growth rate enhancement, quantum dots are formed between the stripes. The size, composition (which determine the bandgap) and density of the quantum dots formed (which determines the amount of gain available for the lasers) are controlled by the stripe width.

- Removal of the SiO$_2$ mask pattern (step 5 in the flowchart):

Following the growth of the active region, the sample is taken out of the reactor to etch away the SiO$_2$ mask pattern. Removal of the SiO$_2$ is necessary to eliminate losses in the fabricated devices due to the formation of non-radiative recombination centers at the semiconductor-dielectric interface.

- Growth of the p-type structure (growth-III):

The sample is again introduced into the reactor for the final growth step. This step involves growth of the GRIN layer, cladding layer and the contact layer on the p-side of the device structures. The layer thickness, composition and doping levels are indicated in Table 11.1.

Figure 11.9 shows the cross-section schematics of the device structures used for fabricating a laser integrated with a passive waveguide and multicolor lasers after the three-step growth process. The SiO$_2$ stripes shown in Figure 11.9b are etched away after the growth of the active region. The active and passive sections of the lasers integrated with waveguides are isolated from each other and only the active region is electrically pumped. For the lasers integrated with passive waveguides, the laser cavity

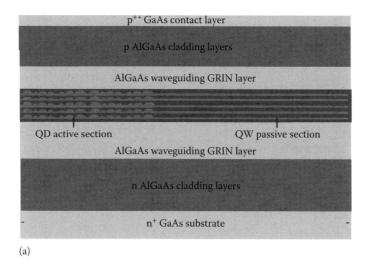

(a)

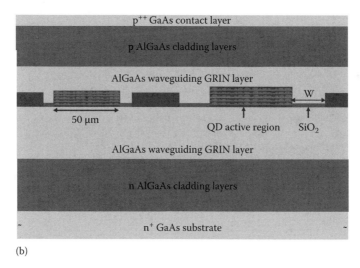

(b)

FIGURE 11.9 Schematic of the cross section of complete device structures for fabrication of (a) a quantum dot laser integrated with a passive waveguide and (b) multicolor lasers.

is defined by cleaved mirror facets formed at one end of the active section and the other end of the passive section; its length is determined by the total length of the active and passive sections $(L_a + L_p)$ as indicated in Figure 11.8.

11.2.5 Device Characteristics

11.2.5.1 Quantum Dot Laser Integrated with a Passive Waveguide

Devices with 2 mm long active (laser) section were cleaved with different lengths of the passive section for determining the losses in the passive section. Figure 11.10 shows the threshold currents and lasing wavelengths for a 2 mm long laser integrated with a passive waveguide of varying lengths. As the length of the passive section increases, due to increase in the total losses in the device, the threshold current increases and the lasing wavelength blueshifts.

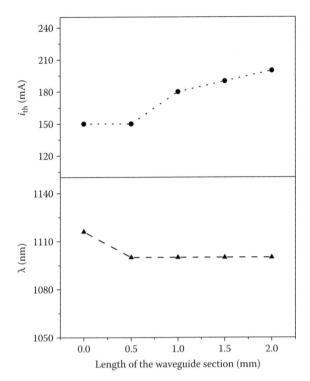

FIGURE 11.10   Threshold current and lasing wavelength of the integrated devices with a 2 mm long active region as a function of the length of the waveguide section. (From Mokkapati, S. et al., *IEEE Photon. Technol. Lett.*, 18(15), 1648, August 1, 2006. With permission. © 2006 IEEE.)

Fitting the differential efficiency of the integrated devices as a function of the length of the passive section with Equation 11.1 (Coldren and Corzine 1995, 446–458) yields a loss of 3 cm⁻¹ in the waveguide (Figure 11.11).

$$\eta_d = \eta_i \frac{\alpha_p L_p + \ln\left(\dfrac{1}{R}\right)}{\left[\alpha_a L_a + \alpha_p L_p + \ln\left(\dfrac{1}{R}\right)\right]} \frac{2(1-R)e^{-\alpha_p L_p}}{\left(1+e^{-\alpha_p L_p}\right)\left(1-R\right)e^{-\alpha_p L_p}} \qquad (11.1)$$

where
   η_d is the differential efficiency of the integrated device
   η_i is the internal efficiency of the lasers fabricated by SAE
   α_p and α_a denote the losses in the passive and active sections, respectively
   L_p and L_a are the lengths of the passive and active sections, respectively
   R is the power reflection coefficient of the cleaved facet

The low losses in the waveguide region are due to a large difference in the bandgap energies of the epitaxial layers in the laser region and the waveguide region. A bandgap energy shift of 200 meV is evident from Figure 11.12, which shows the lasing spectra of lasers fabricated from the selectively grown QDs (that form the active region of the integrated devices) and QWs (that form the passive section of the integrated devices).

11.2.5.2 Multicolor Lasers

The electroluminescence spectra of 3 mm long lasers fabricated from quantum dots grown between SiO₂ stripes of widths (*W*) 5, 10, 15, and 20 μm are shown in Figures 11.13a through d, respectively.

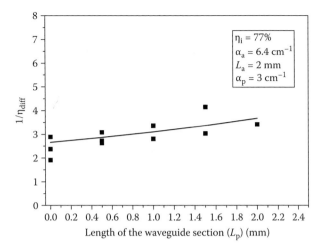

FIGURE 11.11 Inverse differential efficiency of the integrated devices as a function of the length of the passive section, L_p. The solid line shows the fit to the experimental data using Equation 11.1. The parameters used for fitting are also indicated in the figure. (From Mokkapati, S. et al., *IEEE Photon. Technol. Lett.*, 18(15), 1648, August 1, 2006. With permission. © 2006 IEEE.)

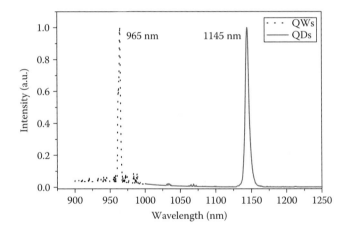

FIGURE 11.12 Lasing spectra of lasers fabricated from selectively grown quantum dots and quantum wells. (From Mokkapati, S. et al., *IEEE Photon. Technol. Lett.*, 18(15), 1648, August 1, 2006. With permission. © 2006 IEEE.)

For a stripe width of 5 μm, the growth rate enhancement is minimal and shallow quantum dots of low density are formed. The quantum dot density is not sufficient for providing enough gain for ground-state lasing. But since excited states have higher degeneracy, they provide higher gain and so devices fabricated from these quantum dots lase from the excited states (Shoji et al. 1997, 188; Bimberg et al. 2000, 235; Liu et al. 2000, 1272; Asryan et al. 2001, 418; Smowton et al. 2001, 2629; Ustinov et al. 2003, 101; Tatebayashi et al. 2005, 053107). The luminescence from the quantum dot ground-state peaks at ~1190 nm at an injection current of 40 mA and luminescence from higher lying energy states appears (as indicated by the dashed arrow in Figure 11.13a). As more current is injected into the device, the luminescence peak from the excited states blueshifts, until the device starts to lase from these states.

When the stripe width is increased to 10 μm, the growth rate enhancement is sufficient to form a high density of quantum dots for the same deposition time. The increase in the quantum dot density now provides enough ground-state gain to overcome the cavity losses and internal optical losses and the

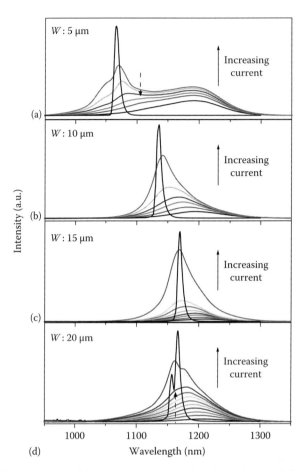

FIGURE 11.13 Electroluminescence and lasing spectra of 3 mm long lasers fabricated from quantum dots grown between pairs of SiO$_2$ stripes of widths (W) (a) 5 μm, (b) 10 μm, (c) 15 μm, and (d) 20 μm. The separation between the SiO$_2$ stripes (O) is 50 μm. Luminescence from quantum dot excited states is indicated by dashed arrows in (a) and (d). (Reused from Mokkapati, S., *Appl. Phys. Lett.*, 90, 171104, 2007. With permission. Copyright 2007 American Institute of Physics.)

lasers fabricated from these dots lase from the ground state (Figure 11.13b). There is no sign of luminescence from quantum dot excited states in the electroluminescence spectra of this device.

By increasing the stripe width to 15 μm, the growth rate enhancement is sufficient to form quantum dots with smaller bandgap energies. The luminescence from these quantum dots peaks at ~1210 nm at an injection current of 40 mA (Figure 11.13c). As the device is pumped harder, the luminescence peak blueshifts and the device starts to lase at ~1170 nm. The electroluminescence spectra of the device do not show any luminescence from higher lying excited states.

By increasing the stripe width to 20 μm, there is no further variation in the quantum dot bandgap energy. The increase in the total amount of material deposited due to enhanced growth rate increases the strain energy, which promotes formation of defects. Once defects appear in the quantum dots, any additional material deposited is preferentially accumulated at these defects; and does not contribute to varying the average size of the optically active quantum dots. So devices fabricated from dots grown between 20 μm wide SiO$_2$ stripes do not lase at longer wavelengths than the devices fabricated from quantum dots grown between 15 μm wide stripes. The devices fabricated from dots grown between 20 μm wide stripes lase at slightly shorter wavelengths than the ones fabricated from quantum dots grown between 15 μm wide stripes (Figure 11.13d), as these devices have to be pumped harder in order to overcome increased losses, which makes them lase from the excited states.

11.3 Ion Implantation–Induced Interdiffusion

11.3.1 Bandgap Tuning of Quantum Dots

Atomic interdiffusion in semiconductor quantum dots leads to compositional changes and alloying of the initially abrupt interfaces between the quantum dot and the barriers which modifies the confining potential and energy levels as illustrated in Figure 11.14 (Malik et al. 1997, 1987; Xu et al. 1998, 3335; Lever 2004, 92). It is well known that high temperature annealing of the III–V compound semiconductors promotes interdiffusion through group III and/or group V vacancy/interstitial diffusion (Kheris et al. 1997, 15813). The magnitude of interdiffusion in as-grown quantum wells/quantum dots is determined by the background vacancy concentration and the concentration gradient. Differential in-plane bandgap energy shifts may be achieved by varying the vacancy concentration across the substrate through ion implantation.

There have been several reports on ion implantation–induced intermixing in quantum dots (Surkova et al. 2001, 6044; Lever et al. 2003, 2053). Energy shifts of up to 280 meV have been reported (Fafard and Allen 1999, 2374; Salem et al. 2005, 241115). Interdiffusion in quantum dots changes both the transition energies and the intersublevel spacing (Perret et al. 2000, 5092). Some authors report that quantum dots retain their 0D density of states even after strong interdiffusion of their potential by intermixing (Fafard and Allen 1999, 2374), while others suggest that interdiffusion creates a 2D system from quantum dots (Babinski et al. 2001, 2576).

Though there are reports on enhanced interdiffusion due to heavy ion implantation, proton implantation is expected to be better than heavy ion implantation (Ji et al. 2003, 1208), as it creates a much higher density of point defects, which is useful for intermixing. Hence results based on proton implantation are presented here. Proton implantation energy (100 keV) has been chosen so that the defect peak coincides with the active region of the devices. An implantation dose of 5×10^{13} cm^{-2} was chosen to maximize the differential bandgap shift based on some annealing studies (Lever 2004, 123). Unimplanted and implanted device structures were annealed at 600°C for 30 min under AsH$_3$ ambient. The annealing conditions were chosen to maximize the photoluminescence recovery from the quantum dots in the active region. Higher temperatures degrade the quantum dot carrier capture, imposing an upper limit on the temperatures that could be used. Lower annealing temperatures necessitate long annealing times to recover the photoluminescence by annealing out the implantation-induced damage.

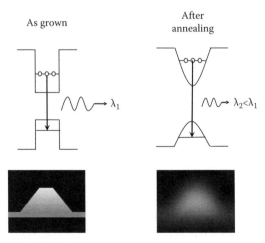

FIGURE 11.14 Schematic illustrating the effect of atomic interdiffusion on quantum dot energy levels; and the interface between quantum dots and the barrier.

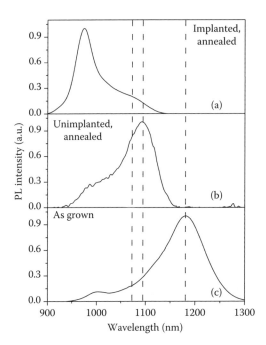

FIGURE 11.15 Normalized room temperature photoluminescence spectra from the quantum dots in (a) laser structure implanted with protons of dose $5 \times 10^{13}\,cm^{-2}$ and annealed at 600°C for 30 min; (b) unimplanted laser structure annealed at 600°C for 30 min; and (c) as-grown laser structure.

Room-temperature photoluminescence spectra from the active region of as-grown, unimplanted, annealed and implanted annealed device structures are shown in Figure 11.15. All the spectra show two distinct peaks, one associated with the quantum dots and one with the wetting layer. The dashed vertical lines in Figure 11.15 give an indication of the relative magnitude of thermally induced shift and implantation-induced differential shift in the quantum dot peak positions (shift in the emission wavelength of quantum dots annealed with and without implantation). The thermal shift in the peak position corresponding to luminescence from quantum dots is 65 meV and the implantation-induced differential shift is only 13 meV. The magnitude of implantation-induced differential energy shift is smaller in quantum dots (in comparison to the differential shifts achieved in QWs) due to much larger thermal shift. It has been reported that the magnitude of thermal interdiffusion in quantum dots is large because of large in-built strain, high In content and large surface area to volume ratio (Heinrichsdorff et al. 1998, 540; Wellmann et al. 1998, 1030; Lever 2004, 89).

Figure 11.16 shows the room temperature photoluminescence spectra from the active region of laser structure implanted with protons and annealed for 30 min, at low (1 W cm^{-2}) and high (15 W cm^{-2}) excitation intensities. The experimentally measured spectrum at high excitation intensity can be fitted well with four Gaussian peaks, centered at 980 nm (1.27 eV), 1018 nm (1.22 eV), 1045 nm (1.19 eV) and 1099 nm (1.13 eV). The integrated intensity of the peak at 1.13 eV is ~0.6 times higher than the integrated intensity of the peak centered at 1.19 eV. Even at very low excitation intensities, the photoluminescence spectrum retains the Gaussian components at 1.19 and 1.13 eV with the same relative intensity. This is an indication that these Gaussian components are due to radiative transitions in ground state of the quantum dots with different mean dimensions. The relative intensity of the integrated emission is an indication of the relative density of quantum dots in each group. The Gaussian component centered at 1.22 eV, present in the photoluminescence spectrum measured at high excitation intensity is not clearly identified in the photoluminescence spectrum measured at low excitation intensity. The presence of this peak only at higher excitation intensities is an indication that it is associated with excited state

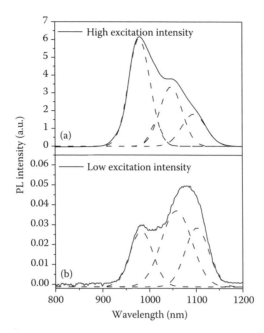

FIGURE 11.16 Room temperature photoluminescence spectra from the active region of laser structure implanted with protons of dose $5 \times 10^{13}\,\text{cm}^{-2}$ and annealed at 600°C for 30 min, measured at (a) high and (b) low excitation intensities. The measured spectra (solid lines) are deconvoluted into several Gaussian components as shown by the dashed lines.

transitions in one of the quantum dot families. This peak is attributed to excited state transitions in quantum dots with larger mean dimensions. The peak centered at the highest energy (1.27 eV) is attributed to radiative transitions in the 2D wetting layer. The relative intensity of the wetting layer peak with respect to total emission from quantum dot ground states increases from 0.91 at low excitation intensity to 2.9 at high excitation intensity. This is because of more states available for carriers in the wetting layer than in quantum dots due to smaller surface coverage of quantum dots. The intensity dependent photoluminescence spectra from the active region of device structures annealed without implantation also show similar behavior. The spectrum at high excitation intensity can be deconvoluted into four Gaussian components, while only three Gaussian components are identified in the spectrum obtained at low excitation intensity. The Gaussian components in the low excitation intensity photoluminescence spectrum are attributed to ground-state transitions from different families of quantum dots differing in their mean dimensions and the wetting layer. The additional peak identified in the high excitation intensity photoluminescence spectrum is attributed to excited state transitions in larger quantum dots.

Assuming a conduction band offset of 60%, the separation between the consecutive energy levels in the conduction band of implanted (or unimplanted), annealed samples is of the order of thermal energy of carriers at room temperature (~26 meV). Reduced energy separation between consecutive energy levels in the conduction band of annealed samples leads to an unconventional behavior in the lasing characteristics of these devices as discussed in Section 11.3.2.

11.3.2 Device Characteristics

11.3.2.1 Lasing Spectra

Figure 11.17 shows the lasing spectra of 1 mm long laser fabricated from the laser structure annealed at 600°C for 30 min. Also shown in the inset is the *L–I* (output light vs. input current) plot of the device,

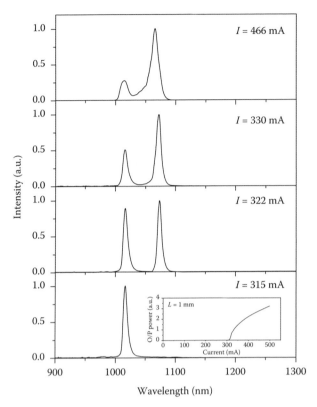

FIGURE 11.17   Lasing spectra at different injection currents for 1 mm long laser fabricated from a laser structure annealed at 600°C for 30 min. The inset shows the output power vs. injected current plot for the device.

showing the lasing threshold. The device has a threshold current of ~310 mA. At an injection current of 315 mA, just above the threshold the lasing spectrum is a single peak at 1017 nm, with a FWHM of 11 meV. As the injection current is increased to 322 mA, the lasing spectrum shows two distinct peaks, located at 1017 nm and 1073 nm with FWHM of 11 and 10 meV, respectively. As the injection current is further increased, the intensity of the peak at the longer wavelength increases with respect to the intensity of the peak at shorter wavelength, and the peaks become broader. At an injection current of 466 mA, the FWHM of the two peaks in the lasing spectrum are 18 and 15 meV, respectively. Lasing spectra of devices fabricated from laser structures implanted with protons and annealed show similar behavior.

Figure 11.18 shows the lasing spectrum of 1 mm long laser fabricated from the annealed device structures, superimposed on the absorption signal (photocurrent spectrum) from the device. The absorption signal below ~870 nm is attributed to transitions in the GaAs barrier or AlGaAs cladding layers. Transitions between ~870 and ~990 nm are associated with wetting layer states and the broad peak at longer wavelengths is due to absorption in quantum dots. Inhomogeneous broadening of the quantum dot size distribution results in a broad quantum dot absorption signal. Absorption in quantum dots is lower in intensity than in the wetting layer because of smaller quantum dot surface coverage. It is interesting to note that the photocurrent spectrum shows continuously increasing absorption signal between the low energy quantum dot signal and the wetting layer related signal. The continuous background absorption has also been observed by others (Toda et al. 1999, 4114; Finley et al. 2001, 073307; Kammerer et al. 2001, 207401). Vasanelli et al. (2002, 216804) showed that the background absorption signal is related to transitions between bound quantum dot and delocalized wetting layer/barrier states.

Comparing the photoluminescence spectrum and the lasing spectrum (Figures 11.15 and 11.17), the lower energy peak in the lasing spectrum can be attributed to quantum dot transitions. The peak at

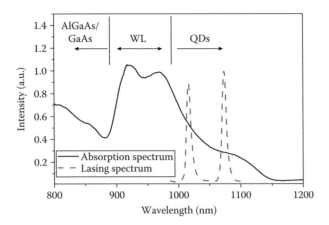

FIGURE 11.18 Photocurrent spectrum with no applied external bias and the lasing spectrum of 1 mm long laser fabricated from devices annealed at 600°C for 30 min.

higher energy in the lasing spectrum is very close to the wetting layer peak in the photoluminescence spectrum, and cannot be attributed to excited quantum dot transitions or the wetting layer transitions without ambiguity. However, comparing the position of the wetting layer peaks in the photocurrent spectra with the lasing spectrum (Figure 11.18), it can be concluded that the peak at shorter wavelengths in the lasing spectrum is also associated with the quantum dot energy levels. The energy of the lasing peak is smaller than the energy of transitions between wetting layer electron and hole states.

The device simultaneously lases from different energy levels associated with the quantum dots. Simultaneous lasing from quantum dot/quantum well ground states and excited states or from quantum dot states and wetting layer states has been observed by several groups (Zory et al. 1986, 16; Karouta et al. 1998, 1474; Lester et al. 1999, 931; Zhukov et al. 1999, 1926), especially for very short cavity lengths. However, all the reports indicate that lasing first starts at longer wavelengths and as higher lying energy levels are progressively filled, these levels may also contribute to lasing at higher injection currents. This is however different to what is observed in the devices studied in this chapter. Lasing first starts at shorter wavelengths and with increasing injection current, another peak at longer wavelengths begins to appear and gains in intensity. This behavior is only observed in short devices. For longer devices ($L = 3$ mm), the lasing spectrum consists of only one peak.

This peculiar behavior can be explained in terms of gain saturation and optical pumping in the active region of the annealed devices. For short cavity lengths, the gain available from the quantum dot ground states is insufficient to overcome the cavity losses. As a result, the devices do not lase from quantum dot ground states. As the injection current is increased to increase the gain in the active region, quantum dot excited states and the wetting layer states are progressively filled. Once the net gain exceeds the cavity losses, the device starts to lase. For short devices, the cavity losses are exceeded only after the higher energy levels (quantum dot excited states) are filled, making them lase initially from these states. As described in Section 11.3.1, the thermal energy of carriers at room temperature is comparable to the energy level separation in the active region of the annealed devices. In such a scenario, higher energy levels can be occupied by carriers even before the lower energy levels are fully saturated. Once the device starts to lase from higher energy levels, the lasing radiation is absorbed in certain regions, creating carriers in un-occupied lower energy levels. This is equivalent to optical pumping in the active region of the devices. Such a phenomenon is only possible in quantum dot devices owing to the inhomogeneous nature of the active region. In quantum well lasers, the homogeneous active region prohibits such a phenomenon. As the injection current is increased, the optical pumping of quantum dot lower energy transitions increases in intensity, until these states start to lase. Once quantum dot lasing starts, its intensity increases with further increase in injection current.

In longer devices, the quantum dot ground states provide enough gain to overcome the cavity losses before excited states are filled with carriers. So the lasing spectra of these devices show only one peak corresponding to the ground-state transitions in quantum dots.

11.3.2.2 Multicolor Lasing

Figure 11.19 shows the lasing wavelength as a function of cavity length for lasers fabricated from device structures with and without proton implantation. Short cavity devices exhibit simultaneous lasing from two states, due to ground-state gain saturation, as described in Section 11.3.2.1. Active region in unimplanted devices with longer cavity lengths $L \geq 2$ mm provides enough ground-state gain to overcome the total cavity loss and the unimplanted devices lase from quantum dot ground states for these cavity lengths.

Figure 11.20 shows the lasing spectra of unimplanted and implanted devices for a cavity length of 2 mm. The lasing spectrum of the implanted device is very broad due to simultaneous lasing from

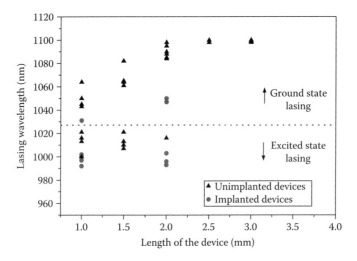

FIGURE 11.19 Lasing wavelength as a function of the cavity length for devices fabricated from laser structures annealed at 600°C for 30 min with and without implantation.

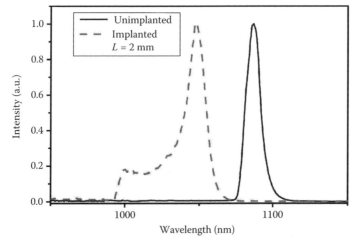

FIGURE 11.20 Normalized lasing spectra for 2 mm long devices fabricated from laser structure annealed at 600°C for 30 min with and without implantation.

two states. In spite of very broad lasing spectra, the shift between the dominant peaks in the lasing spectra is clearly observable. The peaks shown in Figure 11.20 are shifted with respect to each other by ~40 nm. The shift in the lasing wavelength is greater than the implantation-induced differential shift observed in the photoluminescence spectra (Figure 11.15). Enhanced interdiffusion due to implantation also reduces the carrier confinement potential in the active region of the implanted devices, in addition to increasing the photoluminescence peak emission energy (Section 11.3.1). Hence the lasers fabricated from implanted device structures require higher pumping levels to achieve lasing, causing a larger blueshift between the lasing wavelength and photoluminescence wavelength (due to filling up of higher lying energy levels) than in lasers fabricated from unimplanted device structures. Different blueshifts between the lasing spectra and the photoluminescence peaks for the implanted and unimplanted devices result in increased differential shift in the lasing spectra than in the photoluminescence spectra.

11.4 Impurity-Free Vacancy Disordering

11.4.1 Bandgap Tuning of Quantum Dots

As already mentioned in Section 11.3.1, the magnitude of thermally induced interdiffusion in as-grown quantum dots is determined by the background concentration and the concentration gradient of point defects (Kheris et al. 1997, 15813). Differential in-plane bandgap energy shifts can be obtained by annealing the quantum dot sample by varying the vacancy concentration across the substrate. Ion implantation (Section 11.3) is one way to achieve vacancy concentration gradient across the wafer. Another intermixing technique, that has been widely studied is IFVD. IFVD is a technique that controls the degree of disordering in quantum well/quantum dot sample through generation and diffusion of point defects.

The diffusion of group-III/V vacancies is considered to be the mechanism responsible for IFVD (Gillin 2000, 53). The concentration of group III vacancies can be controlled by capping the semiconductor sample with different dielectrics prior to annealing. The solubility of group III atoms (Ga) in the dielectric cap and the strain induced in the semiconductor sample by the dielectric cap during annealing influence the group III vacancy concentration in the sample. Increasing the solubility of Ga in the dielectric cap enhances the number of vacancies created in the sample leading to enhanced interdiffusion. Compressive strain in the semiconductor structure due to the thermal expansion coefficient mismatch between the dielectric and semiconductor materials enhances the diffusion of Ga vacancies into the semiconductor, while a tensile strain traps the vacancies creating large defect clusters (Fu et al. 2002, 3579). The effect of properties of the dielectric capping on interdiffusion has been extensively studied (Kuzuhara et al. 1989, 5833; Burkner et al. 1995, 805; Pepin et al. 1997, 142; Bhattacharya et al. 2000, 4619; Deenapanray et al. 2000, 196; Fu et al. 2000, 837; Fu et al. 2003, 2613; Yu et al. 2003, S458).

A 200 nm thick SiO_2 layer was deposited on half of the substrate, using PECVD. The laser structure was then annealed at 750°C for 30 s to achieve differential bandgap shift between quantum dots in the uncapped regions and regions capped with SiO_2. Annealing conditions were chosen to maximize the differential bandgap shift based on some annealing studies (Lever 2004, 109).

Figure 11.21 shows the room temperature photoluminescence spectra from the active region of an as-grown laser structure and laser structures annealed with and without SiO_2 capping. The photoluminescence spectra of all three samples show two peaks; a peak at shorter wavelength due to the wetting layer and a peak at longer wavelength originating from the quantum dots. The photoluminescence signal from the wetting layer and the quantum dots in the active region of the as-grown laser structure peak at 990 and 1165 nm, respectively. The photoluminescence peaks from the quantum dots in the active region of annealed (with and without SiO_2 capping) device structures are blueshifted with respect to that of the as-grown sample.

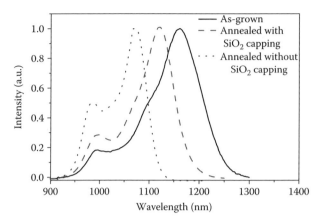

FIGURE 11.21   Room temperature photoluminescence spectra from the active region of as-grown laser structure and laser structures annealed with and without SiO_2 capping.

The photoluminescence signal from the quantum dots in the active region of the device structure annealed without SiO_2 capping peaks at ~1070 nm. The blueshift (~95 nm) of the quantum dot lumines-cence peak from the active region of device structure annealed without SiO_2 capping with respect to the peak from the active region of as-grown device structures is caused by interdiffusion at the quantum dot-barrier interface, due to the background concentration of point defects in the structure. The lumi-nescence from the quantum dots in the active region of device structures annealed with SiO_2 capping peaks at 1120 nm. This peak is redshifted (~50 nm) with respect to the luminescence peak from quantum dots in the active region of device structures annealed without SiO_2 capping and blueshifted (~45 nm) with respect to the luminescence peak from quantum dots in the active region of as-grown device struc-tures. This suggests that capping the laser structure with a layer of SiO_2 prior to annealing reduces the interdiffusion in the active region of the laser structure with respect to that in laser structures without SiO_2 capping. This observation is inconsistent with most of the available reports in literature (Malik et al. 1997, 15813; Bhattacharya et al. 2000, 4619; Deenapanray et al. 2000, 196; Fu et al. 2000, 837; Lever et al. 2004b, 7544).

It is now well understood that capping with SiO_2 results in out-diffusion of Ga from the GaAs sam-ple (Figure 11.22), resulting in increased concentration of Ga vacancies. The difference in the thermal expansion coefficients of SiO_2 and GaAs results in compressive strain in the GaAs layers when annealed at high temperatures. Due to the compressive strain, the Ga vacancies created at the GaAs–SiO_2 inter-face are pushed toward the quantum dot region. Increased concentration of group III vacancies in the vicinity of the quantum dots results in enhanced interdiffusion.

Most of the interdiffusion studies in In(Ga)As/GaAs quantum dot system have been performed on test structures comprising of only quantum dots and the GaAs barrier layers. The structures studied in this work are laser structures that have AlGaAs layers in addition to GaAs barrier and capping layers. There is only one report available to date, which reports on reduced interdiffusion in semiconductor

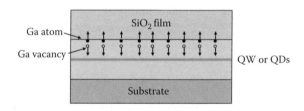

FIGURE 11.22   Schematic showing the mechanism for increased interdiffusion in samples capped with SiO_2.

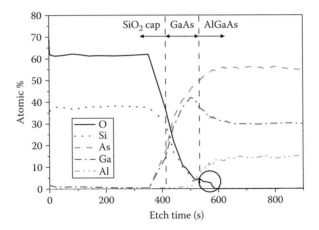

FIGURE 11.23 X-ray photoelectron spectroscopy plot showing the composition of various layers in laser structures annealed with SiO$_2$ capping. SiO$_2$/GaAs and GaAs/AlGaAs interfaces are indicated with dashed vertical lines and the region of the plot indicating the oxygen diffusion into the AlGaAs layers is circled.

samples capped with SiO$_2$, with respect to the uncapped sample (Chamberlin et al. 2002). Cohen et al. (1998, 803) reported a decrease in interdiffusion by depositing a thin layer of Al on the quantum well sample, prior to deposition of the dielectric (Ga$_2$O$_3$) layer. Annealing samples capped with a layer of Ga$_2$O$_3$ enhances interdiffusion, but depositing a layer of Al prior to deposition of the dielectric has been reported to inhibit interdiffusion. The reduction in interdiffusion has been attributed to the oxidation of Al, which injects Ga interstitials from the dielectric cap into the quantum well sample. Hence the concentration of background group III vacancies is reduced, which results in reduced interdiffusion. Chamberlin et al. (2002) observed reduced interdiffusion in samples capped with SiO$_2$ when a layer of oxidized AlGaAs is introduced between the quantum well sample and the dielectric cap. They suggest that the oxide layer either acts to getter the point defects that promote interdiffusion or injects point defects that retard diffusion. These studies however differ from the present study as the Al in the AlGaAs layers in the laser structures is not in direct contact with the dielectric capping. The laser structure has a 100 nm GaAs contact layer (Table 11.1) between the SiO$_2$ cap and the AlGaAs layers. Figure 11.23 shows the x-ray photoelectron spectroscopy measurements of the laser structure annealed with SiO$_2$ capping. The circled region of the plot indicates that oxygen from the SiO$_2$ cap diffuses through the GaAs contact layer into the AlGaAs layers. The diffusion of oxygen through the GaAs contact layer oxidizes the AlGaAs layer that either acts to getter the point defects that promote interdiffusion or injects point defects that retard diffusion (Chamberlin et al. 2002).

11.4.2 Device Characteristics

11.4.2.1 Multicolor Lasing

The electroluminescence spectra, with increasing injection currents up to threshold, from lasers fabricated from device structures annealed with and without SiO$_2$ capping are depicted in Figure 11.24. The cavity length for the two devices is 3.5 mm. It is interesting to note that at lower injection currents, the electroluminescence spectra from both devices are centered at wavelengths longer than the photoluminescence peak wavelengths (Figure 11.21). With increasing injection current, the electroluminescence signal blueshifts and the devices lase at wavelengths slightly shorter than the photoluminescence peak wavelengths. It has been shown that the blueshift of electroluminescence spectra is a consequence of quantum dot size distribution and Fermi–Dirac like carrier distribution in the quantum dot ensemble (Lever et al. 2004a, 1410). At very low injection currents, quantum dots with larger mean heights (and

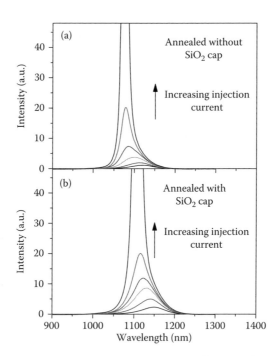

FIGURE 11.24 Electroluminescence spectra for 3.5 mm long lasers fabricated from device structures annealed (a) without and (b) with SiO$_2$ capping.

thus lower lying energy states) are first populated, resulting in electroluminescence spectra redshifted with respect to the photoluminescence spectra. With increasing injection currents, quantum dots with smaller mean heights (and thus higher lying energy states) are progressively filled, leading to the observed blueshift. It can be clearly seen from the electroluminescence spectra that both devices are lasing from quantum dot ground states. Luminescence from quantum dot excited states does not appear in the electroluminescence spectra until the devices lase. Laser fabricated from structure annealed with SiO$_2$ capping lases at 1110 nm, while the laser fabricated from structure annealed without SiO$_2$ capping lases at 1070 nm. The shift between the lasing spectra (~40 nm) is similar to the shift between the ground-state peak photoluminescence emission wavelengths from the active region of the two structures (Figure 11.21).

11.5 Comparison between the Different Bandgap-Tuning Techniques

As discussed in Sections 11.2.2 and 11.2.5.1, a large differential bandgap energy shift between the epitaxial layers forming the active and passive sections of a laser integrated with a passive waveguide is essential for achieving very low passive losses in the integrated devices. The differential bandgap shift is maximized by having quantum dots in the active region and quantum wells in the passive section of the integrated devices. SAE is the only bandgap-tuning technique that can be used to simultaneously grow quantum dots and quantum wells in different regions of a substrate. Even though there are some reports indicating loss of 3D carrier confinement in quantum dots (and thus showing a quantum well like behavior) due to annealing followed by implantation, such studies involve high annealing temperatures and/or longer annealing times (Babinski et al. 2001, 2576). Higher annealing temperatures and/or longer annealing times also cause enhanced interdiffusion (and increase the blueshift in the emission wavelength) in quantum dots in the active region, thereby reducing the differential bandgap shift between the active and passive sections and carrier capture efficiency in the active region of the devices. Reduction in differential bandgap energy shift and carrier capture efficiency in the active region of

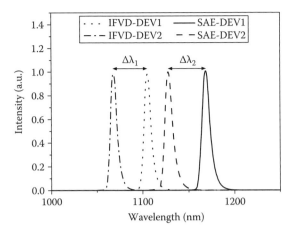

FIGURE 11.25 Normalized lasing spectra of 3 mm long multiple-wavelength lasers fabricated using IFVD and SAE. IFVD-DEV1 and IFVD-DEV2 are devices fabricated from laser structures annealed with SiO_2 capping and without SiO_2 capping, respectively. SAE-DEV1 and SAE-DEV2 are devices fabricated using quantum dots grown selectively in the $50\,\mu$m openings between SiO_2 stripes of width 15 and $10\,\mu$m, respectively.

devices, are both undesirable for optimal device performance. Hence SAE is the preferred technique for integration of a quantum dot laser integrated with a passive waveguide.

While ground-state lasing has been demonstrated in multicolor lasers fabricated using SAE and IFVD (Figures 11.13b and c and 11.24), it is difficult to achieve ground-state lasing from implanted, annealed devices (Figure 11.19). The devices either lase predominantly from the quantum dot excited states ($L < 2$ mm) or start lasing from quantum dot excited states first and then from quantum dot ground states (Section 11.3.2.1). Lasing in implanted devices cannot be achieved for longer cavity lengths due to very high internal losses, possibly due to residual defects. Un-implanted, annealed devices exhibit ground-state lasing only for $L \geq 2$ mm (Figure 11.19). The differential shift in the lasing wavelength of 2 mm long unimplanted, annealed and implanted, annealed devices is $\sim$40 nm (Figure 11.20). This differential shift is comparable to the differential shifts obtained in multicolor lasers fabricated using IFVD ($\Delta\lambda_1 = 35$ nm) or SAE ($\Delta\lambda_2 = 40$ nm) as shown in Figure 11.25. Though similar differential shift in the lasing spectra is obtained using all the techniques, devices fabricated using ion implantation–induced intermixing and IFVD have shorter lasing wavelengths compared to the devices fabricated using SAE. One of the objectives for realizing GaAs-based quantum dot devices is for applications in optical telecommunication systems (Mukai et al. 2000, 3349; Qiu et al. 2001, 3570; Sellin et al. 2001, 1207). In(Ga) As quantum dots have the potential to extend the lasing wavelength of the GaAs based devices to 1.3 μm, which represents the chromatic dispersion minima in optical fibers. Constant efforts for achieving longer wavelength emission from quantum dots have been published (Mukai et al. 2000, 3349; Qiu et al. 2001, 3570; Sellin et al. 2001, 1207). It may not be possible to fabricate quantum dot lasers with longer emission wavelength in the GaAs-based system using ion implantation–induced intermixing or IFVD because the annealing step necessary for initiating interdiffusion blueshifts the quantum dot emission wavelength (Leon et al. 1996, 1888; Mo et al. 1998, 3518; Fafard and Allen 1999, 2374; Jiang et al. 2000, 356; Zhang et al. 2002, 136).

Figures 11.26 and 11.27 compare the inverse differential efficiency and the threshold currents, respectively, as a function of cavity length for the two selectively grown devices that exhibit ground-state lasing. The two selectively grown devices (fabricated from quantum dots grown between SiO_2 stripes of width, W: 10 and 15 μm) have quite similar internal quantum efficiencies and optical losses of $\sim$11 cm^{-1}. The two selectively grown devices also have very similar threshold currents for all cavity lengths examined.

Figure 11.28 shows the inverse differential efficiency vs. cavity length plot for devices fabricated from laser structures annealed with and without SiO_2 capping. The internal quantum efficiency (η_i) for

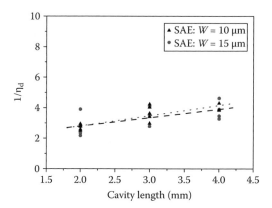

FIGURE 11.26 Inverse differential efficiency vs. cavity length plot for devices fabricated from selectively grown quantum dots (*W*: 10 and 15 μm).

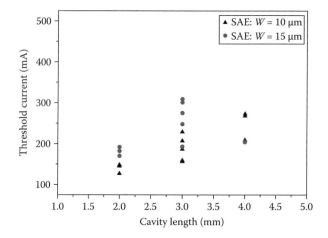

FIGURE 11.27 Threshold current as a function of cavity length for devices fabricated from selectively grown quantum dots grown between SiO_2 stripes of width (*W*) 10 and 15 μm.

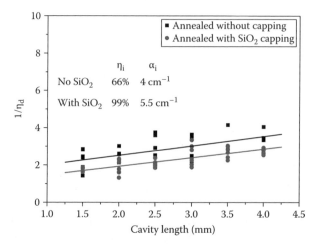

FIGURE 11.28 Inverse differential efficiency vs. cavity length plot for lasers fabricated from device structures annealed with and without SiO_2 capping.

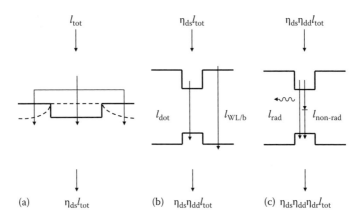

FIGURE 11.29 Schematic illustrating the effect of (a) current spreading, (b) recombination in the wetting layer and barrier, and (c) non-radiative recombination on the internal quantum efficiency of a quantum dot laser. (Adapted from Smowton, P.M. and Blood, P., *IEEE J. Sel. Top. Quantum Electron.*, 3, 491, 1997.) (From Smowton, P. and Blood, P., *IEEE J. Select. Top. Quantam Electron.*, 3(2), 491, April 1997. With permission. © 1997 IEEE.)

devices fabricated from structures annealed with SiO_2 capping is 99% and the internal losses (α_i) are ~5.4 cm^{-1}. Devices fabricated from laser structures annealed without SiO_2 capping have lower internal quantum efficiency (66%) and optical losses (~4 cm^{-1}) than the devices fabricated from laser structures annealed with SiO_2 capping.

Lower internal quantum efficiency for devices annealed without SiO_2 capping compared to the devices annealed with SiO_2 capping can be qualitatively explained in terms of smaller carrier confinement in these devices. Figure 11.29 schematically illustrates the effect of carrier loss mechanisms on the internal quantum efficiency of the quantum dot lasers. The internal differential efficiency η_i can be expressed in terms of η_{ds}, η_{dd} and η_{dr} as (Smowton and Blood 1997, 491)

$$\eta_i = \eta_{ds} \times \eta_{dd} \times \eta_{dr}$$

where

  η_{ds} is associated with lateral current spreading
  η_{dd} is associated with the carrier injection efficiency
  η_{dr} is associated with the radiative efficiency in the quantum dots

η_{ds} depends on the particular device geometry under consideration and can be significantly less than unity for narrow ridge waveguide lasers. η_{ds} cannot account for the differences between devices fabricated from laser structures annealed with and without SiO_2 capping, as both the devices are ridge waveguide lasers with 4 μm wide ridges. η_{dd} represents the fraction of carriers injected into the active region that actually recombine in the quantum dots. Carrier population of the wetting layer/barrier states can significantly reduce the value of η_{dd}. η_{dr} is determined by the fraction of carriers in the quantum dots that recombine radiatively. Non-radiative recombination like Auger recombination process in the quantum dots may significantly reduce its value. Smaller carrier confinement in devices fabricated from laser structures annealed without SiO_2 capping increases the probability of carrier occupancy of the states in wetting layer/barrier. The increased occupancy of the wetting layer and barrier states decreases η_{dd} in devices fabricated from laser structures annealed without SiO_2 capping. It has been shown that increased population of wetting layer and barrier states may also increase the probability of Auger recombination processes involving mixed (wetting layer and quantum dot) states (Sears 2006, 127). The increased probability of such non-radiative processes may result in smaller values of η_{dr}. Thus

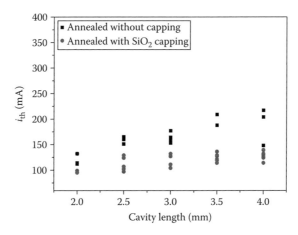

FIGURE 11.30 Threshold currents for devices fabricated from laser structures annealed with and without SiO₂ capping for various cavity lengths.

the reduction in carrier confinement and separation between consecutive energy levels in the active region of devices fabricated from laser structures annealed without capping leads to reduced internal quantum efficiency in these devices.

Comparing the threshold currents, devices fabricated from laser structures annealed with SiO₂ capping have lower threshold currents than the devices fabricated from laser structures annealed without SiO₂ capping for all cavity lengths (Figure 11.30). The difference can be explained in terms of lower carrier confinement in the conduction band in the active region of the devices fabricated from laser structures annealed without SiO₂ capping. Lower confinement energy increases the probability of thermal escape of carriers from the quantum dot ground states to higher energy states (excited states/wetting layer states/barrier states). Loss of carriers from the lasing states reduces the net gain available from the quantum dot active region and increased population of wetting layer/barrier states increases the probability of non-radiative recombination (Deppe et al. 1999, 1238; Sears 2006, 127). Both reduction in the net gain from the lasing states and increased non-radiative recombination lead to an increase in the threshold current of the devices.

Comparing devices fabricated using SAE and IFVD, the devices fabricated using SAE have higher losses and lower internal quantum efficiency than devices fabricated using IFVD, which are a consequence of the quality of regrown interfaces. Imperfections at the interfaces may result in increased scattering of radiation and reduced efficiency of the devices. Increased optical losses in these devices also result in higher threshold currents compared to devices fabricated from laser structures annealed with SiO₂ capping.

Figure 11.31 shows the $L-I$ plots at 10°C for 3 mm long lasers fabricated from device structures grown using SAE and device structures grown on unpatterned substrates and annealed. Annealed devices have lower slope efficiencies ($\eta_d = 2.4\%$) compared to devices fabricated using SAE ($\eta_d = 38\%$). Also the annealed devices have higher threshold currents ($I_{th} = 340\,\text{mA}$) compared to the devices fabricated using SAE ($I_{th} = 135\,\text{mA}$). The lower slope efficiencies and increased threshold currents in annealed devices can be attributed to reduced carrier confinement potentials and inter-level spacing in these devices, which result in loss of carriers from quantum dot ground states to higher energy levels, reducing the net gain available from lower energy levels at a given carrier injection and increasing the possibility of non-radiative recombination rates involving carriers in higher energy levels. However, ion implantation–induced intermixing in carefully optimized device structures such as the ones that use AlGaAs barriers to increase the carrier confinement in the active region (Zaitsev et al. 1997, 4219) may result in multicolor lasers with improved characteristics.

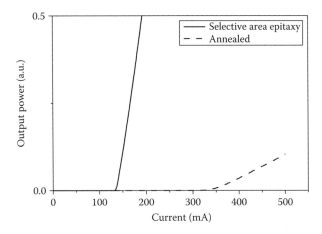

FIGURE 11.31 Output power (*L*) vs. injected current (*I*) plots (at 10°C) for 3 mm long lasers fabricated using device structures grown by SAE and device structures annealed at 600°C for 30 min.

11.6 Summary

This chapter was aimed at giving an overview of bandgap-tuning techniques that can be used for achieving integration of quantum dot based optoelectronic devices. The basic principles involved in SAE, ion implantation–induced intermixing, and IFVD have been discussed.

The local growth parameters have been controlled during MOCVD growth of InGaAs quantum dots by using a SiO_2 mask pattern. The thickness of epitaxial layers deposited across the substrate is determined by the local mask dimensions. Nucleation of high-density, defect-free quantum dots only in certain regions of the substrate and growth of quantum wells on the rest of the substrate were achieved by careful optimization of mask dimensions and the growth parameters. This growth scheme is ideal for a single-step growth of quantum dots and quantum wells for the fabrication of a quantum dot laser integrated with a passive waveguide. In regions where quantum dots were formed, their bandgap energy was tuned by controlling the local mask dimensions. Bandgap tuning of quantum dots is a convolution of two effects, namely variation in average dot size and variation in dot composition. The dot sizes vary because of variation in growth parameters like coverage and deposition rate across the substrate, while the dot composition varies because of differential incorporation of In and Ga into the quantum dots. This growth scheme is also suitable for growing the active region for fabrication of multiple-wavelength quantum dot lasers.

The ability to grow quantum dots and quantum wells simultaneously in a single growth step has been used to fabricate a quantum dot laser integrated with a passive (quantum well) waveguide. The active and passive sections of the integrated devices are electrically isolated from each other. Only the active section is electrically pumped and the cavity length is determined by the sum of the lengths of the active and passive sections. For a fixed length of the active section, as the length of the passive section increases, the lasing wavelength of the integrated devices blueshifts and the threshold current increases due to increase in total losses in the devices. The integrated devices have very low passive losses (3 cm^{-1}). Such low losses are attributed to a large differential shift in the bandgap energies of the active and passive sections of the integrated devices.

The ability to tune the bandgap energy of quantum dots has been used to demonstrate multicolor quantum dot lasers. The lasing wavelength was tuned over ~35 nm, while maintaining ground-state lasing. It has also been demonstrated that if the mask dimensions are not optimized, either too little or too much material may be deposited during the growth of the active region leading to lasing from quantum dot excited states and poor device performance.

Multicolor lasers have been fabricated using post-growth bandgap-tuning techniques of ion implantation–induced intermixing and IFVD. For devices fabricated using ion implantation–induced intermixing, the implantation and annealing conditions used in this study resulted in relatively small implantation-induced differential intermixing compared to the thermal interdiffusion. A differential shift of ~40 nm has been demonstrated in the lasing wavelength of unimplanted and implanted devices. The devices fabricated using ion implantation–induced intermixing exhibit simultaneous lasing from two states for short cavity lengths. Lasing commences at higher energy level first and at higher injection currents an additional peak at lower energy is observed. This behavior is contrary to what has been reported earlier for quantum well and quantum dot lasers and has been explained in terms of lower carrier confinement potential and reduced inter-level spacing in the active region of the annealed (with and without implantation) devices.

Laser structures were annealed with and without a 200 nm thick SiO_2 capping layer to achieve differential bandgap tuning in the active region. A differential shift of ~40 nm was obtained in the photoluminescence emission signal from the structures annealed with and without SiO_2 capping. Device structures annealed without SiO_2 capping exhibited enhanced interdiffusion in the active region compared to device structures annealed with SiO_2 capping. This observation contradicts the results in literature where several groups have reported an increase in interdiffusion in structures capped with SiO_2. Reduced interdiffusion in the active region of device structures capped with SiO_2 is a consequence of oxidation of high Al containing cladding layers. Oxygen from the SiO_2 cap diffuses through the GaAs cap and oxidizes the Al in the AlGaAs layers. The oxide layer either acts to getter the point defects that promote interdiffusion or injects point defects that retard diffusion or compensate for the background Ga vacancy concentration and result in reduced interdiffusion. Ground-state lasing has been demonstrated in lasers fabricated from both device structures.

Devices fabricated using ion implantation–induced intermixing have higher threshold currents and lower differential efficiencies compared to devices fabricated using SAE because of weaker carrier confinement potential in the annealed quantum dots in the active region. The devices fabricated using SAE have threshold currents and internal quantum efficiency comparable to devices fabricated from laser structures annealed without SiO_2 capping. Devices fabricated from laser structures annealed with SiO_2 capping have significantly lower threshold currents and higher internal quantum efficiency than the devices fabricated using SAE. Poor performance of devices fabricated using SAE in comparison to devices fabricated from laser structures annealed with SiO_2 capping is attributed to the quality of regrown interfaces. Imperfections at the regrown interfaces lead to an increase in scattering of radiation, which increases the internal optical losses. Increased optical losses lead to higher threshold currents and lower internal quantum efficiency in devices fabricated using SAE.

Each of the techniques studied in this chapter (SAE, ion implantation–induced intermixing, and IFVD) has its own advantages and disadvantages for integration of quantum dot–based optoelectronic devices. SAE is the technique of choice for fabrication of a quantum dot laser integrated with a low-loss passive waveguide. It is the only bandgap-tuning technique that can be used to realize both quantum dots and quantum wells simultaneously on the same substrate, essential for maximizing the differential bandgap shift/minimizing the passive losses.

IFVD results in multicolor lasers with good performance characteristics and offers ease of fabrication. However, the differential bandgap shift that can be obtained using this technique is highly sensitive to the quality of the oxide used and lacks reproducibility. While ion implantation–induced intermixing is a highly reproducible technique, the quality of the integrated devices fabricated using ion implantation–induced intermixing is degraded due to residual defects. The device performance could however be improved to certain extent by careful design of the device structures. SAE is a technique suitable for fabrication of multicolor lasers emitting at longer wavelengths. However, it involves intensive processing. Hence, the bandgap-tuning technique has to be chosen depending on the specific requirements.

Acknowledgments

Thanks are due to Dr. Manuela Buda and Michael Aggett for fruitful discussions and expert technical advice, respectively. We wish to acknowledge Dr. Vince Craig for his expert advice on AFM and Katie McBean and Prof. Matthew Phillips for help with CL measurements. The Australian Research Council is gratefully acknowledged for their financial support. The Australian National Fabrication Facility funded under the Australian Government's National Cooperative Research Infrastructure Strategy is acknowledged for providing access to the facilities used in this work.

References

Asryan, L. V., M. Grundmann, N. N. Ledentsov, O. Stier, R. A. Suris, and D. Bimberg. 2001. Effect of excited state transitions on the threshold characteristics of a quantum dot laser. *IEEE Journal of Quantum Electronics* 37: 418.

Babinski, A., J. Jasinski, R. Bozek, A. Szepielow, and J. M. Baranowski. 2001. Rapid thermal annealing of InAs/GaAs quantum dots under a GaAs proximity cap. *Applied Physics Letters* 79: 2576–2578.

Bhattacharya, D., A. S. Helmy, A. C. Bryce, E. A. Avrutin, and J. H. Marsh. 2000. Selective control of self-organized In0.5Ga0.5As/GaAs quantum dot properties: Quantum dot intermixing. *Journal of Applied Physics* 88: 4619–4622.

Bimberg, D., M. Grundmann, F. Heinrichsdorff, N. N. Ledentsov, V. M. Ustinov, A. F. Tsatsulnikov, P. S. Kopév, and Z. I. Alferov. 2000. Quantum dot lasers: Breakthrough in optoelectronics. *Thin Solid Films* 367: 235.

Buda, M., J. Hay, H. H. Tan, J. Wong-Leung, and C. Jagadish. 2003. Low loss, thin p-clad 980-nm InGaAs semiconductor laser diodes with an asymmetric structure design. *IEEE Journal of Quantum Electronics* 39: 625–633.

Burkner, S., M. Maier, E. C. Larkins, W. Rothemund, E. P. O'Reilly, and J. D. Ralston. 1995. Process parameter dependence of impurity-free interdiffusion in GaAs/AlGaAs multiple quantum wells. *Journal of Electronic Materials* 24: 805–812.

Chamberlin, D. R., S. A. McHugo, D. Burak, D. Burke, T. Oseentowski, and S. J. Rosner. 2002. Injection of point defects by oxidation of AlGaAs. *Materials Research Symposium Proceedings*, paper F13.4, Warrendale, PA, Vol. 719.

Cockerill, T. M., D. V. Forbes, H. Han, B. A. Turkot, J. A. Dantzig, I. M. Robertson, and J. J. Coleman. 1994. Wavelength tuning in strained layer InGaAs-GaAs-AlGaAs quantum well lasers by selective-area MOCVD. *Journal of Electronic Materials* 23: 115–119.

Cohen, R. M., G. Li, C. Jagadish, P. T. Burke, and M. Gal. 1998. Native defect engineering of interdiffusion using thermally grown oxides of GaAs. *Applied Physics Letters* 72: 2850–2852.

Coldren, L. A. and S. W. Corzine. 1995. *Diode Lasers and Photonic Circuits*. New York: John Wiley & Sons.

Coleman, J. J. 1997. Metalorganic chemical vapor deposition for optoelectronic devices. *Proceedings of the IEEE* 85: 2358–2366.

Deenapanray, P. N. K., H. H. Tan, L. Fu, and C. Jagadish. 2000. Influence of low-temperature chemical vapor deposited SiO$_2$ capping layer porosity on GaAs/AlGaAs quantum well intermixing. *Electrochemical and Solid-State Letters* 3: 196–199.

Deppe, D. G., D. L. Huffaker, S. Csutak, Z. Zou, G. Park, and O. B. Shchekin. 1999. Spontaneous emission and threshold characteristics of 1.3-μm InGaAs-GaAs quantum-dot GaAs-based lasers. *IEEE Journal of Quantum Electronics* 35: 1238–1246.

El-Emawy, A. A., S. Birudavolu, S. Huang, H. Xu, and D. L. Huffaker. 2003. Selective surface migration for defect-free quantum dot ensembles using metal organic chemical vapor deposition. *Journal of Crystal Growth* 255: 213–219.

Fafard, S. and C. N. Allen. 1999. Intermixing in quantum-dot ensembles with sharp adjustable shells. *Applied Physics Letters* 75: 2374–2376.

Finley, J. J., A. D. Ashmore, A. Lemaitre, D. J. Mowbray, M. S. Skolnick, I. E. Itskevich, P. A. Maksym, M. Hopkinson, and T. F. Krauss. 2001. Charged and neutral exciton complexes in individual self-assembled In(Ga)As quantum dots. *Physical Review B* 63: 073307.

Fu, L., P. N. K. Deenapanray, H. H. Tan, C. Jagadish, L. V. Dao, and M. Gal. 2000. Quality of silica capping layer and its influence on quantum-well intermixing. *Applied Physics Letters* 76: 837–839.

Fu, L., J. Wong-Leung, P. N. K. Deenapanray, H. H. Tan, C. Jagadish, B. Gong, R. N. Lamb, R. M. Cohen, and W. Reichert. 2002. Suppression of interdiffusion in GaAs/AlGaAs quantum-well structure capped with dielectric films by deposition of gallium oxide. *Journal of Applied Physics* 92:3579–3583.

Fu, L., P. Lever, H. H. Tan, C. Jagadish, P. Reece, and M. Gal. 2003. Suppression of interdiffusion in InGaAs/GaAs quantum dots using dielectric layer of titanium dioxide. *Applied Physics Letters* 82: 2613–2615.

Gillin, W. P. 2000. Interdiffusion mechanisms in III–V materials, in E. Herbert Li (Ed.), *Semiconductor Quantum Wells Intermixing*, pp. 53–84. Amsterdam, the Netherlands: Gordon and Breach Science Publishers.

Gordeev, N. Y., W. K. Tan, A. C. Bryce, and J. H. Marsh. 2007. Broad-area InAs/GaAs quantum dot lasers incorporating intermixed passive waveguide. *Electronic Letters* 43: 29–30.

Heinrichsdorff, F., M. Grundmann, O. Stier, A. Krost, and D. Bimberg. 1998. Influence of In/Ga intermixing on the optical properties of InGaAs/GaAs quantum dots. *Journal of Crystal Growth* 195: 540–545.

Ji, Y., W. Lu, G. Chen, X. Chen, and Q. Wang. 2003. InAs/GaAs quantum dot intermixing induced by proton implantation. *Journal of Applied Physics* 93: 1208–1211.

Jiang, W. H., H. Z. Xu, X. L. Ye, J. Wu, D. Ding, J. B. Liang, and Z. G. Wang. 2000. Annealing effect on the surface morphology and photoluminescence of InGaAs/GaAs quantum dots grown by molecular beam epitaxy. *Journal of Crystal Growth* 212: 356–359.

Kammerer, C., G. Cassabois, C. Voisin, C. Delalande, P. Roussignol, and J. M. Gérard. 2001. Photoluminescence up-conversion in single self-assembled InAs/GaAs quantum dots. *Physical Review Letters* 87: 207401.

Karouta, F., E. Smalbrugge, W. C. V. der Vleuten, S. Gaillard, and G. A. Acket. 1998. Fabrication of short GaAs wet-etched mirror lasers and their complex spectral behaviour. *IEEE Journal of Quantum Electronics* 34: 1474–1479.

Kheris, O. M., W. P. Gillin, and K. P. Homewood. 1997. Interdiffusion: A probe of vacancy diffusion in III-V materials. *Physical Review B* 55: 15813.

Kuzuhara, M., T. Nozaki, and T. Kamejima. 1989. Characterisation of Ga out-diffusion from GaAs into SiO_xN_y films during thermal annealing. *Journal of Applied Physics* 66: 5833–5836.

Leon, R., Y. Kim, and C. Jagadish. 1996. Effects of interdiffusion on the luminescence of InGaAs/GaAs quantum dots. *Applied Physics Letters* 69: 1888–1890.

Lester, L. F., A. Stintz, T. C. Newell, E. A. Pease, B. A. Fuchs, and K. J. Malloy. 1999. Optical characteristics of 1.24-μm InAs quantum-dot laser diodes. *IEEE Photonics Technology Letters* 11: 931–933.

Lever, P. 2004. Interdiffusion and MOVPE growth of self-assembled InGaAs QD structures and devices. PhD thesis, Australian National University, Canberra, Australia.

Lever, P., H. H. Tan, C. Jagadish, P. Reece, and M. Gal. 2003. Proton-irradiation-induced intermixing of InGaAs quantum dots. *Applied Physics Letters* 82: 2053–2055.

Lever, P., M. Buda, H. H. Tan, and C. Jagadish. 2004a. Investigation of the blueshift in the electroluminescence spectra from MOCVD grown InGaAs quantum dots. *IEEE Journal of Quantum Electronics* 40: 1410–1416.

Lever, P., H. H. Tan, C. Jagadish, P. Reece, and M. Gal. 2004b. Impurity free vacancy disordering of InGaAs quantum dots. *Journal of Applied Physics* 96: 7544–7548.

Liu, G. T., H. L. Stintz, T. C. Newell, A. L. Gray, P. M. Varangis, K. J. Malloy, and L. F. Lester. 2000. The influence of quantum-well composition on the performance of quantum dot lasers using InAs/InGaAs dots-in-a-well (DWELL) structures. *IEEE Journal of Quantum Electronics* 36: 1272.

Malik, S., C. Roberts, R. Murray, and M. Pate. 1997. Tuning self-assembled InAs quantum dots by rapid thermal annealing. *Applied Physics Letters* 71: 1987.

Miller, S. E. 1969. Integrated optics: An introduction. *Bell System Technical Journal* 48: 2059–2069.

Mo, Q. W., T. W. Fan, Q. Gong, J. Wu, and Z. G. Wang. 1998. Effects of annealing on self-organized InAs quantum islands on GaAs(100). *Applied Physics Letters* 73: 3518–3520.

Mokkapati, S. 2005. Controlling the properties of InGaAs quantum dots by selective-area epitaxy. *Applied Physics Letters* 86: 113102.

Mokkapati, S. 2007. Multiple wavelength InGaAs quantum dot lasers using selective area epitaxy. *Applied Physics Letters* 90: 171104.

Mokkapati, S., H.H. Tan, and C. Jagadish. 2006. Integration of an InGaAs quantum dot laser with a low-loss passive waveguide using selective area epitaxy. *IEEE Photonics Technology Letters* 18(15): 1648–1650.

Mokkapati, S., J. Wong-Leung, H.H. Tan, C. Jagadish, K.E. McBean, and M.R. Phillips. 2008. Tuning the bandgap of InAs quantum dots by selective-area MOCVD. *Journal of Physics D: Applied Physics* 41: 085104.

Mukai, K., Y. Nakata, K. Otsubo, M. Sugawara, and N. Yokoyama. 2000. High characteristic temperature of near-1.3-μm InGaAs/GaAs quantum-dot lasers at room temperature. *Applied Physics Letters* 76: 3349–3351.

Pepin, A., C. Vieu, M. Schneider, H. Launois, and Y. Nissim. 1997. Evidence of stress dependence in SiO_2/Si_3N_4 encapsulation-based layer disordering of GaAs/AlGaAs quantum well heterostructures. *Journal of Vacuum Science and Technology B* 15: 142–153.

Perret, N., D. Morris, L. Franchomme-Fossè, R. Côté, S. Fafard, V. Aimex, and J. Beauvais. 2000. Origin of the inhomogeneous broadening and alloy intermixing in InAs/GaAs self-assembled quantum dots. *Physical Review B* 62: 5092–5099.

Qiu, Y., P. Gogna, S. Forouhar, A. Stintz, and L. F. Lester. 2001. High-performance InAs quantum-dot lasers near 1.3 μm. *Applied Physics Letters* 79: 3570–3572.

Salem, B., V. Aimez, D. Morris, A. Turala, P. Regreny, and M. Gendry. 2005. Bandgap tuning of InAs/InP quantum sticks using low-energy ion-implantation-induced intermixing. *Applied Physics Letters* 87: 241115–241123.

Sears, K. 2006. Growth and characterization of self-assembled InAs/GaAs quantum dots and optoelectronic devices. PhD thesis, Australian National University, Canberra, Australia.

Sellin, R. L., C. Ribbat, M. Grundmann, N. N. Ledentsov, and D. Bimberg. 2001. Close-to-ideal device characteristics of high-power InGaAs/GaAs quantum dot lasers. *Applied Physics Letters* 78: 1207–1209.

Shiramine, K.-I., T. Itoh, S. Muto, T. Kozaki, and S. Sato. 2002. Adatom migration in Stranski-Krastanow growth of InAs quantum dots. *Journal of Crystal Growth* 242: 332–338.

Shoji, H., Y. Nakata, K. Mukai, Y. Sugiyama, M. Sugawara, N. Yokoyama, and H. Ishikawa. 1997. Lasing characteristics of self-formed quantum-dot lasers with multistacked dot layer. *IEEE Journal of Selected Topics in Quantum Electronics* 3: 188–195.

Smowton, P. M. and P. Blood. 1997. The differential efficiency of quantum-well lasers. *IEEE Journal of Selected Topics in Quantum Electronics* 3(2): 491–498.

Smowton, P. M., E. Herrmann, Y. Ning, H. D. Summers, P. Blood, and M. Hopkinson. 2001. Optical mode loss and gain of multiple-layer quantum-dot lasers. *Applied Physics Letters* 78: 2629–2631.

Surkova, T., A. Patanè, L. Eaves, M. Henini, A. Polimeni, A. P. Knights, and C. Jeynes. 2001. Indium interdiffusion in annealed and implanted InAs/(AlGa)As self-assembled quantum dots. *Journal of Applied Physics* 89: 6044–6047.

Tatebayashi, J., N. Hatori, M. Ishida, H. Ebe, M. Sugawara, Y. Arakawa, H. Sudo, and A. Kuramata. 2005. 1.28 μm lasing from stacked InAs/GaAs quantum dots with low-temperature-grown AlGaAs cladding layer by metalorganic chemical vapor deposition. *Applied Physics Letters* 86: 053107.

Toda, Y., O. Moriwaka, M. Nishioka, and Y. Arakawa. 1999. Efficient carrier relaxation mechanism in InGaAs/GaAs self-assembled quantum dots based on the existence of continuum states. *Physical Review Letters* 82: 4114.

Ustinov, V. M., A. E. Zhukov, A. Y. Egorov, and N. A. Maleev. 2003. *Quantum Dot Lasers*, 1st edn. New York: Oxford Science Publications.

Vasanelli, A., R. Ferreira, and G. Bastard. 2002. Continuous absorption background and decoherence in quantum dots. *Physical Review Letters* 89: 216804.

Wellmann, P. J., W. V. Schoenfeld, J. M. Garcia, and P. M. Petroff. 1998. Tuning of electronic states in self-assembled InAs quantum dots using an ion implantation technique. *Journal of Electronic Materials* 27: 1030–1033.

Xu, S. J., X. C. Wang, S. J. Chua, C. H. Wang, W. J. Fan, J. Jiang, and X. G. Xie. 1998. Effects of rapid thermal annealing on structure and luminescence of self-assembled InAs/GaAs quantum dots. *Applied Physics Letters* 72: 3335.

Yu, J. S., J. D. Song, Y. T. Lee, and H. Lim. 2003. Effects of the thickness of dielectric capping layer and the distance of quantum wells from the sample surface on the intermixing of In0.2Ga0.8As/GaAs multiple quantum well structures by impurity-free vacancy disordering. *Journal of Korean Physical Society* 42: S458–S461.

Zaitsev, S. V., N. Y. Gordeev, V. I. Kopchatov, V. M. Ustinov, A. E. Zhukov, A. Y. Egorov, N. N. ledentsov, M. V. Maximov, P. S. Kop'ev, A. O. Kosogov, and Z. I. Alferov. 1997. Vertically coupled quantum dot lasers: First device oriented structures with high internal quantum efficiency. *Japanese Journal of Applied Physics* 36: 4219–4220.

Zhang, Y. C., Z. G. Wang, B. Xu, F. Q. Liu, Y. H. Chen, and P. Dowd. 2002. Influence of strain on annealing effects of In(Ga)As quantum dots. *Journal of Crystal Growth* 244: 136–141.

Zhukov, A. E., A. R. Kovsh, N. A. Maleev, S. S. Mikhrin, V. M. Ustinov, A. F. Tsatsul'ikov, M. V. Maximov, B. V. Volovik, D. A. Bedarev, Y. M. Shernyakov, P. S. Kopév, Z. I. Alferov, N. N. Ledentsov, and D. Bimberg. 1999. Long-wavelength lasing from multiply stacked InAs/InGaAs quantum dots on GaAs substrates. *Applied Physics Letters* 75: 1926–1928.

Zory, P. S., A. R. Reisinger, R. G. Watres, L. J. Mawst, C. A. Zmudzinski, M. A. Emanuel, M. E. Givens, and J. J. Coleman. 1986. Anomalous temperature dependence of threshold for thin quantum well AlGaAs diode lasers. *Applied Physics Letters* 19: 16–18.

12

Infrared Physics of Quantum Dots

Manijeh Razeghi
Bijan Movaghar

12.1 Introduction

The field of quantum dot (QD) physics is now a vast subject, with many excellent reviews and original papers [1–8] and even books. In this chapter, I will not try to cover "everything" but will stay focused and concentrate on the field where our group here at Northwestern has an established world-class expertise, which is that of infrared (IR) physics. As we proceed, the reader will be referred to the other special features associated with QDs and their applications. There are many different ways one can make a QD [1–8]. A QD being defined as a three-dimensional domain with nanosize geometry [1–8]. We are here concerned with semiconductor QD, and furthermore those made using the Stranski–Krastanow (SK) self-assembly method [1–8]. Let me briefly recall how this works. When one grows a thin film of one material on top of another one with MBE or MOCVD, for example, i.e., atomic layer-by-layer deposition, and there is lattice mismatch, there is strain energy built into the interface. If the lattice mismatch exceeds a critical amount, typically 4%, the systems prefer to release the strain energy by forming nanosize islands that can exhibit a high degree of regularity. The strained thin film is called the wetting layer (WL). The precise diameter and height of these islands or QD depends on the materials, WL thickness, and growth conditions (temperature and deposition rate) used. Using atomic force microscopy (AFM), one can scan the QD surface to determine the geometry and distribution. A typical AFM picture of QD using InAs as the WL grown on top of InAlAs is shown in Figure 12.1 [9,10]. Density and QD geometry are very important for light-sensing applications as we shall see later.

FIGURE 12.1   Atomic force microscopy imaging of the InAs/InAlAs QDs used in our recent devices.

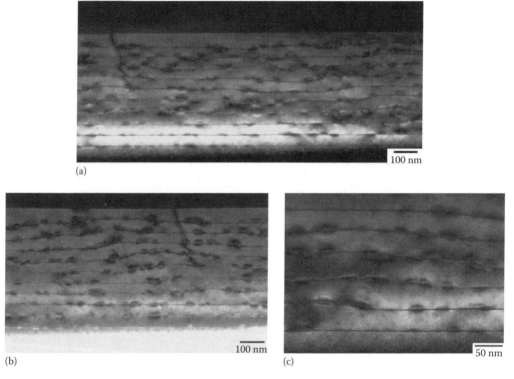

FIGURE 12.2   Transmission electron microscope (TEM) images of the midwavelength-IR–QDIP device with InAs/QDs on GaAs/InAlAs/InP and 10/30 nm two-step barrier growth. (a) Bright-field image shows overall structure; (b) dark-field image shows overall structure; (c) magnified (200) dark-field image of the first few layers.

AFM images of the QDs in our device are shown in Figure 12.1. The dot density is low, ~1 × 10⁹ cm⁻². From AFM imaging experiments, the dots are approximately 6 nm in height and 60 nm in diameter, though these sizes may be overestimated due to tip effects. These dots are larger, however, than those in our previous reported devices [9].

Note that the QD images in Figure 12.1 are AFM scans. Recent TEM pictures have given clear evidence of a WL of InAs of 1 nm thickness on top of InAlAs, but have not produced clear evidence of well-formed QD layers, so the final structure characterization is still a matter of investigation and the work on improving dot formation is in progress.

The reader is also referred to, as an example, the important work of Chakrabarti et al. [11], who have grown InAs QD on GaAs reaching densities of 5×10^{10} cm⁻² for the QD multilayers, and in another realization, using InAs on AlAs, they have made superlattice QD in the well structures, with QD densities near 10^{12} cm^{-2}.

The first question is what is special about QD? The answer is that by growing arrays of QD, one has arrived at one of nanotechnology's prime targets, which was to make three-dimensionally confined systems, which exhibit quantized energy levels. From the particle in a box problem, we recall that energy scales as the inverse effective mass, so the low carrier mass of III–V compounds makes the energy resolution more effective in particular for electron doping. If we can put charge into the QD we can therefore (1) absorb light at well-defined wavelengths, (2) emit light, and (3) emit a charge out of the QD into the barrier layer, and thus generate wavelength-selective photocurrents; see Figures 12.3 and 12.4 for typical quantum-dot–infrared-photodetector (QDIP) device structures. If one is planning to make wavelength-selective photodetectors,

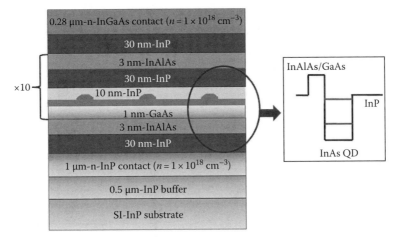

FIGURE 12.3 Illustrates the midwavelength-IR–QDIP device structure of Figure 12.2 grown with two-step barrier growth and InAlAs CBLs. The 10 nm InP capping layer was grown at the same temperature as the QD growth temperature 440°C. The 30 nm InP barriers and 3 nm InAlAs CBL were grown at the higher temperature of 590°C. Inset shows the schematic conduction band alignment.

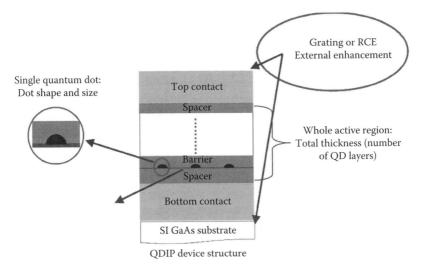

FIGURE 12.4 The typical QDIP device structure used for photodetection. The electron doping (Si) is in the wetting layer or in the barrier layer.

the QDIP, one would regrow a barrier layer on top of the QD layer, regrow the same structure again N-times, and include doped contact layers as shown in Figures 12.3 and 12.4.

12.2 The Quantum Mechanics of Quantum Dots

The energy levels of the localized states inside the QD depend on the geometry and the materials. A typical example with two distinct sphere section geometries and with height H over diameter $2R$, $H/2R$ ratios are shown in Figure 12.5.

The oscillator strength f_{ij}, which is a measure of the absorption strength of a particular optical dipole transition, is denoted by {a, b, c, etc.} in Figure 12.5. Here Ψ_i are the bound QD wavefunctions, E_i are the energy levels, and m^* is the effective mass so that

$$f_{fi} = \frac{2m^*}{\hbar^2}\left(E_f - E_i\right)\left|\left\langle \Psi_f \left| x \right| \Psi_i \right\rangle\right|^2 \tag{12.1}$$

And the absorption efficiency of a transition is measured by

$$\alpha_{fi} \propto \left(n_i - n_f\right)f_{fi} \tag{12.2}$$

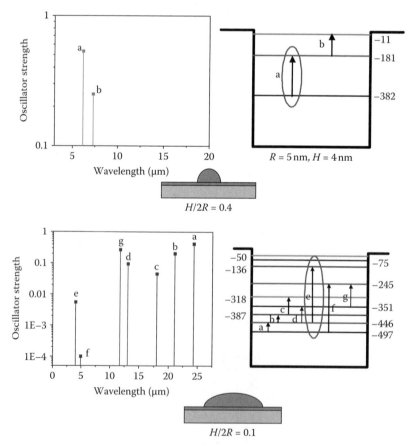

FIGURE 12.5 Calculation of energy levels and oscillator strengths for InGaAs dots in InGaP barriers. The dots have different aspect ratios with radius R and fixed height H.

The reader will note by looking at Figure 12.5 the extreme sensitivity of the energy level structure on the QD size and geometry. One also notes that for the more realistic QD size and geometry shown in Figure 12.5, the number of bound levels in the QD is very high indeed. In Figure 12.5, the energy levels and wavefunctions were evaluated under the assumption that the QD is embedded in a larger cylinder [7,12], of which we know the exact eigenstates analytically. These states are then chosen as the basis states for the QD wavefunctions. It is further assumed that the cylinder that represents the barrier layer has its own effective mass. The advantage of this technique is that one does not have to worry about the matching wavefunctions at the interface boundaries. One has to solve an $M \times M$ matrix eigenvalue problem with a basis that is as large as possible but typically 700 wavefunctions. The QD eigenstates, Ψ, are then linear combinations of the large cylinder eigenstates as

$$\Psi_m(E_s) = \sum_{n,l} a_{n,l}(E_s) J_m(K_{mn}\rho) e^{im\phi} \sin\left(\frac{l\pi z}{L}\right) \tag{12.3}$$

where
the $a_{n,l}(E)$ are the admixture coefficients
E_s are the new eigenvalues
J_m are the Bessel functions with ρ denoting the base radial coordinate of the cylinder
m is the eigenvalue of angular momentum in the plane
l is the index of the confined eigenstates in the z-direction within a cylinder of height L

The energy levels of the basis set are defined by the eigenvalues of $J_m(RK_{mn}) = 0$, where R is the big cylinder radius. The total energy of an eigenstate in the big cylinder with effective mass m^* is given by

$$E_{n,m,l} = \frac{\hbar^2}{2m^*}\left[\{K_{mn}\}^2 + \frac{\pi^2 l^2}{L^2} \right] \tag{12.4}$$

The large cylinder is the barrier layer, and the small dot cylinder is embedded in it, and can have a different effective mass. The change in mass is assumed to be abrupt. But if we know the local variation and strain distribution as discussed in Ref. [7] for example, it is in principle a simple matter to upgrade the calculation. Note that due to the rotational symmetry of the "ideal lens," we can assume that the exact eigenstates can be classified using the angular momentum index m (m states > 0 are twofold degenerate $= +, -$), and this considerably simplifies the problem. An optical transition with s-polarized light is only allowed between states differing in "m" by one unit.

Another way is to compute the wavefunction directly as the solution of the one-band effective mass Schrödinger equation and allow the effective mass to vary as one crosses from one material into the other, and in principle also with energy, as prescribed by the effective one-band Kane Hamiltonian; see Figures 12.6 and 12.7. We have applied this technique using the FEMLAB commercial package to the system shown in Figure 12.7 [10,13]. This now consists of a "hybrid QD in the well" system with the InAs QD grown on top of an InAlAs barrier. The QD are capped by InGaAs quantum well (QW), which is itself covered again with an InAlAs barrier to form the repeat unit, see Figure 12.6.

The novelty of this device, quite apart from the dot in the well structure, which in itself is nowadays not unusual [14,15], is the fact that the band offset between the QD and the QW layer on top is very shallow, ~130 meV. Since the electron effective mass, m^*, of InAs is very small ~$0.043m_0$, this makes the hybrid QDWIP, a system with a single three-dimensional localized bound level inside the QD. The localized level is bound relative to the QW band, and with the parameters shown the bound state energy is only ~34 meV for a 0.6 nm WL.

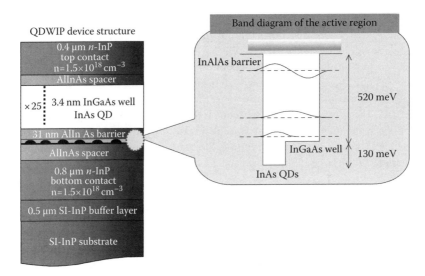

FIGURE 12.6 The "hybrid QD in a well device" structure called "QDWIP" [13]. The Si doping is in the InGaAs QW.

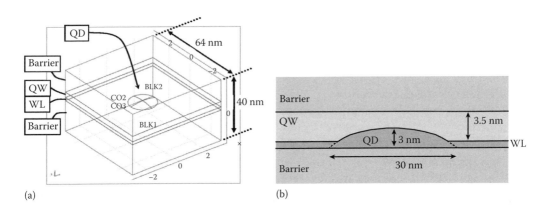

FIGURE 12.7 The QD geometry in the QDWIP. The left image (a) is a schematic picture of the geometry for the energy level calculation, and the right (b) is a cross-sectional drawing around the QD.

The computed wavefunctions and energies are shown in Figure 12.8 and Table 12.1 for a number of different assumed WL (InAs) thicknesses.

In this chapter, we used COMSOL Multiphysics®, a finite element method software package, to calculate the energy levels in our QDWIP structure. We note that the second excited state is already delocalized inside the region adopted for the computation. Whereas Figures 12.2 and 12.3 are examples of conventional QD and QDIP structures, the QDWIP structure shown in Figure 12.6 is a more unusual "hybrid" QDWIP structure because the Si dopant electrons added to the QW, will, at the dopant densities of interest for photodetection ~10^{18} cm^{-3}, occupy both the QD states and the higher QW band; see Figure 12.9. This figure is a schematic representation of what one expects the density of states to look like in a hybrid with $T = 0$ Fermi level in the impurity band.

These two examples shown in Figures 12.4 and 12.6 allow us to discuss a number of established scenarios of QD photodetection. In the past, many workers in this field have used the 8-band k.p method or ab initio methods, including strain for maximum accuracy [16]. The 8-band k.p method is based on the logic that the QD perturbation is an additional source of band mixing and that this mixing

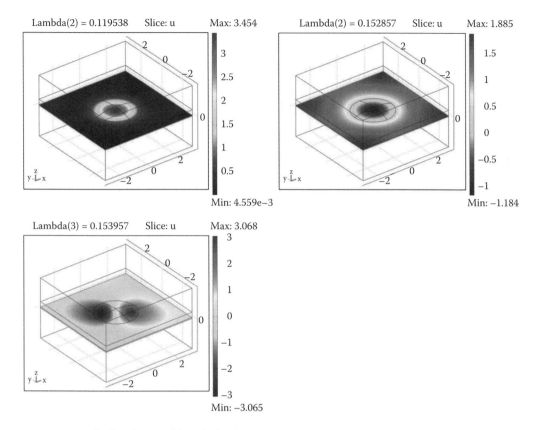

FIGURE 12.8 The distribution of the calculated wavefunction for the first three eigenstates in the QDWIP for the WL thickness 0.6 nm.

therefore takes into account the effective mass corrections caused by the presence of the QD potential. Working with one effective mass band only can therefore lead to errors, but these errors can be small compared to the inaccuracies in the size and shape of the QD. The full 8-band k.p method is appropriate when one is sure about the geometry of the QD. When methods are complex and fully numerical, it is not easy to build in the inevitable statistical size and shape variations.

TABLE 12.1 QD Binding Energy for the Different WL Thicknesses

WL Thickness (nm)	QD Binding Energy (meV)
0.6	33.3
0.8	28.3
1.0	23.4

12.3 Quantum Dots as Detectors of Long-Wavelength Light: Why Use QD for Photodetection?

The question now is this: what is the advantage of using QD arrays and multilayers for detecting light as shown in Figures 12.4 and 12.6? One advantage over the conventional bulk semiconductor is the sharp wavelength selectivity in contrast to the cut-off at the band gap. Remember that the density of states here is a sum of delta functions. Unfortunately one cannot completely select the wavelength because one has no way at present of choosing the exact geometry of the QD. The other advantage is that one can detect long-wavelength radiation down to 20 μm, which is difficult to do with bulk semiconductors if one wants to keep the dark current and noise level low [1–8]. Since noise grows with dark current (shot noise and generation–recombination (GR) noise), and conductivity (Johnson noise), it is essential to keep the dark currents and conductivities low.

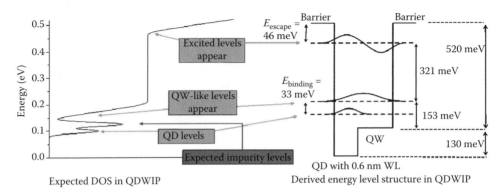

Expected DOS in QDWIP   Derived energy level structure in QDWIP

FIGURE 12.9 A schematic representation of the QDWIP density of states [13]. The QD layer only contributes 10^{10} cm^{-2} shallow bound states compared to 10^{12} cm^{-2} from the Si dopant and 10^{14} cm^{-2} from the QW subband. The QD potentials and dopants will enhance the strong Anderson localization at the lower band edge that will now extend into the first QW "subband" up to the mobility edge which is at about 5×10^{12} cm^{-2}.

Wavelength selectivity with more control over the dark current can also be achieved using the so-called superlattice QWIP technology [17]. This technology is very successful at low temperatures up to 80 K, but again the dark current (noise) level tends to be too high as one goes up in temperature [17–19]. This is now where the QD and thus QDIP comes into play. The QD absorption can, as mentioned above, be wavelength selective, the dark current can in principle be designed to be lower because the QD electron density is lower than the total ground state band occupation density in the QWIP. But one has to be careful, if one reduces the number of QD, one also reduces the number of absorbers and therefore the photocurrent. The full advantage of not having the entire space filled by QD comes into play when we consider the consequences a low filling factor has on capture/recombination and thus on the gain. Once the carrier is photoexcited out of the QD, and if the mobility is high in the barrier layer, and there is no impediment to reinjection of charge at the contacts, the carrier can go around the circuit many times before it is captured again by a QD as shown schematically in Figure 12.10. This enhances the photocurrent yield per absorbed photon: the photoconductive gain.

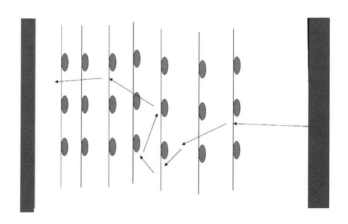

FIGURE 12.10 Illustrates a carrier trajectory through the QD multilayer of a QDIP, showing that in principle the carrier can cross without being captured. Note that in contrast in the QDWIP of Figure 12.6 or QWIP, the carrier has to cross QW/QD regions and will have a much higher capture probability and thus a much lower gain $g < 1$.

12.3.1 Transport in the Growth Direction: The QDIP

The left and right rectangles in Figure 12.10 are the contacts. The carrier is far more likely to be able to cross the QD multilayer without being recaptured in a QDIP device as shown than in a QDWIP or simple QWIP multilayer of Figure 12.6 where the crossing of QW regions is inevitable. In the QDIP, the thin WL is not an effective trap. One can see how there can be a positive trade-off between not filling the entire layers with absorbers, which lower the photo-response, but increasing the photocurrent per carrier because of low capture rates, which is beneficial. The high gain, which is defined as the ratio of capture/recombination time to barrier-transit time, is possible when the mobility in the barrier layer is high, there are few traps, and the reinjection from the electrodes is efficient. The formal theory for the gain is developed in Refs. [20,21] for the QDIP, and the QWIP, where the reader can find all the mathematical details he needs. Basically for QDIPs, what we do is to solve the diffusion equation in the presence of random traps that we assume are spheres of finite radius.

12.3.2 The Diffusion–Recombination Equation: Gain and Photocurrent Relaxation

Consider an electron moving in a band with effective mobility μ and diffusivity D in the presence of an applied field F_x, where $\mu = eD/kT$, in the presence of a distribution of spherical QD [20]. The motion, assumed for simplicity to be isotropic, can be described by the diffusion equation with recombination centers. The trapping and release can be included in the definition of an effective diffusivity $D = D(T,V)$.

$$\frac{\delta n(r,t)}{\delta t} = G_g(r,t) + D\nabla^2 n(r,t) + \frac{\delta n(r,t)}{\delta x}\mu_b F_x - \sum_i n(r,t)V(r - r_i) \qquad (12.5)$$

where
  $n(r, t)$ is the band occupation density at time "t" at point "r"
  $V(r - r_i)$ is the capture rate due to the ith QD at r_i

The generation rate is denoted by G_g. The external field is F_x, and μ_b is the band mobility. For a single carrier created in the band at time $t = 0$ at $r = 0$, $G_g = \delta(r, t)$, the solution of the above equation can be thought of as the Green function of a Schrödinger equation with $V(r)$ being a scattering potential. If the QD were uniformly distributed on a lattice, we would have a standard band structure problem to solve with $V(\vec{r}) = \sum_i V(\vec{r} - \vec{r}_i)$ acting as the periodic potential. For spherical symmetric trapping rates the problem has been solved by Zhao, Ghosh, and Huber in the so-called average t-matrix approximation [22] which is good enough for three-dimensional motion. Even though our QD are in general not spherical, we will assume for the present purpose that they are, and choose the radius of the QD so that it occupies the same volume as the true QD. The full result with various spherical trapping potential models is given in Ref. [20]. For traps of range R_t with capture rate strength V_t, the full result of the trapping rate C_{be} for an isotropic three-dimensional system with trap (QD) concentration N_d and radius R_t is

$$C_{be} = \frac{1}{\tau} = N_d\left\{4\pi DR_t\right\}\left[1 - \left(\frac{D}{V_t R_t^2}\right)^{1/2}\tanh\left\{\frac{V_t R_t^2}{D}\right\}^{1/2}\right] \qquad (12.6)$$

$$V_t(r) = V_t \to r < R_t; \quad V_t = 0; r > R_t$$

The effect of the external electric bias field in three-dimensional motion is not important until the field reaches very high values at which drift dominates over diffusion, when $qFa/kT \geq 1$ where a is the lattice constant; see Refs. [23,24]. This is not the case in the present system. One of the results of the theory is that the time decay of the photoconduction process is basically exponential. Nearly exponential decay has indeed been observed with QD layers using femtosecond transient THz absorption spectroscopy [25–27]. The low-field time decay of the band photocurrent $I_{QD}(t)$ due to QD capture is thus

$$I_{QD}(t) \sim qN_b^0 \mu_b(V)F \frac{1}{\left[1+\left(\dfrac{\mu F}{v_s}\right)^2\right]^{1/2}} \exp\left[-tN_d\left\{4\pi DR_t\right\}\left\{1-\left(\frac{D}{V_tR_t^2}\right)^{1/2}\right\}\tanh\left\{\frac{V_tR_t^2}{D}\right\}^{1/2}\right] \quad (12.7)$$

where

    v_s is the saturated velocity in the barrier continuum
    N_b is the free carrier density
    q is the magnitude of the charge

Other random recombination or trapping processes, if any, would also usually be exponential, and would have to be combined with the QD trapping time to give a total lifetime. In a fast photocurrent relaxation experiment, one would expect the QD trapping and WL trapping to constitute the fastest trapping, or initial decay events. The WL energy state is normally very shallow and carriers de-trap quickly with temperature, so the carrier is back in action again until it really recombines into the deep QD levels in the QDIP in the longer time domain. The gain calculations in [20] show that in the ideal situation, one can get a gain of up to 500 but a number near 50 is often more realistic. We shall see later how the gain enters the performance of the QDIP.

Going back to the problem of QD density, we note that one can reduce the dark current without reducing the QD density and thus keep the absorbers density high, as shown by Bhattacharya et al. [28]. They used a very neat trick by building-in a resonant tunnel barrier into the escape process. Since the escape is resonant with the photoexcitation step, the dark current also has the same narrow limited window at its disposal. The resonant tunnel step limits the spectral width of course too, which is useful if one is making a wavelength-selective system. Temperature, however, strongly detunes the resonance, and the QDIP in Ref. [28] remains therefore, despite this innovation, a low temperature detector. Including a narrow adjacent well, as in the work of Ref. [29], could also do the trick, and this technology would not restrict the spectral region as much. We are studying these interesting innovations ourselves in our group presently.

Finally there are two more potential advantages of using QD. The first one is the fact that the photoexcited state in the QD (see Figure 12.5) can have a longer lifetime than the corresponding photoexcited state in the QW. The reason is that there are far fewer states below the excited state in a QD than there are in a QW. Whereas in the QW the recombination lifetime is never much more than a few ps, in a QD, if the next level down is a long way in energy compared to an optic phonon, the lifetime can be up to 2 orders of magnitude longer than in a QW excited state. This is true in the example shown in the top part of Figure 12.5. The small more round QD in Figure 12.5 is therefore a much better system for QDIP application. Unfortunately, this is not so for the more realistic QD depicted in the lower part of Figure 12.5 made using InP technology. With low aspect ratios and large diameters, there are many eigenstates in the QD, and the recombination is fast again. But with GaAs technology, Chakrabarti et al. [11] have made smaller QD with higher aspect ratios, which should, in principle, exhibit about an order of magnitude longer lifetimes than in the QWIPs. The final, and perhaps the most important, advantage of the QDIP and the QDWIP over the QWIP is the fact that normal incidence light can be absorbed in a QDIP with much higher efficiency than in a QWIP. In a QWIP, the normal incidence light can only be absorbed in a z-intersubband transition when there are breaks in the plane (x, y) to the (z) symmetry. This occurs at impurity dopant sites, defects, and at the interfaces both because of roughness and because of the Bloch bands. The absorption efficiency relative to that of z-polarized light is, in a good quality layer, less than 5% as shown by the work of Gunapala et al. [30]. In a QDIP, the symmetry is broken as a result of the QD potentials which cause both light scattering, i.e., polarization mixing, and QD interlevel transitions. Again one has to be careful here, because if the QD are very flat, and have low aspect ratios, the relevant oscillator strengths for the photodetection process can be very low, and not much is gained by using QD; see Figure 12.5.

In summary, we can conclude that the QDIP technology really comes into its element for photodetection, when the following conditions are satisfied:

1. There is a high density of QD, if possible >10^{11} cm^{-2}, with good uniformity.
2. The QD geometry is such that there are only a few bound states in the QD with the excited state *close to the continuum*, and linked to the ground state with a reasonably high oscillator strength ($f > 0.1$) in normal incidence. This also lengthens the recapture/recombination time and helps the gain.
3. If possible one should be able to engineer the escape barrier to filter (reduce) the dark current as in Ref. [11] or [29]. Filtering will also narrow down the photocurrent spectrum. Thermal detuning can be taken care-off by design.
4. The mobility in the barrier layer should be high, few traps low 1/*f* noise, to maximize the gain.

Adding dark current blocking layers (CBLs) can help when the noise is dominated by non-GR processes such as hot carrier noise (optic phonon emission, intervalley crossing, and intersubband avalanching).

12.3.3 Modeling the QDIP Performance

Now let us express the above in mathematical language. Let us start with absorption. The absorption coefficient, α_{eg}, for the selected transition "ground 'g' to excited 'e'" is given by

$$\alpha_{eg} = \frac{\hbar\omega n_r N_d}{\varepsilon\varepsilon_0 c}\frac{2\pi}{\hbar}\left|\langle g|qx|e\rangle\right|^2 \frac{(n_g - n_e)\Gamma}{\hbar^2(\omega - \omega_{ge})^2 + \Gamma^2} \tag{12.8}$$

$$\alpha_{eg} = \frac{\pi\hbar n_r N_d q^2(n_g - n_e)\Gamma}{m^*\varepsilon\varepsilon_0 c[\Gamma^2 + \hbar^2(\omega - \omega_{eg})^2]} f_{eg} \tag{12.9}$$

where
  Γ is the transition linewidth
  N_d is the QD density
  q is the magnitude of the electric charge
  c is the velocity of light
  n_r is the refractive index in the medium
  ω is the light frequency
  ε is the permittivity
  n_{eg} are the occupation probabilities
  f_{eg} is the corresponding oscillator strength

This expression has to be integrated over a Gaussian distribution of transition frequencies caused by the QD size distribution.

Once excited to "e," the carrier can recombine back down again in the lower QD levels with rate v_{eg} or it can tunnel and thermally escape into the continuum with rate $v_0 \exp[-E_{eff}/kT]$, so that the photocurrent is given by the number of excited photoelectrons in "e" multiplied by the rate of emission into the continuum divided by the capture rate from the barrier continuum into the QDs excited level "e" C_{be}. After some straightforward algebra the responsivity, R_{eg}, becomes

$$R_{eg} = g \frac{q(1 - \exp(-\alpha L))}{\hbar \omega} \left\{ \frac{\nu_0 e^{-E_{eff}/kT}}{\nu_{eg} + \nu_0 e^{-E_{eff}/kT}} \right\} \tag{12.10}$$

$$g = \frac{\mu F}{L C_{be}} \tag{12.11}$$

where α is the effective absorption coefficient, $\alpha_{eg} = \alpha \dfrac{L}{L_a}$ is the active layer absorption coefficient, i.e.,

where L_a is the width of the active region and L the total device length; where also C_{be} is the capture/recombination rate from the continuum to the ground state, and $E_{eff}(F, T)$ is an effective activation energy which characterizes the tunnel/thermal escape rate from the excited state into the continuum; μ_b is the band mobility and F the applied field; L is the length of the device. The capture rate, C_{be}, is only rate determining for recombination from the continuum if the re-escape rate out of the QD is slower than the next recombination step back down. If it is faster than the capture process, it is itself only a trapping process and contributes to the lowering of the overall mobility. In this case, the C_{be} in (12.11) should be replaced by the recombination rate from "e" to the ground level "g"; see Ref. [20c] for an example where this happens at high enough temperatures. We have for simplicity assumed that there are only two levels in the QD. The generalization to allow the multilevel structure is of course straightforward. The internal quantum efficiency (QE) is defined as

$$\eta = \alpha_{eg} L_a \left\{ \frac{\nu_0 e^{-E_{eff}/kT}}{\nu_{eg} + \nu_0 e^{-E_{eff}/kT}} \right\} \tag{12.12}$$

From Equation 12.10 it immediately follows that for a good QDIP responsivity R, the absorption coefficient should be high and the recombination rate ν_{eg} should be slow. The absorption is high when the density is high, and the oscillator strength in normal incidence is high. The excited state lifetime is long when the phonon bottleneck is operating, i.e., when there are very few levels to recombine with, in other words, when the QD size is small and the aspect ratio is high; see Figure 12.5 and the QD in Ref. [11]. From Equation 12.12 it also follows that the gain should be as high as possible. This implies high barrier layer mobility and slow capture/recombination rates. All these features point to having preferably small QD with high aspect ratios, hemispherical QD with 10–15 nm diameters are ideal. A typical photoactive spectrum for the device in Figure 12.3 is shown in Figure 12.11.

One can see that the photoaction spectrum in Figure 12.11 is not as narrow as one would have expected it to be with a single QD shape. The inevitable size distribution that depends on material and growth conditions gives the spectrum a Gaussian inhomogeneous broadening. But photodetection also needs low noise, and that means low dark currents, and low conductivity. Now consider the calculation of the dark current. In evaluating the dark current, we need to take into account two contributions: (1) I_b, the term which is due to carriers excited from occupied levels to the barrier continuum. This term is characterized by the density of states, ρ, with Fermi function, $f(E)$, and continuum band edge, E_c.

$$I_b = qA \int_{E_c} dE \rho(E) f(E) \mu F \tag{12.13}$$

(2) I_{QD}, the contribution from the occupied ground state of the QD to the continuum via the QD excited states; see Figure 12.12.

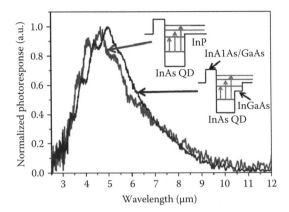

FIGURE 12.11 Normalized spectral response measured by FTIR at 77 K and a bias of 0.4 V for the device with well-formed QD of Figure 12.2. The lighter curve represents the response from an InAs/GaAs/InAlAs/InP-QDIP. The darker curve represents the spectral response from a device with the same structure except that 3 nm InGaAs capping layers have been added on top of the InAs layers.

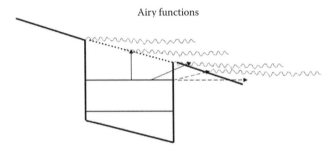

Airy functions

FIGURE 12.12 Illustrates schematically the excitation sum for the escape of the carriers from a QD excited state to the continuum states that are in principle Airy-like functions. A similar escape sum process also applies to the lower levels.

The sum of the escape rates in Figure 12.12 can be conveniently expressed as [7]

$$I_{QD} = \frac{AqN_d}{C_{be}} v_d(F,T) \left[\sum_s \left(\frac{v_s g_s^0 f_s}{1 + \frac{W_{sc} f_s}{C_{be}}} \left\{ \frac{e^{-E_{sc}/kT} - e^{-\varsigma E_{sc}^{3/2}/qFa} e^{-\varsigma E_{sc}^{1/2}} e^{qFa/kT}}{1 - e^{-\varsigma E_{sc}^{1/2}} e^{qFa/kT}} \right\} \right) \right] \qquad (12.14)$$

$$\varsigma = a \left(\frac{2m_e^\star}{\hbar^2} \right)^{1/2} \qquad (12.15)$$

where
  A is the area
  f_s is the Fermi function for QD energy level "s"
  g_s is the degeneracy of the level
  v_{sc} are the escape attempt frequencies
  E_{sc} is the escape energy difference from "s" to the continuum "c"
  a is the lattice constant
  v_d is the drift velocity in the band
  W_{sc} are themselves the combination escape rates from "s to c" summed over the Airy states

For example from "e to c" we have

$$W_{ec} = v_{ec}e^{-E_{ec}(F)/kT} = v_{ec}g_e \left\{ \frac{e^{-E_{ec}/kT} - e^{-\varsigma E_{ec}^{3/2}/qFa}e^{-\varsigma E_{ec}^{1/2}}e^{qFa/kT}}{1 - e^{-\varsigma E_{ec}^{1/2}}e^{qFa/kT}} \right\} \tag{12.16}$$

Normally, the ratio in the denominator of Equation 12.15 is negligible compared to 1. Finally, we must remember that the total dark current, I_D, contributes to the GR noise $I_n(GR)$ via

$$I_n^2(GR) = 4qg\Delta v I_D \tag{12.17}$$

where Δv is the noise bandwidth.

The calculation is performed using a density of states model as shown in Figure 12.9. The QWIP and QDWIP differ basically only via the energy that separates the subband excited state to the continuum. The additional InAs WL in the QDWIP is the main difference to what would be the simple InGaAs-QWIP. The QD layer is sparse and shallow and has almost no effect on the energy spectrum in this technology.

As shown in Figure 12.13, the dark current modeling deviates from the experimental data at low temperature and bias. This indicates that we have an unaccounted source of current leakage, possibly via defects. The contribution from these defects only dominates at low temperature and bias, and the modeled tunneling and band contributions dominate at high temperature and bias. However, the defect-related leakage could still produce the significant excess $1/f$ noise signal that we have observed in our measurements [9,13].

The final figure of merit for a photodetector is the detectivity defined as [1,10]

$$D^* = \frac{R\sqrt{A\Delta v}}{I_n} \tag{12.18}$$

where A is the area, and is measured in units of Jones. An example of the behavior of detectivity and dark current with bias for the device of Figure 12.3 is shown in Figures 12.14 and 12.15.

The noise entering Equation 12.18 is the actual measured noise. One has to be careful, because the measured noise in a given temperature and bias range is not necessarily dominated by the GR noise of Equation 12.17. The noise analysis in QDIPs and QWIPs is not a trivial exercise, and has to be done with great care each time a device is assessed. If the noise, as is often the case, is mainly due to GR, then it follows from Equations 12.18 and 12.17, that the performance is enhanced only by the square root of the gain.

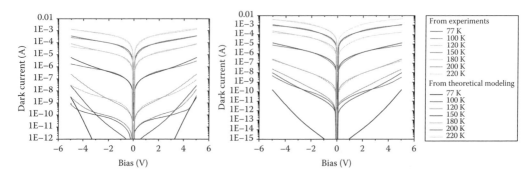

FIGURE 12.13 Fitting of dark current in a QWIP (left) and a QDWIP (right) in the hybrid QDWELL InAs/InGaAs technology of Figure 12.6. Solid lines are the calculated curves and dashed lines are the experimentally obtained dark current curves. The QD contribution to the overall density of states is insignificant; see Figure 12.9.

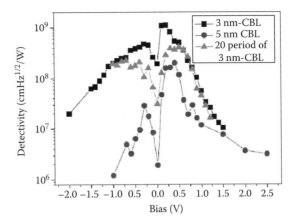

FIGURE 12.14 Comparison of detectivities for the device with well-formed QD in Figure 12.2 at 77 K and with InAlAs CBL. The photospectrum is given in Figure 12.11.

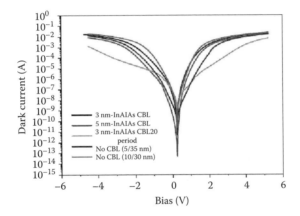

FIGURE 12.15 The dark current in the device of Figure 12.3 at 77 K with well-formed QD shown in the TEM picture Figure 12.2 and with InAlAs CBLs of differing widths and repeat period. The aim is to lower the dark current.

12.3.4 Photodetection: Discussion

Now given these formulae, we can assess the quantities that matter for photodetection using QD. Start with the responsivity R given by Equation 12.10 for the QDIPs. Clearly as mentioned already, it is desirable to have high absorption in the photoactive layer which means high densities of QD with high normal incidence oscillator strengths. Thus small hemispherical and densely packed QD are optimum. The QD layers fabricated by Chakrabarti et al. [11] with InAs/GaAs technology have a very good geometry with ~17 nm width and 8 nm height but the density is only moderately good at 5×10^{10} cm^{-2}. The highest density QD system so far seems to be the superlattice QD in the well structure fabricated with GaAs technology by Chakrabarti et al. [11]. These authors have achieved the remarkable density ~10^{12} cm^{-2} so that with two electrons in the ground state one has as many absorbers as in a QW with 10^{18} cm^{-3}, and the bonus of normal incidence response; but here the motion is in a miniband and the noise turns out to be higher. In QDIP technology, the gain is in general high in QDIPS ($g \gg 1$) because the WL is normally too thin to act as a serious trap, and the effective capture rate from QDs is much smaller than in QWIPs where the gain, g, is <1. The higher gain is the main commonly observed feature in the various QDIP technologies. It can allow a mediocre absorption coefficient of $\alpha \sim 10^2$ cm^{-1} to still give reasonable values of R [7]. In the quantum dot in the well (QDWELL) technologies [14,15], the gain is comparable to that of the QWIP [17,18] and in general $g < 1$ as one would expect. The benefit of using the QDWELL is twofold: (1) it makes a QWIP but with higher normal incidence absorption without having to use a grating and (2) the QW layer can be used to tune the absorption wavelength and possibly makes multicolor cameras. The QD is also expected to lower the dark current when compared to the QWIP. Adding the QW layer as in the QDWIP hybrid of Figure 12.6 gives one an extra engineering tool to adjust the barrier height and the transition wavelength. This is not possible with pure QD self-assembly. Combining QD technology with QW allows one to extend the fabrication design to also allow for a resonant tunneling escape window for high temperature operation as done in Ref. [28]. One can lower the dark by adding resonant sidewells which have a smaller effect on the responsivity as done in Refs. [29,30]. Simply adding barriers or reducing the doping is in general not particularly helpful as it also reduces the responsivity. It is useful when the dark current barrier-free device is not GR-noise limited. Then the addition of an extra barrier can reduce non-GR noise sources. In practice one often finds that the peak responsivity occurs at biases that are high for best focal plane array (FPA) operation. Here, the strategy of using a cascade photovoltaic QDWIP structures with the QDWIP photoactive units built as in Figure 12.6 could and should be tried. The challenge is to arrive at a good-enough material quality. This has been done successfully using QWIPs by Giorgetta et al. [31], the so-called "QCD."

To come back to the normal QDWIP made at Northwestern by our group, we should point out that the advantage here is the fact that with high doping, the majority of carriers are in the QW bands (see Figure 12.9), and they all contribute to the absorption as in a QWIP. The difference is that now the QD corrugation (1) acts to improve the light scattering, (2) enhances the Anderson localization of the QW band states so that the electrons in the ground levels are mostly in localized states below a pseudo-mobility edge, with not too large localization lengths, i.e., comparable to the decay lengths of the 10^{18} cm^{-3} hydrogenic dopant levels in the QW layer [30]. Even with temperatures up to 150 K, a substantial portion of carriers will still be in localized states, and these can absorb normal incidence light with an absorption coefficient of up to 10^3 cm^{-1}. This is still an order of magnitude smaller compared to z-polarized light absorption in QWIPs, but it is possible to image with a noise equivalent difference temperature (NEDT) of 600 mK at 130 K as shown in Figure 12.16. It is interesting that the NEDT actually decreases with temperature at first; this is unique to the QDWIP and not seen in the QWIP. Better light coupling scenarios should enhance the QDWIP performance considerably. This work is in progress.

In Figure 12.16, one can see how the hybrid device recently made in our group at NWU [9,13] using the shallow QDWELL technology of Figure 12.6 performs as a camera which is the final aim of the exercise.

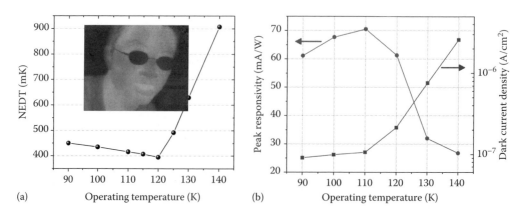

FIGURE 12.16 (a) NEDT as a function of operating temperature with the inset showing a representative FPA image taken at an operating temperature of 120 K for the hybrid device of Figure 12.6. Remember that in this device, the QD are essentially part of the 1 nm InAs wetting layer; (b) the peak responsivity and dark current density of the FPA as a function of operating temperature.

The NEDT is the measure of FPA performance and is defined as the noise equivalent difference temperature and is shown in Figure 12.16 as a function of operating temperature. The NEDT gives us the limit of resolution of temperature differences when viewing an object. It is noteworthy that the performance of the camera actually increases first with temperature (lower NEDT) and then goes up above 120 K [9].

12.4 Some Other Important Applications of QD

12.4.1 The Lateral QD Photodetector: n- and p-Doping

In the lateral photoconduction mode shown in Figure 12.17, the idea is to excite the carrier from a QD layer into the fast transporting channel adjacent to the photoactive layer. This will maximize the gain. The gain is high because the recombination back down to the photoactive layer can be engineered to be very slow. These devices have been made both with electron doping [32] and even recently with hole Be doping of InAs [33]. The responsivity at normal incidence is indeed very high reaching values of ~16 A/W (p-doping at 10 K), and 20 A/W (n-doping at 60 K), which is much larger than in the conventional z-QDIP systems. Hole doping QW or QD photodetectors are very rare, but a hole QWIP had previously been reported by Levine et al. [17]. The advantage of hole doping is that the matrix element for intersubband normal incidence absorption is less restrictive than for electron doping (see Levine's review [17]). The disadvantage is that the hole mass is high, so that tunneling out of the excited state into the continuum is more difficult. The problem with lateral-QDIP technology is how to restrict the dark current, because the charge is sitting in the photoactive layer and can move in the plane. Electrons can be excited thermally out of the QD into the quasi-free QW bands. The barrier is therefore the binding energy of the highest occupied level in the QD with respect to the QW band. One can envisage further improvements of this technology but these are not simple to implement. One way is to reduce the lateral dark current by enhancing the in-plane localization of the QW bands by alloying and defect control. This should work well with the heavier mass GaN technologies. Novel masking techniques could be introduced, such as the use of patched selected area doping in the plane. Lateral structuring of QW/QD layers is challenging and could constitute the next step forward. This technology could prove most valuable in making these very high responsivity systems work efficiently also as photodetectors.

12.4.2 QD Semiconductor Hole-Mediated Ferromagnetism

Mn doping of InAs QD layers embedded in GaAs has raised the ferromagnetic Curie temperature, T_c, of the multilayer stack to 343 K higher than the 150 K obtained in 15 nm thick Mn doped layers [34], and the 80 K of the bulk material. The Mn incorporation releases a weakly bound hole, which is antiferromagnetically coupled to its Mn spin, and which can transfer the antiferromagnetic spin information from Mn to Mn "magnets" and in this way generate ferromagnetic alignment of the much higher Mn (5/2) spins. The exciting feature is that the carrier (hole) confinement appears to help the hole-mediated ferromagnetic coupling of the Mn spins by increasing the local stay probability on the Mn spins. This discovery on its own is already very interesting. The next step forward is to

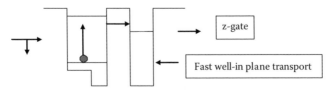

FIGURE 12.17 Schematic diagram of the concept of the lateral photoconductor. The photoexcited charge leaves the QD/QW complex and ends up in the fast lateral channel from which it is difficult to recombine back again. The uncanceled charge draws in more charge until there is recombination.

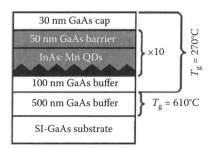

FIGURE 12.18   Schematic heterostructure of a typical InAs; Mn QD multilayer sample grown by low temperature MBE. (Reprinted from Holub, M. et al., *Appl. Phys. Lett.*, 85, 973, August 9, 2004. With permission. American Institute of Physics.)

study the dark and photo-magneto-transport of these devices that may even have serious potential for magneto-IR applications. One should start by measuring the in-plane and z-dark-magneto-resistance, and indeed even the photo-magnetoresistances. The above lateral-QDIP technology in Figure 12.17 should be extended to include Mn hole doping instead of Be hole doping. Since the doping in semiconducting magnets is high (5%), one would want to focus for optical applications, in the lateral devices, on the lowest doping that still gives ferromagnetism and also insulating transport behavior [35]. In the conventional z-QDIP functionality, the physics of ferromagnetism and transport is quite different, and the discussion is beyond the scope of this chapter. Selected area doping could be used to generate Mn doped ferromagnetic "areas" large enough to sustain ferromagnetism, which would lead to the coexistence of ferromagnetism and low in-plane conductivity. But fabricating such devices is a challenging program (Figure 12.18).

12.4.3 Photoluminescence from Quantum Dots

When light can be absorbed, it can also, by the reverse process, be emitted. The rate at which an electron in the excited state "e" returns back to the ground state "g" and emits a photon of energy is given, in the dipole approximation, by the usual golden rule expression

$$W_{eg}(L) = \frac{2\pi}{\hbar}\left|\left\langle e\left|\vec{\varepsilon}_p \cdot \vec{r}\right|g\right\rangle\right|^2 |qE_L|^2 \, \delta\left(E_e - E_g - \hbar\omega\right) \tag{12.19}$$

where $\vec{\varepsilon}_p$ is the polarization vector of the light emitted, q is the magnitude of the charge, and the electric field E_L is the stimulating field that causes the transition, and in the absence of all directly stimulating fields, there is always the stimulations caused by the quantum fluctuations of the vacuum. Inside a box of volume L^3, the amplitude of the vacuum field is given by

$$E_L(\omega_n) = \sqrt{\frac{\hbar\omega_n}{2\varepsilon_0 L^3}} \tag{12.20}$$

The total rate induced by the vacuum fluctuations is referred to as the spontaneous emission rate and, and in an isotropic system, is given in all directions by

$$W_{eg}(sp) = \frac{q^2\omega^3 n_r r_{ge}^2}{3\pi c^3 \hbar\varepsilon_0}$$

$$r_{ge}^2 = \left|x_{ge}\right|^2 + \left|y_{ge}\right|^2 + \left|z_{ge}\right|^2 \tag{12.21}$$

$$x_{ge} = \left\langle g\left|x\right|e\right\rangle$$

In a QW of GaAs, the transition rate between the first two subband levels with energy difference ~0.01 eV is typically ~10^5 Hz which is substantial, though slow, compared to phonon relaxation rates of 10^9–10^{12} Hz, given that it involves only the quantum fluctuations of the vacuum. One should also note that the expression does not depend on the linewidth of the transition, because the light source which is causing the transition encompasses all frequencies allowed in the cavity. Cavity size enters the spontaneous decay rate as shown in Ref. [36a]. An alternative expression that exhibits the relation to absorption is given by the spontaneous rate per unit volume and energy interval (Hz cm^{-3}/eV)

$$dW_{eg}(sp,\omega)/d(\hbar\omega) = \frac{\pi q^2 \omega}{n_r c \varepsilon_0} \frac{8\pi n_r^2 (\hbar\omega)^2}{h^3 c^2} \frac{2}{\Omega} \sum_{i,j} \left| \left\langle i \middle| \vec{\varepsilon}_p \cdot \vec{r} \middle| j \right\rangle \right|^2 \delta\left(E_j - E_i - \hbar\omega\right) f_j \left(1 - f_i\right) \qquad (12.22)$$

where Ω is the volume, and the sums now extend over the allowed thermally averaged transitions. This expression reduces to the previous one for a single transition at $T = 0$. The corresponding simulated rate per energy is (in Hz cm^{-3}/eV)

$$dW_{eg}(sp,\omega)/d(\hbar\omega) = \frac{\pi q^2 \omega}{n_r c \varepsilon_0} \frac{8\pi n_r^2 (\hbar\omega)^2}{h^3 c^2} \frac{2}{\Omega} \sum_{i,j} \left| \left\langle i \middle| \vec{\varepsilon}_p \cdot \vec{r} \middle| j \right\rangle \right|^2 \delta\left(E_j - E_i - \hbar\omega\right)\left(f_j - f_i\right) \qquad (12.23)$$

This depends on the number of excited levels, f_j, which participate in the transition. The distribution functions, f_i, are not necessarily the equilibrium values, they are the steady-state values.

Thus sources of stimulation that enter the total, $E_L(\omega)$, are the actual photons present in the cavity [36a] and include in particular also the photons that are in the blackbody spectrum of the material at finite temperatures, provided the system can be assumed to emit as a blackbody of course. If not, one has to take the actual thermal emission spectrum with Equation 12.23. If the cavity is closed for photon escape, and the main source of photon reabsorption is the transition itself, then the spontaneously emitted photons can be recycled back to stimulate the transition again. The stimulated rate scales as the total number of photons in the cavity, so with sufficient number of photons in the cavity, the self-stimulated emission rate can be made to exceed the spontaneous rate and we can make lasers. To make a laser with quantized electronic energy levels only, we need at least three levels so that the lower level in the photonic transition can relax rapidly and does not reabsorb effectively. In this way we can achieve inversion, which means that we have more electrons in the upper lasing levels than in the lower lasing levels [36b,36c].

Let us go back to the QD system and first note that if IR light can be absorbed, it can also be emitted as shown above, so one should be able to engineer long-wavelength partially tunable PL systems, and be able to make lasers as well. In principle, both intersubband and interband light can be emitted. To understand interband emission in a QD, we need to study the energy levels of electrons and holes. The field of QD-PL is a vast one, and the reader is referred to the original literature and reviews [37a,37b]. We will here consider examples from our own work. Figure 12.19 illustrates the energy level diagram for the shallow-dot hybrid device of Figure 12.6.

Light emission rate will depend on the charge state of the QD, and this is exciting because charges can be added and removed from QD by current, heat, and light pulses and in this way the PL can be manipulated. In the particular class of shallow QD hybrid devices considered in Figure 12.6, the QD transition has not yet been observed, only the QW transitions are observed. We now know why. Recent TEM pictures show that only the WL remains in the final QDWIP structures with current growth parameters in agreement with the above PL analysis shown in Figure 12.20. In order to observe the QD-PL, one has to have a sufficient number of excitons in the QD at room temperature. Since in the QDWIP we only have 10^{10} cm^{-2} QD, and very low QD electron binding energies, it is perhaps not surprising that no QD-PL is observed at room temperature. We need to lower the measurement temperature. Another factor that influences the QD formation is the ripening time. The QD has to be given enough time to form before

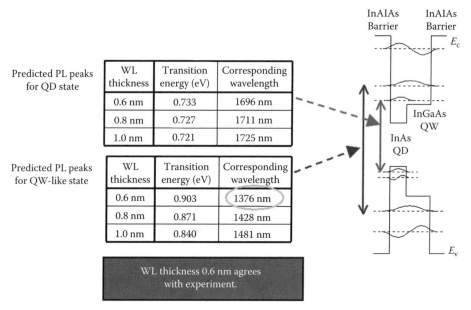

Predicted PL peaks for QD state	WL thickness	Transition energy (eV)	Corresponding wavelength
	0.6 nm	0.733	1696 nm
	0.8 nm	0.727	1711 nm
	1.0 nm	0.721	1725 nm

Predicted PL peaks for QW-like state	WL thickness	Transition energy (eV)	Corresponding wavelength
	0.6 nm	0.903	1376 nm
	0.8 nm	0.871	1428 nm
	1.0 nm	0.840	1481 nm

WL thickness 0.6 nm agrees with experiment.

FIGURE 12.19 Calculations of the electron and hole energy level structure of the QDWIP hybrid device using the AFM parameters for the QD.

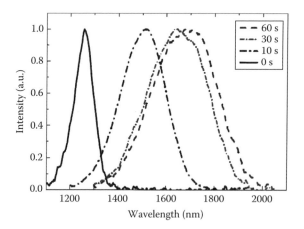

FIGURE 12.20 Room temperature PL of InAs QD on InP. The figure illustrates the ripening time effect using InAs/InP. With time the wavelength shifts to lower energies, which is a way of monitoring the QD formation process.

a clear QD signal can be seen. This effect is shown using InAs QD on InP growth from our own group in Figure 12.20. Work to improve QD formation in the hybrid devices is in progress. The hybrid devices are important because as they stand they give a good photodetecting performance. They constitute at present effectively QWIPs with a better normal incidence light coupling and lower dark currents.

12.4.4 QD Lasers

A very important application of semiconductor QD is in the area of long-wavelength laser physics. Most QD lasers are, however, made as interband lasers rather than intersubband lasers. The latter are made using the quantum cascade laser (QCL) technology [38a,38b], which allows the lower laser level electrons to escape very effectively into the next module. The field of interband QD lasers is very vast and successful and we will refer the reader to the original literature [38a,38b]. The interband laser normally involves double injection of electrons and holes from "n and p" doped layers. The carriers then get trapped inside the QD where the pair can recombine radiatively with the above-mentioned probabilities. Since the carrier-trapping process is not trivial, one should use the rigorous method of Ref. [20] to calculate the steady-state electron and hole distribution and recombination in the QD. The advantages are the long wavelengths and the relatively low threshold currents, high output power, and single mode option. These lasers are promising for fiber optical communication systems [38b].

12.4.5 The QD as a Memory

Semiconductor QD can be grown undoped, they can be embedded in nanopillars, and the charge can be injected from outside, even using nano-injectors. Charge can therefore be stored and retrieved from QD just like floating memory. In a doped system, in which the electron has been photoexcited out of the QD into another channel as in Figure 12.17, the stored, effectively positive charge, can be used to draw in more charge and act as an amplifier. This mechanism is exploited in the lateral photodetector of Figure 12.17. Another way is to except an electron-hole pair near a QD, and arrange it for one sign to trap in the QD. Present-day nanotechnology can make single electron devices too. Single QD charge and spin storage centers have been made [39a–39c] with QD. However, it is clear that single electron technology at the moment is a low temperature technology. The problem of signal to noise can, maybe in the future, be overcome at higher temperatures using neural networking and parallelism. Another way forward is to put QD in nanowires or nanopillars in such a way that the charge current going through the narrow constriction cannot avoid "talking to the state" of charging of the QD via the Coulomb interaction, spin state of the collision, and occupation of the QD. Similarly for light emission, where n-injected electrons and p-holes have to meet in the constriction where they can emit light.

12.4.6 Optical Integration Using Self-Assembled QD

We have seen that the self-assembled semiconductor QD can be a reasonably efficient source of band-to-band light and lasers [40] in the important 1–2 micron region with InAs/GaAs technology but also in the 400–600 nm region with GaN/AlN technology [41]. In this chapter, we are not including the important class of CdSe-type QDs and other chemically synthesized systems. The self-assembled QD can act as a single electrically driven light emitter that operates on a nanoscale, and in principle one can detect light sources down to a single photon in this energy range. Built into a nanopillar or nanowires [42] with n- and p-injection, the QD, is acting as a source of light that can be built into optic fibers technology or liquid crystals, or the very important photonic crystals. Photonic crystals [43] are like ordinary crystals in the sense that they are constituted by Bloch periodic structures with changes in the refractive index playing the role of the atomic or molecular potentials. These periodic changes are now on the scale of the wavelength of the light and they give rise to bandstructure for photonic propagation with all the exciting features such as the potentially low group velocities, and "stop bands" (band gaps) for light propagation. When nanolight sources are built into photonic crystals, one can place them in regions where the light amplitude is strongly enhanced by constructive interference with Bragg scattering playing the role of cavity mirrors. Light emitted from the QD can in this way be amplified, and in some cases the wavelength can be tuned, if the photonic crystal is heated [43] for example, and there are changes in its lattice structures and thus the bandstructures and optical band gaps. This can also be done with liquid crystalline mirrors, which of course can be made to change their transmittance by changes in temperature, and this is used in temperature and chemical sensing technology. The electrically driven nanopillar/QD light source is ideal also as a buried nanolight source for situations where the photonic crystal is a structure-responsive colloidal glass particle gel that can be made to change their shapes when decorated or attached with selective chemical receptors and exposed to analytes. When the lattice structure changes, then so does the transmittance or reflectance, and one has a way of monitoring contamination, in principle with spatial resolution. Light seeps through where the analyte has "deformed" the local network. The responsive gels are, for example, currently being developed for glucose sensing and are built into contact lenses, which sense and monitor the glucose contained in eye fluid [44]. There are of course many potential and beautiful applications for light sources which are built on nanoscales, and which can thus be integrated in electro-optical circuits which themselves have nanoscale components. But these developments are complex and they still need a lot of work to be done on them. The point is that the nano-emission can be amplified by cavity stimulation and, as we shall now see, with "surface plasmons." This is another and most exciting way of amplifying the photonic emission probability, and one of the most exciting developments is indeed the surface plasmon amplifier [41,45]. The plasmon source can be a simple gold or silver particle in the size range 10–100 nm with the plasmon resonance tuned to the same wavelength as the QD-PL emission wavelength itself [41,45]; see Figure 12.21. The tuning is done by changing the size and geometry [45,46] of the particle. Recently, it has been possible to tune the plasmon resonance geometries to reach into the IR at wavelengths or several microns [46] while keeping a relatively sharp resonance profile. The beauty of the plasmon resonator is that it is a simple nano-object, and that it produces at resonance an optical electrical field near itself which can be several orders of magnitude larger than the optic field which excites it. The surface plasmon resonance proximity has been demonstrated to enhance the PL of QD layers by an order of magnitude in many different instances and scenarios [41]. The plasmon electric field has, by design, exactly the right frequency, and thus acts to stimulate the emission, replacing what would normally only be the vacuum fluctuations or blackbody sources as shown by Equation 12.22 (see [36a]).

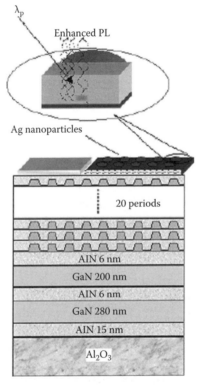

FIGURE 12.21 Schematic of GaN QD–based plasmonic nanostructures. (Reprinted from Neogi, A. and Morkoc, H., *Nanotechnology*, 15, 1252, September 2004. With permission. Institute of Physics Publishing.)

Again the interested reader should research the literature himself by Googling this exciting and topical subject on the Web. The reference list given here only represents a small fraction of the important papers and beautiful work that has been done in this field.

With PL surface plasmon enhancement, one has yet another powerful tool for optical integration of QD, not only by enhancing the PL but also by transporting the light in the form of localized surface charge oscillations and excitons both via the QD array itself (excitons) and via a line of tuned metal particles [47]. These arrays act as effective "photon-hopping" nanoscale light antennae and transport the light up to distances of ~500 nm. The photon is turned into charge oscillations that hop from particle to particle, moving more slowly than light speed; eventually, it can be turned back into light (space photon) again. This most exciting phenomenon is unfortunately limited in its scope by the lifetime of the excitation that decays via nonradiative and radiative decay processes in 10^{-9}–10^{-12} s. In information transfer, the challenges is to replenish the information by again using QD light-emitting nano-circuitry such as GaN QD in nanopillars [42].

Thus we have the exciting reality and prospect of making light on a nanoscale with QD, enhancing the PL on a nanoscale, Raman shifting the wavelength on a nanoscale [48] with plasmons, transporting it with metal particle nanowire waveguides [47], and re-amplifying the signal, switching, directing, slowing down the light, and amplifying with photonic crystals, reflecting from self-assembled liquid crystal mirrors, and finally detecting on a nanoscale with high-gain amplified QD or QWIP technology as discussed in the lateral photodetector of Section 12.4.1. Remember that here, a single "activated" QD (charged) can in principle trigger a large enough current by drawing in more charge into an adjacent high mobility channel. If one then combines these achievements, with the powerful responsive photonic crystal gels [44], liquid crystal technologies, the surface plasmon particles, the selective bio-receptors, and microfluids, one has the elements of an integrated hard and soft "lab" on the chip technology. But there is a lot more work to be done, integrating components, controlling lifetimes and loss.

12.5 Conclusion

In this chapter, we have tried to give the reader an idea of some important recent developments in the field of QD-IR physics. We have naturally focused on our own field of expertise which is IR-detection physics. The field of QD is too vast to be reviewed in a single chapter and the reader is encouraged to study the original literature and other reviews dealing with special topics that can be "Googled" nowadays with no problem.

The IR camera application of QD still needs a lot more work to be done on it to make it competitive with thermal Type II and CdMgTe technologies [1]. One has to remember however that broadband thermal sensing was never the main objective. It was always realized that thermal sensing was not the ideal application for QD. The niche is really in sharp wavelength and then multi-wavelength, resolved detection, and also in sensors and especially integration in photonic nano-circuitry. The main problems in improving the QE of the traditional QDIP are in raising the QD density and improving the normal incidence geometry. Thus high densities of $>10^{12}$ cm^{-2} and high aspect ratios of QD are still the target, even though the "superlattice QD in the well" work of Chakrabarti et al. [11] comes pretty close to the ideal situation. But here, controlling the dark current and noise level has proved difficult. New growth techniques and designs would have to be devised to improve on that. This is why researchers at CQD in NWU decided to try another line of attack: the hybrid QW/QD system shown in Figure 12.6 [13]. This design allows one to keep the high absorber density of the QWIP because most carriers are in the QW-like subbands, but it also keeps the QD potential corrugation, which allows a more efficient normal incident light coupling. The QD potentials seriously increase the tendency of band-edge states to be strongly Anderson localized. Localization breaks the planar to z-symmetry of the traditional QWIP, which is detrimental for normal incidence response. But the other important fact is that the InAs/QD WL complex also increases the barrier height, and thus lowers the dark current, while keeping the detection wavelength in the designed range. The simplest and most important improvement which one can undertake, on the present QDWIP, is to build in a corrugated back reflection mirror and maybe put in some top electrode cavity effects.

In the review of other applications, we have given the readers only a taste of what can be done with QD. They can judge the technology for themselves and see how exciting the long-term prospects are for nanoscale electro-optical-bio integration of QD physics. They can also see the hurdles on the way, which are in the first place fabricational and with targets such as this: what is the smallest photodetector/emitter/laser one can make for a particular wavelength and signal strength?

References

1. P. Martyniuk, Krishna, S., and Rogalski, A., *J. Appl. Phys.* **104**, 034314 (2008).
2. V. A. Shchukin and D. Bimberg, *Rev. Mod. Phys.* **71**, 1121 (1999).
3. V. Ryzhii, *Semicond. Sci. Technol.* **11**, 759 (1996).
4. J. Phillips, *J. Appl. Phys.* **91**, 4590 (2002).
5. M. Razeghi, H. Lim, K. Mi, S. Tsao, J. Szafraniec, W. Zhang, and B. Movaghar, *Nanotechnology* **16**, 219–229 (2005).
6. E.-T. Kim, A. Madhukar, Z. Ye, and J. C. Campbell, *Appl. Phys. Lett.* **84**, 3277 (2004).
7. H. Lim, B. Movaghar, S. Tsao, T. Sills, J. Szafraniec, W. Zhang, and M. Razeghi, *Phys. Rev. B* **72**, 085332 (2005).
8. S. Chakrabarti, A. D. Stiff-Roberts, P. Bhattacharya, S. Gunapala, S. Rafol, S. Bandara, and S. W. Kennerley, *IEEE Photonics Technol. Lett.* **16**, 1361 (2004).
9. S. Tsao, T. Yamanaka, S. Abdollahi Pour, I.-K. Park, B. Movaghar, and M. Razeghi, *Proc. SPIE* **7224**, 7224ov (2009).
10. H. Lim, PhD thesis, Northwestern University, Evanston, IL (2007).
11. S. Chakrabarti, A. D. Stiff-Roberts, P. Bhattacharya, and S. Kennerly, *J. Vac. Sci. Technol. B*, **22**, 1071 (2004).
12. M. Califano and P. Harrison, *J. Appl. Phys.* **91**, 389 (2002).
13. H. Lim, S. Tsao, W. Zhang, and M. Razeghi, *Appl. Phys. Lett.* **90**, 131112 (2007).
14. L. Hoeglund, P. O. Holtz, H. Petterson, C. Asplund, Q. Wang, S. Almqvist, S. Smuk, E. Petrini, and J. Y. Andersson, *Appl. Phys. Lett.* **93**, 103501 (2008).
15. P. Aivaliotis, E. A. Zibik, L. R. Wilson, J. W. Cockburn, M. Hopkinson, and N. Q. Vinh, *Appl. Phys. Lett.* **92**, 023501 (2008).
16a. D. Wood and A. Zunger, *Phys. Rev. B* **53**, 7949 (1996).
16b. H. Fu and A. Zunger, *Phys. Rev. Lett.* **80**, 5397 (1998).
17. B. F. Levine, *J. Appl. Phys.* **74**, R1 (1993).
18. A. Rogalski, *J. Appl. Phys.* **93**, 4356 (2003).
19a. O. O. Cellek and C. Besikci, *Semicond. Sci. Technol.* **19**, 183 (2004).
19b. S. Memis, O. O. Cellek, U. Bostanci, M. Tomak, and C. Besikci, *Turk. J. Phys.* **30**, 335 (2006).
20a. H. Lim, B. Movaghar, S. Tsao, M. Taguchi, W. Zhang, A. A. Quivy, and M. Razeghi, *Phys. Rev. B* **74**, 205321 (2006).
20b. Z. Ye, J. C. Campbell, Z. Chen, E.-T. Kim, and A. Madhukar, *Appl. Phys. Lett.* **83**, 1234 (2003).
20c. X. Lu, J. Vaillancourt, and M. J. Meisner, *Appl. Phys. Lett.* **91**, 051115 (2007).
21. B. Movaghar, S. Tsao, A. A. Pour, T. Yamanaka, and M. Razeghi, *Phys. Rev. B* **17**, 115320 (2008).
22. K. K. Ghosh, L. H. Zhao, and D. L. Huber, *Phys. Rev. B* **25**, 3851 (1982).
23. B. Movaghar, *Semicond. Sci. Technol.* **4**, 95 (1989).
24. P. Grassberger and I. Procaccia, *Phys. Rev. A* **26**, 3686 (1982).
25. D. Turchinovich, K. Pierz, and P. Uhd Jepsen, *Phys. Status Solidi C* **0**, 1556 (2003).
26. D. G. Cooke, F. A. Hegmann, Y. Mazur, W. Q. Ma, X. Wang et al., *Appl. Phys. Lett.* **85**, 3839 (2004).
27. T. Mueller, F. F. Schrey, G. Strasser, and K. Unterrainer, *Appl. Phys. Lett.* **83**, 3572 (2003).

28. P. Bhattacharya, H. H. Su, S. Chrakrabarti, G. Ariyawansa, and A. G. U. Perera, *Appl. Phys. Lett.* **86**, 191106 (2005).

29. K. K. Choi, B. F. Levine, C. G. Bethea, J. Walker, and R. J. Malik, *Phys. Rev. Lett.* **59**, 2459 (1987).

30. S. D. Gunapala, K. M. S. V. Bandara, B. F. Levine, G. Sarusi, J. S. Park, T. L. Lin, W. T. Pike, and J. K. Liu, *Appl. Phys. Lett.* **64**, 3431 (1994).

31. D. R. Giorgetta, E. Baumann, R. Theron, M. L. Pelloton, D. Hofstetter, M. Fischer, and J. Faist, *Appl. Phys. Lett.* **92**, 121101 (2006).

32. L. Chu, A. Zrenner, and G. Abstreiter, *Appl. Phys. Lett.* **79**, 2249 (2001).

33. S. W. Lee and K. Hirakawa, *Nanotechnology* **17**, 3866–3868 (2006).

34. M. Holub, S. Chakrabarti, S. Fathpour, P. Bhattacharya, Y. Lei, and S. Ghosh, *Appl. Phys. Lett.* **85**, 973 (2004).

35. L. F. Arsenault, B. Movaghar, P. Desjardins, and A. Yelon, *Phys. Rev. B* **78**, 075202 (2008).

36a. G. S. Solomon, M. Pelton, and Y. Yamamoto, *Phys. Rev. Lett.* **86**, 3903 (2001).

36b. J. Faist, F. Capasso, D. L. Sivco, C. Sirtori, A. L. Hutchinson, and A. Y. Cho, *Science* **264**, 553–556 (1994).

36c. Y. Bai, S. R. Darvish, S. Slivken, P. Sung, J. Nguyen, A. Evans, W. Zhang, and M. Razeghi, *Appl. Phys. Lett.* **91**, 141123 (2007).

37a. Y. D. Wang, S. J. Chua, S. Tripathy, M. J. Sander, P. Chen, and C. G. Fonstad, *Appl. Phys. Lett.* **86**, 07197 (2005).

37b. M. Bissirri, G. Baldossarri, H. von Hoegersthal, A. S. Bhatti, M. Capizzi, A. Frona, P. Frijeri, and S. Franchi, *Phys. Rev. B* **62**, 4642 (2000).

38a. V. M. Ustinov, A. E. Zhukov, A. Kovsh, S. Mikhrin, N. A. Maleev, B. V. Volovik, Y. G. Musikhin, Y. M. Shernyakov, E. Yu Kondat'eva, M. V. Maximov, A. F. Tsasulnikov, N. N. Lendentsov, Zh I Alferov, J. A. Lott, and D. Bimberg, *Nanotechnology* **11**, 397–400 (2000).

38b. N. F. Masse, I. P. Marko, A. R. Adams, S. J. Sweeney, *J. Mater. Sci. Mater. Electron.* **20**, S272–S276 (2009).

39a. N. Kroutvar, Y. Ducommun, J. J. Finley, M. Bichler, and G. Abstreiter, *Appl. Phys. Lett.* **83**, 443 (2003).

39b. C. Balocco, A. M. Song, and M. Missous, *Appl. Phys. Lett.* **85**, 5911 (2004).

39c. A. J. Williamson, L. W. Wang, and A. Zunger, *Phys. Rev. B* **62**, 12963 (2000).

40. H. Cao, H. Deng, H. Ling, C. Liu, V. A. Smagley, A. Gray, L. F. Lester, P. G. Eliseev, and M. Osinki, *Appl. Phys. Lett.* **86**, 203117(2005).

41. A. Neogi and H. Morkoc, *Nanotechnology* **15**, 1252 (2004).

42. T. Taliercio, S. Rousset, P. Lefebvre, T. Bretagnon, T. Guillet, B. Gil, D. Peyrade, Y. Chen, N. Grandjean, and F. Demangeot, *Superlattices Microstruct.* **36**, 783 (2004).

43. D. Eglund, A. Faraon, I. Fushman, J. Vokovic, N. Stolz, and P. Petroff, *Appl. Phys. Lett.* **90**, 213110 (2007) and see also *Nat. Lett.*, December 6, 2007, *Nature* 06234.

44. V. L. Alexeev, S. Das, D. N. Finegold, and S. A. Asher, *Clin. Chem.* **50**(12), 2353–2360 (2004).

45. T. R. Jensen, G. Schatz, and R. P. Van Duyne, *J. Phys. Chem. B* **103**, 2394 (1999).

46. R. Bukasov and R. S. Parry, *Nano Lett.* **7**(5), 1113 (2007).

47. M. Brongesma, J. Hartrman, and H. A. Atwater, *Phys. Rev. B* **62**, R16356 (2000).

48. K. Kneipp, H. Kneipp, I. Itzkhan, R. R. Dasari, and M. Field, *J. Phys. Condens. Matter* **14**, R597–R624 (2002).

13

III-Nitride Nanotechnology

Manijeh Razeghi
Ryan McClintock

13.1 Introduction

III-nitrides, long regarded as a scientific curiosity, have now earned a most respectful place in the science and technology of compound semiconductors and optoelectronic devices. Although the first AlN, GaN, and InN compounds were synthesized as early as 1907,[1] 1910,[2] and 1932,[3] respectively, little significant progress was reported until the end of the 1960s. In the late 1960s and early 1970s, the advent of modern epitaxial growth techniques led to a resurgent interest in III-nitrides.[4–7] However, it was not until the advent in the 1980s of the low-temperature GaN buffer,[8,9] and the realization of effective p-type doping in the 1990s[10] that the field really took off. Nowadays, the area of III-nitride research is well established and many commercial devices are available based on III-nitrides. This is pushing research into more and more creative areas in an attempt to create new discoveries. One of the interesting niche areas of III-nitride research that has seen progress in the past several years is the area of nanostructures.

III-nitride nanostructures offer unique properties that have the potential to be utilized to improve the properties of the existing devices. There is currently a very high level of commercial and scientific interest in gallium nitride–based one-dimensional (1D) structures (nanowires, nanotube or nanorods) because of the promise for use as ultraviolet (UV) or blue emitters, detectors, laser diodes (LDs), high-speed field-effect transistors, and high-temperature and radiation-resistant electronic devices.[11–17] Though two-dimensional (2D) structure GaN materials have been synthesized successfully for several years, investigations of 1D and zero-dimensional (0D) GaN materials are limited due to difficulties associated with their synthesis.

In spite of a very high dislocation density inherent in III-nitrides, largely due to the lack of lattice-matched substrates, blue LDs have been successfully demonstrated and commercialized.[18,19] However, in order to enhance the properties of these LDs, and to realize green LDs, it is imperative to decrease the dimensions of the active layer structure.[20–22] For this reason, quantum well (QW) and quantum dot (QD) structures have been pursued. Currently LDs with QWs in the active region have been commercialized;

however, the effort to enhance the quality of them still persists. On the other hand, QDs are attractive for their linear and nonlinear optical properties including the following: (1) A perfect population inversion can be achieved due to the discrete nature of energy levels, (2) Energy level quantization is accompanied by the concentration of bulk oscillator into single spectral lines and small volumes, and (3) Disappearance of the temperature dependant broadening mechanism (that dominates room temperature (RT) QW) in QDs as an immediate consequence of electronic state confinement.

13.1.1 Quantum Dots and Quantum Wires

Introduction of QWs in the early 1970s was a turning point in the direction of research on electronic structures.[23] The motion of electrons in a QW structure is bound in two directions if the thickness of the QW layer is on the order of the de Broglie wavelength. In the 1980s the interest of researchers shifted toward structures with further reduced dimensionality: 1D confinement (quantum wires)[24] and 0D confinement (QDs). Localization of carriers in all three dimensions breaks down the classical band structure of a continuous dispersion of energy as a function of momentum. Unlike QWs and quantum wires, the energy level structure of QDs is discrete. The study of nanostructures opens a new chapter in fundamental physics.

In order to demonstrate useful QD-based devices several requirements need to be fulfilled.[25] (1) Sufficiently deep localizing potential for observation of 0D confinement. (2) High density of QDs and a high filling factor to yield a meaningful interaction volume. (3) The QD size should not exceed a lower and an upper size limit dictated by the population of energy levels. The minimum QD size is the one that ensures existence of at least one energy level of an electron or a hole or both: in the GaN/AlN system this value is about 1 nm. The maximum size of a QD is determined by the transition between bulk-like behavior to the crystal and QD-like confined states. The general rule is that the QD should be on the order of the exitonic Bohr radius. The exitonic Bohr radius for GaN is reported to be on the order of 10 nm in the literature.[26,27] (4) The QDs must be uniform. The uniformity issue arises when device performance relies on the integrated gain in a narrow energy range, such as a QD laser. This imposes a constraint on the fluctuation in the QD size and shape. Large size and shape dispersion values results in variation in the position of energy levels resulting in inhomogeneous broadening. (5) Finally, as a general requirement, the density of defects in a QD material should be as low as possible in order to minimize the non-radiative processes.

13.1.2 Overview of Nanostructure Fabrication in III-Nitrides

QD-based structures pose great potential. For this reason, different methods have been utilized to realize QDs on III-nitride devices. Chemically synthesized GaN nanopowders have been studied using high-resolution transmission electron microscopy (HRTEM).[28] Moreover, epitaxially grown GaN dots have been demonstrated both in the Stranski–Krastanow (SK) growth mode[29] and by using a small amount of a surfactant.[30] The former produces self-assembling nanoscale dots on a very thin wetting layer due to lattice mismatch between AlN and GaN, while the latter uses Si as a surfactant for subsequent growth of three-dimensional (3D) material.

Although some improvements have been made in the growth of self-assembled QDs, applying this method to active layer of LDs still requires more study. This is largely due to poor control over the QD uniformity. To solve this problem, selective growth of dot structures has been proposed as an alternative method.[31–33] In selective area growth, InGaN QD structures are grown on SiO_2/GaN/sapphire substrate patterned by either conventional photolithography[33] or by focused ion beam irradiation.[32] By this method extremely uniform dots with hexagonal pyramidal shape have been achieved.[31] It has been suggested that InGaN QD structures are formed at the top of these pyramids.

One drawback of patterned regrowth is that it typically suffers from small filling factors with QDs only being present at the top of the pyramids, and the bulk of the growth being dominated by more traditional QW behavior on the sidewalls of the pyramids. An alternative to this is the use of electron-beam lithography (EBL) to directly create QDs in the III-nitride material by physically etching away material. This technique suffers from etch-induced damage, but gives excellent control of the period, size, and spacing of the QDs.

Another interesting area is the fabrication of nanotubes in III-nitrides. Especially since the discovery of carbon nanotubes by Ijima in 1991,[34] there has been increasing interest in fabrication of 1D nano-materials. Thus far, the fabrication of 1D structure GaN materials has been reported by following four methods: (1) Using carbon nanotubes as the template, which results in the growth of GaN nanorods or nanowires with a diameter similar to that of carbon nanotubes.[35,36] (2) Using the aluminum oxide as the template which is prepared by anodization of aluminum plate and has well-ordered pores. GaN nanowires have been fabricated in nanopores by the capillary effect.[37–40] (3) Self-assembly growth of GaN nanowires on the substrate (graphite, water-cooled Cu plate, Si, or sapphire) with direct growth or aid of metallic catalysts (Fe, Ni, In, or Au) which generated liquid nanoclusters that served as the reactive sites for confining and directing the growth of crystalline GaN nanowires.[41–44] (4) VLS growth on sapphire substrate using metal organic chemical vapor deposition (MOCVD) in which trimethylgal-lium (TMGa) and ammonia are used as the source materials.[45]

In this chapter, we review the III-nitride-based nanostructure work that has been conducted at the Center for Quantum Device. This includes self-assembled InGaN QDs; InGaN, GaN, and AlGaN fabricated by EBL; GaN nanowires grown by VLS; and, nanotubular GaN grown in a porous aluminum oxide template. We also discuss some of the nanostructure-based devices that have been demonstrated.

13.2 III-Nitride Nanostructure Fabrication

A variety of different methods are available to realize the fabrication of nanostructures in III-nitrides. Post-growth patterning of a bulk layer and transferring the pattern to the semiconductor layer via etch-ing was one of the earliest implemented methods.[46] The initial patterning can be achieved via a number of different techniques, with EBL being the most popular. However, due to the difficulties in etching of III-nitride structures, this method usually suffers from low lateral resolution and etch-induced dam-ages. Another method is selective growth on a patterned template; the surface of the underlying semi-conductor is patterned with a mask (SiO_2). Nucleation then takes place on the areas not covered by the mask resulting in hexagonal pyramids in the III-nitride material system.[31] This type of technique can also be used in 3D patterned nanoporous aluminum membranes to realize high-aspect-ratio quantum wires. Another useful method of formation of nanostructures is the self-organization phenomena on crystal surfaces by using strain relaxation to form ordered arrays of quantum wires and QDs.[47]

13.2.1 Growth of Self-Assembled Quantum Dots

The growth of self-assembled growth of QDs is a well-established technique that is widely used in other material systems. By controlling the growth conditions, both MBE and MOCVD can be used for the growth of uniform arrays of 3D QDs.[48–52] The growth of QDs using III-nitride compounds is a relatively new development. Self-assembled QDs in the III-nitride material system are generally grown either by using an "anti-surfactant" or by taking advantage of the lattice mismatch between InGaN/GaN or GaN/AlGaN heterostructures. An example of InGaN QDs grown using this method is shown in Figure 13.1.

In the GaN/AlGaN system, adding a third element which acts as an anti-surfactant such as TESi to the substrate material (AlGaN) can decrease the surface free energy and force 3D growth in this sys-tem.[30] In this case, the shape and the density of QDs can be controlled by the Si dose, the growth tem-perature, the growth duration, and the Al content in the underlying AlGaN layer.[30,53,54]

The SK growth mode has also been studied in the GaN/AlN material system. This is driven by the fact that there exists ~2.5% lattice mismatch between GaN and AlN. The minimum number of MLs for formation of 3D islands by MOCVD appears to be ~4. Similarly, it has been shown that 3D islands form above a critical thickness of 3 MLs by plasma-assisted MBE.[55] Using this method, high density of dots

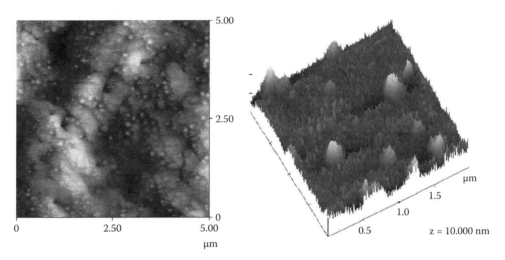

FIGURE 13.1   Image of typical InGaN QDs grown on a GaN template. (Left) 5 μm × 5 μm field of QDs, (Right) detail 3D view of QDs (the height scale is 10 nm).

($\sim 10^{10}$ cm^{-2}) have been achieved by MOCVD.[56] For the GaN/AlN material system, usually the constraint is to grow more than 3 MLs for the island formation and less than 8 MLs to avoid the plastic relaxation of the stress.[57]

The V/III ratio (N/Ga ratio) is another important factor in determining the growth mode. It has been found that growth of GaN on AlN in a N-rich environment results in the formation of GaN QDs.[55] On the other hand, the 2D/3D transition can be completely inhibited by performing the growth in a Ga-rich environment. This behavior is attributed to formation of a stable Ga film, about 2 ML thick, which acts as a "self-surfactant."[58]

13.2.2 Direct Etching of Quantum Structures

One of the easiest methods that can be used to realize very small features in III-nitrides is EBL. First a bulk layer is grown, and then a resist layer is spun on top and patterned via EBL. Finally, the pattern is transferred into the III-nitride layer via dry etching. Using this technique, very-well-ordered arrays of uniform-size QDs can be generated. Additionally, the size of QDs and the array geometry can easily be varied across the wafer. This has the potential to allow the use of quantum size effects to monolithically integrate different color detectors and/or emitters on a single wafer.

In order to fabricate QDs for blue emission, the starting material is a 150 nm thick $In_{0.2}Ga_{0.8}N$ layer grown on a standard 2 μm thick GaN template layer. This layer is grown via conventional low-pressure MOCVD using TMGa, TMIn, and NH, which were used as the gallium, indium, and nitrogen precursors, respectively. Standard bulk GaN or AlGaN can also be grown and patterned depending on the wavelength range(s) of interest.

This material then needs to be patterned with QDs. However, it is not possible to directly pattern an e-beam resist layer on top of this material due to the relatively high resistivity of the sapphire substrate and (In)GaN material. The high resistivity causes excessive charging at the surface degrading the resolution and requiring very low beam current to allow partial discharging while writing the image. In order to avoid these problems, a very thin metal layer is first deposited on top of the samples. This layer acts as a shunt to carry away the e-beam current and prevent charging allowing for higher resolution features and higher probe currents and thus shorter exposure times. A standard HMDS resist primer was then spun on top of the metallized sample followed by a 100 nm layer of PMMA 950K C6 electron-beam resist. The materials were then soft baked on a hot plate at 180°C for 90 s. Electron beam exposure

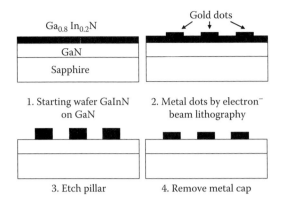

FIGURE 13.2 Schematic diagram of $Ga_{0.8}In_{0.2}N$ QD arrays fabrication process.

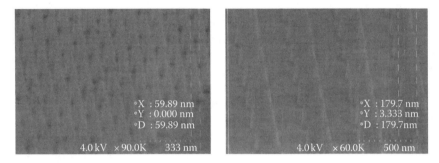

FIGURE 13.3 (Left) SEM image of ~60 nm diameter GaInN pillars with the density of 200 nm center-to-center spacing. (Right) 179 nm GaN pillars with the density of 500 nm center-to-center spacing.

was performed on a Leica LION LV-1 lithography system at dose ranging from 200 to 700 μC cm^{-2}. The exposure pattern was designed to produces dots with dimensions that varied from 300 nm down to as small as 50 nm. Following exposure MIBK:IPA(2:1) developer was used and the patterns were hard baked on a hot plate at 100°C for 90 s. A 100 nm thick gold etch-mask layer was then deposited on the patterned e-beam resist layer, and acetone was used to perform liftoff. This gold metal mask was then used to transfer the pattern into the InGaN layer using a Plasma-Therm SLR series electron-cyclotron resonance–reactive ion etcher (ECR-RIE) with $SiCl_4$:Ar plasma. The metal mask can then be removed by using an aqueous gold-etch solution (8 g KI + 2 g I_2 + 80 mL H_2O). The basic steps in this fabrication process are illustrated in Figure 13.2.

After patterning, the QD arrays were inspected via scanning electron microscopy. Good etching uniformity was achieved with sharp sidewalls. Figure 13.3 depicts arrays of different size of $Ga_{0.8}In_{0.2}N$ and GaN pillars (left and right, respectively). The $Ga_{0.8}In_{0.2}N$ QD array has a 200 nm center-to-center spacing. The GaN QD density is of 500 nm center-to-center spacing and the minimum size of GaN QDs we obtained is about 50 nm.

Photoluminescence (PL) was measured from the $Ga_{0.8}In_{0.2}N$ in order to study the optical properties of these dots. The PL spectra were recorded at room temperature and at 12 K using a cryostat, monochromator, and a 244 nm frequency-doubled Ar$^+$ ion laser. For comparison, an un-etched piece of the $Ga_{0.8}In_{0.2}N$ bulk sample was also characterized. Figure 13.3 shows the PL of QD arrays and $Ga_{0.8}In_{0.2}N$ bulk material at the temperature of 300 and 12 K; Figure 13.4 shows the temperature PL peak intensity of QD array and $Ga_{0.8}In_{0.2}N$ bulk material.

It is very clear in Figure 13.4 that the intensity of PL from QD array is much stronger than that from the $Ga_{0.8}In_{0.2}N$ bulk material. The full-width-at-half-maximum (FWHM) of QDs is also 79 nm

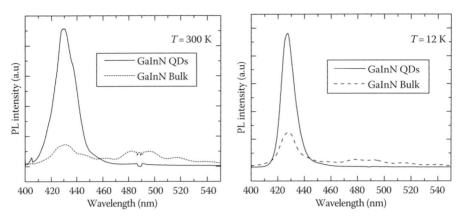

FIGURE 13.4   PL of GaInN QDs and bulk material at (Left) room temperature and (Right) at 12 K.

compared to the FWHM of the bulk $Ga_{0.8}In_{0.2}N$ which is about 101 nm (at 12 K). This represents a significant improvement in the luminous efficiency since the patterning effectively reduces the volume of InGaN in the QD array by a factor of ~19 times. With the decrease of the size of QDs, the number of total dislocations will be decreased in each dot: the dislocation density in nitride material is about 10^9 cm^{-2}, and with the dimension of dot decreased to less than 100 nm, the dots have a high probability of being dislocation free. This reduces the effects of dislocation-related non-radiative recombination. A strong broad peak at 450–500 nm can be seen from room temperature to 12 K PL in the bulk $Ga_{0.8}In_{0.2}N$ sample in Figure 13.4, but the peak disappears at both 12 K and at room temperature in the PL spectra of the QD array sample. The broad yellow peak is considered to be dislocation-related emission. With the decrease of dislocation number in the QD arrays, the dislocation-related emission is decreased. With the decrease of the size of the $Ga_{0.8}In_{0.2}N$ QDs, stain effects will also be decreased.

There is no clear blueshift observed in the PL spectra, irrespective of temperature. This is probably because the size of the carrier localization in the $Ga_{0.8}In_{0.2}N$ material and the quantum confinement are not sufficient to see quantum size effects. The excitonic Bohr radius for GaN is reported to be on the order of 10 nm in the literature,[26,27] whereas the QDs fabricated here are ~60 nm in diameter and 150 nm in thickness. In order to see real quantum confinement effects, the dot size will need to be reduced further. However, compared with bulk $Ga_{0.8}In_{0.2}N$ material, the QD array does give a much stronger PL emission due to the reduction of dislocation and possible reduction of piezoelectric field in these QD arrays. Smaller features than are traditionally allowable via direct patterning can be realized via controlling the etch process to create quantum wires, as shown in Figure 13.5. In this case, the nanowire diameter is determined not by the minimum feature size of the EBL system, but by the careful control of the etching to reduce the diameter. These and more conventional QD arrays can be applied to the fabrication of blue and UV light emitters. They also can be used to nanostructure the active region of UV photodetectors as discussed latter.

13.2.3 Vapor–Liquid–Solid Catalytic Growth of Nanowires

One attractive class of the nanostructures that offers 1D confinement is nanowires. Nanowires are structures with 2D lengths below ~100 nm, and the other dimensions on the order of several microns. Nanowires can be synthesized by a variety of methods; one of the most interesting is the vapor–liquid–solid (VLS) catalytic synthesis technique.[59–61] To grow a nanowire by VLS, a catalyst such as Ni, Au, In, or Fe is used to initiate nucleation of nanowires and help control the growth.[45,62] The pattering of the nanostructured catalysis can be accomplished lithographically. However, well-ordered arrays can

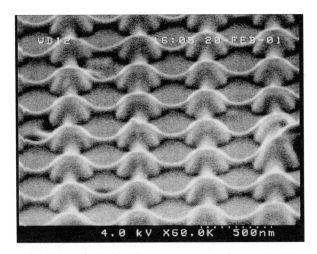

FIGURE 13.5 InGaN quantum wires fabricated using e-beam lithography and dry etching. The small quantum wires are fabricated by precisely controlling the etch process to leave narrow connecting strips of InGaN.

also be obtained by depositing a thin layer of the catalyst metal on the wafer and then heating it above its melting point for the formation of nanoclusters. Upon the start of the growth, these nanoclusters capture the gaseous reactants until they become supersaturated. After supersaturation occurs crystals in the form of nanowires grow against the substrate. The pattern density, array geometry, and nanowire diameter are dependent upon the diameter of nanoclusters, and can be controlled by changing the thickness of the catalyst layer and the annealing. It is also possible to use gallium to self-catalyze the growth of GaN nanowires.[63] However, the stability of these drops is low which makes them short-lived and limits the length of the nanowires resulting in more of QDs.

MOCVD growth of GaN nanowires has been recently investigated by a number of groups. Kuykendall et al. performed atmospheric pressure MOCVD growth of GaN nanowires resulting in the formation of randomly oriented GaN nanowires with triangular cross sections.[45] Su et al. have reported on the MOCVD growth of GaN and AlN nanowires using indium as a solvent agent. The nanowires were grown on mesoporous molecular sieves (MCM-41) consisting of hexagonal arrays of silicate material with nanoscale pores to facilitate yield and control of nucleation.[64] Kipshidze et al. used pulsed MOCVD growth to obtain well-oriented nanowires perpendicular to the substrate, ~100 nm in diameter.[65]

In our experiments with nanowire growth, we have used a nickel catalyst. Before growth, c-plane sapphire substrates were coated 10–30 Å thick nickel. These Ni-coated substrates were then loaded into the growth chamber and was heated for 5 min at 1050°C under a stream of nitrogen. Pulsed-MOCVD growth was then used to grow the nanowires. Ammonia flow was set to 75 sccm, while TMGa flow was 4 µmol/min. Pulse durations were 2 s for TMGa and 1 s for NH_3 with a 2 s interruption between pulses.

Various samples were grown at several different growth temperatures. It follows that the growth temperature has a strong effect on the nanowire growth dynamics. At low growth temperatures (~720°C), nanoclusters, slightly bigger than the original Ni nanoclusters, form on the surface (Figure 13.6a). Raising the growth temperature to ~920°C results in the VLS growth of GaN nanowires (Figure 13.6b). Observation of Ni nanoclusters at the tip of the nanowires is a testament to the VLS growth of the GaN nanowires. At higher temperatures (~1020°C), wide nano-columns, a few hundred nanometers in diameter, form on the surface (Figure 13.6c). Ni nanoclusters cannot be seen on top of these features, which could be indicative of the growth of GaN crystals around the original nanoclusters.

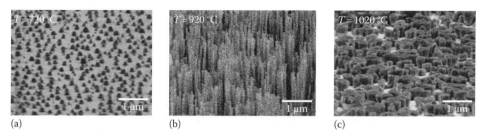

FIGURE 13.6 Evolution of GaN nanowire morphology as a function of growth temperature: (a) $T = 720°C$, (b) $T = 920°C$, and (c) $T = 1020°C$.

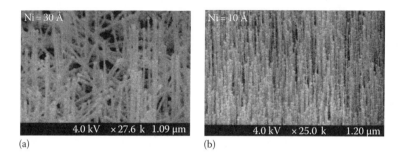

FIGURE 13.7 (a) SEM image of GaN nanowires grown via VLS technique using nickel as a catalyst with an initial thickness of 30 Å, and (b) SEM image of nanowires with an initial catalyst thickness of 10 Å resulting in highly ordered high-density GaN nanowires.

The directionality and the width of GaN nanowires are dependent upon the thickness of the initial nickel layer. A sample with a 10 Å nickel layer possessed very uniform straight nanowires with an average diameter of ~30 nm, while the nanowires of a sample with a 30 Å thick nickel layer had average diameters of ~90–100 nm and not all the nanowires were perpendicular to the surface of the substrate (Figure 13.7a and b). The height of the nanowires depends primarily on the growth duration: 500 series of pulses resulted in 2–2.5 μm tall nanowires. The density of nanowires for sample A is roughly 2×10^{10} cm^{-2}.

These III-nitride-based nanowires can potentially be incorporated into the future generation of optoelectronic devices, such as UV light emitters, to improve their efficiency. If the width of the nanowires can be limited to below the Bohr radius, quantum confinement effects should be observable. By using different initial metal thicknesses, it should be possible to simultaneously grow quantum wires of different sizes allowing for monolithic integration of several different wavelength detectors and/or emitters.

13.2.4 Growth of GaN on Anodic Aluminum Oxide Nanoporous Substrates

Sapphire, or aluminum oxide (Al_2O_3), is the most commonly used template choice for the growth of III-nitride materials. For planar semiconductor growth, commercially available wafers are generally cut from Czochralski-grown boules. However, the anodic growth of aluminum oxides from a raw surface of aluminum metal is a well-understood industrial process used to passivate the surface of many commercial aluminum parts. This anodic coating generally forms with a nanoporous structure that is commercially used to capture dye particles, and then sealed allowing for the realization of hard colored anodized coatings on aluminum parts.

However, more recently this same anodic aluminum oxide growth technique has been used to produce free-standing nanoporous membrane filters. A commercially available aluminum oxide membrane

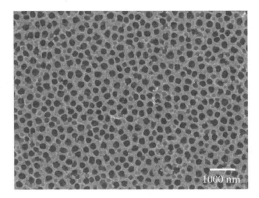

FIGURE 13.8   Scanning electron microscope image of surface morphology of aluminum oxide template before growth.

filter (Whatman Co., U.K.) was utilized as the template for the growth of III-nitride nanostructures. This membrane filter has a precise, nondeformable honeycomb pore distribution. The pore size in this membrane filter is about 200–250 nm and the thickness of membrane filter is 50–60 μm. A scanning electron micrograph of the template structure is shown in Figure 13.8. The crystal structure of the aluminum oxide membrane is found to be amorphous by x-ray diffraction.

The GaN nanotubular material is fabricated on the pores of this template by using a low-pressure MOCVD reactor. Trimethylgallium and ammonia are used as the group III and group V sources, respectively. Hydrogen was used as the carrier gas. In order to obtain good quality material in the deep pores, it is first necessary to nitridize the nanoporous template before starting the growth. Obtaining good coverage then requires a low growth rate.

In order to investigate the GaN growth inside of the nanoporous template a piece of the template was sectioned to allow viewing in a scanning electron microscope. Figure 13.9 shows a cross-sectional view of the aluminum oxide membrane after the deposition of GaN. The inner wall of the template can be seen to be covered by a thin GaN layer. This layer is composed on many fine GaN clusters of approximately 20 nm diameter. The total thickness of deposited GaN layer was estimated to be ~50 nm. Energy dispersive x-ray spectroscopy (EDS) was used to confirm the stoichiometry of this GaN layer.

After the deposition, the as-grown sample was wafer bonded to a silicon substrate to facilitate handling. The sample is then immersed in a solution of phosphoric acid and chromic acid to remove the polycrystalline aluminum oxide membrane. This solution does not attack the GaN nanotubes or the

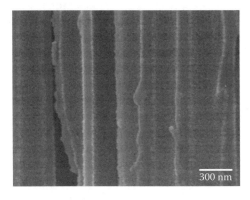

FIGURE 13.9   Scanning electron micrograph of the GaN layer deposited on the inner wall of the aluminum oxide template.

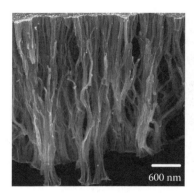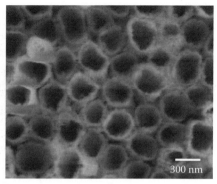

FIGURE 13.10 Scanning electron microscope image of free-standing GaN nanotubes after removal of aluminum oxide template: (Left) Cross-sectional view, (Right) View of the top surface of the hollow nanotubes.

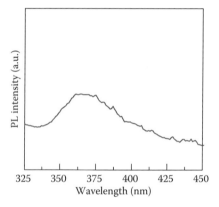

FIGURE 13.11 Room temperature PL spectrum of the GaN nanotubular material.

silicon carrier wafer. Figure 13.10 shows the free-standing GaN nanotubes after removal of the aluminum oxide template. The GaN nanotubes formed throughout the entire length of the nanoporous aluminum oxide template; the total length of the GaN nanotube was estimated to be about 50 μm. We observe that the GaN nanotubes are hollow with an outside diameter of ~250 nm, similar to the original aluminum oxide membrane.

The PL of the GaN nanotubes was measured at room temperature using a 244 nm frequency-doubled Ar ion laser. The PL spectrum for the specimen shown in Figure 13.11 consists of one broad emission between 340 and 400 nm centered at ~370 nm. This corresponds to the expected emission from bulk GaN (~365 nm, 3.4 eV).[4] The excitonic Bohr radius for GaN is reported to be 11 nm in the literature.[26,27] Thus, the diameter of the GaN nanotubes is larger than the excitonic Bohr radius, and thus a blueshift due to quantum size effects is not expected. Similar observations have been made in reports by other investigators.[37,39]

In conclusion, GaN nanotubular material can be fabricated through MOCVD growth on an aluminum oxide membrane. Gallium nitride is deposited on the inner wall of the nanoporous membrane. It is estimated that GaN is nucleated randomly on the inner wall of pores and then grows to form particulates. The GaN nanotubular material thus consists of a tube of many fine GaN particulates with size of 15–30 nm. After growth the aluminum oxide membrane can be chemically removed to yield high-aspect-ratio nanotubes. The outside diameter of these GaN nanotubes is approximately 200–250 nm (dependent on the membrane geometry), and the wall thickness is about 40–50 nm. However, due to the large size, no clear blueshift is observed in the PL spectrum.

13.3 III-Nitride Devices Based on Quantum Structures

13.3.1 Green Quantum Dot–Based LEDs

The development of high-efficiency green LEDs is a crucial part of developing solid-state white lighting to replace current inefficient (white) light sources among many other applications like full color displays.[66] However, while blue and violet LEDs are well understood, with increasing indium content, moving into the green, a significant decrease in performance is observed in QW-based III-nitride LEDs.[67] One of the reasons is that the large lattice mismatch between high indium content InGaN and GaN induces the generation of dislocations in the active region.[67,68] This occurs even if growth is performed on a low dislocation density template.[69] As a better alternative, the use of QDs is being investigated as a means to overcome these limitations in higher-indium-content devices.

Room temperature blue emission from QDs has been reported extensively in the litterature.[70,71] The strong luminescence from these QDs is attributed to exitonic transitions.[72] Blue QDs are being integrated into LEDs.[73] However, there are few reports of InGaN QDs in the green spectral region operating at room temperature.[74] As such, it is useful to study the formation of InGaN QDs for green emission, and compare their optical properties with those of InGaN QWs.

The InGaN layers are grown on a 2 μm GaN template on (00.1) sapphire (Al_2O_3) substrate using a horizontal-flow low-pressure MOCVD reactor. By carefully controlling the temperature, deposition thickness, and the V/III ratio, it is possible to control the formation and optical quality of the InGaN QDs.

Atomic force microscopy (AFM) is used to study the effects of temperature on QD formation. Figure 13.12a through c displays the (1 μm × 1 μm) AFM images of 4.8 monolayer (ML) InGaN grown at 734°C, 679°C, and 633°C, respectively. At lower temperatures, smaller InGaN disks are observed. The decrease in disk size is attributed to decreased adatom mobility. Based on this, we have determined 633°C to be the ideal temperature in our reactor for the growth of green-emitting QDs.

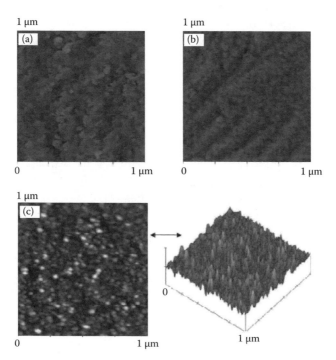

FIGURE 13.12   AFM (1 μm × 1 μm) images of 4.8 ML thick InGaN grown at (a) $T = 734$°C, (b) $T = 679$°C, and (c) $T = 633$°C.

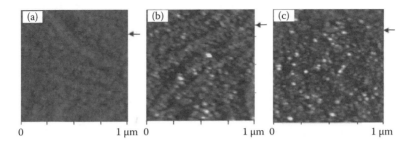

FIGURE 13.13   AFM (1 μm × 1 μm) images of (a) 2.4 ML, (b) 3.6 ML, and (c) 4.8 ML InGaN grown at a temperature of 633°C.

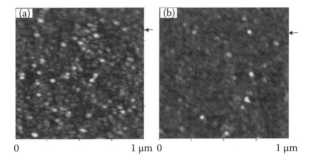

FIGURE 13.14   AFM (1 μm × 1 μm) images of 4.8 ML InGaN grown at T = 633°C with V/III ratios of (a) 13,880 and (b) 13,570.

We have also studied the effects of layer thickness on the formation of QDs. Figure 13.13a through c displays the (1 μm × 1 μm) AFM images of 2.4, 3.6, and 4.8 ML thick InGaN grown at 633°C, respectively. Based on this we have determined the critical thickness for dot formation, at 633°C, to be ~3.3 ML. Below this critical thickness, no QDs are formed, but above this critical thickness, increasing the deposited material results in increase in the density of the QDs at the expense of uniformity.

We have also studied the effects of the V/III ratio on QD formation. Figure 13.14a and b displays the (1 μm × 1 μm) AFM images of 4.8 ML thick InGaN grown at 633°C with V/III ratios being 13,880 and 13,570, respectively. The higher V/III ratio results in larger QDs. However, attempts to increase the V/III ratio further result in less-uniform QDs. For the higher V/III ratio sample, the average size of the QDs is 6 Å in height and 40 nm in diameter, with a density of 5×10^{10} cm^{-2}.

Based off of the optimized QD parameters, a pair of QD and QW structures were grown to allow for comparison of the performance. QW and QD structures consisted of 3 MQW (3 nm InGaN/7 nm GaN) and 3.6 ML InGaN QDs, respectively. However, despite the differing thicknesses, the PL intensities are comparable, as seen in Figure 13.15. The QDs have a 117 meV FWHM which is narrower than the 165 meV for the QWs at 558 nm. This FWHM is smaller than other published values.[74–76]

With increasing excitation power, as seen in the inset of Figure 13.15, the FWHM of the QDs decrease significantly whereas that of the QW remains constant. The lower initial FWHM and decrease of FWHM with increasing excitation power suggest that the QDs have fewer dislocations than QWs since the active layer relaxes elastically by QD formation, and avoids dislocation generation.

13.3.2 GaN Nanostructured Photodiodes

The use of nanostructures in traditional devices brings out new physics that can help to improve their performance up to unknown limits. In nanopillars, quantum confinement effects in the lateral direction are only revealed when the lateral dimensions are lower than 50 nm.[77] For larger mesoscopic systems,

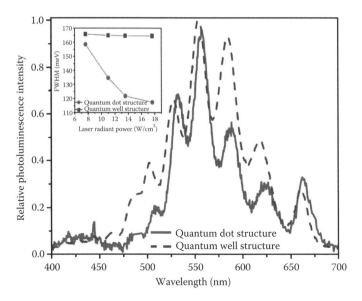

FIGURE 13.15 Relative PL intensity versus wavelength at room temperature for QD and QW structures. Inset displays FWHM dependence on incident laser radiant power for QDs and QWs structures.

quantum effects are not observable but the surface states also make their performance depart significantly from that found in bulk-based devices.[78,79] In particular, light–matter interaction and carrier transport are severely affected in those systems. These effects can be utilized to realize improved UV photodetectors based upon GaN nanostructures.

Samples were grown by low-pressure MOCVD on transparent AlN templates on double-side polished c-plane sapphire substrates (to allow for back-illumination). The basic structure consists of a *p-i-n* GaN junction (285, 200, and 200 nm) with carrier concentrations of 5×10^{17} cm^{-3}, 4×10^{18} for the *p*-type and *n*-type regions, respectively. The *i*-region was undoped with a residual carrier concentration estimated to be ~5×10^{16} cm^{-3}.

Nanopillars were formed in this structure using electron-beam lithography. The etch depth was ~500 nm and the resulting nanopillar diameters were ~200 nm with a packing density of about 0.01. A scanning electron micrograph of the patterned nanopillars is shown in Figure 13.16.

Polyimide was used for planarization, to help passivate the nanopillars, and to allow for contact deposition. HD Microsystem's polyimide was spin-applied to the sample and baked with a mean thickness

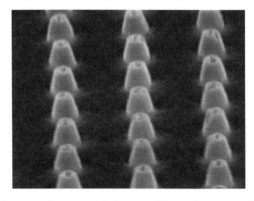

FIGURE 13.16 Scanning electron micrograph of the nanopillar surface array after polyimide deposition and etchback.

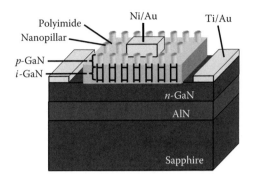

FIGURE 13.17 Schematic diagram of the processed device structure showing the nanopillars, polyimide planarization, and the metal contacts.

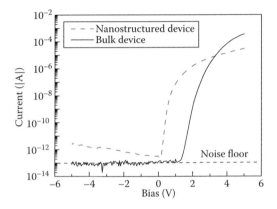

FIGURE 13.18 Current–voltage characteristics of GaN nanostructured *p-i-n* diodes with 625 μm² total device area as compared to 625 μm² bulk diodes.

of around 1.9 μm. Oxygen plasma was then used to etchback the polyimide and reveal the tops of the nanopillars. The etchback process revealed the top 100 nm of the nanopillars approximately. Standard lithography was then used to define circular contacts on the nanostructured surface. The rest of the polyimide that surrounded these contacts was removed by using oxygen plasma, and a bottom contact was deposited. Figure 13.17 illustrates the device scheme after processing. For comparison purposes, bulk devices were also fabricated from the same wafer, following the procedure described elsewhere.[80]

I–V measurements were performed (Figure 13.18). The resulting *I–V* curves of the nanostructured diodes show strong rectifying behavior. However, unlike bulk diodes, which have currents under the noise floor of the system, increasing leakage currents with reverse bias are observed in the picoampere range. This leakage current observed in *p-i-n* diodes at low voltages has been previously attributed to the damage induced by the dry etching on the sidewalls of the mesa structures.[81] In a nanostructured device the ratio of sidewall area to mesa area is significantly larger than in a commercial device, thus the belief that sidewall etch damage is the primary mechanism responsible of the leakage observed in these nanostructured devices. The average leakage current per nanopillar obtained at −5 V is ~5 fA, which corresponds to a current density of ~16 μA/cm².

It is also noticeable in Figure 13.18 that a considerable reduction of the turn-on voltage in the nanostructured diodes takes place. Furthermore, the ideality factor obtained from the fitting of the *I–V* curves under low forward biases decreases from 2.5–3.0, in the bulk diodes, to 1.3–1.6, in the nanostructured diodes. The improvement of the ideality factor as well as the reduction of the turn-on voltage in the nanostructured devices are unexpected. These effects have been attributed to etching-related effects

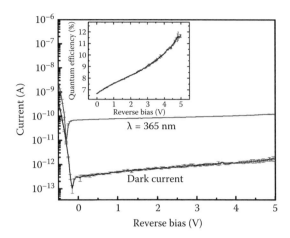

FIGURE 13.19 Reverse current of a 625 μm² diode in darkness and under illumination at 365 nm. Inset: calculated quantum efficiency as a function of bias.

since the top contact now contacts not only the top surface but the sidewall of the 100 nm etched-back *p*-GaN region as well.[82]

Photocurrent measurements under back-illumination were performed at 365 nm using a Xe lamp and a monochromator (Figure 13.19). In this configuration, the light is absorbed in the GaN after passing through the sapphire substrate and the AlN buffer layer with minimal losses. As shown in the inset of Figure 13.19, the optical response increased significantly with reverse bias: a quantum efficiency of 7% was calculated at 0 V, increasing up to 11.6% at 5 V. On the other hand, the quantum efficiency of the bulk diodes was about 28% at 0 V, and increased up to 30% at 5 V. The increase in the bulk devices is attributed to the additional broadening of the depletion region. However, despite experiencing the same depletion region broadening, efficiency increases more than 1.5 times faster between 0 and 5 V reverse bias in nanostructured diodes than in bulk diodes.

Two loss mechanisms can account for the reduction of efficiency in nanostructured diodes: surface recombination and the existence of dead layers due to a low lateral collection efficiency of carriers into the nanopillar volume. The effect of surface recombination is difficult to predict, although it is reasonable to think that it should be fairly insensitive to the bias voltage. Thus, the faster increase of the photocurrent with bias in nanostructured devices compared to bulk diodes seems to confirm the lateral collection efficiency as a limiting factor. Therefore, we can expect a higher response from the reduction of the separation between nanopillars, which provides better lateral collection efficiency besides increasing the packing density. In fact, the distance between nanopillars is in our case about 800 nm, which is still slightly higher than most of the hole diffusion lengths reported in n-type GaN.[83,84] Moreover, the linear behavior of the optical response with the excitation power allows us to rule out other effects like photoconductive gain.

In summary, the fabrication of GaN *p-i-n* photodiodes with embedded nanopillars is reported. Their *I–V* characteristics show an improvement of the ideality factor compared to bulk diodes fabricated on the same material. The enhanced tunneling current through the p-type barrier contact is believed to be responsible of this behavior. Increasing quantum efficiencies with reverse bias were found due to the enhancement of the lateral collection efficiency.

Acknowledgments

The authors of this chapter gratefully acknowledge the many contributions of C. Bayram, W. Jung, P. Kung, K. Mi, J. L. Pau, and A. Yasan over the years in developing the study of III-Nitride nanostructures at the Center for Quantum Devices, Department of Electrical Engineering and Computer Science, Northwestern University, Evanston, IL.

References

1. F. Fichter, *Z. Anorg. Chem.* 54, 322 (1907).
2. F. Fischer and F. Schröter, *Berichte der Deutschen Chemischen Gesellschaft* 43, 1465 (1910).
3. V.C. Johnson, J.B. Parsons, and M.C. Crew, *J. Phys. Chem.* 36, 2588 (1932).
4. H.P. Maruska and J.J. Tietjen, *Appl. Phys. Lett.* 15, 327 (1969).
5. H.M. Manasevit, F.M. Erdmann, and W.I. Simpson, *J. Electrochem. Soc.* 118, 1864 (1971).
6. I. Akasaki et al., MITI report in Japanese only (1974). I. Akasaki, and I. Hayashi, *Ind. Sci. Technol.* 17, 48 (1976).
7. S. Yoshida, S. Misawa, and A. Itoh, *Appl. Phys. Lett.* 26, 461 (1975).
8. S. Yoshida, S. Misawa, and S. Gonda, *Appl. Phys. Lett.* 42, 427 (1983).
9. H. Amano, N. Sawaki, I. Akasaki, and Y. Toyoda, *Appl. Phys. Lett.* 48, 353 (1986).
10. H. Amano, M. Kito, K. Hiramatsu, and I. Akasaki, *Jpn. J. Appl. Phys.* 28, L2112 (1989).
11. H. Morkoc and S.N. Mohammad, *Science* 267, 51 (1995).
12. F.A. Ponce and D.P. Bour, *Nature* 386, 351 (1997).
13. G. Fasol, *Nature* 272, 1751 (1996).
14. S.N. Mohammad and H. Morkoc, *Prog. Quantum Electron.* 20, 361 (1996).
15. S. Nakamura, *Science* 281, 956 (1998).
16. J.C. Zolper, R.J. Shul, A.G. Baca, R.G. Wilson, S.J. Pearton, and R.A. Stall, *Appl. Phys. Lett.* 68, 2273 (1996).
17. Q. Chen, M.A. Khan, J.W. Wang, C.J. Sun, M.S. Shur, and H. Park, *Appl. Phys. Lett.* 69, 794 (1996).
18. S. Nakamura, M. Senoh, S. Nagahara, N. Iwasa, T. Yamada, T. Matsushita, H. Kiyoku, and Y. Sugimoto, *Appl. Phys. Lett.* 68, 2105 (1996).
19. P. Kung, A. Saxler, D. Walker, A. Rybaltowski, X. Zhang, J. Diaz, and M. Razeghi, *MRS Internet J. Nitride Semicond. Res.* 3, 1 (1998).
20. S. Tanaka, S. Iwai, and Y. Aoyagi, *Appl. Phys. Lett.* 69, 4096 (1996).
21. K. Tachibana, T. Someya, S. Ishida, and Y. Arakawa, *Appl. Phys. Lett.* 76, 3212 (2000).
22. P. Ramvall, S. Tanaka, S. Nomura, P. Riblet, and Y. Aoyagi, *Appl. Phys. Lett.* 73, 1104 (1998).
23. R. Dingle, W. Wiegmann, and C.H. Henry, *Phys. Rev. Lett.* 33, 827 (1974).
24. P.M. Petroff, A.C. Gossard, R.A. Logan, and W. Wiegmann, *Appl. Phys. Lett.* 41, 635 (1982).
25. D. Bimberg, M. Grundmann, and N.N. Ledentsov, *Quantum Dot Heterostructures*, John Wiley & Sons, New York (1999).
26. X. Xie, Y. Quian, W. Wang, S. Zhang, and Y. Heng, *Science* 272, 1926 (1996).
27. X. Shan, X.C. Xie, and J.J. Song, *Appl. Phys. Lett.* 67, 2512 (1995).
28. K. Gonsalves, S. Rangarajan, G. Carlson, J. Kumar, K. Yang, M. Benaissa, and M. José-Yacamán, *Appl. Phys. Lett.* 71, 2175 (1997).
29. B. Daudin, F. Widmann, G. Feuillet, Y. Samson, M. Arley, and J. L. Rouviere, *Phys. Rev. B* 56, R7069 (1997).
30. S. Tanaka, S. Iwai, and Y. Aoyagi, *Appl. Phys. Lett.* 69, 4096 (1996).
31. K. Tachibana, T. Someya, S. Ishida, and Y. Arakawa, *Appl. Phys. Lett.* 76, 3212 (2000).
32. J. Wang, M. Nozaki, M. Lachab, Y. Ishikawa, R.S. Qhalid Fareed, T. Wang, M. Hao, and S. Sakai, *Appl. Phys. Lett.* 75, 950 (1999).
33. D. Kapolnek, S. Keller, R.D. Underwood, S.P. DenBaars, and U.K. Mishra, *J. Cryst. Growth* 189/190, 83 (1998).
34. S. Ijima, *Nature* 354, 56 (1991).
35. W.Q. Han, S.S. Fan, Q.Q. Li, and Y.D. Hu, *Science* 277, 1287 (1997).

36. C.C. Chen, C.C. Yeh, C.H. Liang, C.C. Lee, C.H. Chen, M.Y. Yu, H.L. Liu, L.C. Chen, Y.S. Lin, K.J. Ma, and K.H.J. Chen, *Phys. Chem. Solids* 62, 1577 (2001).
37. J. Zhang, X.S. Peng, X.F. Wang, Y.W. Wang, and L.D. Zhang, *Chem. Phys. Lett.* 345, 372 (2001).
38. G.S. Cheng, L.D. Zhang, Y. Zhu, G.T. Fei, L. Li, C.M. Mo, and Y.Q. Mao, *Appl. Phys. Lett.* 75, 2455 (1999).
39. G.S. Cheng, L.D. Zhang, S.H. Chen, Y. Li, L. Li, X.G. Zhu, G.T. Fei, and Y.Q. Mao, *J. Mater. Res.* 15, 347 (2000).
40. G.S. Cheng, S.H. Chen, X.G. Zhu, Y.Q. Mao, and L.D. Zhang, *Mater. Sci. Eng. A* 286, 165 (2000).
41. H.Y. Peng, X.T. Zhou, N. Wang, Y.F. Zheng, L.S. Liao, W.S. Shi, C.S. Lee, and S.T. Lee, *Chem. Phys. Lett.* 327, 263 (2000).
42. H.Y. Peng, N. Wang, X.T. Zhou, Y.F. Zheng, C.S. Lee, and S.T. Lee, *Chem. Phys. Lett.* 359, 241 (2002).
43. M. He, I. Minus, P. Zhou, S.N. Mohammed, J.B. Halpern, R. Jacobs, W.L. Sarney, L. Salamanca-Riba, and R.D. Vispute, *Appl. Phys. Lett.* 77, 3731 (2000).
44. S.Y. Bae, H.W. Seo, J. Park, H. Yang, and B. Kim, *Chem. Phys. Lett.* 376, 445 (2003).
45. T. Kuykendall, P. Pauzauskie, S. Lee, Y. Zhang, J. Goldberger, and P. Yang, *Nano Lett.* 3, 1063 (2003).
46. M.A. Reed, R.T. Bate, K. Bradshaw, W.M. Duncan, W.M. Frensley, J.W. Lee, and H.D. Smith, *J. Vac. Sci. Technol. B* 4, 358 (1986).
47. V.A. Shchukin, N.N. Ledentsov, P.S. Kop'ev, and D. Bimberg, *Phys. Rev. Lett.* 75, 2968 (1995).
48. K. Mukai, N. Ohtsuka, M. Sugawara, and S. Yamazaki, *Jpn. J. Appl. Phys.* 33, L1710 (1994).
49. J. Oshinowo, M. Nishioka, S. Ishida, and Y. Arakawa, *Appl. Phys. Lett.* 65, 1421 (1994).
50. A. Heinrichsdorff, A. Krost, M. Grundmann, D. Bimberg, A.O. Kosogov, and P. Werner, *Appl. Phys. Lett.* 68, 3284 (1996).
51. R. Leon, S. Fafard, D. Leonard, J.L. Merz, and P.M. Petroff, *Appl. Phys. Lett.* 67, 521 (1995).
52. A. Ponchet, A. Le Corre, H. L'Haridon, B. Lambart, and S. Salaün, *Appl. Phys. Lett.* 67, 1850 (1995).
53. X.Q. Shen, S. Tanaka, S. Iwai, and Y. Aoyagi, *Appl. Phys. Lett.* 72, 344 (1998).
54. P. Ramvall, P. Riblet, S. Nomura, and Y. Aoyagi, *J. Appl. Phys.* 87, 3883 (2000).
55. F. Widmann, J. Simon, B. Daudin, G. Feuillet, J.L. Rouvière, N.T. Pelekanos, and G. Fisherman, *Phys. Rev. B* 58, R15989 (1998).
56. M. Miyamura, K. Tachibana, and Y. Arakawa, *Appl. Phys. Lett.* 80, 3937 (2002).
57. N. Gogneau, D. Jalabert, E. Monroy, T. Shibata, M. Tanaka, and B. Daudin, *J. Appl. Phys.* 94, 2254 (2003).
58. G. Mula, C. Adelmann, S. Moehl, J. Oullier, and B. Daudin, *Phys. Rev. B* 64, 5406 (2001).
59. Y. Wu and P. Yang, *J. Am. Chem. Soc.* 123, 3165 (2001).
60. D. Wang, F. Qian, C. Yang, Z. Zhong, and C.M. Lieber, *Nano Lett.* 4, 871 (2004).
61. S. Han, W. Jin, T. Tang, C. Li, D. Zhang, X. Liu, J. Han, and C. Zhou, *J. Mater. Res.* 18, 245 (2003).
62. J. Zhang and L. Zhang, *J. Vac. Sci. Technol. B* 21, 2415 (2003).
63. C.-W. Hu, A. Bell, F.A. Ponce, D.J. Smith, and I.S.T. Tsong, *Appl. Phys. Lett.* 81, 3236 (2002).
64. J. Su, G. Cui, M. Gherasimova, H. Tsukamoto, J. Han, D. Ciuparu, S. Lim, L. Pfefferle, Y. He, A.V. Nurmikko, C. Broadbridge, and A. Lehman, *Appl. Phys. Lett.* 86, 013105 (2005).
65. G. Kipshidze, B. Yavich, A. Chandolu, J. Yun, V. Kuryatkov, I. Ahmad, D. Aurongzeb, M. Holtz, and H. Temkin, *Appl. Phys. Lett.* 86, 033104 (2005).
66. E.D. Jones, *Light Emitting Diodes for General Illumination*, OIDA, Washington, DC, 2001.
67. Y. Cho, S.K. Lee, H.S. Kwack, J.Y. Kim, K.S. Lim, H.M. Kim, T.W. Kang, S.N. Lee, M.S. Seon, O.H. Nam, and Y.J. Park, *Appl. Phys. Lett.* 83, 2578 (2003).
68. F.A. Ponce, S. Srinivasan, A. Bell, L. Geng, R. Liu, M. Stevens, J. Cai, H. Omiya, H. Marui, and S. Tanaka, *Phys. Status Solidi* 240, 273 (2003).
69. T. Kozaki, H. Matsumura, Y. Sugimoto, S. Nagahama, and T. Mukai, *Proc. SPIE* 6133, 613306 (2006).
70. K. Tachibana, T. Someya, and Y. Arakawa, *Appl. Phys. Lett.* 74, 383 (1999).
71. B. Damilano, N. Grandjean, S. Dalmasso, and J. Massies, *Appl. Phys. Lett.* 75, 3751 (1999).

72. O. Moriwaki, T. Someya, K. Tachibana, S. Ishida, and Y. Arakawa, *Appl. Phys. Lett.* 76, 2361 (2000).

73. Y.K. Su, S.J. Chang, L.W. Ji, C.S. Chang, L.W. Wu, W.C. Lai, T.H. Fang, and K.T. Lam, *Semicond. Sci. Technol.* 19, 389 (2004).

74. S. Choi, J. Jang, S. Yi, J. Kim, and W. Jung, *Proc. SPIE* 6479, 64791F (2007).

75. H. Zhou, S.J. Chua, K. Zang, L.S. Wang, S. Tripathy, N. Yakovlev, and O. Thomas, *J. Cryst. Growth* 298, 511 (2007).

76. Y. Wang, X.J. Pei, Z.G. Xing, L.W. Guo, H.Q. Jia, H. Chen, and J.M. Zhou, *J. Appl. Phys.* 101, 033509 (2007).

77. A. Gin, B. Movaghar, M. Razeghi, and G.J. Brown, *Nanotechnology* 16, 1814 (2005).

78. N. Malkova and C. Z. Ning, *Phys. Rev. B* 74, 155308 (2006).

79. Z.-M. Liao, K.-J. Liu, J.-M. Zhang, J. Xu, and D.-P. Yu, *Phys. Lett. A* 367, 207 (2007).

80. J.L. Pau, R. McClintock, K. Minder, C. Bayram, P. Kung, and M. Razeghi, *Appl. Phys. Lett.* 91, 041104 (2007).

81. C. Pernot, A. Hirano, H. Amano, and I. Akasaki, *Jpn. J. Appl. Phys.* 37, L1202 (1998).

82. J.L. Pau, C. Bayram, P. Giedraitis, R. McClintock, and M. Razeghi, *Appl. Phys. Lett.* 93, 221104 (2008).

83. R.J. Kaplar, S.R. Kurtz, and D.D. Koleske, *Appl. Phys. Lett.* 85, 5436 (2004).

84. K. Kamakura, T. Makimoto, N. Kobayashi, T. Hashizume, T. Fukui, and H. Hasegawa, *Appl. Phys. Lett.* 86, 052105 (2005).

VII

Integrated Photonic Technologies: Planar Lightwave Circuits

14

Microphotonic Devices and Circuits in Nanoengineered Polymers

Louay Eldada

Polymers are a compelling choice as the base material for integrated microphotonic devices and circuits as, when properly nanoengineered, they can offer simultaneously high performance, compactness, tunability, environmental stability, mechanical flexibility, ease of fabrication, high yields, and low cost.[1,2]

14.1 Nanoengineered Optical Polymers

Polymers for integrated microphotonics have been under development since the early 1980s. The initial focus was on electro-optic materials. Electro-optic materials, however, have intrinsic reliability issues that have yet to be overcome. The focus shifted in the early 1990s to thermo-optic polymers that can achieve high performance with high reliability, resulting in the early 2000s in commercial dynamic components, and in the mid-2000s in commercial routing and switching modules and subsystems.[3]

Optical polymers were engineered in many laboratories worldwide and some are available commercially.[4] Classes of polymers used in PICs include acrylates, polyimides, polycarbonates (PCs), and olefins (e.g., cyclobutene). Some polymers, such as most polyimides and PCs, are not photosensitive and are typically processed using photoresist patterning and RIE etching. These polymers have most of the issues of the silica-on-silicon technology in terms of roughness-induced and stress-induced scattering loss and polarization dependence. Other polymers are photosensitive and as such are directly photopatternable, resulting in low roughness and low stress waveguides. The full cycle time for the fabrication of a multilayer waveguiding circuit on a wafer is about 30 min, giving polymers a 10- to 1000-fold advantage in throughput over other PLC materials. Furthermore, this technology uses low-cost materials and low-cost/low-maintenance capital equipment (e.g., spin-coater and UV lamp vs. CVD growth system).

The production of commercially viable polymeric optical components is a complex task because optical polymers need to simultaneously meet a large number of properties. Table 14.1 summarizes (in no

TABLE 14.1   The Top 50 List of Properties of Optical Polymers

1. Low absorption loss	26. Isotropy
2. Low wavelength-dependent loss (WDL)	27. Homogeneity
3. Low polarization-dependent loss (PDL)	28. Low level of impurities
4. Low polarization-mode dispersion (PMD)	29. Optimal molecular weight distribution
5. Low chromatic dispersion (CD)	30. Optimal viscosity
6. Low birefringence	31. Low volatility
7. Low stress	32. Optimal photosensitivity
8. Low fiber pigtail loss	33. Optimal cure speed
9. Closeness to refractive index of silica fiber	34. Full curability
10. Stability of refractive index	35. Thermoset material
11. Variable refractive index difference (Δn)	36. High crosslink density
12. Variable refractive index profile	37. Low shrinkage
13. Low refractive index dispersion ($dn/d\lambda$)	38. Optimal free volume
14. Large TO or EO coefficient (dn/dT or r_{33}) if needed	39. Contrast in patterning
15. Linearity of refractive index with temperature	40. Optimal surface energy
16. Optimal thermal conductivity	41. Processability for high film quality
17. Thermal stability	42. Process latitude in patterning
18. No phase transition in operating range	43. Patternability with low scattering loss
19. Stability with humidity	44. Machinability (cleaving, dicing, polishing)
20. Hydrophobicity	45. Mechanical robustness
21. Stability with optical power	46. Flexibility
22. Adhesion (to substrates, self, electrodes)	47. Stability of mechanical integrity
23. Compatibility with electrode patterning	48. Volume manufacturability
24. Low reactivity to acids, bases, solvents	49. Manufacturability with repeatable properties
25. Stability with oxygen	50. Long shelf life

particular order) the 50 main properties that a polymer needs to meet in order to be a viable optical polymer with potential for use in commercial products.

The nanostructure of materials has been recognized by leading developers of optical polymers as a fundamental determinant of their properties long before nanoscale science and engineering was identified as a distinct field of study. The recent emergence of new tools and techniques for the measurement, characterization, and control of nanoscale features has allowed further refinement and better control of material properties. The key properties of high-performance optical polymers, mainly fluoroacrylates, are described in this chapter.

Optical polymers can be nanoengineered to meet all the requirements of Table 14.1. Some of the most promising polymers are hyperbranched fluoroacrylates (Figure 14.1). The main approaches used to meet 14 of the 50 required properties of Table 14.1 are listed in Table 14.2.

Optical polymers can be highly transparent, with absorption loss values around or below 0.1 dB/cm at all the key communication wavelengths (840, 1310, and 1550 nm). In order to reduce absorption loss at telecom wavelengths such as the 1260–1360 nm and 1450–1610 nm windows, halogenated polymers were synthesized such as the one whose IR spectrum is shown in Figure 14.2, demonstrating high transmission at the main communication bands.

Figure 14.3a shows the insertion loss at 1550 nm of a straight waveguide fabricated in 80% fluorinated acrylates, and cut back to different lengths. The plot shows about 0.1 dB/cm propagation loss (slope) and about 0.3 dB fiber coupling loss for both sides of the chip (y intercept). Figure 14.3b depicts the insertion loss of a 3 cm long straight waveguide tested between 1500 and 1570 nm wavelength, exhibiting low wavelength-dependent loss (WDL) in this spectral window.

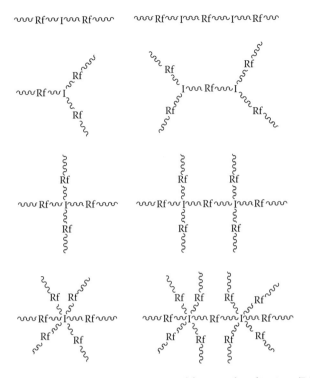

FIGURE 14.1 Hyperbranched fluoroacrylates nanoengineered for optical applications (Rf: fluoroalkyl group).

TABLE 14.2 Some Properties of Optical Polymers and Approaches to Achieve Them

Property	Approach to Achieve Property
Low absorption at 800–1600 nm	Halogenation
Low birefringence	Isotropy, homogeneity
Low PDL	Isotropy, homogeneity
Low shrinkage	High molecular weight
High stability	Molecular design
Fast curing	One end pre-fixed
High crosslink density	One end pre-fixed
Low volatility	High molecular weight
Controlled viscosity	Molecular weight, molecular structure
High dn/dT for thermo-optics	High CTE, low T_g
Mechanical properties	Nanoadditives, molecular structure
High-temperature properties	Low T_g, nanoadditives
Barrier properties	Molecular structure, nanoadditives
Processability for high film quality	Molecular weight, nanoadditives

The PDL (PDL = loss_{TE} – loss_{TM}) varies with processing conditions. The TE loss measured in planar waveguides can be higher than the TM loss when the vertical walls of the core have a higher degree of roughness than the horizontal boundaries, and it can be lower when the vertical evanescent tails overlap with an absorptive substrate or superstrate. These single-mode waveguides are well optimized by having minimal edge roughness and a well-confining material stack, and as a result have PDL values below 0.01 dB/cm.

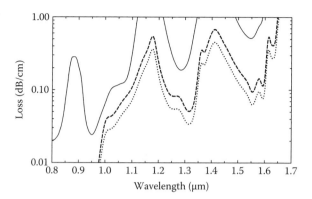

FIGURE 14.2 Absorption loss spectra for three high-performance acrylic optical polymers: solid line is for no halogenation, dashed line is for 70% halogenation, dotted line is for 80% halogenation.

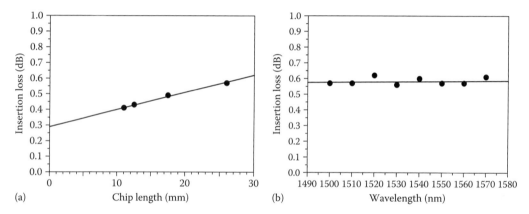

FIGURE 14.3 (a) Results of a cut-back experiment on a polymer waveguide, revealing about 0.1 dB/cm propagation loss and 0.3 dB total fiber coupling loss, and (b) insertion loss of a polymer waveguide between 1500 and 1570 nm wavelength.

Three-dimensionally cross-linked polymers undergo little molecular orientation during processing, and as a result have a birefringence ($n_{TE} - n_{TM}$) value that is extremely low (on the order of 10^{-6}). Devices fabricated in these polymers exhibit low PDL, and low PMD value in devices based on these materials is on the order of 0.01 ps.

Polymer waveguides can achieve a high refractive index contrast (Δn) in a buried channel, which is preferable to the high Δn that can be achieved in air-clad semiconductor waveguides. Buried channel waveguides are needed to avoid excessive loss and polarization dependence when metal electrodes are deposited on a photonic circuit for actuation. The refractive index of polymers varies between 1.28 and 1.72 at 1550 nm. The value of Δn is controlled by material composition and is continuously tunable between 0% and 30%. A Δn of 30% is 7.5 times larger than the maximum Δn of 4% achievable in silica,[5] and it permits the use of small radii of curvature that enable high-density compact waveguiding structures. Table 14.3 shows various index contrast levels achievable in high-performance optical polymers and the resulting core size, fiber coupling loss, minimum bend radius, and relative chip size.

A high index contrast results in many advantages, including shorter devices (therefore lower absorption loss), more confining structures (therefore lower radiation loss), and higher performance devices (Mach–Zehnder interferometers [MZI] have larger bandwidth; ring resonators have larger free spectral range [FSR]; photonic crystals have improved directivity; and amplifiers have higher gain efficiency,

TABLE 14.3 Some Refractive Index Contrast Levels Achievable in High-Performance Optical Polymers and the Resulting Core Size, Fiber Coupling Loss, Minimum Bend Radius, and Relative Chip Size

Δn	Standard	Medium	Medium High	High	Very High	Ultrahigh	Highest
Δn value	0.0075	0.01125	0.015	0.0225	0.045	0.15	0.45
$\Delta n/n$	0.5%	0.75%	1%	1.5%	3%	10%	30%
Core size[a]	$7 \times 7\,\mu m^2$	$6 \times 6\,\mu m^2$	$4.5 \times 4.5\,\mu m^2$	$3 \times 3\,\mu m^2$	$2 \times 2\,\mu m^2$	$1 \times 1\,\mu m^2$	$0.5 \times 0.5\,\mu m^2$
Coupling loss[b]	0.0077 dB	0.23 dB	0.56 dB	1.6 dB	4.2 dB	8.0 dB	9.6 dB
Bend radius[c]	15 mm	6 mm	4 mm	2 mm	0.5 mm	0.1 mm	0.01 mm
Chip size[d]	●	●	•	•	·	·	

[a]Largest core size for square-cross-section single-mode buried channel waveguide.
[b]Lowest possible coupling loss to Corning SMF28 fiber for both sides of chip.
[c]Smallest radius of curvature for negligible bend radiation loss.
[d]Relative dimensions.

lower power dissipation, and lower noise figure).[6] The disadvantages include a larger roughness to core size ratio resulting in higher scattering loss, and a small mode-field diameter resulting in higher fiber coupling loss. The improvements obtained at high index contrasts are very desirable, but one has to deal with the increased scattering loss and fiber coupling loss. Approaches to deal with the high scattering loss include the use of a graded index, or the elimination of roughness by reflowing and coalescing the material when feasible. Approaches to deal with the high fiber coupling loss include the use of adiabatic tapers, or simply having a high level of integration that makes the loss decrease due to having a large number of elements on a short chip larger that the loss increase due to the higher fiber coupling loss. One technically possible (but expensive and slightly bulky) solution that makes the increase in both types of loss essentially a nonissue would be the integration of waveguide amplifiers.

The unique combination of large thermo-optic coefficient and low thermal conductivity makes polymers ideal materials for thermo-optic devices such as optical switches, variable attenuators, and tunable filters. The thermo-optic effect is the change of refractive index, n, with temperature, T, and is commonly referred to as dn/dT. For an amorphous polymer, the refractive index change is predominantly due to its density change. Therefore, in order to increase the thermo-optic effect, polymers were designed with high coefficient of thermal expansion (CTE). In addition, these polymers have a T_g well below the low end of telecom operating temperature specification ($-40°C$) and a large free volume. Figure 14.4 illustrates the refractive index change of one of these polymers with temperature. The measurement was performed with a thin polymer film using an Abbe refractometer. This polymer has a

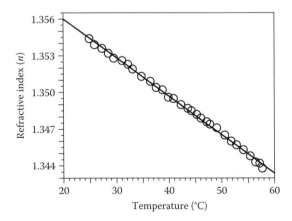

FIGURE 14.4 Refractive index change with temperature for a fluorinated acrylic polymer.

thermo-optic coefficient of about $-3.2 \times 10^{-4}/°C$. This dn/dT is 32 times larger than that of silica, and 3–5 times larger than that of common optical polymers such as polymethylmethacrylate (PMMA) and PC. A proportionate decrease in power consumption results in thermo-optic switches and tunable devices. For a tunable Bragg grating filter made with this polymer, this thermo-optic coefficient translates into a $d\lambda/dT$ of about -3.7 nm/°C, allowing the tunable filter to cover the entire ITU C band (1528–1565 nm) with 100°C temperature range.[7]

Organic materials can have large electro-optic coefficients (r_{33}), enabling low-voltage modulators, fast switches, and PMD compensators. The major barrier for the commercialization of electro-optical polymer–based components has been reliability. Adding a high concentration of nonlinear, optical, organic chromophore molecules to a polymer and poling (aligning) these molecules with a strong electric field during fabrication are necessary to achieve a large electro-optic effect in that polymer. Using such a concentration, however, results in electrostatic interaction between the highly polar molecules, thereby causing relaxation or reorientation, which in turn reduces the electro-optic coefficient over time. This commonly misunderstood and widely published degradation phenomenon is behind the perception of poor reliability in polymers. Isotropic thermo-optic polymers have none of the directional asymmetry that causes the instability of electro-optic polymers—and therefore do not suffer from this reliability problem.

The environmental stability of optical polymers is an important issue because most polymers do not have properties that are adequate for operation in communication environments. Figure 14.5 shows the results of thermal gravimetric analysis (TGA) measurements on nanoengineered thermo-optic polymers, revealing very high thermal stability. The ramp TGA measurement (Figure 14.5a) shows that stability is maintained up to 400°C in nitrogen and 370°C in air, and the isothermal TGA measurement (Figure 14.5b) shows stability in air at 175°C.

A key characteristic for practical applications is the thermal stability of the optical properties of polymeric photonic components since organic materials may be subject to yellowing upon thermal aging due to oxidation. The presence of hydrogen in a polymer allows the formation of H-Halogen elimination products, which result in carbon double bonds, which are subject to oxidation. Fortunately, the absorbing species from thermal decomposition are centered near the blue region of the spectrum, and the thermal stability is high at the datacom wavelength of 840 nm and even greater at the telecom wavelengths of 1310 and 1550 nm. Heating is not caused only by the environment and by electrical heating electrodes on the chip, it can also be the result of high optical power propagating in the waveguides. Components were subjected to highly accelerated aging tests, the results are shown in Figure 14.6. The waveguides exhibit high resistance to heat from the environment (5000 h at 175°C) and from guided 1550 nm optical power (6000 h at 1.5 W). If the optical power would have been slightly higher at 2 W, the silica fiber used

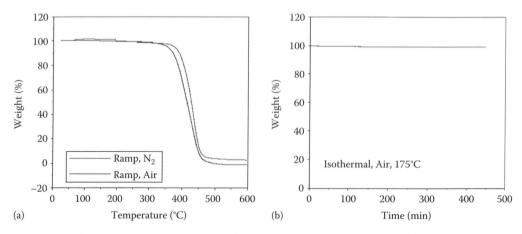

FIGURE 14.5 Ramp TGA results (a) showing stability up to 400°C in nitrogen and 370°C in air, and isothermal TGA results (b) showing stability in air at 175°C.

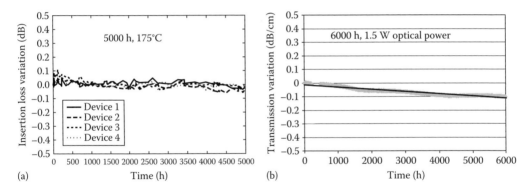

FIGURE 14.6 Transmission of polymer waveguides held at (a) 175°C for 5000 h and (b) 1.5 W optical power for 6000 h.

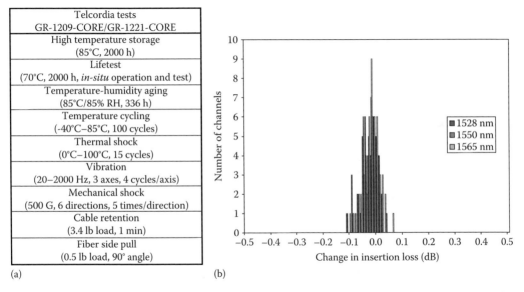

FIGURE 14.7 Telcordia 1209/1221 tests (a), and results obtained from 11 four-channel polymer VOA parts that were subjected to these tests (b). The variation in insertion loss is within about ±0.1 dB, with the Telcordia-defined pass criterion being ±0.5 dB; hence, these components passed the qualification testing with a large margin.

to pigtail the devices would have fused. The power level of 1.5 W in a $7 \times 7 \, \mu m^2$ cross section is $100 \times$ more intense than the optical power density on the surface of the Sun.

Dynamic optical components based on custom-nanoengineered thermo-optic polymers have passed Telcordia 1209/1221 qualification tests (Figure 14.7).[8] Subsystems based on these components, such as variable multiplexers (VMUX) and reconfigurable optical add/drop multiplexers (ROADM), have met the requirements of Telcordia GR-1312 and GR-63 protocols.[9]

14.2 Polymeric Waveguide Fabrication

Polymer films can easily be formed in a wide range of thicknesses (10 nm to 1 mm) on virtually any substrate (e.g., glass, silicon, glass-filled epoxy printed circuit boards, and flexible plastic films) by spin coating, spray coating, slot coating, doctor blading (tape casting), dipping, or other wet film formation technique. Robotized equipment has been developed to high-volume waveguide fabrication on silicon wafers (Figure 14.8).

FIGURE 14.8   Robotic system for the fabrication of polymeric waveguides on 6 in. silicon wafers.

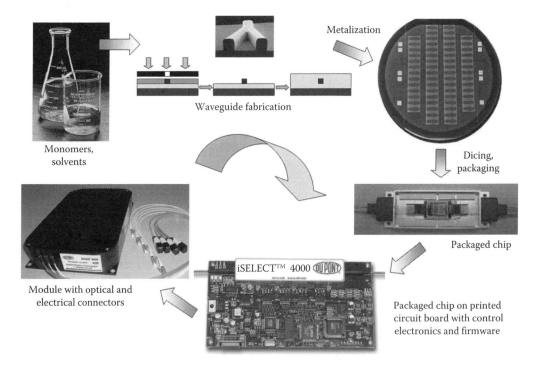

FIGURE 14.9   Polymer-on-silicon photonic IC component manufacturing.

Some polymers, such as most polyimides and PCs, are not photosensitive and are typically processed using photoresist patterning and reactive ion etching (RIE). These polymers have most of the issues of the silica-on-silicon technology in terms of roughness-induced and stress-induced scattering loss and polarization dependence. Other polymers are photosensitive and as such are directly photo-patternable (Figure 14.9), resulting in low roughness and low stress waveguides. The full cycle time for the fabrication of a wafer with a multilayer waveguiding circuit based on photosensitive polymers is about 30 min, giving these polymers a 10- to 1000-fold advantage in throughput over other PLC materials. Furthermore, this technology uses low-cost materials and low-cost/low-maintenance capital equipment (e.g., spin-coater and UV lamp vs. CVD growth system).

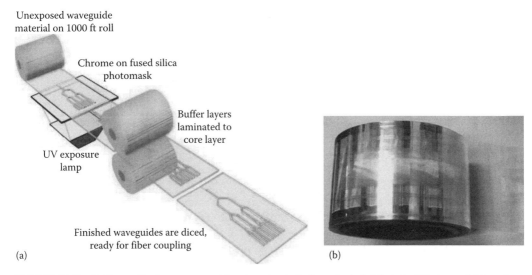

Unexposed waveguide
material on 1000 ft roll

Chrome on fused silica
photomask

Buffer layers
laminated to
core layer

UV exposure
lamp

Finished waveguides are diced,
ready for fiber coupling

(a)

(b)

FIGURE 14.10 Roll-to-roll polymer waveguide patterning techniques include (a) photodiffusion and (b) mechanical stamping.

High-speed patterning techniques can also be used with organic materials, including the following:

- Molding (injection molding, compression molding, etc.). This low-cost technique typically uses soft (e.g., silicone) tooling.
- Hot and soft embossing. This technique uses hard (e.g., metal) tooling, and achieves high uniformity over large areas.
- Photodiffusion combined with reel-to-reel processing. Polyguide® technology[10] (Figure 14.10a) uses this process for the production of polymer optical interconnect strips.
- Reel-to-reel stamping. This technique was developed for the rapid, low-cost, mass production of photonic circuits. Such a manufacturing technique allows us to address high-volume markets such as the consumer optoelectronics market. Figure 14.10b shows a roll of plastic with thousands of $10\,cm \times 10\,cm$ organic photonic circuits stamped at a rate of $20\,m/min$. Each circuit includes thousands of waveguides for optical scanning applications.

Polymeric materials can be nanoengineered to meet all the requirements of optical applications. Waveguides based on these materials offer high performance, compactness, tunability, environmental stability, mechanical flexibility, ease of fabrication, high yields, and low cost.

14.3 Polymeric Microphotonic Devices and Circuits

A healthy photonics industry will be built on standardized component platforms that can be manufactured in high volume at low cost. These platforms will deliver performance:cost ratios that scale exponentially with each technology generation. Polymeric materials provide a versatile set of properties and processes that meet this need for both hybrid and monolithic photonic circuits. Polymeric devices have been fabricated and tested for both passive and active functions. Photonic circuits constructed of polymeric components are being manufactured and deployed in the field. During the next decade, polymeric materials will continue to be a critical ingredient in commercial photonic circuits.

The ability to efficiently guide and control an optical signal carrier will drive the technology for the next decade, and will enable a significant reduction in the cost of optical components. A realistic

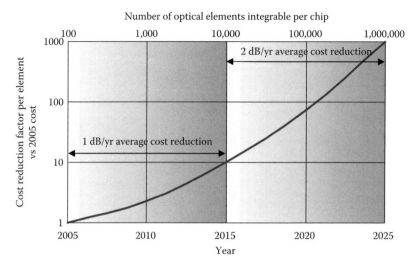

FIGURE 14.11 Integration and cost reduction roadmap for polymer optical device integration.

roadmap for optoelectronic integration is defined by (1) the industry acceptance timeline, colliding with (2) the technical feasibility timeline, and (3) the value driver at all points. These criteria were considered in the generation of the polymer optical device integration roadmap of Figure 14.11, where the ratio of discrete to integrated component cost is plotted vs. the number of functions integrated on a chip between the years 2005 and 2025. From 2005 to 2015, optoelectronic integration is expected to result in about 1 dB (20%) cost reduction vs. discrete components per year, resulting in 2015 in 10 dB cost reduction (10% of cost of discretes). From 2015 to 2025, about 2 dB (40%) cost reduction is expected per year, resulting in 2025 in an additional 20 dB cost reduction, for a total of 30 dB (0.1% of cost of discretes) in two decades.[1]

The market drivers for polymers in optoelectronics vary across application areas. Table 14.4 summarizes these drivers. The great potential of integrated optical waveguiding devices has been recognized for decades, with the most notable advantages being cost reduction, performance improvement, and size reduction—the well-known "cheaper, better, smaller." Additional advantages include interface simplification and reduction of packaging complexity, weight reduction, reduction of power dissipation, reliability improvement, ease of scalability, ease of customization, ease of automation, yield improvement, throughput increase, reduction in need for manual labor, and faster time to market.[11]

14.3.1 Optical Interconnects

Optical interconnects have applications in the following:

- Short-reach data links
  - Local area networks (LAN) interconnects in enterprise: include user-to-user/cabinet-to-cabinet (intra-office), floor-to-floor (intra-building), and building-to-building (intra-campus) interconnects.
  - Control system interconnects: include intra-vehicle/institution/home interconnects, vehicles include all vehicles in the automotive and aerospace industries, and institutions include medical institutions and factories.
  - Video/gaming/education on demand (VoD/GoD/EoD) interconnects in the digital home: include interconnects for high-definition video on demand, gaming on demand, and education on demand.

TABLE 14.4 Market Drivers for Polymer Optical Devices

Industry	Application	Product Examples	Requirements	What Polymers Offer	Technology Needed	Timing of Market Need
Telecom	Optical signal routing, switching, and power level control (long-haul, metro)	ROADM	Low cost (capex and opex for OEM and carrier)	Lowest power consumption dynamic thermo-optic components	Cost-effective "pick-and-place" assembly technology	Now
		Optical crossconnect (OXC)	High reliability	Practical hybrid integration enabling complex functionality PLCs today in material with state-of-the-art performance		
		Protection switches	Small size			
		Variable optical attenuator (VOA)	Low power dissipation			
		Variable multiplexer (VMUX)	Ease of scalability			
		Dynamic channel equalizer (DCE)	Simple fiber management			
		Dynamic gain equalizer (DGE)				
	Optical signal distribution (access, FTTx)	Splitters	Low capex for OEM and carrier	Low-cost, high-volume replication techniques (stamping, etc.)	Low-cost fiber arrays	Now
			Low opex for OEM and carrier			
		Athermal mux/demux	High reliability	Athermal AWG with substrate of proper CTE		
			Athermal passive behavior			
Datacom (computer, automotive, aerospace, security)	Short-reach data links for entertainment and control systems in digital home and vehicles	Plastic optical fiber (POF)	Ease of splicing and connecting	Multimode graded index POF meeting all requirements	Ultralow loss to compete with glass for long distances, ultralow cost to compete with copper for short distances	Now
			Lightweight	With transparency windows at 850, 670, and 530 nm, it is compatible with silicon or polymer photodetectors, and with silica or polymer waveguide circuits		
			Low bending loss			
			Resiliency to mechanical impact			

(continued)

TABLE 14.4 (continued)   Market Drivers for Polymer Optical Devices

Industry	Application	Product Examples	Requirements	What Polymers Offer	Technology Needed	Timing of Market Need
	Computing	Optical interconnects on backplanes/boards/MCM for workstations and servers, and eventually personal computers	Low cost Low loss Ease of connecting PCB lamination (temperature and pressure) compatibility	Easy to manufacture and cost-effective large-area optics (larger than common photomasks)	Dust-resistant connectors	5–10 years
	Computing	On-chip optical interconnects for workstations and servers, and eventually personal computers	Low cost Low loss Ease of connecting Lightweight Low bending loss CMOS compatibility	Ease of coating on CMOS chips Processability with reduced temperature excursions Rapid and cost-effective manufacturing	Compact and inexpensive integrated transceivers, amplifiers, etc.	20–30 years
Display, imaging, scanning, learning	Image capture and display	Displays and cameras in consumer products (cell phones, PDAs, portable multimedia players, electronic newspapers, etc.) Signage Light collection waveguide arrays in optical scanners	Low cost (eventually enabling omnipresent displays, disposable displays, etc.) Lightweight Flexibility Impact resistance Wide viewing angle	Flexibility Impact resistance	Long lifetime	Now
Military, medical	Sensing, monitoring	OEIC in your shirt	Sensing fabrics Lightweight	Polymer sensing fibers can be woven into comfortable, lightweight clothing	Further development of sensing fibers	5–10 years

FIGURE 14.12   Polyguide strips linking transmit/receive boards through MT connectors.

- Compute platform
  - Off-chip interconnects for servers and workstations, eventually personal computers: include interconnects on backplanes, boards, and multi-chip modules (MCMs).
  - On-chip interconnects: include chip-to-chip and intra-chip interconnects.

Short-reach data links are needed today for LAN in enterprise environments, for control systems in hospitals, for intra-vehicular communications in the automotive and aerospace industries, and for entertainment in the digital home. POF is used for these short-reach applications that require simplicity and low cost. This multimode graded index transmission medium exhibits ease of splicing and connecting, lightweight, low bending loss, and resiliency to mechanical impact. With transparency windows at 530, 670, and 850 nm, it is compatible with silicon or polymer photodetectors, and silica or polymer waveguide circuits.

Polymeric waveguides are broadly accepted as the most promising low-cost electronics-friendly media for high-speed optical interconnects.[12] They offer CMOS compatibility (conformal signal routing, simple planarization, minimal temperature excursion, low temperature dissipation for dynamic/active functions, etc.), manufacturability (rapid process, low cost, compatibility with 265°C Pb-free wave soldering, etc.), and adaptability (all elemental optical functions achievable, widely tunable numerical aperture, possibility of post-fabrication trimming/rework, possibility of 3D routing through processes such as two-photon polymerization, etc.). Mature low-cost manufacturing-friendly polymer optical interconnect technologies exist today (Figure 14.12), and are awaiting market pull. Planar polymer optical interconnects have been produced for the last two decades on backplanes for board-to-board interconnects, on boards for MCM-to-MCM interconnects, and on MCMs for chip-to-chip interconnects.[13] Planar polymer optical interconnects on backplanes, boards, and MCMs will be used in workstations and servers in 5–10 years. However, chip-level interconnects will not be needed until 20–30 years from now, when ultracompact, high-performance, and low-cost optoelectronic integrated circuits (OEICs) (e.g., transceivers, amplifiers) become available.

14.3.2 Couplers and Taps

The need for tapping off a controlled amount of optical power from a carrier signal is common in optical networks, particularly at points where the quality of signal (QoS) needs to be assessed in terms of a variety of characteristics including optical power level, spectral content, and optical signal-to-noise ratio (OSNR). Polymeric optical power couplers/taps can be used for this purpose in hybrid circuits.

14.3.3 Splitters and Combiners

The explosion of the FTTx market, initially in Japan, then in the United States and Europe, and now in Korea and China, has generated the need for a large number of optical power splitters, typically 1×8 to 1×32 splitters. Polymeric splitters can be produced inexpensively in large volumes using high-throughput processes such as stamping.

14.3.4 Multiplexers and Demultiplexers

The large thermo-optic coefficient of polymers contributes to a high sensitivity to ambient temperature in interferometric devices such as arrayed waveguide grating (AWG) multiplexers/demultiplexers (mux/demux), which are generally better done in low thermo-optic coefficient materials such as silica (given that they are static devices and as such do not take advantage of the low power actuation). However, athermal AWGs can be easily produced in polymer by matching the CTEs of the substrate, waveguide films, and superstrate, thereby canceling the effect of temperature on the interferometric device. The superstrate thickness can be adjusted to tune its effective CTE, thereby tuning the AWG thermal dependence to achieve athermal performance. Polymeric athermal AWGs exhibiting 1 pm/°C stability have been demonstrated.[14] This type of device is needed for FTTx signal distribution in wavelength division multiplexing passive optical networks (WDM PON).

14.3.5 Polarization Controllers

Polymeric (mainly polyimide) half-wave plates are commonly inserted in the waveguide array of AWGs (typically made in silica) to reduce their polarization dependence.[15] However, since polyimide is a hygroscopic material, hermetically sealed packages should be used to achieved high stability in performance.

14.3.6 Switches

Thermo-optic $N \times N$ switches can be interferometric switches based on directional couplers or MZI, or they can be digital optical switches (DOS) based on X junctions or Y junctions.[16] The most widely used switch design is the Y-junction-based DOS (Y-DOS), because of its simplicity, and its insensitivity to applied electrical power, wavelength, polarization, ambient temperature, and dimensional variation. The insensitivity to applied electrical power is what enables the digital behavior. The building block in a Y-DOS is a small-angle (typically 0.1°) 1×2 splitter with heaters on its arms, as shown in the schematic diagram of Figure 14.13a. These splitters can be connected with bends and crossings to form $M \times N$ switching matrices. Each 1×2 splitter relies on the adiabatic evolution of the mode profile in its two waveguides into the mode of the ON guide (the guide with the higher effective refractive index) when the OFF guide is heated to reduce its index, as shown in the computer simulation of Figure 14.13b.

The device is considered to have switched once it reaches the desired isolation value, which occurs at some level of electrical power dissipation in the electrodes, beyond which power level the device maintains the isolation, resulting in its well-known "digital" behavior (Figure 14.14). A typical maximum time for restoration and restructuring of multiple circuit connections for SONET is 50 ms. The measured response time for these thermo-optic switches is approximately 3 ms, a value that is adequate for system restoration.

DOSs in a 1×2^N configuration can be fabricated with $2^N - 1$ 1×2's, and one electrode at each stage needs to be heated to perform the switching. For instance, a 1×4 switch can be built with three 1×2's, as shown in Figure 14.15a, where the upper electrode in the first stage and the lower electrode in the second stage are powered to switch the light from port 1 to port 3′.

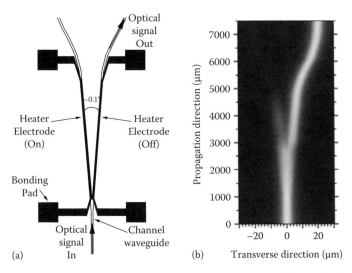

FIGURE 14.13 (a) Schematic diagram of a 1×2 Y-DOS and (b) computer simulation of such a device where the left arm is heated, allowing the light to exit the right arm.

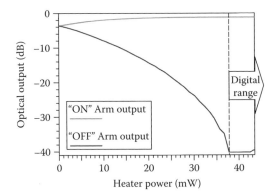

FIGURE 14.14 Operational characteristics of a Y-branch-based 1×2 digital optical switch.

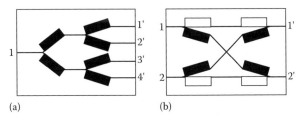

FIGURE 14.15 Schematic layouts of (a) a 1×4 DOS and (b) a 2×2 (cross-bar) DOS. The dark electrodes are powered. The 1×2 is shown switching from port 1 to port $3'$ and the 2×2 is in the bar state.

DOSs in a strictly non-blocking $N \times N$ configuration can be fabricated with $2N(N-1)$ 1×2's (Table 14.5).[6] Figure 14.15b shows a 2×2 (or cross-bar) DOS built with four 1×2's. This switch is operated in the bar state by powering the four inner electrodes, while powering the four outer electrodes results in the cross state.

An eight-channel intelligent optical crossconnect (iOXC) subsystem on a chip achieves 8×8 strictly non-blocking switching with power monitoring and power balancing. The chip architecture

(Figure 14.16) includes 112 1×2 switches that make up a strictly non-blocking 8×8 switch matrix, eight optical power taps and eight integrated photodiodes for per-channel power monitoring, and eight variable optical attenuators (VOAs) for power level control. The photocurrents generated by the photodiodes are used by a feedback electronic circuit to control the VOAs, thereby achieving closed-loop automatic power balancing on all the channels. An 8×8 iOXC is shown in Figure 14.17 at three different stages of production.

The fiber-to-fiber insertion loss of the single-chip iOXC of Figure 14.17, between 1528 and 1610 nm wavelength, is 4 dB (including 5% tapped power). The PDL at minimum insertion loss from any port to any port is 0.2 dB, the PMD is 0.01 ps, and the CD is 0.1 ps/nm. The switch isolation (or extinction) is 45 dB, and the crosstalk from any port to any port is 50 dB.[17]

The small value of dn/dT and limited index contrast in silica PLCs has kept 16×16 switches from being implemented with 1×2 DOS switches. Planar 16×16 switches implemented to date in silica have been based on MZIs.[18] The first DOS-based 16×16 switch was implemented in polymer.[19] This switching matrix consists of 480 1×2 switches, interlinked with 704 S-bends that intersect at 227 locations to provide a strictly non-blocking connectivity. Figure 14.18 shows the 16×16 switch both schematically (a) and to scale (b), as well as a photograph of the heaters in a fabricated device (c).

In the characterization of the switching matrix, the performance of the DOS and crossing building blocks were first measured. In a fully functional 16×16 switch, each path is defined by heating 8 heaters, thus 128 heaters are continuously being used. Since the crosstalk is not limited by the switches, an extinction of 15 dB per 1×2 stage already yields sufficient effective extinction due to the concatenation of the eight switching/combining stages, and taking into account the additional crosstalk due to the crossings. At a power dissipation of 50 mW per DOS, the insertion loss is 6 dB and the extinction is 30 dB, mainly limited by the crosstalk due to crossings in the design. The total nominal power consumption is 6.4 W.

Whereas the 16×16 switch shows the largest fabricated strictly non-blocking N×N switch, significantly larger switching matrices have been designed and are in early development stages. Using the highest Δn achievable in polymers, the first planar 1024×1024 OXC was designed.[20] The most compact design measured 4.2×6.3 cm², allowing 4 chips to fit on a standard 6 in. wafer.

TABLE 14.5 The Number of 1×2 Switches Needed in Planar Strictly Non-Blocking N×N Switches and the Degree of Maturity of the Different Switch Sizes

N	Number of 1×2's
2	4
4	24
8	112
16	480
32	1,984
64	8,064
128	32,512
256	130,560
512	523,264
1024	2,095,104

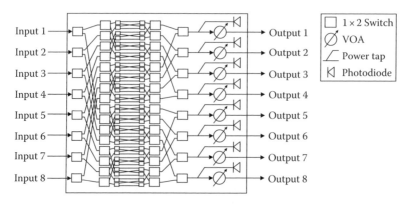

FIGURE 14.16 Chip architecture of an 8×8 iOXC with power monitoring and automatic power balancing.

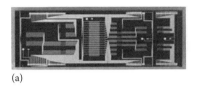
(a)

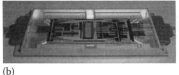
(b)

(c)

FIGURE 14.17 An 8×8 iOXC at three production stages: (a) chip, (b) unsealed package, and (c) sealed package on control PCB.

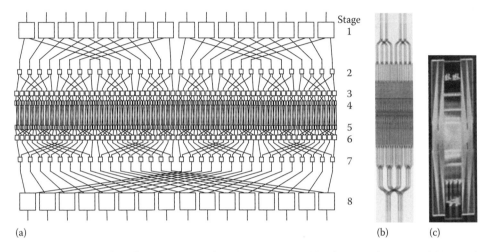

(a)   (b)   (c)

FIGURE 14.18 Design of the first 16×16 DOS-based switch matrix. (a) Schematic diagram and (b) CAD layout of the recursive tree structure that makes up the waveguiding circuit, and (c) electrode layer of a fabricated device.

14.3.7 VOAs, Tunable Couplers, and Variable Taps

VOAs, tunable couplers, and variable taps are analog devices, and as such do not benefit from digital designs and can be based on any switching principle including interferometry, modal transition, or mode confinement. A common interferometric approach uses an MZI, where heat can be applied to at least one of the two arms to induce a phase shift between the signals propagating in them before they recombine, thereby controlling the level of optical power exiting the output guide. Figure 14.19a shows a simulation of this device when power is applied to thermally induce a π phase difference between the optical signals in the two arms, causing the signals at recombination to form an asymmetric mode that radiates into the cladding, since it is not supported by the single-mode output waveguide, resulting in full attenuation. Figure 14.19b shows the operational characteristics of a polymer MZI VOA exhibiting very low power consumption of about 1.4 mW for 30 dB attenuation.[21]

14.3.8 Tunable Filters

Tunable filters based on Bragg gratings have been produced in planar polymers by a variety of techniques such as molding, embossing, stamping, e-beam writing, and photochemical processes.[7] The first three techniques produce surface relief gratings while the last two can produce either relief gratings or bulk index gratings across the waveguide core. Photochemical fabrication processes for bulk index gratings utilize two-beam interference to induce an index modulation. When tunable adding/dropping of signals is desired, a hybrid device of the type shown in Figure 14.20a or an integrated device of the type shown in Figure 14.20b can act as a thermally tunable ROADM, also known as Tunable OADM

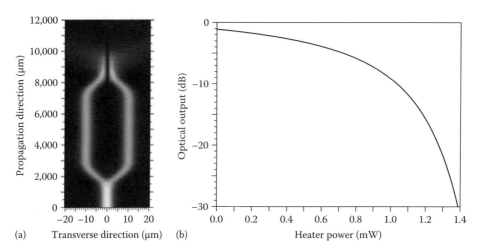

FIGURE 14.19   (a) Computer simulation of an MZI VOA where heat is used to induce a π phase difference between the interferometer arms for full attenuation, and (b) attenuation curve of such a VOA.

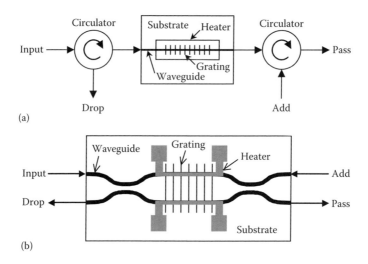

FIGURE 14.20   Hybrid (a) and monolithic (b) TOADM configurations based on tunable gratings.

(TOADM).[22] In the hybrid approach, the chip consists of a grating produced in an integrated single-mode polymeric waveguide with a deposited thin film heater. This chip, combined with two three-port optical circulators, forms a TOADM. In this design, the only critical requirement for the die is to have a grating with the desired spectral response; the circulators provide the add/drop functions. In the integrated approach, an MZI with a grating across its arms forms the TOADM. In this design, the couplers must be fabricated carefully for 3 dB operation in order to achieve high-port isolation. Tuning of the grating is achieved by powering a heater that is fabricated in the proximity of the grating. The large dn/dT in the polymer allows tuning across the entire C band (1528–1565 nm) with a temperature excursion of about 100°C (Figure 14.21).

14.3.9 Chromatic Dispersion Compensators

Ring-resonator-based all-pass filters with phase shifters can be used to achieve tunable CD compensators.[5] The ease of tuning the refractive index of polymers makes them a natural choice for this device.

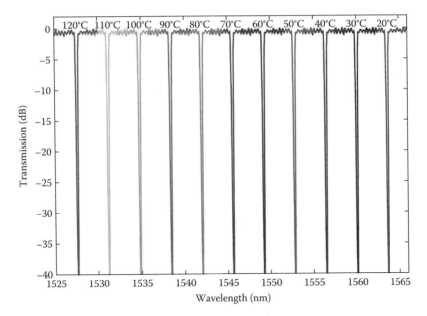

FIGURE 14.21 Transmission spectra for a polymer Bragg grating filter tuned across the C band between 1528 and 1565 nm with a 100°C temperature excursion.

Polymers can be used in ring resonators as cladding materials on semiconductor core to provide easy tuning. They can also be used as core with air cladding, since the Δn with air is high enough for tight confinement and large FSR, but not as high as in semiconductors where the core is too small and the scattering loss too high.

14.3.10 Modulators

High-speed, low-voltage modulators have been the main focus of the R&D in most optical polymer groups for the last two decades. The major barrier for the commercialization of polymer modulators has been reliability. Adding a high concentration of nonlinear, optical, organic chromophore molecules to a polymer and poling (aligning) these molecules with a strong electric field during fabrication are necessary to achieve a large electro-optic effect in that polymer. Using such a concentration, however, results in electrostatic interaction between the highly polar molecules, thereby causing relaxation or reorientation, which in turn reduces the electro-optic coefficient over time. This commonly misunderstood and widely published degradation phenomenon is behind the perception of poor reliability in polymers. Isotropic thermo-optic polymers have none of the directional asymmetry that causes the instability of electro-optic polymers—and therefore do not suffer from this reliability problem.

14.3.11 Ultrafast VOAs, Switches, and PMD Compensators

The high-speed nature of electro-optic actuation makes electro-optic polymers a viable option for ultrafast VOAs that can be used for high-speed transient suppression, as well as for subcarrier modulation. The latter application can be used to add wavelength and channel identification markers, thereby enabling additional functionality in the optical control plane. Ultrafast switches can also be fabricated with electro-optic polymers for high-speed applications such as packet switching. In addition, these materials can be used in PMD compensators, which require high-speed elements that can react to the fast and random changes in PMD.

14.3.12 Active Devices

Active polymer optical components include amplifiers, sources, and detectors. An off-chip optical source (optical bus) architecture alleviates much of the thermal management required for source heat-sinking. Methods must be found for the effective application and patterning of a variety of polymeric materials, since each device is now based on a unique material. Optical amplifiers and index doping can be achieved with semiconductor quantum dot attachment on specific moieties. Polymeric photovoltaic elements were used to achieve photodetectors in the visible range for short-reach data links, as well as solar cells. A high-value application for polymeric materials would be an optical solder equivalent for fiber attach.

Polymeric materials can accomplish the full array of functions needed in optical communications, and can be processed at temperatures that are compatible with CMOS-integrated circuits. Passive dynamic thermo-optic polymeric photonic components with cutting-edge performance and proven reliability are widely available commercially and are used extensively, whereas passive dynamic electro-optic and active polymeric photonics are less mature and do not yet deliver high performance. Over the next decade, polymeric materials will continue to be an indispensable ingredient in integrated photonic circuits and will play a key role in meeting market needs for low-cost, large-volume photonic components.

References

1. L. Eldada, Organics in optoelectronics, in *Microphotonics: Hardware for the Information Age*, Ed. L. Kimerling, MIT, Cambridge, MA (2005).
2. L. Eldada and L.W. Shacklette, Advances in polymer integrated optics, *IEEE J. Sel. Top. Quantum Electron.* **6**, 54 (2000).
3. L. Eldada, Advances in fluoropolymer-based optoelectronics for data transmission, routing and processing, fluoropolymer 2006: Current frontiers and future trends, *Proc. FP* **4**, 24 (2006).
4. L. Eldada, Nanoengineered polymers for photonic integrated circuits, *Proc. SPIE* **5931**, 16 (2005).
5. C.K. Madsen, E.J. Laskowski, J. Bailey, M.A. Cappuzzo, S. Chandrasekhar, L.T. Gomez, A. Griffin, P. Oswald, and L.W. Stulz, Compact integrated tunable dispersion compensators, *Proc. LEOS* **15**, 570 (2002).
6. L. Eldada, Toward the optoelectronic ULSI: Drivers and barriers, *Proc. SPIE* **5363**, 1–15 (2004).
7. L. Eldada, R. Blomquist, M. Maxfield, D. Pant, G. Boudoughian, C. Poga, and R.A. Norwood, Thermo-optic planar polymer Bragg grating OADM's with broad tuning range, *Photonics Technol. Lett.* **11**, 448 (1999).
8. L. Eldada, Telcordia qualification and beyond: Reliability of today's polymer photonic components, *Proc. SPIE* **5724**, 96–106 (2005).
9. L. Eldada, ROADM architectures and technologies for agile optical networks, optoelectronic integrated circuits, *Proc. SPIE* **6476**, 5 (2007).
10. L. Eldada, Organics in optoelectronics: Advances and roadmap, *Proc. SPIE* **6124**, 32 (2006).
11. L. Eldada, Photonic integrated circuits, in *Encyclopedia of Optical Engineering*, Ed. R. Driggers, Marcel Dekker, New York (2003).
12. L. Eldada, Advances in polymer optical interconnects, *Proceedings of the IEEE LEOS Annual Meeting*, Sydney, Australia, Vol. 18, pp. 361–362 (2005).
13. L. Eldada, Polymer optical interconnects, in *Future Trends in Microelectronics: Off the Beaten Path*, Eds. S. Luryi, J. Xu, and A. Zaslavsky, John Wiley & Sons, New York, p. 451 (1999).
14. A. Yeniay, R. Gao, K. Takayama, R. Gao, and A. Garito, Ultra-low-loss polymer waveguides, *J. Lightw. Technol.* **22**, 154 (2004).
15. Y. Inoue, H. Takahashi, S. Ando, T. Sawada, A. Himeno, and M. Kawachi, Elimination of polarization sensitivity in silica-based wavelength division multiplexer using a polyimide half-wave plate, *J. Lightw. Technol.* **15**, 1947 (1997).

16. N. Keil, H.H. Yao, and C. Zawadzki, Integrated optical switching devices for telecommunications made on plastics, *Proc. Plastics Telecommun.* **8**, 11 (1998).

17. L. Eldada, R. Gerhardt, J. Fujita, T. Izuhara, A. Radojevic, D. Pant, F. Wang, and C. Xu, Intelligent optical cross-connect subsystem on a chip, *Proceedings of the Optical Fiber Communication/National Fiber Optic Engineers Conference,* Anaheim, CA, p. 139 (2005).

18. T. Goh, M. Yasu, K. Hattori, A. Himeno, M. Okuno, and Y. Ohmori, Low-loss high-extinction-ratio silica-based strictly nonblocking 16×16 thermooptic matrix switch, *Photonics Technol. Lett.* **10**, 810 (1998).

19. F. L. W. Rabbering, J. F. P. van Nunen, and L. Eldada, Polymeric 16×16 digital optical switch matrix, *Proc. ECOC* **27**, PD-78 (2001).

20. J. Fujita, T. Izuhara, A. Radojevic, R. Gerhardt, and L. Eldada, Ultrahigh index contrast planar polymeric strictly non-blocking 1024×1024 cross-connect switch matrix, *Proc. IPR* **24**, IThC3 (2004).

21. N. S. Lagali, J. F. P. van Nunen, D. Pant, and L. Eldada, Ultra-low power and high dynamic range variable optical attenuator array, *Proc. ECOC* **27**, 430 (2001).

22. L. Eldada, Optical add/drop multiplexing architecture for metro area networks, Optoelectronics & Optical Communications, *SPIE Newsroom*, DOI: 10.1117/2.1200801.0950 (2008). http://spie.org/documents/Newsroom/Imported/0950/0950-2008-01-22.pdf

15

Silicon Photonics Waveguides and Modulators

G. Z. Mashanovich

F. Y. Gardes

M. M. Milosevic

C. E. Png

G. T. Reed

15.1 Introduction

The origins of silicon photonics can be found in the 1980s, notably the pioneering work of Soref et al. (e.g., [1,2]). However, the world as far as silicon as an optical material was concerned was different then than it is today. Silicon was not regarded as an optical material, and it was the application of silicon on insulator (SOI) substrates to photonics technology that began to change things. There can be no doubt that silicon is the dominant semiconductor material for electronic applications. It is an inexpensive, well-understood material, with a high-quality native oxide, and excellent thermal and electrical properties. It also provides strong optical confinement for the telecommunication wavelengths when incorporated into waveguiding structures such as SOI. Seemingly, this would make a fully integrated silicon optoelectronic telecommunications chip an obvious choice. However, because of its indirect bandgap, silicon is a poor light-emitting material. Hence, a fully integrated monolithic light-emitting device cannot be fabricated in conventional ways and, therefore, researchers have studied a variety of ways to modify the structure of silicon in order to force it to emit light [3]. To compound the issue, silicon is a centrosymmetric crystal such that the Pockels effect is nonexistent. Hence, switching and modulation have to be achieved by other means. The global research effort was modest at first, but has increased dramatically in recent years.

The purpose of this chapter is to introduce silicon photonics to the reader. It is impossible to discuss the entire field within a single chapter. For this, the reader is referred to texts on the subject (e.g., [4,5]). Therefore, this chapter will concentrate on the basic building block of optical circuits in silicon photonics, the optical waveguide, as well as a device that has undergone dramatic development in recent years, the silicon optical modulator. These devices are discussed in the context of the dramatic progress that has been made in recent years.

15.2 Monolithic Integrated Devices

One of the frequently posed targets of silicon photonics is the monolithically integrated superchip. The basic premise of such a superchip is that it incorporates all of the necessary components to detect, route, convert, multiplex, encode, reroute, demultiplex, amplify, or even create optical signals that could subsequently be interrogated by electrical circuitry. In Soref's early vision of such a chip, optical fibers were proposed as a means to butt-couple to waveguides on the chip. The fibers were to be supported by high-precision v-grooves etched into the silicon substrate [6]. It is interesting to note that passive alignment of fibers to waveguides on a silicon photonic circuit remains a challenge today, partly due to the more recent trend to smaller cross-sectional dimensions of waveguides and devices. The optical signals would then either be detected by a SiGe or Ge photodiode for electronic interrogation or passed to other devices for optical processing by means of waveguide-based devices. Due to losses and/or penalties in other parts of the system, the light could also be amplified, or perhaps reshaped and reencoded via an optical modulator before retransmission. However, Soref's early vision envisaged not a monolithic chip, but hybrid integration of a range of materials to perform the individual functions. Today, many authors envisage a truly monolithic superchip, although recent work pioneered by Bowers et al. (e.g., [7,8]) in the development of III–V lasers integrated onto a silicon waveguide platform have reinvigorated the "hybrid" view of the integrated chip, or at least the hybrid laser. Other device visions remain predominantly in the monolithic arena, although silicon-compatible compounds such as SiGe (or perhaps even pure germanium) are also commonly utilized for detector work, and hence broader application of such materials is a natural next step. With recent work in Raman amplification and lasers [9,10], and fast modulators [11,12], we can now look forward to monolithic versions of the optical superchip. Furthermore, the vision can also include traditional electronic components to control, drive, and time these functions, typically in CMOS, but also the opportunity of other forms of electronic and photonic integration are also envisaged, such as, for example SiGe-based BiCMOS, with Si-based photonics. Optical and electronic integration has taken place only in a very simple way to date (e.g., [13]), and remains a challenge. However, recent initiatives such as the DARPA EPIC program [14], and the MURI program in the United States, and numerous European projects, notably HELIOS, will also begin to address these issues.

15.3 Waveguide Development

One of the most critical components of any integrated optical system is the optical waveguide itself. A significant amount of research on the planar waveguide was undertaken in the late 1960s and through the 1970s and 1980s [15–22]. In the mid-1980s Soref et al. demonstrated single-crystal silicon waveguides [1,2]. While these first devices were fabricated using highly doped silicon substrates, other substrate configurations were employed subsequently, such as silicon on sapphire (SOS) [23] and SOI [24,25].

In the late 1980s and early 1990s, silicon waveguides were beginning to reap the benefits of the development of separation by implantated oxygen (SIMOX) substrates, as well as bond and etch-back SOI (BESOI) [26,27]. While originally developed for the semiconductor industry to prevent latch-up [28], these materials were an obvious choice as waveguiding substrate materials. Several early assessments of SIMOX and BESOI as waveguiding systems were made [24,29–32], and some of the initial work yielded very large losses [33], but the loss was rapidly improved to respectable levels [34]. In 1989, Davies et al. [31] measured a loss of 4 dB/cm for optical waveguides fabricated in SIMOX. Multiple-layer waveguiding structures using SIMOX technology were also demonstrated [25,32]. Kurdi et al. predicted a loss for a 0.2 μm thick planar waveguide with a 0.5 μm buried oxide (BOX) thickness to have a loss of less than 1 dB/cm [35]. At the University of Surrey research was also underway to reduce the initial high loss. Initially, Weiss et al. [33] determined losses as high as 30 dB/cm from a 2 μm thick planar waveguide.

Further efforts were made by Rickman et al. [34] to reduce the loss by investigating the thickness of the BOX layer. Their results showed that a BOX layer thickness of greater than 0.4 μm was necessary to prevent loss due to substrate coupling for a silicon layer of several microns.

Reduction of the optical loss progressed quickly. In 1991, Schmidtchen et al. [36] reported a loss of 0.4 dB/cm for a silicon rib waveguide. By 1994, this loss was reduced to a level indistinguishable from pure silicon, as reported by Rickman et al. [37] for TE polarized light for a wavelength of 1.5 μm. This measurement demonstrated that silicon was not only a viable waveguiding material, but that the propagation loss was not going to be a serious issue in the development of the technology.

In these early experiments, the majority of the work was conducted on relatively large waveguides, of the order of several microns in cross-sectional dimensions (e.g., [38–40]). However, there were also preliminary reports of very small waveguides being measured. Typically the loss was very high as discussed above (e.g., [33]), due in part to insufficiently well-confined waveguides, and/or significant surface roughness. It is also interesting to note there were views expressed at the time (early to mid-1990s) that very small waveguides were unlikely to be useful, due in part to the high loss, but also due to the fact that coupling from optical fibers was very lossy. It is therefore interesting to note that current research is actively targeting miniaturization to micro- and nanophotonic circuits. The coupling of light to these small waveguides remains an issue today, especially for very small, submicron, waveguides [41–45].

In most applications, a single-mode optical waveguide is required. In order to achieve the necessary criteria of single-mode behavior in planar SOI–based waveguides, the thickness of the overlayer needs to be on the order of a few hundred nanometers. While some devices are based around planar waveguides, the vast majority of devices are based upon the various types of three-dimensional (3D) channel waveguides, namely the rib and strip waveguides.

15.3.1 Rib Waveguides

The three most common basic channel waveguide structures are the rib waveguide (Figure 15.1a), the strip waveguide (Figure 15.1b), and the buried waveguide (Figure 15.1c). In this section, design guidelines for single-mode and birefringence-free small rib waveguides are given, while in Section 15.3.2, strip waveguides are analyzed.

Unlike silica waveguides, which can be relatively easily designed to be single mode, SOI waveguides with dimensions larger than a few hundred nanometers in cross section will support multiple modes. Such multimode waveguides are usually undesirable in photonic circuits as their operation can be seriously compromised by the presence of multiple modes.

It has been shown that large rib waveguides in SOI (Figure 15.2) could be designed such as to be monomodal [46]. These waveguides have been studied extensively by a number of researchers (e.g., [47–50]) to find single-mode behavior and low-loss propagation. Large rib waveguides are interesting because they are multi-micron in cross-sectional dimensions (of the order of 5 μm) facilitating low-loss

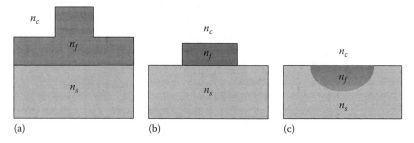

FIGURE 15.1 Three main types of channel waveguides: (a) rib, (b) strip, and (c) buried. (From Reed, G.T. (Ed.), *Silicon Photonics: State of the Art*, John Wiley & Sons, Chichester, U.K., 2008. With permission.)

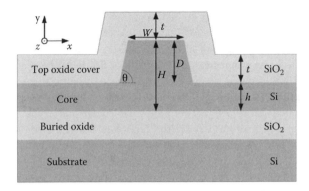

FIGURE 15.2   Rib waveguide in SOI.

coupling to and from optical fibers. Soref et al. [46] first proposed a simple expression for the single-mode condition (SMC) of such waveguides:

$$\frac{W}{H} \leq 0.3 + \frac{r}{\sqrt{1-r^2}} \quad \text{(for } 0.5 \leq r < 1\text{)} \tag{15.1}$$

where

r is the ratio of slab height to overall rib height

W/H is the ratio of waveguide width to overall rib height (Figure 15.2)

Their analysis of the waveguides was limited to shallow-etched ribs ($r > 0.5$) and the waveguide dimensions were assumed to be larger than the operating wavelength. The analysis was based on the assumption that high-order vertical modes (i.e., modes other than the fundamental mode) confined under the rib were coupled to the outer slab region during propagation, therefore yielding high propagation losses for the higher-order modes. Thus the waveguides behave as single-mode waveguides, as all other modes are lost.

Other authors have also considered the SMC for large waveguides, and produced similar expressions (e.g., [47,48]). However, those expressions are not adequate for relatively small rib waveguides, such as the one shown in Figure 15.3.

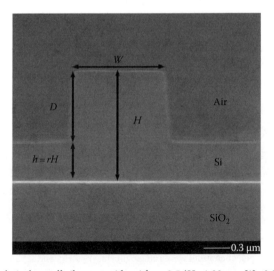

FIGURE 15.3   SEM of a relatively small rib waveguide with $r < 0.5$ ($H = 1.20\,\mu m$, $W = 0.98\,\mu m$, $D = 0.76\,\mu m$). (From Chan, S.P. et al., *Electron. Lett.*, 41, 528, 2005. With permission.)

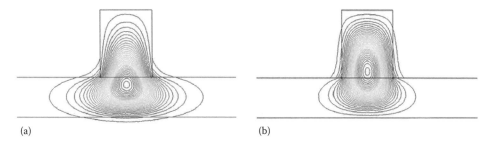

FIGURE 15.4   (a) TE and (b) TM mode shapes for a rib waveguide in SOI with $H = 1.35\,\mu$m, $D = 0.85\,\mu$m, and $W = 0.70\,\mu$m.

The increased polarization dependence in small waveguides is derived from the increasingly differing mode shapes of the TE and TM modes (Figure 15.4). The question arises as to whether the prime concern is to maintain similar losses for the TE and TM modes or to provide similar propagation constants in order to maintain similar phase performance for interferometric-based devices, because it is not generally possible to maintain both.

In this section, we analyze both the SMC and the zero-birefringence condition (ZBC) for a rib waveguide with $H = 1.35\,\mu$m in the presence of oxide stress. Chan et al. [51] have already produced equations to predict single mode and polarization independence for relatively small rib waveguides:

$$\frac{W}{H} \le 0.05 + \frac{(0.94 + 0.25H)r}{\sqrt{1-r^2}} \quad \text{for } 0 \le r \le 0.5 \text{ and } 1.0 \le H \le 1.5 \tag{15.2}$$

$$D_{\min}[\mu\text{m}] = 0.06 + 0.556H[\mu\text{m}] \tag{15.3}$$

Equation 15.2 defines the quasi-TM single-mode boundary, and hence provides guidance on the geometrical limitations to retain single-mode behavior, while Equation 15.3 defines the minimum etch depth, $D_{\min}$, required to obtain polarization-independent propagation. The guidelines of Equations 15.2 and 15.3 are for waveguides with a top cladding of air. However, an upper cladding layer is often deposited to reduce the influence of surface contamination, to passivate the surface, or to provide electrical isolation. For SOI, this layer is usually SiO_2 (Figure 15.5), although nitride and polymer layers are also used in some cases. Therefore, the single-mode and birefringence-free conditions also need to be determined for such a cladding, not only because oxide cladding has a different refractive index than air, but

FIGURE 15.5   SEM of a directional coupler made of two rib waveguides with oxide cover.

also because it causes stress in the waveguide structure, therefore changing effective refractive indexes for TE and TM polarizations [52].

Modal birefringence is defined here as the difference between the effective indexes of the two orthogonally polarized modes, the horizontally polarized mode (quasi-TE), and the vertically polarized mode (quasi-TM): $\Delta N_{eff} = N_{eff}^{TE} - N_{eff}^{TM}$. To achieve zero-birefringence (ZBR) in rib waveguides, an optimization of waveguide dimensions is necessary [53]. As we are investigating a most common situation where there is an oxide upper cladding, the total birefringence is, however, the sum of the geometrical birefringence and the stress-induced birefringence [53]. The structure we analyze in this section, at the operating wavelength of $\lambda = 1.55\,\mu m$, is a rib waveguide with the height of $H = 1.35\,\mu m$, and variable-waveguide rib width W, etch depth D, top oxide cover thickness t, and sidewall angle θ, as shown in Figure 15.2. The BOX layer is $1\,\mu m$ in thickness, while the top oxide cladding used in these simulations is $0.1-3\,\mu m$ thick. These waveguide dimensions are chosen as we have previously reported optical filters based on such waveguides [54,55]. The strain in the upper SiO_2 layer produces a stress distribution within and near the Si rib, which in turn causes a change of the refractive index in both materials due to the photoelastic effect. For these simulations, we used a thickness of the upper cladding film on the rib sidewalls that is 70% of that on the top of the rib, a result obtained by a scanning electron microscope (SEM) investigation of the fabricated waveguides. In this section, we present the effect of stress on the single-mode/multimode boundary and the effect of stress on the polarization-independent curve. The simulations were performed by using two-dimensional (2D) finite element method (FEM) modeling and verified by 3D semi- and full-vectorial beam propagation method (BPM) modeling. The method used was to calculate the effective indices for the fundamental and the first two higher-order modes, for both polarizations, over a range of waveguide dimensions. The single-mode cutoff condition has been evaluated by determining when the first mode of higher order than the fundamental mode begins to propagate.

The calculation window size was chosen to be large enough to minimize the influence of the edge effects on the stress distribution in the vicinity of the rib waveguide. Because of the sharp waveguide corners, stress is inhomogeneous and anisotropic [56]. It is assumed that the difference between the operating (20°C) and reference (1000°C) temperatures is $\Delta T = -980\,K$ [53]. The stress field in the waveguide and Maxwell's equations are solved by FEM, in a nonuniform mesh of triangular elements. For our numerical computations, approximately 40,000 elements were used, together with higher-order shape functions of the Lagrange type.

Figure 15.6 shows the influence of waveguide rib width on the waveguide birefringence for a given value of the etch depth. We assumed vertical rib sidewalls and, for a given top oxide thickness, for

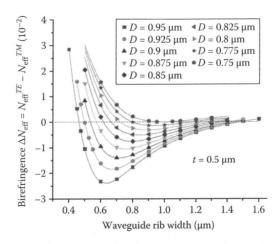

FIGURE 15.6 Birefringence as a function of waveguide rib width and etch depth for the top oxide thickness of $t = 0.5\,\mu m$. (From Milosevic, M.M. et al., *J. Lightw. Technol.*, 26, 1840, 2008. With permission.)

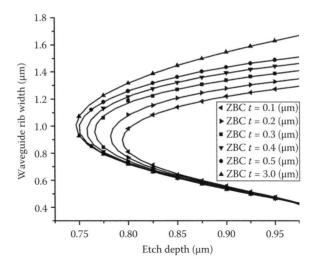

FIGURE 15.7   ZBC as a function of waveguide rib width and etch depth for different values of top oxide thickness. (From Mashanovich, G.Z. et al., *Semicond. Sci. Technol.*, 23, 064002, 2008. With permission.)

each etch depth, a series of ΔN_{eff} points was calculated. The results show that for smaller etch depths ($D < 0.7\,\mu m$) the birefringence is always positive, that is, the propagating phase change of quasi-TE and quasi-TM polarization modes cannot be equalized by changing values of waveguide width and top oxide cover. Deeper-etched devices, on the contrary, show a tendency for two specific waveguide widths for which the ZBC is fulfilled. This means that by using stress, it is possible to produce birefringence-free waveguides for two different waveguide widths when a relatively deep etch depth is employed. The increase of the top oxide cover thickness, t, allows the use of shallower etch depths at which both the modes will be guided through the waveguide with similar propagation constants. For larger t, the family of ΔN_{eff} curves becomes deeper and wider.

The ZBR curves for different values of top oxide cover thickness as a function of waveguide rib width and etch depth is presented in Figure 15.7. It can be seen that the increase of the top oxide cover thickness shifts the ZBR curves to smaller values of etch depth, which is consistent with the results from Figure 15.6. Furthermore, considering constant etch depth, the ZBC can be achieved for larger values of W, while the smaller values of W approach a common asymptote for different etch depths. For the upper part of the ZBR curves, that is for wider ribs, when increasing the oxide thickness the birefringence will become negative and therefore to compensate, and return it to zero again, the width of the rib needs to be increased in order to reduce the influence of stress on the rib. On the other hand, for the lower part of the ZBR curves, that is for narrower ribs, the mode resides mainly in the slab region and therefore the increase of the oxide thickness will not affect the birefringence significantly, hence the common asymptote. The influence and importance of using stress engineering is obvious. The parabolas in Figure 15.7 move to the left with respect to the ZBC curve for a rib waveguide with an air cover ($t=0$). At large top oxide thicknesses ($t>0.8\,\mu m$) a saturation of ZBR curves is observed.

It is interesting to see what happens to the ZBR curve when the sidewall angle is taken into account (Figure 15.8). Here, four examples are presented: one with vertical rib sidewalls ($\theta = 90°$) and three with the slanted sidewalls, $\theta \in (85°, 80°, 75°)$, values that were measured on a series of fabricated rib waveguides. It can be seen that the ZBR curve shifts to smaller values of W. When θ is decreased, for constant W (width of the top of the rib), the width of the bottom of the rib will increase. In other words, the effective width of the rib will increase, and therefore for constant oxide thickness, the ZBC will not be fulfilled. To obtain $\Delta N_{eff}=0$, the effective rib width needs to be decreased, which means that the ZBC

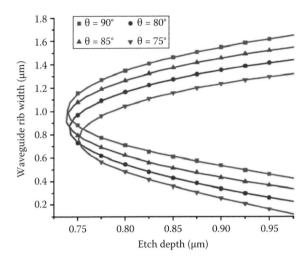

FIGURE 15.8 ZBC as a function of waveguide rib width and etch depth for different values of ridge sidewall angle. (From Milosevic, M.M. et al., *J. Lightw. Technol.*, 26, 1840, 2008. With permission.)

will be fulfilled for narrower ribs. That is why the ZBR curve shifts down in Figure 15.8. When θ is decreased below 80°, the ZBR curve moves to the right, that is, to larger etch depths D. We can conclude that the stress-induced birefringence for shallow-etched rib waveguides has less influence on the total birefringence. Geometrical birefringence is then dominant and that is why the ZBR curve moves to the right, similar to the case when stress effects are not considered. Fabricated waveguides typically have sidewall angles in the 8°–10° range. In that case, the ZBR curve moves down maintaining the values of etch depth. If we assume this range of sidewall angles, the following equation describes the influence of the sidewall on the waveguide width, which maintains the ZBC:

$$W\,[\mu m] = -0.474 + 0.952 \times \theta\,[rad], \quad \text{for } 80° \leq \theta \leq 90° \tag{15.4}$$

It is worth noting that the ZB curves will be different for differently deposited oxides, as the stress level for each will also be different.

In common with our previous work (without stress-induced effects) [51,58] we also found that satisfying the SMC for the TM mode is more restrictive than for the TE mode. The area below the SMC for quasi-TM mode defines the global condition for single-mode behavior. According to our simulations, for given waveguide parameters, there is at least one intersection point between the SMC line for quasi-TM mode and the ZBR curve. A part of the ZBR parabola between these intersection points defines waveguide parameters for which both ZBC and SMC are satisfied.

Figure 15.9 describes two different cases of the waveguide with vertical sidewalls. The first (Figure 15.9a) is the waveguide without an upper oxide layer, while the second (Figure 15.9b) represents the waveguide with an upper oxide cover of 0.3 μm in thickness. In the first case (Figure 15.9a), a similar result was obtained to that in [51], as expected. Both the SMC and ZBC are satisfied and the quasi-TM curve is more critical for the single-mode/multimode boundary. In the case when an oxide layer is deposited, the situation is different (Figure 15.9b). As mentioned previously, the ZBR curve moves to the left, but the single-mode curves are also changed. The single-mode quasi-TM curve limits the boundary between single-mode and multimode behavior, but the stress effect will shift the quasi-TE and quasi-TM curves downward. Now, we have rather limited range of the locus where both SMC and ZBC can be satisfied. However, if the sidewall angle is, for example, 80° rather than 90°, both SMC and ZBC can be satisfied for a larger range of rib waveguide widths and etch depths.

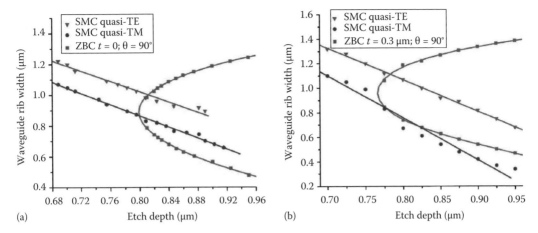

FIGURE 15.9 Single-mode and ZBC as a function of waveguide rib width and etch depth for top oxide thickness of (a) $t = 0$ and (b) $t = 0.3\,\mu m$ ($\theta = 90°$). (From Milosevic, M.M. et al., *J. Lightw. Technol.*, 26, 1840, 2008. With permission.)

15.3.2 Strip Waveguides

SOI is a platform that offers very high refractive index contrast and consequently strong light confinement. This allows shrinking of the waveguide core to submicron dimensions. The core size for single-mode propagation at the 1.3–1.5 μm telecommunications wavelengths is a few hundred nanometers. Such extreme light confinement also allows the minimum bending radius to be reduced to the micron range, and therefore offers the possibility of realization of ultradense photonic circuits, which can further decrease the cost of silicon photonics. It has been shown that small waveguides with small bending radii can improve the characteristics of photonic devices, for example optical modulators and filters [59–61]. Furthermore, such small waveguides, or photonic wires as they are usually called, can realize an ultrahigh optical power density, which can be as much as 1000 times that in a conventional single-mode fiber, enhancing nonlinear optical effects [62].

Unlike rib waveguides that can be monomodal even for cross-sectional dimensions of several microns, strip waveguide (Figure 15.10) dimensions must be significantly smaller than 1 μm to suppress propagation of higher-order modes. For example, by simulating the fundamental mode and the first two higher-order modes for the waveguide with height of 220 nm and top oxide cover of 200 nm, the waveguide width that corresponds to the single-mode/multimode boundary was found to be $W = 500$ nm (Figure 15.11).

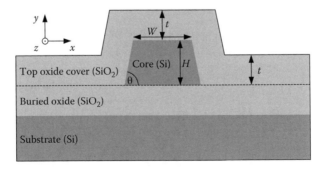

FIGURE 15.10 SOI strip waveguide: t is the top oxide layer thickness, H is waveguide height, W is waveguide width, and θ is the sidewall angle.

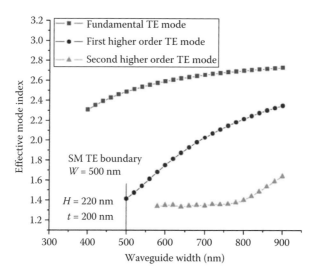

FIGURE 15.11 SMC for TE mode as a function of waveguide width for top oxide cover thickness of $t = 200$ nm ($H = 220$ nm, $\theta = 90°$, and $\lambda = 1.55\,\mu$m). Single-mode/multimode boundary for TE mode was found at $W = 500$ nm.

Alto gave an approximate expression for the SMC for silicon strip waveguides: $W \times H < 0.13\,\mu$m^2 [63], where W and H are the width and height of the strip waveguide, respectively. This condition agrees reasonably well with our 2D simulations that gave an approximation for the boundary between the single-mode and the multimode regime:

$$W \leq -1.405H + 0.746 \tag{15.5}$$

As an example, in order to prevent propagation of higher-order modes, a 0.3 μm high waveguide should not be wider than ~0.35 μm. Clearly such a restriction has implications for fabrication resolution, which is why some authors suggest that aiming for polarization independence in strip waveguides is impractical, and polarization diversity must be employed.

We can expect that the propagation constants for TE and TM polarization modes will be similar for a square cross section of the strip waveguide, and that polarization independence can be achieved. We analyzed a strip waveguide with a BOX layer thickness of 2 μm, a waveguide height of $H = 340$ nm, a top oxide cover of $t = 160$ nm, and vertical sidewalls. For small values of the waveguide width, the birefringence is negative and it increases with increasing width (Figure 15.12). A waveguide width of ~350 nm corresponds to the polarization independence condition. The birefringence curve goes to saturation for high values of waveguide width (Figure 15.12).

Figure 15.13 shows field profiles for TE and TM modes in a 300×300 nm strip waveguide. The TE field profile is characterized by much higher field intensity at the sidewalls, while the TM mode has a relatively small amplitude at the sidewalls but much higher amplitude at the top and bottom interfaces (Figure 15.13). This helps to understand why propagation loss is typically higher for TE modes than for TM modes [64]. In particular, roughness of the sidewalls is typically higher than roughness of the top and bottom interfaces and hence causes additional propagation loss for TE mode as its field is much higher at the sidewalls. The roughness of the sidewalls of the waveguide is a result of an imperfect resist pattern that can then be further deteriorated by the plasma etching. Therefore, lithography and etching are the key processes in achieving smoothly etched designs [65].

Propagation loss of TE modes is typically of the order of 3 dB/cm at the wavelength of 1550 nm (e.g., 3.6 dB/cm for 220×445 nm waveguides in [64], 2.8 dB/cm for 200×400 nm waveguides in [65], and 2.4 dB/cm for 220×500 nm in [66]). These figures could be decreased further by sacrificial oxidation of the sidewalls, which smooths the sidewalls [67]. Lee et al. used both oxidation smoothing and

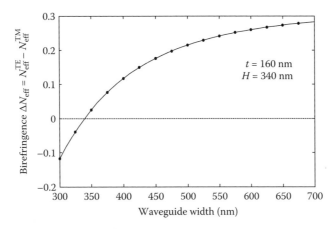

FIGURE 15.12 Birefringence as a function of waveguide width for top oxide cover of $t = 160$ nm at an operating wavelength of $\lambda = 1.55\,\mu$m and waveguide height of $H = 340$ nm. It can be seen that ZBR can be achieved for waveguide width of ~350 nm.

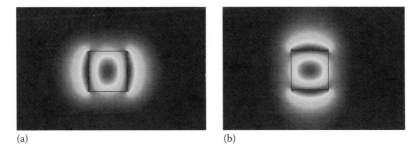

(a)   (b)

FIGURE 15.13 Field profiles for (a) TE and (b) TM polarization in a small strip silicon waveguide.

anisotropic etching and measured propagation loss of only 0.8 dB/cm for single-mode strip waveguides that had a width of 500 nm. Vlasov and McNab also measured losses per 90° bend for 220 × 445 nm strip waveguides and obtained 0.005 dB/turn, 0.013 dB/turn, and 0.086 dB/turn for bending radius of 5, 2, and 1 μm, respectively [64]. Bogaerts et al. measured 0.016 dB/turn for 3 μm bend radius [61].

In passing, we note that it can be seen in Figure 15.13b, that there is a large highly localized electric field distribution near the waveguide surface of small strip SOI waveguides. Therefore, shrinking silicon waveguides to these small dimensions opens another application area, as these waveguides could be used as evanescent field sensors for biological or chemical applications [68].

15.3.3 Other Types of Silicon Waveguides

There are also other types of waveguide geometries and materials that are interesting for silicon photonics. For example, photonic crystal (PhC) waveguides (Figure 15.14) and slot waveguides (Figure 15.15) enable propagation of light in regions with low index of refraction. PhCs are periodic structures composed of alternating regions of dielectric materials with high and low index of refraction, with periodicity on the order of the wavelength. This periodicity induces a high reflectivity in a spectral range for which light cannot propagate in the crystal. The reflectivity occurs for all directions of light incident on the crystal, thus giving a complete bandgap. The periodicity of the PhC can be in one, two, or three dimensions. Due to their waveguiding mechanisms, PhC waveguides can guide light in cores with cross-sectional dimensions smaller than 0.15 μm [69]. In recent years, there has been a growing

FIGURE 15.14 SEM of a PhC waveguide (without the oxide cladding) and the tapered access waveguide. (From White, T.P. et al., *Opt. Express*, 16, 17076, 2008. With permission.)

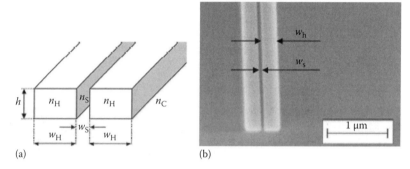

(a)   (b)

FIGURE 15.15 (a) Schematic of the slot waveguide. (b) Top-view SEM picture of the slot waveguide before deposition of the SiO_2 upper cladding. (From Xu, Q. et al., *Opt. Lett.*, 29, 1626, 2004. With permission.)

interest in the realization of PhCs or photonic bandgap structures as optical components and circuits (e.g., [70,72]) mainly due to their compact size.

Slot waveguides can confine light in the low-index slot (typically air or SiO_2) that is embedded between two silicon waveguides (Figure 15.15). Modes with strong field intensity at the two interfaces of the slot are formed and the overlap of the evanescent tail of the modes in the slot give strong light confinement in the low-index region [71,73]. These structures are therefore of interest in sensing or nonlinear applications.

In addition to the near-infrared, the mid-infrared spectral region is very interesting as the practical realization of optoelectronic devices operating in this wavelength range offers potential applications in a wide variety of areas, including optical sensing and environmental monitoring, free-space communications, biomedical and thermal imaging, and infrared counter-measures. Silicon has two low-loss transmission windows, one from 1.2 to 6.6 μm and the other from 24 to 100 μm [74]. However, SiO_2 is very lossy in the 2.6–2.9 μm range and beyond 3.6 μm [75], which means that SOI cannot be used in the mid- and far-infrared. Several silicon-based waveguide structures have been proposed for the mid-infrared [76,77] including suspended waveguides, SiGe or Ge on Si waveguides, slot and hollow waveguides, silicon on porous silicon (SOPS) (Figure 15.16) or freestanding silicon waveguides (Figure 15.17). The latter two structures have been extensively investigated recently at the National University of Singapore and University of Surrey (e.g., [52,78,79]). The waveguides have been fabricated by using focused proton beam irradiation and electrochemical etching. The SOPS waveguides were fabricated by single energy beam writing, while the freestanding waveguides were fabricated by using two beam energies, one to fabricate the waveguides and the other to fabricate the supporting pillars. These waveguides can be used in the mid-infrared region as the cladding is either porous silicon or air. The surface roughness caused by the etching process is the main contribution to the propagation loss and a post-oxidation process is used to reduce the surface roughness and to achieve ~1 dB/cm propagation loss [80].

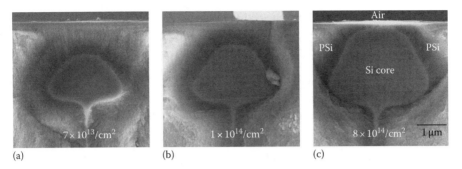

FIGURE 15.16 Cross-sectional SEM of the waveguides irradiated with a dose of (a) 7×10^{13}, (b) 1×10^{14}, and (c) 8×10^{14} cm^{-2}. (From Teo, E.J. et al., *Opt. Express*, 16, 573, 2008.)

FIGURE 15.17 Sideview of a freestanding waveguide with supporting pillar. (From Mashanovich, G.Z. et al., *Semicond. Sci. Technol.*, 23, 064002, 2008. With permission.)

15.4 Silicon Optical Modulators

This section discusses the basis of optical modulation in silicon circuits, together with a discussion of the devices that have appeared in the scientific literature.

Without a strong linear electro-optical effect, silicon was left uninvestigated as a switching/modulating material for years. However, in 1987 Soref and Bennett [81] investigated the potential modulation mechanisms available in silicon. They showed that electric field–based effects are very small but that an injection of free carriers (the plasma dispersion effect) has potential as a modulation mechanism. They studied results in the scientific literature to evaluate the change in refractive index, Δn, due to experimentally produced absorption curves for a wide range of electron and hole densities, over a range of wavelengths. In particular, they focused on the communications wavelengths of 1.3 and 1.55 μm. Interestingly, the changes in refractive index results were in good agreement with the classical Drude–Lorentz model, but only for electrons. For holes they noted a $(\Delta N)^{0.8}$ dependence. As well as quantifying the changes in refractive index with carrier density, they also quantified the changes in absorption [81].

They produced the following extremely useful empirical expressions, which are now used almost universally to evaluate changes due to injection or depletion of carriers in silicon:

At $\lambda_0 = 1.55\,\mu m$

$$\Delta n = \Delta n_e + \Delta n_h = -[8.8 \times 10^{-22}\,\Delta N_e + 8.5 \times 10^{-18}(\Delta N_h)^{0.8}] \tag{15.6}$$

$$\Delta\alpha = \Delta\alpha_e + \Delta\alpha_h = 8.5 \times 10^{-18}\,\Delta N_e + 6.0 \times 10^{-18}\,\Delta N_h \tag{15.7}$$

where

     Δn_e is the change in refractive index resulting from change in free electron carrier concentrations
     Δn_h is the change in refractive index resulting from change in free hole carrier concentrations
     $\Delta\alpha_e$ is the change in absorption resulting from change in free electron carrier concentrations
     $\Delta\alpha_h$ is the change in absorption resulting from change in free hole carrier concentrations

Equation 15.6 implies a change in refractive index in excess of 1.6×10^{-3} for a change in electron and hole density of $5 \times 10^{17}\,cm^{-3}$, the latter being relatively easy to achieve. Thus a realistic modulation mechanism was confirmed in silicon, and the challenge for designers became one of implementing this modulation mechanism both efficiently and at high speed. However, the plasma dispersion effect is not an ideal modulation mechanism because it is derived from a change in the absorption spectrum of silicon via a Kramers–Kronig coupling, and hence any change in refractive index must also suffer an associated change in absorption. For a similar change in electron and hole density of $5 \times 10^{17}\,cm^{-3}$, Equation 15.7 implies a change in absorption coefficient of $\Delta\alpha = 7.25\,cm^{-1}$, or $31.5\,dB/cm$. The latter sounds like an enormous loss, but modulators are typically of the order of $500\,\mu m$ or less in length, so the active loss is of the order $1.5\,dB$. However, this result also shows that the plasma dispersion effect has significant potential as an absorption-based modulator, or a variable optical attenuator (VOA). Similar equations were determined for $\lambda_0 = 1.3\,\mu m$. Thus silicon now had a method to produce relatively fast switches and modulators thereby increasing the likelihood of fully integrated monolithic superchip.

The earliest designs were based upon silicon guiding layers fabricated on doped silicon substrates (to form the lower waveguide boundary), the latter having a reduced refractive index via the plasma dispersion effect. Later SOI waveguiding structures became more popular due to the possibility of much stronger optical confinement. The first plasma dispersion modulator in silicon was proposed by Soref et al. in 1987 [82]. This *p+-n-n+* modulator was based on a single-mode silicon rib waveguide. It was found by modeling such that the interaction length of the modulator required for a π-radian phase shift was less than $1\,mm$. The corresponding loss was less than $1\,dB$ at $\lambda = 1.3\,\mu m$ for both orthogonal polarization modes. The authors noted that to first approximation, the modulator was polarization independent. It is interesting to note the buried "block" of SiO_2 below the waveguide, acting as the lower waveguide boundary. This is the early start of using BOX to fabricate low-loss high-contrast refractive index structures. SOI wafers are used almost universally today in silicon photonics, and are the basis of silicon photonic circuits. In the late 1990s, the reduction in device dimensions was a popular means of improving device performance. The work of Ang et al. [83] proved to be significant when in 1999, *pin* modulator devices were proposed and modeled to operate above $500\,MHz$. These were similar in design to previous devices of Hewitt et al. [84,85] and Tang et al. [86,87], but smaller. The modulators with interaction lengths of $500\,\mu m$ were based upon transverse *pin* structures in an optical rib waveguide with a silicon thickness of $0.98\,\mu m$. In that paper, the authors reported a theoretical *npp* device, with doping concentrations of $10^{20}\,cm^{-3}$, which required a drive current of just $0.29\,mA$ to achieve a π-radian phase shift, the lowest reported then. The corresponding current density was $112\,A/cm^2$. The response time was modeled to be $0.81\,ns$ (~$430\,MHz$). Those findings represented a significant improvement in device performance from those previously reported in the existing literature. Indeed, at the time of this work [88], more and more researchers were recognizing the benefits of reducing the silicon overlayer thickness, not only in terms of modulator performance, but optical circuit performance in general.

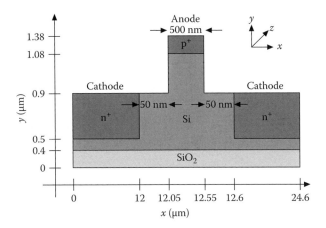

FIGURE 15.18 Proposed three-terminal rib waveguide device based on SOI. (From Png, C.E. et al., *J. Lightw. Technol.*, 22(6), 1573, 2004.)

Png et al. [89], later improved the work of Ang et al. [88] by modeling devices of similar geometry (Figure 15.18), but with improved performance [90,91]. In particular a series of devices were modeled with bandwidths ranging from 70 MHz to in excess of 1 GHz. The devices were based around a rib waveguide, approximately 1 μm in height and between 0.5 and 0.75 μm wide. A feature of these devices was the optimized doping profile in the n^+ regions to optimize injection efficiency. Png et al. [89] also reported the technique of pre-emphasis on critical device rise and fall times to increase device speed, improving a device based on Figure 15.18 from 95 MHz to 5.8 GHz. Using such a scheme, a class of devices with nominal operating speeds of 1 GHz could theoretically be switched in excess of 40 GHz [92].

In 2003, Irace et al. [93] modeled a 1.4 GHz operating bandwidth for a two-terminal Bragg reflector rib waveguide (Figure 15.19). This is essentially a follow up work of Cutolo et al. [94], in a smaller waveguide. The device had a waveguide height and rib width of 1 μm, an etch depth of 100 nm, and an interaction length of 3000 μm. These dimensions are a reduction from the original waveguide height and rib width of 3 μm, and etch depth of 450 nm, while the interaction length remains comparable (3200 μm). Thus, it can be seen that by reducing the silicon overlayer thickness and rib width, the device bandwidth increased from MHz operation to the GHz regime. Irace et al. [93] attributed this improvement in bandwidth to two factors: decreasing and optimizing the device dimensions, and applying a pre-bias to the

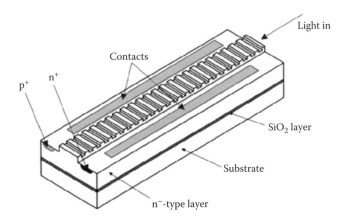

FIGURE 15.19 Schematic of the proposed 1 GHz Bragg modulator. (From Irace, A. et al., *Electron. Lett.*, 39(2), 232, 2003. With permission.)

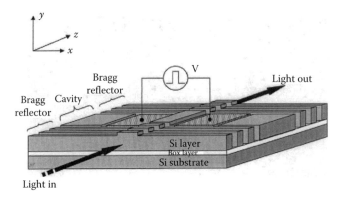

FIGURE 15.20 Fabry–Perot microcavity with high-reflectivity Bragg reflectors in SOI waveguide. (From Barrios, C.A. et al., *J. Lightw. Technol.*, 21(4), 1089, 2003. With permission.)

modulator to an "off" level (0.6 V) just below the turn-on voltage (0.8 V), which allowed faster movement of the injected carriers.

Longer devices potentially hinder high integration levels and hence increase cost. Consequently, Barrios et al. [95] modeled a modulator based on a Fabry–Perot microcavity with Bragg reflectors, as shown in Figure 15.20. The microcavity facilitates confinement of the optical field in a small region, and the transmission of the device near its resonance is highly sensitive to small index changes in the cavity. Thus the device required a low concentration of injected carriers to switch to a nonresonant position. The rib width and silicon thickness were 1.5 μm, and the etch depth was 0.45 μm. The 20 μm long device was predicted to require a dc power of the order of 25 μW at an operating wavelength of 1.55 μm, to achieve 31 MHz operating bandwidth with transmittance of 86% and a modulation depth of 80%. The authors also noted the merits of using trench isolation reported by Hewitt et al. [84] and implemented this feature in their proposed modulator device [95].

The same authors proposed another modulator that was only 250 nm in height [95]. The device, shown in Figure 15.21, consisted of two uniformly doped contacts, *p*+ and *n*+ at a level of 10^{19} cm^{-3} at a distance of 200 nm from the rib edge. The device slab thickness was 50 nm. The authors predicted that in order to induce a refractive index change of 10^{-3}, a dc power of 1.53 μW/μm was required. The switching time for this device was estimated to be 1.29 ns. The dc power is one of the lowest predicted to date, and this device reinforces the merit of reducing device dimensions.

In 2004, researchers from the Intel Corporation experimentally demonstrated a silicon-based optical modulator with a bandwidth that exceeded 1 GHz for the first time [96]. This was a major milestone in

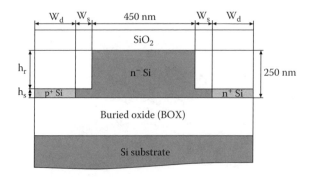

FIGURE 15.21 Proposed rib waveguide with an integrated lateral *pin* diode. (From Barrios, C.A. et al., *J. Lightw. Technol.*, 21(4), 1089, 2003. With permission.)

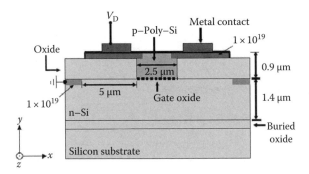

FIGURE 15.22 Schematic diagram of the silicon-based optical modulator demonstrated experimentally to exceed 1 GHz bandwidth fabricated using standard CMOS processing techniques. (From Liu, A. et al., *Nature*, 427(6975), 615, 2004. With permission.)

silicon photonics and attracted huge media attention. Figure 15.22 shows a schematic of the reported device. This device operates by the free carrier effect and bears a close resemblance to a complimentary metal-oxide-semiconductor (CMOS) transistor. The device structure consists of *n*-type crystalline silicon with an upper "rib" of *p*-type polysilicon. The *n*-type and *p*-type regions are separated by a thin insulating oxide layer. Upon application of a positive voltage to the *p*-type polysilicon, charge carriers accumulate at the oxide interface, changing the refractive index distribution in the device. This in turn induces a phase shift in the optical wave propagating through the device.

The bandwidth of the device (a single 2.5 mm long phase modulator) was characterized in two ways in an integrated asymmetric Mach–Zehnder interferometer (MZI). The first technique was to drive the device with a 0.18 V rms sinusoidal source at the wavelength of 1.558 μm, using lensed fibers for coupling into and out of the device. Figure 15.23a shows the normalized optical response of the MZI as a function of frequency (photoreceiver output voltage divided by on-chip drive voltage). Clearly the 3 dB bandwidth exceeds 1 GHz. The second test was the application of a 3.5 V digital pulse pattern with a dc bias of 3 V. A 1 Gb/s pseudorandom bit sequence was applied to the device and a high-bandwidth photoreceiver was used for detecting the transmitted optical signal. Figure 15.23b shows the optical signal faithfully reproducing the 1 Gb/s electrical data stream. However, the on-chip loss for this device was rather high, at ~6.7 dB, and the device was also highly polarization dependent due to the horizontal gate oxide. Phase modulation efficiency for TE polarization was larger than TM polarization by a factor of 7.

The authors also suggested several methods that may improve the device performance even further. The first is replacing the p-type polysilicon with single-crystal silicon, where the latter was expected

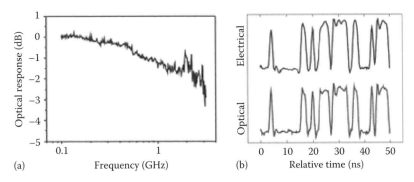

FIGURE 15.23 (a) Optical response of the MZI modulator (Figure 15.22) as a function of frequency. (b) Pseudorandom bit sequence of a silicon MZI containing a single 2.5 mm long MOS capacitor device (Figure 15.22) in one arm at a data bit rate of 1 Gb/s. (From Liu, A. et al., *Nature*, 427(6975), 615, 2004. With permission.)

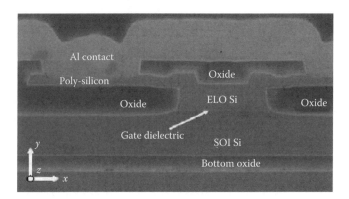

FIGURE 15.24   SEM cross section of the MOS modulator. (From Liao, L. et al., *Opt. Express*, 13(18), 3129, 2005. With permission.)

to reduce on-chip loss by ~5 dB. Another suggestion is the reduction of the device dimensions as the capacitance is reduced through such a shrinkage. They also suggested using a graded doping profile in the vertical direction such that higher doping densities exist in the areas close to the gate oxide and lower doping concentrations in the rest of the waveguide. Furthermore, the authors also reported that their modeling predicted that the device could be scaled to operate at 10 GHz, offering very significant improvement of silicon photonic devices.

A year later, Liao et al. [97] reported an improved version of the MOS optical modulator. Figure 15.24 shows an SEM cross-sectional image of the phase shifter. This modulator is smaller than the previous one, and comprises a 1.0 μm *n*-type doped crystalline Si (the Si layer of the SOI wafer) on the bottom and a 0.55 μm *p*-type doped crystalline Si on the top, with a 10.5 nm gate dielectric, a multilayer stack of silicon dioxide and nitride, sandwiched between them. In the first version of the device, the waveguide cross section was 2.5 μm × 2.3 μm and the top Si layer was poly-silicon (poly-Si), which is significantly more lossy than crystalline Si due to defects and grain boundaries [97]. In the improved device the poly-silicon was replaced by crystalline silicon via epitaxial lateral overgrowth (ELO), and the doping concentration was higher. In this smaller version of the phase shifter the mode-charge interaction is much stronger, which has, according to the authors, improved the $V_\pi L_\pi$ coefficient by 50%. The authors reported an intrinsic data at 10 Gb/s for the modulator with an extinction ratio (ER) of 3.8 dB. Data transmission measurements suggested bandwidth ranging from 6 GHz (ER of 4.5 dB) to 10 GHz. The authors explained that the 6 GHz limitation is due to the driver design and wire bonding, which decreased the cutoff frequency.

Also utilizing an MZI, Gan et al. [98] proposed a device based on a Si–SiO$_2$ high-index contrast waveguide modulator. This device is based on a split-ridge waveguide that operates under forward-biased conditions and shows corner frequencies of up to 24 GHz. Figure 15.25a shows the RF electrodes of the device and Figure 15.25b shows the actual device structure where to obtain single-mode operation, the dimensions are $d_1 = 100$ nm, $d_2 = 350$ nm, $d_3 = 2$ μm, $d_4 = 100$ nm, $h = 550$ nm, $w = 1$ μm. The *pin* junction is forward biased with a dc voltage of 2 V and modulated with a voltage swing of 1 V. By varying the recombination lifetime inside the waveguide, simulations presented in this paper show that, with a carrier lifetime in the intrinsic region, on the order of ps, which could be set by lifetime doping or ion bombardment, as well as operation in the saturation region where carriers travel at saturation velocity close to $v_s \sim 10^7$ cm/s, a phase shifter with a modulation frequency up to 24 GHz should be possible.

As the trend toward minimizing the real estate of devices continued, Xu et al. [99] reported a ring resonator using the waveguide structure proposed by Barrios et al. [95]. The diameter of the device was 12 μm, which according to the authors was at that time three orders of magnitude smaller than previously demonstrated. To use the ring resonator as a modulator it is necessary to operate at a single wavelength. Typically ring resonator modulators are operated between regions of high throughput and low throughput in the resonator response. One way to achieve such switching is via changes in

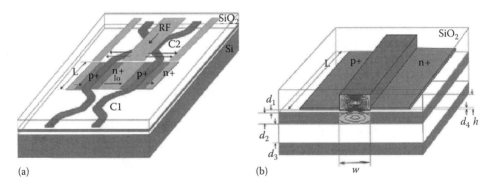

FIGURE 15.25 (a) Externally biased MZ modulator with coplanar RF feeder. (b) Split-ridge waveguide pin-phase modulator with single-mode intensity profile. (From Gan, F. and Kartner, F.X., *IEEE Photonics Technol. Lett.*, 17(5), 1007, 2005. With permission.)

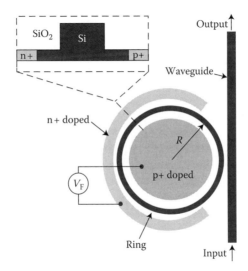

FIGURE 15.26 Schematic of the ring resonator structure. (From Xu, Q. et al., *Nature*, 435(7040), 325, 2005. With permission.)

effective index in the ring either changed through the plasma dispersion effect or changed thermally. Figure 15.26 shows a schematic of the device, where the waveguide structure and electrical structure is the one proposed by Barrios et al. [95] and is based on carrier injection. In this paper, the authors reported a drive voltage of 0.3 V for dc modulation. During modulation, the ring resonator modulator is operated with a peak-to-peak voltage of 3.3 V and shows a significant modulation depth at 0.4 Gb/s as the ER at resonant frequency is above 15 dB. The authors also reported a data rate of 1.5 Gb/s when the ring resonator is operated with a peak-to-peak voltage of 6.9 V. Figure 15.27a shows an SEM image of the ring resonator. The waveguide is a rib structure, where the waveguide width is 450 nm and the separation in the coupling region is 200 nm.

In early 2007, Xu et al. demonstrated an improvement of the previous device [100]. For this device, shown in Figure 15.28, the 5 μm ring is formed by silicon near strip waveguides with a height of 200 nm and width of 450 nm on top of a 50 nm thick slab layer. The distance between the ring and the straight waveguide is around 200 nm. Furthermore, compared to the previous design (Figure 15.26) an additional n^+ doped region is added outside of the straight waveguide to form a nearly closed loop *pin* junction, and the distance between the doped regions and the edge of the ring resonators and straight

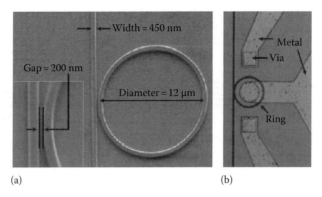

(a)   (b)

FIGURE 15.27 (a) SEM of the ring resonator. (b) Top-view microscope image of the ring resonator. (From Xu, Q. et al., *Nature*, 435(7040), 325, 2005. With permission.)

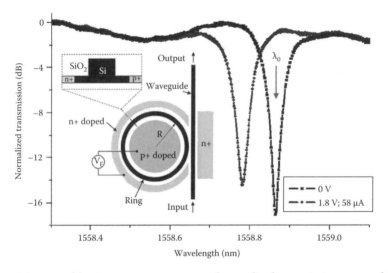

FIGURE 15.28 Schematic of the ring resonator structure and normalized transmission spectra of the modulator proposed in [100]. (From Xu, Q. et al., *Opt. Express.*, 15, 430, 2007. With permission.)

waveguides is reduced to ~300 nm. Using a pre-emphasis (shown in Figure 15.29a and b) NRZ signal, they demonstrated the possibility of decreasing further the rise time and fall time of the ring modulator, hence increasing the bandwidth of the modulator to 12.5 Gb/s with an ER of the signal of around 9 dB. A later result was obtained using the same principle with a bandwidth of 18 Gb/s and an ER of 3 dB [101]. This modulation method follows the same principle proposed previously by Png et al. [92].

Later in 2007, Green et al. [102] from IBM demonstrated a Mach–Zehnder modulator (MZM) based on the plasma dispersion effect and the injection of carriers in a nanowire rib waveguide. The cross section is shown in Figure 15.30. The waveguide is 220 nm high and 550 nm wide with a 185 nm etch depth. The slab was implanted to a concentration of 10^{20} cm^{-3} to form the *p*- and *n*-type resistive contact. In order to achieve high-speed modulation, a pre-emphasized electrical drive shown in Figure 15.31a was applied to the device. The method was similar to that used in [100].

The results obtained were a measured $V_\pi L_\pi$ figure of merit of 0.36 V · mm, with a data rate of 10 Gb/s for a length of 200 microns shown in Figure 15.32b. Furthermore, the RF power consumption was 51 mW. This device provided further evidence that a pre-emphasized method was the way toward the possibility to achieve low-power, high-efficiency, and high-speed modulation using injection type modulator.

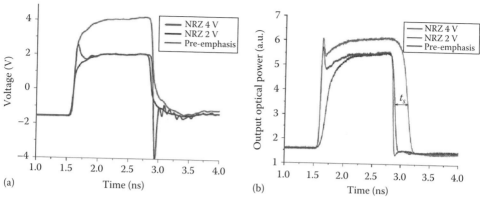

FIGURE 15.29 (a) Square-wave driving signals with (pre-emphasis) and without (NRZ 2–4 V) the pre-emphasis. (b) The output optical power when the modulator is driven by voltage signals shown in (a) (From Xu, Q., et al., *Opt. Express*, 15, 430, 2007. With permission.)

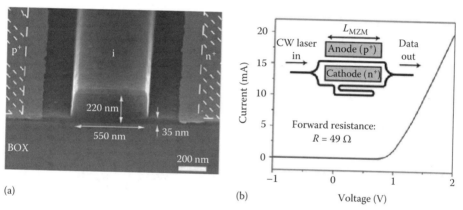

FIGURE 15.30 (a) Cross-sectional SEM image of the SOI p^+-i-n^+ diode nanophotonic rib waveguide used. (b) Electrical I–V trace taken for a modulator with $L_{MZM} = 100\,\mu m$, illustrating a low forward resistance of 49 Ω. (From Green, W.M.J. et al., *Opt. Express*, 15(25), 17106, 2007. With permission.)

Following a new approach and based on the Intel's silicon Raman laser [103], Jones et al. reported a modulator using the Raman effect [104], in which they demonstrated lossless optical modulation in a silicon waveguide. The device is shown in Figure 15.33, as a schematic and in Figure 15.34, as a front-view SEM picture. To achieve net Raman gain in a silicon waveguide, high pump intensities are required and two-photon-initiated free carriers have to be removed from the intrinsic region of the waveguide. To achieve continuous-wave (CW) Raman gain, a reverse-biased *pin* device is embedded in a silicon waveguide to sweep out the two-photon-induced free carriers. The electrical structure of the proposed device is illustrated in Figure 15.33, which shows the position of the *pin* diode inside the waveguide. The doping concentration for the *p*-type and *n*-type region is 1×10^{20} cm^{-3} and to increase the interaction length, hence achieving a larger total Raman gain, the waveguide formed an S-shaped curve. The total length of the waveguide was then 4.8 cm with a bend radius of 400 μm. In order to characterize the dc gain measurements, a CW pump laser emitting around 1548.3 nm was amplified using two EDFAs to a maximum output power of 3 W. The probe laser situated at the Stokes wavelength of 1684 nm was a 2 mW external-cavity tuneable diode laser. For this device, the authors reported a net gain of 3.5 dB for a power pump of 945 mW and showed that a modulation of 5 dB of the net gain should be achievable with a 0–8 V reverse bias voltage swing. A maximum bandwidth of 80 MHz was reported for a reverse bias

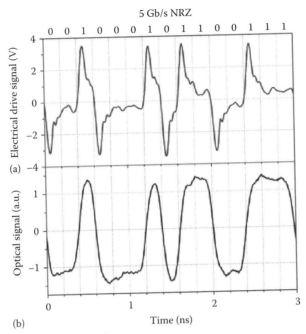

FIGURE 15.31 (a) Pre-emphasized electrical drive signal at 5 Gb/s. Transitions between 0 and 1 bits have large amplitude, while consecutive 0's or 1's are suppressed. (b) Corresponding optical signal at the output of a $L_{MZM} = 100\,\mu m$ modulator. (From Green, W.M.J. et al., *Opt. Express*, 15(25), 17106, 2007. With permission.)

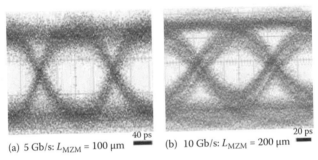

FIGURE 15.32 Eyeline diagrams of NRZ optical data signals (PRBS 2^7-1) produced by several ultracompact MZM devices. (a) $L_{MZM} = 100\,\mu m$ modulator operating at 5 Gb/s. (b) $L_{MZM} = 200\,\mu m$ modulator operating at 10 Gb/s (From Green, W.M.J. et al., *Opt. Express*, 15(25), 17106, 2007. With permission.)

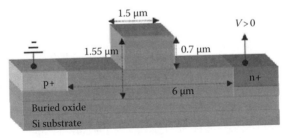

FIGURE 15.33 Schematic of the Raman device, showing the electrical structure and dimensions of the waveguide. (From Jones, R. et al., *Opt. Express*, 13(2), 519, 2005. With permission.)

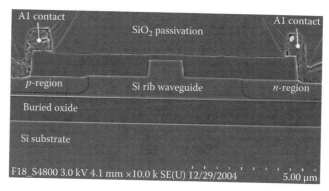

FIGURE 15.34 SEM image of the waveguide and the *pin* electrical structure used in the experiment. (From Jones, R. et al., *Opt. Express*, 13(5), 1716, 2005. With permission.)

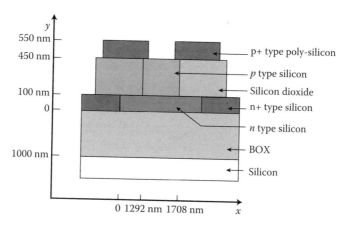

FIGURE 15.35 Schematic of a four-terminal depletion-type modulator. (From Gardes, F.Y. et al., *Opt. Express*, 13(22), 8845, 2005. With permission.)

of 0.5–3.5 V swing, with a pump power of 283 mW where the modulation depth was 0.3 dB. The authors stated that the modulation depth could be improved by using the index change due to the free carriers in an interferometer such as an MZI and offered the possibility of a lossless planar waveguide switch or modulator based in silicon.

In order to increase the bandwidth further, a sub-micrometer modulator based on the depletion of a *pn* junction was proposed in 2005 by Gardes et al. [105]. In common with the MOS capacitor [96], the depletion-type phase shifter is not limited by the minority carrier recombination lifetime and is based on the principle of removing carriers from the junction area when applying a reverse bias. Figure 15.35 shows a four-terminal asymmetric *pn* structure, where the concentration of *n*-type doping is much higher than the concentration of *p*-type doping. The reason for such a structure is first to minimize the optical losses induced by the *n*-type doping and secondly to enhance the depletion overlap between the optical mode and the *p*-type region, in order to induce a better phase shift to length ratio.

The carrier concentration variation in this kind of device is not uniform, as can be seen in the predictions of the refractive index change in the waveguide shown in Figure 15.36, and arises on both sides of the junction over a width of around 200 nm. One way to optimize the device is by increasing the overlap between the optical mode and the *p*-type depleted region. The main advantage of using depletion is obviously the very fast response time, simulated to be 7 ps for this modulator. This corresponds to an intrinsic bandwidth of approximately 50 GHz. The device proposed in [105] is 2.5 mm long and operates with a reverse bias swing of 5 volts in a push–pull configuration as part of a MZI.

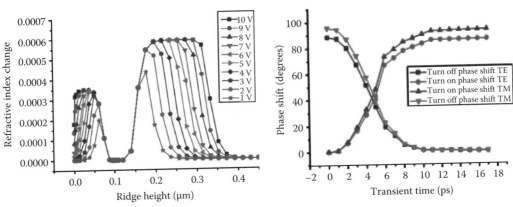

FIGURE 15.36 Left: Variation of the refractive index in the waveguide. Right: Rise and fall time for TE and TM.

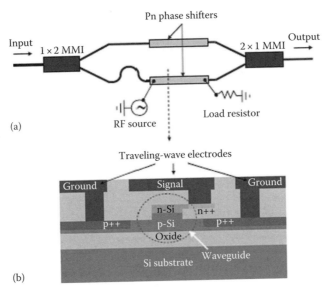

FIGURE 15.37 (a) Top view of the proposed MZI modulator. (b) Cross-sectional view of the modulator. (From Liu, A. et al., *Opt. Express*, 15(2), 660, 2007. With permission.)

In 2007, following the same principle, Liu et al. demonstrated a *pn* junction–based silicon optical modulator [106]. Figures 15.37 and 15.38 show the schematic of the modulator as well as an SEM picture of the modulator cross section.

The modulator shown in Figure 15.37 comprises a *p*-type doped crystalline silicon rib waveguide having a rib width of ~0.6 μm and a rib height of ~0.5 μm with an *n*-type doped silicon cap layer ~1.8 μm wide. This 0.1 μm thick cap layer is formed using a nonselective epitaxial silicon growth process and is used for the formation of the *n*-doped electrical contact. The *p*-doping concentration is about 1.5×10^{17} cm^{-3}, and the *n*-doping concentration varies from around 3×10^{18} cm^{-3} near the top of the cap layer to 1.5×10^{17} cm^{-3} at the *pn* junction. A good ohmic contact is ensured between the silicon and the metal contacts by two slab regions, which are situated 1 μm away from both sides of the rib edge, and a cap layer region at about 0.3 μm away from the rib edge. Those three regions are heavily doped with a dopant concentration of 1×10^{20} cm^{-3}.

Figure 15.39 shows the phase shift against voltage of the phase shifter for three different lengths that results in a modulation efficiency, $V_\pi L$, (where V_π is the bias voltage required for π phase shift and *L* is

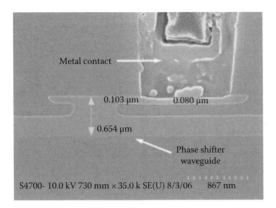

FIGURE 15.38 SEM of the cross section of the optical modulator in [106]. (From Liu, A. et al., *Opt. Express*, 15(2), 660, 2007. With permission.)

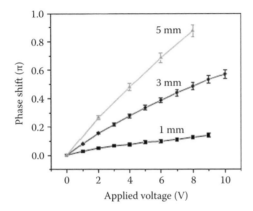

FIGURE 15.39 Phase shift against reverse bias for different length of the active area of the phase shifter. (From Liu, A. et al., *Opt. Express*, 15(2), 660, 2007. With permission.)

the device length) of approximately $4\,V\cdot cm$. The modulation bandwidth was also measured on a 1 mm long phase shifter, and Figure 15.40a shows that the MZI modulator has a 3 dB roll-off frequency of ~20 GHz. Figure 15.40b shows the eye diagram at a bit rate of 30 Gb/s, when using a pseudo-random bit sequence with $2^{31} - 1$ pattern length as the RF source. In October 2007, Liao et al. [107] demonstrated a 30 GHz, 40 Gb/s modulator based on the same structure but with a traveling-wave termination reduced to 14 Ω. The frequency response and eye diagram is shown Figure 15.41. To date, this is the fastest reported experimental optical modulator in SOI based on the plasma dispersion effect, although the modulation depth was only of the order of 1 dB.

The general trend in silicon photonics is design of devices with sub-micrometer waveguides to achieve good phase and power efficiency, as well as to improve the speed and reduce the real estate of devices. The stage, where integration of photonic devices and electronics circuits on the same chip is becoming a major milestone in silicon photonics.

In early 2006, Gunn et al. [108] demonstrated modulation in both MZI and ring resonator with a data rate up to 10 Gb/s. The authors stated that the modulator drivers were integrated on the chip, but did not provide any details about the electronics or the technology used to change the effective index of the mode in the waveguide. The information provided in [108] indicated that the waveguides were 500 nm wide and had a cross-sectional area of $0.1\,\mu m^2$. Figure 15.42 shows a top view of the MZI, where the dashed lines show the optical waveguide structure buried in the silicon and the driver input pads on the

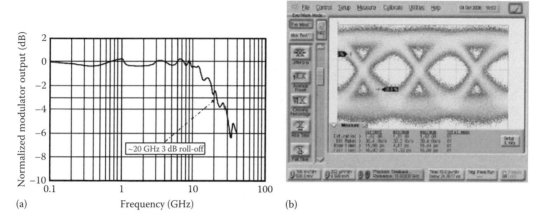

(a) Frequency (GHz)   (b)

FIGURE 15.40   (a) Response of the modulator as a function of the RF frequency for a 1 mm active area. (b) Optical eye diagram of the modulator with a 1 mm long active area. (From Liu, A. et al., *Opt. Express*, 15(2), 660, 2007. With permission.)

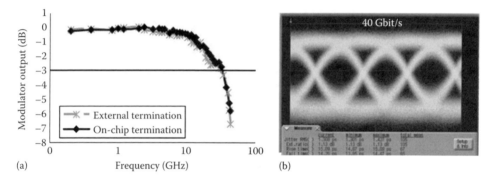

(a) Frequency (GHz)   (b)

FIGURE 15.41   (a) Response of the modulator as a function of the RF frequency for a 1 mm active area. (b) Optical eye diagram of the modulator with a 1 mm long active area. (From Liao, L. et al., *Electron. Lett.*, 43(22), 1196, 2007. With permission.)

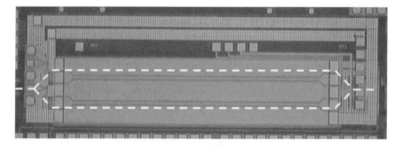

FIGURE 15.42   Photograph of the MZI. (From Gunn, C., *Micro. IEEE*, 26(2), 58, 2006. With permission.)

left, supply rails across the top and termination pads on the right. Figure 15.43 shows an eye diagram, when the modulator is operated at 10 Gb/s and the frequency response that shows a 3 dB cutoff frequency at 10 GHz. The authors also stated that the typical ER were 5 dB when the modulator was driven at 2.5 V.

As mentioned above, Gunn et al. [108] also demonstrated a ring resonator modulator. The top view of the metal contact is shown in Figure 15.44. The ring was used as a tuneable notch filter where the wavelength response is shown in Figure 15.45 and centered between 1524 and 1525 nm. The ring radius was

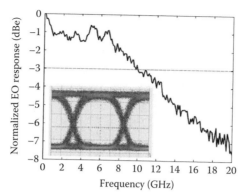

FIGURE 15.43 Silicon optical modulator 10 Gb/s eye diagram and frequency response. (From Gunn, C., *Micro. IEEE*, 26(2), 58, 2006. With permission.)

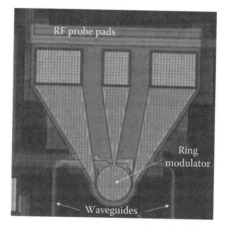

FIGURE 15.44 Photograph of a ring resonator modulator with RF probe pads. (From Gunn, C., *Micro. IEEE*, 26(2), 58, 2006. With permission.)

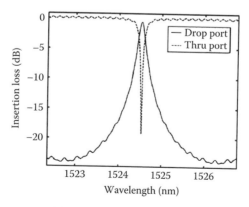

FIGURE 15.45 Wavelength response of the drop port and through port of the ring resonator. (From Gunn, C., *Micro. IEEE*, 26(2), 58, 2006. With permission.)

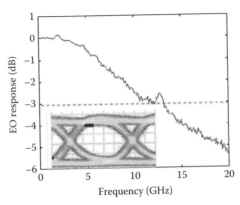

FIGURE 15.46   Resonator modulator 10 Gb/s eye diagram and frequency response. (From Gunn, C., *Micro. IEEE*, 26(2), 58, 2006. With permission.)

30 μm and was a major improvement in terms of real estate compared to the proposed MZI modulators that occupy approximately 2 mm². Figure 15.46 shows the eye diagram of the ring resonator modulator operated at 10 Gb/s and the frequency response that shows a cutoff frequency of around 10 GHz.

In a more recent publication [109] relating to the device proposed by Gunn et al., Huang et al. described the effect used in the optical modulator to be based on the free-carrier plasma dispersion. The transducer was a reverse-biased lateral *pin* diode, where modulation was obtained when majority carriers are swept in and out of the optical mode by an electric field. The modulator has a length of 2 mm and a performance of about 50/mm/V/arm. As high-speed modulation was based on majority carriers, the device was entirely limited by RLC parasitics. The diode was driven using a traveling-wave electrode design, where the diode junction is designed as part of the microwave transmission line. The geometry of the traveling-wave electrode was chosen such that the electrical group velocity and the optical group velocity were approximately matched. The transmission line itself had an impedance of below 25 Ω, but when loaded with the diode, the total system achieves 25 Ω.

Using different methods for the modulation, optical modulators in silicon have seen their bandwidth increased to reach multi-GHz frequencies. In order to simplify fabrication, one requirement for a waveguide, as well as for a modulator, is to retain polarization independence in any state of operation and to be as small as possible. In 2006, Gardes et al. [110] proposed a way to obtain polarization independence and to improve the efficiency of an optical modulator using a V-shaped *pn* junction (Figure 15.47) based on the natural etch angle of silicon, which is 54.7°. This modulator was compared to a flat-junction depletion-type modulator of the same size and doping concentration (Figure 15.48).

The simulations undertaken were to determine the conditions for polarization independence of the waveguide during modulation. Figures 15.49 and 15.50 show the phase shift achieved for different junction depths and voltages. Figure 15.49 shows the simulation resulting from the V-shaped junction and Figure 15.50 shows similar results for the flat junction. In order to achieve polarization independence during modulation, the optimal positioning of the junction inside the waveguide has to be determined. Figures 15.49 and 15.50 show the junction depth for both devices where polarization independence is achieved. For the V-shaped junction the polarization independence during modulation is achieved when the top of the V-shaped junction is situated at a depth of 0.87 μm. The flat-junction modulator achieves polarization independence with a junction situated at 0.5 μm from the bottom of the waveguide. This is situated at the rib–slab interface, which can facilitate fabrication.

The second parameter to extract from those simulations is that the maximum phase shift achieved at polarization independence for both modulators is different. For a waveguide length of 5 mm the V-shaped junction achieves a phase shift of 365° ($L_\pi V_\pi = 2.5$ V·cm) whereas for the flat junction the maximum phase shift is 290° ($L_\pi V_\pi = 3.1$ V·cm). These results show that the V-shaped *pn* junction can

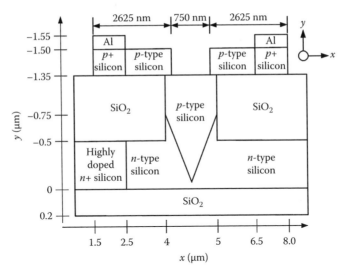

FIGURE 15.47 V-shaped *pn* junction optical modulator in silicon technology. (From Gardes, F.Y. et al., *Opt. Express*, 15(9), 5879, 2007. With permission.)

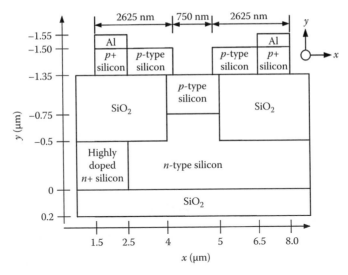

FIGURE 15.48 Flat-shape *pn* junction optical modulator in silicon technology. (From Gardes, F.Y. et al., *Opt. Express*, 15(9), 5879, 2007. With permission.)

achieve the same efficiency as the modulator proposed in [105] without the problems involved with submicron waveguides. The next step is to determine the bandwidth of both modulators. This is done by calculating the phase shift variation with transient time and is shown in Figure 15.51 and Figure 15.52 for TE and TM polarization and for the V- and flat junctions, respectively.

The transients show that the rise and fall time are similar for both types of junctions. For this type of doping, the V-shaped junction has a rise time of 13 ps and a fall time of 23 ps for both TE and TM. For the flat junction the rise time is 12 ps and the fall time is 21 ps for both TE and TM. In both cases, this corresponds to an intrinsic bandwidth in excess of 15 GHz.

Based on the principle of the depletion of a *pn* junction, Marris-Morini et al. [111] demonstrated in 2007 a low-loss (5 dB insertion loss) lateral-depletion modulator. The modulator is inserted in a rib waveguide

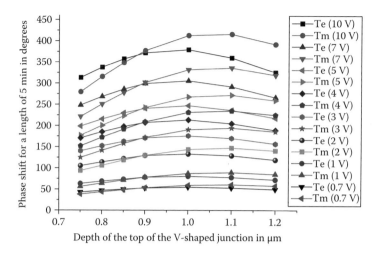

FIGURE 15.49   Phase shift modulation for TE and TM using a V-shaped *pn* junction. (From Gardes, F.Y. et al., *Opt. Express*, 15(9), 5879, 2007. With permission.)

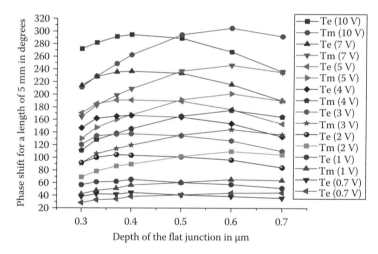

FIGURE 15.50   Phase shift modulation for TE and TM using a flat *pn* junction. (From Gardes, F.Y. et al., *Opt. Express*, 15(9), 5879, 2007. With permission.)

400 nm high, 660 nm wide, and with 100 nm etch depth. Highly doped regions of boron and phosphorus are situated on both sides of the rib waveguide and the junction is placed vertically in the centre of the waveguide, which ease the fabrication process and suppress the need of a top contact. The measured efficiency, $V_\pi L_\pi$, is 5.6 V-cm and the frequency performance of the modulator evaluated by applying an ac signal added to a dc bias has a 3 dB cutoff frequency of 1.4 GHz at 1.55 μm with a 4 mm long phase shifter.

In early 2008, Marris-Morini et al. [112] demonstrated a similar device with improved performance, the device cross section and top view are shown in Figure 15.53. For this device the efficiency, $V_\pi L_\pi$, is 5 V·cm and the normalized optical response of the modulator is plotted in Figure 15.54 for a dc bias of −5 V. The amplitude of the ac signal used for the measurement was 2.8 V peak to peak and a 3 dB cutoff frequency of ~10 GHz was measured.

In the previous devices, interferometers and high-quality resonators were typically used to modulate the light, due to a relatively weak plasma dispersion effect. Thin quantum-well structures made from III/V semiconductors such as GaAs or InP exhibit a much stronger quantum-confined Stark effect (QCSE)

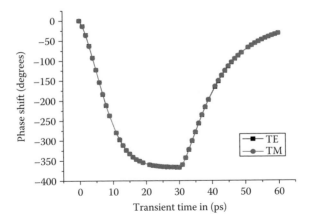

FIGURE 15.51 Phase shift against transient time for TE and TM for the V-shaped junction. (From Gardes, F.Y. et al., *Opt. Express*, 15(9), 5879, 2007. With permission.)

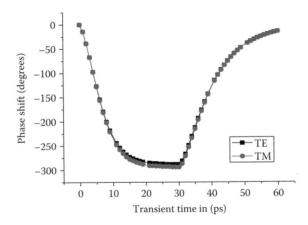

FIGURE 15.52 Phase shift against transient time for TE and TM for the flat junction. (From Gardes, F.Y. et al., *Opt. Express*, 15(9), 5879, 2007. With permission.)

mechanism [113], which allows modulator structures with optical path length in micrometers [114,115]. The QCSE, at room temperature, in thin germanium quantum-well structures grown on silicon (Figure 15.55) has also been reported recently [116]. The QCSE in this structure is comparable to that in III/V materials, and the authors claim that the process is CMOS compatible although fabrication complexity is undoubtedly increased. QCSE devices have typical length of the order of 100 microns [117] and do not require carrier injection, as they are typically operated in reverse bias. The resulting low power dissipation allows large arrays of devices at high data rates, and devices with operation bandwidths larger than 50 GHz have been reported in other materials [117]. This is very encouraging for high-speed, low-power, highly integrated electroabsorption (EA) modulators in silicon photonics, although much more development is required.

Pushing the trend in scaling devices even further, low power and level of high integration have been achieved in silicon by using PhC MZI modulator [118]. PhCs are a class of artificial optical materials with periodic dielectric properties, which result in unusual optical properties. A PhC waveguide is formed by introducing a line defect into a 2D slab [119,120]. The dispersion of PhC optical waveguides offers the possibility to enhance the propagation constant by a factor larger than 100 [118,120]. Therefore the active length of a rib or strip waveguide can be reduced by the same factor, to achieve the phase shift needed for modulation. The device shown in Figure 15.56 is proposed by Jiang et al. [118],

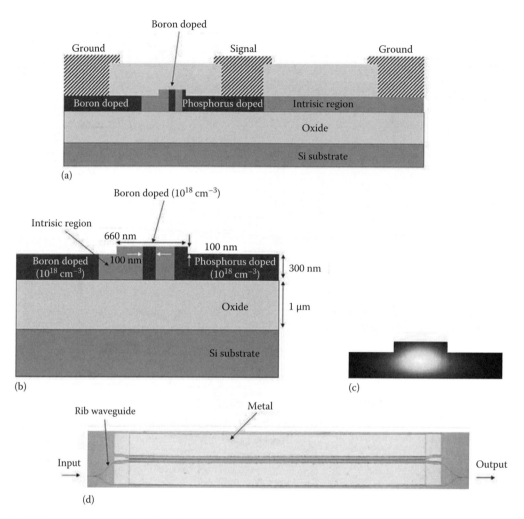

FIGURE 15.53 Cross section of the phase shifter structure integrated into a rib SOI waveguide: (a) Global view of the device with coplanar waveguide electrodes. (b) Phase shifter structure. (c) Intensity profile of the optical mode. (d) Optical microscope view of the modulator. (From Marris-Morini, D. et al., *Opt. Express*, 16(1), 334, 2008. With permission.)

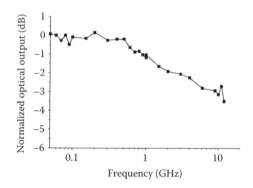

FIGURE 15.54 Normalized optical response of the modulator. (From Marris-Morini, D. et al., *Opt. Express*, 16(1), 334, 2008. With permission.)

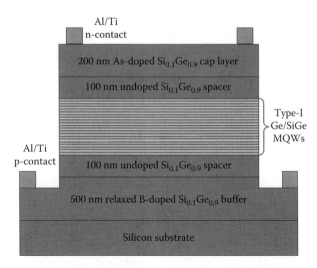

FIGURE 15.55 Schematic diagram of a *pin* structure of strained Ge/Si multiple quantum wells grown on silicon on relaxed SiGe buffer. (From Kuo, Y.-H. et al., *Nature*, 437(7063), 1334, 2005. With permission.)

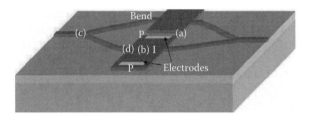

FIGURE 15.56 Schematic of the MZI PhC modulator. (a) Electrodes, (b) PhCs, (c) rib waveguide y-junction, and (d) rib-to-PhC junction. (From Yongqiang, J. et al., *Appl. Phys. Lett.*, 87(22), 221105-1, 2005. With permission.)

and is modulated using a current in the range of μA and a maximum voltage of 7.5 mV, for an active length of 80 μm.

Gu et al. [121] followed up on the results obtained by Jiang et al. [118] and reported high-speed operation using a PhC-based MZI having similar dimensions (lattice constant $a = 400$ nm, hole diameter $d = 220$ nm, overlayer thickness $t = 260$ nm, and interaction length = 80 μm) (Figure 15.57). The active element is made up of a *pin* diode (Figure 15.58) in one arm of the MZI (Figure 15.59) with both *p*-type and *n*-type concentration of 5×10^{17} cm^{-3}. The authors reported bit rates of 2 Mb/s and 1 Gb/s (Figure 15.60) with modulation depths of 85% and 20%, respectively. Maximum dc modulation depth of 93% was obtained with an injection current of 7.1 mA. Interestingly, this level of injection current is similar to those reported by Tang et al. [87] which has multi-micron device dimensions. Gu et al. [121] reported that the device could be further improved by design optimization, intrinsic region reduction, and fabrication to enhance matching and to reduce drive voltage.

Silicon has been limited as an optical material for decades because of a lack of or limited active optical properties. In a paper published in 2006, Jacobsen et al. [122] demonstrated a significant linear electro-optic effect induced in silicon by breaking the crystal symmetry. The inversion symmetry of silicon crystal prohibits the existence of a linear electro-optic effect, hence, by applying an asymmetric strain in the waveguide, the symmetry can be broken. Figure 15.61 shows the principle used for the proposed device [122], where a straining layer is deposited on the top of the waveguide. The proposed structure used a deposited silicon nitride layer (Si_3N_4) as a straining layer [123]. The amorphous Si_3N_4

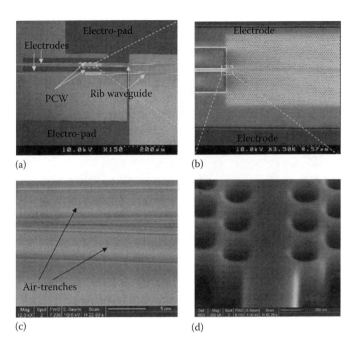

FIGURE 15.57 SEM micrograph of the silicon PhC modulator with overlayer thickness of $t = 215$ nm: (a) Overview picture of the modulator, (b) PhC waveguide with two electrodes, (c) Y-junction, and (d) magnified PhC waveguide based on a triangular lattice with lattice constant $a = 400$ nm and hole diameter $d = 210$ nm. (From Yongqiang, J. et al., *Appl. Phys. Lett.*, 87(22), 221105-1, 2005. With permission.)

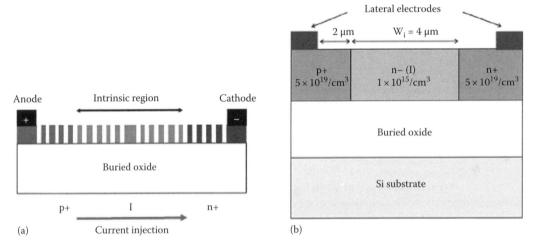

FIGURE 15.58 Cross-sectional schematic of the active *pin* diode (top), 2D model used in the electrical simulation (bottom). (From Gu, L. et al., *Appl. Phys. Lett.*, 90(7), 071105-3, 2007. With permission.)

is pre-compressively strained and hence tries to expand the structure underneath in both horizontal directions, thus creating the asymmetry.

It is theoretically predicted [124] that the material nonlinearity [125] is enhanced linearly with the group index. By using PhC waveguides, an enlarged group index can be obtained, hence increasing the material nonlinearity. The proposed device has been designed to achieve values above 230 for the group index. Figure 15.62 shows the use of PhC waveguides inserted in an MZI.

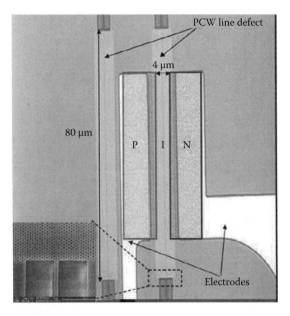

FIGURE 15.59 Top view of the *pin* diode implemented into the PhC-based MZI. (From Gu, L. et al., *Appl. Phys. Lett.*, 90(7), 071105-3, 2007. With permission.)

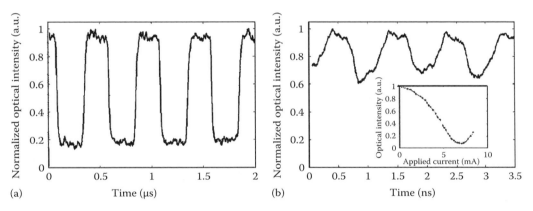

FIGURE 15.60 Normalized optical intensity output at 2 Mb/s (a) and 1 Gb/s (b). The inset shows the modulator optical output intensity against applied current. (From Gu, L. et al., *Appl. Phys. Lett.*, 90(7), 071105-3, 2007. With permission.)

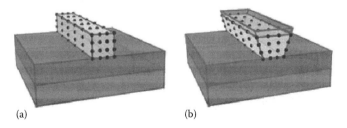

FIGURE 15.61 Applying strain to crystalline silicon. (a) Waveguide fabricated in the top layer of an SOI wafer. (b) The same waveguide with a straining layer deposited on top. The straining layer breaks the inversion symmetry and induces a linear electro-optic effect. (From Jacobsen, R.S. et al., *Nature*, 441(7090), 199, 2006. With permission.)

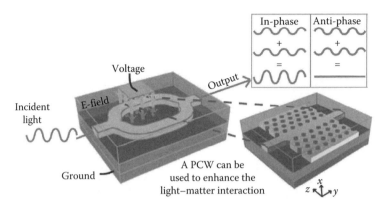

FIGURE 15.62　Diagram of an MZM. Incident light is split into two waveguides. The output amplitude depends on the phase difference at recombination. As shown at top right, in-phase recombination gives "1" bit output while anti-phase recombines. (From Jacobsen, R.S., et al., *Nature*, 441(7090), 199, 2006. With permission.)

Although the enhanced nonlinearity can be increased above $800\,\mathrm{pm}\cdot\mathrm{V}^{-1}$ for a specific wavelength and a specific PhC waveguide [122], it is important to make a fair comparison between different nonlinear materials. To do that, one must compare the material nonlinearity measured in [122], which is determined to be approximately $15\,\mathrm{pm}\cdot\mathrm{V}^{-1}$ and the commonly applied nonlinear material, $LiNbO_3$. In $LiNbO_3$, the largest tensor component [126] is approximately $360\,\mathrm{pm}\cdot\mathrm{V}^{-1}$. However, the authors stated that improvements in the material nonlinearity could be obtained if the silicon forming the waveguide could be deformed more freely by using, for example, nanowires. The advantages of using strain-induced nonlinearities in a modulator are fairly obvious as the speed of such a device would not be limited by charge mobility or charge recombination. Furthermore, only an electric field is required and no current is needed, which is not the case in devices where changing the carrier concentration often requires a large ac current.

In 2007, Liu et al. [127] presented a design of monolithically integrated GeSi EA modulators and photodetectors based on the Franz–Keldysh effect. The GeSi composition was chosen for optimal performance around 1550 nm. The proposed modulator device was butt-coupled to a Si (core)/SiO_2 (cladding) high-index contrast waveguides, and is predicted to have a 3 dB bandwidth of >50 GHz and an ER of 10 dB. The advantage of this design is that using the same device structure, a waveguide-coupled photodetector can also be integrated to the waveguide. The photodetector proposed by Liu et al. has a predicted responsivity of >1 A/W and a 3 dB bandwidth of >35 GHz.

As shown in Figure 15.63, the GeSi EA modulator and the photodetector are based on the same structure, potentially allowing efficient monolithic process integration. Both structures are based on a vertical Si/$Ge_{0.9925}Si_{0.0075}$/Si *pin* diode with a doping level of $2\times10^{19}\,\mathrm{cm}^{-3}$ in n^+ and p^+ Si, and their height (H) and width (W) can be designed to obtain optimal device performance. The only difference in the dimensions of the GeSi EA modulator and the photodetector is that the latter is longer than the former ($L_2 > L_1$) to increase the absorption. The drawbacks of such a structure are the difficulty to obtain efficient butt-coupling to minimize losses due to the impedance mismatch between the strip waveguide and the modulator as well as to fabricate the vertical Si/$Ge_{0.9925}Si_{0.0075}$/Si *pin* diodes.

The question lies herein: is it possible to increase the operating speed of silicon optical devices further? Perhaps utilizing other mechanisms? Recently, Hocberg et al. [128] exploited the Kerr phenomenon by inducing optical nonlinearities in the waveguide material using an intense modulation beam. The authors overcame the relatively weak ultrafast nonlinearities in silicon by cladding a silicon waveguide with a specially engineered nonlinear optical polymer (Figure 15.64) based around an MZI (Figure 15.65).

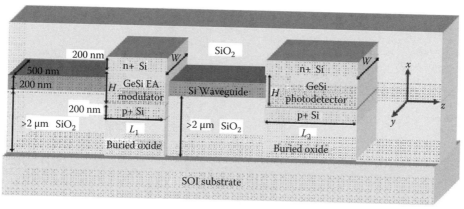

FIGURE 15.63 Structure of the proposed monolithically integrated GeSi electroabsorption modulators and photodetectors. (From Liu, J. et al., *Opt. Express*, 15, 623, 2007. With permission.)

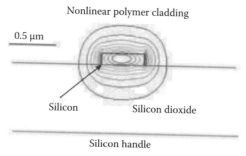

FIGURE 15.64 Mode pattern of the optical signal in the silicon waveguide clad with a specially engineered nonlinear polymer. Contours are drawn in 10% increments of power. (From Hochberg, M. et al., *Nat. Mater.*, 5(9), 703, 2006. With permission.)

The source signal is introduced to the MZI by a beam splitter and in one of the arms, a modulating gate signal is introduced by a 3 dB coupler. The nonlinear Kerr effect allows the gate signal to induce a phase shift in the source signal, which in turn allows the shifted source signal to interfere with the optical signal at the reference arm of the MZI, thereby causing an intensity modulation of the source signal. According to the authors, the demonstrated modulation frequency of 10 GHz is limited by existing measurement equipment and they showed via indirect evidence (spectral measurements) that the device can function into the terahertz range. A key point to note is that the authors took advantage of efficient light confinement in the silicon waveguides and the highly nonlinear properties of optical polymer. However, it remains to be seen if such hybrid processes can be fully integrated into a CMOS process.

From the above, it is clear that silicon optical modulators utilizing the plasma dispersion effect are capable of reaching multiple gigahertz speed using changes in the optical properties of the material via electrical signals. This is a huge improvement during a period when device operating speeds improved from 20 MHz in 1988, to 1 GHz in 2004, and later to 30 GHz and 40 Gb/s in 2007 [107].

In the last few years processing power has increased tremendously and nowadays common desktop computer are able to achieve few billions of floating points operation per second (FLOPs). Recently Intel demonstrated an 80 core processor capable of delivering more than 1 TFLOPs [129], paving the way toward what tomorrow's computing platforms could look like. Such a processing capability necessitates interconnects able to deliver information in excess of a terabit per second (Tb/s). In this regard silicon photonic circuits may be the technology of choice, primarily because of the potential attraction of

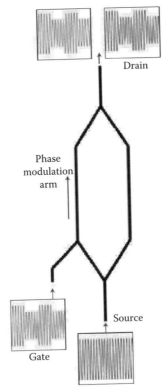

FIGURE 15.65 The gate signal has its intensity modulation transferred to the source signal via nonlinear phase modulation in one arm of the MZI. (From Hochberg, M. et al., *Nat. Mater.*, 5(9), 703, 2006. With permission.)

integration of photonic functionality with electronics in a cost-effective manner, and also because SOI substrates have proved successful for high-volume processing of very low-loss waveguides, of the order of 0.1 dB/cm. Furthermore, multiplexing of 40 Gb/s signals would enable Tb/s signals to be achieved, hence satisfying the need for tomorrow's multicore applications.

15.5 Conclusion

This chapter has attempted to provide an overview of research in silicon photonic waveguides and modulators. In providing such a summary it is impossible to be comprehensive, as a surprising range and depth of work was carried out, not only in single-crystal silicon, but also in related technologies. However, it is hoped that the reader has obtained a sense of the history of the formative years of silicon photonics and a summary of recent developments in the two areas. It is noteworthy that silicon photonics is now progressing very quickly. While the progress made in the past was significant, things are moving much more quickly now, partly due to the increased interest in silicon photonics, as well as the significantly increased resources worldwide. Consequently, silicon photonics can look forward to an intense period of advanced study.

Among several platforms considered for silicon photonics, SOI has become by far the most popular platform. In the early days, SOI waveguides had rather high propagation loss, which has been reduced to the order of 0.1 dB/cm owing to improvements in design and fabrication. At first silicon waveguides had also rather large dimensions in order to reduce the insertion loss when coupled to/from an optical fiber. The current trend is, however, to move to smaller device dimensions for improved cost, efficiency, and device performance. CMOS compatibility is another reason for this move and therefore small strip waveguides are becoming very interesting building blocks for silicon photonic circuits. However, these waveguides have higher propagation losses of 2–3 dB/cm. Furthermore, it is difficult to achieve polarization independence in devices based on these small waveguides. As the cross section of these waveguides is much smaller than the cross section of an optical fiber, coupling to/from fibers remains a very important issue. Mid- and long-wave infrared photonic applications are becoming more interesting and therefore new waveguiding structures suitable for those wavelength ranges need to be found. In this chapter, proton-beam-written structures have been considered.

It is clear from the foregoing discussion that silicon optical modulators have undergone a significant transformation since the early work of the 1980s. Device dimensions have been reduced to allow operation at the market standard and above up to 40 Gb/s, and work is ongoing that predicts improvement of the data rate by at least another order of magnitude in the future. This is already a significant achievement in a material that does not intrinsically exhibit a significant electric field–based modulation mechanism. While these devices still have ground to make up on modulators in other material technologies (power consumption, ER), the possibility of producing all-silicon optical circuits and subsequently integrating them with electronic circuits is a hugely attractive vision. This will not only increase the application areas of silicon but will also enhance the electronic circuits and the bandwidth required in forthcoming multi-core applications. Hence, silicon photonics is enabling the emergence of silicon as a candidate to become the photonics material of the future, and perhaps the platform upon which the next technological revolution will be built.

Thus significant effort is now focusing on the intra-chip interconnection of future multi-cores of the microprocessor, as well as interconnection of component parts of high-performance computers. The components now exist in silicon photonics to suggest that such interconnection architectures are viable. Of course, significant challenges remain, and further device improvements are required, but it is noteworthy that silicon has progressed almost from the status of a non-optical material to a candidate technology to enable future advancement of computing infrastructure. The next decade will be interesting for those working in silicon photonics. Will the technology become mainstream, and part of every computing application? The possibility certainly exists, and developments in the last 5 years suggest that the probability of success is high.

Acknowledgments

The authors would like to acknowledge the approval from Wiley to use material from *Silicon Photonics: The State of the Art*, Ed. G.T. Reed, John Wiley & Sons, Chichester, U.K., 2008. Also acknowledged is funding from EPSRC (UK Silicon Photonics programme) and The Royal Society.

References

1. Soref, R.A. and J.P. Lorenzo, Single-crystal silicon—A new material for 1.3 and 1.6 μm integrated-optical components. *Electron. Lett.*, **21**: 953–954, October 10, 1985.
2. Soref, R.A. and J.P. Lorenzo, All-silicon active and passive guided-wave components for λ = 1.3 and 1.6 μm. *IEEE J. Quantum Electron.*, **QE-22**: 873–879, June 1986.
3. Iyer, S.S. and Y.-H. Xie, Light emission from silicon. *Science*, **260**(5104): 40–46, 1993.
4. Reed, G.T. and A.P. Knights, *Silicon Photonics: An Introduction*, John Wiley & Sons, Chichester, U.K., 2004.
5. Reed, G.T. (Ed.), *Silicon Photonics: State of the Art*, John Wiley & Sons, Chichester, U.K., 2008.
6. Soref, R.A., Silicon-based optoelectronics. *Proc. IEEE*, **81**(12): 1687–1706, 1993.
7. Fang, A.W., H. Park, O. Cohen, R. Jones, M.J. Paniccia, and J.E. Bowers, Electrically pumped hybrid AlGaInAs-silicon evanescent laser. *Opt. Express*, **14**: 9203–9210, October 2006.
8. Fang, A.W., B.R. Koch, R. Jones, E. Lively, D. Liang, Y.-H. Kuo, and J.E. Bowers, A distributed Bragg reflector silicon evanescent laser. *IEEE Photon. Technol. Lett.*, **20**(20): 1667–1669, October 2008.
9. Boyraz, O. and B. Jalali, Demonstration of a silicon Raman laser. *Opt. Express*, **12**(21): 5269–5273, 2004.

10. Rong, H., A. Liu, R. Jones, O. Cohen, D. Hak, R. Nicolaescu, A. Fang, and M. Paniccia, An all-silicon Raman laser. *Nature*, **433**: 292–294, 2005.

11. http://techresearch.intel.com/articles/Tera-Scale/1419.htm

12. Manipatruni, S., Q. Xu, B.S. Schmidt, J. Shakya, and M. Lipson, High speed carrier injection 18 Gb/s silicon micro-ring electro-optic modulator, in *LEOS 2007*, Lake Buena Vista, FL, October 21–25, 2007.

13. Gunn, C., Integration, in *Silicon Photonics: The State of the Art*, John Wiley & Sons, Chichester, U.K., 2008.

14. http://www.darpa.mil/mto/programs/epic/

15. Polky, J.N. and G.L. Mitchell, Metal-clad planar dielectric waveguide for integrated optics. *J. Opt. Soc. Am.*, 64(3): 274–279, 1974.

16. Nezval, J., WKB approximation for optical modes in a periodic planar waveguide. *Opt. Commun.*, **42**(5): 320–322, 1982.

17. de Ruiter, H.M., Limits on the propagation constants of planar optical waveguide modes. *Appl. Opt.*, **20**(5): 731–732, 1981.

18. Tamir, T., Leaky waves in planar optical waveguides. *Nouvelle Revue d'Optique*, **6**(5): 273–284, 1975.

19. Payne, F.P., Generalized transverse resonance model for planar optical waveguides, in *Tenth European Conference Optical Communication* (ECOC '84), Stuttgart, Austria, 1984.

20. Kawachi, M., M. Yasu, and T. Edahiro, Fabrication of SiO_2-TiO_2 glass planar optical waveguides by flame hydrolysis deposition. *Electron. Lett.*, **19**(15): 583–584, 1983.

21. Hanaizumi, O., M. Miyagi, and S. Kawakami, Low radiation loss y-junctions in planar dielectric optical waveguides. *Opt. Commun.*, **51**(4): 236–238, 1984.

22. Jerominek, H., Z. Opilski, and J. Kadziela, Some elements of integrated optics circuits based on planar gradient glass waveguides. *Opt. Appl.*, **13**(2): 159–168, 1983.

23. Albares, D.J. and R.A. Soref, Silicon-on-sapphire waveguides, in *Proc. SPIE: Integrated Optical Circuit Engineering IV*, Bellingham, WA, 1987, vol. 704, pp. 24–25.

24. Cortesi, E., F. Namavar, and R.A. Soref, Novel silicon-on-insulator structures for silicon waveguides, in *1989 IEEE SOS/SOI Technology Conference* (Cat. No.89CH2796-1), Stateline, NV, October 3–5, 1989, IEEE.

25. Namavar, F., E. Cortesi, R.A. Soref, and P. Sioshansi, On the formation of thick and multiple layer SIMOX structures and their applications, in *Ion Beam Processing of Advanced Electronic Materials Symposium*, San Diego, CA, April 25–27, 1989, Mater. Res. Soc.

26. Reed, G.T., L. Jinhua, C.K. Tang, L. Chenglu, P.L.F. Hemment, and A.G. Rickman, Silicon on insulator optical waveguides formed by direct wafer bonding. *Mat. Sci. Eng. B Solid*, **B15**(2): 156–159, 1992.

27. Evans, A.F., D.G. Hall, and W.P. Maszara, Propagation loss measurements in silicon-on-insulator optical waveguides formed by the bond-and-etchback process. *Appl. Phys. Lett.*, **59**(14): 1667–1669, 1991.

28. Colinge, J.-P., *Silicon on Insulator Technology: Materials to VLSI*, 2nd ed., Kluwer Academic Publishers, Norwell, MA, 1997.

29. Mohd Kassim, N., H.P. Ho, T.M. Benson, and D.E. Davies, Assessment of SIMOX material by optical waveguide losses, in *20th European Solid State Device Research Conference* (ESSDERC 90), Nottingham, U.K., September 10–13, 1990, Adam Hilger.

30. Weiss, B.L. and G.T. Reed, The transmission properties of optical waveguides in SIMOX structures. *Opt. Quantum Electron.*, **23**(8): 1061–1065, 1991.

31. Davies, D.E., M. Burnham, T.M. Benson, N.M Kassim, and M. Seifouri, Optical waveguides and SIMOX characterization, in *1989 IEEE SOS/SOI Technology Conference* (Cat. No.89CH2796-1), Stateline, NV, October 3–5, 1989, IEEE.

32. Soref, R.A., E. Cortesi, F. Namavar, and L. Friedman, Vertically integrated silicon-on-insulator waveguides. *IEEE Photonic. Tech. Lett.*, **3**(1): 22–24, 1991.

33. Weiss, B.L., Reed, G.T., Toh, S.K., Soref, R.A., and F. Namavar, Optical waveguides in SIMOX structures. *IEEE Photonic. Tech. Lett.*, **3**(1): 19–21, 1991.

34. Rickman, A., G.T. Reed, B.L. Weiss, F. Namavar, Low-loss planar optical waveguides fabricated in SIMOX material. *IEEE Photonic. Tech. Lett.*, **4**(6): 633–635, 1992.

35. Kurdi, B.N. and D.G. Hall, Optical waveguides in oxygen-implanted buried-oxide silicon-on-insulator structures. *Opt. Lett.*, **13**(2): 175–177, 1988.

36. Schmidtchen, J., A. Splett, B. Schuppert, and K. Petermann, Low loss integrated-optical rib-waveguides in SOI, in *1991 IEEE International SOI Conference*, Vail Valley, CO, October 1–3, 1991, IEEE, Piscataway, NJ, 1992.

37. Rickman, A.G., G.T. Reed, and F. Namavar, Silicon-on-insulator optical rib waveguide loss and mode characteristics. *J. Lightw. Technol.*, **12**(10): 1771–1776, 1994.

38. Rickman, A.G. and G.T. Reed, Silicon-on-insulator optical rib waveguides: Loss, mode characteristics, bends and y-junctions. *IEE Proc. Optoelectron.*, **141**(6): 391–393, 1994.

39. Tang, C.K., A.K. Kewell, G.T. Reed, A.G. Rickman, and F. Namavar, Development of a library of low-loss silicon-on-insulator optoelectronic devices. *IEE Proc. Optoelectron.*, **143**(5): 312–315, 1996.

40. Rickman, A.G., G.T. Reed, and F. Namavar, Silicon-on-insulator optical rib waveguide circuits for fibre optic sensors. *Proc. SPIE: Distributed and Multiplexed Fiber Optic Sensors III*, Bellingham, WA, September 8–9, 1993, vol. 2071, pp. 190–196.

41. Sure, A., T. Dillon, J. Murakowski, C. Lin, D. Pustai, and D.W. Prather, Fabrication and characterization of three-dimensional silicon tapers. *Opt. Express*, **11**(26): 3555–3561, 2003.

42. Masanovic, G.Z., V.M.N. Passaro, and G.T. Reed, Dual grating-assisted directional coupling between fibers and thin semiconductor waveguides. *IEEE Photonic Technol. Lett.*, **15**(10): 1395–1397, 2003.

43. Almeida, V.R., R.R. Panepucci, and M. Lipson, Nanotaper for compact mode conversion. *Opt. Lett.*, **28**(15): 1302–1304, 2003.

44. Shoji, T., T. Tsuchizawa, T. Watanabe, K. Yamada, and H. Morita, Low loss mode size converter from 0.3 mm² Si wire waveguides to singlemode fibres. *Electron. Lett.*, **38**(25): 1669–1670, 2002.

45. Taillaert D., F. Van Laere, M. Ayre, W. Bogaerts, D. Van Thourhout, P. Bienstman, and R. Baets, Grating couplers for coupling between optical fibers and nanophotonic waveguides. *Jpn. J. Appl. Phys.*, **145**: 6071–6077, 2006.

46. Soref, R.A., J. Schmidtchen, and K. Petermann, Large single-mode rib waveguides in Ge-Si and Si-on-SiO₂. *IEEE J. Quantum Electron.*, **27**: 1971–1974, 1991.

47. Pogossian, S.P., L. Vescan, and A. Vonsovici, The single-mode condition for semiconductor rib waveguides with large cross section. *J. Lightw. Technol.*, **16**: 1851–1853, 1998.

48. Powell, O., Single-mode condition for silicon rib waveguides. *J. Lightw. Technol.*, **20**: 1851–1855, 2002.

49. Vivien, L., S. Laval, B. Dumont, S. Lardenois, A. Koster, and E. Cassan, Polarization-independent single-mode rib waveguides on silicon-on-insulator for telecommunication wavelengths. *Opt. Commun.*, **210**: 43–49, 2002.

50. Lousteau, J., D. Furniss, A. Seddon, T.M. Benson, A. Vukovic, and P. Sewell, The single-mode condition for silicon-on-insulator optical rib waveguides with large cross section. *J. Lightw. Technol.*, **22**: 1923–1929, 2004.

51. Chan, S.P., V.M.N. Passaro, and G.T. Reed, Single mode condition and zero birefringence in small SOI waveguides. *Electron. Lett.*, **41**: 528–529, 2005.

52. Mashanovich, G.Z., M. Milosevic, P. Matavulj, B. Timotijevic, S. Stankovic, P.Y. Yang, E.J. Teo, M.B.H. Breese, A.A. Bettiol, and G.T. Reed, Silicon photonic waveguides for different wavelength regions. *Semicond. Sci. Technol.* (invited, special issue on *Silicon Photonics*), **23**: 064002, 2008.

53. Ye, W.N., D.-X. Xu, S. Janz, P. Cheben, M.-J. Picard, B. Lamontagne, and N.G. Tarr, Birefringence control using stress engineering in silicon-on-insulator (SOI) waveguides. *J. Lightw. Technol.*, **23**(3): 1308–1318, March 2005.

54. Headley, W.R., G.T. Reed, M. Pannicia, A. Liu, and S. Howe, Polarization-independent optical race-track resonators using rib waveguides in silicon-on-insulator. *Appl. Phys. Lett.*, **85**(23): 5523–5525, December 2004.

55. Timotijevic, B.D., F.Y. Gardes, W.R. Headley, G.T. Reed, M.J. Paniccia, O. Kohen, D. Hak, and G.Z. Masanovic, Multi-stage racetrack resonator filters in silicon-on-insulator. *J. Opt. A Pure Appl. Opt.*, **8**: S473–S476, 2006.

56. Milosevic, M.M., P.S. Matavulj, and G.Z. Mashanovich, Stress-induced characteristics of silicon-on-insulator rib waveguides, in *Proceedings of the 15th Telecommunication Forum TELFOR*, Belgrade, Serbia, November 2007, pp. 401–404.

57. Milosevic, M.M., P.S. Matavulj, B.D. Timotijevic, G.T. Reed, and G.Z. Mashanovich, Design rules for single mode and polarization independent silicon-on-insulator rib waveguides using stress engineering. *J. Lightw. Technol.*, **26**: 1840–1846, 2008.

58. Timotijevic, B.D., Auto-regressive optical filters in silicon-on-insulator waveguides, PhD dissertation, Advanced Technology Institute, University of Surrey, Guildford, U.K., 2007.

59. Gardes, F.Y., G.T. Reed, N.G. Emerson, and C.E. Png, A sub-micron depletion-type photonic modulator in silicon on insulator. *Opt. Express*, **13**: 8845–8854, 2005.

60. Reed, G.T., W.R. Headley, F.Y. Gardes, B.D. Timotijevic, S.P. Chan, and G.Z. Mashanovich, Characteristics of rib waveguide racetrack resonators in SOI in *Proc. SPIE*, **6183**: 61830G, 2006.

61. Bogaerts, W. et al., Compact wavelength-selective functions in silicon-on-insulator photonic wires. *IEEE J. Sel. Top. Quantum Eletron.*, **12**: 1394–1401, 2006.

62. Yamada, K., H. Fukuda, T. Watanabe, T. Tsuchizawa, T. Shoji, and S.-I. Itabashi, Functional photonic devices on silicon wire waveguide, in *Proc. LEOS 2nd Group IV Photonics*, Antwerp, Belgium, September 2005, pp. 186–188.

63. Alto, T., *Microphotonic Silicon Waveguide Components*, VTT, Finland, 2004 (ISBN 951-38-6422-7).

64. Vlasov, Y.A. and S.J. McNab, Losses in single-mode silicon-on-insulator strip waveguides and bend. *Opt. Express*, **12**: 1622–1627, 2004.

65. Tsuchizawa, T., K. Yamada, H. Fukuda, T. Watanabe, J.-I. Takahashi, M. Takahashi, T. Shoji, E. Tamechika, S.-I. Itabashi, and H. Morita, Microphotonics devices based on silicon microfabrication technology. *IEEE J. Sel. Top. Quantum Electron.*, **11**: 232–240, 2005.

66. Dumon, P., W. Bogaerts, V. Wiaux, J. Wouters, S. Beckx, J. Van Campenhout, D. Taillaert, B. Luyssaert, P. Bienstman, D. Van Thourhout, R. Baets, Low-loss SOI photonic wires and ring resonators fabricated with deep UV lithography. *IEEE Photon. Technol. Lett.*, **16**: 1328–1330, 2004.

67. Tsuchizawa, T., T. Watanabe, E. Tamechika, T. Shoji, K. Yamada, J. Takahashi, S. Uchiyama, S. Itabashi, and H. Morita, Fabrication and evaluation of submicron-square Si-wire waveguides with spot-size converter, in *Proc. of the 15th Annual IEEE LEOS Meeting*, Glasgow, U.K., 2002, pp. 287–288.

68. Veldhuis, G.J., O. Parriaux, H.J.W.M. Hoekstra, and P.V. Lambeck, Sensitivity enhancement in evanescent optical waveguide sensors. *J. Lightw. Technol.*, **18**: 677–682, 2000.

69. Lipson, M., Guiding, modulating, and emitting light on silicon—Challenges and opportunities. *J. Lightw. Technol.*, **23**: 4222–4238, 2005.

70. Beggs, D.M., T.P. White, L. O'Faolain, and T. Krauss, Ultracompact and low power optical switch based on silicon photonic crystals. *Opt. Lett.*, **33**: 147–149, 2008.

71. Almeida, V.R., Q. Xu, C.A. Barrios, and M. Lipson, Guiding and confining light in void nanostructures. *Opt. Lett.*, **29**: 1209–1211, 2004.

72. White, T.P., L. O'Faolain, J. Li, L.C. Andreani, and T.F. Krauss, Silica-embedded silicon photonic crystal waveguides. *Opt. Express*, **16**: 17076–17081, 2008.

73. Xu, Q., V.R. Almeida, R.R. Panepucci, and M. Lipson, Experimental demonstration of guiding and confining light in nanometer-size low-refractive-index material. *Opt. Lett.*, **29**: 1626–1628, 2004.

74. Hawkins, G.J., Spectral characteristics of infrared optical materials and filters, PhD thesis, University of Reading, U.K., 1998.

75. Palik, E.D., *Handbook of Optical Constants of Solids*, vol. 1, Academic, New York, 1985.

76. Soref, R.A., S.J. Emelett, and W.R. Buchwald, Silicon waveguide components for the long-wave infrared region. *J. Opt. A Pure Appl. Opt.*, **8**: 840–848, 2006.

77. Mashanovich, G.Z., S. Stankovic, P.Y. Yang, E.J. Teo, F. Dell'Olio, V.M.N. Passaro, A.A. Bettiol, M.B.H. Breese, and G.T. Reed, Silicon waveguides for mid-infrared wavelength region, in *Photonics West 2008*, San Jose, CA, January 2008, p. 6898-25.

78. Yang, P.Y., G.Z. Mashanovich, I. Gomez-Morilla, W.R. Headley, G.T. Reed, E.J. Teo, D.J. Blackwood, M.B.H. Breese, and A.A. Bettiol, Free standing waveguides in silicon. *Appl. Phys. Lett.*, **90**: 241109, 2007.

79. Teo, E.J., A.A. Bettiol, M.B.H. Breese, P.Y. Yang, G.Z. Mashanovich, W.R. Headley, G.T. Reed, and D.J. Blackwood, Three-dimensional fabrication of silicon waveguides with porous silicon cladding. *Opt. Express*, **16**: 573–578, 2008.

80. Teo, E.J., A.A. Bettiol, P. Yang, M.B.H. Breese, B.Q. Xiong, G.Z. Mashanovich, W.R. Headley, and G.T. Reed, Fabrication of low loss silicon-on-oxidized porous silicon strip waveguides using focused proton beam irradiation. *Opt. Lett.*, **34**(5): 659–661, 2009.

81. Soref, R.A. and B.R. Bennett, Electrooptical effects in silicon. *IEEE J. Quantum Electron.*, **QE-23**(1): 123–129, 1987.

82. Soref, R.A. and B.R. Bennett, Kramers–Kronig analysis of electro-optical switching in silicon, in *Proc. SPIE: Integrated Optical Circuit Engineering IV*, Cambridge, MA, Sept. 16–17, 1986. 1987.

83. Ang, T.W. et al., *Integrated Optics in Unibond for Greater Flexibility*, Electrochem. Soc, Seattle, WA, 1999.

84. Hewitt, P.D. and G.T. Reed, Improving the response of optical phase modulators in SOI by computer simulation. *J. Lightw. Technol.*, **18**(3): 443–450, 2000.

85. Hewitt, P.D. and G.T. Reed, *Multi micron dimension optical p-i-n modulators in silicon-on-insulator*, SPIE-Int. Soc. Opt. Eng, San Jose, CA, 1999.

86. Tang, C.K., G.T. Reed, A.J. Wilson, and A.G. Rickman, Low-loss, single-mode, optical phase modulator in SIMOX material. *J. Lightw. Technol.*, **12**(8): 1394–1400, 1994.

87. Tang, C.K. and G.T. Reed, Highly efficient optical phase modulator in SOI waveguides. *Electron. Lett.*, **31**(6): 451–452, 1995.

88. Ang, T.W. et al., 0.15 dB/cm loss in unibond SOI waveguides. *Electron. Lett.*, **35**(12): 977–978, 1999.

89. Png, C.E. et al., Optical phase modulators for MHz and GHz modulation in silicon-on-insulator (SOI). *J. Lightw. Technol.*, **22**(6): 1573–1582, 2004.

90. Atta, R.M.H. et al., *Fabrication of a Highly Efficient Optical Modulator Based on Silicon-on-Insulator*, The International Society for Optical Engineering, Maspalonas, Gran Canaria, Spain, 2003.

91. Png, C.E. et al., *Development of Small Silicon Modulators in Silicon-on-Insulator (SOI)*, The International Society for Optical Engineering, San Jose, CA, 2003.

92. Png, C.E., Silicon-on-insulator phase modulators, PhD thesis, University of Surrey, Guildford, U.K., 2004.

93. Irace, A., G. Breglio, and A. Cutolo, All-silicon optoelectronic modulator with 1 GHz switching capability. *Electron. Lett.*, **39**(2): 232–233, 2003.

94. Cutolo, A. et al., An electrically controlled Bragg reflector integrated in a rib silicon on insulator waveguide. *Appl. Phys. Lett.*, **71**(2): 199–201, 1997.

95. Barrios, C.A., V.R. de Almeida, and M. Lipson, Low-power-consumption short-length and high-modulation-depth silicon electrooptic modulator. *J. Lightw. Technol.*, **21**(4): 1089–1098, 2003.

96. Liu, A. et al., A high-speed silicon optical modulator based on a metal-oxide-semiconductor capacitor. *Nature*, **427**(6975): 615–618, 2004.

97. Liao, L. et al., High speed silicon Mach-Zehnder modulator. *Opt. Express*, **13**(8): 3129–3135, 2005.

98. Gan, F. and F.X. Kartner, High-speed silicon electrooptic modulator design. *IEEE Photonics Technol. Lett.*, **17**(5): 1007–1009, 2005.

99. Xu, Q. et al., Micrometre-scale silicon electro-optic modulator. *Nature*, **435**(7040): 325–327, 2005.

100. Xu, Q. et al., 12.5 Gbit/s carrier-injection-based silicon micro-ring silicon modulators. *Opt. Express*, **15**: 430–436, 2007.

101. Manipatruni, S. et al., High speed carrier injection 18 Gb/s silicon micro-ring electro-optic modulator, in *Lasers and Electro-Optics Society, 2007 (LEOS 2007). The 20th Annual Meeting of the IEEE*, Lake Buena Vista, FL, 2007.

102. Green, W.M.J. et al., Ultra-compact, low RF power, 10 Gb/s silicon Mach–Zehnder modulator. *Opt. Express*, **15**(25): 17106–17113, 2007.

103. Jones, R. et al., Net continuous wave optical gain in a low loss silicon-on-insulator waveguide by stimulated Raman scattering. *Opt. Express*, **13**(2): 519–525, 2005.

104. Jones, R. et al., Lossless optical modulation in a silicon waveguide using stimulated Raman scattering. *Opt. Express*, **13**(5), 1716–1723, 2005.

105. Gardes, F.Y. et al., A sub-micron depletion-type photonic modulator in silicon on insulator. *Opt. Express*, **13**(22): 8845–8854, 2005.

106. Liu, A. et al., High-speed optical modulation based on carrier depletion in a silicon waveguide. *Opt. Express*, **15**(2): 660–668, 2007.

107. Liao, L. et al., 40 Gbit/s silicon optical modulator for high-speed applications. *Electron. Lett.*, **43**(22): 1196–1197, 2007.

108. Gunn, C., CMOS Photonics for high-speed interconnects. *Micro. IEEE*, **26**(2): 58–66, 2006.

109. Huang, A. et al., A 10 Gb/s photonic modulator and WDM MUX/DEMUX integrated with electronics in 0.13 um SOI CMOS, in *Solid-State Circuits Conference, 2006. ISSCC 2006. Digest of Technical Papers. IEEE International*, San Francisco, CA, 2006.

110. Gardes, F.Y. et al., Micrometer size polarisation independent depletion-type photonic modulator in silicon on insulator. *Opt. Express*, **15**(9): 5879–5884, 2007.

111. Marris-Morini, D. et al., Low loss optical modulator in a silicon waveguide based on a carrier depletion horizontal structure, in *4th IEEE International Conference on Group IV Photonics, 2007*, Tokyo, Japan, 2007.

112. Marris-Morini, D. et al., Low loss and high speed silicon optical modulator based on a lateral carrier depletion structure. *Opt. Express*, **16**(1): 334–339, 2008.

113. Miller, D.A.B. et al., Band-edge electroabsorption in quantum well structures: The quantum-confined Stark effect. *Phys. Rev. Lett.*, **53**(22): 2173–2176, 1984.

114. Arad, U. et al., Development of a large high-performance 2-D array of GaAs–AlGaAs multiple quantum-well modulators. *IEEE Photonics Technol. Lett.*, **15**(11): 1531–1533, 2003.

115. Chin-Pang, L. et al., Design, fabrication and characterization of normal-incidence 1.56 microns multiple-quantum-well asymmetric Fabry–Perot modulators for passive picocells. *IEICE Trans. Electron.*, **E86-C**(7): 1281–1289, 2003.

116. Kuo, Y.-H. et al., Strong quantum-confined Stark effect in germanium quantum-well structures on silicon. *Nature*, **437**(7063): 1334–1336, 2005.

117. Lewen, R. et al., Segmented transmission-line electroabsorption modulators. *J. Lightw. Technol.*, **22**(1): 172–179, 2004.

118. Yongqiang, J. et al., 80-micron interaction length silicon photonic crystal waveguide modulator. *Appl. Phys. Lett.*, **87**(22): 221105–221107, 2005.

119. Johnson, S.G. et al., Linear waveguides in photonic-crystal slabs. *Phys. Rev. B (Condensed Matter)*, **62**(12): 8212–8222, 2000.

120. Soljacic, M. et al., Photonic-crystal slow-light enhancement of nonlinear phase sensitivity. *J. Opt. Soc. Am. B (Opt. Phys.)*, **19**(9): 2052–2059, 2002.

121. Gu, L. et al., High speed silicon photonic crystal waveguide modulator for low voltage operation. *Appl. Phys. Lett.*, **90**(7): 071105-3, 2007.

122. Jacobsen, R.S. et al., Strained silicon as a new electro-optic material. *Nature*, **441**(7090): 199–202, 2006.

123. Madou, J.M., *Fundamentals of Microfabrication*, CRC Press, Boca Raton, FL, 2002.

124. Soljacic, M. and J.D. Joannopoulos, Enhancement of nonlinear effects using photonic crystals. *Nat. Mater.*, **3**(4): 211–219, 2004.

125. Butcher, P.N. and D. Cotter, *The Elements of Nonlinear Optics 5*, Cambridge University Press, Cambridge, U.K., 1990.
126. Li, G.L. and P.K.L. Yu, Optical intensity modulators for digital and analog applications. *J. Lightw. Technol.*, **21**(9): 2010–2030, 2003.
127. Liu, J. et al., Design of monolithically integrated GeSi electro-absorption modulators and photodetectors on a SOI platform. *Opt. Express*, **15**: 623–628, 2007.
128. Hochberg, M. et al., Terahertz all-optical modulation in a silicon-polymer hybrid system. *Nat. Mater.*, **5**(9): 703–709, 2006.
129. Intel. *Intel Research Advances 'Era Of Tera'*. 2007, available from: http://www.intel.com/pressroom/archive/releases/20070204comp.htm

16

Planar Waveguide Multiplexers/ Demultiplexers in Optical Networks: From Improved Designs to Applications

Sailing He
Jun Song
Jian-jun He
Daoxin Dai

16.1 Introduction

Fiber-optic communication has been growing extensively in recent years. A great number of optical fibers have been laid in all parts of the world, dramatically increasing the capacity and quality of telecommunications. However, traditional optical transmission systems are still inadequate to carry the heavy traffic resulting from the exponential increase of the bandwidth demand. Fortunately, the wavelength division multiplexing (WDM) technique provides an effective and low-cost way to increase the capacity by tens or hundreds of times in an optical transmission system.

Optical multiplexers and demultiplexers are the key components in a WDM fiber-optic communication system. Arrayed waveguide gratings (AWGs) and etched diffraction gratings (EDGs) are two typical multiplexers/demultiplexers based on planar optical waveguides [1–4]. They take the advantages of mature semiconductor manufacturing process and can offer more than 40 channels of dense wavelength division multiplexing (DWDM) with relatively low loss.

With the recent development and growth of the FTTx market, etc., both planar waveguide multiplexers/demultiplexers have already many new applications. Their main potential lies in applications with moderate crosstalk requirements and in monolithic or hybrid integration for more complex devices like multi-wavelength receivers and transmitters, add-drop multiplexers, channel monitors, optical cross connections, etc.

16.1.1 Arrayed Waveguide Gratings

Figure 16.1 shows a schematic representation of an AWG demultiplexer. This device consists of two star couplers, connected by a phase-dispersive waveguide array. The operation principle is as follows. Light propagating in the input waveguide will be coupled into the array through the first star coupler, which is also known as free propagation region (FPR). The arrayed waveguides have different lengths. Specifically, the path length difference between adjacent waveguides is constant. As a consequence, the field distribution at the input aperture will be reproduced at the output aperture. At the central wavelength, the light will focus in the center of the image plane (provided that the input waveguide is centered in the input plane). Different wavelengths of the light experience different phase changes within the arrayed waveguides and will be focused at different outputs.

Several substrates and materials have been used for the construction of an AWG, with different waveguide structures, as shown for instance in Figure 16.2. There are many fabrication technologies for AWGs of different materials, such as silica [5,6], InP [7], polymer [8], and silicon-on-insulator (SOI) [9]. Among them, SiO_2-buried rectangular waveguides are the most popular, which can give a low insertion loss to a standard single-mode fiber (SMF). For a SiO_2-based waveguide, however, the bending radius has to be very large (e.g., of the order of several millimeters) due to the low refractive index contrast Δ. This limits the integration density of SiO_2-based photonic integrated devices. Recently, SOI waveguides (see, e.g., [9,10]) have become very popular for planar optical devices due to their compatibility with the complimentary metal-oxide semiconductor (CMOS) technology. The investigations in the past several years were focused on SOI rib waveguides with a large cross section (in order to obtain a single-mode operation as well as a high coupling efficiency to an SMF). Since there is an ultra-high index contrast Δ between the Si core (~3.455) and the cladding of air (1.0) or SiO_2 (1.445), it is also possible to have an ultra-sharp bending and consequently realize some ultrasmall photonic integrated devices. Some results on Si nanowire (instead of rib) waveguide have been reported (see, e.g., [11–13]). For a Si nanowire waveguide,

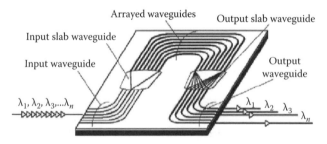

FIGURE 16.1   Schematic configuration for an AWG demultiplexer.

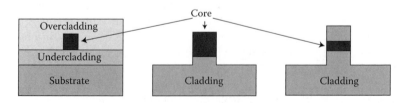

FIGURE 16.2   Structures for optical waveguides used in AWGs: (a) buried waveguide, (b) rib waveguide, and (c) deep-ridge waveguide.

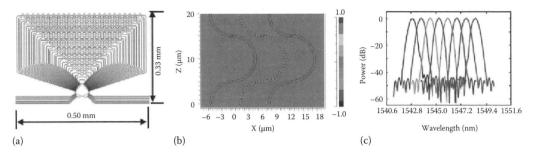

(a) (b) (c)

FIGURE 16.3 (a) AWG layout based on Si nanowire waveguides, (b) light propagation in the microbends, (c) the calculated spectral response. (From Dai, D.X. and He, S., *Opt. Express*, 14(12), 5260, 2006. With permission.)

the scattering loss (per unit of length) due to the roughness of the sidewall is much larger than that for a conventional micrometric waveguide [14]. On the other hand, the reduction of the total size of PLC devices based on Si nanowire waveguides compensates the large scattering loss in terms of the total loss per device. The total loss in a PLC device based on Si nanowire waveguides can be low enough for practical uses. Therefore, it is attractive to develop ultrasmall PLC devices based on Si nanowire waveguides.

As a design example, a novel layout is shown in Figure 16.3a based on Si nanowire waveguides, where the two FPRs are overlapped and a series of microbends are inserted at the middle of the arrayed waveguides [15]. With such a novel design the device size is minimized to only about 0.165 mm². Figure 16.3b shows the light propagation in the microbends of some arrayed waveguides. From this figure, one sees that the coupling between arrayed waveguides is negligible even when the adjacent arrayed waveguides are placed very closely. This is due to the decoupling separation of SOI nanowires of the order of 2 μm. Figure 16.3c shows the calculated spectral response (which does not include the coupling loss between the input/output waveguides and the fibers). In Figure 16.3c, one sees that the calculated crosstalk is smaller than −30 dB. The excess loss due to the inserted microbends in the arrayed waveguides is very small when the bending radius is large enough.

16.1.2 Etched Diffraction Gratings

Compared with an AWG, an EDG (see Figure 16.4) is more compact and potentially has a higher spectral finesse since it can accommodate a larger number of grating facets. These characteristics make an EDG more suitable for high channel density communication systems. However, the application of an EDG demultiplexer is often limited by its large insertion loss associated with imperfectly fabricated grating facets. In order to have a good performance, vertical and smooth grating facets with deep etching depths are required. Recently, significant improvements have been made on chemically assisted ion beam etching (CAIBE), which allows one to fabricate smooth grating sidewalls with vertical angles less than 1°. In this way, one can significantly reduce the excess loss of the reflecting grating (e.g., ~0.5 dB for our layered structure according to the modeling). The improvement in the reliability and reproducibility of the fabrication process, together with the inherent advantage of compactness and the potential

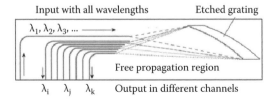

FIGURE 16.4 Schematic diagram of an EDG demultiplexer.

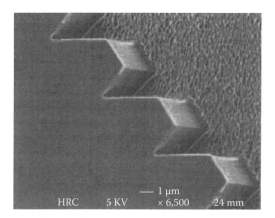

FIGURE 16.5 SEM photograph of a section of the etched grating showing the retroreflecting V-shaped facets. The incident light propagates horizontally from the left side. (From He, J.J. et al., *J. Lightw. Technol.*, 16, 631, 1998. With permission.)

for a large number of closely spaced channels, makes grating-based demultiplexers very attractive for dense WDM applications.

An EDG based on a Rowland mounting is illustrated in Figure 16.4. The field propagating from an input waveguide to the FPR is diffracted by each grating facet. It is then refocused onto an imaging curve and guided into the corresponding output waveguides according to the wavelengths. The grating of an EDG demultiplexer is usually coated with a metal (e.g., Au) at the backside in order to enhance the reflection efficiency. In order to reduce reflection loss without the additional processing steps required for coating the backside of the grating facets with a reflecting metal coating, a retro-reflecting V-shaped facet [16,17] was used at each grating tooth. A scanning electron micrograph (SEM) photograph of the etched grating is shown in Figure 16.5. In this design light hits each grating facet at about 45° incidences producing total internal reflection. This design was used on some of the chips, while some others on the same wafer had flat facets so that direct comparisons could be made.

EDG demultiplexers have been demonstrated in SiO_2/Si [2], InGaAs/AlGaAs/GaAs [18], and InGaAsP/InP [16]. Recently, a very compact EDG device (see Figure 16.6) based on nanophotonic SOI platform

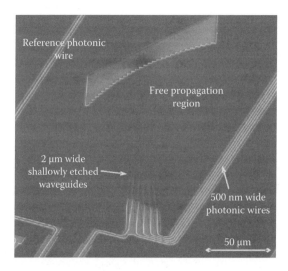

FIGURE 16.6 SEM picture of 1 × 4 EDG demultiplexer using a nanophotonic SOI platform. (From Clemens, P.C. et al., *IEEE Photon. Technol. Lett.*, 4, 886, 1992. With permission.)

has been fabricated with good performances [19]. The demultiplexer is fabricated on a nanophotonic SOI platform [20] using standard wafer-scale CMOS processes including deep-UV lithography. The device has four wavelength channels with a channel spacing of 20 nm and a record-small footprint of $280 \times 150\,\mu m^2$. The on-chip loss is 7.5 dB, and the crosstalk is lower than -30 dB.

16.2 Improved Designs

16.2.1 Passband-Flattened Spectral Responses

For a conventional AWG or EDG demultiplexer, the shape of the spectral response is of Gaussian type and the figure of merit (the ratio of the 3 dB bandwidth to the 33 dB bandwidth) is small. Therefore, the transmission efficiency is sensitive to a slight wavelength shift (caused by, e.g., the laser quality, temperature change, or polarization variation). These drawbacks limit the applications of these planar demultiplexers of Gaussian type in WDM systems. A planar integrated demultiplexer with a flattened spectral response is thus desirable.

Many methods have been introduced to achieve a passband-flattened spectral response of a planar integrated demultiplexer [21–25]. Forming twofold self-image at the end of the input waveguide and improving the design of gratings or arrayed waveguides are two common methods to broaden or flatten the spectral response of a multiplexer/demultiplexer.

16.2.1.1 Forming Twofold Self-Image at the End of the Input Waveguide

To achieve a flattened spectral response, the most common approach is to use a transformer structure such as a parabolically tapered waveguide [26], a multimode interference (MMI) section [27], or a Y-branch [28] at the end of the input waveguide to form a double-peaked field distribution at the input plane.

The method of connecting an MMI section at the end of the input waveguide is simple and effective (see Figure 16.7a). Conventionally, the length of the MMI section was chosen according to the self-image principle [29] so that the twofold self-image was formed at the end of the MMI section (see Figure 16.7b). Such an MMI coupler is not optimal for broadening the spectral response of an AWG/EDG demultiplexer. With a conventional numerical simulation method such as the beam propagation method (BPM), one may have to make a two-dimensional search for the width and length of the MMI section in order to obtain a desired flat-top demultiplexer [30]. The solid line of Figure 16.7c gives the corresponding flattened spectral response calculated with self-image principle. The dotted line of Figure 16.7c indicates the Gaussian spectral response for the same AWG demultiplexer without the MMI coupler. A BPM is used to calculate the spectral response for the designed AWG demultiplexer, and the results are shown in Figure 16.7c with a dashed line. One can see that a larger 3 dB passband can be obtained after optimizing by the BPM method. When carrying out a flat-top design using an MMI at the end of the input waveguide, analytical self-image method can provide the first approximation, and some slight adjustment can be made in the neighborhood of the first approximation by using a more accurate method such as the BPM.

The MMI coupler converts the single-waveguide mode at the input of the coupler into a twofold image. The resulting output field pattern has a "camel-like" shape. The separation between the peaks of the twofold image can be controlled with the MMI width. However, when the separation increases, the depth of the central depression will be enlarged, which increases ripples crosstalk and insertion losses for the center wavelength. Therefore, the flat degree is limited for a uniform MMI coupler. A tapered MMI coupler (Figure 16.8a), instead of a uniform MMI coupler, can be used to obtain a flattened spectral response with low ripples, low insertion losses, and low crosstalk [31]. In Figure 16.8b, the spectral response for different tapered angle is demonstrated. It is easily seen that the spectral response can be improved by adjusting the tapered angle. A low ripple can be obtained when the tapered angle is set to an optimized value.

Moreover, a parabolic MMI section with a reshaping taper connected at the end of the input waveguide has also been presented to obtain a flat passband and sharp transitions [32].

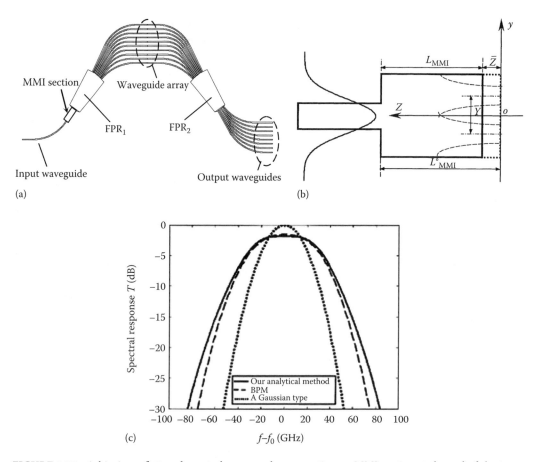

FIGURE 16.7 Achieving a flattened spectral response by connecting an MMI section at the end of the input waveguide. (a) Schematic diagram with an AWG as a design example. (b) Geometrical configuration for the MMI section and the Gaussian beam for the twofold images. (c) Spectral responses. (From Dai, D. et al., *J. Lightw. Technol.*, 20(11), 1957, 2002. With permission.)

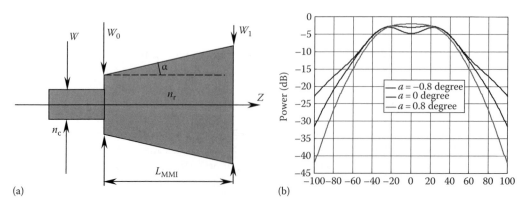

FIGURE 16.8 Use of a tapered MMI coupler to broaden the passband of an AWG. (a) Configuration of a tapered MMI. (b) The spectral response for different tapered angles. (From Dai, D. et al., *Commun.*, 219, 233, 2003. With permission.)

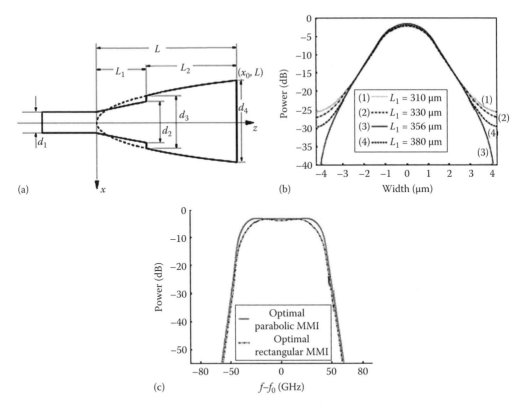

FIGURE 16.9   A flat-topped EDG demultiplexer using a parabolic MMI section with a reshaping taper at the end of the input waveguide. (a) The input structure consisting of a reshaping taper and a parabolic MMI section. (b) The field distribution at the end of the reshaping taper for various lengths of the taper. (c) The spectral responses for a demultiplexer with the optimized parabolic MMI section (solid line), and a demultiplexer with an optimal rectangular MMI section (dash-dotted line). The same reshaping taper is used to excite the MMI section in both cases. (From Song, J. et al., *Opt. Commun.*, 227, 89, 2003. With permission.)

The configuration of the reshaping taper and the parabolic MMI section is shown in Figure 16.9a. Here the taper is used to transfer the Gaussian-type fundamental modal profile into a field distribution with sharp transitions (which then gives a spectral response with sharp transitions). Figure 16.9b gives the simulation results of the field distribution at the end of the taper for various values of the length L_1 of the reshaping taper. There exists an optimal L_1 to give the sharpest transition for the field distribution at the end of the taper. Thus, the optimized taper reshapes the fundamental mode of a standard single-mode waveguide into an excitation field (for the MMI section) with sharper transitions. This will result in sharper transitions (and consequently a much lower crosstalk for adjacent channels) in the spectral response. Figure 16.9c shows the spectral responses for the two structures. The solid line gives the spectral response for the optimized parabolic MMI section, and the dash-dotted line shows the spectral response for an optimized rectangular MMI section. In this figure one sees that the EDG demultiplexer with the optimized parabolic MMI section gives a better performance with a minimal ripple (virtually 0 dB), a very large 1 dB passband.

16.2.1.2 Improving the Design of Gratings or Arrayed Waveguides

As technologies for higher bit rates and higher spectral efficiencies (e.g., 40 Gb/s systems) emerge, the low chromatic dispersion characteristics of (de)multiplexing devices become more important. The

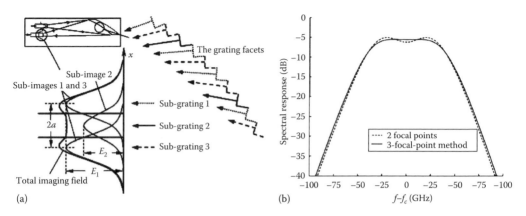

FIGURE 16.10 A three-focal-point method for the optimal design of a flat-top planar waveguide demultiplexer. (a) Geometrical configuration for the three-focal-point method. (b) The flattened spectral response for the designed EDG demultiplexer. (From Shi, Z. and He, S., *IEEE J. Select. Top. Quantum Electron.*, 8(6), 1179, 2002. With permission.)

above-mentioned flattening techniques are usually associated with some degradation in the chromatic dispersion characteristics because of the phase distortion in those input structures. By using some methods of improving the design of gratings/arrayed waveguides, an EDG/AWG can be obtained with both flattened passband and low chromatic dispersion.

For an EDG demultiplexer, a novel design has been presented by dividing the grating into three interleaved sub-gratings with some appropriate ratio of the facet numbers of the sub-gratings [33].

When the sub-gratings are interleaved as shown in Figure 16.10a, the ratio of the peak amplitudes of the sub-images is approximately equal to the ratio of the facet numbers of the corresponding sub-gratings (e.g., when the ratio of the facet numbers of two sub-gratings is 2:1, the ratio of the peak amplitudes of the corresponding sub-images is approximately equal to 2:1). The three-focal-point method is realized by choosing appropriate focal points near the imaging center, which then determines the position of the center of each facet uniquely by fulfilling the condition of the light path difference. The normal to the grating facet is typically along the direction in the middle of the incident ray and the diffracted ray (and the facet angle has a good fabrication tolerance). Thus, incident field with a wavelength different from the designed wavelength would also have three focal points with the same separation near its corresponding imaging centers, and the three-focal-point method is effective over a wide spectrum range. Figure 16.10b shows the ripple level of the spectral response as half-separation a varies for both cases of two and three focal points. When the EDG demultiplexer has only two focal points, the ripple rises quickly when half-separation a increases (see the dashed line in Figure 16.10b). Using the present three-focal-point method, the ripple is always kept at a very low level (see the solid line in Figure 16.10b).

Furthermore, a new passband flattening technique for waveguide grating devices has been proposed and experimentally demonstrated [34]. It uses phase dithering of grating elements to obtain a flat passband with sharp transitions while maintaining single-mode input and output waveguides. The method is similar to the multi-grating method (e.g., three-focal-point method) but more generalized. The technique can be used in both AWG and echelle grating-based devices. The phase dithering method is to introduce an additional phase variation in the grating field distribution so that the resulting field distribution at the output plane (determined by its Fourier transform) is a complex function. The convolution of this complex function with the output waveguide mode profile results in a complex function whose amplitude is as close as possible to a rectangular function. The phase term added at each grating element can be realized by slightly adjusting the waveguide lengths in the case of an AWG or by adjusting the positions of the reflecting facets in the case of an etched diffraction grating. As an example, Figure 16.11a shows the phase variation as a function of the grating facet number of an EDG demultiplexer for

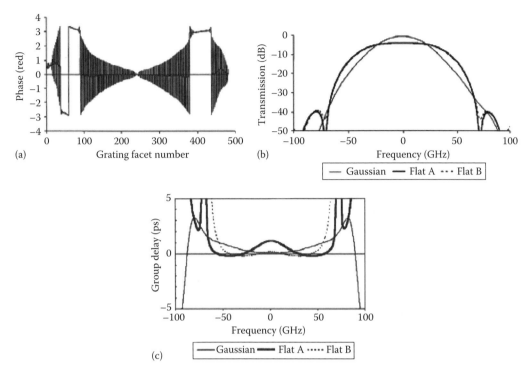

FIGURE 16.11 A phase-dithered EDG demultiplexer with a flat passband and sharp transitions. (a) Phase diagram of a passband-flattened echelle grating. (b) Spectral response of the original Gaussian passband grating (thin solid line) and the flat passband grating (A) corresponding to the phase diagram of (a) (thick solid line). The spectrum of a flat passband grating (B) with improved chromatic characteristics is also shown (dashed line). (c) Group delay responses of the Gaussian passband grating (thin solid line), the flat passband grating (A) corresponding to the phase diagram of (a) (thick solid line) and the flat passband grating (B) with reduced group delay (dashed line). (From He, J.-J., *IEEE J. Select. Top. Quantum Electron.*, 8(6), 1186, 2002. With permission.)

a flat passband design. Note that this phase variation is in addition to the phase difference of $2m\pi$ (where m is the grating order) between two adjacent facets in a standard grating. For the Gaussian passband, this phase variation is zero. For the flat passband grating, the phase variation oscillates around a slowly varying function, with adjacent facets swinging in opposite directions, as shown in the insert. The simulated spectral responses of the demultiplexer with and without the phase adjustments are shown in Figure 16.11b. One can see that, with the phase adjustment, not only is the passband widened and flattened, the slope of the passband transition also becomes sharper. Figure 16.11c gives the group delay responses for the Gaussian and two flat passband gratings with and without the phase adjustments. For a flattened passband design, the phase distribution becomes slightly more asymmetric. The group delay difference over the passband is reduced from 1.37 to 0.38 ps. The group delay difference over the passband for the Gaussian passband grating is 0.62 ps. Therefore, the phase dithering method cannot only flatten the passband, but also reduce the group delay difference. This is in contrast to the MMI or taper-based method in which the passband flattening is typically associated with degradation of the chromatic dispersion property.

However, for the phase-dithering method, the design process using the iterative method is time-consuming, and sometimes it may not converge to a satisfactory spectral response function. Due to the limitation of numerical accuracy, it is sometimes difficult to reduce the sidelobe crosstalk level to below −35 or −40 dB. Compared with the method based on numerical iterations, a design process using an analytic method has been presented using the spatial phase modulation, which is much simpler and less time-consuming [35]. An analytic formula for an EDG demultiplexer is derived using the scalar diffraction theory:

$$E_{out}(x', z') \approx RF^{-1}\left\{W_m \cdot F\left[E_{in}(x)\right]\right\} \tag{16.1}$$

where

 E_{out} is the field distribution along the image plane
 E_{in} is the input field distribution
 R is the reflection coefficient of the grating

The formula means that the image of a conventional EDG demultiplexer can be considered as the result of a Fourier transform of the input field and a weighted inverse Fourier transform with a weight factor W_m (very close to one). When the mth grating facet has a small shift of Δd_m, along the direction perpendicular to its cross-sectional line, weight factor W_m changes as,

$$W_m \approx e^{i\phi_m} \approx e^{-i2k\Delta d_m} \tag{16.2}$$

When phase ϕ_m has a linear profile of u_m, i.e., $\phi_m = 2\pi Au_m$, it results in a transverse shift in the image with a transverse shift distance from the original image center, which is independent of the incident field profile, i.e.,

$$E_{out}(x', z') \approx RF^{-1}\left\{e^{i2\pi Au_m} \cdot F\left[E_{in}(x)\right]\right\} = RE_{in}(x+A) \tag{16.3}$$

When ϕ_m has a nonlinear exponent profile, i.e., $\phi_m = (2\pi Au_m)^P$, the field profile along the image plane also has an approximate transverse shift A, which is also independent of the incident field profile. However, when the profile exponent $P > 1$, the resulting image is distorted, with some sidelobes at the outward side and a smoothly decaying profile at the inward side. Thus, the analytic formula can be characterized by two parameters: a transverse shift distance A and a profile exponent P for the phase modulation.

As a numerical example of a typical SiO_2-etched diffraction grating demultiplexer, Figure 16.12 shows the flattened design of an EDG demultiplexer using the spatial phase modulation method with optimal parameters $A = 4\,\mu m$ and $P = 1.35$. Figure 16.12a shows the actual shifts Δd_m of the grating facets of the optimal design. The spectral response of the central channel for the designed EDG demultiplexer is shown in Figure 16.12b. Using this method, one can obtain a design of planar waveguide demultiplexers with a flat-top and sharp transitions. In Figure 16.12b, one can see that the sidelobe crosstalk level of the designed demultiplexer is always below −40 dB. The dispersion characteristic for the EDG demultiplexer designed with this analytic method of spatial phase modulation is shown in Figure 16.12c. One can see that the designed EDG demultiplexer has a low chromatic dispersion [less than 1.5 ps/nm through out the passband (−50 GHz, 50 GHz)].

16.2.2 Polarization-Insensitive Design

Polarization dependence is an important issue in planar-waveguide-based wavelength (de)multiplexers for DWDM applications. Two main polarization effects will be discussed in this section: polarization-dependent wavelength shift (PDλ) and polarization-dependent loss (PDL).

16.2.2.1 Planar Waveguide Demultiplexers with Low PDλ Design

For a conventional planar waveguide demultiplexer, owing to the difference in the propagation constants of the transverse electric (TE) and transverse magnetic (TM) modes in planar waveguides, the PDλ occurs in the spectral response, which results in a shift in the spectral response peak of each wavelength

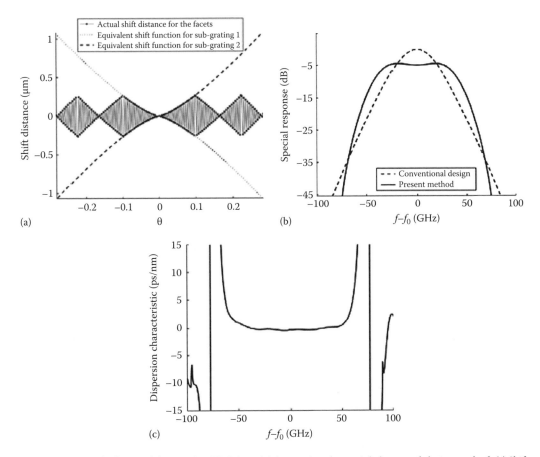

FIGURE 16.12 The flattened design of an EDG demultiplexer using the spatial phase modulation method. (a) Shift distances for grating facets for this design. (b) Spectral responses of the central channel obtained with the optimal design method and the conventional design method. (c) The dispersion characteristic of the designed flattened EDG demultiplexer. (From Shi, Z., *J. Lightw. Technol.*, 21(10), 2314, 2003. With permission.)

channel. This wavelength shift is sensitive to the design of the planar waveguide, and can range from a few tenths of nanometers to a few nanometers. As WDM systems move toward closer and closer channel spacing, even a small polarization-dependent wavelength shift may become a severe problem.

There exist many low PDλ designs applied to planar demultiplexers. The general idea is to compensate the difference of the phase shifts between the two polarizations [36–41]. Usually it is more flexible for an AWG (as compared with an EDG) to achieve low PDλ performance due to more degrees of design freedom (e.g., dimensions of the arrayed waveguides, shapes of the star couplers) to be chosen for an AWG while there is only a simple FPR (a 2D-slab) for an EDG. Two designs will be introduced below to obtain an AWG demultiplexer with low PDλ.

A low PDλ design of AWG demultiplexer based on Si photonic wires has been presented [42]. By optimizing the height and width of the arrayed waveguides, the channel spacing becomes polarization insensitive. To reduce PDλ, different diffraction orders for the TE and TM polarizations are chosen. A non-central input has been used to eliminate PDλ. The Si photonic wire with an air cladding is considered as an example, and the cross section is shown in the inset of Figure 16.13a. When one chooses $w_{co} = 297$ nm and $h_{co} = 362$ nm, a maximal fabrication tolerance for the core width is obtained. By selecting different diffraction order for both polarizations ($m_{TE} = 61$ and $m_{TM} = 72$), one can obtain a low PDλ. The simulated spectral responses are shown in Figure 16.13b, from which one sees that the peaks of the

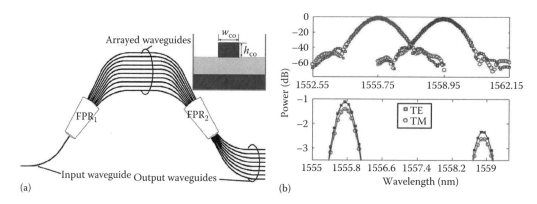

FIGURE 16.13 Design of a low PDλ AWG demultiplexer based on silicon photonic wires. (a) Schematic configuration for an AWG demultiplexer. (b) Simulated spectral responses (TE and TM polarizations) of the designed low AWG for the central and edge channels (top figure), and enlarged view (bottom figure). (From Dai, D. and He, S., *Opt. Lett.*, 31(13), 1988, 2006. With permission.)

spectral responses for the TE and TM polarizations are almost the same. This indicates that both the channel spacing and channel wavelengths are polarization insensitive.

Although a low PDλ can be obtained by introducing two different core widths in the arrayed waveguides, the drawback of this approach is that it introduces coupling loss between different waveguides and the core width difference is limited due to multimode concern. A more effective approach has been presented for eliminating the birefringence of AWGs [43]. Since the birefringence of the slab waveguides of an AWG is generally different from that of the channel waveguides, they can be compensated against each other by appropriately designing the shape of the star couplers. By utilizing the birefringence difference between the slab waveguides and the arrayed waveguides, a polarization-insensitive AWG can be realized without requiring any additional fabrication step or two different channel waveguides. The star couplers are designed according to Rowland circle construction with an oblique incident/diffraction angle θ, similar to the case of EDGs. When θ is not zero, the total path length difference corresponding to adjacent arrayed waveguides is produced in both the slab region and the arrayed waveguide region. Consequently, the diffraction equation for the central wavelength becomes

$$2ns\Delta Ls + na\Delta L = m\lambda c \tag{16.4}$$

In this case, the polarization-dependent wavelength shift becomes $\Delta\lambda = \lambda_{TE} - \lambda_{TM} = [(2\Delta ns\Delta Ls + \Delta na\Delta L)/(2ns\Delta Ls + na\Delta L)]\lambda c$, where $\Delta ns = ns_{(TE)} - ns_{(TM)}$ is the effective index difference (birefringence) of the slab waveguides between the TE and TM modes. The condition for polarization-insensitive operation, i.e., $\Delta\lambda = 0$, becomes

$$2\Delta ns\Delta Ls + \Delta na\Delta L = 0 \tag{16.5}$$

Using geometrical analysis, one can select appropriate parameters to satisfy the zero PDλ condition. Figure 16.14a shows an example design corresponding to a birefringence ratio of 1:3 with the opposite signs between the channel waveguide and the slab waveguide. Figure 16.14b shows the spectral response of the central channel for the birefringence-compensated AWG. The TE and TM responses become completely overlapped. The parameters of the polarization-compensated AWG are determined when both Equations 16.4 and 16.5 are satisfied.

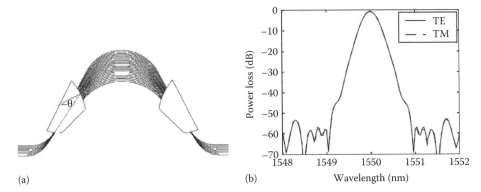

FIGURE 16.14   Birefringence-compensated AWG demultiplexer with angled star couplers. (a) Schematic diagram of a birefringence-compensated AWG with the opposite signs between the channel waveguide and the slab waveguide. (b) Spectral response of the central output waveguide for an AWG with angled star couplers. (From Lang, T.T. et al., *Opt. Express*, 15(23), 15022, 2007. With permission.)

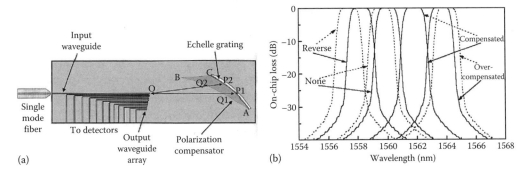

FIGURE 16.15   Integrated polarization compensator for EDG demultiplexers. (a) Schematic diagram of an etched grating waveguide demultiplexer with integrated polarization compensator (region ABC). (b) Simulated channel response functions of InP-based etched grating demultiplexers with different degrees of polarization compensation. (From He, J.-J. et al., *IEEE Photon. Technol. Lett.*, 11, 224, 1999. With permission.)

Both methods mentioned above can be applied for the design of AWGs with a low PDλ. However, for an EDG demultiplexer, there exists a simple FPR (a 2D-slab), and no other design freedom can be used to compensate the birefringence in the only FPR. By integrating a prism-like polarization compensator in the slab waveguide region adjacent to the grating, one can also obtain a low PDλ design for EDG demultiplexers [44]. The same method is also effective for an AWG case. A schematic diagram of the polarization-compensated waveguide demultiplexer based on an etched diffraction grating is presented in Figure 16.15a. The polarization compensator consists of a prism-shaped region, which has different effective indexes (for both TE and TM modes) from those of the surrounding slab waveguide. It can be realized by shallow etching in the same process step as that used to define a single-mode ridge waveguide. Thus, no additional processing steps are required. The shape of this area can be optimized to compensate the birefringence in the FPR. Figure 16.15b illustrates the simulated response function of the demultiplexer with different degrees of polarization compensation: no compensation, compensated, over-compensated, and reverse compensated. When there is no compensation, the channel wavelength for the TM mode is about 0.4 nm shorter than that of TE mode. As compensation increases, the channel passband shifts to longer wavelengths. The wavelength shift for the TM mode is larger than that for the TE mode. The response functions for the TE and TM modes completely overlap in the optimally compensated case.

16.2.2.2 Planar Waveguide Demultiplexers with a Low PDL

For an AWG demultiplexer, PDL mainly results from the polarization-dependent coupling loss between the FPRs and the arrayed waveguides. Therefore, one can easily reduce PDL for an AWG demultiplexer, e.g., by introducing some linear tapers at the ends of the arrayed waveguides [42].

However, for an EDG demultiplexer, the PDL is largely affected by the different diffraction efficiency of the grating for both polarizations. Coating the facets with a metal at the backside of the grating and using retro-reflecting facets [45] are two common methods for reducing the reflection loss and thus improving the diffraction efficiency (see Figure 16.16a and b). For a conventional metallic echelle grating, the metal coating on the shaded facet is the main source of the PDL while it does not contribute to the improvement of the diffraction efficiency [46]. Therefore, the PDL for EDG demultiplexers using retro-reflecting facets is lower than metal-coated one (see Figure 16.17) [47].

It has been shown that a single-side metallic echelle grating (see Figure 16.17c) can have higher diffraction efficiency (especially for the transverse-magnetic (TM) mode) and much lower PDL as compared with a conventional metallic echelle grating [48]. The new groove geometry is suitable for a wide range of grating materials, and is also advantageous over a TIR grating especially for a material with a low refractive index.

Figure 16.18 shows the retro-diffraction efficiency and the PDL of a SiO_2 grating as the incident angle increases for a conventional metallic echelle grating and a single-side metal-coated grating. As the incident angle increases (i.e., the size of the shaded facet increases with respect to the illuminated facet), the influence of the shaded facet in the conventional metallic echelle grating becomes significant rapidly (e.g., PDL exceeds 0.5 dB when the incident angle is 26°). This indicates that the incident angle cannot be large for a conventional metallic echelle grating if PDL criteria are required in the design of a grating device. For an echelle grating with single-side metal-coated grooves, however, the PDL remains less than 0.15 dB even when the incident angle equals 60° (which is large enough for a practical design). Thus, a compact device with very low PDL can be achieved using some single-side metallic echelle grooves.

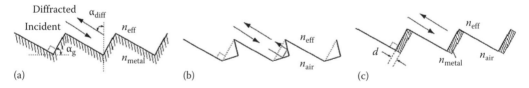

FIGURE 16.16 Geometries for three types of grating grooves. (a) Coating the facets with a metal at the backside of the grating; (b) retro-reflecting facets; and (c) single-side metallic echelle grating.

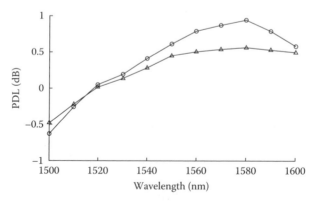

FIGURE 16.17 The PDL of the EDG demultiplexer for different wavelength channels. Circles are for an echelle grating coated with gold. Triangles are for a retro-reflecting grating based on InP material. (From Song, J. et al., *IEEE J. Select. Top. Quantum Electron.*, 11, 224, 2005. With permission.)

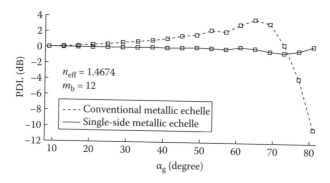

FIGURE 16.18 Retro-diffraction efficiency and PDL for two different types of gratings as incident angle increases. (From Shi, Z. M. et al., *IEEE Photonics Technol. Lett.*, 16(18), 1885, 2004. With permission.)

16.2.3 Other Improved Designs for Planar Waveguide Demultiplexers

16.2.3.1 EDG Demultiplexers with Large Free Spectral Range and Large Grating Facets

For an EDG demultiplexer, one can acquire larger grating facets by selecting a larger diffraction order. This is advantageous in reducing the manufacturing difficulty. In addition, large grating facets in an EDG can reduce the influence of the shaded facets, and then the loss and polarization-dependent loss (PDL) can be reduced. However, the free spectral range (FSR) and the diffraction order of the grating are related with an inverse ratio. Thus, the large diffraction order makes the FSR narrow, and reduces the number of available channels. For a conventional design of the demultiplexer, this is an antinomy that cannot be overcome. The FSR of a diffraction grating is defined as the largest bandwidth of a given order that does not overlap the same bandwidth of an adjacent order. By alternately varying the diffraction order between each adjacent facet, the FSR can overlay the whole diffraction envelop for a special order, but avoid the influence of the adjacent diffraction envelops [49]. When the diffraction order at different facets approximately obeys a Gauss distribution, the extinction ratio between the operated order and adjacent orders can attain a 35 dB or so.

Figure 16.19 shows the field distributions with varied diffraction orders and wavelengths at two different output waveguides. Figure 16.19a and b are based on a conventional design with a fixed diffraction order of 40, and (c) is for the chirped-order design. From Figure 16.19b, one can see that the energy of both wavelengths 1533.4 and 1571.7 nm can be received at the same output waveguide only with a small difference (about 7 dB). Therefore, for a conventional design of the EDG demultiplexer, although the loss is acceptable in some wavelengths, it is not suitable for a WDM application due to too large crosstalk from adjacent envelops. From Figure 16.19c, one can see that the crosstalk from 1533.4 nm is lower than 30 dB and thus the corresponding output waveguide mainly receives the energy from wavelength 1571.7 nm when the chirped-order design is used. This makes the crosstalk from channels in adjacent envelops to those in the operated diffraction envelop acceptable.

16.2.3.2 Improve Channel Uniformity for an Si-Nanowire AWG Demultiplexer

For AWG demultiplexers, it is very important to have good uniformity of channels in some system applications, such as receivers and add/drop scenarios. Otherwise, the signal-to-noise ratio will be degraded for the channel with a large insertion loss after a long-distance transmission. Nonuniformity L_u of an AWG is defined as $L_u = -10 \lg(I_m/I_c)$ (in decibels), where I_m and I_c are the intensities of the marginal channel and the central channel, respectively. The spectral response for a conventional AWG is of quasi-Gaussian shape due to the quasi-Gaussian far field from an individual arrayed waveguide in a conventional AWG demultiplexer. In this case, the marginal channel has a larger insertion loss than the central one, which indicates a fairly large nonuniformity. Many designs have been presented for AWGs with uniform loss properties [50–52]. AWG with uniform loss properties over the entire range of wavelength

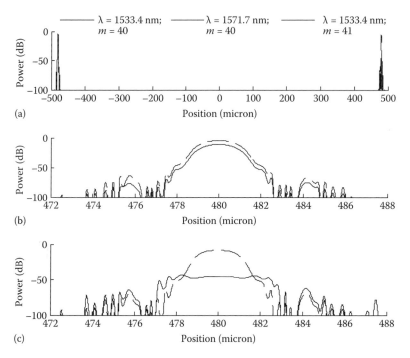

FIGURE 16.19 The field distributions with varied diffraction orders and wavelengths at different positions (a) using the fixed diffraction order; (b) for enlarged figure of (a); (c) using the chirped diffraction-order design. (From Song, J. et al., *IEEE Photonics Technol. Lett.*, 18(24), 2695, 2006. With permission.)

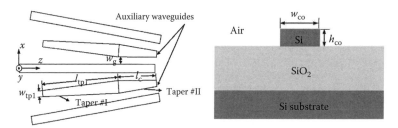

FIGURE 16.20 Schematic configuration of a method for improving channel uniformity. (a) Dual-tapered auxiliary waveguides. (b) Cross section of a Si-nanowire waveguide. (From Sheng, Z. et al., *J. Lightw. Technol.*, 25(10), 3001, 2007. With permission.)

channels have been achieved by combining two light waves with adjacent diffraction orders at its output [53]. Such an AWG employs a configuration in which two output beams with adjacent diffraction orders from the second FPR are combined at off-center wavelengths, which utilizes coherent interference at the output waveguides.

The improvement of the channel uniformity for a Si-nanowire AWG demultiplexer has been carried out by introducing auxiliary waveguides between adjacent arrayed waveguides [54].

The dual-tapered auxiliary waveguide includes two inverse tapers (see Figure 16.20a). In this way, it is easy to optimize the structure for good channel uniformity. The cross section of the Si nanowire is shown in Figure 16.20b. By optimizing all parameters shown in Figure 16.20a and b, the channel uniformity can be improved greatly.

Figure 16.21a shows the simulated spectral responses with optimized dual-tapered auxiliary waveguides. The nonuniformity L_u of the central 12 channels is 0.44 dB (satisfying the requirement of

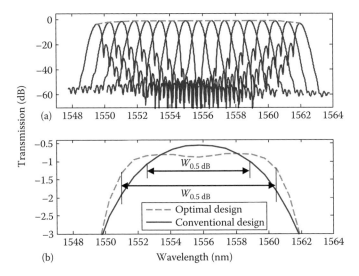

FIGURE 16.21 (a) Simulated spectral responses with optimal parameters for the dual-tapered auxiliary waveguides. (b) Comparison of the envelopes. (From Sheng, Z. et al., *J. Lightw. Technol.*, 25(10), 3001, 2007. With permission.)

$L_u < 0.5\,dB$). The envelope of the spectral responses in the conventional design (without auxiliary waveguides) is also shown in the same figure for comparison. One sees that the conventional design has a quasi-Gaussian-shape envelope, and no more than eight channels are available with the requirement of $L_u < 0.5\,dB$. The 0.5 dB bandwidths ($W_{0.5\,dB}$) for the designs are also shown. The total number of available channels of this AWG demultiplexer (with dual-tapered auxiliary waveguides) is 50% more than that of a conventional one.

16.3 Applications

In the coming years planar waveguide demultiplexers are expected to play an increasingly important role in providing the functionality required in optical networks in a compact form and at steadily decreasing costs. In this section, we describe only three application examples of planar demultiplexers.

16.3.1 Triplexers Based on Cross-Order AWGs

Fiber to the home (FTTH) has been developing rapidly in recent years and will become a major technology for next generation broadband access networks. An FTTH system consists of an optical line termination (OLT), located in an operator's central office, and optical network units (ONUs) located at customers' premises. The triplexer is one of the key components in FTTH systems that employ an analog overlay channel for video broadcasting in addition to bidirectional digital transmissions. At each of the ONUs, a triplexer is used to demultiplex two downstream signals from a single fiber to a digital receiver and an analog receiver and at the same time to couple the upstream signal from a digital transmitter to the same fiber. Similarly, at the OLT, a triplexer is used to combine a digital channel and an analog channel for downstream transmission while receiving an upstream digital signal from the ONUs. Simultaneous transmission on the same fiber is enabled by using different wavelengths for each direction and for each of the digital and analog channels. According to the ITU G.983 standard, the three commonly used wavelengths in passive optical networks (PONs) are 1310, 1490, and 1550 nm, for upstream digital, downstream digital, and downstream analog channels, respectively.

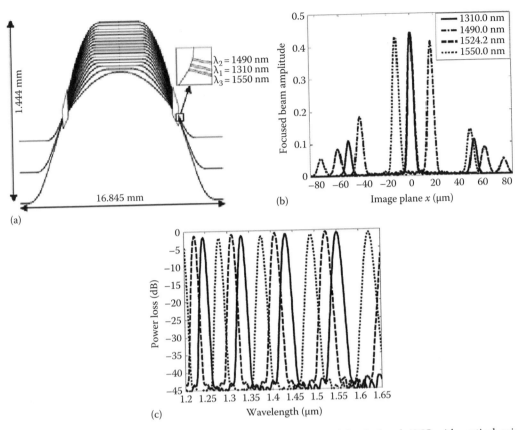

FIGURE 16.22 Triplexers based on cross-order AWGs. (a) Layout of the designed AWG with vertical axis. (b) Simulated field distributions in the focal plane. (c) Simulated spectral responses. (From Lang, T. T. et al., *IEEE Photonics Technol. Lett.*, 18(1), 232, 2006. With permission.)

At present, most commonly used triplexers are based on thin-film filter (TFF) technology, requiring labor-intensive assembly. It is potentially valuable to apply AWGs/EDGs for triplexers due to their small package size, easy integration, high performance, and reliability. Unfortunately, for the triplexer application, the required spectral range is at least 240 nm from the first wavelength (1310 nm) to the third wavelength (1550 nm). To include such a wide wavelength window within a single free-spectral range (FSR) of an AWG/EDG demultiplexer, the diffraction order needs to be very small, and it would be rather difficult or even impossible to realize the mask layout because it requires a very small path length difference between adjacent waveguides/grating grooves.

A novel design [55] using a cross-order waveguide grating has been proposed for FTTH triplexer applications (see Figure 16.22a). The used waveguide structure is a $6 \times 6\,\mu m^2$ SiO_2-based buried channel waveguide. Refractive indexes are assumed to vary linearly with wavelength. The cross-order AWG is designed so that Channel 1 at 1310 nm works at the diffraction order $m = 14$, while the diffraction order for Channels #2 (1490 nm) and #3 (1550 nm) is $m = 12$. As a result, Channel #1, which is relatively far from the other two channels, works at a different diffraction order. The FSR of the AWG is only required to cover Channels #2 and #3 and is much smaller than the full working spectral range. Consequently, the diffraction order, and the path length difference between adjacent waveguides in the AWG, can be increased significantly, making the layout design much easier. In addition, the output waveguides for the three wavelength channels can be arranged closely between each other with almost equal spacing, thus reducing the device size and channel nonuniformity. The simulated output field distributions in the focal plane for the three operating wavelengths as well as the dummy channel are shown in Figure 16.22b. As expected, the peak of channel #1 wavelength (1310 nm) overlaps well with the peak of the dummy channel wavelength (1524.2 nm) at the center of the output plane. The spectral responses of the AWG for the three output waveguides are shown in Figure 16.22c. One can see that the central channel has a response peak at 1310 nm as well as at the dummy channel wavelength of 1524.2 nm. The concept of cross-order grating design for triplexer applications can be applied not only to AWGs but also to EDGs.

16.3.2 Optical Channel Monitors/Receivers Using EDGs

Internet network management has become increasingly important in today's DWDM systems for optical communications. It offers high bandwidth while reducing operational costs. To implement the technology, individual channels must be monitored and controlled dynamically. One of the key devices to achieve this task is optical channel monitor (OCM). High-speed multi-channel receivers are also very important for DWDM networks [56].

A novel InP-based monolithically integrated optical channel monitor/receiver has been proposed [57]. The device, which comprises a flat-field EDG demultiplexer and a slab photodetector array, can monitor or receive signals in 40 wavelength channels separated by 100 GHz. It has a smaller size and a flatter and wider passband than those reported previously. It also features simpler fabrication without the need for output waveguides, thus eliminating issues related to fabrication-sensitive waveguide couplings.

The monolithically integrated optical channel monitor/receiver consists of an EDG demultiplexer and a slab photodetector array. A flat-field EDG design [58] is used in order to minimize detector-size variations and nonuniformity in performance across different channels (see Figure 16.23a). The detector array is aligned on a straight focal line of the grating (instead of an arc in a conventional Rowland circle design). No output waveguides are used. Therefore, the device size is reduced significantly and issues related to fabrication-sensitive waveguide coupling are eliminated. The detector array is fabricated in an active section, which has an active absorption layer and a p-doped InP layer deposited on top of the passive waveguide structure, as schematically shown in Figure 16.23b. In the design, the shapes of the slab waveguide detectors and the vertical layered structure have been optimized to obtain a low crosstalk and a low polarization dependence.

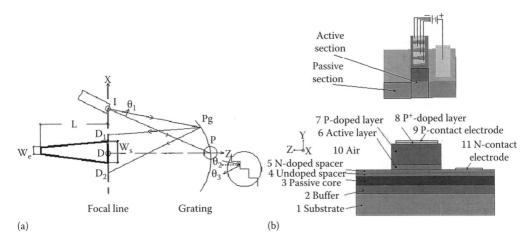

(a) (b)

FIGURE 16.23 InP-based monolithically integrated optical channel monitors/receivers. (a) Schematic of an EDG design using a method of two stigmatic points and the basic shape of the detector. (b) The schematic diagram of the layered structure of the passive section and the active section. (From Wang, L. et al., *J. Lightw. Technol.*, 24(10), 3743, 2007. With permission.)

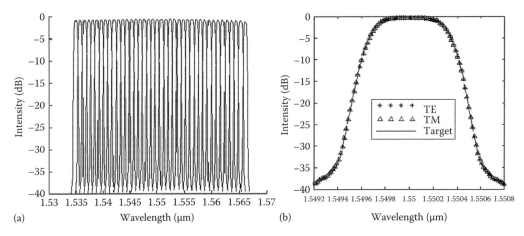

FIGURE 16.24 (a) The spectral responses of the optimized optical channel monitor/receiver for all channels; (b) the spectrum responses of the central channel (around wavelength 1.55 μm) for TE (*) and TM (Δ) polarizations. The solid line is the response curve calculated with some target absorption coefficient. (From Wang, L. et al., *J. Lightw. Technol.*, 24(10), 3743, 2007. With permission.)

Figure 16.24a shows the spectral responses of the optimized optical channel monitor/receiver for all channels. The difference between the TE and TM polarizations are negligible, as shown in Figure 16.24b. The width of the passband at −0.5 dB is about 48 GHz, and the crosstalk between adjacent channels is about −35.5 dB for TE polarization and 34.5 dB for TM polarization.

16.3.3 O-CDMA Encoding and Decoding Using AWGs

Code-division multiple-access (CDMA) communication system allows multiple users to access the network simultaneously by giving unique orthogonal code for each user. Optical CDMA (O-CDMA) has the advantage of using optical processing to perform certain network applications and provides a practically secure communication method with dynamic encoding. A successful spectral phase encoding and decoding operation for an O-CDMA system has been reported in a pair of monolithically integrated InP encoder chips, each consisting of an AWG pair and an eight-channel electro-optic phase shifter array [59].

Figure 16.25a shows the O-CDMA encoder mask layout. Spectral demultiplexing and multiplexing are done by the AWG pair. The phase modulators between the AWG pair give a phase shift corresponding to the O-CDMA code to each demultiplexed spectral channel. The input and output waveguides are chosen for optimal wavelength match of the two AWGs. Figure 16.25b shows the transmission spectrum of an O-CDMA encoder. Figure 16.25c and d shows the packaged chip and an SEM picture of the chip. The fabricated encoder chip was wire-bonded and packaged in a butterfly package for programmable electrical access to the phase shifter arrays. Electro-optical modulation in the phase shifter arrays in the chip gives Walsh-code-based O-CDMA encoding and decoding. The matched-code encoding–decoding operation resulted in nearly error-free performance.

16.4 Conclusion

This chapter has reviewed some recent progresses in planar waveguide multiplexers/demultiplexers, including some improved designs and their applications in optical networks. AWGs and EDGs are two types of key PLC components for constructing flexible and large capacity optical networks. We have described their operation principles. Two design examples for Si-nanowire-based AWGs have also been introduced to meet the urgent demand of the size reduction and scale increase of planar devices.

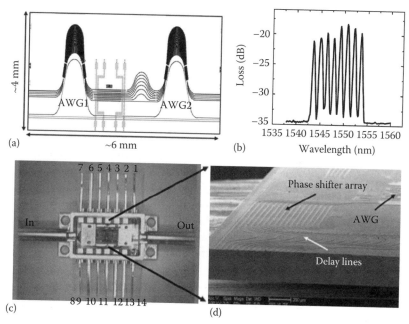

FIGURE 16.25 (a) O-CDMA encoder chip layout; (b) encoder transmission spectrum; (c) packaged O-CDMA encoder chip; and (d) SEM picture of the device. (From Cao, J. et al., *IEEE Photon. Technol. Lett.*, 24(15), 2602, 2006. With permission.)

Some designs for achieving flattened spectral responses, polarization independence, and good channel uniformity have been reviewed in details in this chapter. As examples, three applications of planar demultiplexers have been introduced in the chapter. For AWGs and EDGs, their main potential lies in the applications with moderate crosstalk requirements and in monolithic or hybrid integration in more complex devices like multi-wavelength receivers and transmitters, add-drop filters and optical cross-connects.

Acknowledgment

This work was partially supported by the National Science Foundation of China (No. 60688401). Coauthors of the work used in this review chapter are also acknowledged.

References

1. M. K. Smit and C. Van Dam, PHASAR-based WDM-devices: Principles, design and applications, *IEEE J. Sel. Top. Quantum Electron.*, 2(2): 236–250, 1996.
2. P. C. Clemens, R. Marz, A. Reichelt, and H. W. Schneider, Flat-field spectrograph in SiO$_2$/Si, *IEEE Photonics Technol. Lett.*, 4(8): 886–887, 1992.
3. E. Gini, W. Hunziker, and H. Melchior, Polarization independent InP WDM multiplexer/demultiplexer module, *IEEE J. Lightw. Technol.*, 16(4): 625–630, 1996.
4. Z. J. Sun, K. A. McGreer, and J. N. Broughton, Demultiplexer with 120 channels and 0.29-nm channel spacing, *IEEE Photonics Technol. Lett.*, 10(1): 90–92, 1998.
5. M. Smit, New focusing and dispersive planar component based on an optical phased array, *Electron. Lett.*, 24: 385–386, 1988.
6. C. Dragone, Efficient N × N star coupler based on Fourier optics, *Electron. Lett.*, 24: 942–944, 1988.

7. M. Zirngibl, C. Dragone, and C. H. Joyner, Demonstration of a 15 × 15 arrayed waveguide multiplexer on InP, *IEEE Photonics Technol. Lett.*, 4(11): 1250–1253, 1992.

8. Y. Hida, Y. Inoue, and S. Imamura, Polymeric arrayed-waveguide grating operating around 1.3 micrometers, *Electron. Lett.*, 30: 959–960, 1994.

9. M. Kawachi, Silica waveguides on silicon and their application to integrated-optic components, *Opt. Quantum Electron.*, 22: 391–416, 1990.

10. Y. L. Lin and S. L. Tsao, Improved design of a 64*64 arrayed waveguide grating based on silicon-on-insulator substrate, *Optoelectron., IEEE Proc.*, 153(2): 57–62, 2006.

11. W. Bogaerts, R. Baets, P. Dumon, V. Wiaux, S. Beckx, D. Taillaert, B. Luyssaert, J. Van Campenhout, P. Bienstman, and D. Van Thourhout, Nanophotonic waveguides in silicon-on-insulator fabricated with CMOS technology, *J. Lightw. Technol.*, 23: 401–412, 2005.

12. T. Tsuchizawa, K. Yamada, H. Fukuda, T. Watanabe, J. Takahashi, M. Takahashi, T. Shoji, E. Tamechika, S. Itabashi, and H. Morita, Microphotonics devices based on silicon microfabrication technology, *IEEE J. Sel. Top. Quantum Electron.*, 11: 232–240, 2005.

13. D. Dai, and S. He, Characteristic analysis of nano silicon rectangular waveguides for planar lightwave circuits of high integration, *Appl. Opt.*, 45: 4941–4946, 2006.

14. Y. Hibino, Recent advances in high-density and large-scale AWG multi/demultiplexers with higher index-contrast silica-based PLCs, *IEEE J. Sel. Top. Quantum Electron.*, 8: 1090–1101, 2002.

15. D.X. Dai and S. He, Novel ultracompact Si-nanowire-based arrayed waveguide grating with micro-bends, *Opt. Express*, 14(12): 5260–5265, 2006.

16. J. J. He, B. Lamontagne, A. Delage, L. E. Erickson, M. Davies, and E. S. Koteles, Monolithic integrated wavelength demultiplexer based on a waveguide Rowland circle grating in InGaAsP/InP, *J. Lightw. Technol.*, 16: 631–638, 1998.

17. J. Song, J. J. He, and S. He, Fast analysis method for polarization-dependent performance of a concave diffraction grating with total-internal-reflection facets, *J. Opt. Soc. Am. A*, 22(9): 1947–1951, 2005.

18. M. Fallahi, K. A. McGreer, A. Delâge, I. M. Templeton, F. Chatenoud, and R. Barber, Grating demultiplexer integrated with MSM detector array in InGaAs/AlGaAs/GaAs for WDM, *IEEE Photonics Technol. Lett.*, 5: 794–797, 1993.

19. J. B. Brouckaert, W. Bogaerts, P. Dumon, D. V. Thourhout, and R. Baets, Planar concave grating demultiplexer fabricated on a nanophotonic silicon-on-insulator platform, *IEEE Photonics Technol. Lett.*, 25(5): 1269–1271, 2007.

20. S. Janz, A. Balakrishnan, S. Charbonneau, D. X. Xu, M. Pearson, M. Packirisamy, B. Lamontagne, P. A. Krug, M. Gao, L. Erickson, K. Dossou, A. Delage, M. Cloutier, and P. Cheben, Planar waveguide echelle gratings in silica-on-silicon, *IEEE Photonics Technol. Lett.*, 16(2): 503–505, 2004.

21. K. Okamoto, R. Takiguchi, and Y. Ohmori, Eight-channel flat spectral response arrayed-waveguide multiplexer with asymmetrical Mach–Zehnder filters, *IEEE Photonics Technol. Lett.*, 8(3): 373–374, 1996.

22. K. Okamoto and H. Yamada, Arrayed-wave-guide grating multiplexer with flat spectral response, *Opt. Lett.*, 20(1): 43–45, 1995.

23. M. Kohtoku, H. Takahashi, I. Kitoh, I. Shibata, Y. Inoue, and Y. Hibino, Low-loss flat-top passband arrayed waveguide gratings realised by first-order mode assistance method, *Electron. Lett.*, 32(15): 792–794, 2002.

24. T. Kitoh, Y. Inoue, M. Itoh, M. Kotoku, and Y. Hibino, Low chromatic-dispersion flat-top arrayed waveguide grating filter, *Electron. Lett.*, 38(15): 1116–1118, 2003.

25. H. C. Lu and W. S. Wang, Cyclic arrayed waveguide grating devices with flat-top passband and uniform spectral response, *IEEE Photonics Technol. Lett.*, 20(1–4): 3–5, 2008.

26. K. Okamoto and A. Sugita, Flat spectral response arrayed-waveguide grating multiplexer with parabolic waveguide horns, *Electron. Lett.*, 32(18): 1661–1662, 1996.

27. J. B. D. Soole, M. R. Amersfoort, H. P. Leblanc et al., Use of multimode interference couplers to broaden the passband of wavelength-dispersive integrated WDM filters, *IEEE Photonics Technol. Lett.*, 8(10): 1340–1342, 1996.

28. C. Dragone, Frequency routing device having a wide and substantially. Flat passband, U.S. Patent 5 488 680, Jan. 30, 1996.

29. L. B. Soldano and E. C. M. Pennings, Optical multi-mode interference devices based self-image: Principles and applications, *J. Lightw. Technol.*, 13(4): 615–627, 1995.

30. D. Dai, S. Liu, S. He, and Q. Zhou, Optimal design of an MMI coupler for broadening the spectral response of an AWG demultiplexer, *J. Lightw. Technol.*, 20(11): 1957–1961, 2002.

31. D. Dai, W. Mei, and S. He, Use of a tapered MMI coupler to broaden the passband of an AWG, *Opt. Commun.*, 219: 233–239, 2003.

32. J. Song, D. Q. Pang, and S. He, A planar waveguide demultiplexer with a flat passband, sharp transitions and a low chromatic dispersion, *Opt. Commun.*, 227: 89–97, 2003.

33. Z. Shi and S. He, A three-focal-point method for the optimal design of a flat-top planar waveguide demultiplexer, *IEEE J. Sel. Top. Quantum Electron.*, 8(6): 1179–1185, 2002.

34. J.-J. He, Phase-dithered waveguide grating with flat passband and sharp transitions, *IEEE J. Sel. Top. Quantum Electron.*, 8(6): 1186–1193, 2002.

35. Z. Shi, J.-J. He, and S. He, An analytic method for designing passband flattened DWDM demultiplexers using spatial phase modulation, *J. Lightw. Technol.*, 21(10): 2314–2321, 2003.

36. H. Takahashi, Y. Hibino, and I. Nishi, Polarization-insensitive arrayed-wave-guide grating wavelength multiplexer on silicon, *Opt. Lett.*, 17(7): 499–501, 1992.

37. S. Suzuki, Y. Inoue, and Y. Ohmori, Polarization-insensitive arrayed-wave-guide grating multiplexer with SiO_2-on-SiO_2 structure, *Electron. Lett.*, 30(8): 642–643, 1994.

38. Y. Inoue, H. Takahashi, S. Ando, T. Sawada, A. Himeno, and M. Kawachi, Elimination of polarization sensitivity in silica-based wavelength division multiplexer using a polyimide half waveplate, *J. Lightw. Technol.*, 15(10): 1947–1957, 1997.

39. J. Kobayashi, Y. Inoue, T. Matsuura, and T. Maruno, Tunable and polarization-insensitive arrayed waveguide grating multiplexer fabricated from fluorinated polyimides, *IEICE Trans. Electron.*, E81C(7): 1020–1026, 1998.

40. C. K. Nadler, E. K. Wildermuth, M. Lanker, W. Hunziker, and H. Melchior, Polarization insensitive, low-loss, low-crosstalk wavelength multiplexer modules, *IEEE J. Sel. Top. Quantum Electron.*, 5(5): 1407–1412, 1999.

41. S. R. Park, J. Jeong, O. Beom-Hoan, S.-G. Park, E.-H. Lee, and S. G. Lee, Design and fabrication of polarization-insensitive hybrid solgel arrayed waveguide gratings, *Opt. Lett.*, 28(6): 381–383, 2003.

42. D. Dai and S. He, Design of a polarization-insensitive arrayed wave-guide grating demultiplexer based on silicon photonic wires, *Opt. Lett.*, 31(13): 1988–1990, 2006.

43. T. T. Lang, J. J. He, J. G. Kuang, and S. He, Birefringence compensated AWG demultiplexer with angled star couplers, *Opt. Express*, 15(23): 15022–15028, 2007.

44. J.-J. He, E. S. Koteles, B. Lamontagne, L. Erickson, A. Delage, and M. Davies, Integrated polarization compensator for WDM waveguide demultiplexers, *IEEE Photonics Technol. Lett.*, 11: 224–226, 1999.

45. S. Y. Sadov and K. A. McGreer, Polarization dependence of diffraction gratings that have total internal reflection facets, *J. Opt. Soc. Am. A*, 17: 1590–1594, 2000.

46. J. Song, N. Zhu, and S. He, Effects of surface roughness on the performance of an etched diffraction grating demultiplexer, *J. Opt. Soc. Am. A*, 23(3): 646–650, 2006.

47. J. Song, J. J. He, and S. He, Polarization performance analysis of etched diffraction grating demultiplexer using boundary element method, *IEEE J. Sel. Top. Quantum Electron.*, 11: 224–231, 2005.

48. Z. M. Shi, J. J. He, and S. He, Waveguide echelle grating with low polarization-dependent loss using single-side metal-coated grooves, *IEEE Photonics Technol. Lett.*, 16(8): 1885–1887, 2004.

49. J. Song, N. Zhu, J. J. He, and S. He, Etched diffraction grating demultiplexers with large free-spectral range and large grating facets, *IEEE Photonics Technol. Lett.*, 18(24): 2695–2697, 2006.

50. K. Okamoto, T. Hasegawa, O. Ishida, A. Himeno, and Y. Ohmori, 32×32 arrayed-waveguide grating multiplexer with uniform loss and cyclic frequency characteristics, *Electron. Lett.*, 32(22): 1865–1866, 1997.

51. S. Kamei, A. Kaneko, K. Tanaka, and T. Kitagawa, Scaling limitation of $N \times N$ signal interconnection in uniform-loss and cyclic-frequency arrayed-waveguide grating, *Electron. Lett.*, 36(18): 1578–1580, 2000.

52. S. Kamei, A. Ishii, M. Itoh, T. Shibata, Y. Inoue, and T. Kitagawa, 64×64-channel uniform-loss and cyclic-frequency arrayed-waveguide grating router module, *Electron. Lett.*, 39(1): 83–84, 2003.

53. K. Takiguchi, K. Okamoto, and A. Sugita, Arrayed-waveguide grating with uniform loss properties over the entire range of wavelength channels, *Opt. Lett.*, 31(4): 459–461, 2006.

54. Z. Sheng, D. X. Dai, and S. He, Improve channel uniformity of an Si-nanowire AWG demultiplexer by using dual-tapered auxiliary waveguides, *J. Lightw. Technol.*, 25(10): 3001–3007, 2007.

55. T. T. Lang, J. J. He, J. G. Kuang, and S. He, Cross-order arrayed waveguide grating design for triplexers in fiber access networks, *IEEE Photonics Technol. Lett.*, 18 (1): 232–234, 2006.

56. V. I. Tolstikhin, A. Densmore, K. Pimenov, Y. Logvin, F. Wu, S. Laframboise, and S. Grabtchak, Monolithically integrated optical channel monitor for DWDM transmission systems, *J. Lightw. Technol.*, 22(1): 146–153, 2004.

57. L. Wang, J. J. He, J. Song, and S. He, Design and optimization of a novel InP-based monolithically integrated optical channel monitor, *J. Lightw. Technol.*, 24(10): 3743–3750, 2007.

58. H. Wen, S. He, Z. Sheng, Z. Shi, and J.-J. He, Design of a Flat-field EDG wavelength demultiplxer using a two stigmatic points method, *Acta Photonica Sin.*, 32(3): 308–310, March 2003.

59. J. Cao, R. G. Broeke, N. K. Fontaine, C. Ji, Y. Du, N. Chubun, K. Aihara, Anh-Vu Pham, F. Olsson, S. Lourdudoss, and S. J. Ben Yoo, Demonstration of spectral phase O-CDMA encoding and decoding in monolithically integrated arrayed-waveguide-grating-based encoder, *IEEE Photonics Technol. Lett.*, 24(15): 2602–2604, 2006.

VIII

Integrated Photonic Technologies: Optical-Printed Circuit Boards

17

Optical-Printed Circuit Board and VLSI Photonics

El-Hang Lee
Hyun-Shik Lee

17.1 Introduction

In the introduction of this book, we introduced the concept of optical-printed circuit board (O-PCB) and that of VLSI photonics. Figure 17.1 shows a conceptual schematic diagram of the O-PCB and VLSI photonics. It is a view of a large green planar board on which micro/nanoscale optical wires and devices of diverse functions are interconnected and integrated to form optical and photonic circuits. In this chapter, we describe the process of fabricating O-PCB as a platform for VLSI photonics, and describe technological approaches that have been taken toward that goal.

In generic terms, as mentioned in the introduction, technical approaches toward these goals can be made in three stages. In the first stage, we design and fabricate optical micro/nanowires on a board or on a chip. In the second stage, we integrate microscale photonic devices with micro-photonic wires. And, finally, in the third stage, we integrate nanoscale photonic devices with nano-photonic wires.

17.2 Design and Fabrication of O-PCB

An O-PCB module is made of light sources, waveguides, functional devices, and photodetectors interconnected on a board, hard or flexible. In our study, prefabricated optical waveguides, with 45° mirrors at both ends for vertical lightwave coupling between the light source and the waveguide on the one end and between the waveguide and the photodetector on the other end, are embedded in an electrical-printed circuit board (E-PCB).

The waveguides are made of polymer materials although in principle they can be made by other desired materials. The polymer wires are chosen here for the purpose of low cost, ease of fabrication, and high volume processing. The polymer wires are fabricated by way of thermal or ultraviolet (UV) imprinting technique. In our study, an array of 12 channel polymer waveguides (or an array of four

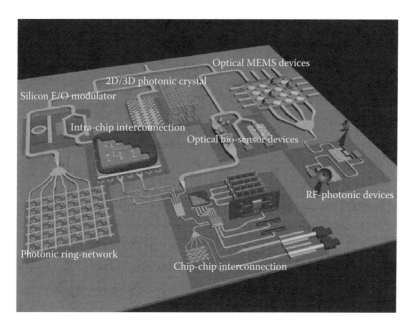

FIGURE 17.1   A conceptual diagram of an O-PCB on which VLSI photonic chips of diverse functions and dimensions are interconnected and integrated by optical wires.

channel polymer waveguides) of rectangle cross section is imprinted using UV embossing technique. First, a silicon mold, formed with 12 channel etched grooves, is fabricated and the waveguides are imprinted using this mold in such a way that both ends of each of the 12 waveguides are cut at 45°. The 45° slopes of the waveguide ends are coated with metal film for efficient lightwave coupling between the light source, waveguide, and photodetector. In our study, a 12-channel vertically coupled surface emitting laser (VCSEL) is used as the light source.

The 45° waveguide end mirrors are known to be fabricated by several methods, which include polishing, microtome, diamond saw, and x-ray [1–4]. In our case, using polymer materials for waveguides for the purpose of low cost and high volume production [5], the 45° waveguide end mirrors are prefabricated in a single imprinting step, using a 12-channel silicon waveguide mold with 45° slope formed at each end of the waveguide groove. A typical cross-sectional size of the waveguide is $50 \times 50 \, \mu m^2$ and its length is 7 cm. The waveguides are placed at a pitch of $250 \, \mu m$ to match the optical fiber sizes.

We have been able to achieve a chip-to-chip optical interconnection on O-PCB, using the 12-polymer waveguide channels at the rate of 2.5, 5, or 10 Gbps for each channel. The inter-chip module, which includes electrical-to-optical (E-O) and optical-to-electrical (O-E) converter units and conventional electronic integrated circuit (IC) for data processing, is attached to the O-PCB with solder balls. Solder ball bonding is designed to maintain the alignment between the optical waveguides and the electrical circuits within $10 \, \mu m$ error.

17.3 Design Consideration for Optical Waveguides

The two most essential rules that govern the VLSI microelectronics are the "scaling rules" for device miniaturization and "Moore's Law" for integration [6]. In VLSI photonics, however, no scaling rules have yet been established. Scaling rules can be useful as design rules. We have introduced the concept and made attempts to establish a design rule, or a scaling rule, for photonic wires [7–9]. For this, we

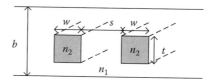

FIGURE 17.2 A simple model of optical interconnection between two wires.

calculated the densities of the optical wires that we can integrate using the concept of minimal separation length determined by the cross talk caused by the evanescent wave interaction between adjacent wires and using the normal waveguide theory.

In order to minimize the waveguide sizes and to maximize the integration density of the optical wires, we propose a simple model for the optical wires, as shown in Figure 17.2.

Here, W and t are the width and the height of the wires, respectively, and b and s are the thickness of the optical layer and the separation between two adjacent waveguides, respectively. And, the n_2 and the n_1 are the refractive indices of the core and of the cladding of the optical waveguide, respectively.

To calculate the optical wire density, we introduce the concept of the minimum separation, which is defined as the distance to guarantee a certain amount of cross talk between an input power P_{in} and an output power P_{out}. We then calculate the minimum separation to guarantee −20 dB cross talk after 100 mm propagation along the wire, which is defined as S_m for the separation distance between two wires. For this calculation, the height of the waveguide is assumed to be the same as the wire width w. The analysis is based on a 2D representation of the optical wire, which can be obtained from actual 3D physical optical wire by the effective index method (EIM). From this model, the density of the optical wire is theoretically analyzed without losing generality [8].

We characterize the coupling coefficient k that describes and indicates the coupling between optical wires for fundamental transverse electric (TE) mode. We can express the coupling coefficient in the following expression, using the normal mode theory [10,11]:

$$\kappa = \frac{2pq^2e^{-ps}}{\beta(d+(2/p))(q^2+p^2)} \tag{17.1}$$

Here
β is the propagation constant
d is the width of waveguide
s is the separation between waveguides

The p and q are expressed as

$$p = k_0\sqrt{n_e^2 - n_{cl}^2}, \quad q = k_0\sqrt{n_c^2 - n_e^2} \tag{17.2}$$

Here, k_0 is the wave vector, $2\pi/\lambda$, and n_e, n_c, and n_{cl} are the effective refractive index, the refractive indices of the optical wire core and cladding, respectively.

In the model of two parallel straight optical wires, we set the input amplitude of one optical wire as A and that of the other wire as B. Then, the initial condition of amplitudes is $A(0) = 1$, $B(0) = 0$, and A and B are expressed by

$$\begin{bmatrix} A(z) \\ B(z) \end{bmatrix} = \begin{bmatrix} \cos(\kappa z) \\ -j\sin(\kappa z) \end{bmatrix} \tag{17.3}$$

And the intensity of each optical wire is expressed by

$$\begin{bmatrix} P_A(z) \\ P_B(z) \end{bmatrix} = \begin{bmatrix} AA^* \\ BB^* \end{bmatrix} = \begin{bmatrix} \cos^2(\kappa z) \\ \sin^2(\kappa z) \end{bmatrix} \tag{17.4}$$

We define the minimum separation to guarantee −20 dB cross talk after 100 mm propagation among the optical wire. We then find the optical wire separation to make $P_B(z)$ be 0.01 (−20 dB). We have the combination of two equations (17.1) and (17.4) and calculate the minimum separation between waveguides as follows:

$$\frac{2pq^2 e^{-ps}}{\beta(d + (2/p))(q^2 + p^2)} = \frac{\sin^{-1}(\sqrt{0.01})}{100,000} = 10^{-6}\pi \tag{17.5}$$

We calculate the minimum separation distance between the optical wires with respect to the wire width for a family of core–cladding refractive index differences, 0.003, 0.005, 0.01, 0.015, and 0.02. Figure 17.3 shows that the minimum separation distance decreases as the waveguide width increases and as the index difference increases.

   The results indicate that, as the wire width increases, the separation between two wires decreases. However, the density of the optical wire depends not only on the wire separation but also on the wire width. From the results of Figure 17.3, we can calculate the density of the optical wires with respect to the wire width, as shown in Figure 17.4. We obtain, for example, a maximum density of 66 wires per millimeter when the waveguide width is 4 μm and the index difference is 0.02. In the case when the refractive index difference of the model is 0.01, as another example, we obtain the maximum density of 49 wires per millimeter for the waveguide width of 5 μm. In the case the refractive index difference is 0.005, we obtain the maximum density of 37 wires per millimeter for the waveguide width of 7 μm. This calculation result means that the higher the index difference, the higher number of wires can be integrated for a given set of waveguide width.

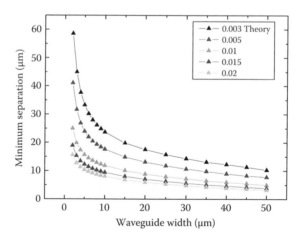

FIGURE 17.3   Minimum separation distance with respect to the wire width for a family of core–cladding refractive index differences.

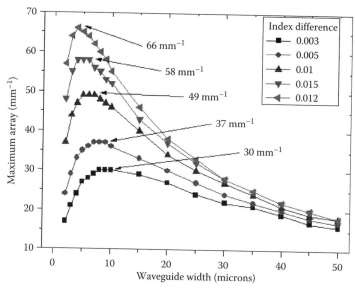

FIGURE 17.4 Calculated densities of the waveguides for various waveguide widths and for various core–cladding refractive index differences.

17.4 Design and Fabrication of Polymer Optical Waveguides

Using the above design rules, we can design and fabricate polymer optical waveguides of any integration density on an O-PCB. However, for practical considerations, the waveguide in our study is designed and fabricated to be in the range of 30–50 μm in width, 30–50 μm in height, and 7–10 cm in length. The angle of the waveguide end mirror facets is set at 45°. This waveguide and the 45° mirrors were fabricated simultaneously by UV imprinting technique using specially fabricated silicon mold [12,13]. The core and cladding materials are hybridized materials, specially designed and formulated by hybridization of organic and inorganic precursors using sol-gel process [5]. An array of 4-channel waveguides, or an array of 12-channel waveguides, was fabricated by imprinting technique. The detailed description of the waveguide fabrication is given elsewhere, and so is not repeated here [5]. The fabricated polymer waveguides have shown data rates up to 10 Gbps with jitter in 10 s of picoseconds. This suggests that the integrated polymer wires can offer not only an enhanced data transmission rate but also clock synchronization in high-speed electronic digital systems for possible applications in optical on-chip clock distribution and optical on-chip interconnection.

17.5 Considerations for Optimization of Optical Interconnection Coupling Efficiency

The most important factor in optical interconnection is to increase the coupling efficiency and to reduce the optical coupling loss. The parameters that would affect the coupling efficiency include the cross-sectional shape and dimension of waveguide, the refractive index contrast between the waveguide core and the cladding, the angle of the mirror facet, the distance between the light source and the waveguide, and the divergence angle of the light source. We analyze the coupling efficiency to minimize the optical loss of the optical interconnects. We first analyze the coupling between the light source and the waveguide and the coupling between the waveguide and the photodetector (PD). The coupling efficiency of the optical interconnects is analyzed using the ray-tracing method, and is optimized in terms of geometric variations, materials properties, and packaging processes.

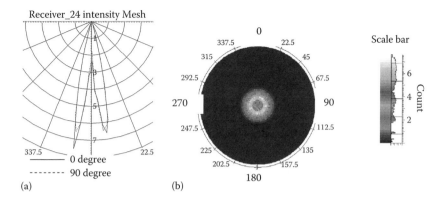

FIGURE 17.5   Modeling of a multimode VCSEL using the ray-tracing simulation method. (a) Radiant intensity distribution of a multimode VCSEL showing 15° angular divergence (where the solid line indicates 0° and the dotted line indicates 90° in the longitudinal direction). (b) Longitudinal angular radiation patterns of a multimode VCSEL having 15° angular divergence.

We use for this study a multimode 850 nm wavelength VCSEL light source, which we assume, based on the ray model, to have a multimode having two Gaussian peak distributions. Figure 17.5a shows the intensity distribution pattern of the multimode VCSEL showing a 15° angular divergence. This graph is calculated by using the ray-tracing method for 0° and 90° longitudinal direction. For the longitudinal direction, the mode shape of the VCSEL having 15° angular divergence has an axial symmetry, as shown in Figure 17.5b. Some 100,000 rays are used for modeling this VCSEL.

Figure 17.6 shows a schematic diagram of a vertically coupled optical interconnection structure. We summarize the parameters used to characterize the optical coupling efficiency. We then conduct simulation study, using the ray-tracing method, to examine how these parameters affect the coupling efficiency.

The operating wavelength of the VCSEL and the PD is 850 nm. The apertures of the VCSEL and the PD are 10 and 80 μm, respectively. The cross-sectional dimension of the waveguide is set to be 50 μm each for both the width and the height. The beam divergence of the light source is varied from 5° to 15° as measured at the full width half maximum (FWHM). The refractive index of the cladding is 1.45 and that of the core is varied from 1.455 to 1.47. The distance between the VCSEL and the waveguide and the distance between the waveguide and the PD is varied from 25 to 100 μm. The angle of the mirror is varied from 42° to 48°. These parameters are summarized in Table 17.1.

At first, we simulate the coupling efficiency using the ray-tracing method for different angles of mirror facets. In this analysis, the distance between the light source and waveguide pair and the waveguide and PD and waveguide is 50 μm and the thickness of upper cladding is 5 μm. The refractive indices of the core and cladding are 1.47 and 1.45, respectively. The divergence angle is 20°. From these results, we obtain the lowest coupling loss of −0.013 dB at 45° mirror angle, as shown in Figure 17.7. For the

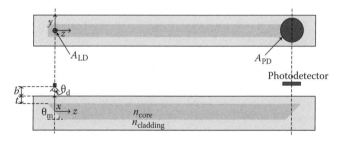

FIGURE 17.6   A schematic diagram of the optical interconnect using vertical coupling scheme.

TABLE 17.1 Parameters Used for the Calculation of the Vertical Optical Coupling between the Light Source and the Waveguide and between the Waveguide and the PD

Parameters	Notations	Values	
Operating wavelength	λ	850 nm	Fixed
Aperture of light source	ALD	10 μm in diameter	Fixed
Aperture of PD	APD	80 μm in diameter	Fixed
Divergence angle	θ_d	10°~30° (FWHM)	Variable
The refractive index of core	n_{core}	1.455~1.47	Variable
The refractive index of cladding	$n_{cladding}$	1.45	Fixed
Index difference	Δn	0.005~0.02	Variable
Thickness of upper cladding	b	0~50 μm	Variable
Distance between VCSEL/PD and waveguide	h	25~100 μm	Variable
Angle of mirror	θ_m	42~58	Variable

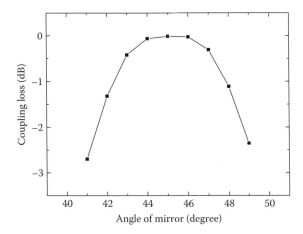

FIGURE 17.7 Coupling loss of the vertical optical interconnection for different mirror angles.

Gaussian beam approximation, it is straightforward to see that the 45° mirror angle achieves the highest coupling efficiency due to the effect of total internal reflection [1–4]. In our study, too, the multi-peak Gaussian beam approximation for the 45° mirror is valid to achieve the highest coupling efficiency, as can be verified from the calculation results.

We then calculate the coupling efficiency for different core materials, having 1.47, 1.46, and 1.445 refractive index, respectively. Here, the distance between the VCSEL and the waveguide pair and the waveguide and the PD pair is 50 μm, respectively, and the thickness of the upper cladding is 5 μm. The divergence angle is 20° and the angle of the mirror facet is 45°. From these results, we learn that the coupling loss does not occur when the index difference exceeds 0.01, as shown in Figure 17.8. An increase of the index contrast ($=\Delta n = n_{core} - n_{cladding}$) results in an increase of the numerical aperture NA ($=n_0 \cdot \sin\theta = n_0\Delta n$) and an increase of the acceptance of the incoming lightwave. Hence, as the index contrast increases, the coupling efficiency also increases, as shown in Figure 17.8.

We can calculate the coupling efficiency when the distance between the VCSEL and the waveguide is varied. Here, the refractive indices of the core and the cladding are 1.47 and 1.45, respectively. The divergence angle is 20° and the angle of the mirror facet is 45°. As the distance between the VCSEL (or PD) and the waveguide decreases, the coupling loss decreases, as shown in Figure 17.9. The reason is that the beam from the source is less diverged when the distance between the VCSEL and waveguide is less.

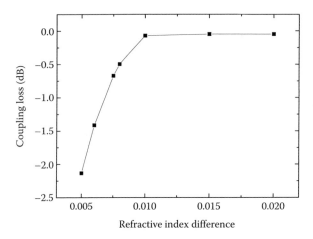

FIGURE 17.8   Coupling loss of the vertical optical interconnection for different values of refractive index differ-ence between the core and the cladding.

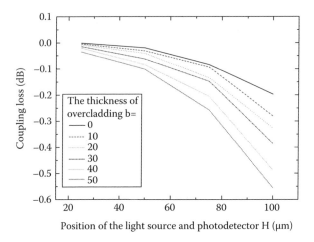

FIGURE 17.9   Coupling loss of the vertical optical interconnection for varied positions of the light source and the PD for a family of waveguides of different cladding thicknesses.

We can also calculate the coupling efficiency for different kinds of divergence angles of the light source, 10°, 20°, and 30°. In this analysis, the distance between the VCSEL and the waveguide pair and the waveguide and the PD pair is set at 50 μm, respectively, and the thickness of upper cladding is set at 5 μm. The refractive indices of the core and cladding are 1.47 and 1.45, respectively. The results show that, as the divergence angle of the light source increases, the coupling loss slightly decreases, as shown in Figure 17.10. The coupling efficiency between the two optical elements is basically determined by the overlap integral of each mode field distribution. The mode field distribution of the incident beam is determined by the beam radius while that of the waveguide is determined by its dimension and refractive index contrast. When a waveguide is fabricated, the radius of the incident beam is the only factor to determine the coupling efficiency. One can denote that the beam radius $W(z)$ of a Gaussian beam pattern of transverse electromagnetic (TEM_{01}) with double peaks is the radius of a circle area which contains 86% of the optical power in its far field pattern. Here, the z-axis denotes the beam propagation direction. The dependence of the beam radius along the z-axis is expressed by the following equation [14]:

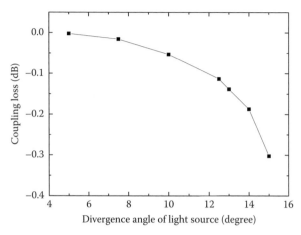

FIGURE 17.10 Coupling loss of the vertical optical interconnection for various divergence angles of the light source.

$$W(z) = W_0 \left[1 + \left(\frac{z}{z_0} \right)^2 \right]^{1/2} \tag{17.6}$$

Here, W_0 is the minimum value of $W(z)$ on the plane $z = 0$. The beam waist increases with z, reaching $\sqrt{2}W_0$ at $z = z_0$. For $z \gg z_0$, the beam waist approximately becomes the product of the divergence angle and the propagation length [14]. Hence, we note that the coupling efficiency decreases when the divergence angle increases and the distance increases, as shown in Figure 17.10.

17.6 Assembly of O-PCB and Test

For assembly of O-PCB, we introduce and discuss a packaging concept where the electrical circuits are placed directly on the same planar surface as that of the optical layer. In this process, we form the electrical circuits directly on the optical layer. The electrical transmission line is formed with a lift-off process. The fabrication process goes as follows. First, on the substrate, a 10 μm thick photoresist (PR) layer is spin-coated and patterned. At the time of the PR patterning, we establish the alignment between the optical layer and the electrical layer using alignment marks, which are already placed at the time of fabricating the optical layer. A 100 nm Cr and a 300 nm Au layer are co-sputtered on a patterned PR. These patterns are used as the transmission lines and the alignment marks. After the PR is removed, the electrical transmission lines are left on the substrate. The VCSEL and PD array operating at 850 nm wavelength are flip-chip bonded on to the substrate. The VCSEL array and the PD array carried Au/ Sn solder ball bumpers of 80 μm diameter and 65 μm height. For bonding, thermal heating is required up to almost 300°C to melt the Au/Sn solder ball. However, because the polymeric materials can suffer degradation and deformation at this temperature, we maintain the substrate temperature below 100°C, although the chip side is heated up to 280°C to melt the Au/Sn solder.

Now, in the study of the dependency of the coupling efficiency on the divergence angle of the light source, we use the nominal values provided by the commercialized light sources and PDs. From the simulation study, we learn that the divergence angle of the light source is a dominant factor affecting the coupling loss. In order to reduce the coupling loss, the divergence angle should be reduced. One of the ways to collimate the light from the VCSEL light source to the waveguide is the use of micro-lenses [15,16]. Micro-lenses have been known to improve the coupling efficiency and the alignment tolerance. Micro-lenses are normally formed by drops of ink-jet drops, where the ink-jet volume of a drop is usually about several tens of picoliters. We can control the size of the micro-lenses by controlling the number of drops. We have been able to achieve the variation of the height and diameter of the micro-lenses formed on the glass substrate with increasing number of drops, normally from 1 drop to 7 drops. Here, the VCSEL has a divergence angle of 15° and the lens dimension is 148 μm in diameter and 20 μm in height. The droplet of the UV curable resin on the aperture automatically forms a lens shape due to the effect of surface tension. Using the micro-lenses thus formed, the divergence angle of the VCSEL output could be decreased from 15° to 13° and the emitted multimode light output from the VCSEL is reduced to a single mode [15]. This mode conversion is caused by the phenomenon where the emitted output from the VCSEL is reflected back by the micro-lens and is refocused in the VCSEL cavity. This study successfully indicates that micro-lenses using micro-droplets of ink-jets can help focus the light effectively for high efficiency light coupling for optical interconnection.

The propagation loss of the waveguide thus fabricated was −0.32 dB/cm as measured by the butt-coupling method and the insertion loss was measured to be −1.55 dB at each 45° facet end mirrors. The total loss of the interconnection, including the propagation loss along the 7 cm waveguide, the vertical coupling loss, and the insertion loss, is −5.34 dB on average with its variation within −1 dB. The transmission characteristics of the UV-cured polymer waveguides, having a square cross-sectional shape of 50 μm both in the width and height and the total length of 70 mm, showed eye patterns of up to 10 Gbps without any significant distortion.

In the case of flexible board, the transmission loss of the optical wires remained nearly constant until the radius of curvature reached below 2 mm, below which the transmission loss rose rapidly to 10s of dB to the point of little or of no utility. This means that a flexible O-PCB board carrying polymer wires can sustain bending down to the radius of curvature to about 2 mm. We have been able to measure the transmission capacity of the bent polymer waveguide up to 10 Gbps without significant distortion by using eye diagram [5]. The successful design and fabrication of micro-optical polymer wires on an O-PCB board, either hard or flexible, thus provided us with a platform to allow integration of functional devices with the wires to form a functional O-PCB. In the section below we briefly describe some of the attempts that have been made to design and characterize the basic functional photonic devices for the purpose of integrating building blocks of VLSI photonic chips to form functional O-PCBs.

17.7 VLSI Photonics on O-PCB

In order to demonstrate the use of O-PCB as a platform for VLSI integration, one must conduct the design and experimentation on the integration of photonic devices with the optical wires on an O-PCB. The electronic, photonic, and biosensing chips and components that we have been preparing for integration include memory chips, central processing unit (CPU) chips, photonic crystal chips, micro-ring resonator chips, plasmonic chips, and other metamaterial chips. We have designed and characterized photonic crystal devices such as 2D cascaded directional couplers, multimode interference devices, and triplexer devices. We have also investigated an effective way of increasing the coupling efficiency of a racetrack micro-resonator waveguide [17–20]. The efforts to fabricate and integrate the above building block devices for applications for VLSI photonics and O-PCB are continuing.

17.8 Summary and Conclusion

In this chapter, we have given an overview of our study on the theory, design, and fabrication of what we call O-PCB as a platform for VLSI photonic ICs. We have presented a detailed procedure in terms of how to implement the concept of O-PCB in the form of a practical prototype. We have identified the need for scaling rules for integration of wires and have presented the detailed procedure of calculating the wire density. We presented the procedure of fabricating an array of polymer wires using silicon mold and showed the procedure of assembly and packaging a set of O-PCB. We calculated and measured the coupling efficiencies in terms of the divergence angle of the light sources, the losses along the waveguide, and the coupling efficiencies between the light source and the waveguide pair and between the waveguide and the PD pair. We also presented ways of improving the coupling efficiencies by using micro-lenses. We also presented means of integrating electrical circuits and optical circuits on the same planar surface. Finally, we presented some examples of integrating chip devices on the O-PCB board and proved that the O-PCB can be made useful. We also presented some examples of using plasmonic wires and dielectric wires to transmit lightwaves through subwavelength nanowires with reduced loss. This study is intended to find ways to design and implement nanowires for nanoscale photonic integration on O-PCB. Although O-PCB will find its way toward application-specific structures, we have presented the O-PCB here as a platform for VLSI photonic integration. VLSI micro-photonics is a new field of study that is being pursued in search of new functional possibilities in miniaturized and integrated photonic chips, reaching down to the micron, submicron, nano, and quantum scale on an O-PCB. This is very much like the case where the electronic chips are integrated on the electrical PCBs for a variety of specific applications. VLSI micro-photonics is pursued in the twenty-first century with an expectation that it would significantly improve the information technology of the future. As the demand rises for increased information capacity, the demand for increased functionality of photonic devices is also expected to rise. In order to meet this challenge, exploration of new materials, new designs, and new functions for miniaturization and VLSI integration of micro/nanodevices is further required. This chapter is a step toward that goal and we have attempted to explore the very first step in that direction.

Acknowledgments

This work has been supported by the Korea Science and Engineering Foundation (KOSEF) through the Grant for the Integrated Photonics Technology Research Center (R11-2003-022) at the Optics and Photonics Elite Research Academy (OPERA), Inha University, Incheon, South Korea.

References

1. R. Yoshimura, M. Hikita, M. Usui, S. Tomaru, and S. Imamura, Polymeric optical waveguide films with 45 mirrors formed with a 90° V-shaped diamond blade, *Electronics Letters*, 33, 1311 (1997).
2. J.-S. Kim and J.-J. Kim, Fabrication of multimode polymeric waveguides and micromirrors using deep X-ray lithography, *IEEE Photonics Technology Letters*, 16, 798 (2004).
3. B. S. Rho, S. Kang, H. S. Cho, H. H. Park, S. W. Ha, and B. H. Rhee, PCB-compatible optical interconnection using 45°-ended connection rods and via holed waveguides, *Journal of Lightwave Technology*, 22, 2128 (2004).
4. Y. Liu, L. Lin, C. Choi, B. Bihari, and R. T. Chen, Optoelectronic integration of polymer waveguide array and metal-semiconductor-metal photodetector through micromirror couplers, *IEEE Photonics Technology Letters*, 13, 355 (2001).
5. Y. K. Kwon, J. K. Han, J. M. Lee, Y. S. Ko, J. H. Oh, H.-S. Lee, and E.-H. Lee, Organic–inorganic hybrid materials for flexible optical waveguide applications, *Journal of Materials of Chemistry*, 18, 579 (2008).
6. G. Moore, Cramming more components onto integrated circuits, *Electronics*, 38, 114–117 (1965).
7. E. H. Lee, S.G. Lee, and O. Beom Hoan, Microphotonics: physics, technology, and an outlook toward the 21st century, *APOC 2001, SPIE Proceedings*, 4580, 263 (2001).
8. E.-H. Lee, S.-G. Lee, O. Beom Hoan, M.-Y. Jeong, K.-H. Kim, and S.-H. Song, Fabrication and integration of VLSI micro/nano-photonic circuit board, *Microelectronic Engineering*, 83, 1767 (2006).
9. H.-S. Lee, S.-M. An, Y. Kim, D.-G. Kim, J.-K. Kang, Y.-W. Choi, S.-G. Lee, O. Beom Hoan, and E.-H. Lee, Fabrication of a 2.5 Gbps × 4 channel optical micro-module for O-PCB application, *Microelectronic Engineering*, 83, 1347–1351 (2006).
10. K. Okamoto, *Fundamentals of Optical Waveguides*, Academic Press, London, U.K. (2000).
11. C.-L. Chen, *Fundamentals for Guide-Wave Optics*, Wiley-Interscience, NJ (2006).

12. S. M. An, H. S. Lee, O. Beom Hoan, and E.-H. Lee, Fabrication of 45°-reflector embedded polymeric waveguides by UV embossing for optical printed circuit board (O-PCB), *Proceedings of 31st International Conference on Micro- and Nano-Engineering*, Vienna, Austria, 1, 3m-23 (2005).

13. S. M. An, H. S. Lee, O. Beom Hoan, S. G. Lee, S. G. Park, and E. H. Lee, The effect of the KOH and KOH/IPA etching on the surface roughness of the silicon mold to be used for polymer waveguide imprinting, *Proceeding of SPIE*, 6897, 41, (2008).

14. B. E. A. Saleh and M. C. Teich, *Fundamentals of Photonics*, John Wiley & Sons, New York (1991).

15. S. H. Park, Y. S. Park, H. J. Kim, H. S. Jeon, S. M. Hwang, J. K. Lee, S. H. Nam, B. C. Koh, J. Y. Sohn, and D. S. Kim, Microlensed vertical-cavity surface-emitting laser for stable single fundamental mode operation, *Applied Physics Letters*, 80, 183 (2002).

16. F. Medrer, R. Michalzik, J. Guttmann, H. Huber, B. Lunits, J. Moisel, and D. Wiedenmann, 10 Gb/s data transmission with TO-packaged multimode GaAs VCSELs over 1 m long polymer waveguides for optical backplane applications, *Optics Communications*, 206, 309 (2002).

17. I. Park, H.-S. Lee, H.-J. Kim, K.-M. Moon, S.-G. Lee, O. Beom-Hoan, S.-G. Park, and E.-H. Lee, Photonic crystal power-splitter based on directional coupling, *Optics Express*, 12, 3599–3604 (2004).

18. H.-J. Kim, I. Park, O. Beom-Hoan, S.-G. Park, E.-H. Lee, and S.-G. Lee, Self-imaging phenomena in multi-mode photonic crystal line-defect waveguides: Application to wavelength de-multiplexing, *Optics Express*, 12, 5625–5633 (2004).

19. D.-S. Park, J.-H. Kim, O. Beom-Hoan, S.-G. Park, E.-H. Lee, and S. G. Lee, Optical triplexer based on a photonic crystal structure with position-tuned point defects, *Journal of the Korean Physical Society*, 54, 2269–2273 (2009).

20. H. S. Lee, C.-H. Choi, O. Beom-Hoan, D.-G. Park, B.-G. Kang, S.-H. Kim, S.-G. Lee, and E.-H. Lee, A nonunitary transfer matrix method for practical analysis of racetrack microresonator waveguide, *IEEE Photonics Technology Letters*, 16, 1086–1088 (2004).

IX

Technologies for Emerging Applications: Renewable Energy Generation

18

Nanostructured Copper Indium Gallium Selenide for Thin-Film Photovoltaics

Louay Eldada

The photovoltaic (PV) industry today is dominated by panels made of crystalline silicon, which comprise the majority of the products on the market, typically installed in large ground-based energy farms and small roof-mounted systems. Despite tremendous (and growing) interest in harnessing the sun for energy, an imbalanced refined silicon feedstock supply coupled with inefficient manufacturing and installation processes continue to drive high costs that hamper the widespread adoption of solar electricity.

In recent years, a new generation of solar electric products has emerged from the lab into the global market. Much innovation centers around thin-film solar technologies that use approximately 1% of the active and expensive PV material to convert photons from the sun into electrons. Through a combination of cost advantages and new product applications, thin-film solar is serving as a paradigm shift toward distributed electricity generation at cost parity with other forms of energy. Copper indium gallium selenide (CIGS) has long been the most promising thin-film PV material, used for its high conversion efficiencies in advanced spacecraft applications, but has not had a reliable and rapid manufacturing process that could scale effectively and provide significant amounts of electricity at the point of use. The field-assisted simultaneous synthesis and transfer (FASST®) process is one such manufacturing process, enabling the rapid printing of microscale CIGS films with p-type and n-type nanodomains that are critical for achieving the highest efficiencies possible in this material system.

18.1 Thin-Film Photovoltaics

The most commonly used cost metric for PV is $/Wp (often casually referred to as $/W), measured in terms of manufacturing costs for the peak output of a given module. For example, the average industry cost for silicon PV is approximately $3.50 for every Watt of power a module could generate during peak hours. The goal that most thin-film PV companies are trying to reach is less than $1/Wp; however, this metric is not adequate for representing customer concerns, namely getting the most power from a system over its lifetime. A module with rapid degradation in conversion efficiency over time or with inefficient energy conversion under low light conditions could still have an attractive $/Wp metric. Additionally, the balance of system cost (the additional components required for a complete, energy-generating system) depends on the number of panels and means of installation. On the other hand, the cents/kWh metric, that is, the average cost of the energy (in cents) produced (in kilowatt-hours), takes into account all the power that will be produced by a module over its lifetime spread over the upfront capital costs of installation, thus representing the true value delivered by the module.

Thin-film PVs encompass several different materials, each with their own unique properties. While all offer material advantages, they do differ in terms of conversion efficiency, lifetime expectancy, and manufacturing capabilities. Current market players beyond CIGS include amorphous silicon (α-Si) and cadmium telluride (CdTe). Additional early breakthroughs are being made with dye-sensitized solar cells (DSSC) containing titanium dioxide (TiO_2) nanoparticles coated with dye molecules, and organic photovoltaic (OPV) cells.

- α-Si: the most developed and well understood thin-film material, α-Si converts approximately 5%–8% of solar irradiance into electricity and can be applied to both rigid and flexible substrates, with maximum laboratory efficiency levels achieved at 13%.
- CdTe: with much recent success by manufacturing company First Solar, CdTe modules offer conversion efficiencies in the 6%–10.5% range, with the maximum laboratory levels for small-area cells around 16.5%. Due to the toxicity of input materials such as cadmium, vendors reclaim and recycle modules, and the addressable market is constrained. Further, applications have been limited to rigid substrates.
- CIGS: the most efficient thin-film material, CIGS has achieved 20% efficiency in laboratory settings for small-area cells [1] and 13.5% for large-area modules [2], offering the possibility to compete on par with traditional crystalline silicon in the 12%–18% product range. Applications include both rigid and flexible products.
- DSSC and OPV: the recipients of much current R&D, these two technologies offer very low expected lifetimes (3–5 years) but very cost-effective manufacturing options. Product applications are still in the exploration stage.

Table 18.1 summarizes the conversion efficiency values achieved in production modules and in record cells, for the three most mature thin-film materials and for crystalline silicon.

TABLE 18.1 Conversion Efficiencies Achieved in Production Modules and in Record Cells for Leading Thin-Film Materials (α-Si, CdTe, CIGS) and for Crystalline Silicon (Monocrystalline Si [c-Si] and Multicrystalline Si [mc-Si])

	Cell Material	Production Modules (2008)	Potential (Record Cell)
Si crystal	c-Si	12%–15%	24%
	mc-Si	10%–13%	20%
Thin film	α-Si	5%–8%	13%
	CdTe	6%–10.5%	16.5%
	CIGS	10%–13.5%	20%

18.2 CIGS Photovoltaics

CIGS thin-film PVs offer a multitude of competitive advantages. First and foremost, the relatively low material usage means that $1 worth of silicon, the most expensive part of current PV modules, can be replaced with just $.03 of CIGS materials. Furthermore, the silicon market has been constrained by an imbalance between supply and demand, resulting in manufacturers being tied to a volatile and unpredictable commodity market characterized by wide price fluctuations. CIGS production benefits not only from material advantages, but from potential for improving costs throughout the value chain, leveraging manufacturing maturity in related thin-film technologies such as the electronics and display industries.

In addition to manufacturing advantages, CIGS thin-film solar offers versatile aesthetic choices in product innovation. While the current solar market has recently been under-served with demand outpacing supply, differentiators such as appearance and the ability to adapt products to various market segments will play an increasingly important role as the market approaches equilibrium. Silicon PV companies such as SunPower have already proven that high-end products with a strong aesthetic advantage can attract significant investment and secure market share at a premium. CIGS thin film builds on this idea, offering the highest conversion efficiencies of any thin-film material, with the additional opportunity of being integrated into a wide variety of building and construction materials, appealing to the aesthetics needed in different designs and applications. Semi-transmissive modules are also an option, which can serve the dual function of vision glass and electricity generation.

Among the thin-film PV materials, CIGS is the most promising for cost-effective power generation, offering the highest conversion efficiency and thus providing the maximum power per unit area available. This also means an attractive energy payback time: currently 1.5 years, with expected advances resulting in a payback time around 0.5 years in the near future. Depending on the manufacturing technique, CIGS also offers the capability of monolithically interconnected modules. This means that instead of discrete, individual cells that are packaged together, PV modules can take the form of photovoltaic integrated circuits (PVICs) to simplify the manufacturing process and further reduce costs (see Section 18.4). A similar approach has been taken and proven extremely effective for high throughput and manufacturing scalability of CdTe thin-film modules. While not all CIGS manufacturing processes can utilize PVICs, the FASST reactive transfer process capitalizes on the advantages of this approach while also offering rapid deposition over large areas.

18.3 CIGS by Field-Assisted Simultaneous Synthesis and Transfer Printing

High-performance CIGS is characterized by (1) relatively large grain sizes, (2) an overall copper deficiency compared to the structure of the conventional α-phase copper indium di-selenide ($CuInSe_2$), and (3) a composition lying in the equilibrium $\alpha + \beta$ two-phase domain [3]. This latter characteristic is behind the intra-absorber junction (IAJ) model [4] that describes the formation within individual grains of α domains that are Cu-rich with p-type conductivity and β domains that are Cu-poor with n-type conductivity. The two domain types form nanoscale p-n junction networks. The n-type networks act as preferential electron pathways, while the p-type networks act as preferential hole pathways, allowing positive and negative charges to travel to the contacts in physically separated paths, reducing recombination and improving efficiency.

The conventional method used to synthesize high-performance thin-film CIGS devices is a high-temperature co-evaporation method. The multistep deposition sequences developed to achieve this performance always involve the topotactic transformation of a fairly large grain precursor into very large grain CIGS rather than the direct synthesis of CIGS from condensation of elemental vapors as in molecular beam deposition. This same topotactic transformation is used by the reactive transfer process.

The FASST reactive transfer process utilizes a two-stage reactive transfer printing method relying on chemical reaction between two separate precursor films to form CIGS. A schematic of the reactive

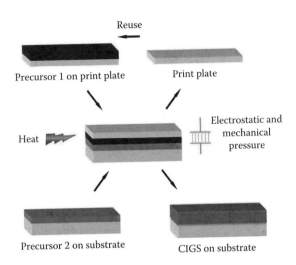

FIGURE 18.1   Schematic of the FASST reactive transfer process.

transfer process is shown in Figure 18.1. In the first stage, two Cu-In-Ga-Se-based precursor layers, forming the chemical basis of CIGS, are deposited onto a substrate and a print plate, respectively. The two separate precursors provide the benefit of independently optimized composition, structure, deposition method, and processing conditions for each precursor. Separating the precursors eliminates prereaction prior to the second-stage reactive transfer process, and facilitates optimized CIGS formation in the second stage. Furthermore, precursors can be deposited at a low substrate temperature enabling lower cost and higher throughput.

In the second stage, these precursors are brought into intimate contact and are rapidly reacted under the pressure in the presence of an applied electrostatic field. The method utilizes physical mechanisms characteristic of rapid thermal processing (RTP) and anodic wafer bonding (AWB), effectively creating a sealed micro-reactor that insures high material utilization, direct control of reaction pressure, and low thermal budget. The rapid thermal transient provides the similarity between the reactive transfer process and RTP. By pulse heating the film through the print plate, the overall thermal budget is significantly reduced, allowing the use of low-cost less thermally stable substrate materials.

Sufficient mechanical pressure can substantially prevent the loss of Se vapor from the reaction zone, thereby achieving highly efficient incorporation of Se into the composition layer. The use of an electrical bias between the print plate and substrate creates between them an attractive force that serves to insure intimate contact between the precursor films on an atomic scale, and can thus be used in conjunction with mechanical pressure to control the total pressure in the reaction zone. This is the resemblance between the reactive transfer process and AWB, a method developed historically to reduce the temperature required to bond two dissimilar materials together.

18.3.1 Nanostructured CIGS Thin Films

Large-grain high-quality CIGS is synthesized from two precursors in 5 min using the reactive transfer process. Figure 18.2 shows a precursor film with its grains having dimensions on the order of a quarter micrometer, and the cross section of a reactive transfer printed CIGS film with its grains being columnar and having dimensions on the order of a micrometer.

Figure 18.3 shows a secondary ion mass spectrometry (SIMS) depth profile of a CIGS thin film processed by reactive transfer. The precursors for the film are made by physical vapor deposition (PVD) methods. The uniform elemental distribution indicates a complete reaction of the precursors, and the

FIGURE 18.2 (a) Precursor film and (b) FASST CIGS cross section.

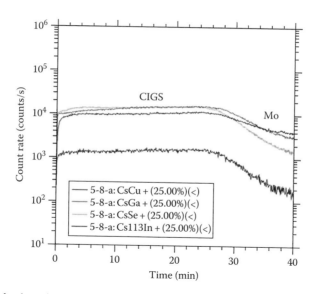

FIGURE 18.3 SIMS depth profile of a CIGS film. This film was formed in 6 min by the FASST process.

x-ray diffraction (XRD) analysis (Figure 18.4) confirms the absence of deleterious phases other than CIGS. All the XRD peaks are indexed based on chalcopyrite-type CIGS and Mo structure. The reactive transfer processed film has a (220/204) preferred orientation. Evidence indicated that the (220/204) oriented films help junction formation and improve solar cell performance [5]. The rapid processing of CIGS formation significantly increases the manufacturing throughput. As described above, this unique processing approach results in a much lower thermal budget as compared to co-evaporation and two-step selenization processes, which are common CIGS manufacturing. The lower thermal budget, removal of selenization process, and high throughput, all contribute to a low-cost process leading to improved manufacturability.

The opportunity to tailor the two precursors independently allows for the use of unconventional, non-vacuum deposition techniques such as die extrusion coating, ultrasonic atomization spraying, pneumatic atomization spraying, inkjet printing, direct writing, and screen printing. These atmospheric-pressure-based deposition tools offer great flexibility and open up entirely new windows for

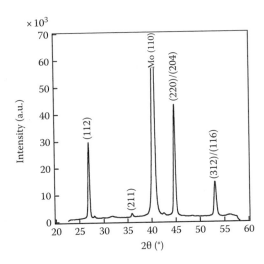

FIGURE 18.4   XRD pattern of a CIGS film fabricated by the FASST process.

materials processing. They also offer a viable means of introducing nanoparticle technology, metal-organic chemistry, and novel reaction paths to produce CIGS. The low capital equipment cost and high throughput capabilities associated with atmospheric pressure processing potentially reduce the manufacturing cost. These materials can be deposited in air at temperatures below 200°C, which will lower the thermal budget [6].

Proprietary inks containing a variety of soluble Cu-, In-, and Ga-multinary selenide materials have been developed. These metal-organic inks are designed to decompose into the desired precursors, and are called metal-organic decomposition (MOD) precursors. The resultant precursors are then used in step one of the reactive transfer process.

For the work described in this chapter, the inks were deposited using an ultrasonic spray head fed by a variable speed liquid pump. A substrate heater mounted on a computer-controlled X-Y motion system allowed for movement of heated substrates under the sprayed stream. The thickness of the sprayed film was controlled by varying the ink concentration, the flow rate through the sprayer, and the number of coats sprayed. Conditions were optimized such that smooth, uniform precursor films were obtained for all of the sprayed inks.

The precursor films were converted to the desired materials through RTP in a controlled atmosphere. The RTP conditions were varied systematically to ascertain the effect of conditions on the film compositions and morphologies obtained. The film compositions were characterized by x-ray fluorescence (XRF), crystalline phases were identified using XRD and film morphology was examined using scanning electron microscopy (SEM).

Binary Cu-Se, In-Se, and Ga-Se materials were developed and used to produce precursors. Various formulations of Cu-Se MOD precursors were also produced, resulting in tunable phase and stoichiometry in as-deposited films. Figure 18.5 shows the crystalline phase of Cu-Se films that can be produced from a single MOD ink, ranging from phase pure $CuSe_2$ to Cu_2Se.

Figure 18.6 compares the cross-sectional morphology of vapor-deposited CuSe films to that of atmospheric MOD CuSe films. The remarkable similarity between grain size, morphology, and density shows the promise of using solution-based precursors as substitutions for vacuum deposition processes. All temperatures used were below 200°C. This result represents a breakthrough for solution-based precursors.

Hybrid CIGS is produced by reactive transfer when one of the precursors is PVD-deposited in vacuum and the other precursor is atmospheric-pressure-deposited from an ink. Figure 18.7 shows cross-sectional and top-view SEM micrographs of such a hybrid CIGS film, revealing high-quality large columnar grains up to 4 µm in size.

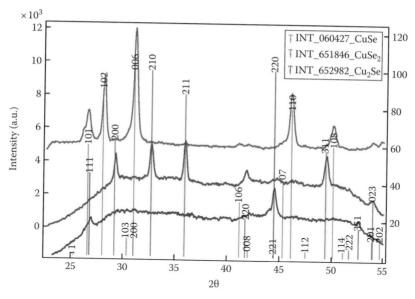

FIGURE 18.5 Range of crystalline Cu-Se phases obtained from a single MOD ink.

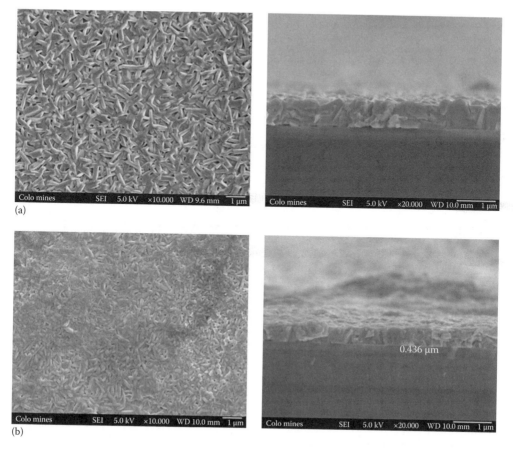

FIGURE 18.6 Top views and cross-sectional SEM views of (a) a PVD-deposited CuSe film, and (b) an atmospherically spray-deposited CuSe film from a solution-based process.

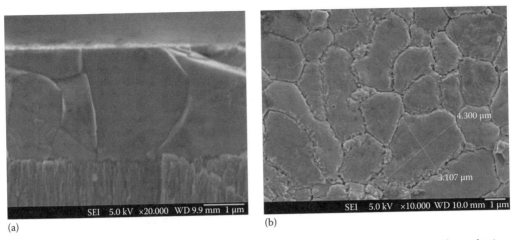

(a)   (b)

FIGURE 18.7   SEM micrographs of (a) the cross section and (b) the top view of a CIGS film synthesized using a non-vacuum-deposited precursor.

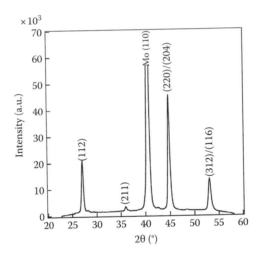

FIGURE 18.8   XRD pattern of a CIGS film by FASST using a non-vacuum deposited precursor.

The XRD pattern for the Figure 18.7 sample is shown in Figure 18.8 and the chalcopyrite CIGS phase is clearly identified. Again, (220/204) textured film is made by the reactive transfer process. The CIGS films are being applied to solar devices.

Various In-Se and Ga-Se precursors were developed and used both individually and in combination to produce In_2Se_3, Ga_2Se_3, and $(In,Ga)_2Se_3$ films. In-containing precursor films can be sprayed on heated, Mo-coated glass substrates and processed by RTP to give In_2Se_3. Figure 18.9 shows XRD scans of thickened In_2Se_3 films processed under these conditions. These In-Se and Ga-Se films were reacted with solution-deposited Cu-Se to produce CIGS absorber layers on Mo/glass substrates. In another approach, a single-source Cu-In-Ga-Se MOD precursor was developed by mixing the binary inks in the proper ratio, then deposited by ultrasonic spray and thermally processed to directly make CIGS.

An important benefit of the reactive transfer process is that there is no constraint on the combination or type of precursors that can be brought together. The only requirement is that all of the elements in the correct stoichiometry must be present on the substrate and print plate prior to the reactive transfer process.

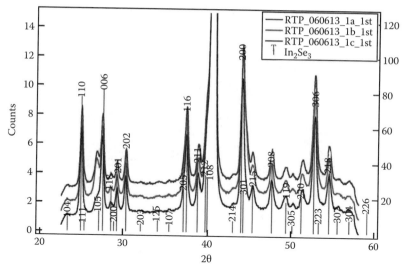

FIGURE 18.9 XRD scans of In$_2$Se$_3$ films of different thicknesses deposited using an In-containing ink. The large peak at ~40° is due to Mo (molybdenum back contact).

The tunability, scalability, high throughput, low thermal budget, and capital equipment cost reduction that atmospheric processing of MOD precursors provides is a promising route to potentially replace vacuum deposition methods for CIGS absorber layer fabrication.

From a manufacturing standpoint, any deposition method, whether it is PVD-based or atmospheric-pressure-based, has relative advantages and disadvantages. While the capital equipment cost for atmospheric-pressure-based systems is lower than that of the corresponding PVD systems, the raw material costs tend to be higher for solution rather than for PVD sources. Therefore, the best choice is ultimately governed by differences in the performance and yield of products manufactured by these two approaches. Better material utilization coupled with the decreasing cost of liquid precursors due to the maturation of the nanotechnology field make atmospheric-pressure-based processing a more attractive alternative.

18.3.2 Solar Cells Based on Nanostructured CIGS Thin Films

Solar cells with a conventional device structure of glass/Mo/CIGS/buffer/TCO were fabricated. The CIGS absorber in these cells was formed by the reactive transfer process with PVD-based precursors. Figure 18.10 shows a cross-sectional schematic of a typical CIGS solar cell. The SEM micrograph in

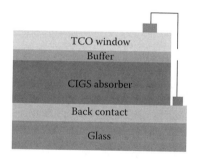

FIGURE 18.10 Cross-sectional schematic of the material stack in a typical CIGS solar cell.

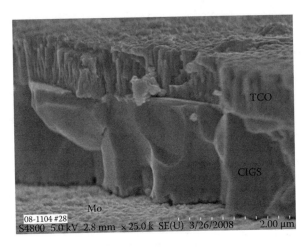

FIGURE 18.11   SEM cross-sectional image of a FASST CIGS device.

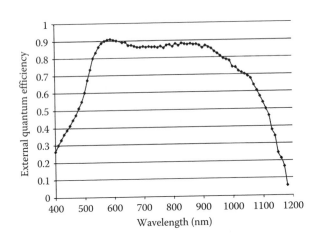

FIGURE 18.12   QE curve of a FASST CIGS solar cell.

Figure 18.11 depicts a cross section of a representative device. As can be seen, high-quality CIGS films with large columnar grains are obtained.

External quantum efficiency (QE) versus wavelength for the device without anti-reflection (AR) coating is shown in Figure 18.12. High QE at wavelengths over 550 nm reveals very good carrier collection and good performance of the CIGS layer. A low QE at short wavelengths indicates the need to further optimize window layers.

Solar cells of over 12% efficiency have been fabricated using the reactive transfer process. An *I–V* curve of a 12.2% efficient device is shown in Figure 18.13. An AR layer was deposited on this device and the efficiency was confirmed by Colorado State University. The composition of the CIGS film in this device as measured by XRF is 21.8% Cu, 21.6% In, 6.3% Ga, and 50.4% Se, which gives a Cu ratio (Cu/(In + Ga)) of 0.78 and Ga ratio (Ga/(In + Ga)) of 0.22. Optimizing Cu and Ga ratios should increase device efficiency by the reactive transfer process. For instance, an open-circuit voltage of 590 mV was obtained by increasing the Ga/(In + Ga) ratio to 0.3.

Analysis of the *J–V* data of the device gave a diode quality factor of about 2, and high saturation current density, which means that the large recombination at the junction region limits the open-circuit voltage and fill factor.

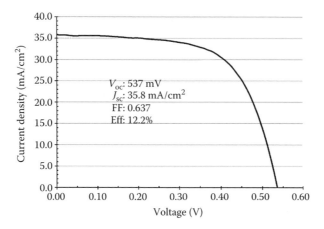

FIGURE 18.13   *J–V* curve of a FASST CIGS solar cell measured by Colorado State University.

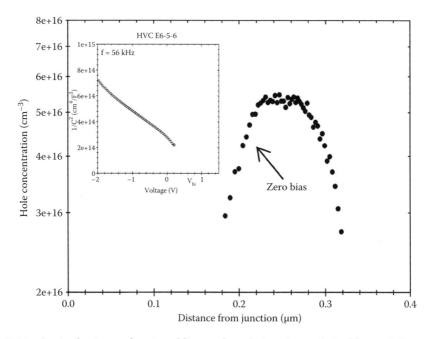

FIGURE 18.14   Carrier density as a function of distance from the junction, as derived from a *C–V* measurement.

Further analysis of the device was carried out by capacitance–voltage (*C–V*) measurement [7]. This measurement gave a hole density of 2.5×10^{16} cm^{-3} and a depletion width of ~0.2 μm. Carrier concentration as a function of distance from the junction, as derived from *C–V* data, is shown in Figure 18.14. The hump in carrier density against distance might be a signature of the measurement responding to deep states near the interface [8,9], and direct measurement of deep-level defects would be needed to verify it. Such states have a detrimental effect on the solar cell efficiency, because they constitute effective recombination paths for forward current opposing the photo-generated current. Such forward currents might also result from enhanced tunneling recombination through these states [10]. This suggests that the interface region, including the CIGS surface termination and post-CIGS treatment, need to be further optimized. Elimination of these states by improving the CIGS surface termination could significantly improve the open-circuit voltage and the fill factor.

18.4 CIGS Monolithically Integrated Modules

A monolithically integrated CIGS module consists of a packaged PVIC. In a PVIC, the thin films are deposited on a substrate and undergo a series of scribe patterning steps that create a monolithically interconnected integrated circuit of cells in series. The integrated approach circumvents the significant cost associated with cell cutting, testing, sorting, tabbing, and stringing processes that are labor intensive. Only thin-film technologies that exhibit large-area composition and thickness uniformity can benefit from this important cost advantage over crystalline-silicon-based modules.

The reactive transfer process prints CIGS directly on a variety of rigid and flexible substrates, including

- Glass
- Metals
- Alloys
- Composites
- Plastics

Soda lime glass (SLG) produced by the float process is the most commonly used substrate material, and has produced some of the best results in terms of both performance and reproducibility. It meets cost, smoothness, and stability criteria, making it well suited for commercial production. One limitation that needs to be addressed in the development of production processes is that SLG starts to soften above 520°C, as it goes through it glass transition temperature (T_g) range of 520°C–600°C. Sodium, an important element to have in the CIGS film for the formation of high-quality large grain crystals, can be transported controllably from the SLG into the CIGS thin film through the back contact, or can be added in controlled amounts.

Flexible substrate materials can be attractive either for manufacturing lightweight flexible products for some applications or for the use of low-cost roll-to-roll deposition processes in manufacturing. Flexible substrate materials on which good performance was achieved include polyimide, titanium, and stainless steel [11,12]. The main drawback of polyimide is low temperature tolerance, since the best commercially available polyimide films can withstand temperatures only up to 450°C; another drawback is high thermal expansion. The main drawback of titanium and steel is their conductivity, requiring an electrically isolating layer in order to allow monolithic integration; the isolation layer needs to be void of defects that cause shunting of the cells. For these flexible substrates, sodium needs to be introduced.

A typical monolithic interconnection scheme involves three main patterns, as illustrated schematically in Figure 18.15. The first pattern (P1) consists of isolation lines cut in the back contact (usually molybdenum) typically using laser ablation, prior to the CIGS deposition. Mechanical scribing is

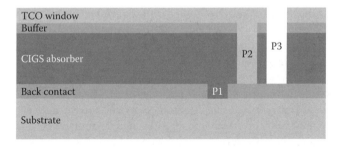

FIGURE 18.15 Cross-sectional schematic of the material stack in a typical CIGS PVIC, showing the three scribe patterns (P1, P2, P3) that achieve the isolation and interconnection between two segments.

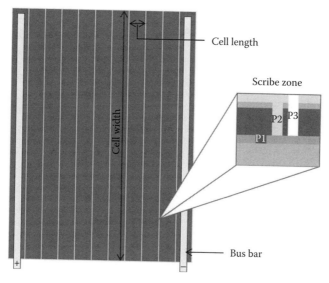

FIGURE 18.16   Schematic of a CIGS PVIC with bus bars. The scribe zones between active segments create a pin-stripe pattern.

typically used for the two subsequent patterning steps (P2 and P3), where the P2 lines are cut through the buffer and the CIGS down to the back contact, then filled with TCO as the interconnect material, and P3 lines are cut through the TCO, buffer, and CIGS down to the back contact to complete the isolation of the cell. The scribe zones between active segments create a pin-stripe pattern, as can be seen in Figure 18.16. The interconnect pattern can be designed to fit various shape and voltage requirements. An additional patterning step known as edge deletion is performed on the module level (as opposed to the segment level), where the thin-film materials are removed from the outer rim of the substrate in order to improve the adhesion to the encapsulant used during lamination. This step completes the fabrication of the PVIC.

The PVIC goes through packaging in order to become a module. Bus bars are attached to avoid resistive losses associated with current collection; these metal stripes can be soldered, welded, or glued to contact areas near the substrate edges, as shown in Figure 18.16. A superstrate is laminated to the substrate using an encapsulant sheet made of a material that is optically transparent and can withstand environmental and UV exposure. The encapsulant, commonly ethylene vinyl acetate (EVA), also serves the purpose of protecting the PVIC from moisture. An edge sealant can additionally be used for further protection from the environment. A junction box is typically integrated with connectors for creating module strings. The junction box must capture and seal the bus bar exit zone, and it often integrates a bypass diode to protect the module string from partial shading losses. Framing finishes the module, but can be omitted for some commercial applications.

18.5 CIGS Photovoltaic Module Reliability Testing and Certification

The reliability of PV modules over time is critically important, along with the initial cost and efficiency, if a PV technology is to make a significant impact in the power generation market, and for it to compete with conventional electricity producing technologies. The reliability of PV modules has progressed significantly in the last several years, as evidenced by warrantees as long as 25 years available on commercial modules.

TABLE 18.2   Key PV Panel Certification Tests per UL and IEC Reliability and Safety Standards

Test	Description
Performance	±10% of specified electrical parameters
Outdoor exposure	60 kWh/m², maintain performance
Thermal cycling	−40°C to +90°C
Damp heat	85°C/85%RH
Humidity freeze	−40°C to +90°C with condensation
Mechanical robustness	Shading hot spot
	Connectors/J-box pull test
	400 lb weight loading
	Steel ball impact (51″, 1.18 lb)
Shock hazard	No leakage current after environmental exposure

Accelerated reliability testing in laboratories includes testing under stressing conditions of

- Temperature
- Voltage
- Current
- Moisture
- Thermal cycling
- UV-visible radiation
- Mechanical impact

Certification testing requirements per UL (Underwriters Laboratories) and IEC (International Electrotechnical Commission) are tabulated in Table 18.2. The main reliability and safety standards for PV panels are UL-1703, IEC-61646, and IEC-61730.

18.6 Applications of CIGS Thin-Film Photovoltaics

The reactive transfer process is compatible with a variety of rigid and flexible substrates including glass, metals, alloys, composites, and plastics. This means that, depending on market needs and demands, a wider variety of product applications are available, from traditional glass modules to building construction materials that incorporate thin-film PVs. By integrating PV into building construction and design, further system-level advantages are available and costs can be reduced even more by leveraging the construction material installation. Moreover, new market applications are possible, as solar power breaks from rooftop installations and ground-mounted energy farms into architectural design and building façades.

Thin-film building-integrated PV (BIPV) is a paradigm shift for solar power, enabling communities and buildings with a solar-powered "skin." Distributed generation through BIPV means that the building is the power plant. This will open the door to a new industry that is the intersection of the multi-trillion U.S. dollar construction and power generation industries.

Aesthetically, the solid dark gray appearance of CIGS modules is usually preferred to the nonuniform bluish appearance of the crystalline silicon modules in BIPV applications. CIGS can offer the appearance of tinted glass on the sides of buildings, or that of slate on rooftop applications.

The performance of CIGS modules at high temperature is superior to that of crystalline silicon modules, making them better suited for hot climates, including harsh desert conditions.

For space applications, CIGS panels offer other advantages since their radiation tolerance is higher than that of crystalline silicon panels [13,14]. Furthermore, lightweight plastic substrates could lead to

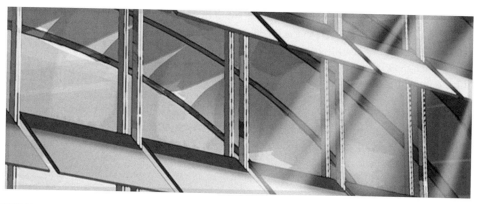

FIGURE 18.17   HelioVolt CIGS BIPV panels used as sunshades.

solar cells with very high specific power (power divided by mass), an important requirement in some space applications [15].

In the near future, CIGS can be incorporated into everyday lives in entirely new ways. Solar skins will become pervasive not only in roofing, curtain walls, and façades, but also in sunshades (Figure 18.17), skylights, atriums, canopies, and pergolas. Further, BIPV will come in a variety of form factors, shapes, colors, and transparencies, and will incorporate dynamic behavior and innovations in lighting, heating, and cooling.

The reactive transfer process produces high-performance CIGS devices rapidly and at low cost, based on harnessing the key elements required for the creation of the highest quality materials. Reactive transfer printing, especially combined with the use of liquid ink precursors, provides a sustainable long-term technology-based cost advantage and opens the doors to further advances in the manufacturing of CIGS thin-film PVs.

References

1. M.A. Contreras, I. Repins, W.K. Metzger, and D. Abou-Ras, Se activity and its effect on Cu(In,Ga)Se$_2$ photovoltaic thin-film materials, *Proceedings of the International Conference on Ternary and Multinary Compounds*, Berlin, Germany, Vol. 16, 2008.
2. M. Powalla, The R&D potential of CIS thin-film solar modules, *Proceedings of the European Photovoltaic Solar Energy Conference*, Dresden, Germany, Vol. 21, p. 1789, 2006.
3. B.J. Stanbery, Copper indium selenides and related materials for photovoltaic devices, *Critical Reviews in Solid State and Materials Sciences*, **27**, 73 (2002).
4. Y. Yan, R. Noufi, K.M. Jones, K. Ramanathan, M.M. Al-Jassim, and B.J. Stanbery, Chemical fluctuation-induced nanodomains in Cu(In,Ga)Se$_2$ films, *Applied Physics Letters*, **87**, 121904 (2005).
5. L. Eldada, F. Adurodija, B. Sang, M. Taylor, A. Lim, J. Taylor, Y. Chang, S. McWilliams, R. Oswald, and B.J. Stanbery, Development of hybrid copper indium gallium selenide photovoltaic devices by the FASST® printing process, *Proceedings of the European Photovoltaic Solar Energy Conference*, Valencia, Spain, Vol. 23, p. 2142, 2008.
6. C. Curtis, M. Hest, A. Miedaner, J. Nekuda, P. Hersh, J. Leisch, and D. Ginley, Spray deposition of high quality CuInSe$_2$ and CdTe films, *Proceedings of the IEEE Photovoltaic Specialist Conference*, San Diego, CA, Vol. 33, p. 1065, 2008.
7. S. Chaisitsak, A. Yamada, and M. Konagai, Preferred orientation control of Cu(In$_{1-x}$Ga$_x$)Se$_2$ ($x \approx 0.28$) thin films and its influence on solar cell characteristics, *Japanese Journal of Applied Physics*, **41**, 507 (2002).

8. P. Mauk, H. Tavakolian, and J. Sites, Interpretation of thin-film polycrystalline solar cell capacitance, *IEEE Transactions on Electron Devices*, **37**, 422 (1990).

9. H. Tavakolian and J. Sites, Effect of interfacial states on open-circuit voltage, *Proceedings of the IEEE Photovoltaic Specialist Conference*, Las Vegas, NV, Vol. 20, p. 1608, 1988.

10. I.L. Repins, B.J. Stanbery, D.L. Young, S.S. Li, W.K. Metzger, C.L. Perkins, W.N. Shafarman, M.E. Beck, L. Chen, V.K. Kapur, D. Tarrant, M.D. Gonzalez, D.G. Jensen, T.J. Anderson, X. Wang, L.L. Kerr, B. Keyes, S. Asher, A. Delahoy, and B Von Roedern, Comparison of device performance and measured transport parameters in widely-varying Cu(In,Ga)(Se,S) solar cells, *Progress in Photovoltaics: Research and Applications*, **14**, 25 (2006).

11. B.M. Basol, V.K. Kapur, A. Halani, and C. Leidholm, Flexible and lightweight copper indium selenide solar cells, *Proceedings of the IEEE Photovoltaic Specialists Conference*, Washington, DC, Vol. 25, p. 157, 1996.

12. M. Hartmann, M. Schmidt, A. Jasenek, H.W. Shock, F. Kessler, K. Herz and M. Powalla, Flexible and lightweight substrates for Cu(In,Ga)Se$_2$ solar cells and modules, *Proceedings of the IEEE Photovoltaic Specialists Conference*, Anchorage, AK, Vol. 28, p. 638, 2000.

13. R.M. Burgess, W.S. Chen, W.E. Devaney, D.H. Doyle, N.P. Kim, and B.J. Stanbery, Electron and proton radiation effects on GaAs and CuInSe$_2$ thin film solar cells, *Proceedings of the IEEE Photovoltaic Specialists Conference*, Las Vegas, NV, Vol. 20, p. 909, 1988.

14. A. Jaseneka, U. Raua, K. Weinerta, I.M. Kötschaua, G. Hannaa, G. Voorwindenb, M. Powallab, H.W. Schocka, and J.H. Wernera, Radiation resistance of Cu(In,Ga)Se2 solar cells under 1-MeV electron irradiation, *Thin Solid Films*, **387**, 228 (2001).

15. N.P. Kim, B.J. Stanbery, R.M. Burgess, R.A. Mickelsen, R.W. McClelland, B.D. King, and R.P. Gale, High specific power (AlGaAs)GaAs/CuInSe$_2$ tandem junction solar cells for space applications, *Proceedings of the Energy Conversion Engineering Conference*, Washington, DC, Vol. 2, p. 779, 1989.

19

High-Efficiency Intermediate Band Solar Cells Implemented with Quantum Dots

Elisa Antolín
Antonio Martí
Antonio Luque

The production of photovoltaic (PV) energy in a massive way could contribute to the reduction of CO_2 emissions and to palliate the effects of the global climatic change. The amount of energy from the Sun that reaches the Earth's surface each day is huge. But the areal density of this flux is relatively low, which makes its exploitation inefficient and expensive. This is the ultimate reason why until now PV only contributes with 0.15% to the production of the world electricity [1]. Even though the module cost of presently commercialized technology, based on crystalline-silicon cells, has declined from about 70 US$/$W_p$ in 1976 to about 3.5 US$/$W_p$ in 2003 [2], there is a wide consensus that it will not be able to reach the threshold required for PV technology to penetrate the global energy context, which has been estimated in about 0.35 US$/$W_p$ [3]. From the definition of PV cost, there are two main paths to achieve its reduction: one is to lower the areal cost and the other, to increase the conversion efficiency (the efficiency of present crystalline-Si technology reaches 24% in the lab and industrially the usual figure is around 15%). Unfortunately, these two strategies appear to be mutually exclusive, since high-efficiency approaches often involve the fabrication of more sophisticated and expensive cells. On the other hand, the apparent equivalence between lowering the areal production cost of PV systems and increasing their efficiency is only valid if we consider an unlimited dedicated area. But in practical applications, a technology that requires less exploitation land while rendering the same final electricity price will be a better energetic solution. In this respect, the development of high-concentration techniques is expected to be a determining factor to counteract the high cost of new and sophisticated devices. The idea is to build an optical system that intercepts a given area of solar flux and focuses the light power on a solar cell of much smaller area [4]. The areal cost of an optical system based on lenses or mirrors is notably lower than that of semiconductor devices. Then, if a sufficiently high concentration ratio (i.e., the ratio between optical system area and cell area) is achieved, the impact on the system cost of using high-efficiency solar cells can be minimized, leading to an effective decrease of the areal cost/efficiency ratio. Besides, the use of concentration is very positive from the point of view of energetic cost and sustainability, since the amount of semiconductor material used can be drastically reduced. Finally, it has also an impact on the cell efficiency, which is increased in absolute terms under high light power density. For all these reasons, the application of high-concentration techniques has opened the path for the development of new high-efficiency PV technologies. Within this context, different approaches are being researched, including the application of nanotechnology. In order to design devices capable to substantially surpass the efficiency of present solar cells, it is important to analyze the fundamental reasons that limit their performance.

19.1 Intrinsic Limitations of Single-Gap Solar Cells

The conversion efficiency of conventional, or single-gap, solar cells is fundamentally limited by the fact that they only use a part of the solar spectrum efficiently. They are made of a semiconductor material, which is characterized by an energy bandgap (E_G). Figure 19.1a shows the energy-band diagram of such a semiconductor and the different photon absorption cases that take place in it. A photon of energy equal to E_G (represented by photon (1) in the picture), can be absorbed by an electron at the valence band (VB), which will use its energy to promote to the conduction band (CB), creating an electron–hole pair. On the contrary, photons of lower energy, as photon (2), will not be absorbed. We have included in Figure 19.1b the standard spectrum of the solar flux as it reaches the Earth's surface and have marked the bandgap energy of some common semiconductors. It can be clearly seen that this condition leaves a relevant part of the solar spectrum unexploited.

Let us now consider the case of photon (3), which energy is higher than E_G. This photon can be absorbed in the semiconductor, but because within the bands the electronic states form a continuum, the produced carriers will easily migrate to lower energy states, transferring the energy excess to the lattice in form of heat (phonon emission). In any conventional semiconductor, carriers will relax by this *thermalization* process to the bandgap edges in sub-picoseconds time and the remaining collectable energy from photon (3) will be the same as that in the case of photon (1). We face then a trade-off: if we choose a semiconductor of high E_G, such as GaAs, few photons will be absorbed, whereas if a low-bandgap material as Ge is chosen, we will collect more carriers, but a greater part of their energy will be lost by thermalization.

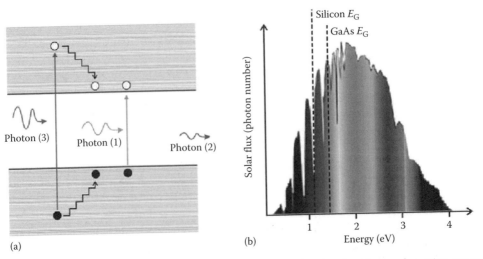

FIGURE 19.1 (a) Simplified band diagram of a semiconductor showing the photon absorption processes described in the text and (b) standard AM1.5G solar spectrum (the bandgaps of some semiconductors commonly used to fabricate solar cells are marked).

In a solar cell the generated carriers have to be extracted before they recombine in order to collect the energy derived from photon absorption. This is achieved through the implementation of selective contacts, which in the case of a conventional solar cell is done by fabricating a pn-junction or diode. Then, the extraction of power ($P = V \times I$) is governed by the current–voltage characteristic of the diode under illumination. The maximum current that can be extracted (the short-circuit current, I_{SC}) will be more or less proportional to the number of photons absorbed, while the maximum output voltage (the open-circuit voltage, V_{OC}) will never be allowed to exceed the value of the bandgap (divided by the electron charge, e). This is so because the output voltage corresponds to the split μ_{CV} between the quasi-Fermi level of electrons, ε_{Fe}, and the quasi-Fermi level of holes, ε_{Fh} (again divided by e). A voltage exceeding the bandgap would then imply a population inversion and would release stimulated emission. Therefore, it is concluded that the trade-off found for the bandgap width can be translated to the association of a low current with a large bandgap and a low voltage with a narrow bandgap.

Of course, within this framework, the actual performance of a particular solar cell will be determined by different factors. Particularly critical is the amount of recombination, which directly affects the output voltage. In the context of high-efficiency PV it is of importance to know the theoretical limit of PV conversion of a single-gap solar cell, regardless of possible technology-related nonidealities. The obtained value can be seen as an efficiency ceiling, which of course cannot be reached in practical devices, but it allows us to estimate the room of improvement by technological perfection of single-gap solar cells. On the other hand, the arguments developed for the single-gap solar cell can be useful to understand the operation of novel PV concepts.

The efficiency limit can be calculated using detailed balance arguments, following the model first proposed by Shockley and Queisser [5] by considering a cell with ohmic contacts, full absorption (no reflexion and quantum efficiency QE = 1 for any photon energy > E_G), and infinite carrier mobility. As the possible losses in this cell are reduced to a minimum, still it must be realized that the recombination cannot be fully annihilated. There is an unavoidable loss derived from radiative recombination, because light emission is the detailed balance counterpart of light absorption: if the first is not possible at a given wavelength, the second will be impossible as well. In fact, not the whole radiative recombination that takes place in the semiconductor material can be regarded as losses, since a great part of the photons emitted will be reabsorbed creating new electron–hole pairs, a mechanism called photon recycling, but this is also taken into account in the Shockley–Queisser model [6]. Therefore, the condition for a cell to behave close to ideal is not that the dominant carrier lifetime is very high, but that the lifetime associated to non-radiative processes is much higher than the radiative lifetime, τ_{rad}. As long as this condition is fulfilled and the thickness of the cell assures complete absorption, the absolute value of τ_{rad} is not so significant, because the absorption coefficient, which determines the photon recycling, is inverse proportional to this value [7].

Therefore, the loss of an ideal, fully radiative cell includes only the photons that are emitted by the cell to its surroundings. From this point of view, the current density extracted from the cell can be expressed as

$$J \left[\text{A/cm}^2 \right] = e \left(\text{absorbed photon flux} - \text{emitted photon flux} \right) \tag{19.1}$$

Equation 19.1 can be used to calculate the power that can be extracted from a cell of a given bandgap if we take into account that the emitted photon flux depends on the quasi-Fermi level split, and thereby, on the voltage. When Equation 19.1 is solved, the photon absorption will increase proportionally to the concentration ratio assumed, while the reemission will not be so much enlarged, resulting in a higher V_{OC}. For this reason, to compute the absolute limiting efficiency of solar cells, the theoretical maximum concentration ratio is assumed (46050X, the one providing isotropic illumination on the cell with the radiance of the sun's photosphere and the cell surrounded by a medium of refraction index one).

The absolute efficiency limit for the optimal bandgap, which is about 1.1 eV (~ silicon), is 40.7% [5]. If one-sun illumination is considered (no concentration) the value drops to about 31%. The efficiency of commercially available solar cells rarely surpasses 20%, although the record efficiencies of single-gap cells, even under one-sun illumination, are higher (24.8% Si-cell, 24.6% GaAs cell [8]). One possible approach to improve the production efficiencies is to continue to research on technological improvements and minimize the gap between average production efficiency and the best experimental cell performance, and the gap between these champion experimental cells and theoretical efficiency. But the room for improvement following this approach is limited. Therefore, it is important to push up the theoretical efficiency ceiling through the development of new devices not subject to the Shockley–Queisser limit.

A logical consequence of the argumentations given is that if we are able to implement more than one energy threshold in a PV device, it may be possible for this device to make a better use of the solar spectrum and surpass the Shockley–Queisser limit. The simplest way would be to combine semiconductor materials of different bandgaps. But this cannot be done straightforward, just by fabricating a single pn-junction using two materials of bandgap E_{G1} and E_{G2}, so that $E_{G1} < E_{G2}$. Although it is true that the semiconductor with bandgap E_{G1} can absorb the photons that are wasted in the other semiconductor, the quasi-Fermi level split is limited by the lower bandgap and the resulting efficiency is subject to the Shockley–Queisser limit of bandgap E_{G1} [9]. Hence, we are compelled to reflect on what is exactly implied by "implementing different energy thresholds": first, the existence of electronic states that enable optical transitions of diverse energy, and second, that some paths for thermalization of carriers are blocked in order to extract efficiently the energy gained in the different absorptions.

Both requirements for implementation of multiple energy thresholds are fulfilled in a *multi-junction solar cell* (MJC), consisting of a stack of pn-junctions with different bandgaps [10]. The subcells are arranged from top to bottom with decreasing bandgap, and each of them absorbs the corresponding part of the solar spectrum. The usual implementation is an expitaxially grown, monolithic, two-terminal device that integrates tunnel junctions between the subcells. The tunnel junctions connect the VB of one subcell to the CB of the next, blocking the pass of carriers between subcells along the CB or VB and, thus, avoiding the thermalization of part of their energy. In this way, the quasi-Fermi level splits of the subcells are added following a series connection scheme. The increase in efficiencies at a range of about one absolute percent per year seen in MJCs in the 1999–2006 time frame [11] backs the strategy of going to different cell designs with a higher theoretical efficiency ceiling. So far, the best efficiencies achieved in this kind of devices are 40.7% for a metamorphic triple-junction GaInP/GaInAs/Ge cell operating at 240 suns [12] and 40.8% at 326 suns for an inverted metamorphic triple-junction GaInP/$In_{.04}Ga_{.96}As/In_{.37}Ga_{.63}As$ cell [13].

19.2 Why Quantum Dots?

The properties of planary grown quantum dot (QD) arrays embedded in a higher bandgap semiconductor, such as self-assembled In(Ga)As dots with GaAs as host material, introduce an alternative way for engineering multiple energy thresholds in a PV device in a more compact design. The three-dimensional quantum confinement produces a series of discrete levels within the wider semiconductor bandgap, offering the possibility of introducing new electronic populations and transitions of different energy besides those associated to the CB and VB of the host semiconductor. As in the case of one-dimensional (quantum wells, QW) or two-dimensional (quantum wire, QWI) confinement, band-to-band recombination (of confined electrons with confined holes, also called *excitonic* recombination) is of dominant radiative nature. But in addition to that, a QD has a δ-like density of confined states. This means that there are gaps with zero density of states (DOS) that separate the levels from each other and also from the continuous states of the bulk semiconductor bands (see Figure 19.2). If these gaps are larger than the energy of optical phonons, carrier relaxation from one state to a lower state by interaction with a single phonon is ruled out due to energy nonconservation. Thus, relaxation of excited carriers to the ground

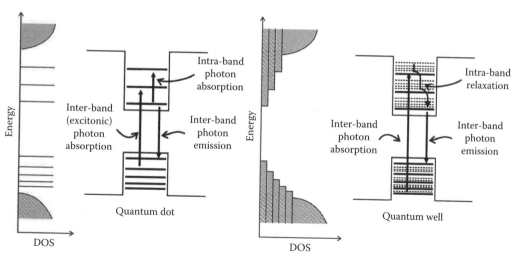

FIGURE 19.2　(Left) Density of states (DOS) and band structure of a quantum dot (QD) and (right) a quantum well (QW). Some of the numerous possible transitions are shown. A remarkable fact is that optical intra-band transitions under normal incidence are allowed for electrons confined in the QD, but not for electrons confined in the QW, and also that fast relaxation of confined electrons has a higher probability in the QW.

state is only possible through less probable carrier–carrier scattering and simultaneous multiphonon processes that comply with energy and momentum conservation. This phenomenon, usually termed *phonon-bottleneck effect* (PBE) [14–16], can be used to prevent fast thermalization of carriers as required in the PV context if suitable devices are designed based on QDs. Such an application would not be possible with quantum structures of lower dimensionality because these structures do not introduce a true zero DOS between the confined energy levels and the bulk semiconductor bands. Consequently, while in QWs an electron in one sub-band can fall to another in sub-picoseconds time, in the case of self-assembled InGaAs/GaAs QDs relaxation times ranging from 1 ns to 10 ps have been reported [17]. Nevertheless, the existence of an effective relaxation blockage in real QD materials as theoretically predicted is quite a controversial point in current research. Several experimental works report about its existence, but others present data that deny it (see [18] for a literature review). It seems that in real QDs, and especially in real QD arrays, some mechanisms can appear that introduce paths for fast relaxation, such as inelastic multiphonon processes bound to inhomogeneity of the QD ensemble [19], second-order electron–phonon interaction (coupling of optical and acoustical phonons) [20], coupling to defect states [21], electron–electron scattering enhancement by coupling to an electron plasma (such as the wetting layer) [22] or to an *n*-type δ-doping region [23], and electron–hole scattering [24,25]. However, it must be considered that in this moment most of the QD applications intend to exploit excitonic luminescence. In this context, special emphasis is put in identifying and promoting the mechanisms that enable a fast relaxation of carriers to the ground state of the QDs and a relaxation time in the order of 10 or 100 ps is often regarded as the proof of a violation of a strictly defined PBE. Contrarily, in the present context, attention has to be paid to mitigate these mechanisms in order to achieve a relaxation time as long as possible. For the PV application that will be discussed here, it is sought that the relaxation time is comparable with the radiative lifetime, and this has been found to be possible in some grown InGaAs/GaAs QDs even at room temperature [17].

With respect to the multiple absorption in QDs, it has been shown that QDs enable the absorption of photons both in excitonic and intra-band transitions (see Figure 19.2). It is of particular importance for PV applications that in the case of confined electron intra-band transitions light can be also absorbed in normal incidence [26–28], and not only when it propagates parallel to the growth plane, as it is the case of QWs [29]. The absorption strength between confined levels is found to be high even in the case of intra-band transitions (10^4 cm^{-1}, [26]), whereas for transitions between confined states and continuous states of the bulk semiconductor bands, the absorption can be weaker. In any case, it must be taken into account that the density of dots is rather low when compared to the atomic density of bulk semiconductors (in current In(Ga)As/GaAs QD arrays it is typically $\leq 10^{17}$ cm^{-3}). Then, even if the dots are characterized by high absorption coefficients, the resulting volumetric absorption coefficient in the transitions only allowed inside the dots can be much lower.

19.3 The Quantum Dot Intermediate Band Solar Cell

The intermediate band solar cell (IBSC) [30,31] is a device with potential for exploiting the properties of QDs to achieve high PV conversion efficiencies. In this cell, the confined states introduced by a QD array are used to form an intermediate band (IB) located between the CB and VB of the host semiconductor [32]. The IB is used to enhance the photocurrent with respect to a single-gap cell. As depicted in Figure 19.3, photons of energy higher than the host semiconductor bandgap E_G (photon 3) can be absorbed as usual by electrons in a VB $\rightarrow$ CB transition, but photons of energy lower can now also be absorbed, either by electrons located in the VB that promote to the IB (photon 1) or by electrons located at the IB that promote to the CB (photon 2). To fabricate the PV device, the QD material has to be sandwiched between two conventional semiconductors of opposite doping as represented in Figure 19.4. These layers, usually called *emitters*, act as selective contacts for electrons and holes, disconnecting the IB from the external circuit. This means that the sub-bandgap photons contribute to the photocurrent by a two-step absorption mechanism mediated by the IB. It must be remarked that this two-step absorption differs from a three particle collision (simultaneous absorption of two photons by a single electron located in the VB), which is a process of low probability. The existence of both sub-bandgap photon absorptions and the production of photocurrent out of the two-step mechanism constitute one of the two key operating principles of the IBSC. If both transitions are operative in the device, the partial bandgaps created by the IB (labeled E_L and E_H in Figure 19.3) represent the extra absorption thresholds required to exceed the Shockley–Queisser limit.

Under operating conditions, the electronic population in the IB material can be modeled using three quasi-Fermi levels that are associated to the three electronic gasses: ε_{Fe} for electrons in the CB, ε_{Fh} for

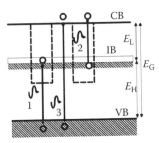

FIGURE 19.3 Band structure of a QD IB material, showing the possible absorption processes. The bandgaps (E_L, E_H, and E_G) are also labeled. (Reprinted from Martí, A. et al., *Phys. Rev. Lett.*, 97(24), 247701, 2006. With permission. Copyright 2006 by the American Physical Society.)

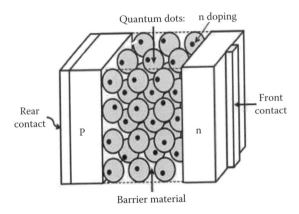

FIGURE 19.4 Elements of a QD-IBSC.

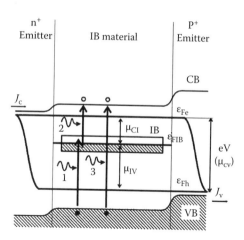

FIGURE 19.5 Band diagram of an IBSC under operation conditions showing the three quasi-Fermi levels (ε_{Fe}, ε_{Fh}, and ε_{FIB}) and the corresponding quasi-Fermi level splits (μ_{CI}, μ_{IV}, and μ_{CV}).

holes in the VB and ε_{FIB} for carriers in the IB (see band diagram in Figure 19.5). The existence of these three distinguished quasi-Fermi levels assumes that carrier relaxation within bands is a much faster process than carrier generation-recombination between bands. With respect to the role of the emitters as selective contacts we notice that, because of them, only the quasi-Fermi levels of electrons and holes are continuous through the device. Their gradient determines the electron and hole current density flow and the split between them (μ_{CV}, the difference between ε_{Fe} at the electron contact and ε_{Fh} at the hole contact) divided by e equals the output voltage of the cell. In this way, the voltage is only limited by the fundamental bandgap of the semiconductor used and not by the position of the IB.

The preservation of an output voltage equivalent to that of a single-gap solar cell is the second fundamental requirement for the IBSC to operate as a high-efficiency converter. It must be noted also that the position of ε_{FIB} is not directly determined by the external voltage. When the cell is working, ε_{FIB} has to be located in the IB or at least very close to it. Then, the band is semi-filled with electrons and both kinds of sub-bandgap photon absorption are simultaneously possible because the IB has both empty states to receive electrons from the VB as electrons to be pumped to the CB. The half-occupancy of the IB could be in principle induced by the respective absorption rates in transitions VB → IB and IB → CB, but in the ideal IBSC model it is assumed that ε_{FIB} is always pinned at the IB in order to make the half-filling of the IB independent of external parameters, such as illumination and biasing. This implies that under equilibrium conditions, the Fermi level is located at the IB and that the DOS of this band is sufficiently high as to provide both positive and negative charge without significant displacement of ε_{FIB} under operation. When the IB material is implemented with QDs, the Fermi level can be located at the IB using δ-doping [34]. Introducing a shallow dopant species in a concentration equal to the density of dots results in the half-occupation of the ground states (these states can host two electrons due to spin degeneracy). It is also found that the density of the dots can be kept below one order of magnitude under the DOS at the CB and VB still providing a clamping of the ε_{FIB} at the IB within 1 kT even when the cell is operated up to 1000 suns [35].

In the general case, there is a nonzero split μ_{CI} between ε_{Fe} and ε_{FIB}, and a nonzero split μ_{IV} between ε_{FIB} and ε_{Fh}, which are related to the external voltage through the equation

$$\mu_{CV} = \mu_{CI} + \mu_{IV}. \tag{19.2}$$

The ideal current-voltage curves plotted in Figure 19.6 illustrate the potential of the IBSC concept. In this device, the constrain that relates photocurrent and voltage in a single-gap solar cell is broken,

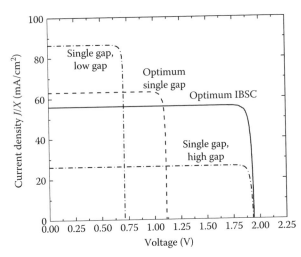

FIGURE 19.6 Calculated current–voltage characteristic of an ideal IBSC with optimized bandgap configuration compared to the characteristic of an ideal optimized single-gap cell. The curves corresponding to single-gap cells with low bandgap (equivalent to the E_L of the IBSC) and high bandgap (equivalent to the E_G of the IBSC) are also included.

enabling the confluence of a high short-circuit current and a high open-circuit voltage. To calculate the efficiency of the IBSC, two inherent constrains have to be taken into account: the relation between quasi-Fermi level splits given by Equation 19.2 and the fact that no current can be extracted from the IB. The latest can be expressed as

$$J_{VI} = J_{IC}, \tag{19.3}$$

where J_{XY} stands for the net carrier generation (multiplied by the electron charge) between bands X and Y. Adding these constrains to Equation 19.1, the efficiency limit of the IBSC can be obtained basing on detailed balance arguments analogously to the Shockley–Queisser calculation for the single-gap cell [5]. In the case of an ideal IBSC (only radiative recombination, ohmic contacts, infinite mobility in all bands and ε_{FIB} clamped to the IB, full absorptivity and optimal absorption coefficient spectral distribution) the efficiency limit is 63.2% when the device works under maximal concentration [30] (see Figure 19.7). This value is obtained for an optimized bandgap configuration of $E_G = 1.95$, $E_H = 1.24$, and $E_L = 0.71$ eV. It surpasses the Shockley–Queisser limit of 40.7% for single gap solar cells and is similar to the efficiency limit of a triple-junction cell. Furthermore, the efficiency ceiling of the IBSC can be pushed up if a multi-junction device is constructed using different IBSCs as subcells. In this respect, it has been demonstrated that the design of a series-connected IBSC double-junction cell, which requires only one tunnel junction, has an efficiency limit of 72.5% [36], to be compared with the 73.3% obtained for a series-connected six-junction cell (with five tunnel junctions). Brown et al. [37] have also calculated a slightly lower efficiency (71.7%) considering the insertion of two IBs in a single junction IBSC. The later approach shows the advantage of not needing the implementation of any tunnel junction, although it seems difficult at this state of the research that the half-filling of both IBs and the respective absorption rates can be sufficiently controlled under operation.

Let us now consider the properties that the QD array has to exhibit in order to fulfill the model of the IBSC with the characteristics summarized in Figures 19.3 and 19.4. In the first place, we have to pay attention to the energy level distribution within the QD material. The material depicted in Figure 19.3 represents one possible configuration in which the IB arises from confinement in the CB (type I QD band structure). The approach has been used to realize quantum dot intermediate band solar cell (QD-IBSC) prototypes, using the InAs/GaAs [38], and the InGaAs/GaAs [39] material systems. It would also be possible in principle to engineer an IB out of hole confined levels in a type II QD array, such as

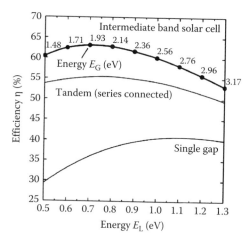

FIGURE 19.7 Efficiency limit for an IBSC with and for a series-connected double-junction cell, in both cases vs. the lowest bandgap E_L, and for a cell with a single bandgap (in this case the x-axis quantifies the total bandgap E_G). On the curve corresponding to the IBSC, the values of the total bandgap (E_G) for maximum efficiency are also indicated. (From Luque, A. and Martí, A., *Phys. Rev. Lett.*, 78, 5014, 1997.)

in a GaSb/GaAs QD material. This material system has indeed also been used to fabricate QD-IBSC prototypes [40] and the use of a type II broken-gap heterostructure, such as an InAs/GaSb QD super-lattice, has also been proposed as a material candidate [41]. One fundamental aspect to be considered here is that to preserve the voltage, both quasi-Fermi level splits μ_{CI} and μ_{IV} have to be nonzero. In this respect, the split related to the band suffering the confinement (μ_{CI} in Figure 19.3) is the most problem-atic. A nonnegligible split indicates that the relaxation of carriers to the ground state is slow, or in other words, that the process can be seen as a recombination, rather than as a thermalization (carrier lifetime <100 fs). As in a single-gap solar cell, the ideal case is that the lifetime is determined by radiative recom-bination (~2 ns for CB–VB recombination in doped GaAs), because this recombination is unavoidable and if another mechanism dominates, it will be necessarily characterized by a shorter lifetime. To this end, it is needed that the confined states are well separated in order to profit from an effective blockage of carrier relaxation to the ground state by the PBE. It must be remembered that this effect relies on the fact that the separation between states does not match the optical phonon energy (approx. 32 meV in GaAs). In this respect, confinement of electrons is preferred to confinement of holes, because the higher effective mass of holes leads to a higher number of confined states for the same confinement potential. For this reason, and also because it is a well-characterized material, In(Ga)As/GaAs QDs arrays have been chosen in most cases to develop the first QD-IBSC prototypes, in spite of the fact that the bandgap configuration is far from the optimal one. This will be the material system that will be mostly discussed throughout this text.

As a matter of fact, even if dominant radiative recombination is assured, it is still preferred that the number of excited dot levels is as low as possible. They introduce extra recombination paths, whose effects are not expected to be counteracted by the extra light absorption that they may introduce. As a conse-quence, a trade-off appears in the design of the QD-IBSC. To optimize the production of photocurrent big dots are required, for they exhibit a wider gap between IB and CB (assuming a type I QD structure). But big dots also host an increased number of excited levels. Further research will be needed on the depen-dence of the electronic structure of QDs with their shape, stoichiometry, and density in order to solve this problem. However, it has been proposed that the use of concentrated light can be an efficient tool to minimize the effect of extra levels and make the IBSC tolerant to the use of relatively large QDs [42].

In order to maximize the total quasi-Fermi level split, the absence of a VB offset in the type I band structure is also desirable. Unfortunately, that is not the case of the InAs/GaAs QD system, in which

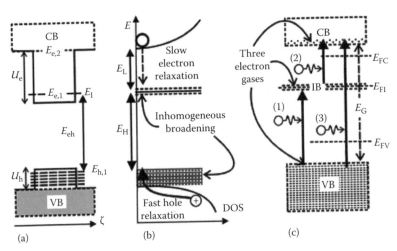

FIGURE 19.8 (a) Band structure of a single QD along a line ζ crossing its centre. (b) DOS of the QD/barrier system, illustrating the slow electron relaxation and the effect of the VB offset, where relaxation of holes is expected to be fast. (c) Resulting simplifies band diagram of the IB material. E_{FC}, E_{FI}, and E_{FV} label here the three quasi-Fermi levels associated to the three bands. (From Cuadra, L. et al., Phonon bottleneck effect and photon absorption in self-ordered quantum dot intermediate band solar cells, in *Proceedings of the 19th European Photovoltaic Solar Energy Conference*, Paris, France, 2004, WIP-Renewable Energies and ETA: Munich, Germany/Florence, Italy, pp. 250–254. With permission.)

a VB offset of approx. 200 meV is present. The resulting band diagram is shown in Figure 19.8. Because of the excessive number of levels in the VB confinement potential, holes are expected to thermalize almost instantaneously. That results in a lower effective bandgap, an inherent and unavoidable cause of voltage loss. Furthermore, in the situation of having both types of carriers spatially confined in the same volume enhances the probability of electron–hole scattering (the mechanism by which an electron in the CB transfers its energy to a hole in an Auger-like process; it is relaxed to the QD ground state, while the excited hole losses the energy by thermalization). This mechanism has been identified as one of the main causes of phonon-bottleneck breakdown in practical (In,Ga)As/(Ga,Al)As QDs even at low injection levels [24,25]. Type I QD material combinations with zero VB offset, such as $InAs_{0.9}N_{0.1}/GaAs_{0.98}Sb_{0.02}$, have been proposed by Levy et al. [44] for IBSC fabrication, although they have not been implemented yet. With respect to the electron confined levels, Figure 19.8 illustrates the most favorable case, in which only one excited level $E_{e,2}$ is present, located as close as possible to the CB. The effect of this level is twofold: it provides the condition for a PBE and enhances the absorption probability in the IB → CB transition. The absorption enhancement is based on the fact that in QDs, the optical transition rates depend on the dipole matrix and thereby on the overlapping of the initial and final states involved [45] (see also Figure 19.9).

Regarding the electrical properties of the IB, the IBSC model is in principle not very demanding. The formation of minibands out of confined states has proven to be possible in highly ordered and size homogeneous $In_xGa_{1-x}As/GaAs$ QD arrays [46]. But it is not clear that the formation of actual minibands is strictly required. The IB does not have

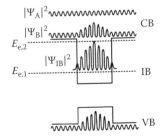

FIGURE 19.9 Band diagram of an In(Ga)As/GaAs QD showing the shape of the different electronic wavefunctions. Confined levels are characterized by a more localized wavefunction, while the levels in the CB and VB are delocalized. The overlap between wavefunctions determines the absorption strength in a given transition. In that way, the transition from the localized Ψ_{IB} to the partially localized Ψ_B (resulting from coupling of the excited level $E_{e,2}$ and the lower states of the CB) is more probable than transitions between Ψ_{IB} and fully delocalized states of the CB (Ψ_A).

to carry the current flow and, therefore, its mobility can be low. It seems at this point of the research that a lower bound for the mobility may only be found in practical devices related to the need of spatial redistribution of carriers along the IB to maximize sub-bandgap photon absorption. Then, the formation of a miniband will have to be rather subordinated to the control of the phononic dispersions and the maintenance of an effective PBE.

On the contrary, the optical properties of the QDs seem to be critical for the overall performance. The relative strength and spectral dependence of the three possible optical transitions are a sensitive aspect of the IBSC design. It is quite intuitive that the efficiency of the device would sink if part of the energy of the photons gets lost because they are absorbed in a transition of lower energy than possible. The most certain way to achieve optimal performance in an IBSC would be then to ensure the selectivity of the absorption coefficients (α_{VC}, α_{VI}, and α_{IC}) associated respectively to the VB $\rightarrow$ CB, VB $\rightarrow$ IB, and IB $\rightarrow$ CB transitions. This implies that α_{IC} should be zero in the range where α_{VI} is nonzero (photon energies $\geq E_H$) and α_{VI} should behave analogously with respect to α_{VC} (see graph in Figure 19.8, left). This is the configuration used in the ideal case for efficiency limit calculations. But it seems difficult to achieve such a complete selectivity in a real IB material. It has been found that an IB material where $\alpha_{IC} \ll \alpha_{VI} \ll \alpha_{VC}$ (Figure 19.10, right) would also lead to optimal efficiencies [47], in particular when assisted by light confinement techniques for the weaker transitions [48]. We find that this condition is in agreement with the absorption properties found in QDs, at least in the In(Ga)As/GaAs material system. Looking back at Figure 19.9 and remembering that the optical transition rates depend on the overlapping of the initial and final states involved, we find that the excitonic transition VB $\rightarrow$ IB has to be stronger than the transition IB $\rightarrow$ CB. The transition associated to the fundamental bandgap is far less problematic in this context. It is reasonably expected to be stronger than the sub-bandgap ones because it is present in the barrier material of the QD stack and also in the emitters. It can be concluded then that the step-like absorption coefficient distribution in In(Ga)As/GaAs QDs is promising for the implementation of the IBSC concept. Logically, the performance of the device will be compromised if the values of α_{IC} and α_{VI} are too low to enable sufficient sub-bandgap photon absorption. This is in fact a problem that has been found in practical QD-IBSC prototypes and that will be discussed in more detail.

To finalize with the description of the IBSC and the way to implement it, it is of interest mentioning that an alternative to QDs is also being considered as IB material. It is often said the QDs "mimic" atoms, or that they are an "engineered form of atoms and molecules." Thinking the other way round, in the IBSC context impurity atoms can be seen as the minimal size limit of QDs. In fact, it is not new at all that an atom of a foreign species can introduce discrete levels in the bandgap of a semiconductor,

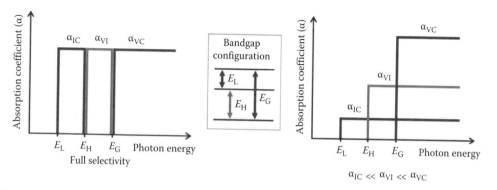

FIGURE 19.10   Diagrams showing the two possible absorption coefficient spectral distributions that result in an efficient photon-sorting in the IBSC. α_{XY} represents the absorption coefficient associated to transitions between bands X and Y. The diagram on the left shows the ideal case of full selectivity and the one on the right the step-distribution, which is thought to be close to the absorption properties in InAs/GaAs QD-IBSCs. The assumed bandgap configuration is included for reference.

sometimes quite distant from both the conduction and VBs (the so-called *deep levels*). Then, a bulk IB material could be synthesized by doping of a semiconductor material with appropriated deep-level impurities. A major complication of this approach is the expected appearance of a high non-radiative recombination through the IB, because isolated impurities in a semiconductor do not show a PBE. On the contrary, they act as "carrier traps," promoting non-radiative relaxation of electrons through multiphonon emission [49]. Mainly because of this fact, the research on practical IBSC devices based on bulk materials is currently not as advanced as the QD-IBSC research. However, it is worth mentioning that a method for suppression of the non-radiative recombination in bulk IB materials has been recently proposed [50] and also that Yu et al. [51,52] and Lucena et al. [53] have reported evidence of an IB in the optical characterization of bulk materials synthesized by different methods.

19.4 Characterization of QD-IBSC Prototypes

In the last years, there have been several attempts to implement the QD-IBSC concept in practical devices. Some of these prototypes have been used to test experimentally the principles of the IBSC model with positive results. However, to date, it has not been possible yet to fabricate a device that actually develops the PV conversion potential of the IBSC model. In this section we will review the characterization of state-of-the-art QD-IBSCs, trying to identify the main factors that limit practical efficiencies and to outline strategies for overcoming these limitations.

The experimental data presented in this chapter have been obtained with different QD-IBSC prototypes. A generic device structure is sketched in Figure 19.11, but the particular growth and structure details can be found in the respective references included in the text. Unless something else is stated, the QD-IBSC prototypes contain 10 layers of self-assembled δ-doped InAs/GaAs QDs, grown by molecular-beam epitaxy using the Stranski–Krastanov [45] growth mode, and have GaAs emitters. The density of the dots is in the order of $1-4 \times 10^{10}$ cm^{-2}. To study the performance of the prototypes, they are compared with GaAs reference cells fabricated with the same technology. Their structure differs from the QD cell only in that they do not contain a QD array. Figure 19.12 (from [54]) shows the current–voltage characteristic under one-sun illumination of a QD-IBSC compared with that of a reference cell. It can be clearly seen from this plot that in state-of-the-art QD-IBSCs none of the expected effects (photocurrent enhancement and voltage preservation) are observed yet. We will discuss both effects separately.

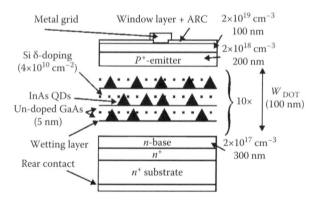

FIGURE 19.11 Example of the structure of an InAs/GaAs QD-IBSC with 10 layers of δ-doped QDs. The thickness and doping densities of the different layers are shown. The QD stack is grown by the Stranski–Krastanov growth mode and, therefore, each QD layer is accompanied by a wetting layer (a very thin QW). (Reprinted from Luque, A. et al., *J. Appl. Phys.*, 99(1), 094503, 2006. With permission. Copyright 2006 American Institute of Physics.)

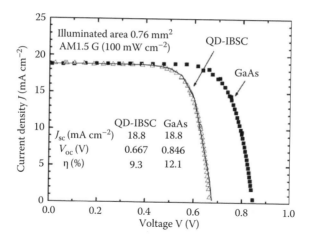

FIGURE 19.12   One-sun illuminated current–voltage characteristics of an InAs/GaAs QD-IBSC (open triangles) and a GaAs reference cell (solid squares). (Reprinted from Luque, A. et al., *J. Appl. Phys.*, 99(1), 094503, 2006. With permission. Copyright 2006 American Institute of Physics.)

19.4.1 Enhancement of the Photocurrent

The fact that the J_{sc} of the QD cell does not surpass the value obtained for the reference cell does not necessarily mean that the QDs are not absorbing light and can be attributed to the reduced thickness of the QD stack. This assessment is corroborated by the quantum efficiencies shown in Figure 19.13 (from [55]). It can be observed that for wavelengths greater than 879 nm, which corresponds to the gap energy of GaAs (1.41 eV), the response of the reference cell decays abruptly, while that of the QD-IBSC sample shows further photocurrent until approximately 1300 nm and then decays. The measurement proves that the QDs are actually contributing to the photocurrent with sub-bandgap absorption, but their input is about two orders of magnitude lower than the GaAs contribution. Then, it is not surprising that the enhancement cannot be reflected in a current–voltage curve. Another negative effect of the reduced QD material thickness that has been identified in the characterization of QD-IBSC prototypes

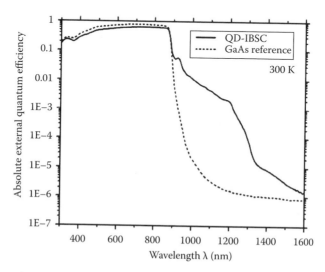

FIGURE 19.13   External quantum efficiency of an InAs/GaAs QD-IBSC sample (solid line). The quantum efficiency of a GaAs cell with identical structure but no QD material is also shown for reference (dashed line). (From Antolín, E. et al., *Thin Solid Films*, 516, 6919, 2008. With permission.)

is the complete immersion of the QDs in the space-charge region of the cell [54]. Attempts have been made to grow thicker QD stacks, but the results obtained for 20 and 50 layers of InAs/GaAs QDs have shown severe emitter degradation with increased QD layers due to strain buildup [56]. This is an inherent problem of the Stranski–Krastanov growth mode: the lattice mismatch between the substrate and the epitaxial layer triggers the QD formation, but it also has the negative effect of accumulating strain along the growth direction. For thick QD stacks the accumulated strain can result in dislocations that spread into the emitter grown on top of the QD material, leading to photocurrent values below the reference cell photocurrent.

To overcome the problem of strain accumulation, new growth methods are being developed that intend to relief vertical strain in In(Ga)As/GaAs QD stacks. Fabrication of QD-IBSCs using strain-balanced materials based on the insertion of P in the GaAs barrier has been reported by Hubbard et al. [57,58], Laghumavarapu et al. [59], Popescu et al. [60], and Alonso-Álvarez et al. [61]. Oshima et al. have also produced strain-balanced QD-IBSCs using GaNAs as barrier material [62]. Figure 19.14 shows an example of the effect of strain balancing in the QD stack growth when compared with a non-balanced sample (from [60]). Although the effect of strain compensation on the material quality is impressive, only Hubbard et al. have reported from five-layer non-δ-doped QD-IBSC samples a photocurrent that equals, and in one case even exceeds [58], the photocurrent of the reference.

It seems a good strategy for the further development of the QD-IBSC technology to continue investigating strain-balanced growth techniques until a defect-free thick QD stack can be inserted into practical devices. Besides, strain balancing is expected to have another positive effect. It seems at this point of the research that for the optimization of the bandgap configuration in practical QD devices, it is required to enlarge as much as possible the value of E_L (the bandgap between IB and CB). This is achieved by growing large QDs. But the growth of larger QDs induces a higher strain accumulation. This can be observed from the comparison between Figures 19.12 and 19.13: the current of the QD-IBSC in Figure 19.12 equals the current of the reference cell, while in Figure 19.13 it is lower, although in both cases the stacks contain 10 layers of dots. This comes from the fact that the samples used in [55] have larger QDs than the samples used in [54]. For this reason, while the QE of Figure 19.13 drops at 1300 nm, the QE corresponding to the samples measured in Figure 19.12 (which can be found in [54]) drops at 1150 nm. An increase of ~130 meV in the value of E_L has been achieved through the growth of larger QDs, but this has also induced a degradation of the emitter. It must be noted also that, taking into account the VB offset and the presence of a wetting layer, which also reduces the effective bandgap [31], the value of E_L

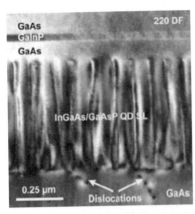
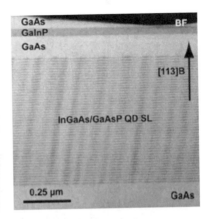

FIGURE 19.14 Left, 220 Dark-field transmission electron microscopy (TEM) image showing the generation of misfit dislocations in a 50-period (In,Ga)As/Ga(As,P) (0.8% P) QD superlattice grown on (113)B GaAs without strain balance. Right, bright-field TEM image showing highly uniform 50-period (In,Ga)As/Ga(As,P) (14% P) QD superlattice containing no misfit dislocations grown on (113)B GaAs with strain balance. (Reprinted from Popescu, V. et al., *Phys. Rev. B*, 78, 205321, 2008. With permission. American Physical Society.)

is rather low even in the case of Figure 19.13: from the position of the photocurrent edge and discounting a bandgap narrowing of ~200 meV, it is deduced that $E_L \sim 0.26$ eV. In this respect, it is expected that the developing of strain-compensation techniques will enable the growth not only of thicker stacks, but also of larger dots for that may enable a better bandgap configuration. In this context we are expecting that the other known drawback of large dots for IBSC implementation, the appearance of extra confined levels, can be counteracted by the use of high concentration as exposed in [42].

19.4.2 Voltage Preservation

Taking into account that even In(Ga)As/Ga(P)As QD-IBSCs, which do not show a severe emitter degradation also present a low V_{OC}, most of the voltage loss can be attributed to a narrowing of the effective bandgap E_G by the presence of the VB offset and a wetting layer. Nevertheless, it is of interest to evaluate the impact that the recombination properties of the IB have on the output voltage and, thereby, to test the principles of the IBSC on practical devices. Because of the controversy that exists around the existence of an effective PBE in practical QD arrays, one of the first questions to be addressed in this context is the possibility of inducing a nonnegligible split between the quasi-Fermi levels associated to the CB and the IB. This possibility has been demonstrated in the work presented in Ref. [38] through the analysis of the electroluminescence spectrum of InAs/GaAs QD-IBSCs (i.e., the quasi-Fermi level split is achieved under forward bias) (see Figure 19.15). Furthermore, it has been possible to determine the carrier lifetimes associated to the CB → IB and the IB → VB transitions through curve fitting to experimental current density, quantum efficiency, and electroluminescence emission spectrum data using the IBSC model [54]. The low lifetime values obtained of 0.5 and 40 ps, respectively, have been attributed to non-radiative channels associated to localized defects in the QD stack and point out again the necessity for better material quality through strain-compensation and improved growth techniques.

However, the direct observation of absorption in the IB → CB transition and the demonstration of photocurrent production due to two-photon absorption remained elusive by standard characterization methods. We will see that this demonstration is of the outmost importance because, although we have presented the production of photocurrent out of two (and not one) sub-bandgap photon absorption and the voltage preservation as two independent principles of the IBSC operation, it can be shown that they are directly related [63]. In this respect, QE measurements as the one showed in Figure 19.13 do not

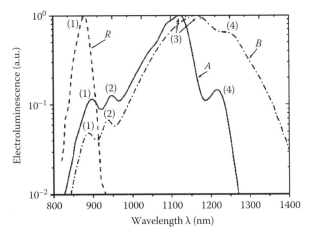

FIGURE 19.15 Electroluminescence of InAs/GaAs QD-IBSCs. "R" labels the GaAs reference cell, "A" a sample with 10 layers of QDs and δ-doping layers, and "B" a sample with 10 layers QDs but without δ-doped layers. The numbers "1" to "4" identify the different peaks. (Reprinted from Luque, A. et al., *Appl. Phys. Lett.*, 87(8), 083505-3, 2005. With permission. Copyright 2005 American Institute of Physics.)

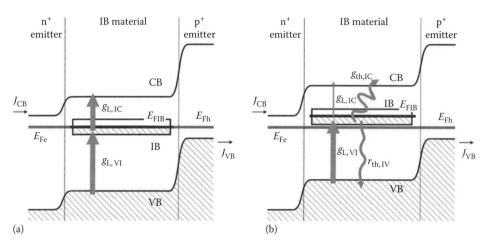

FIGURE 19.16 Two possible band diagrams for an IBSC when it is short circuited and illuminated only with sub-bandgap photons. Light generation (g_L), thermal generation (g_{th}), and thermal recombination (r_{th}) processes are represented, as well as the position of the quasi-Fermi levels. In (a) the two light generation rates are equal, and in (b) the light generation rate between VB and IB is much higher than that between IB and CB. (From Antolín, E. et al., Demonstration and analysis of the photocurrent produced by absorption of two sub-bandgap photons in a quantum dot intermediate band solar cell, in *Proceedings of the 23rd European Photovoltaic Solar Energy Conference*, Valencia, Spain, 2008. With permission.)

constitute a sufficient proof-of-the-concept, because they allow different interpretations. The extraction of photocurrent below the GaAs bandgap edge in Figure 19.13 can indeed result from the two situations depicted in Figure 19.16a and b. Both represent the band diagram of an IBSC short circuited and illuminated only with sub-bandgap photons. In Figure 19.16a, the generation rates in the VB → IB and the IB → CB transitions are equal, and the photocurrent extracted results from two-photon absorption. Figure 19.16b represents the case in which the generation rate associated to the VB → IB transition is higher than that of the IB → CB transition. Given the difference in absorption strength between both transitions discussed in the previous section, this will be generally the case if the IB is implemented with InAs/GaAs QDs and no light-trapping techniques are used (more so if we take into account that the high photon energy considered, between 0.95 and 1.41 eV, is much larger than E_L). In this situation ε_{FIB} can be higher than ε_{Fe}, as shown in Figure 19.16b. The negative quasi-Fermi split will necessarily cause a reaction of the system in order to reestablish the carrier population at equilibrium. In the same way that in any solar cell a positive $\varepsilon_{Fe} - \varepsilon_{Fh}$ split implies the occurrence of recombination, here the negative split between ε_{FIB} and ε_{Fe} will release thermal generation between the corresponding bands. This thermal escape of carriers from the dots to the CB is indeed the most plausible explanation for the sub-bandgap photocurrent that is observed in Figure 19.10.

The effect of thermal escape on the output voltage of the cell can be understood by means of a circuital model. A general equivalent circuit of an IBSC has been presented in [64]. In Figure 19.17a, we include a simplified circuit model of the IBSC where, for example, Auger processes and resistive losses have been neglected, but which will serve the purpose of carrying out this analysis. The current generators labeled $J_{L,XY}$ represent the photogeneration between bands X and Y, while the diodes J_{0YX} represent the recombination between the same bands. As the net carrier flow J_{XY} equals ($J_{L,XY} - J_{0YX}$), the structure of the circuit results from the direct application of Equations 19.2 and 19.3. Starting from this circuit we can draw the equivalent circuit for the QE measurement in the sub-bandgap range. It is shown in Figure 19.17b, where it has been assumed that the sub-bandgap photocurrent measured has no significant contribution of $J_{L,IC}$. Since in a QE measurement the cell is short circuited, J_{0CV} is not conducting and can be neglected. Regarding the other two recombination diodes, their voltage biases have to compensate

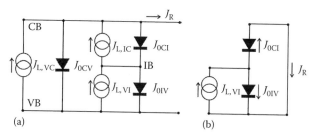

FIGURE 19.17   Equivalent circuit of the QD-IBSC operation under the different conditions explained in the text: (a) general equivalent circuit, (b) under sub-bandgap continuous illumination (neglecting transitions from the IB to the CB) and short circuit.

each other. This means that when a photocurrent J_R is extracted, J_{0IV} is forward biased and J_{0CI} is reverse biased with the same absolute voltage ($V_{IV} = -V_{CI}$). The reverse bias of J_{0CI} represents the negative $\varepsilon_{FIB} - \varepsilon_{Fe}$ split in Figure 19.16b, while the thermal escape between IB and CB corresponds to the inverse saturation current of that diode. The split $\varepsilon_{FIB} - \varepsilon_{Fh}$, which determines the recombination rate between IB and VB, has its circuit counterpart on the forward bias of J_{0IV}. If the short-circuit condition is removed, the extraction of sub-bandgap photocurrent through J_{0CI} profiting from thermal escape implies a voltage loss. From the previous argumentation we realize that it is possible to increase the photocurrent of the cell by means of thermal escape, but voltage preservation requires that the absorption of a second photon is feasible to promote electrons from the IB to the CB. From a more fundamental point of view, it can be demonstrated that it is impossible for a pseudo-IBSC that relays on the absorption of one sub-bandgap photon and subsequent thermal escape to have an output voltage larger than the bandgap E_H without violating the second law of thermodynamics [65,66]. A process identified as an alternative to the absorption of a second photon that would enable voltage preservation in cells where sub-bandgap absorption is only possible through VB → IB transitions is impact ionization (an inverse Auger mechanism) [67]. It is unknown for the moment to what extent this effect is present in current QD-IBSCs. As a matter of fact, the exploitation of impact-ionization processes is the aim of another high-efficiency PV concept, the multiple-exciton-generation (MEG) solar cell [68], which is currently also under development using a different kind of QDs.

It can be seen from the argumentation exposed that the feasibility of the second photon absorption process in QD devices is indispensable to corroborate that this approach fits into the theoretical model of the IBSC. Because of the low value of E_L in existing QD-IBSC prototypes, the saturation current of J_{0CI} can be very high at room temperature even if only radiative processes are taken into account, not limiting the extraction of J_R for practical values of $J_{L,VI}$ [69]. Therefore, it is necessary to lower the operating temperature of the devices in order to detect any contribution of IB → CB absorption ($J_{L,IC}$) to the photocurrent. Specific experiments have been carried out using a two-light-sources frequency-modulated photocurrent setup, in which it has been possible for temperatures lower than 80 K to extract photocurrent produced by two-photon absorption from InAs/GaAs QD-IBSCs [33]. The experiments were performed in short-circuit conditions to assure that the measured photocurrent has a PV nature (in contraposition to photoconductance effects). The dependence of this photocurrent has been found to be in agreement with the IBSC circuital model [63]. However, the photocurrent detected is low (tens of nA/cm^{-2}), which probably indicates that the low absorption of QDs in the IB → CB transition constitute an important limitation for the development of InAs/GaAs QD-IBSCs.

19.4.3 Boosting the Efficiency of QD-IBSCs

It is concluded that the main principles of the IBSC theoretical model have been demonstrated in practical QD-IBSC devices, but that their effects are jeopardized by different factors that prevent the achievement of high efficiencies. As we have seen, some of these factors have been already identified in the characterization of InAs/GaAs QD-IBSCs. In some cases, the mitigation of different problems leads to a trade-off in the design of the QD-IBSC. We can now summarize some strategies that can be adopted to overcome the drawbacks of the current technology as follows:

- *Modification of the material composition.* In this respect, two different objectives can be pursued, which are not mutually exclusive [44]. The first is the fabrication of a QD material system with more optimum bandgaps. The second, the improvement of strain-balanced QD growth techniques [57–62], which is expected to have a significant impact on the global performance of the device for many reasons. The strain accumulation induces structural defects in the QD material grown and subsequently enhances the non-radiative recombination, preventing the realization in practice of the QD theoretical properties, in particular the existence of an effective PBE at room temperature. Furthermore, strain-balance techniques can enable the growth of thicker QD stacks to enhance sub-bandgap photon absorption and larger QDs to increase the value of E_L. The later is required for two reasons: to improve the exploitation of the solar spectrum and to inhibit the thermal escape from the IB to the CB for voltage preservation.

- *Use of concentrated light.* While large dots are desirable for their wider E_L value, small dots host less excited levels and, therefore, introduce less possible recombination paths. The use of high concentration has been proposed as a practical solution contributing to this trade-off in the short-term basing on the prediction of a saturation of the extra recombination paths for high injection levels [42]. Moreover, the bandgap of GaAs is too low to really profit from the insertion of an IB if the cell is operated at one sun [42]. Consequently, unless major changes in the material composition are made or the cell is inserted in a multi-junction configuration [36], the use of concentration is mandatory.

- *Application of specific light-trapping techniques for the sub-bandgap energy photon range.* The characterization of prototypes has shown that the values of α_{VI} and α_{IC} are too low to produce a significant enhancement of the photocurrent. One approach to solve this problem is to increase the density of dots. But this strategy would bring us again to a trade-off with strain accumulation. It is not clear yet if the use of strain-balanced techniques can allow the growth of stacks as thick as to obtain a sufficient absorption in the VB → IB transition. What seems to be clear at this point of the research is that the required absorption in the IB → CB cannot be achieved in stacks of practical thickness. One possibility to circumvent this limitation is to enhance the effective optical path for IR photons in the QD stacks. This can be done by implementing diffraction structures that deflect the light of the relevant wavelength almost parallel to the surface [70].

19.5 Summary

The performance of conventional single-gap solar cells is fundamentally limited to a conversion efficiency ceiling of 40.7% (the Shockley–Queisser limit) because they cannot use a relevant part of the solar spectrum. The utilization of semiconductor QDs constitutes a means for enhancing the conversion potential of solar cells over the Shockley–Queisser limit. In particular, the fabrication of an IBSC using QD arrays is proposed. This concept has a theoretical efficiency limit of 63.2%. In this chapter, the properties required in the QD arrays to implement the IBSC theoretical model have been discussed. The characterization of state-of-the-art QD-IBSCs, mainly based on the InAs/GaAs QD material system, has been reviewed. It has been concluded that the key principles of the IBSC theoretical model (the production of photocurrent due to two sub-bandgap photon absorption and the description of the electronic population by three separated quasi-Fermi levels as required for voltage preservation) have proven to be operative in the QD-IBSCs characterized. However, the performance of current QD-IBSC prototypes does not comply with the predictions of the IBSC theoretical model. Different factors affecting the efficiency of practical devices have been identified, such as the existence of a VB offset, strain accumulation in the QD material, existence of excited levels in the QDs, low sub-bandgap absorption, and reduced energy gap between the IB and the CB. Based on these conclusions, strategies have been proposed to boost the efficiency of future QD-IBSCs prototypes, such as the use of concentrated illumination and sub-bandgap light trapping, as well as improvements in the QD material composition. The later includes the implementation of QD materials with negligible VB offset and the further development of strain-balanced growth methods with the aim of growing larger QDs and thicker QD stacks.

Acknowledgments

This work has been supported by the project IBPOWER (www.ies.upm.es/ibpower), funded by the European Commission under Contract No. 211640, the Spanish Plan for R&D Consolider-GENESIS FV (CSD2006-00004), and the Comunidad de Madrid Program NUMANCIA (S-0505/ENE/000310). The authors would like to acknowledge Prof. C. R. Stanley and Dr. C. D. Farmer from the University of Glasgow for the fabrication of many of the devices analyzed in this work and for valuable discussions.

References

1. Kammen, D.M., *The Rise of Renewable Energy*. Scientific American, pp. 84–93, September 2006.
2. Lewis, N.S. et al., *Basic Research Needs for Solar Energy Utilization*, MD, April 18–21, 2005.
3. Luque, A. and A. Martí, Non-conventional photovoltaic technology: A need to reach goals. In A. Martí and A. Luque (Eds.), *Next Generation Photovoltaics: High Efficiency through Full Spectrum Utilization*, pp. 1–18, Institute of Physics Publishing: Bristol, U.K., 2003.
4. Luque, A., *Solar Cells and Optics for Photovoltaic Concentration, Non-Imagining Optics and Static Concentration*, Adam Hilguer: Bristol, U.K., 1989.
5. Shockley, W. and H.J. Queisser, Detailed balance limit of efficiency of p-n junction solar cells. *J. Appl. Phys.*, 1961, **32**(3): 510–519.
6. Martí, A., J.L. Balenzategui, and R.F. Reyna, Photon recycling and Shockley's diode equation. *J. Appl. Phys.*, 1997, **82**(8): 4067–4075.
7. Roosbroeck, W.V. and W. Shockley, Photon-radiative recombination of electron and holes in germanium. *Phys. Rev.*, 1954, **94**(6): 1558–1560.
8. Green, M.A. et al., Tables of solar cell record efficiency. *Prog. Photovoltaics Res. Appl.*, 2002, **10**: 55–61.
9. Wolf, M., Limitations and possibilities for improvements of photovoltaic solar energy converters. Part I. Considerations for earth's surface operation. *Proc. IRE*, 1960, **48**(7): 1246–1263. United States.
10. Jackson, E.D., Areas for improvement of the solar energy converter. In *Transactions of the Conference on the Use of Solar Energy*, Tucson, AZ, 1955, University of Arizona Press, Tucson, AZ, 1958, Vol. 5, pp. 122–126.
11. King, R.R. et al., Raising the efficiency ceiling with multijunction III–V concentrator photovoltaics. In *Proceedings of the 23rd European Photovoltaic Solar Energy Conference*, Valencia, Spain, 2008, WIP Renewable Energies, Munich, Germany, pp. 24–29.
12. King, R.R. et al., 40% efficient metamorphic GaInP/GaInAs/Ge multijunction solar cell. *Appl. Phys. Lett.*, 2007, **90**(18): 183516.
13. Geisz, J.F. et al., 40.8% efficient inverted triple-junction solar cell with two independently metamorphic junctions. *Appl. Phys. Lett.*, 2008, **93**(12): 123505.
14. Bockelmann, U. and G. Bastard, Phonon scattering and energy relaxation in two-, one- and zero-dimensional electron gasses. *Phys. Rev. B*, 1990, **42**(14): 8947–8951.
15. Benisty, H., C.M. Sotomayor-Torres, and C. Weisbuch, Intrinsic mechanism for the poor luminescence properties of quantum-box systems. *Phys. Rev. B*, 1991, **44**(19): 10945–10947.

16. Benisty, H., Reduced electron-phonon relaxation in quantum-box systems: Theoretical analysis. *Phys. Rev. B*, 1995, **51**(19): 3281–13293.

17. Mukai, K. et al., Phonon bottleneck in self-formed $In_xGa_{1-x}As/GaAs$ quantum dots by electroluminescence and time-resolved photoluminescence. *Phys. Rev. B*, 1996, **54**(8): R5243–R5246.

18. Nozik, A.J., Spectroscopy and hot electron relaxation dynamics in semiconductor quantum wells and dots. *Annu. Rev. Phys. Chem.*, 2001, **52**: 193–231.

19. Heitz, R. et al., Energy relaxation by multiphonon processes in InAs/GaAs quantum dots. *Phys. Rev. B*, 1997, **56**(16): 10435–10445.

20. Inoshita, T. and H. Sakaki, Electron relaxation in a quantum dot: Significance of multiphonon processes. *Phys. Rev. B*, 1992, **46**(11): 7260–7263.

21. Li, X.-Q. and Y. Arakawa, Ultrafast energy relaxation in quantum dots through defect states: A lattice-relaxation approach. *Phys. Rev. B*, 1997, **56**(16): 10423–10427.

22. Bockelmann, U. and T. Egeler, Electron relaxation in quantum dots by means of Auger processes. *Phys. Rev. B*, 1992, **46**(23): 15574–15577.

23. Fekete, D. et al., Temperature dependence of the coupling between n-type δ–doping region and quantum dot assemblies. *J. Appl. Phys.*, 2006, **99**: 034304.

24. Vurgaftman, I. and J. Singh, Effect of spectral broadening and electron–hole scattering on carrier relaxation in GaAs quantum dots. *Appl. Phys. Lett.*, 1994, **64**(2): 232–234.

25. Sosnowski, T.S. et al., Rapid carrier relaxation in In0.4Ga0.6As/GaAs quantum dots characterized by differential transmission spectroscopy. *Phys. Rev. B*, 1998, **57**(16): R9423.

26. Li, S.S. and J.B. Xia, Intraband optical absorption in semiconductor coupled quantum dots. *Phys. Rev. B*, 1997, **55**(23): 15434–15437.

27. Phillips, J. et al., Intersubband absorption and photoluminescence in Si-doped self-organized InAs/GaAlAs quantum dots. *J. Vac. Sci. Technol. B*, 1997, **16**(3): 1243–1346.

28. Pan, D. et al., Normal incident infrared absorption from InGaAs/GaAs quantum dot superlattice. *Electron. Lett.*, 1996, **32**(18): 1726–1727.

29. Loehr, J.P. and M.O. Manasreh, In M.O. Manasreh (Ed.), *Theoretical Modeling of the Intersubband Transitions in III-V Semiconductor Multiple Quantum Wells*, in *Semiconductor Quantum Wells and Superlattices for Long-Wavelength Infrared Detectors*, Artech House: Boston, MA/London, 1993.

30. Luque, A. and A. Martí, Increasing the efficiency of ideal solar cells by photon induced transitions at intermediate levels. *Phys. Rev. Lett.*, 1997, **78**(26): 5014–5017.

31. Martí, A., L. Cuadra, and A. Luque, Intermediate band solar cells. In A. Martí and A. Luque (Eds.), *Next Generation Photovoltaics: High Efficiency through Full Spectrum Utilization*, pp. 140–162, Institute of Physics Publishing: Bristol, U.K., 2003.

32. Martí, A., L. Cuadra, and A. Luque, Quantum dot intermediate band solar cell. In *Conference Record of the 28th IEEE Photovoltaics Specialists Conference*, Anchorage, AK, 2000, pp. 940–943.

33. Martí, A. et al., Production of photocurrent due to intermediate-to-conduction-band transitions: A demonstration of a key operating principle of the intermediate-band solar cell. *Phys. Rev. Lett.*, 2006, **97**(24): 247701.

34. Martí, A., L. Cuadra, and A. Luque, Partial filling of a quantum dot intermediate band for solar cells. *IEEE Trans. Electron. Devices*, 2001, **48**(10): 2394–2399.

35. Martí, A., L. Cuadra, and A. Luque, Quantum dot analysis of the space charge region of intermediate band solar cell. In *Proceedings of the 199th Electrochemical Society Meeting*, Washington, DC, 2001, The Electrochemical Society: Pennington, NJ, pp. 46–60.

36. Antolín, E., A. Martí, and A. Luque, Energy conversion efficiency limit of series connected intermediate band solar cells. In *Proceedings of the 21st European Photovoltaic Solar Energy Conference*, Dresden, Germany, 2006, pp. 412–415.

37. Brown, A.S., M.A. Green, and R.P. Corkish, Limiting efficiency for a multi-band solar cell containing three and four bands. *Physica E*, 2002, **14**: 121–125.

38. Luque, A. et al., Experimental analysis of the quasi-Fermi level split in quantum dot intermediate-band solar cells. *Appl. Phys. Lett.*, 2005, **87**(8): 083505-1–083505-3.
39. Norman, A.G. et al., InGaAs/GaAs QD superlattices: MOVPE growth, structural and optical characterization, and application in intermediate-band solar cells. In *31st IEEE Photovoltaic Specialists Conference*, Orlando, FL, 2005, IEEE: New York, pp. 43–48.
40. Laghumavarapu, R.B. et al., GaSb/GaAs type II quantum dot solar cells for enhanced infrared spectral response. *Appl. Phys. Lett.*, 2007, **90**(17): 173125.
41. Cuadra, L., A. Martí, and A. Luque, Type II broken band heterostructure quantum dot to obtain a material for the intermediate band solar cell. *Phys. E Low Dimen. Syst. Nanostruct.*, 2002, **14**(1–2): 162–165.
42. Martí, A. et al., Elements of the design and analysis of quantum-dot intermediate band solar cells. *Thin Solid Films*, 2007, **516**: 6716–6722.
43. Cuadra, L. et al., Phonon bottleneck effect and photon absorption in self-ordered quantum dot intermediate band solar cells. In *Proceedings of the 19th European Photovoltaic Solar Energy Conference*, Paris, France, 2004, WIP-Renewable Energies and ETA: Munich, Germany/Florence, Italy, pp. 250–254.
44. Levy, M.Y. et al., Quantum dot intermediate band solar cell material systems with negligible valence band offsets. In *31th IEEE Photovoltaics Specialists Conference*, Lake Buena Vista, FL, 2005, pp. 90–93.
45. Sugawara, M., Self-assembled InGaAs/GaAs quantum dots. *Semiconductors and Semimetals*, Vol. 60, Academic Press: New York, 1999.
46. Song, H.Z. et al., In-plane photocurrent of self-assembled $In_xGa_{1-x}As$/GaAs(311)B quantum dot arrays. *Phys. Rev. B*, 2001, **64**(8): 085303.
47. Cuadra, L., A. Martí, and A. Luque, Influence of the overlap between the absorption coefficients on the efficiency of the intermediate band solar cell. *IEEE Trans. Electron. Devices*, 2004, **51**(6): 1002–1007.
48. Martí, A. et al., Light management issues in intermediate band solar cells. In *MRS Proceedings, Spring Meeting 2008*, San Francisco, CA, Vol. 1101E, pp. KK06–02.
49. Lang, D.V. and C.H. Henry, Nonradiative recombination at deep levels in GaAs and GaP by lattice-relaxation multiphonon emission. *Phys. Rev. Lett.*, 1975, **35**(32): 1525.
50. Luque, A. et al., Intermediate bands versus levels in non-radiative recombination. *Phys. B Condens. Matter*, 2006, **382**(2): 320–327.
51. Yu, K.M. et al., Diluted II–VI oxide semiconductors with multiple band gaps. *Phys. Rev. Lett.*, 2003, **91**(24): 246403–246404.
52. Yu, K.M. et al., Multiband GaNAsP quaternary alloys. *Appl. Phys. Lett.*, 2006, **88**(9): 092110–092113.
53. Lucena, R. et al., Synthesis and spectral properties of nanocrystalline V-substituted In_2S_3, a novel material for more efficient use of solar radiation. *Chem. Mater.*, 2008, **20**(16): 5125–5127.
54. Luque, A. et al., Operation of the intermediate band solar cell under nonideal space charge region conditions and half filling of the intermediate band. *J. Appl. Phys.*, 2006, **99**(1): 094503.
55. Antolín, E. et al., Low temperature characterization of the photocurrent produced by two-photon transitions in a quantum dot intermediate band solar cell. *Thin Solid Films*, 2008, **516**: 6919–6923.
56. Martí, A. et al., Emitter degradation in quantum dot intermediate band solar cells. *Appl. Phys. Lett.*, 2007, **90**: 233510.
57. Hubbard, S.M. et al., Effect of strain compensation on quantum dot enhanced GaAs solar cells. *Appl. Phys. Lett.*, 2008, **92**(12): 123512.
58. Hubbard, S.M. et al., Short circuit current enhancement of GaAs solar cells using strain compensated InAs quantum dots. In *33rd IEEE Photovoltaics Specialists Conference*, San Diego, CA, 2008, DOI: 10.119/PVSC.2008.4922600.

59. Laghumavarapu, R.B. et al., Improved device performance of InAs/GaAs quantum dot solar cells with strain compensation layers. *Appl. Phys. Lett.*, 2007, **91**: 243115.
60. Popescu, V. et al., Theoretical and experimental examination of the intermediate-band concept for strain-balanced (In,Ga)As/Ga(As,P) quantum dot solar cells. *Phys. Rev. B*, 2008, **78**: 205321.
61. Alonso-Álvarez, D. et al., Carrier recombination effects in strain compensated quantum dot stacks embedded in solar cells. *Appl. Phys. Lett.*, 2008, **93**(12): 123114.
62. Oshima, R., A. Takata, and Y. Okada, Strain-compensated InAs/GaNAs quantum dots for use in high-efficiency solar cells. *Appl. Phys. Lett.*, 2008, **93**: 083111.
63. Antolín, E. et al., Demonstration and analysis of the photocurrent produced by absorption of two sub-bandgap photons in a quantum dot intermediate band solar cell. In *Proceedings of the 23rd European Photovoltaic Solar Energy Conference*, Valencia, Spain, 2008, WIP Renewable Energies, Munich, Germany, pp. 5–10.
64. Luque, A. et al., General equivalent circuit for intermediate band devices: potentials, currents and electroluminescence. *J. Appl. Phys.*, 2004, **96**(1): 903–909.
65. Luque, A., A. Martí, and L. Cuadra, Thermodynamic consistency of sub-bandgap absorbing solar cell proposals. *IEEE Trans. Electron. Devices*, 2001, **48**(9): 2118–2124.
66. Luque, A., A. Martí, and L. Cuadra, Thermodynamics of solar energy conversion in novel structures. *Physica E*, 2002, **14**(1–2): 107–114.
67. Luque, A., A. Martí, and L. Cuadra, Impact-ionization-assisted intermediate band solar cell. *IEEE Trans. Electron. Devices*, 2003, **50**(2): 447–454.
68. Ellingson, R.J. et al., Highly efficient multiple exciton generation in colloidal PbSe and PbS quantum dots. *Nano Lett.*, 2005, **5**(5): 865–871.
69. Martí, A., C.R. Stanley, and A. Luque, Intermediate band solar cells (IBSC) using nanotechnology. In T. Soga (Ed.), *Nanostructured Materials for Solar Energy Conversion*, Chap. 17, 2006: Elsevier, Amsterdam, the Netherlands/Boston, MA.
70. Tobías, I., A. Luque, and A. Martí, Light intensity enhancement by diffracting structures in solar cells. *J. Appl. Phys.*, 2008, **104**(3): 034502.

Technologies for Emerging Applications: Photonic DNA Computing

20

Nanoscale Information Technology Based on Photonic DNA Computing

Yusuke Ogura
Jun Tanida

20.1 Introduction: Nanoscale Information Technology Using Photonics

Many novel methodologies have recently been developed for nanoscale science and engineering. Advances in nanotechnology have a strong impact on various fields of research; for example, materials, processing, measurement, and fabrication. Photonics-based methods are surely promising, and a lot of effort has been made to utilize the superior properties of photonics, such as the capability in making interaction with various materials. One of the most exciting approaches is nanophotonics, which is based on near-field optics, nonlinear optics, plasmonics, and other interesting phenomena relating to light.[1,2] The essential idea of nanophotonics is the confinement of light into a nanoscale volume to achieve the resolution of the order of a nanometer. The applications of nanophotonics include molecular or bio-sensing by use of nanoscale probes, fine-structure fabrication, and various imaging by surface plasmon-enhanced spectroscopy.[3–8]

Nanoscale science and technology should also be combined with information technology. It is certain that the nano-world contains much information and that the effective use of the information promotes revolutionary changes in a human's life, as information technology has led to many technical innovations so far. Nature offers a variety of matters and phenomena that give us hints for realizing high-performance information technology. Information technology inspired by nature, which is referred to as natural computing, is actively studied and is expected to be effective for nanoscale information technology due to its potential. A typical example of natural computing is information photonics, including the idea of optical computing. Information photonics is a paradigm for manipulating information based on the effective use of photonics, often combined with electronics, biotechnology, and so on.[9,10] Various interesting techniques and systems have been proposed relating to, for example, optical interconnection, digital parallel processing, optical pattern recognition, and optical security.[11–26] Achievements in information photonics research are expected to be applied to develop nanoscale information technology using photonics. However, the diffraction limit of light, by which the spot size is restricted as typically around μm for visible light, prevents the straightforward use of photonics at the nanoscale level. Although this restriction can be overcome with nanophotonics, it is generally difficult to utilize the spatial parallelism of photonics effectively. Some new ideas are necessary to develop valuable nanoscale information technology by the use of photonics; and the utilization of nanoscale mechanisms in nature becomes a promising solution.

DNA is as an attractive information carrier, and nanoscale information technology using the nature of DNA, usually called DNA computing, has been paid some attention.[27,28] The activity and the evolution of living beings are regulated by information stored in DNA. This fact indicates that DNA is an excellent carrier of information and is suitable for long-term preservation. Many promising applications have been proposed in the scope of DNA computing, for example, gene analysis, nanofabrication, nano machines, dense memory, and synthetic biology.[29–45] Although a simple strategy is adopted to mimic a biological system utilizing DNA information, the complexity of the biological system prevents us from taking the simple strategy.

A scheme of nanoscale information technology that combines photonics and DNA is being studied.[46–51] This is a new parallel method for the effective manipulation of information, by making light and DNA work cooperatively as an information carrier or a manipulation method of information. Encoding information into DNA molecules enables the simultaneous control of a large amount of information based on reaction parallelism. Information is transferred from the macro-world to the DNA through light, and the DNA works depending on the light signal. High information density is achievable because the DNA molecules are small and react depending on their base sequences. The use of DNA ensures accessibility to material, phenomena, and other things in the nano-world.

In this chapter, we describe photonic DNA computing, which is an instance of nanoscale information technology using photonics. In Section 20.2, the concept of photonic DNA computing is shown. In Section 20.3, parallel optical tweezers for manipulating DNA clusters, DNA vessels attached to microscopic beads, is described. In Section 20.4, a photonic method for controlling DNA molecules in local volumes is shown. In Section 20.5, an addressing method toward memory using photonics and DNA is introduced and some preliminary results are shown. In Section 20.6, we summarize this chapter.

20.2 Photonic DNA Computing

In conventional DNA computing, information is encoded as the form, the sequence, the length, or another attribute of DNA molecules, which are referred to as information DNA in this chapter, and which is processed through a sequence of DNA reactions. The reactions proceed autonomously and in parallel by mechanisms that are incorporated in the DNA intentionally. Watson–Crick base pairing is an important property, in particular, to control the DNA for processing. The DNA reactions are induced in a spatially uniform manner in the whole of a DNA solution.

In contrast, photonic DNA computing, which is on the basis of photonics and DNA computing, utilizes the local control of information DNA by changing light distributions. The concept of photonic DNA computing is illustrated in Figure 20.1. A solution containing information DNA is placed on a substrate. The solution is irradiated with multiple beams to control the DNA within the individual beams. The information is broadcasted to the DNA, and the DNA behaves by following the program, which is previously built into the DNA. For example, the reaction can be controlled by local temperature change induced by laser irradiation. This means that the solution is considered to be divided into multiple small spaces (subspaces) virtually with a micro-scale size corresponding to each of the irradiation beam spots. This method enables us to access particular information DNA molecules in the solution and manipulate them selectively. This is a major difference between photonic DNA computing and conventional DNA computing.

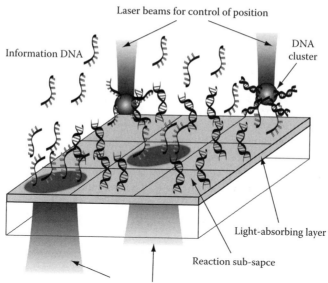

FIGURE 20.1 The basic concept of photonic DNA computing.

Many DNA molecules react in parallel within the size of a diffraction-limited light spot. For example, there exists tens of thousands of DNA molecules in a $(1\,\mu m)^3$ of a $100\,\mu M$ solution, which is a typical concentration of DNA experiments. This strategy offers a manipulation method of information with the size smaller than the light spot generated with propagating light in a diffraction-limited system because the DNA molecules work independently even if they are under the control of the same beam. Information DNAs are transported to another subspace by fabricating DNA clusters, which are vessels of DNA attached to microscopic beads. To keep information DNA at appropriate positions, they can be attached to and detached from the substrate by hybridization and denaturation with the DNA strands that are bound to the substrate.

In photonic DNA computing, light and DNA must work cooperatively to deal with information. Light and photonics techniques are useful for global processing on a microscale, whereas DNA and DNA technologies are adequate for local processing on a nanoscale. The idea of photonic DNA computing can offer novel information techniques that bridge over information systems between the microscale and the macroscale. The scheme provides many features that are useful for realizing information systems, as follows:

1. Massive parallelism based on spatial parallelism of photonics and reaction parallelism of molecules.
2. Unique information processing by combining global processing using photonics and local processing using DNA.
3. Large capacity and dense information systems.
4. Flexible and programmable processing at the nanoscale level.
5. Compatibility to electronics or other systems by bridging with photonics.
6. Communication capability between the microscale and the nanoscale.
7. Capability to obtain and control molecular information such as genetic information, which includes ambiguity.
8. Tolerance for alignment and package of optical systems as the benefit of the autonomous and regulated action of DNA.

Photonic methods for parallel manipulation of information DNA are indispensable to realize the concept of photonic DNA computing. We have developed several methods including a parallel optical-manipulation method for controlling the position of the DNA individually, and a method for controlling the reaction of the DNA in the local space by laser irradiation.

20.3 Parallel Optical Tweezers for Manipulating DNA Clusters

Optical manipulation is a method for dealing with a microscopic object without contact using a radiation pressure force induced by interaction between the light and the object. Since Ashkin demonstrated the single-beam optical trap of a dielectric bead in 1986,[52] optical manipulation has been used as a valuable tool in various fields. By attaching molecules to a microbead, the molecules can be manipulated, and the method can be applied to the measurement of a force powered by motor molecules such as kinesin, the measurement of the elasticity of DNA strands, and others.[53–56]

Parallel or flexible manipulations of objects require the generation of proper distributions of a light field. Holographic optical tweezers, which uses a spatial light modulator for indirect modulation of a light source, is a good solution. This technique is applied to the kinetically locked-in colloidal transportation, the processing of carbon nanotubes, and so on.[57,58]

We have developed another method that is based on the direct modulation of light sources on a vertical-cavity surface-emitting laser (VCSEL) array.[59–62] VCSELs are semiconductor lasers with a unique structure different from that of the edge-emitting lasers,[63] and they can be arranged at a period of several tens or a few hundred micrometers. The emission intensities of the individual pixels of the VCSEL array are independently controlled by electronics at the maximum modulation rate of more than 10 GHz. With the VCSEL array, arbitrary spot array patterns are generated on demand.

The conceptual illustration of the VCSEL array optical manipulation is shown in Figure 20.2.

A flexible manipulation for micro-objects is achievable by the control of the spatial and the temporal intensity distribution generated by the VCSEL array sources. The method is, in particular, useful for the parallel manipulation of objects with compact hardware and simple control software. A same system configuration provides various modes of manipulation.

Direct control of the position of DNA molecules in parallel is difficult because of the scale-gap between light spots and DNA molecules. Indirect manipulation by fabricating DNA clusters with microscopic beads is an alternative method for the purpose. The construction of a DNA cluster is shown in Figure 20.3.

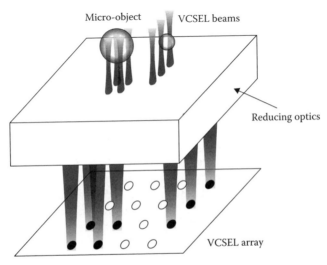

FIGURE 20.2 The scheme of VCSEL array optical manipulation.

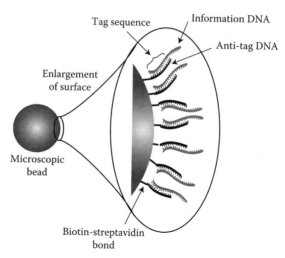

FIGURE 20.3 The composition of a DNA cluster.

DNA molecules with particular sequences, which are referred to as anti-tag DNA, are modified with biotin at the 5′-end, and are bound to the surfaces of the beads by biotin-streptavidin binding.

The sequence of information DNA includes a tag sequence, which is complementary to the anti-tag DNA, and the information DNA is attached to the beads by a hybridization of the tag sequence and the anti-tag sequence. A vessel of information DNA on the bead is referred to as a DNA cluster. Approximately 10^4 DNA molecules are bound to a single bead 10 μm in diameter. This composition makes it easy not only to construct DNA clusters, but also to destroy them by letting information DNA free through general reactions of DNA.

Figure 20.4 shows the experimental setup to demonstrate the parallel transportation of DNA clusters. The VCSEL array (NTT Photonics Laboratory, wavelength of 854 ± 5 nm, maximum output power of more than 3 mW, aperture of 15 μm-diameter, and pixel pitch of 250 μm) has 8 × 8 VCSEL pixels, after

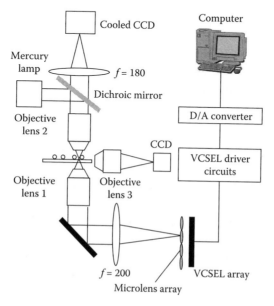

FIGURE 20.4 Experimental system of VCSEL array optical manipulation.

which a micro-lens array (focal length of 720 μm and lens pitch of 250 μm) was set to increase the light efficiency. Objective lens 1 (OLYMPUS, LUMPlan FI 60× W/IR, NA = 0.90) was a water-immersible, long working-distance objective lens and was used as a focusing lens.

Objective lens 2 (OLYMPUS, LUMPlan, 100 × NA = 1.0) was used for observation from above and objective lens 3 (Mitsutoyo, MPlan ApoSL 50×, NA = 0.42) was used for observation from the side. The beam spacing period, the maximum intensity, and the beam diameter on the sample plane were 3.75 μm, approximately 1.1 mW per pixel, and 2.6 μm, respectively. The sample was observed through a cooled CCD (Nippon Roper, CoolSNAP fx) and a CCD. The beads used to fabricate DNA clusters in this experiment were polystyrene particles of 6 μm in diameter. Fluorescence molecules (Molecular Probes: Alexa Fluor 546) were attached to information DNA strands for observation.

Figure 20.5 shows the fluorescence images before and after the simultaneous transportation of two DNA clusters by changing the emission patterns of VCSELs. One or two pixels of the VCSEL array were assigned to capture the beads and the assigned pixels were switched sequentially. This result demonstrated that the DNA clusters were transported toward different directions in response to the emitting pixels simultaneously. The average velocity was 0.38 μm/s.

Multiple DNA clusters can be stacked by using multiple VCSEL beams entering from the bottom. A procedure for stacking DNA clusters is shown in Figure 20.6a. A DNA cluster is levitated by irradiating with multiple beams (2 × 2 beams in the figure) from below because the beams at the edge of the DNA cluster create a strong force for levitation. Then, if the next DNA cluster is irradiated, the second DNA cluster is also levitated and reaches just below the first DNA cluster. By repeating this operation, some DNA clusters can be stacked vertically.

A sequence of images observed from the horizontal side in the experiment of stacking is shown in Figure 20.6b. The DNA clusters are indicated by dotted circles to show their positions clearly. The result demonstrates that five DNA clusters can be stacked vertically. This manipulation is useful in, for example, temporarily storing information DNA and increasing the density of information DNA within a small area.

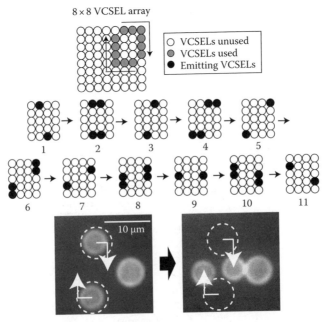

FIGURE 20.5 Experimental result of parallel transportation of two DNA clusters. Upper; a sequence of the emission pattern of the VCSEL array, and lower; fluorescence images before (left) and after (right) transportation.

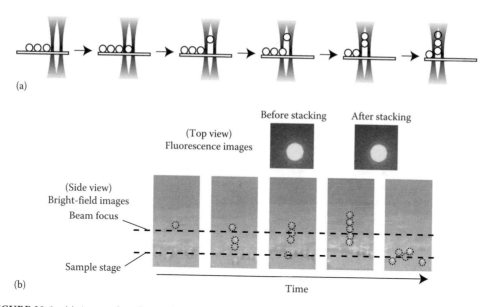

(a)

(b)

FIGURE 20.6 (a) A procedure for stacking multiple DNA clusters vertically, and (b) the experimental result by stacking DNA clusters.

20.4 Photonic Control of DNA Molecules in Local Volumes

Temperature is an essential parameter for the control of various reactions of DNA molecules. Light-thermal conversion induced by irradiation with a laser beam is applicable to control the temperature of a DNA solution.[64] The scheme of a method for controlling reaction of DNAs by irradiating with a laser beam is shown in Figure 20.7. A substrate is coated with a sort of material that absorbs light and a DNA solution is put on it. Irradiation of the substrate with a focused laser beam heats up the surface of the substrate due to light absorption. The thermal energy transfers to the solution on the substrate, and the temperature of the solution around the irradiated area increases. The temperature of the solution inside a small volume can be controlled by changing the power of the beam used. This method offers a local control of temperature because various light distributions can be easily generated and modulated at a micro-scale using appropriate light sources and devices. When focusing a laser beam using an objective lens with a high numerical aperture, we can control a reaction of the DNAs selectively at a microscale.

A glass substrate is coated with titanylphthalocyanine of $0.15\,\mu m$ thickness as a layer for absorbing light, then gold is deposited on the substrate to attach the DNA to the substrate. The anti-tag DNAs that were modified with thiol at the 5′-end were bound to the substrate. The information DNAs labeled with fluorescence molecules were immobilized to the substrate by hybridization with the anti-tag DNAs.

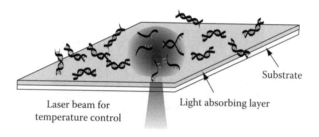

Laser beam for
temperature control

Light absorbing layer

Substrate

FIGURE 20.7 A method for controlling the temperature of a solution in a micrometer scale.

Then, we irradiated the substrate with a focused laser beam, and scanned the beam in the shape of the letter "T" by moving the sample stage. Figure 20.8 shows a fluorescence image captured after the above procedure. The fluorescent intensity of the area where the beam passed is lower than that of the other area. The result shows that the denaturation of the information DNAs can be controlled by irradiating with a laser beam in an approximately 5 μm resolution.

We also performed experiments for the reaction on a bead to investigate the reaction of DNA clusters.

In the experiment, a hairpin DNA was used because the hairpin DNA has two stable states as shown in Figure 20.9. A hairpin DNA is a DNA molecule that has a self-complementary part in its sequence and can form hairpin-like structure. The hairpin DNA has two stable structures at a low temperature: an open state and a close state. The hairpin DNA maintains its open form with a specific DNA molecule, which contains a complementary sequence of the stem part of

30 μm

FIGURE 20.8 Experimental result of detachment of information DNA from a substrate.

the hairpin DNA. If the temperature of the solution is decreased gradually, hairpin DNA and linear DNA molecules hybridize with each other (open state). In contrast, if the temperature of the solution is decreased rapidly, the hairpin DNA forms the hairpin formation and does not hybridize with the linear DNA (close state). This is usable to express two states of information, such as binary digits "1" and "0". In this case, the linear DNA can be considered as the information DNA.

Writing and erasing of information DNA on beads were studied experimentally. We prepared a solution containing DNA clusters constructing hairpin DNA and fluorescence-labeled information DNA, and placed it on a substrate. The sample was irradiated from below with a beam that was generated from a semiconductor laser of a wavelength of 854 nm and focused with an objective lens (Olympus Corp., LUMPlan Fl 60×). A fluorescence microscope with a cooled CCD was used for observation.

A DNA cluster was irradiated with a laser beam of 5 mW for 10 s, and the fluorescence images captured before and after irradiation are shown in Figure 20.10. If information DNA (linear DNA) is attached to

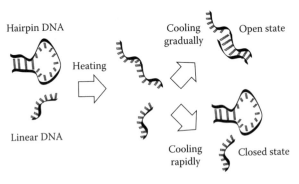

FIGURE 20.9   Behavior of hairpin DNA and linear DNA. Linear DNA can be considered as information DNA.

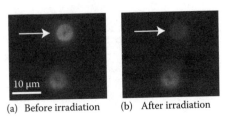

(a) Before irradiation   (b) After irradiation

FIGURE 20.10   (a and b) Denaturation of DNA on a bead by laser irradiation.

a bead, the fluorescence power observed around the bead is high because fluorescence molecules are concentrated on the bead as shown in Figure 20.10a. After denaturing, the information DNA dispersed in the solution, and the intensity around the bead decreases (Figure 20.10b). This result shows that the information DNA was denatured by laser irradiation.

We investigated a suitable irradiating condition for writing information DNA on a bead. The sample solution was prepared by mixing the solution of the beads including the hairpin DNA with the solution of the information DNA. The mixed solution contained enough information DNA for reaction. To perform hybridization of information DNA with linear DNA on a bead, a gradual cooling of the solution is required. A target bead was irradiated with the irradiation schedule shown in Figure 20.11a, and the obtained fluorescence images are shown in Figure 20.11b. Figure 20.11c shows cross sections of fluorescence images along the dotted line as indicated in Figure 20.11b. The background fluorescence intensity is removed to show these figures.

It can be seen that the fluorescence intensity of the irradiated bead increased obviously. This result indicates that the information DNA hybridizes with the hairpin DNA on the bead. We succeeded in

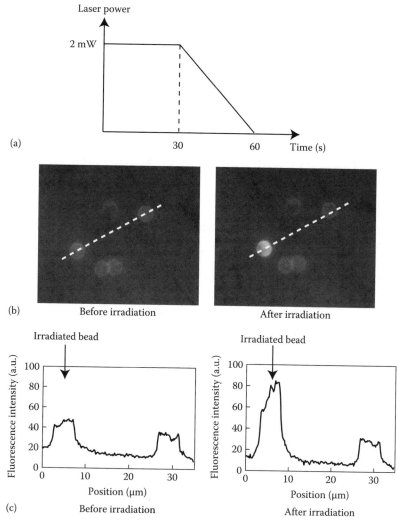

FIGURE 20.11 Experimental results on writing data DNA on a bead. (a) Irradiation schedule, (b) fluorescence images, and (c) cross sections along the dotted line shown in (b).

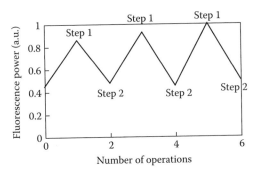

FIGURE 20.12   The experimental result of repetitions of writing and erasing operations on a bead.

writing the information DNA on the bead using the optical technique. The fluorescence intensities of other beads around the target bead did not change, and a hybridization reaction could be controlled with a resolution of no more than 10 μm.

In the next experiment, we repeated writing and erasing operations on a bead. At the beginning, the information DNA molecules were not attached to a bead with the hairpin DNA. The following steps were repeated three times. Step 1: writing with the irradiation schedule shown in Figure 20.11a, step 2: erasing by irradiating with a laser beam of 5 mW for 10 s. Figure 20.12 shows the fluorescence power measured after the individual steps. The fluorescence power increases after step 1 and decreases after step 2. This is an expected result. Note that efficiencies of writing and erasing indicate almost the same values during the three repetitions. We demonstrate that our method was usable to write and to erase from beads repeatedly.

20.5 Addressing by Use of Repeated Hairpin DNA for Memory

A memory is an important functionality to enhance the performance of a photonic DNA computing system. With a memory that uses DNA and photonics, we can store not only digital data originated from various sources but also directly from the molecules themselves. To utilize such information on demand, addressing of the stored information is necessary.

For the purpose, we considered a combination of molecular addressing and spatial addressing.[65] The concept of molecular addressing and spatial addressing is shown in Figure 20.13. The molecular address is determined by what sequence of DNA is used. Let us consider that a memory DNA has a

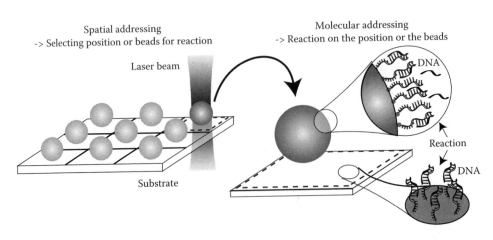

FIGURE 20.13   Operation of spatial addressing and molecular addressing.

part of a sequence for addressing and another part for storing data, and different address sequences are assigned to the individual memory DNAs. When a bank of memory DNAs are mixed with an addressing-DNA having specific sequences, only the memory DNA that contains the complementary sequence to the addressing-DNA in its address part can react with the addressing-DNA. From this reaction, we can distinguish the memory DNA of the target address from the others; namely, we can achieve molecular addressing. On the other hand, the spatial address is determined by the position of the DNA. The local controlling methods using light irradiation described in Sections 20.3 and 20.4 can be applied for spatial addressing. The spatial address is determined by selecting a reaction area or beads. Execution of spatial addressing makes it possible to store and use different information on individual beads.

Molecular addressing is realized utilizing repeated hairpin DNA. Figure 20.14 shows the reaction scheme of the repeated hairpin DNA used. The DNA consists of two layers, each of which consists of a hairpin DNA. Note that multiple kinds of the sequence of hairpin DNA are used for each layer. When $2n$ sequences of hairpin DNA are designed, and n sequences are assigned to each of the two layers, the size of the address space is n^2. On the other hand, when using only one layer, the size of the address space is $2n$. The layered structure is effective for increasing the size of the address space. A first-layer opener can open a corresponding first-layer hairpin DNA. A second-layer opener can open a corresponding second-layer hairpin DNA, only if the first-layer hairpin is open. If the first-layer hairpin is not open, the second-layer opener cannot hybridize with the second-layer hairpin, and further reactions do not occur. After opening the first- and second-layer hairpin DNAs, a new DNA sequence appears and lets us store information.

To demonstrate the fundamental functions of addressing on beads, the reaction of repeated hairpin DNA was verified by some experiments. Two kinds of hairpin DNAs are assigned to each of the layers and the total number of the hairpin DNAs is $2 \times 2 = 4$. Hairpin DNAs A and B are for the first layer and the hairpin DNAs C and D are for the second layer. The individual hairpin DNAs have their own opener DNAs.

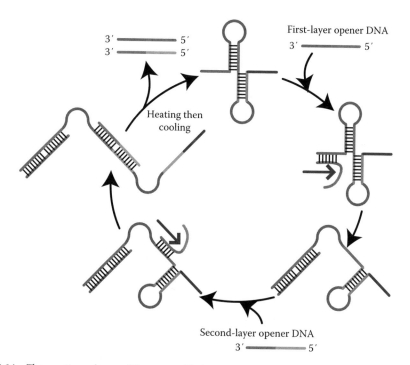

FIGURE 20.14 The reaction scheme of the repeated hairpin DNA.

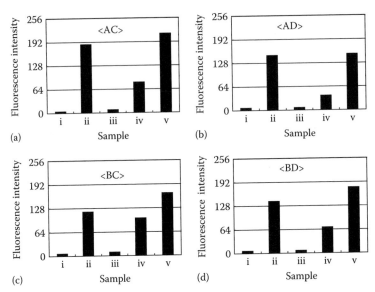

FIGURE 20.15 The fluorescence intensity of the solutions after reactions. The hairpin DNA used are (a) A and C, (b) A and D, (c) B and C, and (d) B and D.

We prepared five solutions (i)–(v) containing beads with the repeated hairpin DNA and the opener DNAs. Fluorescence molecules (Molecular Probes; Alexa Fluor 546) are attached to the opener DNAs. Here the opener DNA is expressed as $O_i^{(f)}$ where i is the layer number (1 or 2) and the superscript (f) denotes that the opener is fluorescence-labeled. The opener in the solutions were the following: (i) no opener, (ii) $O_1^{(f)}$, (iii) $O_2^{(f)}$, (iv) O_1 and $O_2^{(f)}$, (v) $O_1^{(f)}$ and $O_2^{(f)}$. The solutions were put on a substrate after reactions, and were observed through a fluorescence microscope to detect the fluorescence intensity of the beads. The state of the repeated hairpin DNA was determined from a fluorescence image of the sample. If a opener DNA opens a hairpin DNA, the fluorescence intensity of the bead increases because the opener DNA binds to the hairpin DNA.

The fluorescence intensity of each solution is shown in Figure 20.15. The fluorescence intensities of solutions (i) and (iii) are low enough, and the maximum intensity is obtained for solution (v). The middle intensities are obtained for solutions (ii) and (iv). The difference in the intensities of solutions (ii) and (iv) is considered to cause the difference of the reaction efficiency of the first and second layer because the reaction of the second layer must follow the first-layer reaction. Although the result suggests some problems to be solved, it also demonstrates the correct reaction of the repeated hairpin DNA on beads.

In the next experiment, the reaction of the first layer was induced by the methods for transporting DNA clusters and for controlling temperature on a substrate. The schematic diagram of the reaction is shown in Figure 20.16. Opener DNA is attached on the substrate. A DNA cluster is transported to a position by optical tweezers (Figure 20.16a). By irradiating with the laser beam, the opener DNA on the substrate is detached and moves to the hairpin DNA on the bead (Figure 20.16b), then the opener opens the hairpin (Figure 20.16c). By transporting the DNA cluster, the opener DNA on a track of the bead transportation can be transferred from the substrate to the DNA clusters (Figure 20.16d).

Figure 20.17 shows the fluorescence intensities of the DNA cluster and the track of the bead transportation on the substrate. It can be seen that the fluorescence intensity of the DNA cluster increases whereas that of the track decreases. This result shows that the opener DNA moves from the substrate to the bead, and opened the first-layer hairpin DNA, namely, an addressing of the first-layer was demonstrated successfully.

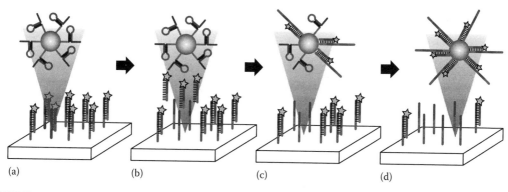

FIGURE 20.16   A sequence of the first-layer reaction on a substrate.

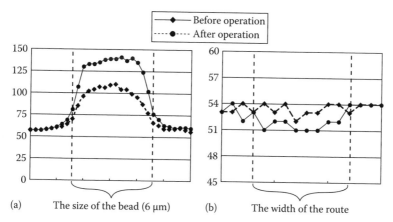

FIGURE 20.17   Fluorescence intensities of (a) the DNA cluster and (b) the track of the bead transportation on the substrate.

20.6 Summary

Photonic DNA computing, which is a parallel computing scheme using photonics and DNA technologies, is promising to realize nanoscale computing. We developed optical techniques that make it possible to achieve a parallel manipulation of DNA for realizing photonic DNA computing. The experimental results demonstrate that the VCSEL array optical manipulation is applicable to parallel transportation and the stacking of DNA clusters. The reactions of DNA can be controlled within a local volume by irradiating with a laser beam. As an application of these techniques, we proposed an addressing method based on the photonic control of repeated hairpin DNA. Experimental results demonstrate that molecular addressing is correctly executed on beads. This functionality is useful in realizing nano-sized memory using photonics.

The applications of photonic DNA computing foresee a secure image memory and nanoprocessors for dealing with molecular information. We started a study on photonic nanoscale automaton as a fundamental processor.[66,67] The research on photonic DNA computing is now at the initial stage, but we believe that a cooperation of the concepts of optical computing and DNA computing have great potential for evolution in the field of information science.

Acknowledgments

The experiments were performed by Dr. Rui Shogenji, Dr. Naoya Tate, Dr. Fumika Sumiyama, Takashi Kawakami, and Taro Beppu. We thank them for their great contributions to this research. This work was supported by Core Research for Evolutional Science and Technology (CREST) of Japan Science and Technology Agency (JST); the Ministry of Education, Culture, Sports, Science and Technology, Japan; a Grant-in-Aid for Scientific Research (A) 18200022, 2006–2008; and Grant-in-Aid for Young Scientists (B) 17760044, 2005–2006.

References

1. V. M. Shalaev and S. Kawata (eds.), *Nanophotonics with Surface Plasmons*, Amsterdam, the Netherlands: Elsevier Science, 2007.
2. M. Ohtsu, K. Kobayashi, T. Kawazoe, T. Yatsui, and M. Naruse, *Principles of Nanophotonics*, London, U.K.: Chapman & Hall, 2008.
3. S. Nie and S. R. Emory, Probing single molecules and single nanoparticles by surface-enhanced Raman scattering, *Science* 275, 1997, 1102–1106.
4. S. Kawata, H.-B. Sun, T. Tanaka, and K. Takada, Finer features for functional microdevices, *Nature* 412, 2001, 697–698.
5. A. J. Haes and R. P. V. Duyne, A unified view of propagating and localized surface plasmon resonance biosensors, *Anal. Bioanal. Chem.* 379, 2004, 920–930.
6. A. D. McFarland, M. A. Young, J. A. Dieringer, and R. P. V. Duyne, Wavelength-scanned surface-enhanced Raman excitation spectroscopy, *J. Phys. Chem. B* 109, 2005, 11279–11285.
7. M. Mrejen, A. Israel, H. Taha, M. Palchan, and A. Lewis, Near-field characterization of extraordinary optical transmission in sub-wavelength aperture arrays, *Opt. Express* 15, 2007, 9129–9138.
8. L. Eurenius, C. Hagglund, E. Olsson, B. Kasemo, and D. Chakarov, Grating formation by metal-nanoparticle-mediated coupling of light into waveguided modes, *Nat. Photonics* 2, 2008, 360–364.
9. G. Barbastathis, A. Krishnamoorthy, and S. C. Esener, Information photonics: Introduction, *Appl. Opt.* 45, 2006, 6315–6317.
10. J. Tanida and Y. Ichioka, Optical computing, in T. G. Brown, K. Creath, H. Kogelnik, M. A. Kriss, J. Schmit and M. J. Weber (eds.), *The Optics Encyclopedia 3*, Berlin: Wiley-VCH, 2003, pp. 1883–1902.
11. B. Javidi, G. Zhang, and J. Li, Encrypted optical memory using double-random phase encoding, *Appl. Opt.* 36, 1997, 1054–1058.
12. J. Tanida, String data alignment by a spatial coding and moire technique, *Opt. Lett.* 24, 1999, 1681–1683.
13. J. Tanida, T. Kumagai, K. Yamada, S. Miyatake, K. Ishida, T. Morimoto, N. Kondou, D. Miyazaki, and Y. Ichioka, Thin observation module by bound optics (TOMBO): Concept and experimental verification, *Appl. Opt.* 40, 2001, 1806–1813.
14. K. Kagawa, K. Nitta, Y. Ogura, J. Tanida, and Y. Ichioka, Optoelectronic parallel-matching architecture: Architecture description, performance estimation, and prototype demonstration, *Appl. Opt.* 40, 2001, 283–298.
15. O. Matoba and B. Javidi, Secure holographic memory by double-random polarization encryption, *Appl. Opt.* 43, 2004, 2915–2919.
16. H. Yamamoto, Y. Hayasaki, and N. Nishida, Secure information display with limited viewing zone by use of multi-color visual cryptography, *Opt. Express* 12, 2004, 1258–1270.
17. P. E. X. Silveira, G. S. Pati, and K. H. Wagner, Optoelectronic implementation of a 256-channel sonar adaptive-array processor, *Appl. Opt.* 43, 2004, 6421–6439.
18. E. Watanabe and K. Kodate, Implementation of a high-speed face recognition system that uses an optical parallel correlator, *Appl. Opt.* 44, 2005, 666–676.

19. D. Miyazaki, K. Shiba, K. Sotsuka, and K. Matsushita, Volumetric display system based on three-dimensional scanning of inclined optical image, *Opt. Express* 14, 2006, 12760–12769.

20. M. Otaka, H. Yamamoto, and Y. Hayasaki, Manually operated low-coherence interferometer for optical information hiding, *Opt. Express* 14, 2006, 9421–9429.

21. T. Nomura and B. Javidi, Object recognition by use of polarimetric phase-shifting digital holography, *Opt. Lett.* 32, 2007, 2146–2148.

22. D. Song, H. Zhang, P. Wen, M. Gross, and S. Esener, Misalignment correction for optical interconnects using vertical cavity semiconductor optical amplifiers, *Appl. Opt.* 46, 2007, 5168–5175.

23. T. Kubota, K. Komai, M. Yamagiwa, and Y. Awatsuji, Moving picture recording and observation of three-dimensional image of femtosecond light pulse propagation, *Opt. Express* 15, 2007, 14348–14354.

24. N. Tate, Y. Ogura, and J. Tanida, Photonic implementation of quantum computation algorithm based on spatial coding, *Opt. Rev.* 14, 2007, 260–265.

25. K. Nitta, O. Matoba, and T. Yoshimura, Parallel processing for multiplication modulo by means of phase modulation, *Appl. Opt.* 47, 2008, 611–616.

26. R. Heming, L.-C. Wittig, P. Dannberg, J. Jahns, E.-B. Kley, and M. Gruber, Efficient planar-integrated free-space optical interconnects fabricated by a combination of binary and analog lithography, *J. Lightw. Technol.* 26, 2008, 2136–2141.

27. L. M. Adleman, Molecular computation of solutions to combinatorial problems, *Science* 266, 1994, 1021–1024.

28. M. H. Garzon and H. Yan (eds.), DNA computing, in *13th International Meeting on DNA Computing (DNA13)*, Memphis, TN, June 4–8, 2007, Revised Selected Papers, New York: Springer-Verlag, New York, 2008.

29. E. Winfree, F. Liu, L. A. Wenzler, and N. C. Seeman, Design and self-assembly of two-dimensional DNA crystals, *Nature* 394, 1998, 539–544.

30. K. Sakamoto, H. Gouzu, K. Komiya, D. Kiga, S. Yokoyama, T. Yokomori, and M. Hagiya, Molecular computation by DNA hairpin formation, *Science* 288, 2000, 1223–1226.

31. C. Mao, T. H. Labean, J. H. Reif, and N. C. Seeman, Logical computation using algorithmic self-assembly of DNA triple crossover molecules, *Nature* 407, 2000, 493–496.

32. B. Yurke, A. J. Turberfield, A. P. Mills Jr., F. C. Simmel, and J. L. Neumann, A DNA-fuelled molecular machine made of DNA, *Nature* 406, 2000, 605–608.

33. N. C. Seeman, DNA in a material world, *Nature* 421, 2003, 427–431.

34. H. Yan, S. H. Park, G. Finkelstein, J. H. Reif, and T. H. LaBean, DNA-templated self-assembly of protein arrays and highly conductive nanowires, *Science* 301, 2003, 1882–1884.

35. Y. Chen, M. Wang, and C. Mao, An autonomous DNA nanomotor powered by DNA enzyme, *Angew. Chem. Int. Ed.* 43, 2004, 3354–3357.

36. J.-S. Shin and N. A. Pierce, A synthetic DNA walker for molecular transport, *J. Am. Chem. Soc.* 126, 2004, 10834–10835.

37. J. Shin and N. A. Pierce, Rewritable memory by controllable nanopatterning of DNA, *Nano Lett.* 4, 2004, 905–909.

38. Y. Benenson, B. Gil, U. Ben-Dor, R. Adar, and E. Shapiro, An autonomous molecular computer for logical control of gene expression, *Nature* 429, 2004, 423–429.

39. A. Kameda, M. Yamamoto, H. Uejima, M. Hagiya, K. Sakamoto, and A. Ohuchi, Hairpin-based state machine and conformational addressing: Design and experiment, *Nat. Comput.* 4, 2005, 103–126.

40. H. Lederman, J. Macdonald, D. Stefanovic, and M. N. Stojanovic, Deoxyribozyme-based three-input logic gates and construction of a molecular full adder, *Biochemistry* 45, 2006, 1194–1199.

41. K. Komiya, K. Sakamoto, A. Kameda, M. Yamamoto, A. Ohuchi, D. Kiga, S. Yokoyama, and M. Hagiya, DNA polymerase programmed with a hairpin DNA incorporates a multiple-instruction architecture into molecular computing, *Biosystems* 83, 2006, 18–25.

42. P. W. K. Rothemund, Folding DNA to create nanoscale shapes and patterns, *Nature* 440, 2006, 297–302.

43. J. Bath and A. J. Turberfield, DNA nanomachines, *Nat. Nanotechnol.* 2, 2007, 275–284.

44. M. Nowacki, V. Vijayan, Y. Zhou, K. Schotanus, T. G. Doak, and L. F. Landweber, RNA-mediated epigenetic programming of a genome-rearrangement pathway, *Nature* 451, 2008, 153–158.

45. P. Yin, H. M. T. Choi, C. R. Calvert, and N. A. Pierce, Programming biomolecular self-assembly pathways, *Nature* 451, 2008, 318–322.

46. J. Tanida, Y. Ogura, and S. Saito, Photonic DNA computing: Concept and implementation, *Proc. SPIE* 6027, 2006, 602724.

47. Y. Ogura, T. Beppu, F. Sumiyama, and J. Tanida, Translation of DNA molecules based on optical control of DNA reactions for photonic DNA computing, *Proc. SPIE* 5931, 2005, 110–120.

48. Y. Ogura, T. Beppu, M. Takinoue, A. Suyama, and J. Tanida, Control of DNA molecules on a microscopic bead using optical techniques for photonic DNA memory, *Lect. Notes Comput. Sci.* 3892, 2006, 213–223.

49. Y. Ogura, T. Beppu, R. Shogenji, and J. Tanida, Photonic translation of DNAs between microscopic beads and a substrate for a photonic DNA memory, *Proc. SPIE* 6327, 2006, 63270V.

50. Y. Ogura, R. Shogenji, S. Saito, and J. Tanida, Evaluation of fundamental characteristics of information systems based on photonic DNA computing, *Lect. Notes Comput. Sci.* 3853, 2006, 192–205.

51. Y. Ogura, T. Nishimura, and J. Tanida, State-transition of DNA nanomachines based on photonic control, *Proc. SPIE* 7039, 2008, 70390K.

52. A. Ashkin, J. M. Dziedzic, J. E. Bjorkholm, and S. Chu, Observation of a single-beam gradient force optical trap for dielectric particles, *Opt. Lett.* 11, 2006, 288–290.

53. S. C. Kuo and M. P. Sheetz, Force of single kinesin molecules measured with optical tweezers, *Science* 260, 1993, 232–234.

54. Y. Arai, R. Yasuda, K. Akashi, Y. Harada, H. Miyata, K. Kinosita Jr., and H. Itoh, Tying a molecular knot with optical tweezers, *Nature* 399, 1999, 446–448.

55. Y. Taniguchi, M. Nishiyama, Y. Ishii, and T. Yanagida, Entropy rectifies the Brownian steps of kinesin, *Nat. Chem. Biol.* 1, 2005, 342–347.

56. U. F. Keyser, B. N. Koeleman, S. V. Dorp, D. Krapf, R. M. M. Smeets, S. G. Lemay, N. H. Dekker, and C. Dekker, Direct force measurements on DNA in a solid-state nanopore, *Nat. Phys.* 2, 2006, 473–477.

57. P. T. Korda, M. B. Taylor, and D. G. Grier, Kinetically locked-in colloidal transport in an array of optical tweezers, *Phys. Rev. Lett.* 89, 2002, 128301.

58. J. Plewa, E. Tanner, D. M. Muech, and D. G. Grier, Processing carbon nanotubes with holographic optical tweezers, *Opt. Express* 12, 2004, 1978–1981.

59. Y. Ogura, K. Kagawa, and J. Tanida, Optical manipulation of microscopic objects by means of vertical-cavity surface-emitting laser array sources, *Appl. Opt.* 40, 2001, 5430–5435.

60. Y. Ogura, N. Shirai, and J. Tanida, Optical levitation and translation of a microscopic particle by use of multiple beams generated by vertical-cavity surface-emitting laser array sources, *Appl. Opt.* 41, 2002, 5645–5654.

61. F. Sumiyama, Y. Ogura, and J. Tanida, Stacking and translation of microscopic particles by means of 2×2 beams emitted from vertical-cavity surface-emitting laser array, *Appl. Phys. Lett.* 82, 2003, 2969–2971.

62. Y. Ogura, T. Kawakami, F. Sumiyama, A. Suyama, and J. Tanida, Parallel translation of DNA clusters by VCSEL array trapping and temperature control with laser illumination, *Lect. Notes Comput. Sci.* 2943, 2004, 10–18.

63. K. Iga, Surface-emitting laser—Its birth and generation of new optoelectronics field, *IEEE J. Sel. Top. Quantum Electron.* 6, 2000, 1201–1215.

64. Y. Ogura, T. Kawakami, F. Sumiyama, S. Irie, A. Suyama, and J. Tanida, Methods for manipulating DNA molecules in a micrometer scale using optical techniques, *Lect. Notes Comput. Sci.* 3384, 2005, 258–267.

65. N. Tate, Y. Horiguchi, Y. Ogura, M. Hagiya, and J. Tanida, Hierarchic molecular addressing for photonic DNA memory, in M. Garzon and H. Yan (eds.), in *Preliminary Proceedings of the 13th International Meeting on DNA Computing*, Memphis, TN, 2007, p. 372.

66. Y. Ogura, T. Nishimura, Y. Horiguchi, and J. Tanida, Concept and primal implementation of photonic nanoscale automaton, in *Technical Digest of International Meeting on Information Photonics 2008*, Hyogo, Japan, 2008, pp. 32–33.

67. H. Sakai, Y. Ogura, and J. Tanida, An implementation of a nanoscale automaton using DNA conformation controlled by optical signals, in *Technical Digest of International Meeting on Information Photonics 2008*, Hyogo, Japan, 2008, pp. 172–173.

Technologies for Emerging Applications: Sensing Applications

21

Evanescent Fiber Bragg Grating Biosensors

Mario Dagenais
Christopher
J. Stanford

21.1 Introduction

Optical fiber sensors [1–5] are devices for which a physical, chemical, biological, or other measurand alters the properties of the light guided in an optical fiber (an intrinsic sensor) or guided to and from an interaction region (extrinsic sensor) by an optical fiber and leads to the generation of an optical signal related to the measurand. This optical signal then ultimately generates a current in a detector. The optical information is conveyed by a change of phase, polarization, frequency, intensity, or a combination thereof. The operation of sensing using phase, polarization, or frequency modulation involves interferometric or grating-based signal processing. The intensity-based sensor is the simpler sensor but is much less sensitive than the interferometric sensor. The relevant measurands can be temperature, stress, strain, pressure, vibration, position, humidity, viscosity, rotation, acceleration, electrical currents, electric and magnetic fields, and bio-attachment molecules. Some attractive reasons for using optical fiber sensors include small sizes, light weight, immunity to electromagnetic interference (EMI), high-temperature performance, environmental ruggedness, large bandwidth, distributive sensing, easy space, time, frequency, and polarization multiplexing. They can also operate over very long distances. Optical fiber sensors are chemically and biologically inert since the basic sensor material, silica, is unaffected by most chemical and many biological agents. Even though strain, temperature, and pressure have been the most widely studied measurands, applications of fiber sensors in biology have been gaining traction in the last several years and several new applications of these sensors have emerged. Fiber Bragg grating (FBG) sensors are sensors based on an inscribed grating at the core of a fiber. As we will see, they can be seen as stable interferometric sensors. Over the last 10 years, because of their ease of fabrication, use, and good sensitivity, they have become one of the most popular topics in optical fiber sensors.

In this chapter, we focus on FBGs and on their applications in bio-sensing. We succinctly present the theory of FBGs, discuss their fabrication and their properties, and discuss their use and how to increase their sensitivity and their selectivity for biological applications. Finally, we look at applications of FBG sensors in biology.

21.2 Evanescent Fiber Bragg Grating Sensors

In 1978, Hill et al. [6] observed the formation of a photogenerated grating in a germanosilicate fiber following the interference of an argon-ion laser with its reflection coming from the end of the fiber. Even though this was an important realization, it remained an oddity since the photogenerated gratings could only operate at a wavelength near 488 nm, the wavelength of the writing beam, therefore drastically limiting the applications. It was not until 1989 [7] that it was realized that Bragg gratings could be inscribed directly in the fiber core from the side using a holographic interferometer illuminated with a coherent UV source operating at 244 nm. This technique was possible because the fiber cladding is transparent to the ultraviolet light whereas the fiber core is highly absorbent to the ultraviolet light. The other big achievement was that the Bragg wavelength could be chosen independently from the laser writing wavelength. Subsequent to this work, considerable interest was generated in FBGs. They became key passive elements in optical fiber telecommunications where they started being used in wavelength-division-multiplexing, fiber laser and amplifier pump reflectors, gain flattening, and dispersion compensation devices. They also started being extensively used in sensing applications for strain, temperature, and pressure measurements, among others. Since the information is usually obtained by detecting a wavelength shift induced by the measurand, the measurement became insensitive to intensity fluctuations. One of the key applications of FBG sensors has become the "fiber-optic smart structures," where FBGs are imbedded into the structure to monitor the strain distribution.

A strong index of refraction change is created when a germanium-doped fiber (germanosilicate fiber) is exposed to UV light close to the absorption peak of a germania-related defect at a wavelength between 240 and 250 nm [8]. A sharp threshold for the generation of a strong grating was observed when the fluence exceeds 450–500 mJ/cm^2. Below this threshold, the index modulation grows linearly with energy density, whereas above the threshold, the index modulation grows non-linearly and then saturates. The gratings generated below the threshold are usually referred to as type I gratings and the ones produced above the threshold are referred to as type II gratings. Damage at the core-cladding interface is usually observed in type II gratings. Very large indices of refraction changes can also be observed in germanium-boron co-doped fibers. Hydrogen loading and flame brushing can be used to further enhance the photosensitivity of germanosilicate fiber. Hydrogen loading is carried out by diffusing hydrogen molecules into the fiber core at high temperatures (several 100's °C) and high pressures (of the order of 1 atm). The increased photosensitivity in the fiber due to flame-brushing is permanent, as opposed to hydrogen-loading where the fiber loses its photosensitivity as the hydrogen diffuses out of the fiber. Another approach of producing strong Bragg gratings is based on exposure with 193 nm ArF excimer laser, without the need for H$_2$ loading.

Three techniques exist for writing gratings: the interferometric technique, the phase mask technique, and the point-by-point technique. In the interferometric technique, the UV beam is split into two beams and then recombine to form an interference pattern. This interference pattern is used to expose a photosensitive fiber and creating a refractive index modulation in the fiber core. The Bragg grating period, Λ, is identical to the period of the interference fringe pattern and is given by

$$\Lambda = \frac{\lambda_w}{2 \sin \varphi},$$
(21.1)

where
    λ_w is the UV wavelength
    φ is the half-angle of the interfering beams

The Bragg condition based on momentum conservation is given by $\lambda_B = 2n\Lambda$, where n is the core effective index. The Bragg grating wavelength can be varied by changing λ_w or by changing the half-angle φ between the intersecting beams. This method offers complete flexibility for producing gratings of different lengths and at different wavelengths. By using curved reflectors, chirped gratings can also be produced. The temporal coherence of the laser has to be such that it is coherent over the length of the grating so that the two beams can interfere with a good contrast ratio. One of the most effective techniques for writing gratings is the phase-mask technique. It uses a diffractive optical element, a phase mask, to spatially modulate the writing beam. The phase mask is a one-dimensional surface-relief structure written in a high quality fused silica substrate transparent to the UV beam. The grating is designed so as to suppress the zero-order diffracted beam to less than a few percent of the transmitted power. This is done by requiring that the grating groove height d is chosen to be

$$d = \frac{\lambda_w}{2(n_w - 1)}.$$ (21.2)

The two diffracted orders (±1 orders) with approximately 40% of the total power interfere and generate the grating (see Figure 21.1). If the period of the phase mask is Λ_{mask}, the period of the photoimprinted index grating is $\Lambda_{mask}/2$. In comparison to the holographic technique, the phase mask approach permits easier alignment of the fiber with reduced stability requirements and allows lower coherence requirements on the laser. A drawback of the phase mask technique is that a separate phase mask is required for each Bragg wavelength. The phase-mask technique also allows index modulation along the fiber length leading to apodization and suppression of side lobes in transmission. Furthermore, it is possible to linearly change the period of the grating leading to a chirped grating. KrF excimer lasers are the most common UV sources used for making Bragg gratings with a phase mask. The low spatial coherence of the laser requires the fiber to be placed in close proximity to the phase mask. A third method for making gratings is based on a point-by-point writing technique. This method is not very efficient but it allows the making of coarse gratings with pitches of order 100 μm that are used for mode converters.

We will now describe the spectral characteristics of a FBG. If the index change is taken to be proportional to the intensity of the writing beam, the grating produced by the interference of two beams has an index profile perturbation that is written in the general case as

$$\delta n(z) = \delta n_0(z) \left[1 + m \cos\left(2\pi \frac{z}{\Lambda_B} \right) \right]$$ (21.3)

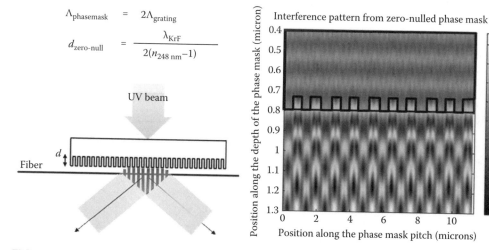

FIGURE 21.1 FBG writing using a phase mask.

where

$\delta n_0(z)$ represents the average index change along the optical axis

$(\delta n_0(z)m)$ is the envelope of the grating modulation

The resulting modal index n_{eff} usually varies along the grating length. The visibility of the UV fringe pattern is given by the parameter m. If the chirping introduced by $\delta n_0(z)$ is neglected by taking it to be constant, then the grating reflectivity is given by [5,7]

$$R(l,\lambda) = \frac{\kappa^2 \sinh^2(sL)}{\Delta k^2 \sinh^2(sL) + s^2 \cosh^2(sL)}, \tag{21.4}$$

where

L is the grating length

κ is the coupling coefficient

$\Delta k = k - \pi/\lambda$ is the detuning wavevector

$k = 2\pi \dfrac{n_0}{\lambda}$ is the propagation constant

$s = \sqrt{\kappa^2 + \Delta k^2}$

For a sinusoidal variation of the index along the fiber axis, the coupling coefficient κ is given by

$$\kappa = \frac{\pi \Delta n \eta(V)}{\lambda}, \tag{21.5}$$

where $\eta(V)$ is the fraction of the power that is contained in the core and depends on the V-parameter of the fiber. An approximate expression for the full width at half-maximum bandwidth of a grating is given by [8]

$$\Delta\lambda = \lambda_B \alpha \sqrt{\left(\frac{\Delta n}{2n_0}\right)^2 + \left(\frac{1}{N}\right)^2}, \tag{21.6}$$

where

N is the number of grating planes

α is ~1 for strong gratings and α is ~0.5 for weak gratings

Counter-propagating waves that interact with a periodic index modulation may be generally expressed in terms of a complex amplitude and phase:

$$E_F(z,t) = F(z)e^{i(\omega t - \beta z)}, \tag{21.7}$$

$$E_B(z,t) = B(z)e^{i(\omega t + \beta z)}. \tag{21.8}$$

Coupled wave equations govern how the field amplitudes behave when light encounters a periodic modulation [8]:

$$\frac{dF(z)}{dz} = i\kappa B(x)e^{-2i\Delta\beta z}, \tag{21.9}$$

$$\frac{dB(z)}{dz} = -i\kappa^* F(x)e^{2i\Delta\beta z}. \tag{21.10}$$

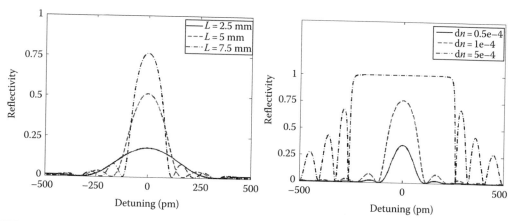

FIGURE 21.2   Reflectivity of grating when the grating length or index step is changed.

FBG spectra may also be simulated utilizing the transfer matrix method, where transmission and reflection matrices can be defined for uniform sections along the grating region. From the coupled mode equations, we may express a transfer matrix as follows [6,8]:

$$\begin{bmatrix} F_0 \\ B_0 \end{bmatrix} = \begin{bmatrix} T_{11} & T_{12} \\ T_{21} & T_{22} \end{bmatrix} \begin{bmatrix} F_L \\ B_L \end{bmatrix}$$

(21.11)

where F_0 and B_0 are the normalized field amplitudes of the forward and backward traveling waves, respectively, at the beginning of the grating. F_L and B_L are the normalized field amplitudes at the end of the grating. F_L and B_0 represent the grating transmission and the grating reflection as a function of the detuning from the center Bragg wavelength. The elements of the transfer matrix are given by

$$\begin{bmatrix} T_{11} & T_{12} \\ T_{21} & T_{22} \end{bmatrix} = \begin{bmatrix} \dfrac{\Delta k \sinh(\kappa L) + i\kappa \cosh(\kappa L)}{i\kappa} e^{-i\beta_0 L} & \dfrac{\kappa \sinh(\kappa L)}{i\kappa} e^{-i\beta_0 L} \\ -\dfrac{\kappa \sinh(\kappa L)}{i\kappa} e^{i\beta_0 L} & \dfrac{-\Delta k \sinh(\kappa L) + i\kappa \cosh(\kappa L)}{i\kappa} e^{i\beta_0 L} \end{bmatrix}$$

(21.12)

Implementing the transfer matrix method in a MATLAB® routine provides insight into the spectral features of practical gratings. Gratings with gradual changes in the grating pitch, grating strength, and uniformity may be simulated and analyzed with the transfer matrix method. Figure 21.2 shows the reflectivity of several gratings when the grating length, L, or the index step is varied.

21.3 Fiber Bragg Grating Sensitivity

FBGs can be very sensitive sensors provided that they are optimized for a particular measurand. For detecting biological species on the surface of the fiber, one needs to etch part of the diameter of the fiber so that the evanescent propagating field on the outside of the fiber can probe the deposited molecules at the surface of the fiber and/or the surrounding liquid. We will now study the sensitivity of the fiber to a change of temperature, strain, and the surrounding index of refraction. The Bragg wavelength ($\lambda_B = 2n_{eff}\Lambda$) will vary according to changes in temperature, length, and outside surrounding index of refraction according to

$$\Delta\lambda_{\text{Bragg}} = 2\left(\Lambda_g \frac{\partial n_{\text{eff}}}{\partial T} + n_{\text{eff}} \frac{\partial \Lambda_g}{\partial T}\right)\Delta T$$

$$+2\left(\Lambda_g \frac{\partial n_{\text{eff}}}{\partial L} + n_{\text{eff}} \frac{\partial \Lambda_g}{\partial L}\right)\Delta L$$

$$+2\left(\Lambda_g \frac{\partial n_{\text{eff}}}{\partial n_{\text{surr}}}\right)\Delta n_{\text{surr}} \tag{21.13}$$

The first term is related to the temperature sensitivity of the fiber due to the thermo-optic effect. The thermo-optic effect represents the change in the refractive index of a material due to the change in temperature. Different materials may exhibit positive or negative thermo-optic constants and some materials (e.g., silica) exhibit both, depending on dopant type and relative concentration. Ferreira et al. and others quote the value of typical germanosilicate fibers as having a thermo-optic constant of 8.5×10^{-6} [9]. The thermo-optic constant of plain germanosilicate is greater than that of the boron co-doped fibers. The concentration of boron and germanium within the core and cladding will determine the properties in each region. The second term is related to the thermal expansion of the fiber. Thermal expansion of the glass fiber also causes shifts in the Bragg wavelength. Most materials expand when the temperature increases; however, there are some exotic materials that contract. Silica, for example, has a thermal expansion constant of 0.5×10^{-6}. This means that every unit of length along the fiber will increase by the same fraction of that length, including between the high and low index regions within the grating. If the fiber expands with an increase in temperature, then the grating pitch increases and the Bragg wavelength redshifts. In terms of temperature dependence, the thermal expansion accounts for only 5.5% of the total effect. Silica is known to have a particularly low constant of thermal expansion (CTE), as compared to common metals and plastics, which expand and contract 20–100 times more with temperature fluctuations. The Bragg wavelength will shift a total of 10 pm/°C at 1550 nm for typical germanosilicate fibers. The third term is related to the stress-optic effect and the Pockels effect and leads to a change of the index of refraction as the fiber expands. The strain-optic tensor relates the refractive index shifts along different crystal directions with an applied strain in each or a combination of directions. For cylindrical fibers, the Pockels coefficients p_{11} and p_{12} are the only relevant tensor components, describing index shifts of two transverse polarizations for tension applied along the z-axis of the fiber. The photoelastic coefficient of the fiber can be written as

$$P = \frac{n^2}{2}\left[p_{12} - \upsilon \times (p_{11} - p_{12})\right], \tag{21.14}$$

where υ is the Poisson's ratio. The fourth term is related to the change of the Bragg wavelength due to strain. The net effect of strain on the fiber is reduced from that expected from elongation alone. Standard experiments have shown that the multiplicative factor M is between 71% and 78% of the expected shift [10]:

$$\Delta\lambda_{\text{Bragg,strain}} = 2\left(n_{\text{eff}} \frac{\partial \Lambda_g}{\partial L}\right)(1 - P)\Delta L, \tag{21.15}$$

$$\Delta\lambda_{\text{Bragg,strain}} = 2 \times M \times \left(n_{\text{eff}} \frac{\partial \Lambda_g}{\partial L}\right)\Delta L. \tag{21.16}$$

The last term is related to the sensitivity of the Bragg wavelength to the index of refraction of the liquid surrounding the fiber.

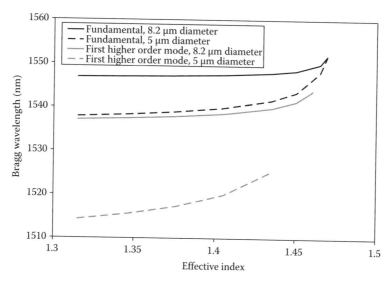

FIGURE 21.3 Plot of the Bragg wavelength versus effective index for fundamental and first higher order modes.

We have modeled the theoretical shifts in effective index with surrounding index changes. As Bragg shifts respond linearly to the change in effective index, the Bragg sensitivity follows according to the sensitivity of the effective index to the change in surrounding solution. Thus, it has been shown that the Bragg wavelength shifts most with an etched fiber where the surrounding index is near to the core index. Furthermore, we may take the limit that the modal confinement goes to 100% in the limit that

$n_{surrounding} \rightarrow n_{core}$. As this occurs, $\frac{\partial n_{eff}}{\partial n_{surr}} \rightarrow 1$, and the ultimate wavelength sensitivity is

$$S = \frac{\Delta\lambda_{Bragg,RI}}{\Delta n_{surrounding}} = 2\Lambda_{grating}\Big/ {riu} \approx 1063 \text{ nm/riu} \qquad (21.17)$$

where "riu" represents the "refractive index unit." While practical limitations of handling and measuring make this sensitivity unrealistic, etched fibers with diameters of 3–5 μm have been shown by us to be very sensitive. We have plotted the Bragg wavelength for various surrounding indices, given that the fiber is a standard etched Corning SMF-28.

The slope varies and represents the Bragg sensitivity to surrounding index (Figure 21.3). For lower indices, higher order modes display greater Bragg sensitivity to index shifts than the fundamental modes at any given fiber diameter. However, as the surrounding index approaches the core index, the most sensitive mode is the fundamental mode.

21.4 Evanescent Wave Fiber Bragg Gratings

In order to measure small changes in composition, it is important for the optical mode to penetrate evanescently into the surrounding solution. Sensors based upon long-period grating-assisted coupling to cladding modes of a fiber have been presented [11]. A method of increased sensitivity to surrounding index by etching the fiber close to the core diameter was presented by Asseh et al. [12]. A sensitivity of 2.66 nm/riu was achieved with the diameter of the fiber etched to 11 μm. The sensitivity was increased to 7.3 nm/riu by etching the fiber to a diameter of 8.3 μm in [13]. Chen et al. demonstrated a sensitivity of 172 nm/riu for the third order mode [14]. Schroeder et al. [15] proposed a method of side-polishing the fiber and achieved a sensitivity of 340 nm/riu. A sensor with an index sensitivity of 10^{-5} was demonstrated by Iadicicco et al. [16,17]. In a recent paper by our group we have demonstrated that the sensitivity can be appreciably increased by etching single mode fibers to diameters as small as 3.4 µm and having the surrounding index close to that of the fiber core [18]. A maximum sensitivity of 1394 nm/riu was demonstrated, a bit higher than our theoretical prediction (see Equation 21.17). Recent work on microstructure gratings, where different parts of a grating are etched to different thicknesses, was reported by Cusano et al. [19]. Also, the field of FBG evanescent wave sensors was recently reviewed by Cusano et al. [20].

The basic approach commonly used in FBG-based sensors is to monitor the shift in wavelength of the returned "Bragg" signal with changes in the measurand. The sensitivity of the sensor depends upon the change in the effective index for the waveguided mode, which is related to the change in the refractive index of the measured solution. For obtaining high sensitivity, it is advisable to etch away the cladding of a single mode fiber and even to etch away part of the core region. The well-known approach for solving for the guided modes of a step-index multi-mode fiber is used to extract the effective index of the modes as a function of the fiber core radius [21]. This is done by solving the Helmholtz equation for the applicable boundary conditions. We have developed a simple graphical solution to determine the effective index of the fiber in the presence of a grating. The Bragg wavelength of a grating is given by $\lambda = 2n_{\text{eff}}\Lambda$, where Λ is the grating pitch, and n_{eff} is the effective index of the propagating mode. As the core of the fiber is etched, the modal effective index changes and the reflected Bragg wavelength shifts. In Figure 21.4, we show the well-known graph representing the normalized effective index b of a mode plotted versus the normalized frequency V for a fiber of core index n_1 and cladding index n_2. b and V are given by

$$b = \frac{n_{\text{eff}}^2 - n_2^2}{n_1^2 - n_2^2} = \frac{\left(\dfrac{\beta}{k}\right)^2 - n_2^2}{n_1^2 - n_2^2} \qquad (21.18)$$

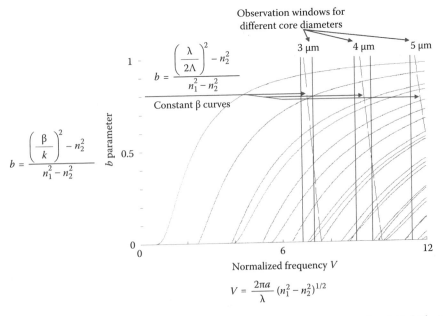

FIGURE 21.4   Normalized propagation versus V of a fiber. Constant grating curves are also plotted. The intersection gives the Bragg wavelength of the grating.

$$V = \frac{2\pi a}{\lambda}(n_1^2 - n_2^2)^{1/2} \tag{21.19}$$

where a is the diameter of the fiber. On the other hand, on the same graph, we can also plot the equation describing b vs. V for the grating. The solution for the normalized index of the different modes will be given by the intersections of the two curves. For the grating, we can write

$$b = \frac{\left(\dfrac{\lambda}{2\Lambda}\right)^2 - n_2^2}{n_1^2 - n_2^2} = \frac{\left(\dfrac{\pi a}{V\Lambda}\right)^2 (n_1^2 - n_2^2) - n_2^2}{n_1^2 - n_2^2} \tag{21.20}$$

The intersections of the curve representing the normalized index of the fiber with the curve representing the normalized index for the grating are the solutions and determine the n_{eff} and consequently the Bragg wavelength. We have plotted the solution for different fiber diameters on the same graph in Figure 21.4. The window regions between 1.5 and 1.6 μm are shown by the two vertical lines associated with different fiber diameters. As the core of the fiber is etched, the Bragg wavelength decreases for the same surrounding index. This property can be used to *in situ* monitor the diameter of the core during the etching process, and very good control of the diameter of the fiber can be achieved. Also the Bragg wavelength changes faster for a same change of the surrounding index for fiber gratings with reduced core, resulting in increased sensitivity for the sensor. This is because the evanescent field penetrates further into the surrounding medium for small diameter fibers. For a constant diameter, the sensitivity also increases as the index of the surrounding medium approaches that of the core of the fiber. In Figure 21.5, we show the b vs. λ dependence for both the fiber propagation modes and the grating, for two different indices of refraction at wavelengths between 1.5 and 1.6 μm. The intersections give the reflected wavelengths for the fundamental and the higher order modes as well. From Figure 21.4, we can see that as we decrease the diameter, the distance between the first- and the second-order solution increases. This results in a faster displacement of the intersection point of the second-order and thus yields higher sensitivity for the higher order modes. In Figure 21.5, a plot of b vs. λ is presented when the surrounding index of refraction is increased (a) and when the grating pitch is increased (b). It can be seen that the normalized index b decreases and the Bragg wavelength moves to a longer wavelength. When the grating pitch is increased, the fiber modes are not altered and the Bragg wavelength moves to a longer wavelength.

An important feature of this theoretical approach is that if we know the index of refraction of the etchant that was used, we can monitor the etching process *in situ* and achieve good accuracy and

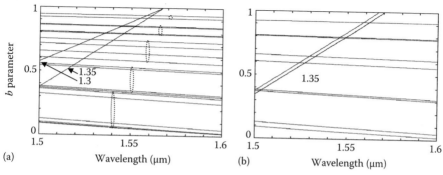

FIGURE 21.5 Graphical solution for the reflected Bragg wavelength. (a) For two different indices of refraction: $n_2 = 1.3$ and 1.35. The circles on the left plot designate the same order modes for the two different index values. (b) For two grating pitches different by 1 nm at $n_2 = 1.35$. We can see that the fiber modes do not change in this case, only the grating curves change.

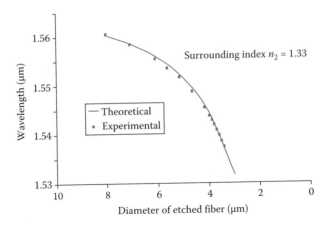

FIGURE 21.6 Shift of the Bragg wavelength as the core of the fiber is etched.

repeatability of the fiber diameter without the use of a microscope and without any interruptions of the process. In Figure 21.6, we present the theoretical and experimental results of the peak shift during etching, which indicates the accuracy of our monitoring technique.

After studying the simulation results for many different cases, we now summarize some of the useful observations we made. As we etch the grating down to the core, the cladding index gradually decreases effectively from 1.4447 to 1.33, which is the estimated etchant index. This big change in the cladding region pushes our fiber into a multimode regime. However, if we continue to etch further to smaller diameters, we can see from Equation 21.19 that there is a critical diameter that will bring our observation window at the cutoff value of $V = 2.405$. Past that point, we restore our single mode operation. Another way would be to immerse our sensor into a higher index liquid. Decreasing index and diameter tend to shift the grating peak to lower wavelengths. The highest sensitivity is achieved when the surrounding medium index is closely matched to the core index. The straight grating lines in the b vs. λ plot shown in Figures 21.4 and 21.5 do not change for varying fiber diameters, as expected from Equation 21.20, but do change for varying indices and grating periods. In Figure 21.5, we have shown how these changes affect our solutions. In the left part of the Figure 21.5, it is clear why the higher order mode sensitivity should be higher than the lower order, while in the right part of the figure, one can see that a change in the grating period, which might be thermally or mechanically induced, does not affect the fiber propagation modes. Therefore in the latter case, the higher order fiber modes, which are almost parallel to the λ-axis, should yield the same amount of peak shift. Therefore we can assume that, when the higher order modes have a wavelength shift proportional to the shift of the fundamental mode, then the shift is due to an index change, while when they shift together by about the same amount as for the fundamental mode, then it implies a change of the period of the grating due to a length change caused by stress or due to a thermal elongation of the grating. We have recently verified these predictions [22] and demonstrated that the sensitivity of the third-order mode is increased by at least a factor of six as compared to that of the fundamental mode for a fiber etched down to a core diameter of 5 μm. These experiments were carried out by first etching down the diameter of a FBG. We used a commercially available JDSU FBG with an unetched Bragg grating wavelength of 1563 nm and a bandwidth of 0.4 nm. For light source, we used a Photonetics broadband erbium-doped fiber amplifier (EDFA) source. An advantest optical spectrum analyzer with a 0.007 nm wavelength resolution was used to monitor the reflected Bragg spectrum. A schematic of the apparatus is shown in Figure 21.7. A J.T. Baker buffered oxide etchant (BOE) 7:1 was used to etch the fiber. A very smooth etched surface is obtained as can be seen from an SEM of an etched fiber (Figure 21.8). The fiber can be etched down to 3.5 μm. However, at such small diameters, we observed an etching-rate dependence on the index of refraction of the FBG, which led to a periodic variation of the fiber diameter and to a broadening of the spectrum and a reduction of the reflected signal. To study the

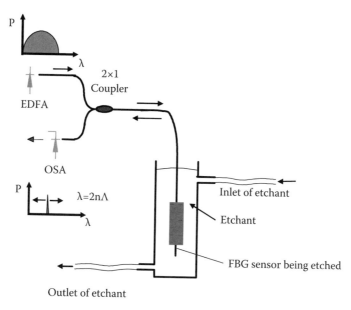

FIGURE 21.7 Experimental setup for *in situ* monitoring of fiber etch.

FIGURE 21.8 Fiber etched to about 6.3 μm diameter. This demonstrates the smoothness of the etching process.

sensitivity of the etched core FBG to a change of the surrounding index of refraction, we etched several fibers to different diameters and we surrounded them with liquids with different indexes of refraction. The sensitivity results are shown in Figures 21.9 and 21.10. When the fiber gratings were etched, a spectral feature appeared in the reflected spectrum of linewidth about 10 pm. This spectral feature was used to monitor the FBG shift for different liquids of different indexes of refraction surrounding the sensor. The center of this spectral feature can be positioned to an accuracy of 1 pm. This leads to an index resolution of 7.2×10^{-7} (see Figure 21.10), an improvement of a factor of 10 over our previously published results [18].

Even though commercial FBGs were used in the previous measurements, we also have the capability for making FBGs in-house. We have used a KrF Lambda PhysiK LPX-250T excimer laser operating at 248 nm to write gratings in Newport's boron co-doped photosensitive fibers. An Ibsen Photonics phase mask with a grating pitch of 1062 nm was used. Fibers are placed on a custom mount that securely holds the fiber in contact mode with the phase mask. Grating writing is monitored in real time with an Agilent tunable laser system. Some of the transmission spectra obtained from strong gratings are shown in Figure 21.11.

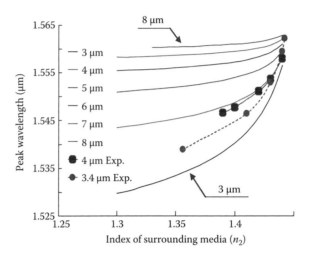

FIGURE 21.9   Change in Bragg wavelength with changing surrounding index for different diameters of fiber.

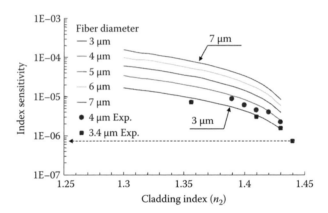

FIGURE 21.10   Measurable change of index for FBGs of different diameters. 0.001 nm resolution of peak wavelength detection is assumed.

21.5 Bio-Sensing with Fiber Bragg Gratings

A biosensor has to be sensitive and selective to the analyte being detected and has to be biocompatible and insensitive to external disturbances such as temperature and pressure. Guided-wave optical biosensors have recently been reviewed [23–27]. A biosensor is a device that consists of a biological sensing element (bio-receptor) integrated with a physical transducer to transform a measurand into an output signal. Different bio-receptors have been used to recognize some biochemical substances, including nucleic acids, antibodies, enzymes, proteins, cells, and tissues. The transduction process of interest to optical biosensors is based on intensity phase and wavelength changes of an optical signal. DNA or RNA hybridization exploits the complementarities of pairs of nucleotides: adenine (A)-thymine (T) and cytosine (C)-guanine (G). Known DNA sequences (probes) can be used to discover specific DNA sequences in biological samples. Antibodies are biomolecules formed by a hundred or so amino acids arranged in a large Y-shaped ordered sequence. The antibody recognizes a specific target, called the antigen. The interaction between an antibody and its antigen is highly specific and the antigen–antibody bond is very

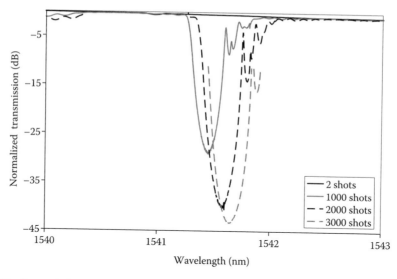

FIGURE 21.11 Transmission spectra during continuous grating writing. The "2 Shots" transmission spectrum is barely noticeable and centered at 1539.95 nm. The rate of Bragg shifting and grating strengthening saturates for longer exposures.

stable. The high specificity of the antigen–antibody reaction allows the antibody to detect a very small amount of antigen in the analyte, even in the presence of a large number of other substances. Antibodies provide immunity against invading pathogens. Enzymes can be used as bio-receptors because of their highly specific binding capabilities and because of their catalytic activity that allows the amplification of the detection process. Different types of sugar molecules can also be made very selective to certain proteins and can be used as detection agent (glycobiology). For instance, Concanavalin A, a lectin protein, can bind selectively to glucose but not to lactose.

For all the bio-experiments, we are looking for a certain biological material in the analyte to bind to an appropriate pre-attached biomaterial on the fiber. The pre-attached biomaterial can be an antibody, a single-strand DNA, or a saccharide molecule depending on whether an antigen–antibody binding, a DNA hybridization, or a protein–sugar recognition experiment is performed. We have therefore developed a model to translate the attachment of biomaterial to the fiber for a material of given thickness and index of refraction and extracted the corresponding shift of the FBG peak [28]. This model also applies to a situation where several biomaterials attach successively to the fiber. In the case where mono or multiple layers are immobilized on the fiber (as in the case of the hybridization of DNA), we first calculate the effective index of the surrounding medium in the presence of an immobilized layer on the surface of the fiber by using a beam-propagation method, as is shown in Figure 21.12. The optical mode of the fiber is calculated for the fiber with the biological/chemical layer added and using a very fine grid (at least 10 points in the immobilized layer) The modal index is then extracted. In order to calculate the effective index of the surrounding medium, the biological/chemical layer is then removed and the mode is calculated again by adding a small perturbation to the index of the surrounding medium till the calculated modal index matches the previously calculated modal index. The change in Bragg wavelength is calculated using the change in effective index in the presence of the biological or chemical layer(s) on the fiber and multiplying it by the sensitivity of the etched-core FBG sensor, which is calculated using the graphical method that we previously described. This approach is schematically represented in Figure 21.12. The Bragg wavelength shift is typically referenced to a solution like water if the buffer solution is water-based or to a solvent if the biological solution is dissolved in a solvent solution. When the sensor is first immersed in a biological solution, there is a shift of the

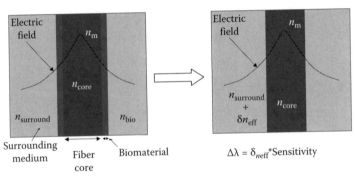

FIGURE 21.12 Schematic of the calculation for the effective index of the surrounding medium in the presence of thin layers of chemical or biological agents on the surface of the fiber.

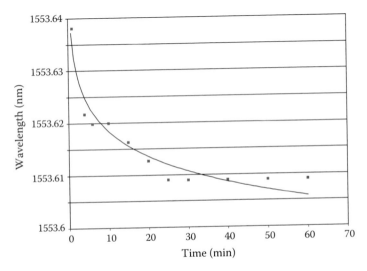

FIGURE 21.13 Real-time monitoring of the hybridization of 20-base probe DNA.

Bragg wavelength due to the change of index going from the water/solvent to the biological solution. Then, there will be a progressive shift of the Bragg wavelength as the biomaterial accumulates on the fiber. At the end of the experiment, when the sensor is put back in the reference water/solvent solution, there will be a shift of the Bragg peak of a very similar size to the initial shift. The final shift of the sensor measured in the reference water/solvent solution is typically related to the amount of biological material that has attached on the fiber. When a water layer forms next to the deposited biomaterial on the fiber, the situation is more complicated and was recently discussed by us in the case of siloxane film formation on silica [29].

DNA hybridization was monitored using an etched-core FBG [30]. A 20-base single-stranded DNA target was used. The peak shift caused by the hybridization of DNA is shown in Figure 21.13. A repeat of the hybridization experiment using non-complementary target DNA led to an expected nonmeasurable wavelength shift.

In situ real-time silanization experiments were performed with 3-Aminopropyltriethoxysilane (APTES) of different concentrations diluted in ethanol [29]. The measurements are shown in Figure 21.14. A shift toward shorter wavelengths was first observed at times shorter than 100 s and it was identified with the formation of a water-layer next to the grating. For longer times, silanization of the fiber surface was observed.

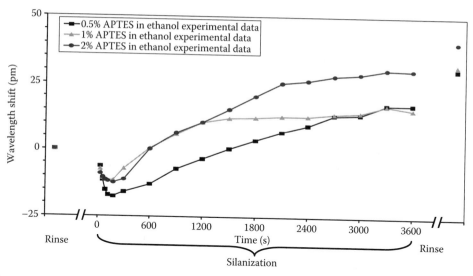

FIGURE 21.14 *In situ* data for silanization with 0.5%, 1%, and 2% APTES concentrations in ethanol.

21.6 Conclusion

Etched core FBG sensors were shown to be very sensitive sensors. Index changes smaller than 1×10^{-6} can be detected with such a sensor. These sensors allow real-time monitoring of the bio-attachment of molecules on the fiber surface. By properly functionalizing the surface of the sensor, it is possible to detect the hybridization of DNA, the attachment of antigens to antibodies, and the detection of glyco-processes involving proteins. Further optimization of the sensor will permit the measurement of even smaller index of refraction changes. We expect that such sensors will find use in medicine, environment monitoring, and in bio-defense.

References

1. B. Culshaw, Optical fiber sensor technologies: Opportunities and—perhaps—pitfalls, *J. Lightw. Technol.* 22, 39 (2004).
2. B. Lee, Review of the present status of optical fiber sensors, *Opt. Fiber Technol.* 9, 57 (2003).
3. A.D. Kersey, M.A. Davis, H.J. Patrick, M. LeBlanc, K.P. Koo, C.G. Askins, M.A. Putnam, and E.J. Friebele, Fiber grating sensors, *J. Lightw. Technol.* 15, 1442 (1997).
4. B. Gholamzadeh and H. Nabovati, Fiber optic sensors, *Proc. World Acad. Sci. Eng. Technol.* 32, 327 (2008).
5. K. Hotate, Fiber sensor technology today, *Jpn. J. Appl. Phys.* 45, 6616 (2006).
6. K.O. Hill and G. Meltz, Fiber Bragg grating technology fundamentals and overview, *J. Lightw. Technol.* 15, 1263 (1997).
7. G. Meltz, W.W. Morey, and W.H. Glenn, Formation of Bragg gratings in optical fibers by a transverse holographic method, *Opt. Lett.* 14, 823 (1989).
8. A. Othonos, Fiber Bragg gratings, *Rev. Sci. Instrum.* 68, 4309 (1997).
9. L.A. Ferreira, F.M. Araujo, J.L. Santos, and F. Farahi, Simultaneous measurement of strain and temperature using interferometrically interrogated fiber Bragg grating sensors, *Opt. Eng.* 39, 2226–2234 (2000).
10. K.O. Hill, Photosensitivity in optical fiber waveguides: Applications to reflection filter fabrication, *Appl. Phys. Lett.* 32, 647 (1978).

11. V. Bhatia and A.M. Vengsarkar, Optical fiber long-period grating sensors, *Opt. Lett.* 9(21), 692–694, 1996.
12. A. Asseh, S. Sandgren, H. Ahlfeldt, B. Sahlgren, R. Stubbe, and G. Edwall, Fiber optical Bragg grating refractometer, *Fiber Integr. Opt.* 17, 51 (1998).
13. N. Chen, B. Yun, and Y. Cui, Cladding mode resonances of eth-eroded fiber Bragg grating for ambient refractive index sensing, *App. Phys. Lett.* 88, 133902 (2006).
14. D.A. Pereira, O. Frazao, and J.L. Santos, Fiber Bragg grating sensing system for simultaneous measurement of salinity and temperature, *Opt. Eng.* 43(2), 299–304 (2004).
15. K. Schroeder, W. Ecke, R. Mueller, R. Willsch, and A. Andreev, A fiber Bragg grating refractometer, *Meas. Sci. Technol.* 12, 757–764 (2001).
16. A. Iadicicco, A. Cusano, A. Cutolo, R. Bernini, and M. Giordano, Thinned fiber Bragg gratings as high sensitivity refractive index sensor, *IEEE Photonics Technol. Lett.* 16, 1149 (2004).
17. A. Iadicicco, A. Cusano, S. Campopiano, A. Cutolo, and M. Giordano, Thinned fiber Bragg gratings as refractive index sensors, *IEEE Sensors J.* 5, 1288 (2005).
18. A.N. Chryssis, S.M. Lee, S.B. Lee, S.S. Saini, and M. Dagenais, High sensitivity evanescent field fiber Bragg grating sensor, *IEEE Photon. Technol. Lett.* 17, 1253 (2005).
19. A. Iadicicco, S. Campopiano, D. Paladino, A. Cutolo, and A. Cusano, Micro-structured fiber Bragg gratings: Optimization of the fabrication process, *Opt. Express* 15, 15011 (2007).
20. A. Cusano, A. Cutolo, and M. Giordano, Fiber Bragg gratings evanescent wave sensors: A view back and recent advancements, in *Sensors: Advancements in Modeling, Design Issues, Fabrication and Practical Applications*, S.C. Mukhopadhyay and Y.-M. Huang, Eds., Springer, Berlin, Heidelberg, Part II, pp. 113–152.
21. J.A. Buck, *Fundamentals of Optical Fibers*, John Wiley & Sons, New York, 2004, 332pp.
22. A.N. Chryssis, S.S. Saini, S.M. Lee, and M. Dagenais, Increased sensitivity and parametric discrimination using higher order modes of etched-core fiber Bragg grating sensors, *IEEE Photonics Technol. Lett.* 18, 178 (2006).
23. V.M.N. Passaro, F. Dell'Olio, B. Casamassima, and F. DeLeonardis, Guided-wave optical biosensors, *Sensors* 7, 508 (2007).
24. C.L. Baird and D.G. Myszka, Current and emerging commercial optical biosensors, *J. Mol. Recognit.* 14, 261 (2001).
25. D.J. Monk and D.R. Walt, Optical fiber-based biosensors, *Anal. Bioanal. Chem.* 379, 931 (2004).
26. X. Fan, I.M. White, H. Zhu, J.D. Suter, and H. Oveys, Overview of novel integrated optical ring resonator bio/chemical sensors, Laser Resonators and Beam Control IX, San Jose, CA, A.V. Kudryashov, A.H. Paxton, and V.S. Ilchenko, Eds., *Proc. SPIE* 6452, 64520M (2007).
27. L.M. Lechuga, Optical biosensors, in *Biosensors and Modern Biospecific Analytical Techniques*, Lo Gorton, Ed., Vol. XLIV, Elsevier, Oxford, U.K., 2005, 635pp.
28. S.S. Saini, C. Stanford, S.M. Lee, J. park, P. DeShong, W.E. Bentley, and M. Dagenais, Monolayer detection of biochemical agents using etched-core fiber Bragg grating sensors, *IEEE Photonics Technol. Lett.* 19, 1341 (2007).
29. C.J. Stanford, M. Dagenais, and P. DeShong, Real-time monitoring of siloxane monolayer film formation on silica using a fiber Bragg grating, *Curr. Anal. Chem.* 4, 356 (2008).
30. A. Chryssis, S.S. Saini, S.M. Lee, H. Yi, W.E. Bentley, and M. Dagenais, Detecting hybridization of DNA by highly sensitive evanescent field etched core fiber Bragg grating sensors, *IEEE J. Sel. Top. Quantum Electron.* 11, 864 (2005).

22

Nano-Injection Photon Detectors for Sensitive, Efficient Infrared Photon Detection and Counting

Hooman Mohseni
Omer G. Memis

22.1 Introduction

A new class of short-wave infrared (SWIR) photon detectors, called the nano-injection photon detectors, has been recently developed to address the main problem of using nanoscale features for light detection: Even though nanoscale sensors offer high sensitivity, their interaction with visible or infrared light is severely limited by their miniscule volume. The nano-injection detectors have highly sensitive nanometer-sized pillars, "nano-injectors," on large, thick absorption layers. These large, thick layers capture incoming light efficiently, and the nano-injectors sense and amplify these signals. Due to the pairing of thick absorbers and nano-injectors, high sensitivity and high efficiency can be simultaneously attained, which can satisfy a growing need for identifying and counting photons in many modern applications.

Measurements on nano-injection detectors clearly reveal their capabilities. Amplification factors reaching 10,000 have already been recorded, together with low dark current densities at room temperature. Noise suppression behavior is observed at amplification factors up to 4000+, which lower the detector noise to values below the theoretical shot noise limit. The devices, when properly surface-treated, show bandwidths exceeding 3 GHz with an impressive time uncertainty (jitter) of 15 ps. Tests over arrays of nano-injection detectors exhibit high uniformity with dark current standard deviation less than 3%. These properties make the nano-injection photon detectors extremely suitable for demanding imaging applications, which require high efficiency, high sensitivity, and high uniformity, in addition to many applications such as nondestructive material inspection, high-speed quantum cryptography, or medical optical imaging.

22.2 Single Photon Detectors: Importance and Applications

Many modern applications, such as medical instruments, imaging systems, and telecommunication devices, have demanding requirements to achieve state-of-the-art performance. However, in many cases, the performance of the system suffers from the limitations imposed by the photodetector. Improving the sensitivity of photodetectors is of the utmost importance in these systems, and the ultimate target for researchers is the detection of a single quantum of light, the photon [1]. It represents a very small amount of energy, less than 10^{-18} J in the infrared spectrum, so detectors specialized for individual photon detection need to have very low noise levels. These detectors are called single photon detectors (SPDs).

An ideal SPD can be used in almost every application that utilizes a regular photodetector. The improved signal-to-noise ratio (SNR) that SPDs present would increase the performance of the system, resulting in deeper medical noninvasive probing, longer range fiber links, more accurate nondestructive parts inspection for aircraft safety, less error-prone satellite communications, or clearer and higher resolution night vision images for military. All existing applications, whether in military, commercial, environmental, or medical, will benefit tremendously.

In parallel to improving existing applications, SPDs are quickly becoming the enabling technology for many emerging applications, which were previously thought impossible. One example of these technologies is quantum computing [2], in which bits can assume many states and perform calculations on these multiple states at once [3]. They promise many orders of improvement in calculation speed and efficiency and can potentially break down all existing barriers of traditional computation.

Quantum cryptography [4] relies on the quantum theory to provide secure communications. The current public–private key encryption schemes rely on the mathematical infeasibility of some inverse calculations, which will be significantly challenged by the paradigm shift to be introduced by quantum computing [5]. Quantum cryptography, on the other hand, will be immune from the effects of this vast change [6].

Another example is quantum ghost imaging [7], where the scenery in a remote location can be reconstructed from correlating the acquired signals, using reflected entangled photon pairs.

22.3 Current Single Photon Detectors

Toward photon detection in visible and infrared bands, many photodetector technologies have been proposed and developed for decades, and among these, higher sensitive detectors with lower noise levels have been adopted and used for single photon detection. The mainstream SPD technologies are photomultiplier tubes (PMTs), PIN detectors, avalanche photodiodes, and superconducting SPDs (SSPDs).

The PMT is a single photon detection technology that relies on photoelectric effect and secondary emission of electrons. They can operate at wavelengths from 200 to 1700 nm [8] at QE of 10%–25% [9]. They have dark count rates of $10-10^5$/s and a bandwidth of hundreds of megahertz.

PIN detectors are small devices with high-speed response and ultralow jitter [10]. They have extremely low dark current values; however, PIN detectors do not provide internal amplification and are usually paired with sophisticated, very-low-noise electronic preamplifiers.

Avalanche photodetectors (APDs), in contrast, are built around an internal multiplication scheme coupled to the detection mechanism [11]. The two common types of APDs are Si and InGaAs/InP based. Si single photon APDs (SPADs) are very established, very capable detectors that cover a wavelength range of 300–1000 nm with a quantum efficiency of ~70% [12]. They have low dark counts on the order of 100/s and can be operated at a rate of 5 MHz.

In contrast to Si SPADs, InGaAs/InP SPADs do not exhibit such a high performance, primarily due to unfavorable electron-to-hole ionization rates and more defect states compared to Si. InGaAs/InP SPAD operate in 1000–1700 nm with QE of ~20%–30% [13]. They have dark count rates of 10^5–10^6/s and maximum operation rates on the order of 10 MHz [13]. Due to their internal positive feedback in avalanche multiplication, the gain tends to destabilize at higher values and increases amplitude uncertainty (i.e., noise) [14].

SSPDs rely on one of the two mechanisms: transition edge detection [15] or current-assisted hot spot formation [16]. In transition edge detection, a very small superconducting device is kept at just below the critical temperature, where an incoming photon heats the device slightly and causes a transition from superconducting to resistive operation. In current-assisted hot spot formation, the detector consists of lines of superconducting material. When a photon hits the detector, it causes a local hot spot where superconductivity is suppressed. This hot spot forces current to flow around itself, thereby increasing the current density around the hotspot and forcing it to exceed the critical current density, creating larger hotspots until superconductivity is suppressed in the whole width of the strip. In either type of mechanism, the detection is achieved by observing the resistance of the device. SSPDs can detect wavelengths from 200 to 1700 nm with a QE reaching 10% at SWIR [17,18]. They have dark counts on the order of 10^4/s and they can operate at very fast speeds exceeding 1 GHz.

22.4 SWIR Single Photon Imagers

On the imaging side, image intensifiers (I^2) have been developed for many decades. They typically consist of a detector that converts light to electrons, a gain medium that amplifies the electrons, and a phosphorus screen that converts the amplified electrons back to visible photons [19] (Figure 22.1). There have been five generations of I^2 [20]. The very first one, generation 0 active I^2, utilized infrared light sources to illuminate target to compensate for the poor sensitivity of vacuum-packed photon detectors. Generation 1 I^2 introduced the first passive detection systems by using tri-alkali photocathodes, which achieved an amplification of ~1000 but still needed ambient light, i.e., primarily moonlight.

With the development of microchannel plate (MCP) technology, generation 2 I^2 devices were designed. MCPs consist of microscopic (10 μm) conducting channels of hollow glass, which provide significant secondary electron amplification (~20,000×) to the incoming electrons. The generation 2 I^2 also use multi-alkali photocathodes for improved performance. Generation 3 I^2 devices replaced the multi-alkali photocathodes while improving MCPs, and started using GaAs photocathodes instead. This resulted in gain values of 30,000–50,000 and increased the lifetime substantially to 10,000 h. However, the second and third generation devices lacked the sensitivity to operate in SWIR. The spectral response can be seen in Figure 22.2. In generation 4 I^2, one of the targets besides enhancing MCP performance is increasing the sensitive spectrum up to 1.6 μm. InGaAs-based photocathodes were proposed [21] for detection beyond 1.1 μm.

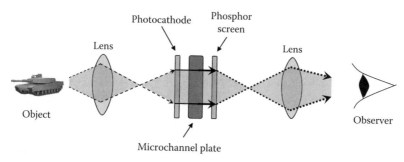

FIGURE 22.1 The schematics of an image intensifier system.

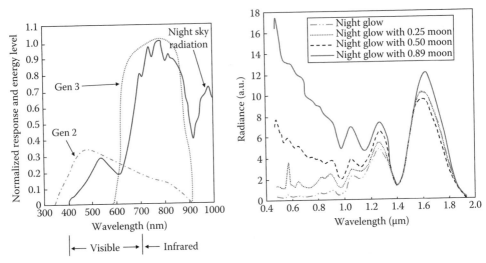

FIGURE 22.2 (Left) The spectral response of second generation and third generation image intensifier systems. (Right) The spectral radiance of night sky, showing significant spectral components in SWIR that is present even in nights without moonlight.

To quantify the noise levels of I² devices, the noise factor is used, which is the ratio of the input SNR to output SNR. The typical noise factor values [22] for I² are 3–5, which corresponds to a current noise of 0.49–0.63 nA/Hz$^{0.5}$ on a 50 Ω system.

To address the wavelength limitations of the above, night vision focal plane arrays (FPA) have been introduced. They have gone through two generations and are currently in their third generation [23]. The first generation FPAs were scanning systems, made of either a single detector or a linear array of detectors such as photoconductive detectors. The detectors were physically scanned to form image, which meant short integration times and the requirement of external multiplexing [23]. Furthermore, common detectors, such as photoconductive HgCdTe detectors, required cooling down to 25 K.

In contrast, the second generation detectors were staring arrays. Instead of scanning 0D or 1D arrays, this generation had full 2D arrays with no mechanical movement [23]. The first versions were photovoltaic HgCdTe arrays with a resolution of 64 × 64, which later improved up to 2000 × 2000 pixels. These arrays had multiplexing onboard, which let them scan electronically and efficiently. They had high fill factors and a lot of signal processing capabilities. They were coupled with advanced readout integrated circuits, with processing options such as antiblooming, subframe imaging, and pixel deselecting [24].

The third generation infrared detectors are also staring systems. They are not as clearly defined as first or second generation, but their primary target is to add more functionality to the detectors in addition to improving existing parameters (e.g., resolution). The definition of third generation detectors by the U.S. Army consists of three groups [25]:

- Multiple color, high performance, large format cooled sensors
- Medium to high performance uncooled imagers
- Very low cost expandable uncooled imagers

Besides these groups, there is other functionality of interest [26,27]. One would be biologically inspired systems with integrated temporal and spatial processing, curved sensors for less aberrations, or wide field of view. Another would be hyperspectral imaging, where an imager can simultaneously see different bands (or colors). On-FPA processing is also a desirable functionality with capabilities such as motion detection or edge enhancement.

22.5 A Biological Single Photon Detector: The Human Eye

As researchers have found almost half a century ago, the human eye is surprisingly sensitive [28] and capable of detecting a few photons with orders of magnitude higher SNR compared to existing SPDs [29]. To see the reasons for such sensitivity, the detection mechanism was studied at the molecular level and some essential concepts were established for ultralow light level detection (Figure 22.3). First, it was identified that the specialized geometry of a rod cell and the arrangement of photosensitive rhodopsin (Rh) molecules in the cell provide large absorption volumes that increase the chance of absorption. Another important property is the inherent gain mechanism of the rod cells, triggered by Rh molecules. A low-noise gain mechanism in the detection stage increases the overall SNR, as it makes the signal less susceptive to other biological, acoustic, or electronic noise sources present after the initial detection.

As illustrated in Figure 22.4, the rod cell consists of two main sections [29]. The small section, or the inner segment, is responsible for the cell's metabolism and other functions. The large section, on the other hand, mainly consists of structures rich with the photosensitive molecule Rh. This section is specialized toward the detection of incoming light. The rest of the cell structure includes the axon and dendrites, which are responsible for receiving and transmitting electrical pulses to the neighboring cells.

Rod cells that are responsible for the high sensitivity of human visual cells are made of inner and outer segments, among which a steady current flows due to Na^+ and K^+ transport in dark. This flow is facilitated by the ion channels, and these channels are held open by the binding of cyclic guanosine monophosphate (cGMP).

Figure 22.5 shows the overall detection mechanism in the rod cells schematically. Outer segment is very rich in photosensitive Rh molecule, which is a strong absorber of photons with peak sensitivity in blue-green spectrum. Upon light reception, Rh is triggered with structural changes and acts as a catalyst. Activated Rh triggers a chain of reactions that lead to the destruction of chemical messenger cGMP.

When Rh is activated by photon absorption, active Rh starts catalyzing the activation of transducin T. Activated transducin T* activates cGMP phosphodiesterase (PDE). When activated, PDE hydrolyzes cGMP and lowers its concentration. Finally, the transport channels for Na^+ and K^+ respond quickly to any change in cGMP concentration by closing the ion channels, and significantly changing the current passing through the rod cell and creating an electrical pulse until the cGMP concentration is restored.

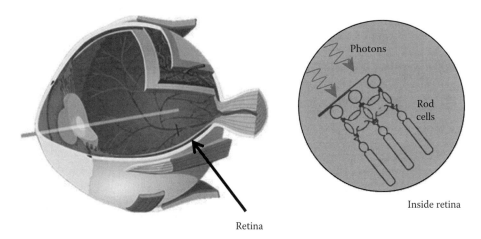

FIGURE 22.3 The cross section of the eye, showing the sensor layer—the retina and the rod cells within.

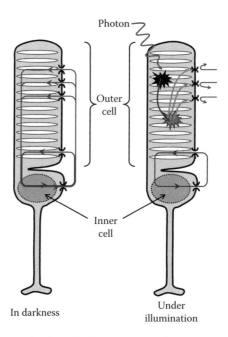

FIGURE 22.4 The schematics of a rod cell. Under darkness, an ion flow occurs within the cell, and the absorption of light prevents this flow, leading to a detectable electrical signal.

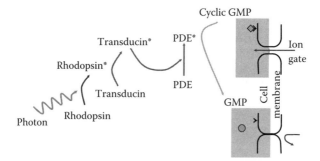

FIGURE 22.5 The mechanism of photon detection and molecular amplification in a rod cell. Absorbing volume and small nano-injectors.

22.5.1 Important Properties of the Rod Cell and Implications

The significance of the detection mechanism of human visual system is that it can provide both high efficiency and high sensitivity at room temperature, a condition that is very difficult to achieve in conventional SPDs. Simply put, the energy of a single photon in the visible or short infrared is extremely small, less than one atto Joule, and the only reliable way of sensing this small energy is to use a very small sensing volume, for example a quantum dot. However, the wavelength of light is significantly larger than such a sensor, and hence the interaction between the photon and the sensor, or quantum efficiency, is extremely small. Any attempt to enhance the efficiency by increasing the volume would simply reduce the sensitivity. Rod cell's detection mechanism resolves this conflict by using a micron-scale absorbing volume, the outer cell, and nanoscale sensing elements, or the ion channels.

Another important property of the rod cell is the inherent amplification. The detection mechanism involves many molecules that act as catalysts and individually affect hundreds to thousands of

TABLE 22.1 Comparison of the Parameters of Rod Cells, Cone Cells, PMTs, and CCDs

	Quantum Efficiency (%)	Dark Counts (s⁻¹)	Integration Time (ms)	Pixel Area (μm²)
Rod	25	0.012	300	3–4
Cone	5	>1000	50	3–4
PMT	25	10	$<10^{-5}$	$<10^{8}$
CCD	75	1	100	~60

molecules. At the final step, the closing of a single ion channel results in the prevention of thousands ions from entering the cell. As a consequence, the chain reaction that begins with activated Rh results in a molecular amplification that has an effective gain exceeding several millions. This is of critical importance, as the internal amplification ensures that the detected signal is not buried under the noise of the proceeding sensory stages.

Table 22.1 illustrates some of the parameters of the rod cell, and how it compares [29] to cone cells, which are responsible for color vision, PMTs, and charge-coupled devices (CCDs).

22.6 Nano-Injection-Based Single Photon Detectors

The concepts that were learned from human visual system were the following:

1. Coupling a micron-scale absorbing volume with nanoscale sensing elements for improved sensitivity.
2. Having a significant internal amplification to boost the miniscule photon energy to a detectable signal level above the system noise floor.

We incorporated these principles in a novel semiconductor platform, called the nano-injection photon detector. Structurally, we made the device analogous to the rod cells: The detector features large regions, which absorb and channel the photoexcited carriers to nano-injectors where the control of the amplified flow of carriers is maintained similar to the ion channels.

We have also engineered the band structure such that it would provide an internal amplification. The amplification method (i.e., nano-injection) is designed to have stability based on an internal negative feedback.

The structure of the nano-injection photon detector is highlighted in Figure 22.6, including large absorbing volume and small nano-injectors.

As highlighted in Figure 22.6, the detector features nano-injectors on a large InGaAs absorption layer. The device is based on InP/GaAsSb/InGaAs material system and therefore has type-II band alignment (Figure 22.7), with the GaAsSb layer acting as a barrier for electrons and a trap for holes. As detailed in Ref. [30], the most tested active layer structure consists of 1000 nm $In_{0.53}Ga_{0.47}As$

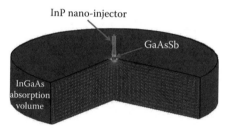

FIGURE 22.6 The device geometry of a nano-injection photon detector.

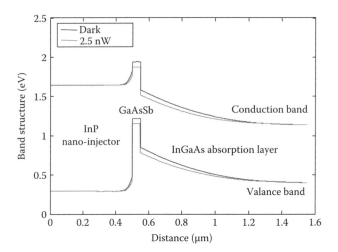

FIGURE 22.7   The band structure of a nano-injection photon detector at the central axis.

(n-doped), 50 nm GaAs$_{0.51}$Sb$_{0.49}$ (p-doped), and 500 nm InP (n-doped), from bottom to up. Five variations on this structure have been designed, with slightly different doping levels, compositions, and
thicknesses.

The band structure across the central axis of the device exhibits a distinct type-II band alignment
shape when properly biased with 0.5–1 V (Figure 22.7). The conduction band incorporates a GaAsSb
barrier to limit the flow of the electrons from InP side to InGaAs side. The valance band, on the other
hand, has a well structure, bound by the higher potential of InP and InGaAs layers (Figure 22.8).

Due to the doping level and work function of each layer, the device generates an internal electric
field in InGaAs region, which gets stronger when the device is biased correctly. Upon absorption, photons generate electron–hole pairs in the large absorption region. The electrons and holes are separated
by the internal electric field of the device. Holes are attracted to the nano-injector, which presents a

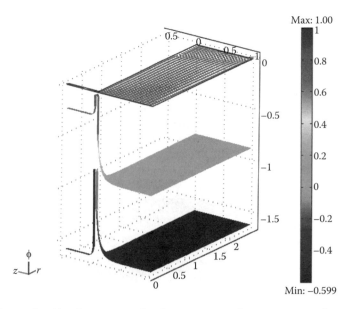

FIGURE 22.8   The simulated band structure in the 2D cross section of a nano-injection detector.

potential trap for holes due to the type-II band alignment. A single photo-generated hole in the absorption region is equivalent to a charge density of 1.4×10^{-3} C/m³. However, when trapped inside the 50 nm high and 100 nm wide diameter nano-injector, the same hole creates an effective charge density of more than 400 C/m³. Therefore, the impact of the hole increases by more than five orders of magnitude. Equivalently, the small volume of the trap represents an ultralow capacitance, and hence the entrapment of a single hole leads to a large change of potential and produces an amplified electron injection, similar to a single electron transistor (SET). Our detailed simulations show that a single hole can alter the potential barrier by more than 52 mV. This is significantly higher than the thermal fluctuation energy of carriers at 300 K, and hence a high SNR is possible even at room temperature.

Device relaxation is achieved by thermal recombination of the trapped holes. The holes that are trapped in the GaAsSb well will eventually recombine, relaxing the bands and restoring the current to the low dark current level.

As the multiplication mechanism is purely applied to one carrier, the amplification noise can be very small in nano-injection detectors. One possible explanation for this low-noise behavior of the device is the negative potential feedback mechanism in the device [31]. Even though the voltage of the barrier is mainly controlled by hole flux, the injected electrons also play an important role in the regulation of barrier layer voltage. Compared to holes, which are trapped in the GaAsSb barrier for a relatively long time, electrons have a very short but finite transit time through the barrier. During the transit time, they lower the local potential and increase the barrier height. The increase in barrier height opposes the flow of electrons and reduces the transmission probability.

22.7 The Performance of Nano-Injection Single Photon Detectors

22.7.1 Low-Frequency Measurements

The dark current, photo-response, and spectral noise power of the devices were measured at room temperature. At each data step, the dark current and spectral noise power measurements were taken simultaneously, quickly followed by photocurrent measurements. During photocurrent measurements, the devices were illuminated with a CW laser at 1550 nm, which was focused into a spot of approximately 5 μm in diameter, located 5 μm away from the injector of the device under test. The signal was amplified using a trans-impedance amplifier and recorded by a high-accuracy multimeter and a spectrum analyzer.

For accurate laser power calibration, a commercial PIN detector was placed inside the setup as a separate experiment and its response was measured to accurately quantify the laser power reaching the sample.

Fabricated unpassivated devices were tested using a computerized setup. The measured dark current showed good agreement with the simulation results, despite not being fit to any parameters (Figure 22.9). 30 μm devices with 10 μm nano-injector showed dark current values around 1 μA, whereas devices with 500 nm size nano-injector had less than 45 nA of dark current (Figure 22.10). The DC current measurements, when coupled with the optical gain measurements, yielded a unity gain dark current density of less than 900 nA/cm² at 1 V.

The uniformity of arrays of nano-injection detectors was also evaluated. Dark current measurements over tens of devices revealed high uniformity, with a standard deviation less than 3%.

Since detection and amplification mechanisms are tightly coupled in the nano-injection detector, the components of responsivity had to be identified. Responsivity, R, is

$$R = \frac{q}{E_{\mathrm{ph}}}\left(1-r\right)\left(1-e^{-\alpha L}\right)\eta_{\mathrm{c}}M$$

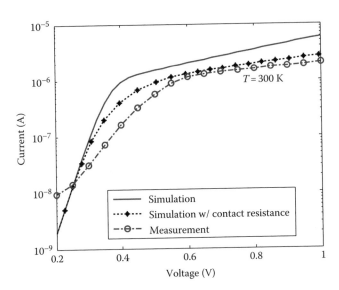

FIGURE 22.9 The current–voltage plot for simulation and measurements (10 μm diameter).

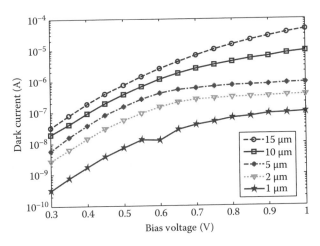

FIGURE 22.10 The current–voltage plots for different-sized devices at room temperature.

where

q is electron charge

E_{ph} is photon energy (~0.8 eV here)

r is reflection from the air–semiconductor boundary (~30%)

α is InGaAs absorption coefficient (~1 μm^{-1} at $\lambda = 1.55$ μm)

L is the path length in InGaAs absorption layer (1 μm)

η_c is carrier collection efficiency

M is the internal electrical gain

Among these factors, collection efficiency and gain are the unknowns and they could not be separately identified in the measurements. In the experiments, the responsivity was measured using the dark and photocurrent measurements with a calibrated laser source. Then using the relation above, the product, $\eta_c M$, was calculated and the product of gain and collection efficiency was defined as the "effective optical–electrical conversion factor" ($\eta_c M$) or "optical gain." In other words, optical-to-electrical conversion factor corresponds to the number of electrons flowing through the device per absorbed photon. The simulation model predicts that the collection efficiency for our device is around 70%. However, this number was not incorporated into further calculations, as it is not a physical or measured value. Instead, the conservative approach of equating the gain to the optical-to-electrical conversion efficiency was taken: The conversion factor is always smaller than the internal gain, since the collection efficiency is always less than unity.

Measurements indicated that the gain increases with the bias (Figure 22.11): Beyond ~1 V, a stable gain of more than 10,000 was measured for 30 μm devices with 10 μm diameter nano-injector (Figure 22.12). For small devices, optical gain values less than larger devices were recorded. Still, devices with 1 μm injectors exhibited respectable gain values around 1000 at modest voltages less than 1 V.

The gains of existing avalanche-based detectors show an exponential relation to the bias, and hence controllable gain values are limited to several hundreds. In contrast, the nano-injection devices show more than an order of magnitude higher stable gain, and much better tolerance to variations in voltage bias.

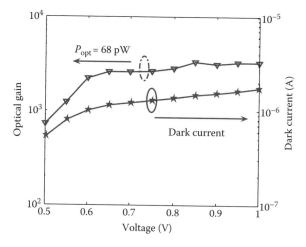

FIGURE 22.11 The current–voltage and optical gain–voltage plots for a nano-injection detector at room temperature.

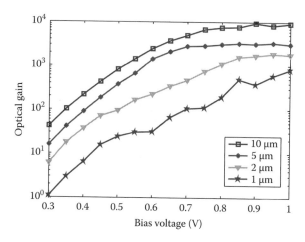

FIGURE 22.12 The optical gain plots for different-sized nano-injection detectors at room temperature.

Similar to Ref. [32], the spectral noise power after amplification in unpassivated devices was measured with a spectrum analyzer around 1.5 kHz, which is beyond the $1/f$ noise knee but lower than the measured bandwidth of the device of about 4 kHz. The measured spectral power was compared to predicted spectral noise density due to Poissonian shot noise with amplification ($2qM^2I_{int}\Delta f$), where a noise suppression phenomenon similar to the Fano effect was observed.

Going beyond the classical limit for shot noise, devices with sub-Poissonian (suppressed) shot noise have been demonstrated. This phenomenon, called Fano effect [33], is shown to result from temporal correlation mechanisms influencing particle flow, such as Coulomb blockade [34] or Pauli exclusion principle [35]. The Fano factor has been shown to assume values as low as 0.4 by imposing temporal correlations of charged carriers [36]. The strength of shot noise suppression (or enhancement) is quantified by the Fano factor [37] γ as

$$\gamma = \frac{I_n^2}{I_{shot}^2} = \frac{I_n^2}{2qI_{DC}\Delta f}$$

where
  I_n is the measured standard deviation of current, or current noise
  I_{shot} is the Poissonian shot noise
  q is electron charge
  I_{DC} is the average value of current
  Δf is the bandwidth

Note that Fano devices generally do not have an associated gain, and thus the right-hand side of this equation is only valid for such devices. For devices with internal amplification, the denominator of the right-hand side needs to be modified into $2qM^2I_{DC}\Delta f$. This is because of the amplification that applies to both the signal and the noise, and hence the noise power needs to be scaled by M^2 to conserve SNR.

Measurements on devices with 10 μm injectors revealed that the active area extended to 7–8 μm away from the injector, resulting in a total active diameter of ~25 μm. The devices showed high, stable optical–electrical conversion factors around 3000 at 0.7 V and 5000 at around 1 V, indicating an internal gain that can extend beyond 5000. The measured Fano factor was $F \sim 0.55$ at such high gain values (Figure 22.13). Since the exact value of the internal gain was not known, a conservative approach was taken, and the optical–electrical conversion factor was used as $\eta_c = 1$. This would always lead to an overestimation of the Fano factor. For instance, assuming a collection efficiency of 70% yields a Fano factor $F \sim 0.39$, which would mean the Poissonian shot noise is further suppressed in reality.

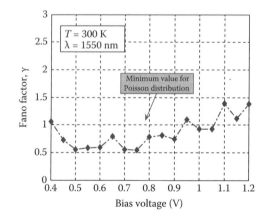

FIGURE 22.13 The variation of noise versus voltage for a 10 μm diameter detector.

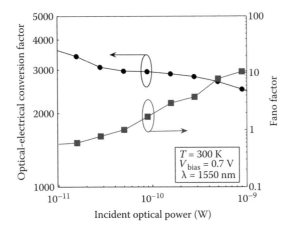

FIGURE 22.14 The behavior of optical gain and Fano factor versus incident optical power.

However, using optical–electrical conversion factor as gain becomes a poor assumption at high optical intensities. As the incident optical power was increased, the measurements showed an overall decrease in optical–electrical conversion factor (Figure 22.14), which is a result of decreasing collection efficiencies. Low collection efficiency is mainly due to carrier shielding and device saturation at high optical intensities. As the holes are collected from the large absorption region and drawn into the small GaAsSb volume, the hole current density shows a rapid increase in the radial direction toward the center. The high density of holes at the center of the device repels the incoming hole flux through Coulomb forces and introduces carrier shielding. The collection efficiency, η_c, drops more as the carrier shielding becomes more pronounced at stronger illumination. The effect of lowered collection efficiency manifests itself as low optical–electrical conversion factors and an overall increase in Fano factor at high intensities.

The noise-equivalent power (NEP) of the devices was measured as $4.5\,\text{fW/Hz}^{0.5}$ at room temperature without any gating, using the relation

$$P_{\text{NEP}} = \frac{I_{n,\text{meas}}}{\eta_c M}$$

which equates $I_{n,\text{meas}}$ (the measured spectral current noise in $\text{A/Hz}^{0.5}$) to the photocurrent that gives a unity SNR. The change in NEP versus bias voltage is plotted in Figure 22.15. Here, we see that the

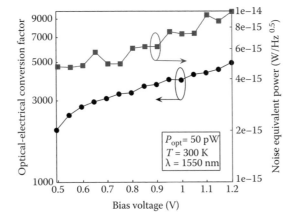

FIGURE 22.15 The optical gain and NEP versus bias voltage at room temperature.

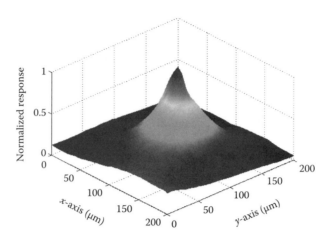

FIGURE 22.16 The spatial sensitivity map of an unpassivated 10 μm nano-injection detector.

optimum bias voltage for this detector is around 0.6–0.7 V, which is in agreement with the expectations from the band lineup and built-in potentials. Below 0.5 V, the electron injection through the InP-GaAsSb junction is the bottleneck of the transport across the device. When biased slightly above this critical point (0.5 V), the transport through this junction starts to increase rapidly. This bias point is where the device exhibits a low electron injection (i.e., low dark current, and low noise) with high sensitivity, which makes this bias more favorable than higher voltages. Hence, the lowest NEP is achieved around these bias voltages. Comparing these values with a survey for detectors operating at 1.55 μm wavelength[13], we see that uncooled APDs can provide similar NEP values around a few $fW/Hz^{0.5}$ (1–2.2 $fW/Hz^{0.5}$). PIN detectors, which do not provide internal amplification, can achieve NEP on the order of a $W/Hz^{0.5}$ with cooling.

The spatial response of the unpassivated devices was measured using a surface scanning beam with ~1.5 μm diameter, and 10 nm step resolution. Despite such a high gain, the device shows a very uniform spatial response, as shown in Figure 22.16. We believe that the low internal electric field in our devices is the main reason for the observed uniformity. The measured response decreases rapidly beyond a radius of about 8 μm, in complete agreement with the simulation model. This property suggests that 2D arrays of such detectors might not necessarily need pixel isolation methods such as ion implantation or mesa etching.

The spatial response of passivated devices was also evaluated. In stark contrast to unpassivated devices, surface-passivated ones showed a much larger spatial response, exceeding an area of 100 μm by 100 μm (Figure 22.17).

22.7.2 Bandwidth Measurements

The measured bandwidth of the unpassivated devices was around 3–4 kHz, much different compared to the lifetime in the GaAsSb trap layer (~1 ns with the device doping levels), which we believe will be the ultimate constraint on bandwidth. We have attributed this difference to the existence of surface traps with long recombination lifetime.

When the devices were passivated, drastically different behavior was observed. The gain decreased significantly to values around 10 and the spatial response extended to beyond 100 μm when the devices were not confined by hard etching the trap layer. However, the bandwidth of these devices exceeded 3 GHz. Risetime values of 200 ps were measured. Further simulations confirmed that the InAlAs layer had more influence in passivated devices than unpassivated ones, and that it was responsible for the observed behavior.

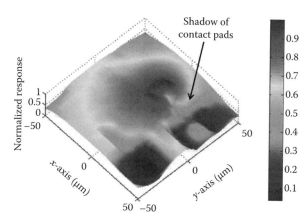

FIGURE 22.17   The spatial sensitivity map of a passivated 5 μm nano-injection detector.

22.7.3 High-Speed Measurements

To evaluate the high-speed transient response of the nano-injection detectors, the devices were tested in a custom-designed setup capable of calibrated infrared illumination, visible/infrared imaging, and motorized scanning. High-frequency RF probes were connected to individual devices and performed measurements. The detectors were biased using a low-noise DC power supply fed through a bias-tee. A femtosecond pulsed laser with <70 fs jitter was attenuated and focused onto the devices using an NIR microscope setup. Equipped with remotely controllable actuators, the setup has the capability of scanning the sample to map the spatial dependence of parameters. The RF signal coming from the devices was extracted using the bias-tee and amplified by an LNA with 2.4 dB noise figure, 2.5 GHz nominal bandwidth and 9 ps jitter. The amplified signal was then acquired by a high-speed sampling oscilloscope, Agilent 86100C (jitter: 1.7 ps), which directly measured delay, rise-time, and jitter. The timing signal was generated by a low-jitter (1.2 ps) PIN detector connected to the second output port of the laser.

The measured devices had a layer structure of 1000 nm InGaAs, 50 nm GaAsSb, 50 nm InAlAs, and 500 nm InP from bottom to top. After dry etching, the samples were passivated with polyimide of thickness 500 nm. The tested devices had an injector radius of 1 μm, which were chosen because of their larger SNR compared to submicron devices. The devices were illuminated with a 300 fs pulsed fiber laser with 20 MHz repetition rate and 1.4 pJ pulse energy after 10 dB of attenuation. The corresponding rise-time at 1 V bias was 200 ps with a jitter of 15 ps. To see the effect of optical saturation, the power was increased to 14 pJ where the pulse shape exhibits compression (Figure 22.18). The compressed response was compared to the results without the LNA in the signal path to verify that the source of compression was indeed the nano-injection detector. Measurements of nominal RMS jitter with increased pulse intensity kept stable around 15 ps, indicating that there was no significant change of the jitter.

Using calibrated attenuators, the pulsed power reaching the sample was reduced significantly. Then, the response was measured with the oscilloscope while the waveform was averaged. Then the number of photons reaching the device per pulse was calculated and the effective number of photons on the device was calculated based on the bandwidth of the devices and the incoming pulse. All efficiency values were assumed to be 1 (Figure 22.19).

This calculation resulted in an overestimation of the number of photons the device experiences. In other words, it places an upper limit on the number of photons that would create the measured response.

The 2 μm pulsed laser spot was scanned using motorized drivers and a map of the amplitude of the response was recorded. Measurements indicate that the active area of the device for pulsed illumination extended to 8–9 μm away from the nano-injectors (Figure 22.20). This was significantly less than the active area with CW illumination, which extended to beyond 100 μm, and we believe that the difference is due to a transient effect on the filling/emptying of surface states on the passivation interface.

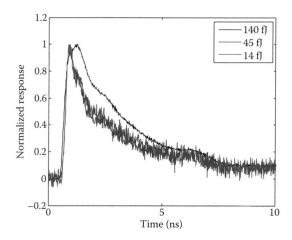

FIGURE 22.18 The normalized pulse response under different pulse energies.

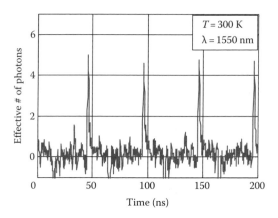

FIGURE 22.19 The transient response of a passivated nano-injection detector under pulse illumination.

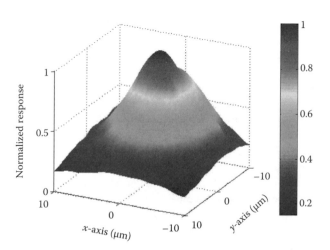

FIGURE 22.20 The spatial sensitivity map of a passivated 5 μm nano-injection detector, under pulsed illumination.

To evaluate the transit time, the 2 μm laser spot was scanned using motorized drivers, and a map of the delay around the device (Figure 22.21) was extracted together with the amplitude of the response [38]. From this map, the effective carrier velocity around the nano-injector was calculated as 7×10^4 m/s. In order to evaluate the results, we used our 3D nonlinear finite-element-method-based simulation model, which has shown good agreement with the measured low-frequency response of the device. The electric field, carrier density, and current density distributions inside the device were calculated. The estimated drift velocity was found to be 4×10^4 m/s. The carrier diffusion velocity was around 2×10^4 m/s in the opposite direction, due to the focalization and compression of holes and the elevated carrier density around the center.

The jitter performance was also evaluated using the measured delay and amplitude maps. A variable illumination spot size was assumed, and the spatial generation of carriers and corresponding transit delays were modeled and analyzed in MATLAB®. The expected jitter was calculated using the probability distribution for carrier arrival times. The jitter was predicted to be 9 ps for a 6 μm spot size, 19 ps for a 15 μm, and 22 ps for a 24 μm diameter (Figure 22.22).

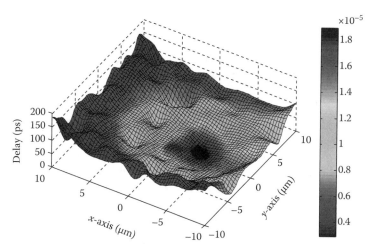

FIGURE 22.21 The spatial delay map of a passivated 5 μm nano-injection detector. The grayscale coding corresponds to the local sensitivity.

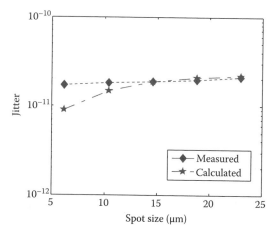

FIGURE 22.22 The behavior of jitter versus spot size. The diamonds show the measured data and the stars show the calculated.

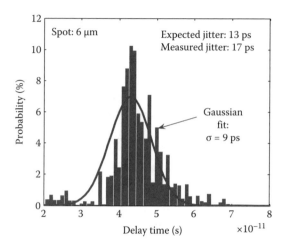

FIGURE 22.23 The expected carrier arrivals at different delay values, based on the measured delay and sensitivity maps.

In parallel with the theorized delay–jitter relation and calculations, the laser was defocused to see and experimentally quantify the effect of spot size on jitter. The spot size was calculated using the deviation from the focal plane and the numerical aperture of the focusing lens. The values showed a monotonically increasing trend with larger spot sizes, ranging from 17 ps for smallest spot sizes to 22 ps for a 24 μm spot diameter (Figure 22.23). The increase is mainly due to the larger optical generation area.

The values obtained from direct jitter measurements (Figure 22.24) matches well with jitter predictions from delay measurements, particularly at larger spot sizes. These results suggest that the jitter is mostly due to transit time, confirming our predictions regarding the stable nature of nano-injection-based amplification in time domain.

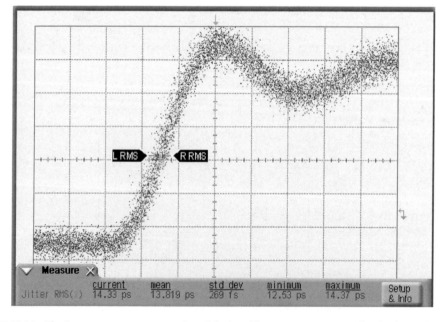

FIGURE 22.24 The jitter measurement on passivated devices. The measurement was taken by the Agilent 86100C high-speed sampling oscilloscope.

For small spot sizes, the calculated values were smaller than the measured values even when the jitter of the measurement setup was considered. We believe this is primarily due to device saturation. We have observed the effects of saturation in pulse shapes (Figure 22.18), which create an apparent increase in delay due to the increase in rise times. A similar increase manifests itself for small spot sizes in the vicinity of the injector, where the pulses saturate the device due to the larger local amplitude response. Hence, a rather flat delay versus distance relationship around the center of the device is observed instead of the expected concave shape, and the jitter values are underestimated in the calculations for small spot sizes. This phenomenon agrees with the overestimated effective carrier velocity around the center of the device (7×10^4 m/s measured versus 2×10^4 m/s expected), which was also due to flat region in spatial delay maps.

22.8 Conclusion

A new type of SWIR photon detectors, called the nano-injection photon detectors, is developed. The nano-injection detector incorporates nanoscale features with large absorption layers for sensitive and efficient light detection. Designed to be a competitor to the mainstream SPDs, the nano-injection detectors rely on the nano-injectors to detect and amplify low-power light signals.

Measurements on nano-injection detectors show optical gain values reaching 10,000, which indicate the very strong internal amplification, with low dark current densities at room temperature. Noise suppression behavior is recorded with Fano suppression factors ~0.55 and amplification factors exceeding 4000. The noise levels, suppressed beyond the theoretical shot noise limit, highlight the negative internal feedback inside the device, stabilizing the current injection. The surface-treated devices show bandwidths exceeding 3 GHz with a very low time uncertainty (jitter) of 15 ps. The source of this jitter was strongly correlated with the variation in transit time inside the device, providing strong evidence toward the very fast nature of nano-injection-based amplification. With high array uniformity (dark current standard deviation less than 3%), the nano-injection photon detectors provide extremely suitable properties for demanding detection and imaging applications, which require high efficiency, high sensitivity, and high uniformity.

References

1. A. Migdall and J. Dowling, *J. Mod. Opt.* **51**, 1265 (2004).
2. D. Deutsch, *Proc. R. Soc. Lond. A* **400**, 97 (1985).
3. T. P. Spiller, *Proc. IEEE* **84**, 1719 (1996).
4. R. J. Hughes, D. M Alde, P. Dyer, G. G. Luther, G. L. Morgan, and M. Schauer, *Contemp. Phys.* **36**, 149 (1995).
5. P. W. Shor, *SIAM J. Sci. Stat. Comput.* **26**, 1484 (1997).
6. C. Elliott, *IEEE Secur. Privacy* **2**(4), 57–61 (2004).
7. T. B. Pittman, Y. H. Shih, D. V. Strekalov, and A. V. Sergienko, *Phys. Rev. A* **52**, R3429 (1995).
8. M. R. Phillips, M. H. Zareie, O. Gelhausen, M. Drago, T. Schmidtling, and W. Richter, *J. Crystal Growth* **269**, 106 (2004).
9. S. Takeuchi, J. Kim, Y. Yamamoto, and H. H. Hogue, *Appl. Phys. Lett.* **74**, 1063 (1999).
10. A. Poloczek, M. Weiss, S. Fedderwitz, A. Stoehr, W. Prost, D. Jaeger, and F. J. Tegude, *The 20th Annual Meeting of the IEEE LEOS 2007*, Lake Buena Vista, FL, p. 180 (2007).
11. S. Cova, M. Ghioni, A. Lotito, I. Rech, and F. Zappa, *J. Mod. Opt.* **51**, 1267 (2004).
12. M. Ghioni, A. Gulinatti, I. Rech, F. Zappa, and S. Cova, *IEEE J. Sel. Top. Quantum Electron.* **13**, 852 (2007).
13. P. L. Voss, K. G. Koprulu, S. K. Choi, S. Dugan, and P. Kumar, *J. Mod. Opt.* **51**, 1369 (2004).
14. R. J. Mcintyre, *IEEE Trans. Electron. Devices* **13**, 164 (1966).

15. B. Cabrera, R. M. Clarke, P. Colling, A. J. Miller, S. Nam, and R. W. Romani, *Appl. Phys. Lett.* **73**, 735 (1998).

16. G. N. Gol'tsman, O. Okunev, G. Chulkova, A. Lipatov, A. Semenov, K. Smirnov, B. Voronov, A. Dzardanov, C. Williams, and R. Sobolewski, *Appl. Phys. Lett.* **79**, 705 (2001).

17. R. Sobolewski, A. Verevkin, G. N. Gol'tsman, A. Lipatov, and K. Wilsher, *IEEE Trans. Appl. Supercond.* **13**, 1151 (2003).

18. A. Korneev, V. Matvienko, O. Minaeva, I. Milostnaya, I. Rubtsova, G. Chulkova, K. Smirnov, V. Voronov, G. Gol'tsman, W. Slysz, A. Pearlman, A. Verevkin, and R. Sobolewski, *IEEE Trans. Appl. Supercond.* **15**, 571 (2005).

19. J. P. Estrera and M. R. Saldana, *Proc. SPIE* **5079**, 212 (2003).

20. A. Rogalski, *Opto-Electron. Rev.* **10**, 111 (2002).

21. L. E. Bourree, D. R. Chasse, P. L. S. Thamban, and R. Glosser, *Proc. SPIE* **4796**, 1 (2003).

22. C. D. Brewer, B. D. Duncan, and E. A. Watson, *Opt. Eng.* **41**, 1577 (2002).

23. A. Rogalski, *Opt. Eng.* **42**, 3498 (2003).

24. T. R. Hoelter, S. M. Petronio, R. J. Carralejo, J. D. Frank, and J. H. Graff, *Proc. SPIE* **3698**, 837 (1999).

25. P. Norton, J. Campbell III, S. Horn, and D. Reago, *Proc. SPIE* **4130**, 226 (2000).

26. A. Rogalski and P. Martyniuk, *Infrared Phys. Technol.* **48**, 39 (2006).

27. G. Sarusi, *Infrared Phys. Technol.* **44**, 439 (2003).

28. S. Hecht, S. Shlaer, and M. H. Pirenne, *J. Gen. Physiol.* **25**, 819 (1942).

29. F. Rieke and D. A. Baylor, *Rev. Mod. Phys.* **70**, 1027 (1998).

30. O. G. Memis, A. Katsnelson, S. C. Kong, H. Mohseni, M. Yan, S. Zhang, T. Hossain, N. Jin, and I. Adesida, *Appl. Phys. Lett.* **91**, 171112 (2007).

31. O. G. Memis, A. Katsnelson, S. C. Kong, H. Mohseni, M. Yan, S. Zhang, T. Hossain, N. Jin, and I. Adesida, *Opt. Express* **16**, 12701 (2008).

32. F. Z. Xie, D. Kuhl, E. H. Bottcher, S. Y. Ren, and D. Bimberg, *J. Appl. Phys.* **73**, 8641 (1993).

33. P. J. Edwards, *Aust. J. Phys.* **53**, 179–192 (2000).

34. G. Kiesslich, H. Sprekeler, A. Wacker, and E. Scholl, *Semicond. Sci. Technol.* **19**, S37–S39 (2004).

35. I. A. Maione, M. Macucci, G. Iannaccone, G. Basso, B. Pellegrini, M. Lazzarino, L. Sorba, and F. Beltram, *Phys. Rev. B* **75**, 125327 (2007).

36. V. Y. Aleshkin, L. Reggiani, N. V. Alkeev, V. E. Lyubchenko, C. N. Ironside, J. M. L. Figueiredo, and C. R. Stanley, *Semicond. Sci. Technol.* **18**, L35–L38 (2003).

37. A. Wacker, E. Schöll, A. Nauen, F. Hohls, R. J. Haug, and G. Kiesslich, *Phys. Status Solidi C* **0**, 1293–1296 (2003).

38. O. G. Memis, A. Katsnelson, H. Mohseni, M. Yan, S. Zhang, T. Hossain, N. Jin, and I. Adesida, *IEEE Electron. Device Lett.* **29**, 867 (2008).

23

Quantum Dot Infrared Photodetectors

Lan Fu
Thomas Vandervelde
Sanjay Krishna

23.1 Introduction

Over the past 10–15 years, there has been significant interest in developing intersubband quantum well or quantum dot infrared photodetectors (QWIPs/QDIPs) for the mid-wave infrared (MWIR) and long-wave infrared (LWIR) regimes (Maimon et al. 1998, Chu et al. 1999, Rappaport et al. 2000, Chen et al. 2001, Liu et al. 2001b, Wang et al. 2001, Raghavan et al. 2002, Stiff-Roberts et al. 2002, Jiang et al. 2003). The 3–25 μm IR regime is of particular interest, due to its wide range of applications, including missile detection and tracking, battlefield imaging and communications, medical diagnostics and treatment, surveillance, and biological/chemical agent identification. Presently, mercury–cadmium–telluride (MCT) detectors continue to demonstrate superior responsivity and specific-detectivity (D^*) when compared to the state-of-the-art QDIPs or QWIPs. Despite their superior performance, difficulties in spatial uniformity, epitaxial growth, and low fabrication yields make the manufacture of large focal plane arrays (FPAs) exceedingly expensive (Sidorov et al. 1997, Phillips 2002). This fact drives scientists and engineers to try to make intersubband systems competitive.

One of the key factors in the performance of the intersubband detectors is the utilization of quantum mechanical confinement of the charge carriers. This confinement enables the possibility of novel high performing devices (Motohisa et al. 1995, Ryzhii 1997). Standard QWIP heterostructures are based on the mature (In,Ga,Al)As/GaAs material system that increases the feasibility of industrial production of low-cost, large-scale, optoelectronic devices. Since many other optoelectronic devices are GaAs based, this also implies easy integration with existing devices and circuits (Shen et al. 2000). The QDIPs, on the other hand, have a number of advantages compared to QWIPs; specifically, they are sensitive to normally incident light. This sensitivity to normally incident photons originates in having the direction of quantum confinement parallel to the electric field of the incident photon. The confinement in QWIPs lies only in the growth direction; therefore, additional efforts must be deployed to alter the trajectory on the normally incident light. QDs are zero-dimensional objects and, therefore, have confinement in three dimensions (Phillips 2002). This implies that QDIPs are sensitive to light incident from any direction, including normal. Additionally, QDIPs have higher optical gain and longer carrier lifetimes; therefore, the efficacy of a carrier providing photocurrent is greater for a QDIP than a QWIP (Horiguchi et al. 1999, Liu et al. 2001a, Kim et al. 2002, Zibik et al. 2004), which in turn could result in a significantly larger D^* (Phillips et al. 1999). Unfortunately, QDIPs also come with some serious drawbacks; specifically, QDIPs have been experimentally shown to have high dark current values (Ryzhii 1996, Ryzhii et al. 2000, Mashade et al. 2003, Nasr and Mashade 2006). It is believed, however, that these high dark current levels are largely due to dot formation issues and problems with subsequent overgrowth degrading the confinement. With lower dark current levels predicted with better three-dimensional confinement of carriers, higher operating temperatures can also be expected, thereby reducing the complexity of associated cryo-coolers. QD detectors with a room temperature operation have already been demonstrated (Lim et al. 2007), and several other groups have shown some promising results in this spectral window (Maimon et al. 1998, Chu et al. 1999, Rappaport et al. 2000, Chen et al. 2001, Liu et al. 2001b, Wang et al. 2001, Raghavan et al. 2002, Stiff-Roberts et al. 2002, Jiang et al. 2003). However, because of the intricate dependence of the operating wavelength on the size and shape of the dot, which in turn depends on the inherently random self-assembly process, presently there are no "dial-in recipes" for obtaining a desired spectral response from QD detectors.

Several research groups, including those of the authors, are investigating detectors based on intersubband transitions in a quantum dots-in-a-well (DWELL) design. While exact material compositions and design for the various layers differ from group to group, the overall design remains constant. The most basic design has the InAs dots placed in an $In_{0.15}Ga_{0.85}As$ well, which in turn is positioned in a GaAs matrix. The DWELL detector is a hybrid design between a conventional QWIP and the QDIP and benefits from the advantages of both of its progenitors. Apart from sensitivity to normally incident light and lower dark currents, the DWELL structure demonstrates a better control over the operating wavelength and nature of the transition (bound-to-bound versus bound-to-continuum) (Krishna et al. 2003b). In the DWELL structure, an intersubband transition occurs when the electrons in the ground state of the QD are photoexcited into a bound state of the QW, as schematically shown in Figure 23.1. The modeling of this structure and its excitations, however, is extremely challenging due to the nature of the potential profile. Preliminary calculations of the electronic states and wavefunctions in the DWELL structure have been published and shown reasonably good agreement between the calculated energy level spacings and the obtained intersubband spectrum (Amtout et al. 2004). In this chapter, we wish to highlight some of the recent results that have been obtained based on the DWELL designs. In Section 23.2, the growth and fabrication of the DWELL detectors and FPAs are discussed. Section 23.3 is devoted to the figures of merit for DWELL FPAs. Section 23.4 discusses some recent progress in the field of multicolor detection, while Section 23.5 will conclude with upcoming advancements in DWELL detectors.

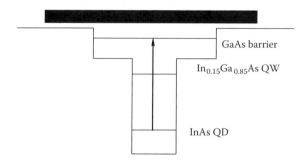

FIGURE 23.1   A schematic for the conduction band diagram of an InAs/InGaAs DWELL detector showing the QD to QW transition for detector applications.

23.2 Growth and Fabrication of DWELL Detectors and Focal Plane Arrays

From the growth point of view, accurate control over the size, uniformity, and density of self-assembled QDs is extremely important for obtaining high-performance QD-based optoelectronic devices. In general, growth of QDs by the technique of metal-organic chemical vapor deposition (MOCVD) is more challenging than that by molecular beam epitaxy (MBE) due to two main reasons: (1) lack of *in situ* monitoring technique. Unlike MBE, MOCVD in general does not have any reliable *in situ* monitoring system that makes it difficult to monitor the two-dimensional to three-dimensional growth transition for the formation of QDs; (2) use of high growth temperature. To suppress the adatom mobility across the surface and thus avoid clustering, dots grown by MBE are typically carried out at around 500°C. However, for MOCVD, the precursor cracking efficiencies are too low at these temperatures to obtain good quality material. Thus, the growth temperatures used by MOCVD are normally 50°C–100°C higher than those used by MBE, making the growth of high-quality QDs and QD stacks (for QDIPs) more challenging by this technique. Therefore, the majority of the work published in the literature on the study of QD devices such as lasers (Wang 2008) and IR photodetectors (Maimon et al. 1998, Chu et al. 1999, Rappaport et al. 2000, Chen et al. 2001, Liu et al. 2001b, Wang et al. 2001, Raghavan et al. 2002, Stiff-Roberts et al. 2002, Jiang et al. 2003) are based on growth using MBE despite the fact that MOCVD is a preferable technique for industrial applications due to its simplicity, scalability, and lower cost of operation. Nevertheless, in recent years, QDIPs by MOCVD with progressive improvements in performances have been reported (Kim et al. 1998, Jiang et al. 2004b, Fu et al. 2005a, Zhang et al. 2005, Szafraniec et al. 2006, Jolley et al. 2007, Lim et al. 2007, Jolley 2008) demonstrating the viability of this technique for future commercial QDIP applications.

For the purposes of this section, we will detail the growth and fabrication of a basic DWELL structure mentioned above by MBE. Other QDIP/DWELL structures grown by both MBE and MOCVD will be discussed later in the chapter.

23.2.1 Growth

The DWELL structures were grown in a V-80 MBE system, with an As_2 cracker source. 2.4 monolayers (MLs) of InAs dots were deposited on the sample, at a rate of 0.053 ML s^{-1}. The dots were Si-doped at a level of 1e^{-}/QD. The cross section of the device shown in Figure 23.2 consists of a 15-stack DWELL heterostructure between two n^{+} doped GaAs contact layers. The DWELL active regions are typically grown at 470°C, while the GaAs barrier was grown at a temperature of 590°C.

High-resolution cross-sectional transmission electron microscopy (TEM) images (Raghavan et al. 2002) has confirmed the absence of threading dislocations in the DWELL structure and revealed that the dot is confined to the top half of the well (Figure 23.3). Room-temperature photoluminescence measurements on these samples with a 2.5 mW, 632.8 nm He–Ne laser and a Ge detector yielded a peak at $\lambda = 1.22$–$1.24\,\mu m$, which is attributed to the interband ground state transition in the dot. Following the analysis by Kim et al. (Kim et al. 2001) and using a 60:40 rule (conduction band: valence band ratio), we can attribute any peak longer than $5.5\,\mu m$ observed in the photocurrent spectrum to a transition from the ground state in the dot to a bound state in the InGaAs well (see Figure 23.1). By increasing the InGaAs well width, the position of the bound state in the well could be lowered. This effect would be expected to result in a redshift in the operating wavelength of the detector.

23.2.2 Processing

Post-growth processing was done using standard contact lithography, plasma-etching, and metallization techniques in a class 100 clean-room environment. Individual $400\,\mu m$ square n-i-n mesas with

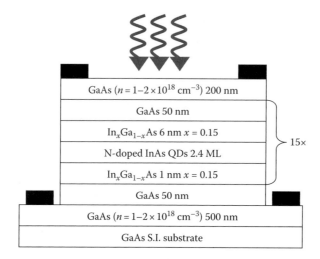

FIGURE 23.2   Diagram of the InAs/InGaAs DWELL after single-pixel processing. Here, the contact layers are shown with ohmic contacts. Between the contact layers lies the active region consisting of repetitions of QDs in quantum wells, nested in GaAs barriers. In this case, the detector is shown as being front illuminated.

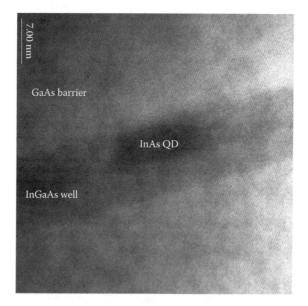

FIGURE 23.3   Cross-sectional TEM image of an InAs/InGaAs DWELL heterostructure.

top pixel apertures, ranging from 25 to 300 μm in diameter, were lithographically defined in the top metal contact for illumination in a front-normal configuration (Raghavan et al. 2002). The contacts were annealed at 400°C using rapid thermal annealing.

The sample used to create the DWELL FPA was also grown by MBE using the already proven single-pixel DWELL structure. In the structure used to create the FPA, the active regions of each pixel consisted of 15 layers of InAs QDs embedded in $In_{0.15}Ga_{0.85}As$ QWs. The pixels are essentially identical to the single-pixel structure shown in Figure 23.2, except that the substrate and bottom GaAs layer are removed and the pixel is flipped by 180°. Following the growth process, the sample was processed into a 320 × 256 array of detectors using standard lithography (each pixel occupies an area of approximately

FIGURE 23.4 Scanning electron microscopy (SEM) images of portions of a 320×256 FPA after mesa etch and contact metal deposition.

5.76×10^{-6} cm^2, or 576 μm^2, and has a 25 μm pitch). Processing included under-bump metallization (UBM) and adding indium bumps at each detector location to facilitate device hybridization to a read-out integrated circuit (ROIC); see Figures 23.4 and 23.5.

23.2.3 Broadband Figures of Merit

While most DWELLs, presently being produced, exhibit multicolor response (Krishna et al. 2003a), it is still valuable to discuss the broadband, unfiltered, figures of merit for a DWELL FPA. Throughout this section, we will discuss these broadband figures of merit by examining how they are derived and detailing new data for the 320×256 DWELL FPA.

Several detector figures of merit can be calculated by measuring mean FPA output and noise versus irradiance. As a standard procedure, the irradiance is provided by operating a calibrated blackbody source at various temperatures. Assuming a properly designed experimental setup, the photon irradiance values (E_q [photons/s-cm^2] and E_e [Watts/cm^2]) at the FPA can be treated as uniform across the array and were calculated using Equations 23.1 and 23.2 (Dereniak and Boreman 1996):

$$E_q = \frac{\pi L_q}{4(f\#)^2 + 1}$$

(23.1)

where

    L_q is photon radiance [photons/cm^2-s-sr-μm]
    $f\#$ is the ratio of the lens focal length to the lens diameter, which is a conveniently defined parameter,

$$L_q = \frac{2c}{\lambda^4 \left(\exp\left(\dfrac{hc}{\lambda kT} \right) - 1 \right)}$$

(23.2)

FIGURE 23.5 SEM image of the DWELL FPA with indium bumps attached. The detector array was hybridized to a commercially available Indigo Systems Corporation ISC9705 ROIC by QmagiQ, Inc. to produce the FPA.

where

  h is Planck's constant (6.626×10^{-34} J-s)
  c is the speed of light (2.998×10^{8} m/s)
  k is Boltzmann's constant (1.381×10^{-23} J/K)
  T is temperature (K)

In general, up to the limit of the ROIC's integration capacitors, the DWELL FPA's output usually has a fairly linear response as a function of irradiance. Figure 23.6 shows an example of the DWELL FPA at detector biases of (V_{DB} ~0.5, 0.75, 1.0, and 1.1 V). The ROIC integration capacitors were full at approximately −0.35 V. Once this output voltage was reached, no further FPA response could be measured.

When the detector array is operated in a photon shot noise dominant regime, the voltage output of DWELL FPA (V_{output}) along with the photon noise voltage (in Equation 23.3) can be utilized to calculate the conversion-gain product, $C_G G$:

$$V_{output} = C_G \left(\eta G E_q A_d T_{int} + \frac{I_{dark} T_{int}}{q} \right) \qquad (23.3)$$

where

  C_G is the conversion gain (volts per electron)
  G is the photoconductive gain
  η is the detector quantum efficiency (electrons per photon)

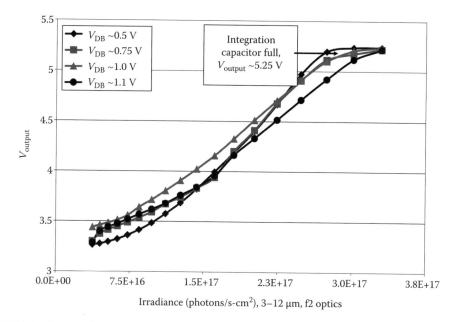

FIGURE 23.6 Output voltage versus irradiance for the DWELL FPA at 77 K.

A_d is the detector area (cm²)
I_{dark} is the detector dark current (A)
q is the electron charge (1.6×10^{-19} C)
T_{int} is the integration time (s)

Squaring Equation 23.3 yields noise variance that can be used to solve for $\eta G E_q A_d T_{int}$. Next, this quantity can be substituted in Equation 23.4,

$$V_{n\text{Photon}} = C_G G \sqrt{2\eta E_q A_d T_{int}} \qquad (23.4)$$

The slope of the resulting equation given by Equation 23.5

$$V_{n\text{Photon}}^2 = 2C_G G(V_{output}) - 2C_G^2 G \frac{I_{dark} T_{int}}{q} \qquad (23.5)$$

corresponds to the $C_G G$ product. This method provides an estimate of the conversion-gain product at the four test biases. For most DWELL arrays, there is an anticipated trend of higher $C_G G$ at higher biases (from the contribution of G) (Campbell and Madhukar 2007). For our example FPA, the estimated $C_G G$ product values for the four detector test biases are shown in Table 23.1.

23.2.3.1 Responsivity

Responsivity is defined as the detector output per unit of radiant input. A higher responsivity is generally desirable, since it is directly related to the sensitivity of the device and is proportional to the detector's quantum efficiency (QE). For a DWELL FPA, responsivity is proportional to the QE, photoconductive gain (G), and conversion gain (C_G) product. For the DWELL FPA, the responsivity was measured at four detector biases ($V_{DB} \sim 0.5, 0.75, 1.0,$ and 1.1 V) by measuring the output voltage versus irradiance. With this data, the peak responsivities can be calculated by using the two equations below:

$$R_v \left[\frac{V}{\text{photon}} \right] = \frac{V_{\text{output}}}{\sum_{\lambda_1}^{\lambda_2} R_n(\lambda) E_e(\lambda, T) \tau_{\text{win}} d\lambda} \tag{23.6}$$

and

$$R_v \left[\frac{V}{W} \right] = \frac{V_{\text{output}}}{\sum_{\lambda_1}^{\lambda_2} R_n(\lambda) E_e(\lambda, T) \tau_{\text{win}} d\lambda} \tag{23.7}$$

TABLE 23.1   Conversion Gain Product ($C_G G$) Estimates at 77 K

V_{DETCOM} (V)	$C_G G$ (V/e⁻)
5.5	1.30×10^{-6}
5.75	2.29×10^{-6}
6.0	2.13×10^{-6}
6.1	1.98×10^{-6}

where

V_{output} is the output voltage (V)
R_n is the normalized spectral response
$d\lambda$ is the wavelength scanning step size from spectral response data (100 nm)

To complete the peak responsivity calculations, the collected spectral response data for the FPA must be used. Table 23.2 details the responsivity values for our example device at 77 K.

23.2.3.2 Noise Equivalent Power

Noise equivalent power (NEP) is a parameter defined as the required optical power incident on a photodetector that produces a signal-to-noise ratio equal to 1. This represents the minimum amount of optical input power that must be exceeded for detection to occur. A low value of NEP is an indicator of good detector performance, indicating a small amount of optical input is detectable. DWELL FPA NEP was calculated using calculated responsivity and Equation 23.8 (Dereniak and Boreman 1996),

$$\text{NEP} = \frac{V_n}{R_v} \tag{23.8}$$

where

V_n is the recorded noise voltage (V_{RMS})
R_v is the voltage responsivity (V/W)

The NEP for our example is plotted against irradiance in Figure 23.7 and the minimum NEP values at each detector test bias are shown in Table 23.3.

23.2.3.3 Noise Equivalent Irradiance

While noise equivalent irradiance (NEI) is simply a shift in units from NEP, it can be useful as a design property. NEI is defined by the number of photons per unit area incident upon a photodetector that produce a signal-to-noise ratio equal to 1 and, therefore, delineates the minimum flux detectable for the device. NEI is calculated with Equation 23.9:

TABLE 23.2   Responsivity Values at 77 K

V_{DB} (V)	R_v (V/photon)	R_v (V/W)
0.5	5.11×10^{-10}	4.32×10^{7}
0.75	1.24×10^{-9}	4.70×10^{7}
1.0	3.81×10^{-9}	3.16×10^{7}
1.1	8.33×10^{-9}	6.05×10^{7}

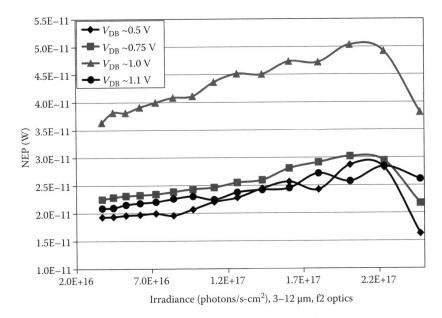

FIGURE 23.7 NEP versus irradiance at 77 K.

$$NEI = \frac{V_n}{R_v A_d} \qquad (23.9)$$

It should be noted that there is another definition of NEI that defined as the irradiance at an f1 input, rather than at the detector. Figure 23.8 shows an example of NEI plotted versus detector irradiance for the four test biases on the same FPA as depicted in Figures 23.6 and 23.7. Minimum NEI values for our example device are listed in Table 23.4.

TABLE 23.3 Minimum Recorded NEP Results at 77 K

V_{DB} (V)	NEP (W)
0.5	1.62×10^{-11}
0.75	2.17×10^{-11}
1.0	3.64×10^{-11}
1.1	2.09×10^{-11}

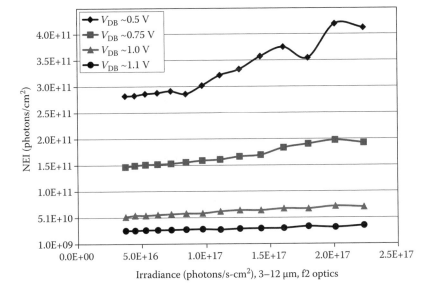

FIGURE 23.8 NEI versus irradiance at 77 K.

23.2.3.4 Detectivity

Inversely proportional to the NEP of a detector is the detectivity, or D^*, (units of cm-(Hz)$^{1/2}$/W). As the name implies, the detectivity is a measure of how little optical irradiance the FPA can detect, and, therefore, having a high detectivity is important. The detectivity for a DWELL FPA is calculated using the NEP calculation results and Equation 23.10 (Bhattacharya et al. 2005) and Equation 23.11 (Dereniak and Boreman 1996),

TABLE 23.4 Minimum Recorded NEI Results at 77 K

V_{DB} (V)	NEI (photons/cm^2)
0.5	2.83×10^{11}
0.75	1.48×10^{11}
1.0	5.24×10^{10}
1.1	2.64×10^{10}

$$\Delta f = \frac{1}{2T_{int}} \tag{23.10}$$

$$D^* = \frac{\sqrt{A_D \Delta f}}{NEP} \tag{23.11}$$

in which Δf is the noise bandwidth (Hz). Detectivities for the same device detailed in the figures and tables above are shown in the Figure 23.9 and Table 23.5.

Under the conditions of background limited infrared performance (BLIP) where photon noise dominates, the theoretical BLIP detectivity may be used to estimate the QE of a photodetector. Theoretical BLIP detectivity is calculated by Equation 23.12 (Dereniak and Boreman 1996),

$$D^*{}_{BLIP} = \frac{\lambda}{2hc}\sqrt{\frac{\eta}{E_q}} \tag{23.12}$$

The BLIP detectivity estimate was made using Equation 23.12 plotted against irradiance at two different values of QE. This plot is compared to the DWELL FPA detectivity values for $V_{DB} \sim 0.5$ V in Figure 23.10. Using this estimation technique, the QE for DWELL FPA detailed in these figures is approximately 0.25%–0.45% at 77 K.

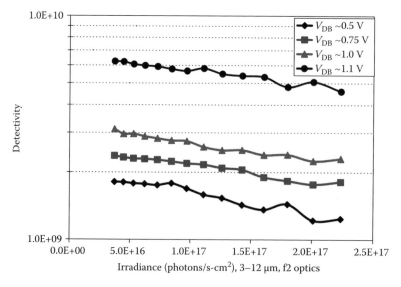

FIGURE 23.9 Detectivity versus irradiance at 77 K.

23.2.3.5 Noise Equivalent Difference in Temperature

The smallest difference in a uniform temperature scene that the FPA can detect is called the noise equivalent difference in temperature (NEDT, units of Kelvin). The smaller NEDT, therefore, the better, since it is a performance measure of the FPA's sensitivity. NEDT is calculated using the voltage output and noise versus irradiance data using the expression below (Stiff-Roberts et al. 2002):

TABLE 23.5 Peak Detectivity Results at 77 K

V_{DB} (V)	D^* (cm Hz$^{1/2}$/W)
0.5	1.81×10^9
0.75	2.37×10^9
1.0	3.11×10^9
1.1	6.25×10^9

$$\text{NEDT} = \frac{\Delta T}{V_s/V_n} \tag{23.13}$$

where
ΔT is the difference in black body temperatures (K)
V_s is the response between two temperatures (V)
V_n is the recorded noise voltage at the lower temperature (V)

Minimum NEDT values are observed just prior to the integration capacitor becoming full, where noise decreased due to a decline in readout noise (Table 23.6). An example NEDT is plotted versus detector output in Figure 23.11 for the same device as the previous figures.

23.3 Recent Advances in QDIP FPA

Large size, highly uniform, and multicolor FPAs are desirable for advanced IR systems requiring high resolution and fast modulation response. Since the first demonstration of a 256×256 InGaAs/InGaP QDIP FPA in 2004 (Jiang 2004a), much effort has been made to improve QDIP FPA performances. DWELLs have become the more and more popular structure of choice due to their flexibility and superior device characteristics demonstrated in single-pixel devices.

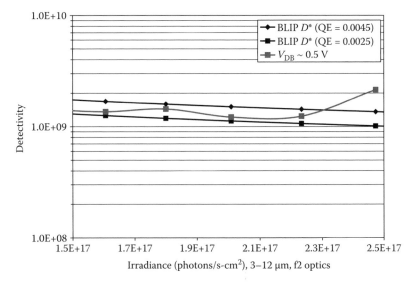

FIGURE 23.10 BLIP detectivity for QE estimation at 77 K.

23.3.1 High Operating Temperature

A high operating temperature is desirable for most of the applications. Cryo-coolers significantly increase the initial equipment cost, increase the long-term operational cost, and decrease the field-based utility of the detector. Recently, the group of Northwest University (NWU) led by Prof. M. Razaghi has demonstrated high operation temperature results for a 4.1 μm single-pixel device (Lim et al. 2007) and a 4 μm FPA (Tsao et al. 2007).

TABLE 23.6 Minimum NEDT Results at 77 K

V_{DB} (V)	NEDT (K)
0.5	0.031
0.75	0.049
1.0	0.064
1.1	0.058

Although the NWU group does not call their structures DWELLs, their devices have the same basic architecture: consisting of a QD in a QW surrounded by barriers. Presently, one of the primary differences is the material system with which they are working. Unlike most other DWELL groups, which use the (In,Al,Ga)As/GaAs material family and grow their material via MBE, the Razeghi group uses (In,Al)As/InP and grows their material via MOCVD (Tsao et al. 2007). A typical example of a DWELL grown by the group is shown schematically in Figure 23.12. A semi-insulating InP substrate is covered by an undoped buffer layer and then a doped contact layer. Then, 25 repetitions of the active region follow. Each repetition consists of 29 nm of InAlAs barriers surrounding an n-doped 3.5 nm InGaAs QW and 1.8 ML InAs to form the QDs. This material composition offers several advantages, most significant of which is the ease of strain balancing in the lattice. With each repetition of the active region, the barrier can compensate for added

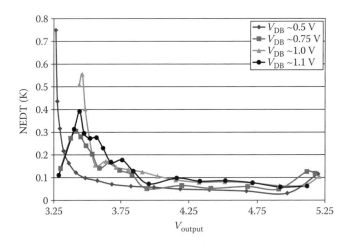

FIGURE 23.11 NEDT versus FPA output voltage at 77 K.

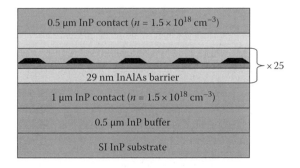

FIGURE 23.12 A schematic for the NWU group DWELL structure. (Reprinted from Lim, H. et al., *Appl. Phys. Lett.*, 90, 131112, 2007. With permission. Copyright 2007 American Institute of Physics.)

strain from the QW and QD. This neutralization of the strain, keeping the average lattice constant equal to the substrate, enables the growth of a more extensive active region.

Although these devices have been operated in a lower wavelength regime and, therefore, a significantly lower dark current regime, their results are notable. Specifically, the 4.1 μm single-pixel DWELL detector achieved a detectivity of 6.7×10^7 cm $Hz^{1/2}/W$ at room temperature, 300 K; with 2.8×10^{11} cm $Hz^{1/2}/W$ at 120 K and 35% QE (Lim et al. 2007). Similarly, the corresponding 320×256 FPA was operational at 200 K and achieved a responsivity of 34 mA/W, a conversion efficiency of 1.1%, and an NEDT of 344 mK at an operating temperature of 120 K (Tsao et al. 2007).

23.3.2 Low NEDT

Recently, a highly effective 640×512 DWELL FPA has been fabricated by Gunapala's group at Jet Propulsion Laboratory (JPL) (Gunapala et al. 2007) based on a careful DWELL design proposed by Krishna's group at University of New Mexico (UNM) (Krishna 2005). For the DWELL photodetectors, each was grown and fabricated in a similar manner to that as was described in Section 23.1. Minor differences are depicted in Figure 23.13 compared with the previous DWELL described in Section 23.1, which include (1) the InGaAs QW's composition being 12% In rather than the standard 15%, and (2) the active region was repeated 30 times rather than the standard 10–15 times, which helps to account for the increase in QE (Krishna 2005).

As discussed above, one of the DWELL structure's primary advantages over typical QDIP structures is the ease of tuning the response by varying the well width (Raghavan et al. 2002, Krishna et al. 2005, Ting et al. 2007), this was aptly demonstrated by the JPL group as seen in Figure 23.14.

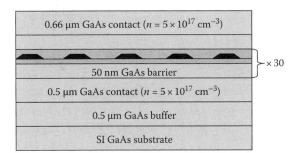

FIGURE 23.13 A schematic for the JPL DWELL structure. (From Krishna, S., *J. Phys. D*, 38, 2142, 2005.)

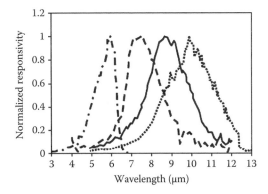

FIGURE 23.14 By varying the InGaAs well width from 5.5 to 11 nm one can select a peak responsivity anywhere from ~5.5 to 10.5 μm. (From Krishna, S., *J. Phys. D*, 38, 2142, 2005.)

One specific contribution that the JPL group added to the design of DWELL has to do with their sensitivity to non-normally incident light (not S-polarized). While one of QDIP and DWELL advantages over QWIPs lies in its sensitivity to S-polarized light, it is also known that DWELLs should have increased sensitivity to non-normally incident light as well. This is believed to be related to the QD's lateral size being dramatically larger than its height (Krishna 2005). This was confirmed by the JPL group and is shown in Figures 23.15 and 23.16. While the S-polarized light is approximately an order of magnitude more responsive than comparable QWIPs, the P-polarized light is five times more responsive than that (see Figure 23.15). These dramatic improvements naturally lead to the inclusion of a grating on the surface of the detector (see Figure 23.16).

Following a process similar to the standard process outlined in Section 23.1, a 640 × 512 DWELL FPA was processed at JPL using this newly enhanced design. Under test, the FPA had an NEDT of 40 mK, which is only twice that of the best NEDTs reported for QWIPs and a QE of 5% for a LWIR response at 60 K (Krishna 2005). This was by far the best LWIR response reported for a DWELL FPA. This FPA additionally was used to image scenes, an example of which is shown in Figure 23.17, where the quality of the image is visually dramatic.

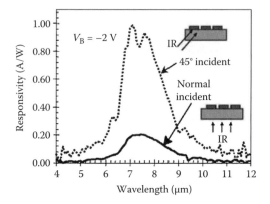

FIGURE 23.15 A plot of wavelength vs. responsivity for a DWELL photodetector under normally incident light and light incident at 45°. (From Krishna, S., *J. Phys. D*, 38, 2142, 2005.)

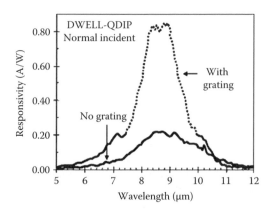

FIGURE 23.16 A plot of wavelength vs. responsivity for two DWELL photodetectors under normally incident light, one with a grating and the other without. (From Krishna, S., *J. Phys. D*, 38, 2142, 2005.)

FIGURE 23.17 Image taken with the first 640×512 pixels QDIP LWIR FPA imaging system with an $f/2$ AR-coated germanium optical assembly. (From Krishna, S., *J. Phys. D*, 38, 2142, 2005.)

23.4 Multicolor Detectors

Two-color or multicolor IR photodetectors are important for future high-performance IR systems. By incorporating two stacks of different InGaAs/GaAs and InAs/GaAs QD structures, multiple photoresponse peaks were reported by Jiang et al. (2005). Compared with QWIPs, this method can be easily incorporated into a FPA since no complicated light-coupling scheme is needed owing to the normal incidence operation in QDIPs. However, due to the extremely sensitive self-organized process of QD formation, the growth of such multiple QD stacks structure with different compositions is highly challenging and difficult to reproduce. The UNM group has been focused on bringing color to the IR spectrum by employing the DWELL structures. Most of the designs, therefore, have two strong responses, one in the MWIR and one in the LWIR. This single bump, multicolor technique provides enhanced utility for the detector in the applications mentioned in the introduction. For example, being able to examine a scene at two wavelengths removes the ambiguity generated by objects having different emissivity. This ability to differentiate objects is also the beginning of spectroscopy, enabling the fine differentiation between chemicals, biological agents, or types of tissues for medical and security applications (Varley et al. 2007).

Using the example, DWELL design detailed in the earlier section on broadband response, the UNM group have had a great deal of success in eliciting a two-color response (Figure 23.18). For this FPA, standard processing was performed, ending with a single indium bump per pixel on the 320×256 array. The detector matrix was then hybridized by a commercial partner (QmagiQ LLC) to an Indigo Systems Corporation ISC9705 readout circuit. After hybridization, the FPA was tested at UNM using CamIRa™ system manufactured by SE-IR Corp. Two-color (MWIR and LWIR) responses were observed from the DWELL FPA at 77 K at a nominal bias voltage ranging from 0.5 to 1.0 V. Larger bias voltages could not be applied due to the saturation of the integration capacitors. The operation of the FPA was evaluated (Shenoi et al. 2008).

The response of the detector existed in two bands. In each of the two bands, ASIO filter lenses were used to spectrally limit the incoming irradiance to 3–5 μm (f2) and 8–12 μm (f2.3), MWIR and LWIR, respectively. At a detector bias of 0.5 V, the integration time for DWELL FPA was 2.37 ms for the MWIR and LWIR responses. All measurements were made at a device temperature of 77 K with a liquid-nitrogen pour-fill Dewar to exclude noise caused by the compressor on a temperature-controllable closed-cycle Dewar (Shenoi et al. 2008).

Figure 23.19 displays the detectivity results for the DWELL FPA. Peak values of 1.46×10^9 (cm² Hz)$^{1/2}$/W and 3.64×10^{10} (cm² Hz)$^{1/2}$/W (LWIR and MWIR, respectively) occurred just before the integration capacitors were fully charged. The observed greater detectivity for the LWIR response compared

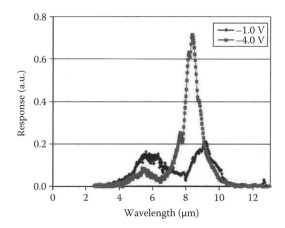

FIGURE 23.18 Multicolor response from a typical DWELL detector. The MWIR and LWIR peaks are probably due to the transitions from a state in the quantum dot to a state in the well and to the continuum, respectively. (Note that the data measured at −1.0 V were scaled by a factor of 5 for readability.) (Reprinted from Varley, E. et al., *Appl. Phys. Lett.*, 91, 081120, 2007. With permission. Copyright 2007 American Institute of Physics.)

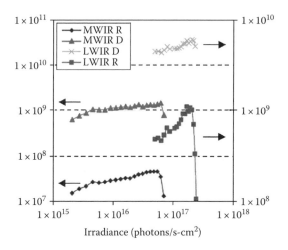

FIGURE 23.19 Peak responsivity (V/W) and detectivity [(cm² Hz)^{1/2}/W] for the MWIR (left axis) and LWIR (right axis) for the DWELL FPA at 77 K. Irradiance levels for MWIR and LWIR are 3–5 μm (f2) and 8–12 μm (f2.3), respectively. (Reprinted from Varley, E. et al., *Appl. Phys. Lett.*, 91, 081120, 2007. With permission. Copyright 2007 American Institute of Physics.)

to the MWIR response may be more an artifact than a fact: this is because LWIR could not be accurately measured in a single pixel at this low bias thus leading to an overestimate of the responsivity. In order to more accurately calculate the figures of merit, spectral response data for the DWELL FPA itself are needed. Therefore, there has been ongoing work to establish the capability to perform spectral response measurements for an entire array (Shenoi et al. 2008).

In addition, the two-color NEDT for the DWELL FPA has also been calculated by using the output voltage (V_0), noise voltage (V_n), and temperature step of the blackbody source (ΔT). Minimum NEDT values of 55 mK (MWIR) and 70 mK (LWIR) were measured for the DWELL FPA, as shown in Figure 23.20 (Shenoi et al. 2008).

By reviewing the two-color response performance measures summarized in Tables 23.7 and 23.8, one can see that the figures of merit calculated tend to be better in the MWIR region, though the difference

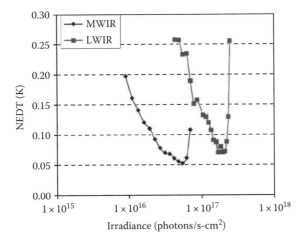

FIGURE 23.20 Noise equivalent temperature difference obtained in the MWIR and LWIR bands at 77 K. Irradiance levels for MWIR and LWIR are 3–5 μm (f2) and 8–12 μm (f2.3), respectively. (Reprinted from Varley, E. et al., *Appl. Phys. Lett.*, 91, 081120, 2007. With permission. Copyright 2007 American Institute of Physics.)

TABLE 23.7 Summary of Two-Color Figures of Merit

V_{DB} (V)	R_v (V/photon)	R_v (V/W)	NEP (W)
	MWIR		
0.5	1.39×10^{-9}	6.76×10^7	1.47×10^{-11}
0.75	2.08×10^{-8}	4.68×10^8	2.37×10^{-12}
1.0	9.44×10^{-9}	4.49×10^7	2.63×10^{-11}
1.1	4.17×10^{-8}	1.54×10^8	7.90×10^{-12}
	LWIR		
0.5	6.30×10^{-10}	7.30×10^7	1.07×10^{-11}
0.75	1.88×10^{-9}	9.35×10^7	1.47×10^{-11}
1.0	4.77×10^{-9}	5.28×10^7	2.56×10^{-11}
1.1	8.48×10^{-9}	7.04×10^7	1.68×10^{-11}

TABLE 23.8 Summary of Two-Color Figures of Merit

V_{DB} (V)	NEI (photons/cm²)	D^* (cm Hz$^{1/2}$/W)	NEDT (K)
	MWIR		
0.5	1.30×10^{11}	2.38×10^9	1.88×10^9
0.75	9.23×10^9	2.54×10^{10}	3.57×10^9
1.0	2.18×10^{10}	4.30×10^9	3.72×10^9
1.1	5.09×10^9	1.65×10^{10}	5.18×10^9
	LWIR		
0.5	1.17×10^{11}	6.12×10^9	1.95×10^9
0.75	1.27×10^{11}	3.63×10^9	1.73×10^9
1.0	4.93×10^{10}	4.41×10^9	2.65×10^9
1.1	2.49×10^{10}	7.76×10^9	3.62×10^9

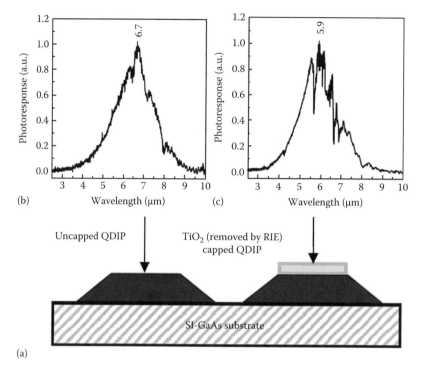

FIGURE 23.21 (a) Schematic of the two-color QDIPS. (b) Spectral photoresponse from the uncapped QDIP. (c) Spectral response from the TiO$_2$-capped QDIP.

between MWIR performance and LWIR performance was not dramatically different at the four detector biases used in testing. This is attributed to the fact that the MWIR and LWIR responses are comparable at these lower detector biases, where bound-to-continuum energy transitions are favored leading to a slightly larger MWIR response. At larger reverse bias, the LWIR response would be expected to become dominant because the probability of carriers tunneling from the bound-to-bound and bound-to-quasi-bound states increases, leading to the increased LWIR response. With the 9705 ROIC, two-color response is noted from the measured spectral response of the FPA, but because of the limitation of biases that can be applied with the 9705 ROIC, the concept of a bias-tunable FPA camera could not be more thoroughly explored.

Another method to achieve multispectral response is through the post-growth quantum dot intermixing (QDI) technique (Leon et al. 1996, Fu et al. 2005b). This technique generates the atomic interdiffusion of the QDs and their surrounding barriers, which will in turn modify the shape of the QD potential, leading to an increased ground state and thus redshifted intersubband transition energy as used by QDIPs. Redshift in wavelength of QDIPs after QDI using simple thermal interdiffusion through annealing has been reported (Hwang et al. 2005, Fu et al. 2006, Aivaliotis et al. 2007). To further realize the multicolor detection, Fu et al (Fu et al. 2008) employed a layer of TiO$_2$ film to suppress thermal interdiffusion (Fu et al. 2003) on selected regions of the sample. By forming a TiO$_2$ capped region and an uncapped region on the same wafer to simultaneously prohibit and generate thermal interdiffusion, the two-color detector displayed a differential wavelength shift of ~0.8 μm with very similar device characteristics (see Figure 23.21).

23.5 Future Advancements and Conclusion

The purpose of this chapter is to give an overview of the important figures of merit for QDIP-based FPAs and review the recent advancements in field. There are a number of new and exciting techniques

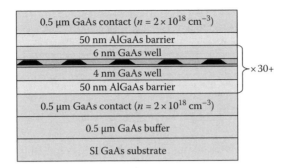

FIGURE 23.22 A schematic for the structure of a low-strain DWELL, specifically a double DWELL or a DDWELL design. Here, a GaAs well acts as a primary well, surrounded by AlGaAs barriers. Inside the GaAs well lies a minimized InGaAs well surrounding the InAs QDs. (Reprinted from Attaluri, R.S. et al., *J. Vac. Sci. Technol. B*, 25, 1186, 2007. With permission. Copyright 2007 American Institute of Physics.)

that are starting to appear in single-pixel devices that will soon be applied to FPAs as well. One of these advancements includes a new DWELL design shown in Figure 23.22 that uses a double well. In this structure, the GaAs barriers are replaced with AlGaAs barriers and the role of the primary well is played by a GaAs layer. Here, the InGaAs layer thicknesses, which constitute the second well, are reduced to a minimum and, therefore, the strain due to lattice mismatch with the GaAs substrate is also minimized. This enables larger stacks to be grown (30–80 repetitions of the active region). FPAs of this design are presently under test at UNM at the time of this chapter being written (Attaluri et al. 2007, unpublished).

Based on a semiempirical estimate, we believe that the energy difference between the ground state in the dot and the conduction band edge of GaAs is around 164 meV (Krishna et al. 2003b). We have modeled the electronic states and wavefunctions in the DWELL structure, using Bessel function expansion (Amtout et al. 2004) and using finite difference method, as plotted in Figure 23.23. There is a reasonably good agreement between the calculated energy level spacings and the obtained intersubband spectrum listed in Table 23.9.

The results confirm that the LWIR response is due to transition from ground state of the quantum dot to ground state of the QW structure while the MWIR response is from the transition to second

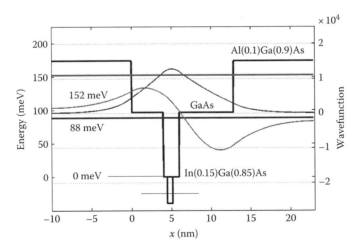

FIGURE 23.23 Model of the double-DWELL photodetectors active region. Note: the portion of the band diagram corresponding to the InAs QD is not depicted here. The ground state and first excited state calculated are solely for the nested QW system. The two modeled states show a close match with the observed transition wavelengths.

TABLE 23.9   Comparison between the
Measured Peak Values for the Double-DWELL
Photodetectors and the FDM Calculated Values

	Measured (µm)	Calculated (µm)
MWIR	5.5	5.6
LWIR	8.3	7.9

bound state, which is very close to the continuum state. From this modeling, the strong quantum-confined Stark effect in asymmetric QDs is apparent, which produces a shift of ~2 µm in the LWIR response for change in the bias polarity. This effect is useful for tunable response from the DWELL detectors.

Another advancement uses a resonant cavity (RC) to increase the number of passes incident photons make through the active region and, therefore, enhance the QE (Shenoi et al. 2007). The RC is formed using a distributed Bragg reflector (DBR) at the bottom of the stack; the natural semiconductor–air interface is all that is used at the top for this design shown in Figure 23.24. This RC-DWELL was designed to enhance the LWIR signal, which does enhance by approximately a factor of 3, as displayed in Figure 23.25 (Shenoi et al. 2007).

Additional experiments are also being performed to enhance the DWELL's functionality, from surface plasmons and photonic crystals as part of the detectors (Vandervelde et al. 2007) to specialized capping materials for the QDs to enhance their functionality directly. In the end, DWELLs are a young technology and their application in FPAs is even more recent. In the coming years of research, one can expect significant improvement in a number of areas: operating temperature, sensitivity (NEDT), QE, and functionality (multicolor).

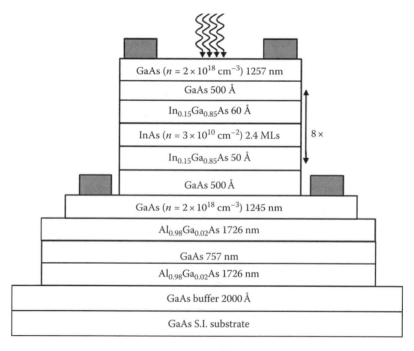

FIGURE 23.24   A schematic of the structure for a processed RC-DWELL. By adding the DBR at the bottom of the stack, the indecent light will make multiple passes through the active region enhancing QE. (From Shenoi, R.V. et al., Plasmon assisted photonic crystal quantum dot sensors, Paper read at *Nanophotonics and Macrophotonics for Space Environments, SPIE Optics Photonics*, San Diego, CA, 2007. With permission.)

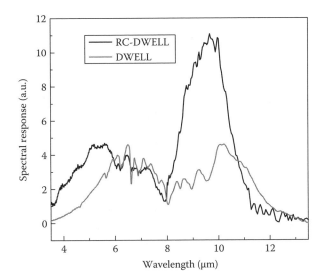

FIGURE 23.25 Spectral response data for the RC-DWELL (upper curve) and the standard DWELL (lower curve) samples. All the spectra were taken at $T = 30$ K at a bias of $V_b = -1.8$ V. (From Shenoi, R.V. et al., Plasmon assisted photonic crystal quantum dot sensors, Paper read at *Nanophotonics and Macrophotonics for Space Environments*, *SPIE Optics Photonics*, San Diego, CA, 2007. With permission.)

Acknowledgments

The authors would like to acknowledge the work and efforts of all of their colleagues who have made this work possible, including Dr. A. Stintz, Dr. A. Amtout, Dr. P. Dowd, G. von Winckel, R. S. Attaluri, S. Annamalai, N. R. Weisse-Bernstein, D. Formann, S. Raghavan, S. J. Lee, J. S. Brown, G. Jolley, Dr. H. H. Tan, and Prof. C. Jagadish. The financial support from AFRL, AFOSR, IC Postdoc Program, NSF, and the Australian Research Council is also gratefully acknowledged.

References

Aivaliotis, P., E. A. Zibik, L. R. Wilson, J. W. Cockburn, M. Hopkinson, and R. J. Airey. 2007. Tuning the photoresponse of quantum dot infrared photodetectors across the 8–12 μm atmospheric window via rapid thermal annealing. *Appl. Phys. Lett.* 91:143502–143504.

Amtout, A., S. Raghavan, P. Rotella, G. von Winckel, A. Stintz, and S. Krishna. 2004. Theoretical modeling and experimental characterization of InAs/InGaAs quantum dots in a well detector. *J. Appl. Phys.* 96:3782–3786.

Attaluri, R. S., J. Shao, K. T. Posani, S. J. Lee, J. S. Brown, A. Stintz, and S. Krishna. 2007. Resonant cavity enhanced InAs/In$_{0.15}$Ga$_{0.85}$As dots-in-a-well quantum dot infrared photodetector. *J. Vac. Sci. Technol. B* 25:1186–1190.

Attaluri, R. S., R. Shenoi, J. Shao, T. E. Vandervelde, A. Stintz, and S. Krishna. Three-color InAs/InGaAs/GaAs dots-in-double-well (DDWELL) infrared photodetector. Unpublished.

Bhattacharya, P., S. Chakrabarti, X. Su, and A. D. Stiff-Roberts. 2005. Research propels quantum dots forward. *Laser Focus World*, May.

Campbell, J. C. and A. Madhukar. 2007. Quantum dot infrared photodetectors. *Proc. IEEE* 95:1815–1827.

Chen, Z. H., O. Baklenov, E. T. Kim, I. Mukhametzhanov, J. Tie, A. Madhukar, Z. Ye, and J. Campbell. 2001. Normal incidence InAs/Al$_x$Ga$_{1-x}$As quantum dot infrared photodetectors with undoped active region. *J. Appl. Phys.* 89:4558–4563.

Chu, L., A. Zrenner, G. Böhm, and G. Abstreiter. 1999. Normal-incident intersubband photocurrent spectroscopy on InAs/GaAs quantum dots. *Appl. Phys. Lett.* 75:3599–3601.

Dereniak, E. L. and G. D. Boreman. 1996. *Infrared Detectors and Systems.* New York: John Wiley & Sons.

Fu, L., P. Lever, H. H. Tan, C. Jagadish, P. Reece, and M. Gal. 2003. Suppression of interdiffusion in InGaAs/GaAs quantum dots using dielectric layer of titanium dioxide. *Appl. Phys. Lett.* 82:2613–2615.

Fu, L., P. Lever, K. Sears, H. H. Tan, and C. Jagadish. 2005a. $In_{0.5}Ga_{0.5}As$/GaAs quantum dot infrared photodetectors grown by metal-organic chemical vapor deposition. *IEEE Electron. Device Lett.* 26:628–630.

Fu, L., P. Lever, H. H. Tan, C. Jagadish, P. Reece, and M. Gal. 2005b. Study of intermixing in InGaAs/(Al) GaAs quantum well and quantum dot structures for optoelectronic/photonic integration. *IEE Proc. Circ. Dev. Syst.* 152:491–496.

Fu, L., H. H. Tan, I. McKerracher, J. Wong-Leung, C. Jagadish, N. Vukmirović, and P. Harrison. 2006. Effect of rapid thermal annealing on device characteristics of InGaAs/GaAs quantum dot infrared photodetectors. *J. Appl. Phys.* 99:114517–114535.

Fu, L., Q. Li, P. Kuffner, G. Jolley, P. Gareso, H. H. Tan, and C. Jagadish. 2008. Two-color InGaAs/GaAs quantum dot infrared photodetectors by selective area interdiffusion. *Appl. Phys. Lett.* 93:013504–013506.

Gunapala, S. D., S. V. Bandara, C. J. Hill, D. Z. Ting, J. K. Liu, S. B. Rafol, E. R. Blazejewski, Jason, M. Mumolo, S. A. Keo, S. Krishna, Y.-C. Chang, and C. A. Shott. 2007. 640×512 pixels long-wavelength infrared (LWIR) quantum-dot infrared photodetector (QDIP) imaging focal plane array. *IEEE J. Quantum Electron.* 43:230–237.

Horiguchi, N., T. Futatsugi, Y. Nakata, N. Yokoyama, T. Mankad, and P. M. Petroff. 1999. Quantum dot infrared photodetector using modulation doped InAs self-assembled quantum dots. *Jpn. J. Appl. Phys.* 38:2559–2561.

Hwang, S. H., J. C. Shin, J. D. Song, W. J. Choi, J. I. Lee, and H. Han. 2005. Detection wavelength tuning of InGaAs/GaAs quantum dot infrared photodetector with thermal treatment. *Microelectron. J.* 36:203–206.

Jiang, L., S. S. Li, N.-T. Yeh, J.-I. Chyi, C. E. Ross, and K. S. Jones. 2003. $In_{0.6}Ga_{0.4}As$/GaAs quantum-dot infrared photodetector with operating temperature up to 260 K. *Appl. Phys. Lett.* 82:1986–1988.

Jiang, J., K. Mi, S. Tsao, W. Zhang, H. Lim, T. O'Sullivan, T. Sills, M. Razeghi, G. J. Brown, and M. Z. Tidrow. 2004a. Demonstration of a 256×256 middle-wavelength infrared focal plane array based on InGaAs/InGaP quantum dot infrared photodetectors. *Appl. Phys. Lett.* 84:2232–2234.

Jiang, J., S. Tsao, T. O'Sullivan, W. Zhang, H. Lim, T. Sills, K. Mi, M. Razeghi, G. J. Brown, and M. Z. Tidrow. 2004b. High detectivity InGaAs/InGaP quantum-dot infrared photodetectors grown by low pressure metalorganic chemical vapor deposition. *Appl. Phys. Lett.* 84:2166–2168.

Jiang, L., S. S. Li, W.-S. Liu, N.-T. Yeh, and J.-I. Chyi. 2005. A two-stack, multi-color $In_{0.5}Ga_{0.5}As$/GaAs and InAs/GaAs quantum dot infrared photodetector for long wavelength infrared detection. *Infrared Phys. Technol.* 46:249–256.

Jolley, G., L. Fu, H. H. Tan, and C. Jagadish. 2007. Influence of quantum well and barrier composition on the spectral behavior of InGaAs quantum dots-in-a-well infrared photodetectors. *Appl. Phys. Lett.* 91:173508–173510.

Jolley, G., L. Fu, H. H. Tan, and C. Jagadish. 2008. Effects of well thickness on the spectral properties of $In_{0.5}Ga_{0.5}As$/GaAs/$Al_{0.2}Ga_{0.8}As$ quantum dots-in-a-well infrared photodetectors. *Appl. Phys. Lett.* 92:193507–193509.

Kim, S., H. Mohseni, M. Erdtmann, E. Michel, C. Jelen, and M. Razeghi. 1998. Growth and characterization of InGaAs/InGaP quantum dots for midinfrared photoconductive detector. *Appl. Phys. Lett.* 73:963–965.

Kim, E.-T., Z. Chen, and A. Madhukar. 2001. Tailoring detection bands of InAs quantum-dot infrared photodetectors using $In_xGa_{1-x}As$ strain-relieving quantum wells. *Appl. Phys. Lett.* 79:3341–3343.

Kim, K., J. Urayama, T. B. Norris, J. Singh, J. Phillips, and P. Bhattacharya. 2002. Gain dynamics and ultrafast spectral hole burning in In(Ga)As self-organized quantum dots. *Appl. Phys. Lett.* 81:670–672.

Krishna, S. 2005. Quantum dots-in-a-well infrared photodetector. *J. Phys. D* 38:2142–2150.

Krishna, S., S. Raghavan, G. von Winckel, P. Rotella, A. Stintz, C. P. Morath, D. Le, and S. W. Kennerly. 2003a. Two color InAs/InGaAs dots-in-a-well detector with background-limited performance at 91 K. *Appl. Phys. Lett.* 82:2574.

Krishna, S., S. Raghavan, G. von Winckel, A. Stintz, G. Ariyawansa, S. G. Matsik, and A. G. U. Perera. 2003b. Three-color (λ_{p1}~3.8 μm, λ_{p2}~8.5 μm, and λ_{p3}~23.2 μm) InAs/InGaAs quantum-dots-in-a-well detector. *Appl. Phys. Lett.* 83:2745–2747.

Krishna, S., D. Forman, S. Annamalai, P. Dowd, P. Varangis, T. Tumolillo Jr., A. Gray, J. Zilko, K. Sun, M. Liu, and J. Campbell. 2005. Demonstration of a 320 × 256 two-color focal plane array using InAs/InGaAs quantum dots in well detectors. *Appl. Phys. Lett.* 86:193501–193503.

Leon, R., Y. Kim, C. Jagadish, M. Gal, J. Zou, and D. J. H. Cockayne. 1996. Effects of interdiffusion on the luminescence of InGaAs/GaAs quantum dots. *Appl. Phys. Lett.* 69:1888–1890.

Lim, H., S. Tsao, W. Zhang, and M. Razeghi. 2007. High-performance InAs quantum-dot infrared photodetectors grown on InP substrate operating at room temperature. *Appl. Phys. Lett.* 90:131112–131114.

Liu, H. C., R. Dudek, A. Shen, E. Dupont, C. Y. Song, Z. R. Wasilewski, and M. Buchanan. 2001a. High absorption (90%) quantum-well infrared photodetectors. *Appl. Phys. Lett.* 79:4237–4239.

Liu, H. C., M. Gao, J. McCaffrey, Z. R. Wasilewski, and S. Fafard. 2001b. Quantum dot infrared photodetectors. *Appl. Phys. Lett.* 78:79–81.

Maimon, S., E. Finkman, G. Bahir, S. E. Schacham, J. M. Garcia, and P. M. Petroff. 1998. Intersublevel transitions in InAs/GaAs quantum dots infrared photodetectors. *Appl. Phys. Lett.* 73:2003–2005.

Mashade, M. B. El, M. Ashry, and A. Nasr. 2003. Theoretical analysis of quantum dot infrared photodetectors. *Semicond. Sci. Technol.* 18:891–900.

Motohisa, J., K. Kumakura, M. Kishida, T. Yamazaki, T. Fukui, H. Hasegawa, and K. Wada. 1995. Fabrication of GaAs/AlGaAs quantum dots by metalorganic vapor phase epitaxy on patterned GaAs substrates. *Jpn. J. Appl. Phys.* 34:1098–1101.

Nasr, A. and M. B. EL Mashade. 2006. Theoretical comparison between quantum well and dot infrared photodetectors. *IEE Proc. Optoelectron.* 153:183–190.

Phillips, J. 2002. Evaluation of the fundamental properties of quantum dot infrared detectors. *J. Appl. Phys.* 91:4590–4594.

Phillips, J., P. Bhattacharya, S. W. Kennerly, D. W. Beekman, and M. Dutta. 1999. Self-assembled InAs-GaAs quantum-dot intersubband detectors. *IEEE J. Quantum Electron.* 35:936–943.

Raghavan, S., P. Rotella, A. Stintz, B. Fuchs, S. Krishna, C. Morath, D. A. Cardimona, and S. W. Kennerly. 2002. High-responsivity, normal-incidence long-wave infrared (~7.2 μm) InAs/In$_{0.15}$Ga$_{0.85}$As dots-in-a-well detector. *Appl. Phys. Lett.* 81:1369–1371.

Rappaport, N., E. Finkman, T. Brunhes, P. Boucaud, S. Sauvage, N. Yam, V. Le Thanh, and D. Bouchier. 2000. Midinfrared photoconductivity of Ge/Si self-assembled quantum dots. *Appl. Phys. Lett.* 77:3224–3226.

Ryzhii, V. 1996. The theory of quantum-dot infrared phototransistors. *Semicond. Sci. Technol.* 11:759–765.

Ryzhii, V. 1997. Characteristics of quantum well infrared photodetectors. *J. Appl. Phys.* 81:6442–6448.

Ryzhii, V., V. Pipa, I. Khmyrova, V. Mitin, and M. Willander. 2000. Dark current quantum dot infrared photodetectors. *Jpn. J. Appl. Phys.* 39:L1283–L1285.

Shen, A., H. C. Liu, M. Buchanan, M. Gao, F. Szmulowicz, G. J. Brown, and J. Ehret. 2000. Progress on optimization of p-type GaAs/AlAs quantum well infrared photodetectors. *J. Vac. Sci. Technol. A,* 18:601–604.

Shenoi, R. V., D. A. Ramirez, Y. Sharma, R. S. Attaluri, J. Rosenberg, O. J. Painter, and S. Krishna. 2007. Plasmon assisted photonic crystal quantum dot sensors. Paper read at *Nanophotonics and Macrophotonics for Space Environments, SPIE Optics Photonics,* San Diego, CA.

Shenoi, R. V., R. S. Attaluri, A. Siroya, J. Shao, Y. D. Sharma, A. Stintz, T. E. Vandervelde, and S. Krishna. 2008. Low-strain InAs/InGaAs/GaAs quantum dots-in-a-well infrared photodetector. *J. Vac. Sci. Technol. B* 26:1136–1139.

Sidorov, Y. G., S. A. Dvoretsky, M. V. Yakushev, N. N. Mikhailov, V. S. Varavin, and V. I. Liberman. 1997. Peculiarities of the MBE growth physics and technology of narrow-gap II–VI compounds. *Thin Solid Films* 306:253–265.

Stiff-Roberts, A. D., S. Krishna, P. Bhattacharya, and S. Kennerly. 2002. Low-bias, high-temperature performance of a normal-incidence InAs/GaAs vertical quantum-dot infrared photodetector with a current-blocking barrier *J. Vac. Sci. Technol. B* 20:1185–1187.

Szafraniec, J., S. Tsao, W. Zhang, H. Lim, M. Taguchi, A. A. Quivy, B. Movaghar, and M. Razeghi. 2006. High-detectivity quantum-dot infrared photodetectors grown by metalorganic chemical-vapor deposition. *Appl. Phys. Lett.* 88:121102–121104.

Ting, D. Z. Y., Y. C. Chang, S. V. Bandara, C. J. Hill, and S. D. Gunapala. 2007. Band structure and impurity effects on optical properties of quantum well and quantum dot infrared photodetectors. *Infrared Phys. Technol.* 50:136–141.

Tsao, S., H. Lim, W. Zhang, and M. Razeghi. 2007. High operating temperature 320×256 middle-wavelength infrared focal plane array imaging based on an InAs/InGaAs/InAlAs/InP quantum dot infrared photodetector. *Appl. Phys. Lett.* 90:201109–201111.

Vandervelde, T. E., J. Shao, A. Stintz, and S. Krishna. 2007. Investigation of shape engineering in InAs quantum dots using various capping materials. Paper read at *MRS Fall Meeting*, Boston, MA.

Varley, E., M. Lenz, S. J. Lee, J. S. Brown, D. A. Ramirez, A. Stintz, and S. Krishna. 2007. Single bump, two-color quantum dot camera. *Appl. Phys. Lett.* 91:081120–081122.

Wang, Z. M., ed. 2008. *Self-Assembled Quantum Dots*. New York: Springer.

Wang, S. Y., S. D. Lin, H. W. Wu, and C. P. Lee. 2001. Low dark current quantum-dot infrared photodetectors with an AlGaAs current blocking layer. *Appl. Phys. Lett.* 78:1023–1025.

Zhang, W., H. Lim, M. Taguchi, S. Tsao, B. Movaghar, and M. Razeghi. 2005. High-detectivity InAs quantum-dot infrared photodetectors grown on InP by metal–organic chemical–vapor deposition. *Appl. Phys. Lett.* 86:191103–191105.

Zibik, E. A., L. R. Wilson, R. P. Green, G. Bastard, R. Ferreira, P. J. Phillips, D. A. Carder, J-P. R. Wells, J. W. Cockburn, M. S. Skolnick, M. J. Steer, and M. Hopkinson. 2004. Intraband relaxation via polaron decay in InAs self-assembled quantum dots. *Phys. Rev. B* 70:161305–161308.

24

Type-II InAs/GaSb Superlattice Photon Detectors and Focal Plane Arrays

Manijeh Razeghi
Binh-Minh Nguyen
Pierre-Yves
   Delaunay

24.1 Introduction

Among the currently developing technologies, the only three that take advantage of low-dimensional properties of quantum mechanics include type-II InAs/GaSb superlattice (SL) photodetectors, the quantum well intersubband photodetectors, and the quantum dot infrared photodetector. Going by dimensionality, the type-II superlattice is still a three-dimensional system, the same as a bulk semiconductor, while the other two systems are so to speak "two- and zero-dimensional systems," respectively, meaning that the confinements of carriers are locally in one or more than one direction. However, compared with the quantum well and the quantum dot, type-II superlattices share the same capability of spatially localizing electrons in a nanoscale space. This is due to the spatial separation of carriers in the constituent layers. Moreover, the physics that underlie the type-II superlattice give us the unique ability to control the interaction between localized electrons by way of varying potential barrier thickness. This gives us, in this system, an additional degree of freedom for designing and creating new "materials." This is not unlike the way Nature created solid crystals from atoms. Having bulk-like properties, type-II superlattices can be exploited using the considerable understanding acquired for bulk semiconductors in the past. This knowledge can be used to optimally design device architectures for photodetecting applications. The close match between theoretical work and experimental developments has allowed type-II superlattice photodetectors to become one of the fastest-growing technologies in infrared detection. During a decade of maturing, the material system has experienced significant progress that has lifted the detector performance to now have comparable levels to the state-of-the-art mercury cadmium telluride (MCT) technology. Whether the superlattice infrared photodetectors can outperform the bulk narrow-gap MCT detectors or not is one of the most important questions for the future of the infrared detection technology.

24.2 Material System and Variants of Type-II Superlattices

24.2.1 The 6.1 Angstrom Family

The 6.1 Angstrom family (Figure 24.1) is an important group in the semiconductor technology. It consists of three members that are closely lattice matched to each other: InAs ($a = 6.0584$ Å), GaSb ($a = 6.0959$ Å), and AlSb ($a = 6.1355$ Å). Since GaSb substrates became available, and the growth of III-antimonide semiconductors was feasible, the 6.1 Å and its compounds have been providing enormous flexibility in designing heterostructures for optical and electronic applications. One great advantage of the family is the small lattice mismatch of the family to GaSb substrates and similar growth windows for the three materials. This enables the growth of high-quality materials with a low density of defects and dislocations. The energy gap of the family and related compounds ranging from 0.41 eV (for InAs) to 1.70 eV (for AlSb) is of particular interest in optoelectronic devices in the short-wavelength infrared (SWIR) and mid-wavelength infrared (MWIR) regimes. Moreover, the heterojunctions between InAs and the other two members benefit of the unique features of the type-II band alignment. On the one hand, the InAs/AlSb interface forms a type-II staggered lineup where the conduction band of InAs is slightly above the

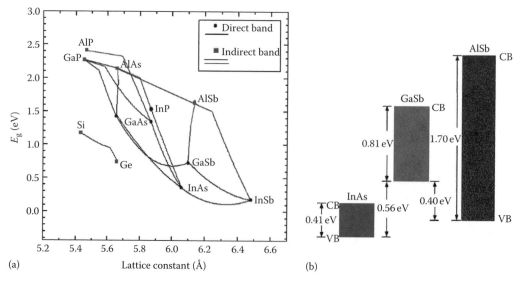

FIGURE 24.1 (a) The energy gap versus the lattice constant of InAs, GaSb, and AlSb compared with other semiconductors. (b) InAs, GaSb, and AlSb energy band lineups.

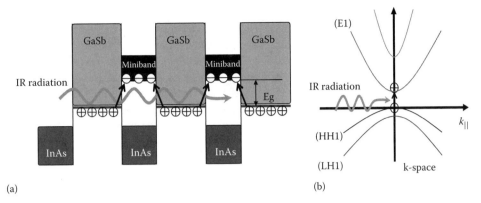

FIGURE 24.2   (a) Spatial band alignment in type-II superlattice and (b) band structure with direct bandgap and absorption process in k-space.

valence band of AlSb. The high energy gap of AlSb leads to an exceptionally large conduction band offset of about 1.45 eV, enabling the realization of very deep quantum wells and very high tunneling barriers. This heterostructure has been widely utilized in high-frequency field-effect transistors (FETs) and resonant interband tunneling diodes (RITDs). On the other hand, the heterojunction between InAs and GaSb leads to the exotic so-called *broken gap* lineup, where the conduction band of InAs is about 0.15 eV lower than the valence band of Gasb. This type of *misaligned* structure is the reason why type-II superlattices have a flexible bandgap engineering capability, and this will be discussed in the following sections.

24.2.2 Type-II InAs/GaSb Superlattice

The idea of type-II InAs/GaSb superlattice was first proposed by Sai-Halasz and Esaki in the 1970s. The superlattice is formed by alternating the InAs and GaSb layers over several periods. This creates a one-dimensional periodic structure, like the periodic atomic chain in naturally occurring crystals (Figure 24.2). The type-II broken gap alignment leads to the separation of electrons and holes into the InAs and GaSb layers, respectively. The charge transfer gives rise to a high local electric field and strong interlayer tunneling of carriers without requiring an external bias or external doping. Large period superlattices behave like semimetals but if the superlattice period is shortened, the quantization effects are enhanced causing a transition from a semimetal to a narrow-gap semiconductor. The resulting energy gaps depend on the layer thicknesses and interface compositions. In reciprocal k-space, the superlattice is a direct bandgap material that enables optical coupling, as shown in Figure 24.2b.

24.2.3 Variants of Sb-Based Superlattices

24.2.3.1 InAs/GaInSb

The first variant of the original type-II binary/binary InAs/GaSb superlattice is the binary/ternary InAs/GaInSb superlattice. The structure was proposed by Fuchs et al. (1997) to compensate for the spatial separation of electrons and holes, in order to enhance the optical matrix element. By adding Indium to the GaSb material, the barrier height is reduced; the electron wavefunction becomes more delocalized, leading to a stronger overlap with the hole wavefunctions and a better transition probability. The GaInSb barrier also allows for one other degree of freedom in controlling the energy gap, i.e., via the change in the Ga/In molar fraction. This however also induces additional experimental difficulties because the growth of ternary GaInSb requires stricter control of growth temperature and III/V flux ratio than the growth of the binary GaSb. It is also more sensitive to the growth parameters, and results in a lower uniformity across a large growth area.

Another issue for ternary superlattices is the positive net strain of the structure when it is grown on a GaSb substrate. As the lattice constant of InSb is much larger than that of GaSb, the GaInSb layer will make the superlattice structure more compressive (the lattice parameter of superlattice is larger than the lattice parameter of the substrate). Moreover, the lowering of the barrier due to the introduction of indium into the GaSb layer tends to reduce the energy gap of the superlattice. When designing for a MWIR cutoff or shorter, the width of the InAs wells must be shortened in order to raise the electron energy level. This then increases the energy gap. The reduction of the InAs thickness will increase the average lattice constant of the superlattice and induce even more strain that can further degrade the material quality.

24.2.3.2 W-Structure

Contrary to the InAs/GaInSb superlattice, the W-structure superlattice (WSL) proposed by Meyer et al. is a design where the delocalization of electron is reduced by the insertion of GaAlSb layers. The structure was first used for quantum well lasers and then applied for infrared photodiodes (Meyer et al. 1995). The schematic diagram of the W-structure is presented in Figure 24.3. The GaAlSb barrier was used to block the electron wavefunction overlap and increase the effective mass of the superlattice. In parallel, the thin GaInSb barrier is used to enhance the electron/hole interaction, as in the case of InAs/GaInSb superlattice. The disadvantages of the W-structure are again the difficulties associated with material growth. Similar to the ternary InAs/GaInSb superlattice, W-structure is based on ternary compounds and is sensitive to the growth condition. Moreover, the switching from the InAs layer to the GaAlSb layer means no common atom at the interface. This adds another degree of complexity in both theoretical calculations and experimental realizations. The large lattice mismatch between InAs and GaAlSb also limits the growth quality.

24.2.3.3 M-Structure

The M-structure superlattice, proposed by (Nguyen et al. 2007a), is constructed by inserting a thin AlSb barrier in the middle of the GaSb layer of a normal type-II binary InAs/GaSb superlattice. Figure 24.4 shows the schematic diagram of the energy band alignment of its constituents. The colored regions

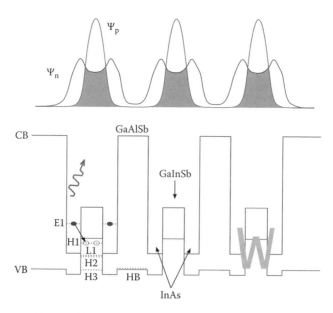

FIGURE 24.3 Schematic diagram of the W-structure superlattice. (a) Spatial wavefunction of electron and hole. (b) Spatial alignment of the lowest conduction band and the highest valence band of the constituent materials.

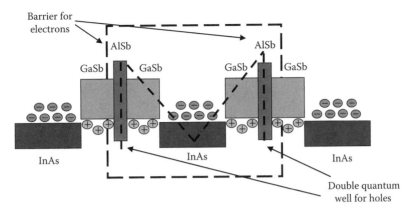

FIGURE 24.4 Schematic diagram of the M-structure. The inserted AlSb layer forms a barrier for electrons in the conduction band and a double quantum well for holes in the valence band. (From Nguyen, B.M. et al., Type-II M structure photodiodes; An alternative material design for mid-wave to long wavelength infrared regimes, in *Quantum Sensing and Nanophotonic Devices IV*, 1st edn., vol. 6479, San Jose, CA: SPIE, 2007c, pp. 64790s–64710s. With permission. Copyright 2007 by SPIE.)

represent the prohibited bandgap of the structure. This AlSb-containing superlattice is named the M-structure. This stands for the shape of the letter M of the band alignment of the AlSb/GaSb/InAs/GaSb/AlSb layers. The alignment allows the structure to acquire the beneficial properties of the well-established binary–binary antimonide-based SL structures used in high-performance infrared detectors. First, it conserves the type-II misalignment of InAs and GaSb, which is capable of eliminating the electron/hole split-off band Auger transition by reducing the resonance between the energy gap and split-off band. Second, as illustrated in the Figure 24.4, AlSb has a large energy bandgap with the conduction band offset higher than that of GaSb and the valence band offset slightly higher than that of InAs. Due to this band offset, AlSb can be utilized as a blocking layer for both electrons in the conduction band and holes in the valence band. As a consequence, the insertion of the AlSb into the GaSb, forming the M-superlattice, will reduce the tunneling probability in long-wavelength infrared (LWIR) and very-long-wavelength infrared (VLWIR) photodiodes because the effective mass of the carriers over the standard binary–binary SL is increased. In comparison with other ternary/quaternary superlattices, such as the WSL or the InAs/GaInSb superlattices, the M-structure is easier to realize because it still keeps the simple form of a binary-only structure, which is less sensitive to the growth temperature and the III–V flux ratio. Finally, the growth of the M-structure is as simple as a standard binary–binary InAs/GaSb structure because there is a small mismatch in the lattice constant between AlSb (6.136 Å) and GaSb (6.097 Å) and the common antimony atom allows for a smooth interface between the two layers. The M-structure does not need as much care as the growth of AlSb on InAs in the WSL where the interface of AlSb and InAs have no common atom and AlSb grown on InAs is highly strained.

24.3 Historic Development of Type-II Superlattice Photodetectors

In 1975, Tsu et al. (1975) first proposed the concept of the semiconductor superlattice using an alternating GaAs/GaAlAs heterostructure. They identified the quantum states and predicted how this would give rise to the phenomenon of negative differential conductance (NDR) in the device. Two years later, the same group predicted important effects for superlattices with type-II alignment between two constituent layers in which the conduction band of one layer lies below the valence band of the other (Sai-Halasz et al. 1977). Taking InAs/GaSb as a typical example for a type-II superlattice, theoretical calculations were carried out to illustrate the energy gap shifting and the semiconductor/semimetal transition when the layer thicknesses are changed. The structure was then proposed by Smith and Mailhiot (1987) for infrared detection applications, exploiting the fact that the cutoff wavelength could be adjusted in the infrared range, and that the effective mass was constant, regardless of the energy gap.

Despite positive theoretical predictions, high-quality growth of this type of materials was not demonstrated until recent advances in the technology of the molecular-beam epitaxy (MBE) technique. The first experimental demonstration of photon detectors comprising superlattices was by Johnson et al. (1996). The first photoconductors grown on GaAs substrate were demonstrated at the Center for Quantum Devices (CQD) at Northwestern University in 1997 (Mohseni et al. 1997). However, type-II superlattices grown on GaAs substrate experienced many difficulties due to the 7% lattice mismatch between the epitaxial layer and the substrate. Since epi-ready GaSb substrates became commercially available in the late 1990s, the growth of type-II superlattice on lattice-matched GaSb have been significantly improved. Defect- and dislocation-free crystals enable the realization of photodiode detectors that require large differential resistance and low leakage. In the perpendicular p-i-n superlattice photodiode configuration, the signal current flows in the growth direction, the size of the detector element could thus be reduced, and multielement devices, such as focal plane arrays became feasible. First high-performance photodiodes using ternary InAs/GaInSb superlattices were demonstrated by F. Fuchs et al. in Germany in 1997 (Fuchs et al. 1997). Since then, efforts have been concentrated in the optimization of photodiode performance and the fabrication of FPAs. In 2003, the first demonstration of an FPA based on type-II superlattice material systems was carried out at the CQD. This provided the proof of concept of this technology. One year later, in 2004, a high-performance FPA with $5\,\mu m$ cutoff was demonstrated in Germany (Cabanski et al. 2004).

Recently, excellent material quality has been achieved with matured growth techniques, allowing for the realization of complex device architectures with higher performance. Vurgaftman et al. (2006) proposed graded gap, and then hybrid graded gap designs for photodiodes based on W-structure superlattice, and have proved that the dark current due to tunneling and recombination processes in LWIR photodiodes can be suppressed. Nguyen et al. (2007b) have applied the great tunability of the M-structure superlattice for the p+-π-M-n+ design that also exhibited an even greater suppression of the tunneling current in LWIR and VLWIR photodiodes. Rodriguez et al. utilized the nBn design for type-II superlattice and proved the concept for MWIR single-element detectors and FPA (Rodriguez et al. 2007). Heterostructure designs with high bandgap contact regions were also independently proposed by Delaunay et al. (2007a) and by Vurgaftman et al. (2006) to avoid the bending of energy bands near the device's sidewall, resulting in a significant reduction of surface leakage even on unpassivated devices. With high-quality material, Nguyen et al. (2007c) have demonstrated a quantum efficiency as high as 50% in topside illumination configuration. Delaunay et al. (2007b) then showed that the same device could achieve a quantum efficiency up to 75% at backside illumination configuration, without antireflective coating. This value is equivalent to that of the state-of-the-art MCT photodetectors. Nguyen et al. (2008) then combined both the optical and electrical optimization schemes into one superlattice structure that exhibited a specific detectivity $8.1 \times 10^{11}\,cm\sqrt{Hz}/W$. The internal noise of the detector was even lower than the noise due to the fluctuation of the radiation background, and the performance was limited by the background radiation, and it was said to be a background-limited infrared photodetector (BLIP).

Side by side with the development of device performance, device processing techniques have been gradually improved, making the fabrication of high-quality single element and FPAs more routine. Researchers at Fraunhofer have commercialized MWIR FPAs based on type-II superlattice, and in 2007, they demonstrated for the first time a MWIR–MWIR two-color type-II superlattice FPAs (Walther et al. 2007). In the LWIR range, Delaunay et al. (2009) realized FPA using M-structure superlattice design, and showed a reduction of dark currents by one order of magnitude compared to FPA using the conventional type-II InAs/GaSb p-i-n photodiode design. The historical progress in the development of this material system is summarized in Figure 24.5.

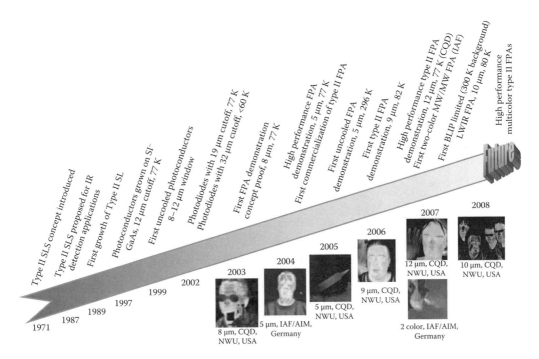

FIGURE 24.5 Historical development of type-II superlattice photodetectors.

24.4 Physics of Type-II InAs/GaSb Superlattice

Theoretical methods that have been developed for type-II superlattices include the k.p-based theory (Grein et al. 1993, 1994), the empirical tight-binding model (ETBM) (Wei and Razeghi 2004), and the pseudo-potential model (Piquini et al. 2008). Compared to other methods, the ETBM exhibits great advantages with the capability to calculate the band structure in the whole Brillouin zone and to precisely describe the atomic layering to the superlattice, taking into account the growth imperfection. The method is also fast and does not require massive and complex numerical calculation.

The main spirit of the ETBM is the decomposition of the wavefunction of the superlattice into the Bloch superposition of localized atomic orbitals:

$$\Psi_{\vec{k}}(\vec{r}) = \sum_{\vec{R}_{SL}} \sum_{\alpha} \sum_{n=1}^{N} \exp\left(i\vec{k} \cdot \left(\vec{R} + \vec{\tau}_n\right)\right) A_n^{\alpha} \varphi_n^{\alpha}\left(\vec{r} - \vec{R} - \vec{\tau}_n\right) \qquad (24.1)$$

where
 N is the total number of atoms in one unit cell

A_n^{α} are constants, and the sum of $\vec{R}_{SL}$ runs through all the unit cells that are involved in the nearest-neighbor interaction, and $\varphi_n^{\alpha}(\vec{r}\text{-}\vec{R}\text{-}\vec{\tau}_n)$ is the α orbital ($\alpha = s, p, s^*$) of the nth atom in the unit cell. By utilizing the formalism of the Schrödinger equation, the band structure calculation problem is reduced to an eigenvalue problem with the form

$$\hat{H}\,\Psi_k(\vec{r}) = E(k)\Psi_k(\vec{r}) \qquad (24.2)$$

where H is the Hamiltonian for the system and $E(k)$ is the energy eigenvalues that are dependent on $\vec{k}$ values in the first Brillouin zone. Using the orthogonal properties of the atomic orbitals, the eigenvalues can be rewritten:

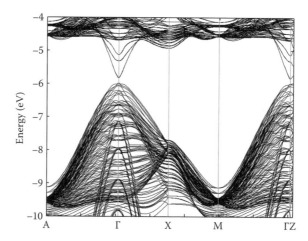

FIGURE 24.6 Calculated band structure for a typical superlattice structure along high-symmetry directions in the entire first Brillouin zone.

$$\sum_{\vec{R}_{SL}} \sum_{\alpha} \sum_{n=1}^{N} \exp\left(i\vec{k}(\vec{R} + \vec{\tau}_n - \vec{R}' + \vec{\tau}'_n)\right)$$

$$\times A_n^{\alpha} \iiint dr^3 \varphi_m^{\beta*}(\vec{r} - \vec{R}' - \vec{\tau}_m)\hat{H}\varphi_n^{\alpha}(\vec{r} - \vec{R} - \vec{\tau}_n) = E \cdot A_m^{\beta} \qquad (24.3)$$

The coefficient A_n^{α} and the energy $E_n(k)$ can be calculated once the Hamiltonian matrix elements are explicitly defined. In the ETBM, the Hamiltonian matrix of a superlattice is constructed from blocks of Hamiltonian matrix for bulk semiconductors, such as GaSb, GaAs, InSb, and InAs, with the same atomic sequence.

Figure 24.6 sketches the band structure $E(\vec{k})$ of a typical type-II superlattice using the ETBM formalism. Energy dispersion can be calculated for hundreds of bands, from the very low valence to the high-conduction continuum at the GaSb's conduction band level. The minima of the conduction band and the maxima of the valence band both occur at the center of the Brillouin zone, indicating that the superlattice is a direct gap material despite the spatial separation of electron/hole in the real space. The energy gap, determined by the difference between the highest valence band and the lowest conduction band of several superlattice designs, was compared to the experimental values extracted from the optical response measurements, and showed very good agreement (Figure 24.7).

From the band structure calculation, the conduction band and valence band limits for all possible superlattice configurations are mapped in Figure 24.8. The evolution of the energy levels is exactly as predicted by the qualitative descriptions in the previous section. The conduction band shows a strong dependence on the InAs thickness, due to the modification of the electron's well width, and also a slight variation with the change of the GaSb layer thickness. This is due to the change of the wavefunction overlap.

24.5 Material Growth and Characterization

The growth of Sb-based type-II superlattice is much more difficult than the growth of bulk semiconductors or simpler heterostructures. A typical device structure consists of hundreds to thousands of sandwiched layers with an equivalent amount of interfaces. Each superlattice period requires strict controls of layer thickness and interface composition in order to achieve an overall smoothness of the whole structure. Moreover, each material has its own growth window, with a small tolerance of growth

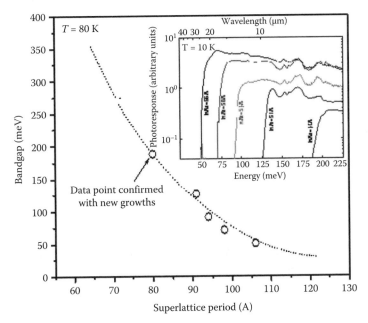

FIGURE 24.7 Comparison between experimental energy gap (extracted from the optical responses in the inset) and theoretical prediction for several type-II superlattice designs.

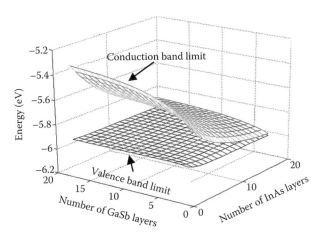

FIGURE 24.8 ETBM calculation for the band edge energy distribution for type-II InAs/GaSb superlattices.

temperature and III/V flux ratio. Since the growth temperature and material flux cannot be changed within the fast switching sequence of the superlattice growth, the growth condition for the superlattice must be a compromise between the growth parameters of the constituent layers. For type-II superlattices, the growth of GaSb and InAs requires high growth temperature, while the InSb interfaces can be melted at above 450°C. A lower growth temperature must be set in order to maintain the sharpness of the interfaces. For such low growth temperature, molecular-beam epitaxy is the best candidate to realize high-quality superlattice since the growth with MOCVD will require high substrate temperature to crack the gaseous products prior to the deposition of materials. Moreover, with MBE, the layer thicknesses can be precisely determined by computer-controlled shutter sequences, which can be actuated with accuracy up to 0.1 s. The ultrahigh vacuum condition of the MBE also allows for high purity and low background material growth.

High-quality superlattice is often characterized by the following:

- Smooth surface with clear atomic steps under atomic force microscopy
- Narrow peaks, high-order diffraction patterns under high-resolution x-ray diffraction
- Narrow photoluminescence peaks at exactly the expected position
- Sharp interface and precise layer thickness as revealed by transmission electron microscopy or scanning tunneling microscopy (Figure 24.9)

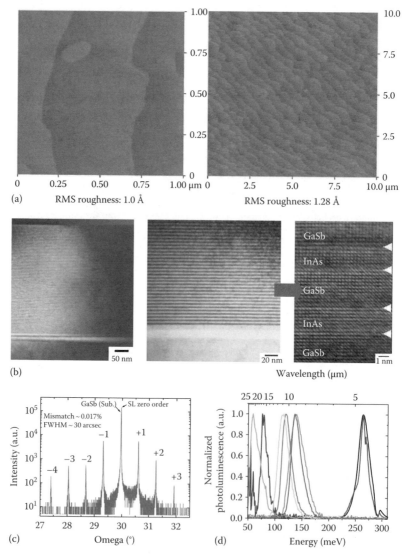

FIGURE 24.9 Routined structural characterization for type-II superlattice using (a) atomic step microscopy. (From Razeghi, M. et al., Type-II superlattice photodetectors for MWIR to VLWIR focal plane arrays, in *Infrared Technology and Applications XXXII*, vol. 6206, 2006, pp. 62060N. With permission. Copyright 2006 by SPIE.) (b) Transmission electron microscopy. (From Razeghi, M. et al., Development of material quality and structures design for high performance Type II InAs/GaSb superlattice photodiodes and focal plane arrays, *in Infrared Spacebcone Remote Sensing and Instrumentation XVI*, vol. 7082, 2008, pp. 708204. With permission. Copyright 2008 by SPIE.) (c) High-resolution x-ray diffraction, and (d) photoluminescence. (From Razeghi, M. et al., Type-II superlattice photodetectors for MWIR to VLWIR focal plane arrays, in *Infrared Technology and Applications XXXII*, vol. 6206, 2006, pp. 62060N. With permission. Copyright 2006 by SPIE.)

24.6 Device Fabrication and Passivations

24.6.1 Single-Element Device for Testing

To characterize the quality and performance of the material, single-element devices are fabricated using standard processing techniques. Basic fabrication steps are presented in Figure 24.10. Diodes with variant sizes undergo the mesa definition process with positive lithography and a combination of dry and wet etching. Then the metal contacts are defined using negative lithography and deposited using electron-beam metal evaporator.

For narrow bandgap material and small-size devices, the surface leakage becomes a limiting factor to the device performance and has been a known processing challenge for years. The surface leakage results from the abrupt termination of the periodic structure at the diodes' side walls. Due to the incomplete bond of surface atoms, a bending of the conduction and valence bands near the surface occurs,

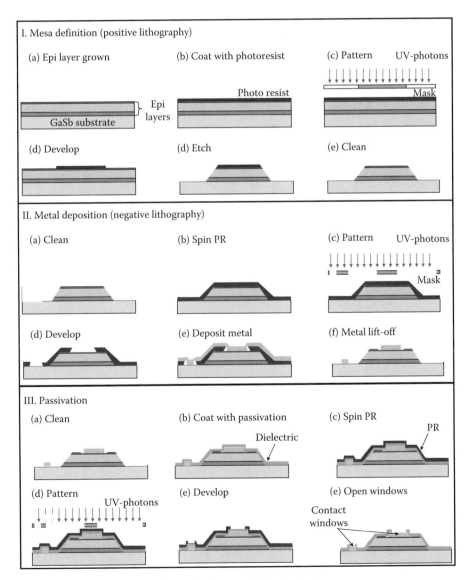

FIGURE 24.10 Basic processing steps for single-element device fabrication.

creating a surface conduction channel along the mesa sidewalls. Moreover, foreign adsorbents and process contaminants can further alter the surface potential and introduce trap levels within the energy gap, leading to more efficient trap-assisted tunneling currents. A good passivation layer is expected to fix the pinning of the Fermi level near the surface and to physically protect the mesa side walls from outside contaminants.

24.6.2 Passivation of Type-II Superlattice

Most semiconductor crystals are described by a band structure that results from the overall electric potential created by the atoms of the material. The discrete atomic energy level bands are replaced by continuous bands because of the periodic repetition of the structure of the crystal. This model, accurate for a bulk material, cannot describe accurately the electrical properties of a semiconductor in the vicinity of its surface, where the periodicity of the structure is broken. In addition, chemical bonds are left dangling at the surface and interact with contaminants, thus modifying the electric potential. The problem is even more complicated for superlattices as the crystal structure is not isotropic and as this semiconductor is made of two different materials. As a result, the nature of the dangling bonds will not be the same on the sidewalls and at the surface of the mesas (Figure 24.11).

The abrupt termination of the crystal and the interaction with surface contaminants is often modeled as a bending of the valence and the conduction band. As the bands bend, the difference of energy with the Fermi level changes (Figure 24.12). This modifies the concentration of carriers in the vicinity of the

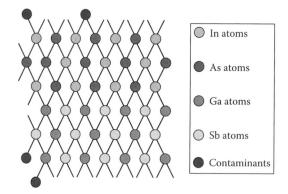

FIGURE 24.11 Dangling bonds at the surface of the semiconductor can be occupied by surface contaminants. Different dangling bonds are available depending on the plan of the surface.

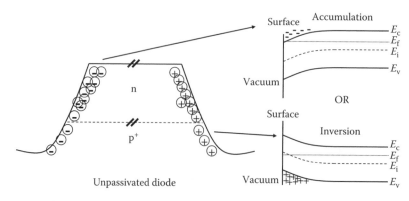

FIGURE 24.12 A schematic depiction of a semiconductor mesa diode with either an accumulated or inverted surface resulting from surface-state-induced band bending.

sidewalls and creates conduction channels that have different I–V characteristics than the bulk material. If the bulk was optimized to minimize the dark current of the device, it is very likely that these conduction channels will degrade the *I–V* of the devices. The presence of contaminants can also create energy traps within the bandgap that increase the trap-assisted component of the dark current. An effective passivation layer has to minimize the band bending at the surface of the devices, while physically protecting the surface from further contamination.

The effect of the band bending is even more important as the bandgap of infrared materials is small. A rough criterion to estimate the effect on the electrical properties of the device consists in calculating the amount of band bending, δ, that will invert the type of the material:

$$\delta_n^{inv} = -\frac{1}{2}\left[E_g + kT \cdot \ln\left(\frac{N_D^2}{N_c N_v}\right)\right] \quad n\,\text{region} \tag{24.4}$$

$$\delta_p^{inv} = \frac{1}{2}\left[E_g + kT \cdot \ln\left(\frac{N_A^2}{N_c N_v}\right)\right] \quad p\,\text{region} \tag{24.5}$$

where
   N_c is the effective density of states for electrons (m^{-3})
   N_v is the effective density of states for holes (m^{-3})
   N_D is the density of donors (m^{-3})
   N_A is the density of acceptors (m^{-3})
   E_g is the bandgap (J)
   k is Boltzmann's constant (m^2 kg s^{-2} K^{-1})
   T is the temperature (K)

The lower the bandgap is, the lower is the inversion potential. Once the p or n side of the photodiode is inverted, a strong electron or hole conduction channel appears along the sidewalls. This channel shorts the diode and drastically increases the dark current and the noise of the photon detectors. Therefore, this paragraph will focus on the description of the three main passivation techniques developed for LWIR type-II superlattice photon detectors:

- Wet chemical passivation
- Dielectric passivation (SiO$_2$ and polyimide)
- Semiconductor overgrowth

24.6.2.1 Wet Chemical Passivation

Unlike popular silicon technology, the native oxide of most III-V compounds is not beneficial as a natural passivation. In the case of GaSb, natural oxides actually create conductive pathways parallel to the surface. The oxidation of GaSb is carried out at relatively low temperatures according to the reaction

$$2GaSb + 3O_2 \rightarrow Ga_2O_3 + Sb_2O_3 \tag{24.6}$$

Wet chemical passivation operates in two steps. First, the solution removes the native oxides at the surface of the semiconductor. Then, new atoms occupy the dangling bonds of the superlattice to prevent reoxidation of the material, surface contamination, and to minimize band bending.

The most common chemical passivations used for type-II superlattices are ammonium sulfide and sodium sulfide based solutions (Zhang et al. 2003, Gin et al. 2004). The solution etches the surface of the material, removing the oxide and, chains of sulfur atoms occupy the dangling bonds. The differential

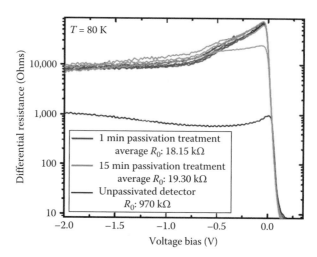

FIGURE 24.13 R_0A increased nearly two orders of magnitude for passivated diodes after passivation treatment in diluted ammonium sulfide solution.

resistance of type-II InAs/GaSb photodiodes improves by one order of magnitude after an exposition of one minute to the solution (Figure 24.13). Wet chemical passivation successfully improves the electrical performance of the superlattice. However, the sulfuric chains easily dissolve into solvents. As a result, this technique is unpractical if the samples need to be processed after passivation. Dielectrics are generally preferred for more complex processes such as the fabrication of focal plane arrays.

24.6.2.2 Dielectric Passivation

The most common passivation scheme used for electronic devices consists in depositing a thin layer of insulator to protect the surface of the semiconductor. The dielectric satisfies the dangling bonds and prevents contaminants from shorting the p-i-n junction. SiO_2 is often chosen for such applications as it is one of the best-known insulators.

An ideal dielectric passivation would be electrically neutral and present a very low concentration of defects at the interface with the semiconductor. Dielectric layers are never completely neutral. Fixed and mobile charges due to defects attract charges with the opposite polarity at the surface of the semiconductor. In the best case, these charges can create a potential that compensates for the natural bending of the bands at the interface semiconductor/dielectric. But, they may also increase the bending and locally invert the superlattice, creating a hole or electron conduction channel that shorts the main p-n junction. A good control of the quality of the dielectric material is very important for the success of the passivation. This is especially important for small bandgap materials that can easily be inverted.

24.6.2.2.1 Silicon Dioxide

While SiO_2 is a reliable passivation in the MWIR (Walther et al. 2005), the situation is more complex at longer wavelengths. Homojunctions passivated with SiO_2 deposited with plasma-enhanced chemical vapor deposition (PECVD) present differential resistances two orders of magnitude lower than unpassivated samples. This is probably because the fixed charges of the dielectric invert the low bandgap superlattice. The electrical performance of the photodiodes can be restored by soaking the sample into water for a few minutes (Delaunay et al. 2007a). This behavior is attributed to compressive stress imparted to the SiO_2 thin film due to the electrostatic interaction of water molecules trapped in pores of the SiO_2 (Robic et al. 1996). However, this is not stable and therefore not suitable as a passivation.

Several groups recently suggested using a double heterostructure with large bandgap p and n contacts to reduce the surface leakage current (Vurgaftman et al. 2006). The contact regions with a cutoff

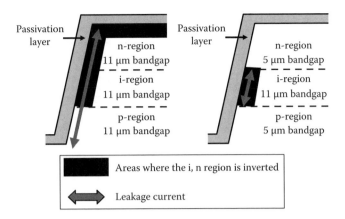

FIGURE 24.14 Sketch of the technique used to suppress the surface leakage in a double heterostructure. This figure represents the case when the n contact region is inverted. (From Delaunay, P.Y. et al., *Appl. Phys. Lett.*, 91(9), 091112, 2007. With permission. American Institute of Physics.)

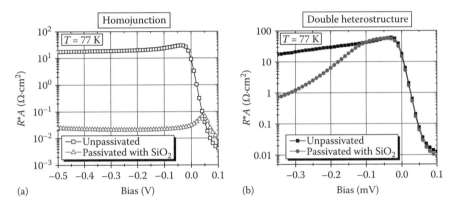

FIGURE 24.15 (a) R_0A of an homojunction before and after SiO_2 passivation. (b) R_0A of a double heterostructure before and after SiO_2 passivation.

wavelength in the MWIR are less sensitive to the presence of surface contaminants and fixed charges. The absorbing region is not changed in order to preserve the optical response of the devices. The high bandgap contacts are not inverted by the fixed charges of the dielectric layer. Thus, even if the LWIR absorbing superlattice is inverted, the electron or hole conduction channel is stopped at the interface with the higher bandgap superlattice (Figure 24.14). The device does not get shorted by the surface conduction channels and the I-V characteristic of the diodes is not affected by the passivation layer.

Double heterostructures passivated with SiO_2 present similar R_0A to the unpassivated superlattice, two orders of magnitudes higher than an homojunction passivated with the same technique (Figure 24.15). And, the surface resistivity of the sidewalls is one order of magnitude higher, indicating a significant reduction of the surface leakage (Table 24.1).

24.6.2.2.2 Polyimide

Conformal polyimide physically passivates and protects the underlying semiconductor from the ambient environment while maintaining or even improving the electrical performance of the device measured prior to passivation. This technique demonstrated nearly bulk-limited performance in type-II InAs/GaSb superlattice photodiodes (Hood et al. 2007). The surface resistivity of polyimide passivated photodiodes is increased at least by one order of magnitude compared to as-processed devices. This

TABLE 24.1 Comparison of Surface Resistivities and R_0A Statistics for Passivated, Unpassivated Homojunctions and Double Heterostructures

Design	Passivation	Surface Resistivity ($\Omega \cdot$ cm)	Average R_0A ($\Omega \cdot$ cm^2)	Standard Deviation R_0A ($\Omega \cdot$ cm^2)
Homojunction	Unpassivated	3.6×10^3	8	6
	SiO$_2$	3.6	0.047	0.029
Double heterostructure	Unpassivated	3.9×10^3	13	2.4
	SiO$_2$	47×10^3	16	2.7

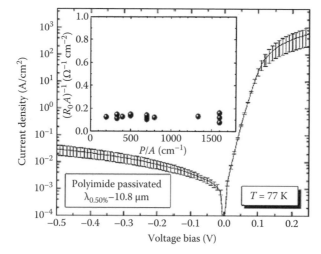

FIGURE 24.16 Average (solid line) and standard deviation (vertical bars) of 10 representative *I–V* curves from polyamide-passivated diodes ranging in size from 25 to 55 µm a side. (From Hood, P. Y. et al., *Appl. Phys. Lett.*, 90, 233513, 2007. American Institute of Physics.)

process has been demonstrated to be compatible with small mesa dimensions and pitches found in typical FPA fabrication processes (Figure 24.16). The polyimide passivation was shown to be stable upon exposure to various ambient conditions as well as over time.

24.6.2.3 Semiconductor Overgrowth

The epitaxial regrowth of single-crystalline semiconductor materials on the sidewall of processed mesas is considered the most ideal, and certainly the most elegant approach to device passivation. Using this approach, it is the band alignment of the epitaxial regrowth material and the underlying superlattice that defines the concentrations of carriers at the surface (Figure 24.17). Additionally, with proper doping of the regrowth layer, the Fermi level can be further defined to reduce the accumulation or inversion of carriers at the diode sidewall. If done properly there is a depletion of electrons at the interface, greatly reducing the likelihood of electrons arriving at the surface and contributing to sidewall leakage current. The concept was demonstrated by several groups using Al$_x$Ga$_{1-x}$As$_y$Sb$_{1-y}$ (Rehm et al. 2005, Hill et al. 2007).

However, the surface of the superlattice needs to be perfectly cleaned and deoxidized prior to semiconductor regrowth in an MBE or MOCVD system. Therefore, it is difficult to achieve high-quality regrowth on processed devices. In addition, as the size of the array increases, the spacing between the mesas is reduced. The high aspect ratio will, in the long term, impact the quality of the regrowth. While semiconductor regrowth is elegant and promising at a single-element level, it is not suitable for focal plane arrays.

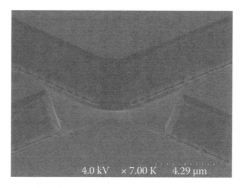

FIGURE 24.17   SEM pictures of FPA mesas after overgrowth.

24.7 Focal Plane Arrays Based on Type-II InAs/GaSb

The ultimate goal of type-II superlattices is to replace HgCdTe as the technology of choice for infrared imaging from the MWIR to the LWIR. Even if some effort has been made to grow the superlattice on GaAs or Si, GaSb remains the only substrate with a lattice mismatch small enough for the growth of a high-quality material. As GaSb is not suitable for regular electronics, the signal processing part of the imager has to be realized on another chip called readout integrated circuit (ROIC). The ROIC, bonded with indium to the array of sensors, extracts the signal from each pixel and processes it into a video signal. The processing of the arrays of photodetectors and the bonding to the ROIC is a very complex process that needs to be controlled carefully in order to preserve the performance of the InAs/GaSb superlattice.

24.7.1 Focal Plane Array Fabrication

The first step of the fabrication consists in defining the array of sensors. Each pixel has to be electrically insulated from its neighbors to avoid compromising the spatial resolution of the imager. An array of mesa is defined using wet and/or dry etching. Dry etching using a BCl_3 plasma is generally preferred because it leads to a higher fill factor. The fill factor is defined as the ratio of the area of a mesa over the area of a unit cell for the ROIC. It quantifies the blind area of the camera. The etching is critical because damaged sidewalls can lead to high surface leakage currents. The photoresist is then removed using a combination of solvents and photoresist stripper. Another photolithography is used to define the top and bottom metal contacts. Ti/Pt/Au gives good ohmic contacts for both electron and hole collection. After removal of the photoresist, the array is passivated with SiO_2 deposited at 260°C using PECVD. Silicon dioxide is preferred to polyimide because it gives more uniform results across a large array. Windows are then opened in the passivation layer using a CF_4-based plasma in an ECR-RIE system (Figure 24.18).

Each mesa has to be connected to its corresponding unit cell on the ROIC. Indium is commonly used for this bonding because it is cheap, soft, and melts at low temperature. It is deposited using electroplating or thermal evaporation. Electroplating can give thick bumps with a high packing density, but the bumps often lack uniformity. The nonuniformity can lead to nonconnected pixels. Thermal evaporation on the other end has very high throughput and uniformity but the packing density is low. Independently of the deposition technique, a thin indium oxide layer forms at the surface of the bumps. This layer is detrimental to the bonding because it prevents a good adhesion between the bump on the ROIC and the bump on the array of sensors. The bumps are generally reflowed above the melting point of indium in a solder flux solution or in an oxygen-free atmosphere. The indium oxide is etched away during the reflow

FIGURE 24.18 Left: SEM picture of a 25 μm × 25 μm FPA pixel. Right: SEM picture of a few FPA pixels.

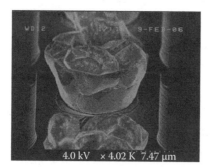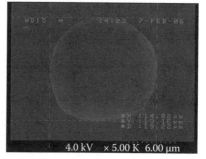

FIGURE 24.19 Left: Image of an indium bump after electroplating. Right, SEM picture of a bump after reflow in a soldering flux solution.

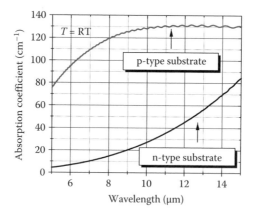

FIGURE 24.20 Absorption coefficients of n-type and p-type substrates measured at room temperature.

and the bumps are reshaped into spheres to minimize surface tension. This shape is optimal for bonding as it concentrates the force in a very small area. Thus, it is more likely to break through an outside oxide layer (Figure 24.19).

The array of sensors is then flip-chip bonded to the ROIC using a high force at room temperature. The indium bumps stick with each other, mechanically and electrically bonding the array to the ROIC. The gap between the two chips is filled with epoxy to improve the mechanical reliability of the FPA.

The p-type and n-type GaSb substrate is not transparent in the infrared because of free electron absorption (Figure 24.20). As the light has to go through the substrate before it reaches the superlattice, it is necessary to remove the substrate in order to maximize the optical response of the array. This is achieved in two steps. First, the GaSb is thinned down to 20–50 μm using a combination of lapping and polishing. The

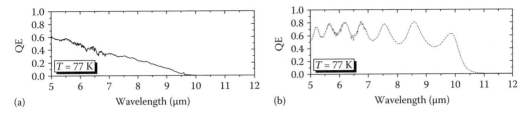

FIGURE 24.21 (a) Quantum efficiency of a sample with a 500 μm thick substrate at 77 K. (b) Quantum efficiency of the same sample after substrate removal at 77 K.

remaining substrate is wet etched with a CrO_3-based solution (Delaunay et al. 2007b). The superlattice is protected during the chemical etching by an etch-stop layer (InAsSb) that was grown prior to the super-lattice. In addition, the thin array of sensors can deform with the ROIC, thus reducing the chance of crack due to thermal expansion. The quantum efficiency is improved by a factor 3 in the LWIR (Figure 24.21).

The FPA is finally wire bonded on a chip carrier and mounted in a testing system that transforms the analog output of the ROIC into a digital video signal.

24.7.2 State-of-the-Art FPAs Based on Superlattices

A lot of progress has been made since the fabrication of the first MWIR and LWIR FPAs based on type-II InAs/GaSb superlattices. In the following paragraph, we will present the state-of-the-art of infrared imaging based on this new technology.

24.7.2.1 MWIR

Infrared imaging using type-II InAs/GaSb superlattices was demonstrated for the first time in the MWIR by Walther et al. (2005). A MWIR camera was later fabricated at the CQD from a type-II super-lattice with a 4.3 μm 50% cutoff wavelength. The array was passivated with polyimide and the substrate was completely removed. The camera was tested both with a narrow-band filter centered around 3.3 μm for radiometric measurement and a lens for imaging characterization. The integration time was set at 23 ms to minimize the noise equivalent temperature difference (NEDT).

At 81 K, the quantum efficiency at 3.3 μm was between 42% and 49%, depending on the bias settings. The operability and the uniformity were respectively 98% and 0.11%. The nonresponsive pixels corresponded to detectors that were not connected (missing indium bumps) or that were saturated by the dark current. After two-point nonuniformity correction, the uniformity went down to 0.0034. The noise increased with the photon irradiance from 10^{13} ph s^{-1} cm^{-2} and the imager approached background-limited performance at 3×10^{13} ph s^{-1} cm^{-2}. At low photon irradiance, the noise and the noise equivalent input were limited by the noise of the IS9705 ROIC from Indigo systems (Figure 24.22). The median NEDT measured with a MWIR lens ($F\# = 2.3$) was measured as low as 10 mK. Those performances are very similar to HgCdTe at this temperature.

The FPA could perform imaging at higher temperatures. Imaging of soldering iron was even obtained at room temperature (Figure 24.23). However, the dark current increased at higher temperature and the integration time had to be reduced. The photosignal was reduced, which led to higher NEDTs. Shot noise started to limit the noise performance of the imager. Therefore, the dark current of type-II superlattice has to be further reduced in order to increase the operating temperature of the camera.

24.7.2.2 LWIR

Recent progress in growth techniques, structure design, and processing has lifted the performances of LWIR superlattice photodetectors to a level where it can compete with existing technologies such as MCT and quantum well infrared photodetectors (QWIP). The introduction of an M-structure barrier

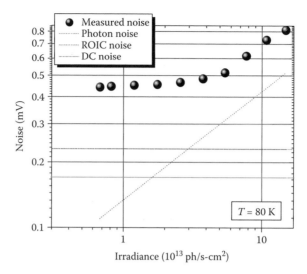

FIGURE 24.22   Variation of the noise of the FPA with the irradiance.

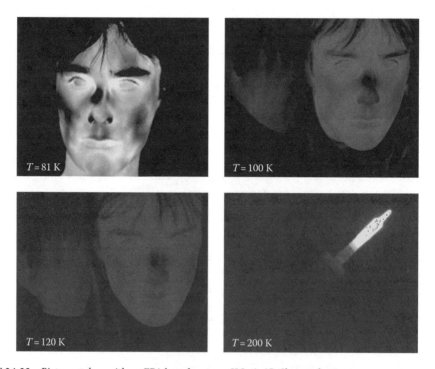

FIGURE 24.23   Pictures taken with an FPA based on type-II InAs/GaSb superlattices.

at the interface of the main p-n junction reduced the dark current by one order of magnitude (Nguyen et al. 2007b). The additional progress made in the FPA fabrication process led to the fabrication of the first BLIP FPA based on this material platform.

The device structure presented in Figure 24.24 includes different characteristics, each of them necessary to maximize the performance of the superlattice FPA. An imager was characterized at 80 K with a 0.11 ms integration time. The median dark current measured with an 81 K cold shield was between 1.1 and 1.6 nA, depending on the bias settings (Figure 24.25).

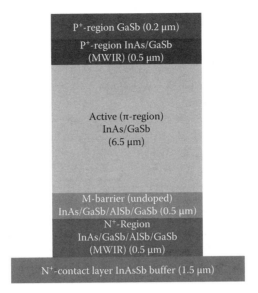

FIGURE 24.24 Structure of type-II superlattice devices. (From Delaunay, P.Y. et al., *IEEE J. Quantum Electron.*, 45(2), 157, 2009. With permission. Institute of Electrical and Electronics Engineers.)

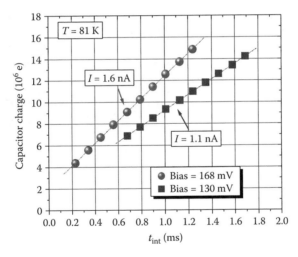

FIGURE 24.25 Variation of the capacitor charge with the integration time at 81 K, under dark conditions. (From Delaunay, P.Y. et al., *IEEE J. Quantum Electron.*, 45(2), 157, 2009. With permission. Institute of Electrical and Electronics Engineers.)

The quantum efficiency at $8\,\mu m$ was 74%, with an operability of 98%. The uniformity of 3.6% was limited by the nonuniformity of the blackbody illumination. The uniformity improved to 0.19% after two-point nonuniformity correction, similar to QWIP-based FPAs. For integration times shorter than 0.5 ms, the noise of the imager was limited by the ROIC. The shot noise became dominant for longer integration times. The FPA approached BLIP at 9×10^{15} ph cm^{-2} s^{-1}.

The NEDT was measured as low as 23 mK, corresponding to an operability of 98% (Figure 24.26). In this configuration, the dark current still corresponds to 5%–10% of the charges accumulated on the capacitors. This means that the dark current needs to be further reduced in order to use the full capacitor charge to store the photoexcited carriers. This would also improve the performance of the imager at lower backgrounds where longer integration times are necessary (Figure 24.27).

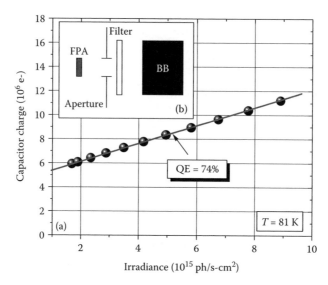

FIGURE 24.26 Variation of the capacitor charge versus the photon irradiance at 8 μm. (From Delaunay, P.Y. et al., *IEEE J. Quantum Electron.*, 45(2), 157, 2009. With permission.)

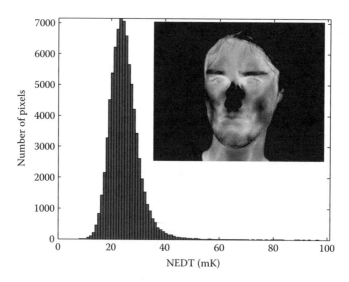

FIGURE 24.27 Histogram of the NEDT at 81 K with a LWIR lens (*F#* = 2.0). The inset is picture of a student taken with the type-II superlattice FPA.

24.7.3 Toward the Third Generation of Type-II InAs/GaSb FPAs

Now that several technologies are able to perform BLIP imaging up to 12 μm, the research effort has shifted toward the demonstration of large and multicolor FPAs. HgCdTe has already demonstrated high-quality two-color arrays (Smith et al. 2003), while QWIP has proven its value for very large format imagers (Gunapala et al. 2005). However, both technologies have flows that either limit the performance of the sensors or increase the cost of a camera. Type-II InAs/GaSb superlattices seem to be a good candidate to solve those issues. The uniformity across a wafer is high enough to allow the fabrication of megapixel cameras and preliminary work has shown its potential for multicolor imaging (Delaunay et al. 2008). However, an effort still has to be made to transfer single-element results to the FPA level.

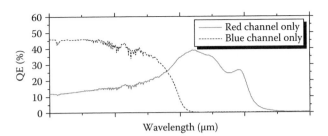

FIGURE 24.28   Optical response of a two-color LWIR/LWIR superlattice device at 77 K.

New processing techniques have to be developed to reduce the pitch of the arrays and improve the performance of the devices (Figure 24.28).

24.8 Type-II Superlattice Detectors for MWIR and LWIR Free-Space Communication

24.8.1 Introduction

Free-space communications (FSC or FSO–free-space optics) is the transmission of data, voice, video, etc., through the atmosphere via a modulated emitter, typically a laser, which is received by a photodetector, typically a photovoltaic or avalanche photodiode device. It has been proclaimed as a solution to creating highly secure, high-bandwidth, point-to-point, and "last-mile" communications links. Such systems can offer up to 100 Gbps (broadband) data rates between two points. Unlike fiber-based optical communication technologies, which rely on expensive glass fibers and corresponding infrastructure for transmissions, FSO operates through the atmosphere. This has the inherent benefit that an optical link can be established *anywhere*, as long as a clear line of sight is present between the source and destination, without the demolition of roads, sidewalks, buildings, etc., needed to lay fiber. Additional advantages of free-space optical networks compared to traditional fiber networks include the security offered by a direct, point-to-point optical link. The data stream/optical link cannot be intercepted without a complete or partial loss in power detected by the receiver (Figure 24.29).

Furthermore, many institutions in urban environments tend to have offices in close proximity to one another (i.e., universities, financial institutions, the media, consulting firms, etc.), but existing high-speed data networks are either nonexistent or are susceptible to eavesdropping on shared networks. A dedicated, non-shared, optical link would provide the needed security and data rates for such a situation. Moreover, since this portion of the electromagnetic spectrum is unlicensed, the link can be established without the need for any FCC (Federal Communications Commission) or governmental licensing, unlike microwave or other radio frequency (RF) technologies, hastening the implementation time and ability to relocate the network when and if necessary.

Realizing the benefits and advantages of this technology, industry has jumped on the opportunity to provide free-space optical communications to end users with a variety of bandwidths, and protocols using "turn-key" communications systems. These systems typically use high-power LEDs or laser diodes operating in the 0.9–1.55 μm, near-infrared range with output powers of ~100 mW per unit. Companies desiring to become involved in this emerging field used existing telecommunications lasers and detectors for quicker integration and time to market. While providing a viable alternative for secure, freely-configurable, high-speed communications, fundamental physical properties limit the performance and practicality of current commercially available systems operating at these wavelengths.

The range of FSC systems is limited by the atmospheric attenuation of the transmitted laser power. Below a finite power level, the receiving detector is unable to adequately process the received radiation under a specified bit error rate (BER). For a useful data link the BER should be 10^{-9} or better, which

FIGURE 24.29   An example application of a point-to-point FSC network in an urban environment.

corresponds to a detector signal-to-noise ratio (S/N) of 6. Commercial, near-infrared systems overcome this limitation by increasing the output power of the emitted signal. This often requires the operation of multiple laser diodes in unison. Companies tend to market this as "redundant transmission," but the observant consumer realizes that this increases the mean time to failure (MTTF) and that the link budget seriously degrades if one of the laser diodes was to fail, thus rendering the long-distance link inoperable, or at least with unacceptably high BERs. An alternative solution is to choose an alternative wavelength where atmospheric attenuation is reduced and longer ranges can be achieved for a given output power. Longer infrared wavelengths are desired due to considerable reductions in atmospheric scattering losses when compared to near-infrared communication links. By moving the operating wavelength into the MWIR or LWIR, enhanced link uptimes and increased range can be achieved due to less susceptibility atmospheric affects.

As with all lasers in the visible region, eye safety must be kept in mind during the design process to keep humans safe from harmful laser radiation. The powers used for FSC are certainly not high enough to cause burns on skin or surrounding structures but for wavelengths less than 1.4 μm only a few milliwatts of laser radiation directed into the eye is enough to severely harm or blind someone by damaging the retina. Though the probability of a well-aligned laser beam entering a human eye is small, the liability is still present for lasers operating below 1.4 μm, which could pose a hazard in the event that any stray reflections arise. Light from eye-safe lasers, however, is absorbed in the fluid within the eye before it reaches the retina to cause damage. Regardless, guaranteeing a perfectly eye-safe system versus one that is "mostly safe" has its benefits to the consumer. MWIR and LWIR lasers, of course, operate within the eye-safe range.

Another benefit of using MWIR/LWIR laser technology is the power and response time offered in a single package. Typically in NIR telecom lasers, the power is limited to a few milliwatts in order avoid nonlinear effects in the fiber-optic cable. Therefore, amplification using erbium-doped fiber amplifiers (EDFA), or similar amplifiers, is typically used to achieve the high powers needed for FSC over extended distances. In the MWIR, lasers such as the quantum cascade laser (QCL) are used widely in spectroscopy applications where high power is required so no external amplification is needed to meet the requirements of an FSC system. Additionally, QCLs can be directly, electrically modulated, eliminating the need for an external modulator, which would increase system cost and complexity.

The combination of room-temperature, continuous-wave (CW) high-power QCLs and high-operating-temperature type-II superlattice photodetectors offers the benefits of MWIR and LWIR systems as well as practical operating conditions.

24.8.2 Free-Space Communication Work at CQD

At the CQD we have already demonstrated the mid-wavelength emitters and detectors that are required to take full advantage of mid-infrared FSC. Since the invention of QCL (Faist et al. 1994), their small size, wide wavelength range (3–160 μm), and high operating temperatures have positioned them well for use in FSC systems. The CQD has led the world in the development of 100% duty-cycle (CW) (Evans et al. 2004) QCLs with high power and high efficiency, which is essential for free-space applications. CW powers over 650 mW at 300 K and nearly 1.2 W at 150 K have been reported (Evans et al. 2007) for mid-infrared QCLs with record efficiencies of nearly 10% at room temperature (300 K) and 20% when slightly cooled to 150 K. In addition, the CQD was the first group to demonstrate high-power CW operation of QCLs at temperatures as high as 90°C, which exceeds the operating temperature requirements for existing telecom and military applications (Evans et al. 2006a). Recent lifetime and reliability studies (Evans et al. 2006b) at the CQD have demonstrated QCLs with 4.6 μm emission wavelengths operating continuously at high powers (>150 mW) for over 18,500 h (over 2 years).

The CQD has also achieved a number of world's firsts in demonstrating photodetectors based on type-II InAs/GaSb superlattice photodiodes, initially proposed by Nobel Laureate Leo Esaki in 1977 (Sai-Halasz et al. 1977). Type-II strained layer superlattices (SLS) have been demonstrated to be a promising alternative to current state-of-the-art MCT photodetector technology throughout the infrared spectrum. The CQD was the first in the world to demonstrate imaging from a room-temperature operating, MWIR focal plane array based on type-II SLS (Wei et al. 2005).

For an initial proof of concept of high-speed operation we used a QCL, operating at room temperature, as a high-speed source of IR radiation at $\lambda \sim 5\,\mu m$ to study the response time of the uncooled type-II devices. The detector used had a cutoff wavelength of ~8 μm. Figure 24.30 shows the current of the QCL as well as the output of the preamplifier versus time. The inset shows the schematic diagram

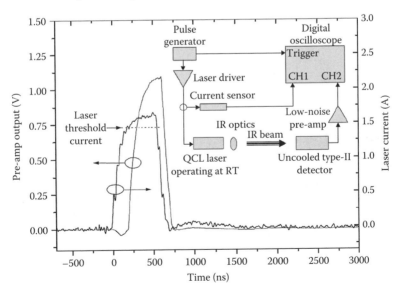

FIGURE 24.30 The current of the QCL laser (right axis) and the amplified output of the detector (left axis) versus time. Inset shows the measurement setup. Note that the laser emits when current is above the threshold level. (From Hood, A. et al., *Conference on Quantum Sensing and Nanophotonic Devices V*, San Jose, CA, 2008. With permission. Copyright 2008 by SPIE.)

of the measurement setup. The laser threshold current marks the current above which the laser starts emitting. Considering the fall time of the preamplifier (~50 ns) and the fall time of the output signal of the preamplifier (~110 ns), the detector response time was about $(110^2 - 50^2)^{1/2} = ~100$ ns using the sum-of-squares approach. Such a detector response time would support data transmission rates of up to 10 Mbps. It should be noted that this laser and detector setup was not optimized for high-speed performance and parasitic delays are believed to be arise from this particular measurement setup when the modulation frequency increases above ~1 MHz. Additionally, this detector was optimized for room-temperature operation at this far extent of the MWIR regime and a detector with a shorter cutoff wavelength, optimized for high-speed data transmission is expected to allow for significantly higher bandwidth.

Following this successful laboratory demonstration, the next step was to take the laser and detector off of the lab bench and test their robustness in a portable demonstration. For simplicity of the prototype, passive cooling was chosen for both the laser and detector modules, which limits their operation to low duty cycles at room temperature. The laser and detector modules, once developed for higher heat loads and data input, could easily be incorporated into any number of application-specific systems, including FSC. The proof-of-concept prototype is shown in Figure 24.31.

The prototype system cost was only $450, without the optical elements, compared to over $10,000 for laboratory-grade components. Furthermore, the prototype is powered using three 9 V batteries (for the transmitter) and 8 AA batteries (for the receiver) which is similar to a battery-powered toy. The laser module weighs approximately 0.9 lbs with batteries and the detector module is slightly heavier at about 2 lbs due to the additional optical power readout display and amplifying circuitry and batteries contained within the module. The laser produces about 2 W peak power and 4 mW average power depending on the duty cycle selected by the driver. Typical operating conditions for the QCL module permit up to 45% duty cycle operation and pulses as short as pulsing at 200 ns at 10 kHz with an effective battery lifetime of 10,000 h for the laser module and 1,000 h for the detector module, which is limited again by the higher power required for the on-board display and amplifying circuits.

This proof-of-concept prototype is designed primarily to demonstrate the robustness of the experimental devices off of the lab bench as well as the potential for creating a portable system using devices grown in our laboratories. The primary drawbacks of the initial prototype are that it is bulky, not very user friendly, and very sensitive to alignment, which meant that the two modules had to be fixed at close range from each other. A more practical system would be capable of more arbitrary alignment and could be implemented with the transmitter and receiver at a useful distance from one another. Additionally, further miniaturization of the electronics, including the amplifying circuit and the readout display can be accomplished so that they fit in a smaller space and require fewer batteries. Ultimately, passively

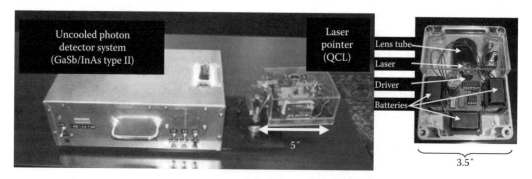

FIGURE 24.31 First proof-of-concept uncooled FSC system developed at the CQD. The system is comprised of an uncooled InAs/GaSb type-II photon detector module with a cutoff wavelength of ~8 μm and a passively cooled QC-laser module emitting at 5.3 μm. (From Hood, A. et al., *Conference on Quantum Sensing and Nanophotonic Devices V*, San Jose, CA, 2008. With permission. Copyright 2008 by SPIE.)

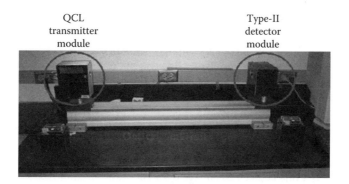

FIGURE 24.32 Prototype modular mid-infrared free-space communication system developed at the CQD setup for laboratory testing on a 3 ft long optical ray. The system is comprised of an uncooled type-II photon detector module and a room-temperature CW QCL transmitting module. (From Hood, A. et al., *Conference on Quantum Sensing and Nanophotonic Devices V*, San Jose, CA, 2008. With permission. Copyright 2008 by SPIE.)

cooled modules should be able to be reduced from shoebox-sized units to pocket-sized devices. All these tasks involve a fair amount of physics, optics, electrical, and mechanical engineering. With these ideas in mind, in conjunction with improvements to the detector and laser epi-materials, a revision of the communications system was developed, and is pictured below in Figure 24.32.

This latest prototype featured active alignment and automatic beam-finding routines, features typically found only in the most advanced commercial systems. Unlike the previous prototype, the new prototype laser module, while slightly larger in size and weight, incorporates active cooling to support CW (100% duty cycle) operation, high laser currents above 1 A DC, and/or higher output power, which all allow for high-speed communications using transmitter direct or encoded (serial) modulation and receiver demodulation. The detector used was one of our typical type-II InAs/GaSb superlattice photodiode designs with a cutoff wavelength of ~5 μm. Single-element detectors fabricated from this material were thoroughly characterized at room temperature and exhibited a specific detectivity, D^*, of ~1×10^9 cm Hz$^{1/2}$ W^{-1}. Operating at a bandwidth of 155 Mbps and a BER of 1×10^{-9} ($S/N = 6$) this corresponds to a sensitivity of −30 dBm. This nonoptimized device compares quite favorably to avalanche photodiodes (APDs) operating at 0.91–1.55 μm with a sensitivity of −45 dBm. Typical performance characteristics for these photodiodes operating at room temperature are shown in Figure 24.33.

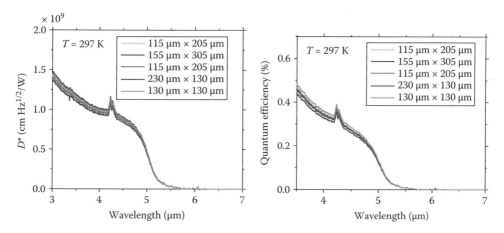

FIGURE 24.33 Device performance characteristics of room-temperature operating MWIR type-II InAs/GaSb superlattice photodiodes. (From Hood, A. et al., *Conference on Quantum Sensing and Nanophotonic Devices V*, San Jose, CA, 2008. With permission. Copyright 2008 by SPIE.)

The receiver prototype consists of a stepper-motor-driven X–Y stage controlled by an 8-bit microcontroller that also handles all of the multiplexing, signal demodulation, user interface, etc. The received laser signal is amplified by a trans-impedance amplifier (TIA) circuit with fixed gain. The modulated signal can be serially decoded and the received data is output to an LCD screen or a standard RS-232 interface. Rough system alignment is done by using a red diode laser on the transmitter side to aim at a target on the receiver. As long as the red laser spot is within a specified target envelope, the automatic beam-finding algorithm programmed into the microcontroller will handle the fine alignment in order to locate and track the infrared communications laser. No collection optics have been integrated to the receiver side yet, but because of the low beam divergence, the laser spot size over ~100 m is still smaller than the detector element size (1 mm × 1 mm). In practice, typically the beam is intentionally expanded or diverges naturally and this is taken advantage of to allow for building sway due to the elements or thermal expansion. The drawback is that the power throughput is significantly decreased as the beam diameter increases, decreasing the effective range of the channel. This broadening of the beam can be mostly avoided with the use of an actively pointed system which will self-align itself.

The QCL used in the transmitter module emits at a wavelength of 4.8 μm and produces over 200 mW of output power at room temperature at a DC drive current of 0.75 A and an operating voltage of 12 V. Active heatsinking is required to stabilize the laser wavelength and dissipate the ~9 W of heat generated during laser operation, which is accomplished using a Peltier cooler and forced air copper heat sink. ZnSe aspheric optics provide a collimated output beam 3 mm in diameter. A rechargeable battery pack provides up to 4 h of continuous operation. Modulation of the output beam is performed by biasing the laser below threshold and coupling in a modulated signal using a DC bias tee.

Figure 24.34 shows an oscilloscope trace of the gate of the laser driver's current driving FET. At a gate voltage of approximately 3.75 V, the current through the FET and through the laser is approximately 400 mA, near the pulsed-mode threshold current density of the laser diode. Above this value, the laser is above threshold, is lasing, and the output power increases with increasing drive current, until thermal rollover. However, some small signal oscillations in the FET's gate voltage as well as the drive current were observed. These are attributed to ground bounce, a deviation from the ground's reference value of 0 V. This undesired effect can be ameliorated by better power supply decoupling of the operational amplifier. The oscillations decrease as the differential increase in current being injected into the ground plane decreases. A more robust driver design could reduce or eliminate these oscillations.

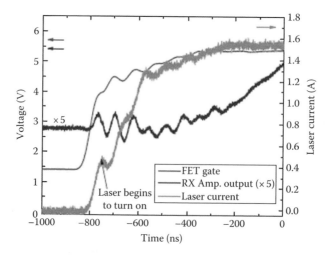

FIGURE 24.34 Traces of the current driving FET's gate voltage, the detector amplifier output (×5), and the laser current versus time.

As expected, the oscillations also modulate the laser's output power, which can be seen in the trace of detector amplifier's output response (blue). The detector's output has been multiplied by a factor of 5 to make it easier to observe small voltage variations in this signal. Though the detector amplifier has a gain-bandwidth product of ~130 MHz, in order to enhance the detector signal to the useful range of the Schmidt trigger, a ~9 kΩ gain resistor acts as the feedback resistor of this TIA. Thus the time response of the amplifier is considerably compromised. However, an improved multistage design with a preamplifier and offset correction could conserve the high-frequency response of the amplifier while still allowing a voltage swing compatible with the signal processing electronics. For serial data transmission with a computer, however, where response times are not expected to be appreciably faster than ~5 μs (corresponding to 200,000 bits per second), this sort of amplifier large signal response is acceptable. For megabit or gigabit speeds, much more attention would have to be paid to the amplifier design as well as the device packaging.

24.8.3 Conclusion

FSC is a proven technology for secure, high-bandwidth communications that can be operated without a license, discretely, and with rapid setup and installation. For fundamentally physical reasons atmospheric attenuation of the beam in adverse conditions such as snow, rain, and fog is expected to decrease as one moves further into the infrared. Additionally, the demonstration of high-performance, room-temperature lasers and detectors in the MWIR show that these devices are ready to be integrated into a practical system. The direct electrical modulation and high output power available in QCL technology allows for more compact and less complex components to power the next generation of free-space optical communications systems.

Acknowledgment

We would like to acknowledge Darin Hoffman and Edward Kwei-wei Huang for their scientific contributions to the work. We would also like to thank Dr. Fenner Milton from the U.S. Army Night Vision and Electronic Sensor Directorate (NVESD), Dr. Meimei Tidrow from the Missile Defense Agency, and Dr. Joe Pellegrino from the NVESD for their collaborations and supports.

References

Cabanski, W.A., Eberhardt, K., Rode, W., Wendler, J.C., Ziegler, J., Fleissner, J., Fuchs, F., Rehm, R.H., Schmitz, J., Schneider, H., and Walther, M., Third-generation focal plane array IR detection modules and applications, in *Infrared Technology and Applications XXX*, 1 edn., vol. 5406, Orlando, FL: SPIE, 2004, pp. 184–192.

Delaunay, P.-Y., Hood, A., Nguyen, B.M., Hoffman, D., Wei, Y., and Razeghi, M., Passivation of type-II InAs/GaSb double heterostructure, *Applied Physics Letters*, 2007a, 91(9), 091112–091113.

Delaunay, P.-Y., Nguyen, B.M., Hofman, D., and Razeghi, M., Substrate removal for high quantum efficiency back side illuminated type-II InAs/GaSb photodetectors, *Applied Physics Letters*, 2007b, 91(23), 231106-1–231106-3.

Delaunay, P.-Y., Nguyen, B.-M., Hoffman, D., Hood, A., Huang, E.K.-W., Razeghi, M., and Tidrow, M.Z., High quantum efficiency two color type-II InAs/GaSb n-i-p-p-i-n photodiodes, *Applied Physics Letters*, 2008, 92(11), 111112–111113.

Delaunay, P.Y., Nguyen, B.M., Hoffman, D., Huang, E., and Razeghi, M., Background limited performance of long wavelength infrared focal plane arrays fabricated from M-structure InAs/GaSb supperlatices, *IEEE Journal of Quantum Electronics*, 2009, 45(2), 157.

Evans, A., Yu, J.S., David, J., Doris, L., Mi, K., Slivken, S., and Razeghi, M., High-temperature, high-power, continuous-wave operation of buried heterostructure quantum-cascade lasers, *Applied Physics Letters*, 2004, 84(3), 314–316.

Evans, A., Nguyen, J., Slivken, S., Yu, J.S., Darvish, S.R., and Razeghi, M., Quantum-cascade lasers operating in continuous-wave mode above 90°C at lambda ~ 5.25 μm, *Applied Physics Letters*, 2006a, 88(5), 051105-1–051105-3.

Evans, A. and Razeghi, M., Reliability of strain-balanced Ga[sub 0.331]In[sub 0.669]As/Al[sub 0.659]In[sub 0.341]As/InP quantum-cascade lasers under continuous-wave room-temperature operation, *Applied Physics Letters*, 2006b, 88(26), 261106-1–261106-3.

Evans, A., Darvish, S.R., Slivken, S., Nguyen, J., Bai, Y., and Razeghi, M., Buried heterostructure quantum cascade lasers with high continuous-wave wall plug efficiency, *Applied Physics Letters*, 2007, 91(7), 071101–071103.

Faist, J., Capasso, F., Sivco, D.L., Sirtori, C., Hutchinson, A.L., and Cho, A.Y., Quantum cascade laser, *Science*, 1994, 264(5158), 553–556.

Fuchs, F., Weimer, U., Pletschen, W., Schmitz, J., Ahlswede, E., Walther, M., Wagner, J., and Koidl, P., High performance InAs/Ga[sub 1-x]In[sub x]Sb superlattice infrared photodiodes, *Applied Physics Letters*, 1997, 71(22), 3251–3253.

Gin, A., Wei, Y., Hood, A., Bajowala, A., Yazdanpanah, V., Razeghi, M., and Tidrow, M., Ammonium sulfide passivation of type-II InAs/GaSb superlattice photodiodes, *Applied Physics Letters*, 2004, 84(12), 2037–2039.

Grein, C., Young, P., Ehrenreich, H., and McGill, T., Auger lifetimes in ideal InGaSb/InAs superlattices, *Journal of Electronic Materials*, 1993, 22(8), 1093–1096.

Grein, C.H., Cruz, H., Flatte, M.E., and Ehrenreich, H., Theoretical performance of very long wavelength InAs/In[sub x]Ga[sub 1 – x]Sb superlattice based infrared detectors, *Applied Physics Letters*, 1994, 65(20), 2530–2532.

Gunapala, S.D., Bandara, S.V., Liu, J.K., Hill, C.J., Rafol, S.B., Mumolo, J.M., Trinh, J.T., Tidrow, M.Z., and LeVan, P.D., 1024 × 1024 pixel mid-wavelength and long-wavelength infrared QWIP focal plane arrays for imaging applications, *Semiconductor Science and Technology*, 2005, 20(5), 473–480.

Hill, C.J., Li, J.V., Mumolo, J.M., Gunapala, S.D., Rhiger, D.R., Kvaas, R.E., and Harris, S.F., MBE grown type-II superlattice photodiodes for MWIR and LWIR imaging applications, in *Infrared Technology and Applications XXXIII*, 1 edn., vol. 6542, Orlando, FL: SPIE, 2007, pp. 654205–654210.

Hood, A., Delaunay, P.-Y., Hoffman, D., Nguyen, B.-M., Wei, Y., Razeghi, M., and Nathan, V., Near bulk-limited R[sub 0]A of long-wavelength infrared type-II InAs/GaSb superlattice photodiodes with polyimide surface passivation, *Applied Physics Letters*, 2007, 90(23), 233513-1–233513-3.

Hood, A., Evans, A., and Razeghi, M., *Conference on Quantum Sensing and Nanophotonic Devices V*, San Jose, CA, 2008.

Johnson, J.L., Samoska, L.A., Gossard, A.C., Merz, J.L., Jack, M.D., Chapman, G.R., Baumgratz, B.A., Kosai, K., and Johnson, S.M., Electrical and optical properties of infrared photodiodes using the InAs/Ga[sub 1 – x]In[sub x]Sb superlattice in heterojunctions with GaSb, *Journal of Applied Physics*, 1996, 80(2), 1116–1127.

Meyer, J.R., Hoffman, C.A., Bartoli, F.J., and Ram-Mohan, L.R., Type-II quantum-well lasers for the mid-wavelength infrared, *Applied Physics Letters*, 1995, 67(6), 757–759.

Mohseni, H., Michel, E., Sandoen, J., Razeghi, M., Mitchel, W., and Brown, G., Growth and characterization of InAs/GaSb photoconductors for long wavelength infrared range, *Applied Physics Letters*, 1997, 71(10), 1403–1405.

Nguyen, B.M., Razeghi, M., Nathan, V., and Brown, G.J., Type-II M structure photodiodes: An alternative material design for mid-wave to long wavelength infrared regimes, in *Quantum Sensing and Nanophotonic Devices IV*, 1 edn., vol. 6479, San Jose, CA: SPIE, 2007a, pp. 64790S–64710S.

Nguyen, B.-M., Hoffman, D., Delaunay, P.-Y., and Razeghi, M., Dark current suppression in type II InAs/GaSb superlattice long wavelength infrared photodiodes with M-structure barrier, *Applied Physics Letters*, 2007b, 91(16), 163511–163513.

Nguyen, B.-M., Hoffman, D., Wei, Y., Delaunay, P.-Y., Hood, A., and Razeghi, M., Very high quantum efficiency in type-II InAs/GaSb superlattice photodiode with cutoff of 12 mu m, *Applied Physics Letters*, 2007c, 90(23), 231108-1–231108-3.

Nguyen, B.-M., Hoffman, D., Huang, E.K.-W., Delaunay, P.-Y., and Razeghi, M., Background limited long wavelength infrared type-II InAs/GaSb superlattice photodiodes operating at 110 K, *Applied Physics Letters*, 2008, 93, 123502.

Piquini, P., Zunger, A., and Magri, R., Pseudopotential calculations of band gaps and band edges of short-period (InAs)[sub n]/(GaSb)[sub m] superlattices with different substrates, layer orientations, and interfacial bonds, *Physical Review B (Condensed Matter and Materials Physics)*, 2008, 77(11), 115314–115316.

Razeghi, M., Nguyen, B.-M., Hoffman, D., Delaunay, P.Y., Huang, E.K.-W., Tidrow, M., and Nathan, V., Development of material quality and structural design for high performance Type II InAs/GaSb superlattice photodiodes and focal plane arrays, in *Infrared Spaceborne Remote Sensing and Instrumentation XVI*, vol. 7082, 2008, SPIE, pp. 708204.

Razeghi, M., Wei, Y., Hood, A., Hoffman, D., Nguyen, B.M., Delaunay, P.Y., Michel, E., and McClintock, R., Type-II superlattice photodetectors for MWIR to VLWIR focal plane arrays, in *Infrared Technology and Applications XXXLL*, vol. 6206, 2006, SPIE, pp. 62060N.

Rehm, R., Walther, M., Fuchs, F., Schmitz, J., and Fleissner, J., Passivation of InAs/(GaIn)Sb short-period superlattice photodiodes with 10 mu m cutoff wavelength by epitaxial overgrowth with Al[sub x]Ga[sub 1 − x]As[sub y]Sb[sub 1 − y], *Applied Physics Letters*, 2005, 86(17), 173501–173503.

Robic, J.Y., Leplan, H., Pauleau, Y., and Rafin, B., Residual stress in silicon dioxide thin films produced by ion-assisted deposition, *Thin Solid Films*, 1996, 290–291, 34–39.

Rodriguez, J.B., Plis, E., Bishop, G., Sharma, Y.D., Kim, H., Dawson, L.R., and Krishna, S., nBn structure based on InAs/GaSb type-II strained layer superlattices, *Applied Physics Letters*, 2007, 91(4), 043514–043516.

Sai-Halasz, G.A., Tsu, R., and Esaki, L., A new semiconductor superlattice, *Applied Physics Letters*, 1977, 30(12), 651–653.

Smith, D.L. and Mailhiot, C., Proposal for strained type II superlattice infrared detectors, *Journal of Applied Physics*, 1987, 62(6), 2545–2548.

Smith, E.P., Pham, L.T., Venzor, G.M., Norton, E., Newton, M., Goetz, P., Randall, V., Pierce, G., Patten, E.A., Coussa, R.A., Kosai, K., Radford, W.A., Edwards, J., Johnson, S.M., Baur, S.T., Roth, J.A., Nosho, B., Jensen, J.E., and Longshore, R.E., Two-color HgCdTe infrared staring focal plane arrays, in *Materials for Infrared Detectors III*, 1 ed., vol. 5209, San Diego, CA: SPIE, 2003, pp. 1–13.

Tsu, R., Chang, L.L., Sai-Halasz, G.A., and Esaki, L., Effects of quantum states on the photocurrent in a "Superlattice", *Physical Review Letters*, 1975, 34(24), 1509.

Vurgaftman, I., Aifer, E.H., Canedy, C.L., Tischler, J.G., Meyer, J.R., Warner, J.H., Jackson, E.M., Hildebrandt, G., and Sullivan, G.J., Graded band gap for dark-current suppression in long-wave infrared W-structured type-II superlattice photodiodes, *Applied Physics Letters*, 2006, 89(12), 121114-1–121114-3.

Walther, M., Schmitz, J., Rehm, R., Kopta, S., Fuchs, F., Fleiner, J., Cabanski, W., and Ziegler, J., Growth of InAs/GaSb short-period superlattices for high-resolution mid-wavelength infrared focal plane array detectors, *Journal of Crystal Growth*, 2005, 278(1–4), 156–161.

Walther, M., Rehm, R., Fleissner, J., Schmitz, J., Ziegler, J., Cabanski, W., and Breiter, R., InAs/GaSb type-II short-period superlattices for advanced single and dual-color focal plane arrays, in *Infrared Technology and Applications XXXIII*, 1 edn., vol. 6542, Orlando, FL: SPIE, 2007, pp. 654206–654208.

Wei, Y. and Razeghi, M., Modeling of type-II InAs/GaSb superlattices using an empirical tight-binding method and interface engineering, *Physical Review B (Condensed Matter and Materials Physics)*, 2004, 69(8), 085316–085317.

Wei, Y., Hood, A., Yau, H., Gin, A., Razeghi, M., Tidrow, M.Z., and Nathan, V., Uncooled operation of type-II InAs/GaSb superlattice photodiodes in the midwavelength infrared range, *Applied Physics Letters*, 2005, 86(23), 233106–233109.

Zhang, X., Li, A.Z., Lin, C., Zheng, Y.L., Xu, G.Y., Qi, M., and Zhang, Y.G., The effects of (NH4)2S passivation treatments on the dark current-voltage characteristics of InGaAsSb PIN detectors, *Journal of Crystal Growth*, 2003, 251(1–4), 782–786.

Index

N